AF560722

NUCLEAR MATERIALS
Issues and Concerns

NUCLEAR MATERIALS
Issues and Concerns

Vol. I

Editor

Dr. Digumarti Bhaskara Rao

M.Sc., M.A., M.A., M.Ed., Ph.D.

Secretary
"Academy of Communication Culture, Education Science and Service"
1-22-10 Srinivasa Nagar
Guntur 522006
Andhra Pradesh
India

2001

DISCOVERY PUBLISHING HOUSE

NEW DELHI—110 002

Published by :
Discovery Publishing House
4831/24, Ansari Road, Prahlad Street
Darya Ganj, New Delhi—110 002 (INDIA)
Phone : 327 92 45
Fax : 91-11-3253475

First Published—2001

ISBN 81-7141-611-X(Set)

Laser Typeset by :
Allied Computers,
Karnal (Haryana)

Printed at:
Arora Offset Press
Laxmi Nagar, Delhi 110 092.

Preface

Nuclear materials have civilian utilisation too along with military utilisation. The dual nature of the utilisation of nuclear materials has originated many problems but has also created several opportunities. If the nuclear materials are used for the production of civilian nuclear energy even to a limited extent, then there will be no energy crisis atleast in sectors like electricity. But, many world countries are developing and or developed sophisticated technologies for arms development at the cost of civilian safety. It is the time now for elimination of the danger of nuclear conflicts from human life and for developing nuclear technologies that work for the welfare of mankind.

This book covers a wide range of vital issues and concerns related to nuclear materials such as nuclear research and development, disarmament and arms control, illegal nuclear traffic and its associated aspects, disposal of radio-active waste/fissile materials, decommissioning and management of surplus nuclear equipments and materials, etc. A special attention to Indian and Pakistan nuclear tests has also been given.

The articles included in this book are selected from the (1) International Conference on Military Conversion and Science" Utilisation/Disposal of the Excess Fissile Weapons Materials: Scientific, Technological and Socio-Economic Aspects", 18–20 March 1996; (2) International Forum on "Illegal Nuclear Traffic: Risks, Safeguards and Countermeasures", 12–13 June 1997; (3) International Forum on "Nuclear and Biological Decommissioning: Management of Global Security Threats", 27–28 June 1997; and (4) Forum of the UNESCO International School of Science for Peace on "Nuclear Disarmament, Safe Disposal of Nuclear Material for new weapons Developments ? Where are the National Laboratories Going ?", 2–4 July 1998 organised by the UNESCO Venice Office and others. The articles have been written by the experts in nuclear issues and concerns.

I am grateful to Prof. Pierre Lasserre, Director, UNESCO Venice Office, for giving me permission to reproduce articles from volumes one, four, five and eight of Science For Peace series. I am thankful to Dr. Rossanna Santesso, Asst. Information Officer, UNESCO Venice Office for readily responding to all of my queries. Special thanks are extended to UNESCO Venice Office, Landau Network-Centro Volta, Italian Ministry of Foreign Affairs, Italian Ministry of Universities and Scientific and Technological Research, ENEA, and ACCESS.

I hope that this book will have an impact on politicians and policy makers, scientists and technologists, economists and educationists and advocates and activists with regard to nuclear issues and concerns.

Dr. Digumarti Bhaskara Rao

Contents

PART–4
PRESENT RESEARCH IN THE FIELD OF NUCLEAR WEAPONS, THE FUTURE OF NATIONAL LABORATORIES AND LAB-TO-LAB COLLABORATION

PART — 1
PERSPECTIVES OF NUCLEAR RESEARCH AND DEVELOPMENT

1

Nuclear Disarmament and Risks of Proliferation

PAOLO COTTA-RAMUSINO

Fifty Years of Nuclear Weapons

More than fifty years have passed since the nuclear bombing of Hiroshima and Nagasaki. During this time we have witnessed the intense yet futile competition between the United States and the Soviet Union regarding the design, construction and deployment of nuclear weapons and their delivery systems. The whole period of the Cold War has been dominated by the nuclear arms race.

The end of the Cold War coincides with the beginning of nuclear disarmament, that is with a dramatic reversal of tendency. But 50 years of nuclear armaments can certainly not be erased instantly and without problems. Nuclear disarmament therefore presents itself as being a long and complicated process, in which there are, and will be, no shortage of problems and contradictory aspects, with periods of good progress which will probably alternate with periods of delay and reversal.

Before talking about nuclear disarmament and its problems, it may be useful to remember some data which in a synthetic way, even if necessarily schematic, can be chosen to represent these 50 years of nuclear arms race.

— Since 1945 about 130000 nuclear warheads have been built—of these the United States have built about 70000 and the USSR more than 55000.

— From the 1960s to the present day two enormous nuclear arsenals have been facing each other, each of which was at all times made up of more than 10000 weapons. The largest number was reached in 1967 by the U.S.A. (32500 weapons) and in 1986 by the USSR (45000 weapons, if not more).

— About 2050 nuclear tests have been carried out, of which 514 in the atmosphere (or in any case not underground).

— The countries in possession of nuclear arms have passed from one (1948) to 8 (today). Of these 8 nuclear countries, 5 declare themselves officially as being so (USA, Russia, Great Britain, France, China) and 3 are *de facto* (Israel, India, Pakistan).

— More than 250 tons of plutonium and 2200 tons of highly enriched uranium (HEU) have been produced for military use.

— Apart from the production of nuclear arms, we must remember the production of the nuclear delivery systems. For example, the United States alone have produced about 67500 nuclear missiles.[1]

— The construction and management of nuclear arms and of the hundreds of thousands of delivery systems has meant a colossal expenditure for humanity. According to an estimate published in the Nov.–Dec. '95 issues of the *Bulletin of the Atomic Scientists*, the United States alone have spent since 1940 the equivalent of about 3900 billion U.S. dollars (current 1995 value) for their nuclear programmes. This is more than 150 times the cost, in 1995 US dollars, of the Manhattan Project. The USSR has probably spent a similar amount which, together with the expenditures of the "minor" nuclear powers, brings the total cost of nuclear weapons to something in the region of nine thousand billion dollars, equivalent to nine times the present annual Italian GNP.

Two points very worthy of separate attention are the loss of human life and the consequences on the environment which have been caused by military nuclear activities from the post-war era to today. In this respect, we should take into account the consequences of nuclear tests (in particular in the Pacific islands, in Nevada, in Kazakhstan), of accidents and contamination occurred in places where plutonium was produced, of the management of radioactive waste. As an example, consider the Mayak/Chelyabinsk-65 plant in Ozersk where more than 26000 sq. km, and more than 400000 people have been contaminated.[2]

Lastly we must mention the specific social consequences which the nuclear arms race has brought about in the ex-USSR, where the military effort has consumed a far higher quota of national resources than that of corresponding Western Countries. A significant example of the economic and social distortion caused by the arms race was the creation of entire cities closed to the outside world and dedicated to the production of fissile material and other products for nuclear weapons. The total population of these cities has been estimated to be over 700000.[3]

The production and maintenance cost of nuclear arms has therefore been very high with regards the use of resources, the loss of human life, the ecological damage caused, the social problems created, and overall the risk of a nuclear catastrophe that has been with us for the last 50 years. On the other hand the military utility of those weapons has been close to zero. After Hiroshima and Nagasaki nuclear weapons have never been used in any of the numerous conflicts which have developed during the last 50 years.

Despite the great variety of nuclear warheads in terms of explosive yield and of range and type of delivery systems, despite the sophisticated strategic concepts that have been developed, no flexibility for the use of nuclear weapons has emerged in the real world.

The belief that even a limited use of those weapons would have caused a global catastrophe has fortunately prevailed. Nuclear weapons therefore represent a unique phenomenon in the history of mankind: never has so much energy been dedicated to the development, the production, and the deployment of weapon-systems which, for about 50 years, have been accumulated in large quantities, without being used.

Plutonium and Uranium for Military Purposes

The first inheritance of the arms race is the enormous quantity of fissile material produced—that is highly enriched uranium (HEU) and plutonium. The fissile material is classified as being weapon-grade if the isotopic composition is above a certain level (93% of U-235 for highly enriched Uranium and 93% of Pu-239 for plutonium).

In order to build even a rudimental pure fission bomb it is not necessary to have only weapon-grade type material on hand.

Also plutonium with a higher percentage of Pu-240 or uranium with a lower percentage of U-235 can be used for the preparation of a bomb.

In this respect, the International Atomic Energy Agency in Vienna (IAEA) defines the *significant quantity*, from the point of view of the possible illegal manufacture of nuclear weapons, as being 8 kg of plutonium and 25 kg of U-235 contained in HEU (enriched with more than 20% of U-235).

In reality, according to a recent estimate of the Natural Resources Defence Council (NRDC) in Washington D.C., a fission bomb could be constructed with an amount between 1 to 6 kg of plutonium and between 3 to 16 kg of U-235 contained in HEU depending on the technical capabilities of the nuclear proliferators and the explosive power of the bomb they intend to produce.

If on one hand the quantity of fissile material needed to construct a fission bomb is quite modest, on the other the quantity of Pu and HEU produced by countries which possess nuclear weapons is extremely vast.

A recent report by the U.S. Department of Energy declares[4] that the same DOE acquired, between 1944 and 1994, 111.4 tons of plutonium for military use, of which 99.5 tons are still in its inventory.

An estimate made by A.S. Diakov[5] claims that the quantity of plutonium produced by the USSR/Russia up to 1995 is about 126 tons.

With regards to enriched Uranium, the same U.S. DOE declared that it had produced a total of 994 tons of HEU, of which the amount destined for nuclear weapons has been estimated as being 730 tons[6], while the corresponding quantity[7] for the USSR/Russia could be anything from 15% to 30% larger than that of the USA.

Following the dismantlement of the nuclear weapons, a considerable part of this fissile material has been, or will be, declared *in excess*: this regards at least 100 tons of plutonium and 700 tons of HEU.

Apart from fissile material connected with military activity, large quantities of plutonium exist in the spent fuel of nuclear reactors; in fact this is the most consistent part of the existing plutonium.

There is also some plutonium of civil origin which has already undergone the process of separation. An estimate made by the C.I.S.A.C. of the American National Academy of Science in 1994 gives the total amount of Plutonium existing in the world in 1992 as being 1100 tons, and foresees that this amount will increase in the year 2000 to 1600–1700 tons.

The problems which concern the enormous quantity of fissile material produced are above all those regarding security. The main obstacle which bars the way to the construction of rudimental nuclear weapons is not the lack of the necessary technological information, but rather the difficulty of obtaining fissile material. The problem therefore is how to avoid that countries or illegal organisations, which may be interested in acquiring nuclear weapons, manage to obtain HEU and plutonium.

The plutonium contained in the spent fuel of nuclear reactors is very difficult to reach since, by definition, it is not separated from the rest of the radioactive waste.

On the basis of the experience accumulated in 50 years of nuclear weapons, it is also reasonable to expect that intact nuclear warheads (in active duty), being protected by military structures, may not easily fell into the hands of potential nuclear proliferators.[8]

In other words, the dismantled warheads, as well as the separated civilian plutonium may form the weakest link in the chain of the control of fissile material.

This problem has been brought to the general public's attention after to the demise of the USSR and the relative political, economical, and organisational difficulties which appeared in the former Soviet republics.

The warheads which must be dismantled have a long road ahead of them before they arrive at their final destination. First of all the warheads must be de-activated (that is, the trigger which is located in the external part of the warhead must be removed). Then they must be transported to the deposits to which they are directed. At this point the warheads must be opened, separating the fissile material contained in a metal container or *pit* from the rest of the warhead (chemical explosive, secondary system in thermonuclear devices, etc.).

Now the *pit*, which contains the plutonium (or the enriched uranium), can be further dismantled only when the final disposal of the fissile material has been decided.

From the security point of view the process of dismantling a warhead therefore presents the following problems:

— The security of the transport of the de-activated warheads to the deposits.

— The accurate checking and registration of all the warheads and pieces of warheads which are dismantled.

— The safe custody of the pits while awaiting the disposal of the fissile material.

— The decisions concerning the disposal of the fissile material and its safe implementation.

It must be emphasised that the large number of warheads to be dismantled will impose a prolonging of the custody phase of the fissile material in the form of *pits*, and that consequently the relative security problems will remain for some considerable amount of time, whatever may be the choices concerning the disposal of the fissile material.

With regards to the enriched uranium, its logical destination will be the dilution with natural or depleted uranium, so to create low enriched uranium which may be used in nuclear reactors.

A similar choice is not feasible for plutonium since the mixing of the different isotopes of plutonium does not eliminate the risk connected to nuclear proliferation. On the other hand, mixing plutonium with other elements (uranium) is a procedure which can be easily reverted by chemical reprocessing.

With regards to the disposal of the plutonium, among the numerous choices proposed and considered in the before-mentioned report of the American National Academy of Science, there are two which emerged as the most worth of consideration:

1. To consider the plutonium the same as waste and to keep it in custody indefinitely after having treated it in such a way as to make access difficult. (For example, vitrifying it together with a highly radioactive material).
2. To use the plutonium for the preparation of MOX fuel (mixed oxides of U and Pu) to be brunt in civil nuclear reactors.

The choice between the two above options will be determined by different factors, and not only by a question of security. The practicability of the second solution, for example, will be determined by the existence of an adequate number of plants which can prepare of MOX, by the characteristics of the nuclear reactors which can use this fuel, by the economic advantages of the whole operation (cost of MOX opposed to the cost of low level enriched uranium fuel), by the political attitudes of the countries which are potentially interested in the use of plutonium as nuclear fuel.

For example, the hostility of public opinion in some Western countries towards the nuclear choice could play a part—just as the belief of the Russian government that "plutonium is a national treasure" could have weight in the opposite direction.

With regards to the time needed for dismantling nuclear warheads, we can easily realise that the speed of this operation most likely will not be very high. It is instructive to compare the following data which refers to the USA[9]:

— Average number of American nuclear warheads produced annually in the years 1959 and 1960: more than 7000.
— American nuclear warheads dismantled in 1969: more than 3000.
— Years in which more than 2000 American nuclear warheads were dismantled per year: 1959, 1964, 1966, 1968, 1969, 1975, 1976.
— Average number of American nuclear warheads dismantled annually in the post cold war era (from 1991 to 1995): 1550.

As far as Russia is concerned, notice that since 1986 it has dismantled nuclear weapons at an initial rate of between 2000 and 3000 per year.

The rate seems to have slowed down now to less than 2000 warheads per year[10].

Illegal Traffic of Nuclear Material

Since the demise of the Soviet Union, western public opinion has been worrying about the possibility of a dramatic increase in nuclear proliferation. Particular anxiety has been caused by:

— The possible creation of new independent nuclear States among the republics of the ex-USSR.
— The prospect of an intense illegal traffic of fissile material or even of whole nuclear warheads.
— The possible migration of large number of technicians and

scientists from the ex-USSR to countries interested in buying nuclear weapons (in Russia there are reportedly about 2000 people who have a detailed knowledge of the design of nuclear weapons and around 3000/5000 experts in the production of fissile material.[11]

Fortunately the present situation is quite a long way from these pessimistic predictions.

More than 6000 tactical nuclear weapons have been reportedly transferred from 14 ex-Soviet Republics to Russia without incidents.

Kazakhstan, Belarus and Ukraine, on whose territory strategic nuclear weapons are still deployed, have signed the Non-Proliferation Treaty as non-nuclear countries, declaring in this way their explicit will to renounce the possession of nuclear weapons.

Up to this moment there is no evidence that significant episodes of recruitment of Russian nuclear scientists have taken place on behalf of countries interested in nuclear proliferation.

Finally, with regard to the episodes of illegal traffic of nuclear material, tens if not hundreds of cases have been reported. Most of these cases have proven, however, to be of dubious authenticity or in any case irrelevant.[12]

Only a few significant episodes of nuclear smuggling have been identified, as are mentioned later. However even in these significant cases the whole picture is not very clear. In particular, even when the fissile material confiscated has been significant and the identity of the thieves established, there is no significant information regarding the possible buyers, whether they be States or criminal organisations.

The following table shows the main episodes of theft of fissile material so far identified.[13] Please note that all the confiscated quantities do not reach the minimum quantities needed to build a single bomb.

1. Podolsk, Russia 9.10.1992 Kg 1.538 HEU (90% enriched)
2. S. Petersburg March 1994 Kg 3.5 HEU
3. Tengen (Baden-Wuetemberg) 10.5.1994 g. 5.6 Pu-239
4. Polyarny (Murmansk) June 1994 Kg 4.5 HEU (20% enriched) (stolen 27.11.93).
5. Vilnius 1994 Kg 2 HEU hidden in 4 tons of Berillium
6. Munich 10.8.1994 g. 560 MOX with g. 363 of Pu-239

7. Prague 14.12.1994 Kg 2.72 HEU (87.7% enriched).

The relatively optimistic evaluation made so far regarding the problem of illegal traffic of nuclear material and the overall effects of the disintegration the USSR on nuclear proliferation, must not however, make us think that the dangers on this front can be considered marginal in the future.

If the situation has been kept (relatively) under control up to now this does not necessarily mean that it will remain so indefinitely, especially in the absence of an adequate international initiative which face up to the numerous problems left to be solved.

With regard to these problems, we must remember some hard facts.

— The quantity of fissile material present in Russia is extremely high and the systems of protection, control and accountability of nuclear material need significant improvement.
— The (illegal) transport of fissile material in quantities which are significant from the point of view of nuclear proliferation can take place with a relatively simple level of precautions, and on the whole fairly easily.
— In the world there exist countries (or criminal groups) which are, in principle, interested in buying nuclear fissile material.

Furthermore in the Russia of today (as in other parts of the world) the illegal or semi-legal traffic of various commodities is a widespread phenomenon which does not seem destined to decrease in the near future. The situation regarding Russia's nuclear activities is further complicated by the difficult economic conditions in which technicians and scientists are forced to live. This contrasts sharply with the relatively privileged conditions, that they enjoyed not too long ago.

Up to now the Russian authorities have shown that they are able to cope with such a difficult and complex situation. The logical reply from the more industrialised countries should have been the development of a comprehensive initiative to support the programmes for the control of the nuclear material in Russia.

Among the steps to be taken, one should enlist specific agreements allowing the drawing of a detailed "nuclear map" of today's nuclear superpowers with a list of all the locations, the quantities, the type of pits, the warheads, the parts of warheads. The separated fissile material, and allowing a wide-ranging system of international inspections.

Another reasonable initiative is the development of scientific, technological, and economic cooperation between scientific institutions of the more industrialised countries and Russian nuclear research and production facilities.

All these initiatives have been, at least partially, carried ahead, but the rhythm and the dimension has been insufficient, as has also been the financial commitment of the more industrialised countries, with respect to the gravity of the situation. The whole affair has been accompanied by political-bureaucratic obstacles of a various nature, in particular with regards to the definition of the agreement concerning the *exchange of data* in order to put together the "nuclear maps", which we mentioned above.

The United States, being by far the Western country which has committed herself more, has in five years (from 1992 to 1996) destined an overall sum of about 530 million dollars to programmes regarding the problems of the nuclear facilities in Russia[14], which is about 1/300th of her current expenditure for her own military nuclear programmes (33 billion dollors in 1995).

The obvious question is this: for global security, is it more important to have an extra submarine with nuclear missiles patrolling the oceans, or rather to develop initiatives which prevent the transfer of fissile material to terrorist groups or countries interested in acquiring atomic weapons ?

Nuclear Proliferation

While discussing about the illegal traffic of nuclear material, we are naturally led to consider the countries which are potentially interested in buying this material. In other words who are the potential proliferators and how high is the probability of a successful nuclear proliferation?

The scientists who participated in the first phase of the construction of the atomic bomb and several American politicians of the immediate post-war era, thought that, in the absence of an international control over atomic energy, nuclear weapons would have spread to many countries, even though no one foresaw the building of a global number of nuclear arms, remotely comparable to the present one.

The number of nuclear countries has instead remained minimal with respect to the initial pessimistic predictions. A true regime of nuclear non-proliferation has been established and has shown notable stability during all these years.

There are five nuclear countries officially recognised as such (USA, Russia, France, Great Britain, China) and there are three *unofficial* nuclear countries (Israel, India, Pakistan).

All or almost all the countries in the world, with the notable exception of the three undeclared countries, have adhered to the Non-Proliferation Treaty (NPT). This treaty was extended indefinitely in May 1995.

Therefore the first element which has contributed strongly to the containment of nuclear proliferation is the NPT Treaty, that together with the treaties which establish nuclear weapons free zones in Latin America (Treaty of Tiatelolco), in Oceania (Rarotonga), and (in the near future) in Africa (Pelindaba), provide a clear framework of international legislation.

Other elements that have contributed to the strengthening of the non-proliferation regime are the significant costs tied to the construction and maintenance of nuclear arms, and the difficulties and costs of the acquisition of fissile material.

These last elements become even more important if compared to the evident uselessness of nuclear weapons. Nuclear weapons have never been used in a conflict since the Second World War. Even the simple presence of nuclear weapons. has shown itself to be a marginal element in the principal conflicts, local or not, which have developed since 1945. Think of Korea, of Vietnam, of the Falklands, of the Middle-East wars etc.

In recent times a strong push towards the strengthening of the non-proliferation regime has been given by the disarmament steps taken by the nuclear powers.

The dramatic reduction of the emphasis given to nuclear weapons by the USA and Russia in the years 1987–94, the unilateral and bilateral initiatives for nuclear disarmament and the relative treaties, the treaty (in preparation) on the total prohibition of nuclear experiments (CTBT) are all elements which have contributed and contribute to diminishing the role of nuclear weapons in international politics.

Lastly the non-proliferation regime was enhanced by the failure or voluntary abandonment of various attempts of nuclear proliferation.

South Africa built 6 fission bombs of the *gun-assembly* type that they successively dismantled. Brazil and Argentina have abandoned their nuclear projects, North Korea has been persuaded to abandon its attempts to acquire fissile material for nuclear weapons, and Iraq has been forced to abandon similar attempts.

We must now consider the opposite type of motivations, that is the ones that can encourage a country to try to acquire nuclear arms, even in violation of the NPT Treaty.

The possession of nuclear weapons has for a long time been associated with a misplaced international prestige.

Possessing nuclear weapons means belonging to an exclusive club of countries, which includes, for example, all the permanent members of the United Nations Security Council.

Another aspect we must consider is that acquiring nuclear weapons can have contrasting effects on nearby countries: it can frighten them or push them to nuclear emulation. We can understand, however, that a country, surrounded by a hostile environment or one perceived as being such, might think of playing the nuclear card. The cases of the clandestine nuclear powers (Israel, India and Pakistan) fit clearly this scheme.

Countries that are today considered to be potential proliferators are some countries of the Middle East. Among these the name which appears most frequently is that of Iran.

In fact the position of Iran on the nuclear problem is not clearly supportive of the non-proliferation regime. Observers, even of different opinions, all agree that Iran could try to acquire a sufficient quantity of fissile material for a few nuclear warheads.[15]

The open or hidden desire to violate the non-proliferation regime may be influenced by the overall attitude of all the nuclear powers, and of the United States and Russia in particular, regarding nuclear disarmament.

If nuclear disarmament proceeds quickly, if the emphasis on nuclear weapons is rapidly reduced until it disappears, then the general attitude of the international public opinion will be less and less prepared to condone nuclear proliferation and the prestige connected to the possession of nuclear arms will fade away.

If instead the disarmament process should slow down, and if the emphasis on the nuclear component of defense will be kept at the present level, if not increased, then the net effect on the non-proliferation regime can be the most unwelcome one.

There are various elements which suggest that the second hypothesis, the most unfavourable one, is not the most unlikely.

Nuclear Strategies after the Cold War

The problem we would like to discuss briefly may be formulated in a very simple way: how many nuclear weapons will be left after the implementation of the present disarmament agreements and what will be the role assigned to the nuclear weapons which remain?

The connection with the problem of nuclear proliferation is clear: if the nuclear powers plan to maintain in the foreseeable future, a high number of nuclear weapons and an aggressive nuclear strategy, then they are also planning to send the wrong message to potential proliferators.

A recent estimate has established that 7500 American nuclear warheads (including the *hedge*) could be allowed after the implementation of Start II (the ratification of which by the Russian Parliament, we must remember, presents many difficulties). An equal number of Russian warheads plus the warheads of minor nuclear countries would bring the total number of warheads to well over 16000. This is the amount of nuclear weapons that are likely to survive the present disarmament agreements.

So there is the serious risk that the emphasis on the role of nuclear weapons will be maintained high by the large number of weapons that will be kept intact in the arsenals of the major powers, by the reluctance of the 'minor' nuclear powers to join the disarmament process and by reluctance of all the nuclear powers to change their ideas and procedures concerning the possible use of nuclear weapons in conflict.

With regards to the nuclear strategy of the major nuclear powers we observe:

— The American *nuclear posture review* of 1995 does not significantly modify the conditions of use for nuclear arms. It explicitly plans for an increase of operational warheads if the political conditions in Russia should deteriorate.

— The American Congress sustains the necessity for a revision of the ABM Treaty and proposes a larger expenditure in the anti-missile defense systems

— Russia has restricted the policy of nuclear *no-first-use* so as to make it practically irrelevant. Russia declares that she will not initiate the use of nuclear weapons, only if the opponent will be a non-nuclear country that does not belong to any military alliance which includes nuclear countries.

— Russia is making the old NATO doctrine its own on the role of nuclear arms to countervail a possible inferiority in the field of conventional weapons. In particular some Russian political and military leaders speaks insistently about a "new" role for tactical nuclear weapons, the removal of which had been announced with separate unilateral initiatives by Presidents Bush and Gorbachev, but which had never been established by treaty.

Lastly, a question of the greatest importance is the problem of the alert-levels of the nuclear delivery systems. Maintaining a good part of the nuclear delivery systems at maximum alert levels was a characteristic of the entire period of the Cold War. This meant that ground-based missiles could be launched not after an enemy attack, but simply after the warning that enemy missiles had been launched and were on their way to their targets. A number of the submarines equipped with nuclear missiles have been kept in continuous navigation, ready to launch their missiles at the shortest notice. Similar levels of alert were kept by nuclear bombers.

After the end of the Cold War, the levels of alert were changed, but only slightly.[17] This is a most relevant fact, since the disarmament process can proceed by two different tracks: by decreasing the number of weapons and by rendering the use of the existing weapons less and less accessible.

Keeping nuclear weapons in the *ready-to-use* condition not only continuously present the risk of a *war by mistake*, but also directly establishes a significant barrier to the further development of nuclear disarmament.

Conclusions

The post Cold-War world is not yet a world free of nuclear weapons, and there are significant elements which oppose the elimination of these weapons of mass destruction.

The role assigned to nuclear weapons by the super powers is still one of great importance. This not only does not eliminate the risk of a global nuclear conflict, but also has a significant effect on the incentives and risks of nuclear proliferation.

The presence of large quantities of fissile material available for nuclear weapons and the problems of the control of this material, could in the future facilitate nuclear proliferation, or the construction of rudimental nuclear weapons by criminal organisations.

Even if the recent progress in the field of nuclear disarmament has certainly been historic, the tasks of the international community with regards to nuclear weapons are far from being finished.

The decisions concerning the procedures and methods for the disposal of the excess-weapon fissile material and in particular of the weapon-grade plutonium, is one the problems to be faced, but it is far from being the only one.

Among the many other problems we have mentioned or schematically discussed, there are the ones concerning the control, the accountability of the existing fissile material, the reduction of nuclear warheads (reaching and going beyond the limits fixed by the current disarmament-agreements), the reduction of the overall emphasis given to nuclear weapons, the reduction of the levels of alert. We also mentioned the involvement of the minor nuclear powers in the disarmament process. All these aspects will play an important role in the future evolution of the non-proliferation regime.

References

1. Bull. Atom. Scient. Nov. Dec. 1995
2. T. Cochran, R.S. Norris, O.A. Bukharin: *Making the Russian Bomb*; Westview Press Boulder Co. (1995)
3. Yomiuri Shimbun 17.11.1991 quoted in M. De Andreis, F. Calogero: *The Soviet Nuclear Legacy*; Oxford University Press, Oxford (1995)
4. US DOE: *Plutonium the First 50 years*, Washington DC (Feb. 1996)
5. A. S. Diakov contribution presented at this conference (Como 18.03.1966)
6. Estimate made by T. Cochran in US *Inventories of Nuclear Weapons and Weapon Usable Fissile Material* NRDC 25.09.1995
7. See T. Cochran, R.S. Norris, O. Bucharin (loc. cit.)

8. "Difficult to access" does not mean "absolutely impossible to access". In particular the custody of nuclear warheads is a subject which deserves some extended discussion.
9. T. Cochran in *US inventories of Nuclear Weapons and Weapon-Usable Fissile Materials* loc. cit
10. See G. Allison, O. Coté, R. Falkenrath, S. Miller *Avoiding Nuclear Anarchy*: MIT Press, 1996
11. R.S. Norris in *Arms Control Today*
12. For a more extended list of the episodes of nuclear smuggling, see the report by the Director of the CIA, J. Deutch, to the Permanent Subcommittee on Investigations of the Committee of the American Senate on Government Affairs (20–3–1996)
13. The table has been compiled on the basis of information reported by T. Cochran (Conference held at Villa Olmo, Como, 05.07.1995), W. Potter (to be published in the Proceedings of the USPID Castiglioncello Conference, 28.09/01.10.1995), G. Allison and others, loc.cit.
14. G. Allison and others, loc. cit.
15. See for example the contribution on Iran in the Proceedings of the *USPID Castiglioncello Conference*, 28–9/01-10-1995
16. See R.S. Norris, in the proceedings of the USPID *Castiglioncello Conference* 28-09/01-10-1995
17. See B.G. Blair, *Global Zero Alert for Nuclear Forces*, Brookings Institution, Washington D.C. (1995)

Address:

PAOLO COTTA-RAMUSINO
Secretary, USPID
Department of Physics
University of Milan
Italy

2

Principal Viewpoint on Nuclear Energy Development in Japan

SADAO KIJIMA

Introduction

Electric power generation by Light Water Reactor has already covered a good deal of the electricity demand in most of developed countries including Japan. Japan considers that the utilisation of Plutonium is also very important in terms of securing the energy resources in the future.

The early part of this paper briefly describes the potential role that Uranium can play as an energy resource. This paper presents Japan's principal viewpoint for Plutonium is also very important in terms of securing the energy resource. This paper also presents Japan's principal viewpoint for Plutonium utilisation taking account of non-proliferation aspects.

THE ROLE OF URANIUM AS AN ENERGY RESOURCE IN THE 21ST CENTURY AND AFTERWARDS

The framework of nuclear energy development programme in Japan has been provided by "The long-term programme for research, development and utilisation of nuclear energy", which Japan Atomic Energy Commission (JAEC) revises about every 5 years. The newest revision was issued in June, 1994 through almost 2 years of debate by a number of people who have various opinions including anti-nuclear. The author itself contributed to the debate, and following quotations from the long-term programme are consistent with its own opinion.

This newest long-term programme describes the importance of nuclear energy as follows:

"World energy consumption has been increasing ever since occurrence of the industrial revolution. In terms of oil consumption, it was only about 100 million tons a year in the middle of the 19th Century, but by 1990 it reached 8,000 million tons. The world population doubled in the last 40 years, but over the same period the world energy consumption increased more than 4 times, which means that per capita energy consumption has also doubled. Energy consumption is closely geared to factors such as level of economic development and life quality. With the advent after Second World War of a society depending on mass production and mass consumption, there has been a sharp increase in energy consumption. For a time after the two oil crises energy consumption did not increase as much as it had before, but after that it resumed its upward trend on account of lower oil princes and other factors, and that trend still continuing."

Figure 1 illustrates the predictions of the global primary energy consumption which have been carried out so far. Summarising these predictions, the global primary energy demand is 13–24 billion tons oil equivalent (TOE) in 2050, and 18~41 TOE in 2100.

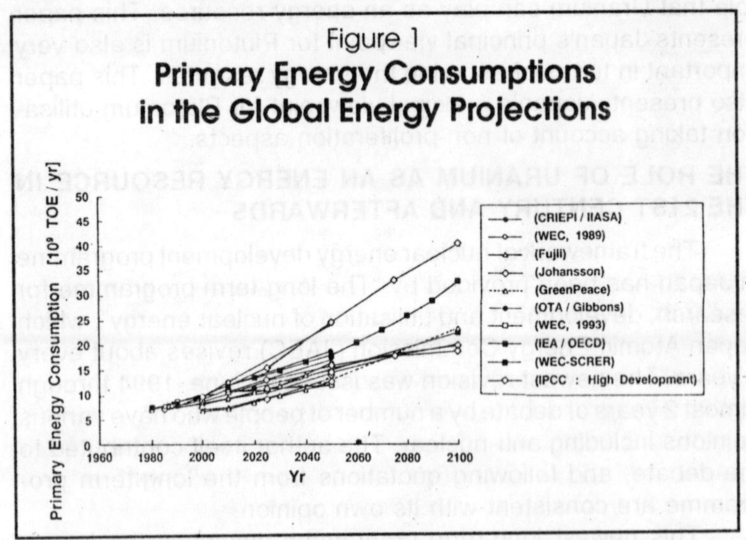

Figure 1
Primary Energy Consumptions in the Global Energy Projections

Figure 2 shows the comparison of the cumulative oil consumption and the deposits. This indicates that one trillion barrels of the present proven oil resources are to be exhausted between

2020 and 2040, and even the 2.2 trillion barrels of the ultimate oil resources are to be exhausted by 2100 at latest.

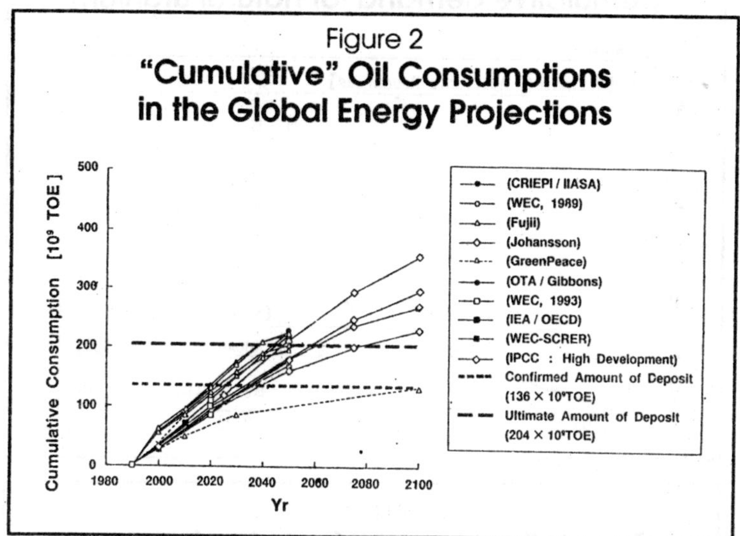

Figure 2
"Cumulative" Oil Consumptions in the Global Energy Projections

It is said that natural gas will also have been exhausted by around 2070, as long as the great deal of new resources are not discovered. These mean that the ultimate available fossil resources except for coal will have been used up by the middle of the 21st century.

Then, we carried out analyses on a long-term scenario that nuclear energy compensates the lack of fossil resources. Figure 3 gives the cumulative demand of natural Uranium corresponding to a model case assuming that WEC's energy outlook until 2020 will keep the same pace after 2020. It is also assumed that all of nuclear energy is covered by conventional LWR. This results also indicate the known natural uranium also will have been used up by around 2050, and the ultimate Uranium resources cannot cover the very long-term demand. Figure 4 illustrates the results of an example analysis which performed to clarify the effects of FBR introduction to the demand reduction of natural Uranium. It is known that the breeding ratio and the out-of-core time of spent fuels are the key factors, these results indicated that the very long-term. Uranium demand can be covered by the ultimate resources in case the breeding ratio is 1.2 and out-of-core time of spent fuel is two years.

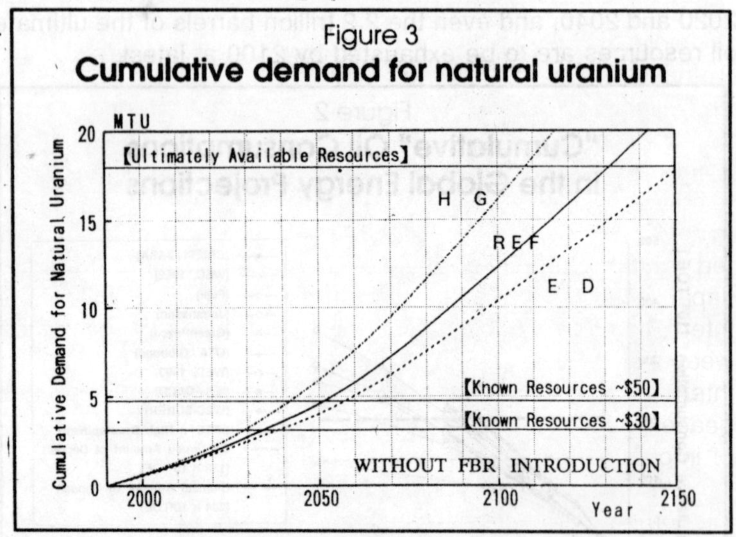

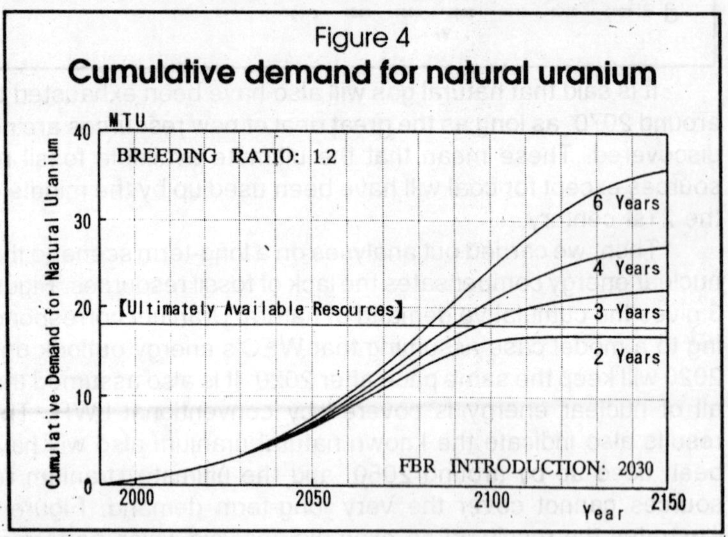

JAPAN'S PRINCIPAL VIEWPOINT FOR PLUTONIUM UTILISATION

Peaceful Use

It is Japan's grand rule that the utilisation of nuclear energy have to strictly limited only for peaceful use. Relating to this principle, the long-term programme of Japan Atomic Energy Commission describes as followings:

"Japan today enjoys the fruits of its past efforts in development and utilisation of nuclear energy with consistent limitation to peaceful purposes, and it will be important in the future to hold on to such posture of peaceful use of nuclear energy and to develop nuclear energy policy appropriate to a nation committed to peaceful use of nuclear energy. That being the case, Japan intends to redouble its efforts with respect to winning international confidence concerning non-proliferation of nuclear weapons, development of technology for peaceful use, acting internationally in a way that befits and advanced nation in peaceful use of nuclear energy and transparency and availability of information.

With respect to obtaining international confidence regarding non-proliferation of nuclear weapons, Japan has worked for maintenance and reinforcement of the system comprised by the Non-Proliferation Treaty and the IAEA safeguards based on it as the important international framework which guarantees compatibility of peaceful use and non-proliferation of nuclear weapons as well as making efforts of its own. At the same time it is important for Japan to clearly express once more its firm determination to limit itself to peaceful use and to show that there is absolutely no possibility of developing nuclear weapons in terms of either systems or actual conditions by refraining from possession of any technology related to nuclear weapons."

Information Disclosure

This programme also emphasises the importance of information disclosure in order to accomplish the above-mentioned objectives as followings:

"The Atomic Energy Basic Law stipulates the requirement to make public the results of research on development and utilisation of nuclear energy in order to ensure that it will be for peaceful purposes, and efforts have always been made for such transparency in the past. Those efforts to make information on nuclear energy public will be continued in the future on the basis of full realisation of the great importance of transparency to ensuring only peaceful use. Furthermore, it is important to always bear in mind the need to formulate nuclear energy policy on the basis of what the Japanese public wants in view of how closely peaceful use of nuclear energy relates to the daily lives of the Japanese people. Accordingly, improved measures will be sought for obtaining understanding and cooperation on the part of the general public through transparency and availability of information so as to reassure people about the safety of nuclear energy."

R&D Programme for Establishment of Fuel Cycle

Based on these principles, Japan is carrying out a series of R&D programmes in order to establish the fuel cycle so that the future energy security can be assured. Our master plan is expressed as follows:

"As a country practically without its own energy resources, it is indispensable for Japan to plan for its energy security on the basis of the future outlook in order to be able to continue to maintain and develop its economic and social activity. Uranium resources are limited just as fossil fuel resources are, and it cannot be denied that if the present situation with use mainly of light water reactors continues, uranium supply-demand relation will be tight by around the middle of the next century. Japan intends to guarantee its future energy security by steadily carrying forward research and development efforts aimed at future commercial commissioning of nuclear fuel recycling Involving reprocessing of spent fuel and recovery of plutonium, uranium, etc. for reuse as nuclear fuel. Another reason for doing so is that recycling of nuclear fuel is also meaningful in terms of sparing resources and the environment and improving management of radioactive waste. Specifically, with the basic idea of first having a long period during which fast breeder reactors, which will make it possible to raise the efficiency of utilisation of uranium resources by leaps and bounds since they produce more nuclear fuel than they consume, are used alongside light water reactors take over as the mainstream of nuclear power generation, research and development will be steadily carried forward in stages on the basis of cooperation between government and the private sector for the purpose of establishing a technological system of nuclear fuel recycling based on fast breeder reactors so as to be able to commission fast breeders commercially by about the year 2030 after first passing through the prototype reactor and the demonstration reactor stages. Furthermore, from the standpoint of establishing the comprehensive technological system concerning use of plutonium that will be needed in the future age of fast breeder reactors and long-term overall improvement of the economic efficiency of nuclear fuel recycling, it is important first to carry out nuclear fuel recycling on a certain scale. Therefore, it is intended to build reprocessing plants on a commercial scale to gain experience in operation and bring about nuclear fuel recycling based on existing light water reactors. Regarding the economic efficiency the nuclear fuel cycle, although at the present time it is estimated that use of MOX fuel by light water reactors will be somewhat more costly than direct disposal of the spent fuel, there is substantially no difference if one considers total power generation cost. Therefore it is intended to work for improvement of economic efficiency from a long-term perspective, including standardisation of fuel specifications".

No Surplus Plutonium

Japan realises that it is inevitable to obtain the worldwide understanding that we never have intention to take any action threatening the nuclear non-proliferation. It is one of counter-measures that Japan possesses no surplus Plutonium. This fact explained in the next chapter.

Plutonium from Weapon

Japan's principle on the treatment of Plutonium from nuclear weapon is as follows: "If, now that the Cold War is over, a shift takes place in international society from emphasis on military concerns to emphasis on the quality of life, nuclear energy, too, can be expected to play an important role in improving the quality of life (peaceful use). At the present time the matter of what to do with the nuclear materials resulting from dismantling of the nuclear weapons left behind by the Cold War is a very important international problem. The basic consideration should be that of preventing those nuclear materials from being used for military purposes again, and the idea has occurred to some that a good way of ensuring that would be to use them as fuel in nuclear power generation. That means we will come across a happening symbolising the current world trends that the technology for peaceful use of nuclear energy. Looking back on the history of nuclear energy, one sees that under the Cold War structure one aspect of nuclear energy was unfortunately maintenance of order in international society by its use as military deterrent, but it is to be earnestly hoped that in the 21st Century nuclear energy will cast off its military yoke to be used solely for the peaceful purpose of supporting the stability and development of international society so as to contribute to the good of mankind, as it should have been from the outset".

"While, on the one hand, progress is being made between the United States and Russia in nuclear disarmament as a result of the end of the Cold War, on the other hand, management and control of nuclear weapons in the ex-Soviet Union has become less unstable, and there is now anxiety about possible new proliferation of nuclear weapons in unstable regions.

Regarding fears of proliferation of nuclear weapons in the ex-Soviet Union, it is important to proceed with nuclear disarmament in a safe and steady manner. As for, particular, the plutonium and other nuclear materials resulting from dismantling of accountancy so as to ensure that they will not be reused for nuclear weapons. Although such nuclear materials should be appropriately disposed of by the countries concerned as their own problem, it is also important that other countries collaborate in devising long-term solutions to that problem. On the basis of awareness of the significance of doing so from the viewpoint of promoting nuclear disarmament and prevention of proliferation of nuclear weapons, Japan intends to contribute to storage and other measures adopted with respect to such nuclear materials by the countries concerned out of consideration of its basic standpoint of promoting utilisation of nuclear energy that is limited to peaceful purposes".

APPROXIMATE ESTIMATION OF SUPPLY AND DEMAND OF PLUTONIUM IN JAPAN

Revised Reflecting the Cancellation of Demonstration ATR

The production of the supply and demand of Plutonium depends on the progress of related nuclear development programmes. Followings are results of approximate estimations of supply and demand of Plutonium in Japan which were carried out based on the development plan as of August 25, 1996. These figures are to be revised in case such as the schedules of related programme change.

All the Plutonium is under IAEA's safeguard, and it is always being confirmed that it is never utilised for any other objective except for the peaceful use. The fraction of fissile is reprocessed Plutonium is 60–70%, and following estimations summarised in Table 1 to 5, were carried out for the fissile Plutonium.

1. Estimation of Supply/Demand of Domestically Reprocessed Pu

1) 1994–1999

Table 1 : Annual Balance

1. Demand	2. Supply
"JOYO" (Experimental FBR) "MONJU" (Prototype FBR) "FUGEN" (Prototype ATR) etc.	Tokai Reprocessing Plant
0.6 ton/yr	0 4 ton/yr

Table 2 : Cumulative Balance

1. Cumulative Domestic Demand (1994–1999)	2. Cumulative Domestic Supply (1994–1999)
"JOYO" (Experimental FBR) "MONJU" (Prototype FBR) "FUGEN" (Prototype ATR) etc.	Tokai Reprocessing Plant and Returned Plutonium from Overseas Reprocessing Plant
0.6 ton	0.4 ton

2) 2000–2010

Table 3 : Annual Balance

1. Demand	2. Supply
"MONJU" etc. 0.6 ton/yr	Rokkasyo Reprocessing Plant 4.8 ton/yr
Demonstration FBR 0.7 ton/yr	Tokai Reprocessing Plant *0.2 ton/yr*
Full MOX ABWR 1.1 ton/yr	
Thermal MOX 2.6 ton/yr	
Total 5 ton/yr	Total 5.0 ton/yr

Table 4 : Cumulative Balance

1. Cumulative Domestic Demand	2. Cumulative Domestic Supply
(2000–2010) "JOYO", "MONJU", "FUGEN", "DFBR" 10–15 ton (*) Full MOX ABWR Thermal MOX 25–30 ton Total 35–45 ton	(2000–2010) Rokkasyo Reprocessing Plant *Tokai Reprocessing Plant* Total 35–45 ton

(*) The amount of Plutonium required to operate reactors for R&D such as "JOYO", "MONJU", and so forth is approximately 15 ton. In case that the Plutonium supply from Rokkasyo reprocessing plan decreases, the period when the domestic supply cannot cover the domestic demand may occur. In such a situation, at least 10 ton of plutonium is supplied from the Rokkasyo plant, and remain is covered by the Plutonium returned from the overseas reprocessing plant.

(2) Plutonium Reprocessed in Overseas Plants

Table 5

1. Cumulative Amount	2. Supply
by around 2010 30 ton.	Basically, the Plutonium reprocessed in the overseas plants are utilised in the full MOX ABWR or LWR, after they are fabricated to MOX fuel in overseas plants and transported. (However, in case the shortage of Plutonium for R&D reactors occurs before the normal operation of Rokkasyo plant, some of them are utilised to compensate it).

CONCLUSION

The development of Fast Breeder Reactors, which is very useful to utilise the energy of Plutonium and can play a significant role to cover the energy demand in the future, is facing to several difficulties. Although France is carrying out its best effort to accomplish the full power operation of Super Phenix I, they had to meet some troubles to overcome in the course of developing it. We also heard that Russian programme is being involved in financial problems. And, FBR, development programme in Japan is also undergoing the difficulties due to secondary sodium leak of prototype FBR "MONJU". However, FBR is a every important option that we should not abandon, and we hope that very developed countries in the field of FBRs will collaborate to establish their utilisation technologies.

Address:

YASUKO KIJIMA
CRIEP
1-6-1 Ohtemachi Chiyoda-ku
Tokyo
Japan

3
Nuclear Power of Russia

O. FAVORSKY
V. KAGRAMANIAN
L. RYABEV

The development of the Russian fuel production and power generation complex is supposed to meet the requirement of the sustainability of power generation in Russia, and to some extent that in Europe, being in compliance with traditions and relations formed thus far. The "Energy Strategy of Russia" adopted by the RF Government in December 1994, defines the ways for the most efficient use of resources and potentials of the complex through its dynamic development and adequate solutions for economic, ecological, safety problems faced. It considers the aforementioned requirement as well. And the development of nuclear power in Russia is an important element to ensure the realisation of this requirement.

At the moment, 9 of the 15 nuclear plants operating in the CIS are located in Russia, with total capacity of 21.242 MWe. By the end of 1995, 29 reactors were operated, including 13 VVER type (6 VVER-440 and 7-VVER-1000); 11 RBMK, 1 fast reactor BN-600 and 4 small graphite moderated, water cooled district heating reactors BN-600 at Bilibino NPP.

All reactors of NPPs are operating reliably at base load, except for variable loading of reactors at Bilibino NPP, which is used to meet both power and heat demand of that outlying area. The thirteen VVER recorded an average annual load factor of 52.7 per cent with a lifetime average to the end 1995 of 65.8 per cent. The eleven RBMKs recorded an average load factor for 1995 of 54.8 and an average lifetime load factor of 71.7 per cent to end 1995. The fast reactor recorded a load factor of 70.4 per cent in 1995 and lifetime figure of 69.4 per cent.

In 1993, the Russian NPPs produced 12.48 per cent of the total amount generated in the country, while in 1994 they produced 11.3 per cent of the total production. To the end of 1995 the Russian units have generated a total of 1713.51 TWh over some 358 reactor-years of operation. The average performance data for all nuclear power plants are presented in Table 1.

Table—1 : Characteristics of Nuclear Power Plants Operating in Russia

General Data	YEAR			
	1991	*1992*	*1993*	*1994*
Installed Capacity (MWe)	20242.0	20242.0	21242.0	21242.0
Electricity Output (Billion kWh)	120.0	119.6	119.2	97.8
Share in Total Output (%)	11.4	11.8	12.7	11.3
Load Factor of Installed Capacity (%)	67.7	67.28	57.22	52.6
Number of Safety System shutdowns Per Unit	1.04	1.32	0.79	0.38
Operational Breakdowns:				
– to be reported according to INES	38.0	31.0	29.0	13.0
– caused by personnel errors	48.0	33.0	48.0	52.0
–Total (incl. other Breakdowns)	172.0	205.0	154.0	127.0

Ensuring safe operation of facilities is the primary goal of nuclear power industry. To achieve this goal, long-term measures are being pursued and periodically reviewed in the light of actual results and the relevant plans for reconstruction and modernisation. At the same time being, the first stage of work has been completed. Information on non-standard operation events at NPPs and their assessment using the IAEA International Nuclear Events Scale (INES) reflects the improvement in performance of operating plants and their comparability with that in countries with developed nuclear power industry. To fully implement these projects, the capital investment of $5.4 Billion is required by the year 2000.

It is interesting to mention also that an idea of the earlier decommissioning of first generation units to improve the Russian nuclear power safety was not justified by the Russian-American joint research carried out in 1994–95 on the alternative options for the development of power industry. The study was done in the frame of the agreement between Russian Prime Minister and US Vice-President.

Normal operation of NPPs does not result in radioactive contamination of environment above the background, irrespective of the distance from the plant. Better ecological impacts of reliably operating NPPs belongs to key arguments, when compared to fossil fuels.

All operational NPPs in Russia received ecological certificates approved by regional environment monitoring authorities. These documents indicating major factors of the plants' environmental impacts are the main tool to ensure ecological safety of units. As a result of measures taken, hazardous environmental impacts of nuclear fuel fabrication facilities are currently estimated as insignificant.

Within the framework of the "Energy Strategy for Russia" during the restructuring of the Russian fuel and energy complex in the next 10 to 15 years steps have been determined to develop nuclear power sector. The necessary level of development by 2010 is 125 billion KWh (22 GWe of installed capacity), i.e. 11 per cent of total domestic electricity generation. Maximum level of power generation at NPPs in 2010 could be 160 billion KWh (28 GWe of installed capacity), i.e. 13 per cent of the total output.

The development of the nuclear power sector is economically justified due to the following considerations:

a) Nuclear power industry is the only sector of the energy complex which does not require mining of natural resources in the near future. In contrast to conventional power production based on fossil fuels, where one of the most important problems is securing of fuel supplies (which means vast capital investment), the nuclear power programme up to 2010 can be carried out using readily available and cheap stocks of uranium, 1.5–2.0 times in excess of demand.

Problems in fuel supply for nuclear power industry may arise only by the year 2030, assuming a sharp increase in the installed capacity. For example, a growth in capacity by 2030, which would provide 30 per cent of total power generated in Russia by NPPs, could result in exhaustion of all uranium stock available within 30–35 years. The outcome could be found if new uranium deposits were discovered in Russia or ex-weapons uranium (WU) was used; another option would be the introduction of closed fuel cycle using civil and ex-weapons plutonium (WPu); or development of nuclear power industry based on thorium. An approximate estimate shows that the total energy potential of ex-weapons uranium to be used in thermal reactors and ex-weapons plutonium to be used in fast reactors could correspond in the long run to an electricity generation of 12–14 trillion KWh, i.e., 12–14 as much as annual output in 1993, and thus save around 3.5 trillion cubic meters of natural gas. That would create a realistic possibility of guaranteeing the export of natural gas to Europe. According to Western estimates by 2000 the export volume of natural gas from Russia should amount to 131 billion cubic meters, by 2010 – to 184 billion cubic meters, and by 2020 – to 204 billion cubic meters.

b) Analysis of nuclear fuel production capacity requirements for various development options in the nuclear power sector up to 2010 indicates that production capacity is also enough to assure any of the options and there will still maintain significant production reserves. Furthermore, average load in work capacity of nuclear fuel fabrication facilities ranges from 10 to 76 per cent.

c) Analysis of the power industry construction companies and regional construction entities assumes that the powerful production infrastructure formed to date, including construction enterprises and companies dealing with installation and assembling, is capable of achieving an annual work level of $120–150 million per unit, which makes the planned programme of nuclear power industry development up to 2010 totally feasible.

The major companies involved in the machine manufacturing for nuclear power industry are: The coordination centre for nuclear power machine construction at Minatom of Russia "Energy Machine Manufacturing Corporation Ltd.", "TENMA" Ltd., MGO "Energomash", "Atomonergomash Ltd.", "Izhorsky Zavod Ltd.", "Podolsk Machine Building Works Ltd.", and others.

Assuming the supply of equipment by main unit manufacturing plants and the maximum development alternative for the nuclear power industry from 1997 to 2010, the average load of "Izhorsky Zavod Ltd" would be 1.38 units per year with an average capacity of 782 MWe per year (29% of production capacity), and that "Atommash Ltd" would be 1.62 units per year with an average capacity of 1.187 MWe per year (40% of production capacity). The load of "Leningrad Metalworks" Ltd. would be 75–100 per cent of production capacity. Units of small capacity require an insignificant part of the main programme in terms of steel and labour, and the programme of their planned commissioning will be secured by the production capacities "Izhorsky Zavod" Ltd. and joint venture "Podolsk Machine Building Works" Ltd.

d) Highly developed R&D facilities in the sector ensure the implementation of the programme of nuclear power industry development up to 2010 and thereafter. The programme is based on new generation reactor projects, supported by a highly developed network of test facilities and centres. Among nuclear reactors for installation at nuclear power plants, priority is given to:

— VVER-640 (Project NP-500), thermal capacity of 1800 MW, the pilot unit to be installed at Sosnovy Bor in Leningrad district;
— VVER-1000 (Projects NP-1000 and NP-1100), thermal capacity of 3000-3300 MW, the pilot unit to be installed at the Novovoronezh nuclear power plant.

A whole range of economic and technical factors will determine the priority within this group for realisation on schedule of intermediate capacity reactor projects based on rich experience gained with VVER-440 and VVER-1000 reactors. This indicates an evolution path for the improvement of nuclear power plants as shown by the world experience. The decrease of reactor capacity and the energy density of the core makes it possible to considerably improve safety without losing competitiveness.

The list of perspective projects includes fast reactors (BN-800) to use civil and ex-weapons plutonium and to burn long-live radio-toxic minor actinides.

To supply energy to outlying regions and industrial districts and cities, a range of projects for small size nuclear power installations (2.5–150 MWh) are planned—the so-called small scale power production.

e) Analysis of the competitiveness of the planned nuclear power plants power generating facilities at the Russian energy market with fossil fuel ones shows the following:

— if capital investment in NPP exceeds that in combined cycle power plants by maximum 1.5 times, and if it is 15–20 per cent higher than in coal-fired plants, then the nuclear options will be the most efficient in all regions of Russia;

— to maintain the competitiveness of NPPs against thermal steam-gas combined cycle power plants (CCPPs), capital investment in nuclear plants should not exceed the corresponding amounts for thermal CCPPs by more than 1.8 times for the North West and Central Russia, by more than 1.9 times for the Northern Caucasus, by more than 1.7 times for the Middle Volga region and by more than 1.6 times for the Urals;

— to maintain the competitiveness of nuclear power plants against coal-fired plants, capital investment in nuclear plants should not exceed the corresponding amounts for coal-fired plants by more than 1.08 times for the Urals, by more than 1.1 times for the Middle Volga, by more than 2.1 times for the North West and Far East and by more than 1.25 times for Central Russia.

As of the end of 1994, the average cost of power generation at operating nuclear power plants was 26 per cent less than the average cost at the seven largest thermal power plants in Russia. The mentioned Russian-American joint research also confirms the results presented above.

Under the present situation of declining production of energy resources (especially of oil and gas) and their inevitable exhaustion in the near future, nuclear power industry can be considered as a guarantee of energy security in Russia and Europe.

Considering the development of the nuclear power industry in this aspect, it is possible to distinguish two main stages:

— maintenance of nuclear power generation capacity in the next 10–15 years at the current level of electricity production by reconstruction and technical upgrading of the units, by completion of the construction of new nuclear power plants and the development and realisation of projects of new generation, high safety nuclear power plants;

— early establishment of the necessary pre-conditions for a significant future increase in the nuclear contribution to the total energy balance of the country and especially for a large-scale development of nuclear power industry after 2010, with a nuclear share in the which produce around 30 per cent to 35 per cent of total power generated in the country, 40 per cent–50 per cent for European part.

The accomplishment of above-referenced works will help resolve the following strategic tasks aimed at power independence for Russia and Europe in the near future:

1) the diversification of national energy sources and saving of fuel resources to ensure the necessary exports (the share of gas used in Russian electricity production is more than 60%, whereas in most Western countries it is between 25% and 30%). Even today it is possible to release for exports the additional energy resources by a higher utilisation of operational capacities of the nuclear power plants available. For example, only by maintaining 1994 electricity production by Russian NPPs at the level of 1993, it would have been quite possible to generate additionally 22 billion KWh of electricity, and consequently save about 5.5 billion cubic meters of natural gas.

The annual saving of natural gas of nuclear power could in 1995 be 25 billion cubic meters of which, at present export prices, amounts to $2 billion. (The whole effect for the entire period utilising nuclear power plants could be 440 billion meters of gas or $35 billion at current world market gas prices.);

2) the establishment of realistic conditions for the increasing of electricity export to European countries. Even in the near future, there is a prospective increase in the export of electricity and in the realisation of investment projects paid back by exports of electricity, for the support of construction of new generation facilities for electricity export, in particular, nuclear power plants. In this connection, great importance is given to the availability of surplus resources primarily in the energy systems directly adjacent to the borders of Russia. The North West energy system is an example where around 50 per cent of electricity is produced by nuclear power plants. The planned decommissioning of units at Leningrad and Kola nuclear power plants, to begin in 2003, should be accompanied by replacement units with enhanced safety standards. Implementation of the projected interconnection Russia-Poland-Germany is extremely significant, especially since it could improve the possibilities for export of electricity from the energy system of Central Russia. The share of nuclear power generation there is at present more than 20 per cent;

3) the solution of the problem of disposition of ex-weapons plutonium and disposal of radioactive waste within the frame-

work of the large-scale development of the nuclear power industry based on fast reactors, and the introduction of a closed fuel cycle.

4) Of late, a keen and growing interest has shown by banks and investment funds to Russian power sector of economy.

Estimates of the investment requirement of the nuclear power industry jointly evaluated by Russian-American researchers indicate that in the period up to 2005 on the most optimistic scenario, development costs for the nuclear power industry will amount to $ 8.83 billion. The greatest part of capital will be required in the period 2000–2005, when it is planned to put into operation 5.5 GWe of new installed capacity nuclear power plants. Safety enhancement constitutes roughly half of the total investment.

Investment projects, selected by Russian and foreign organisations for financing by international credit institutions and by investors, where construction of the units should be completed for their commissioning by 1998.

The completion of the construction of unit 3 at the Kalinin nuclear power plant, for example, is considered as economically justified. Unit 3 of the Kalinin nuclear power plant, with a second-generation VVER-1000 reactor, is 75 per cent complete. According to the most optimistic assessments, to bring this unit into operation will take four years, and roughly $347 million investment is required. This unit can be considered as a candidate for private financing using the BOT principle whereby the unit in question returns to State ownership at the end of its design service life. With a 60:40 ratio of loan to equity, it will be necessary for private investors to invest around $140 million: (Ryabev G.D., "Nuclear Power in Russia: Status and Prospects", Nuclear Europe Worldscan, 1 112, Jan./Feb. 1996).

Investments in the nuclear power industry in Russia are relatively attractive and will have great practical significance for ensuring power independence secure both in Russia and in Europe.

Addresses:

Vladimir Kagramanian
Head of Division
Institute of Physics and
Power Engineering
Moscow
Russian Federation

OLEG FAVORSKY
Head, Branch et Scientific
Problems of Energetics
Russian Academy of Sciences
Moscow
Russian Federation

4

Nuclear Energy

UGO FARINELLI

Just as chemical energy, in addition to being used in explosives, is also used to heat our homes or move our cars, so nuclear energy also has a peaceful utilisation, in nuclear ractors that generate electricity.

The dual nature of nuclear energy has originated many problems but has also created several opportunities.

In 1955, the American President Dwight I. Eisenhower launched the programme "Atoms for Peace", which had the objective of diffusing civil nuclear technologies among the countries that had peaceful intentions, provided they would accept stringent controls to verify that information and materials supplied for civil nuclear applications would not be diverted to military utilisation.

In a great international conference held at the end of August 1955 in Geneva, under the auspices of the United Nations, the Americans unveiled many of the scientific discoveries and of the methods developed for the peaceful use of nuclear energy, which until that moment had been classified as secret because of their possible implication for military applications. They were surprised to find the Russians equally available and open. It was soon clear that in many cases the same routes had been followed and the same results had been reached, while in many other cases the approach had been completely different, and the results obtained in the East were complementary to those from the West. For instance, to study neutron transport and diffusion in a reactor the Russians, with a strong mathematical tradition but still without the electronic computers available in the US, would develop ingenious methods based on analytical calculus, while the Americans would employ numerical methods more congenial to computers.

The surprise and the excitement of the scientists participating in the conference for this reciprocal discovery was such that they may have gone some way beyond the instructions they had originally received, and the exchange of information, soon published and made available to everybody, was quite larger than it had been planned at the start.

In the following fifteen years or so, nuclear energy for peaceful applications had a rapid development, may be even more rapid than it was justified by the maturity and the economic value of this technology. A commercial practice was developed according to which the USA and the USSR did not sell the nuclear fuel, which could be exploited also for military uses, but rented it, so that when it had finished its cycle in the reactor, it was to be returned to the country of origin.

On the wake of the "Atoms for Peace" programme, and in the frame of the United Nations, the International Atomic Energy Agency (IAEA) was founded, with headquarters in Vienna. Also this Agency had, since the beginning, a double mandate. On one side, it was charged with a function of surveillance and safeguard in all the countries that wanted to access to a peaceful nuclear programme, carrying out inspections to all nuclear plants and an accurate book-keeping of all nuclear material, in order to verify that their activities were only directed to peaceful nuclear programme, carrying out inspections to all nuclear plants and an accurate book-keeping of all nuclear material, in order to verify that their activities were only directed to peaceful applications. On the other side, it had the mandate of promoting the peaceful uses of nuclear energy.

In those times (I am referring in particular to the 1960's) there was such an enthusiasm and optimism concerning nuclear energy that today appear unjustified and difficult to understand. One of the fathers of nuclear energy, both military and peaceful, Alvin Weinberg, then director of Oak Ridge National Laboratories in Tennessee, USA, went as far as predicting that nuclear energy would diffuse so much and would be so inexpensive, that it would become unnecessary to measure the individual consumption of electricity ("too cheap to meter"). This prediction did not materialise: when in 1973 the first oil crisis put in doubt the possibility of continuing to rely on oil, only few countries (notably France and Japan) felt like reacting to this crisis by turning to large-scale utilisation of nuclear energy; and when at the end of the 1980's the preoccupation on the stability of the global climate put under suspicion all fossil fuels, nuclear energy was seen as a cure worse than the illness.

Why did this change of attitude occur? For two, interrelated reasons: the increase in investment costs necessary for the construction of nuclear plants, and the mounting fears about their safety. The connection between the two derives from the fact that the more and more stringent requests on the safety features of the plant increase its cost and make their licensing procedure, at all stages of design, construction, commissioning and operation, longer and more uncertain, introducing long dead times in which interests on invested capital run anyway.

If the "Atoms for Peace" programme reflected in some way a guilt complex on the part of some Americans for having developed and used such a deadly weapon, many others, who had no reasons to feel guilty, never forgave the "original sin" of nuclear energy, its military and destructive origin. An anti-nuclear movement was thus born, and grew steadily with time, worried about the safely of civil nuclear plants and of possible releases of radioactivity. This preoccupation was directed more against civil nuclear plants (that until the very serious accident at Chernobyl had made no human casualty nor released dangerous quantities of radioactivity) than against nuclear military applications, that with all the test explosions conducted in the atmosphere until 1967 has released enormous quantities of radioactivity, many times more than the Chernobyl accident itself, and caused other very serious damage to environment and health, both in the USSR and in the USA.

But let us go back to the specific theme of the connection between military and civil nuclear applications. The bomb by itself has practically nothing in common with a nuclear reactor: no more than a bomb based on chemical explosives has in common with the engine of a car, based on a series of chemical micro-explosions in the cylinders.

Where military and civil technology are interconnected in the production of nuclear materials necessary for the weapons. This concerns both the enrichment methods used to produce highly enriched uranium for the nuclear weapons, and how enriched uranium for reactor fuel; and for the production of plutonium, which is obtained by irradiating uranium in a nuclear reactor.

In 1968, after long discussions, the Non-Proliferation Treaty (NPT) was signed; two years later it was ratified and entered in force. This Treaty requires the signatories which do not possess nuclear weapons to engage not to acquire them in the future; while the five countries that at the time of the signature already has such weapons (namely USA, USSR, France, the United Kingdom and China) were engaged to gradually decrease their arsenal and to rely less and less on nuclear weapons for their defence. In jargon, "horizontal proliferation" (i.e. the birth of new nuclear weapon states) is discouraged as well as "vertical proliferation" (the increase in number and destructive power of the arsenals of the five official nuclear weapon states). In addition, nuclear countries pledge not to supply to other countries nuclear weapons, or materials, instruments and knowledge that would facilitate the acquisition or development of peaceful nuclear technology in non-nuclear weapon countries which have signed the NPT. These last will submit their nuclear plants to the safeguards and inspections by the IAEA.

The TNP had a certain success in limiting horizontal proliferation: the situation today is not much worse than it was in 1970; however the non-nuclear weapon states accuse the nuclear states that they have done very little to respect their obligation to reduce vertical proliferation. Only the dramatic changes in the East at the beginning of the 1990's have allowed to see at the horizon a substantial reduction of nuclear arsenals, which however proceeds rather slowly.

Nuclear states have difficulties in convincing the others that nuclear weapons are neither necessary nor useful, while at the same time they show to be very reluctant to renounce to them. For this reason, the majority of non-nuclear weapon countries, and especially developing countries, went in a very controversial mood to the conference that in the Spring of 1995 was convened, after 25 years of NPT, to decide whether for an indefinite time, but the criticisms expressed in the conference are likely to have an echo for a long time.

When Jimmy Carter was President of the United States, he made (in 1977) another step in the direction of making horizontal proliferation even more difficult than the NPT already made it.

Carter had been a nuclear engineer, and knew that he was talking about. His preoccupation was that civil nuclear programmes could open the way to proliferation, and his idea was that to make this impossible, or at least much more difficult, it was necessary (and sufficient) to control two critical points: uranium enrichment and separation of plutonium from irradiated fuel. For the first point, not much more could be done than what was already being done, except avoiding that enrichment plants were built in other countries and putting a stricter control on the leak of information on the relative technology. As for the second point, it was sufficient to prohibit any chemical operation on irradiated fuel, renouncing to recuperate plutonium also for civil utilisation and to reuse the unburned fraction of uranium 235, or to separate the fission products for an easier disposal after vitrification (i.e. the inglobation of radioactive materials into a glass-like structure that cannot be attacked by water or by chemicals present in the ground).

This last part raised very negative reactions from the European allies of the United States, from Japan and also from the Soviet Union. These countries considered that renouncing to recycle irradiated fuel and to recuperate the non-utilised part of the fissile materials was an unacceptable limitation to the energy one could obtain from the available uranium resources, which were being studied especially in France (at the time with the prototype Phénix reactor), in Britain and in the Soviet Union, but also in the United States, Japan, Germany and Italy. Its main advantage is that it allows to exploit uranium so deeply as to make the cost of the procedure energy practically independent from the cost of uranium, and therefore virtually unlimited: known uranium deposits, including those at low concentration of uranium ore, would be sufficient for millions of years, without considering the uranium dissolved in sea water. With fast neutron reactors, nuclear energy becomes equivalent to a renewable source of energy. In addition, at that time German legislation required fission products to be separated from the fuel in order to facilitate their storage them for unlimited periods of time, and therefore made chemical treatment of irradiated fuel practically mandatory.

Who was right. Carter or the others? In retrospective, I think that Carter was right, but for the wrong reasons. He was right in assuming that there was no hurry to exploit completely the nuclear fuel and to turn to fast breeder reactors, and one could wait thirty of forty years before having to worry: but this turned out to be the case because nuclear energy did not develop as much and fast as it was generally predicted (in large part as a consequence of the unpredictable Chernobyl catastrophe), and the uranium market unexpectedly reversed its trend completely, with a fall of prices (in real terms) of a factor of live or more.

On the other hand, it is still doubtful whether civil nuclear power is the main route to proliferation, or if it would be simpler for a country that intended to build nuclear weapons to use a completely distinct and secret programme, with specialised reactors for plutonium production or, better still, to set up a uranium enrichment capability (as the subsequent experience would show, in particular in the case of Iraq).

As a consequence of his policy, Jimmy Carter requested and obtained the organisation, through the International Atomic Energy Agency in Vienna, of a huge exercise of examining and assessing all nuclear technologies, and in particular those connected with the fuel cycle, called the International Nuclear Fuel Cycle Evaluation (INFCE). In 61 meetings of eight working groups made up of experts from all over the world (519 scientists from 46 countries and five international organisations) held in Vienna between the end of 1977 and the beginning of 1980, plus two plenary meetings, all the aspects of nuclear technologies which could be significant from the point of view of proliferation were examined, and this was also an occasion for rethinking all the choices that had been made so far.

The results were not extraordinary, and probably not worth the effort that went into the exercise. Some proposals for new, non-proliferating technologies were presented, like an enrichment scheme from France based on small differences in chemical properties of the different isotopes of uranium, which would not lend itself to military applications (because it would take too long a time to bring the system to equilibrium for high enrichments). Several hypotheses which had been examined and, for one reason or another, had been abandoned, were revived and discussed: among these, the possibility of using thorium, instead of uranium, as the basis of the nuclear fuel cycle, and the idea to use systems based on accelerators rather than self-sustaining reactors (these two concepts were later combined in a proposal brought to the attention of specialists and of public opinion by the Noble prize winner Carlo Rubbia).

The most concrete output of the work of INFCE was a set of practical recommendations on how to store plutonium recovered from spent fuel under international surveillance, and how to control that it would be used for civil applications only, so as to make fuel reprocessing possible without undue proliferation risks.

On the whole the conclusions of INFCE, although written in cold and scientific terms, were closer to the European starting point than to Carter's, and were basically pessimistic on the possibility of finding a technical fix to the problem of proliferation. The message that transpired from it (although not bluntly formulated) was addressed to the politicians: "If you want to find a solution to the problem of nuclear proliferation, it is a political solution you must be looking for. You must remove the motivations for which the countries want to proliferate, you must make the possession of nuclear weapons uninteresting and counterproductive. Otherwise, you may delay of some years the acquisition of nuclear weapons by one country or another, but you won't stop it forever."

References

1. A. Amaldi, U. Farinelli, C. Silvi, "Conversion of weapon-grade nuclear materials to civilian purposes", Workshop on International Security and Disarmament: the Role of the Scientific Academies", Rome, Accademia Nazionale dei Lincei 6–9 June 1989.
2. F. Berkhout, H. Feiveson, "Securing Nuclear Materials in a Changing World", Annual Review of Energy and Environment, 1993, 18: 631–655.
3. Committee on International Security and Arms Control (CISAC) of the National Academy of Sciences, "Management and Disposition of Excess Weapons Plutonium", National Academy Press, Washington D.C. 1994.
4. EIA (Energy Information Administration), US Department of Energy, "World Nuclear Capacity and Fuel Requirements 1993", DOE/EIA-0463(93), Washington D.C. November 30, 1993.
5. U. Farinelli, C. Silvi, "An Update on the Issue of Conversion to Peaceful Uses of Military Nuclear Materials", International Conference on Security in Europe and the Transition Away from Confrontation towards Co-operation, Rome, Accademia Nazionale dei Lincei, 4–7 June 1990.
6. U. Farinelli, C. Silvi, "Conversation to peaceful uses of nuclear

materials", Amaldi Symposium on Science, Technology and International Security, Cambridge, U.K., 8–10 July 1991.
7. Arthur Fisher, Popular Science, April 1990, pp. 68–78.
8. Michael W. Golay, Neil E. Todreas, Scientific American Vol. 262, n. 4, p. 82–89, April 1990.
9. Chuck Hansen, "US Nuclear Weapons—The Secret History", Orion Books, New York 1988.
10. International Nuclear Fuel Cycle Evaluation (INFCE), INFCE Summary Volume, International Atomic Energy Agency, Vienna, 1980.
11. André L. Jaumotte, Alain Michel, "Recycling Military Plutonium: the Academic Community has a Definite Role to Play", 7th Amaldi Conference, Jablonna, Poland, September 1994.
12. Wolfgang K. H. Panofsky, "Safeguarding the Ingredients for Making Nuclear Weapons", Issues in Science and Technology, Spring 1994.
13. Wolfgang K.H. Panofsky, "The Future of Non-Proliferation in Relation to the Nuclear Doctrines of the Nuclear Weapons States", Seventh Internatiional Amaldi Conferenza "How to Reduce Threats of Peace and General Securety", Jablonna, Polonia, Settembre 1994.
14. Michael Renner, "Preparing for Peace", State of the World 1993, p. 139.
15. SIPRI (Stockholm Internatiional Peace Research Institute), "Nuclear Energy and Nuclear Weapon Proliferation", Taylor & Francis, London 1979.
16. SIPRI (Stockholm International Peace Research Institute), "Internationalisation to Prevent the Spread of Nuclear Weapons", Tylor & Francis, London 1980.
17. T.B. Taylor, "Verfified Elimination of Nuclear Warheads", in "Verification on Nuclear Warheads Reductions and Space Reactor Limitations", a cooperation project of the Federation of the American Scientists and the Committee of Soviet Scientists, March 6, 1989.
18. U.S. Congress, Office of Technology Assessment, "Dismantling the Bomb and Managing the Nuclear Materials", OTA, Washington DC, US Government Printing Office, September Alvin M. Weinberg, Annual Review of Energy and Environment, Vol. 19, pp. 15–36, 1994.

Address:

UGO FARINELLI
ENEA
Directorate of Studies and Strategies of Development
00196 Rome
Italy

PART — 2

PROSPECTS FOR DISARMAMENT AND ARMS CONTROL

5

The Nuclear Tests in South Asia: Implications for the Middle East

SHAI FELDMAN

The nuclear tests conducted by India and Pakistan in May created considerable interest in the Middle East. The potential implications of these tests involve their impact on the nuclear policies adopted by the region's states, primarily Iraq, Iran, Egypt, and Israel; the likelihood of applying arms control measures in the region; and America's ability to achieve its objectives in the region, including in the realm of Arab-Israeli peacemaking.

THE NUCLEAR TESTS AND THE MUSLIM STATES

The Indian and Pakistan nuclear tests may affect Muslim states in a number of ways. First, the tests are likely to affect the international debate in a number of these states—primarily Iran—with regard to the size of the investments that should be made in pursuit of nuclear weapons. Those advocating larger investments in nuclear activities may now argue that global trends have been reversed do favour nuclear deterrence and that it would be irresponsible to refrain from the acquisition of such capabilities. These "pro-nuclear" forces will also claim that explicit nuclear capabilities provide their possessors for status of "nuclear powers", and that if the region's states wish to avoid being ignored, they must "go nuclear".

Those favouring the pursuit of nuclear capabilities will also argue that the weak international reaction—failing to compel India and Pakistan to refrain from further testing—indicates that nuclear capabilities may be acquired and possessed at tolerable costs. They will also point out that India and Pakistan can further minimise the costs entailed in their recent testing through a "grand bargain" that would lift the economic sanctions that the United States is applying against them in exchange for signing the Comprehensive Test Ban Treaty (CTBT). Such a deal would leave India and Pakistan with their new status as "nuclear powers" without requiring them to pay substantial long-term costs for this change.

By contrast, those opposed to investing scarce resource in nuclear technology are bound to argue that the nuclear tests have brought about a further deterioration in the economic situation in India and Pakistan, without resulting in any significant improvement in their strategic standing. This is because the Pakistani nuclear tests have cancelled the deterrent effect of the tests conducted by India. Consequently, the nature of the competition and conflict between them—including in Kashmir—will remain unchanged.

It was also speculated that the nuclear tests conducted in South Asia would accelerate nuclear proliferation in the Middle East directly—through Pakistani transfers of fissile material or sensitive nuclear technology to another Muslim state, primarily Iran. Such a development would provide Pakistan's nuclear capability an "Islamic" facet and would transform its capability—a process already celebrated in the mass media—into an "Islamic bomb".

Pakistani spokespersons categorically deny any intention to transfer sensitive materials and technology to another Muslim state. They insist that the nuclear capability they now possess is not an "Islamic bomb", just as the French nuclear deterrent does not comprise a "Christian bomb" and Israel's nuclear potential does not constitute a "Jewish bomb".

Beyond these denials, there are four reasons why Pakistan is unlikely to transfer sensitive nuclear material and technologies to another Muslim state. First, Pakistan developed its nuclear capability in response to India's nuclear programme and as a result of its concern regarding its overall strategic standing vis-a-vis India. Under such circumstances, it is unlikely that Pakistan would wish to exacerbate further its security problems by taking steps that are bound to result in a further intensification of the growing defense relationship between India and Israel.

Second, Pakistan is now hard pressed to persuade the United States to lift the economic sanctions imposed upon it.

Pakistani leaders made various statements to the effect that their people are prepared "to eat grass" if such a sacrifice would be required to sustain their nuclear capability. However, these leaders will very soon have to confront the harsh economic consequences of America's reaction to the nuclear tests and will have to devote much of their energies to minimising such damage. Under these circumstances, it is unlikely that the Pakistani Government would wish to cross another—and much more important—"red line" in U.S. nuclear non-proliferation policy. In Washington's eyes, the transfer of sensitive nuclear materials and technology would mean that in addition to helping breach the nuclear "Club of Five", Pakistan would be providing direct assistance to the efforts of additional states to acquire nuclear weapons. America's reaction to any evidence of such assistance can be expected to be much sharper than its recent reaction to the nuclear tests.

Third, if Pakistan's nuclear capability has an "Islamic" dimension, it is as a "card" in the country's quest for influence in the Muslim world. Yet such a capability can provide Pakistan influence among these states only so long as it continues to enjoy a nuclear monopoly among the Muslim states. The transfer of sensitive nuclear material and technology to another Muslim state would contribute to the loss of such a monopoly, and would erode whatever additional influence Pakistan has gained among the Muslim countries through the recent nuclear tests.

Finally, it is noteworthy that the recent tests have not changed in any way Pakistan's capacity to transfer sensitive nuclear material and technology to another state. Even prior to these tests there was little doubt in the Middle East regarding Pakistan's nuclear capability. The various factors that have dissuaded Pakistan from transferring nuclear materials and technology in the past are also likely to inhibit such transfers in the future. Hence, the recent nuclear tests are unlikely to contribute directly to further nuclear proliferation in the Middle East.

IMPLICATIONS FOR THE ARAB STATES

The nuclear tests conducted in South Asia raised serious concerns among Arab states. First, the weak response by the

United States to these tests was perceived by Arab observers as indicating Washington's tacit support for the emergence of a new strategic balance in Asia, based on India's contribution to containing China's growing power. According to this interpretation, the United States may adopt a similar approach in the Middle East and may support the emergence of a new "balance to terror" in the region, based primarily on Israel's nuclear potential.

Second, Arab observers fear that "the Pakistani bomb" would grant international legitimacy to Israel's nuclear option by making it clear that a Muslim country had become a nuclear power first. In turn, this paves the way for Israel's nuclear option by making it clear that a Muslim country had become a nuclear power first. In turn, this paves the way for Israel's nuclear policy to become more explicit at lower costs than might have been the case prior to Pakistan's nuclear testing.

Third, there is concern among some Arab states—Egypt among them—that the strategic surprise caused by India's nuclear tests to the intelligence communities of the Western countries may strengthen Israel's arguments to the effect that it is impossible to assure detection of activities that violate global arms control agreements. Thus, Israel may use this surprise to illustrate why it must continue to refrain from signing the 1968 Nuclear Non-proliferation Treaty (NPT).

Fourth, it is possible that Egypt and other Arab states might interpret Washington's weak response to the nuclear tests in South Asia as indicating that it is highly unlikely that the United States and other Western countries could even compel Israel to sign the NPT. In turn, this may propel the Arab states to adopt a "more realistic"—and hence more constructive and co-operative-approach toward the prospects of applying arms control and confidence building measures in the Middle East. Alternatively, it may lead these states to conclude that the multilateral Arms Control and Regional Security (ACRS) talks should be permanently abandoned since they are unlikely ever to induce Israel to disarm its nuclear option.

Finally, America's inability to compel India and Pakistan to adopt its priorities in the nuclear realm may be interpreted by the Arab states as another manifestation of the limits of U.S. power and prestige. Thus, Arab observers may argue that if

Washington failed to dissuade Pakistan from reacting to India's nuclear tests, it cannot be expected to be more effective in compelling Israel—a country enjoying considerable domestic support in the United States—to make greater concessions in Israeli-Arab negotiations. In turn, such an inference may lead Arabs to two different reactions: regarding the peace process, possibly resulting suicide bombings; or, the adoption of lower expectations, making more gradual progress in the peace process possible.

WOULD ISRAEL FOLLOW INDIA AND PAKISTAN?

The speculation that Israel might follow the South-Asian example and conduct nuclear tests ignores the significant differences between its strategic circumstances—including its relations with the United States and its approach to global nuclear arms control treaties—and those of India and Pakistan. Analysing the implications of such a step would dissuade Israel from changing its nuclear policy; the costs involved in such action would far out-weigh its benefits.

First, India already faces two nuclear neighbours: China, an explicit nuclear superpower, and Pakistan, an undeclared nuclear state. Similarly, Pakistan already faces an ambiguous nuclear India. By contrast, Israeli does not face an adversary capable of presenting it with an existential threat. While a nuclear Iran might pose such a threat in the future, its military nuclear programme is still in its infancy. Clearly, Israel would not wish to take steps that would be bound to encourage Iran to accelerate its nuclear efforts.

Instead, Israel is devoting its energy to persuading the United States to exert pressures that would prevent the proliferation of mass destruction weapons to Iran. Similarly, it is attempting to persuade the members of the United Nations Security Council—including the United States—to refrain from closing the "nuclear file" of the inspection and monitoring regime that the International Atomic Energy Commission (IAEA) and the United Nations Special Commission (UNSCOM) are applying in Iraq.

Hence, Israel would wish to avoid any step that might provide legitimacy to the nuclear efforts of Iran and Iraq and would bring about a complete halt in U.S.-Israeli cooperation in preventing the acquisition of nuclear weapons by the two countries.

Second, Israel's approach to global nuclear non-proliferation efforts is quite different from India's. New Delhi always rejected the discriminatory nature of the NPT—the division it institutionalised between the world's nuclear "have's" and "have not's". By contrast, Israel never defied the NPT regime, nor did it reject its premises—it merely pointed out that the political conditions in the Middle East to not yet allow it to join the Treaty. It also emphasised that the verification measures applied by the IAEA are insufficient to prevent states from violating the stipulations of the Treaty and from developing nuclear weapons clandestinely, as was demonstrated in the case of Iraq.

Third, U.S.-Israeli relations are very different from Washington's approach to India and Pakistan. First, in contrast to the latter two countries, Israel is the recipient of $ 3 billion in annual direct U.S. economic and military assistance. Existing legislation compels the Clinton Administration to end all direct assistance to any state that conducts a nuclear test.

In addition, the array of Israeli responses to the emerging strategic threats—primarily the proliferation of ballistic missiles in the Middle East—all require close defense ties between Israel and the United States. The efforts to prevent the transfer of missile technology from Russia to Iran require close coordination between Jerusalem and Washington; Israeli deterrence against existential threats requires that the United States continue to regard Israel as a "special case" in its nuclear non-proliferation policy; the central pillar of Israel's "active defense" efforts—the Arrow missile defense system—is funded largely by the Ballistic Missile Defense Organisation (BMDO) of the U.S. Department of Defense; the efficacy of Israeli "passive defense" efforts depends on its ability to obtain instant warning of missile launching—a capability that only the U.S. spy satellite network can provide; finally, the U.S. provides Israel the means of acting of fensively against strategic threats—for example, F-15-I long-range strike aircraft that can be employed against ballistic missile launchers and facilities where mass destruction weapons are being produced.

In short, Israeli nuclear testing would result in an immediate and complete halt to the close defense ties it enjoys wlth the United States. As a result, Israel's ability to confront the strategic threats it now faces will be eroded. Hence, Israel is unlikely to follow the examples of India and Pakistan.

Summary

The analysis provided here points to a number of conclusions regarding the implications of the nuclear tests conducted recently by India and Pakistan:

- The influence of the nuclear tests conducted in South Asia on the internal debates in Arab states and Iran with regard to the efforts that should be invested in the nuclear realm is unclear. Recent developments may strengthen the case of both those advocating greater investments in this realm and those arguing the opposite.
- Pakistan is highly unlikely to transfer sensitive nuclear material and technology to another Muslim state. Hence, the nuclear tests recently conducted in South Asia are unlikely to accelerate the process of nuclear proliferation in the Middle East.
- The Arab states have good reasons to suspect that the nuclear tests conducted by India and Pakistan as well as the weak Western response to these tests, will strengthen the arguments that comprise the basis of Israel nuclear policy. They may further conclude from this response that the United States would not compel Israel to accept its preferences in the realms of arms control and the Middle East Peace process.
- Israel will not follow the examples of India and Pakistan; it will continue to refrain from nuclear testing. This is due to the significant differences between Israel's strategic circumstances—including its relations with the United States and its approach to global nuclear arms control treaties—and those of India and Pakistan.

Address:

SHAI FELDMAN
Director
The Jafee Centre for Strategic Studies
Tel Aviv University
Tel Aviv 69978
Israel

6
On Possible New NWFZs in Asia

JAN PRAWITZ

INTRODUCTION

The Arms Control and Non-Proliferation Background

Approaching the year 2000, the Cold War nuclear arms race has definitely reversed its direction into disarmament which is the main nuclear arms trend today despite various "disturbances". This process proceeds generally along two paths: enhancing the non-proliferation régime and pursuing nuclear disarmament.

The most basic of all attempts to raise barriers against the nuclear arms race, the 1968 Non-Proliferation Treaty (NPT), has over the years, within the limits of its letter, developed into a great success.

- In 1999, almost all states of the world have become parties to the treaty including the five established nuclear-weapon states. Only six states are non-parties (August 1st, 1999), but three of them are parts of nuclear-weapon-free zones[1]. The remaining three are the well-known threshold states of India, Israel and Pakistan. With the important exception of five recognised nuclear-weapon states with a treaty right to have nuclear weapons, possession of nuclear weapons could thus be considered illegitimate.
- Almost unnoticed, the NPT parties decided in 1985 that the treaty should be implemented "under any circumstances", i.e. also in wartime[2].
- In 1995 after 25 years in force, the NPT parties decided to extend the treaty's duration indefinitely[3].

* This Work was supported by **The Swedish Agency for Civil Emergency Planning**. Views and opinions expressed in this paper are those of the author and do not imply the expression of any position on the part of the Swedish Institute of International Affairs or the The Swedish Agency for Civil Emergency Planning.

- On the other hand, the NPT permits stationing and deployment of nuclear weapons controlled by the five nuclear-weapon states in the territories of any NPT party and at sea. In large areas of the world, however, included in nuclear-weapon-free zones, such stationing and deployment is prohibited. Today, five nuclear-weapon-free zones have been established, covering more than half of the world's land-mass, including 99 per cent of the Southern Hemisphere land areas, while excluding most sea areas, and encompassing 112 states (out of some 195) and 17 other territories.
- At long last, dramatic nuclear disarmament (NPT, Art. VI) has been agreed beginning in 1987. Today, agreed measures include the INF (1987), the START (1991) and the START II (1993) agreements[4]. All land-based missiles with ranges between 500 and 5500 kilometers are now eliminated and a verification régime is in operation. Implementation of the START and START II treaties have been rescheduled to be finalised by the year 2007 leaving about 3500 strategic nuclear weapons each in the possession of Russia and the USA[5]. The three smaller nuclear-weapon powers are not participating in these disarmament projects.
- Most important is the withdrawal of sub-strategic nuclear weapons from theaters of deployment and from ships, unilaterally declared in the fall of 1991 by the USA and by the USSR later confirmed by the Russian Federation. Many of those weapons are now being dismantled, particularly in the USA, others will be kept in centrally located storages. These most important measures are based on mutual agreement but got the form of unilateral though coordinated declarations and are thus not legally binding nor do they have a permanent duration in force.[6]
- Beside these reduction agreements on nuclear hardware, the nuclear weapon powers have also agreed to institute a number of software and nuclear confidence- and security-building measures, such as pre-notification of missile launchings, establishment of nuclear risk reduction centres and communication lines, lowering the alert of their remaining strategic nuclear forces, de-targeting weapons aiming at each other, and cooperation in development of measures

for the improvement of the security and safety of nuclear weapons. Recently, the most important Comprehensive Nuclear Test Ban Treat (CTBT) was adopted by the UN General Assembly in September 1996[7]. The CTBT prohibits also nuclear explosions for peaceful purposes.

- Security guarantees extended to the many non-nuclear-weapon states agreeing to renounce their option to acquire nuclear weapons have been an important component of the non-proliferation régime. Coordinated with the adoption of the NPT in June 1968, the UN Security Council passed a resolution outlining rules for assisting non-nuclear -weapon states parties to the NPT subject to attack or threat of attack by nuclear weapons[8]. Later, all nuclear-weapon powers extended unilateral "negative"[9] assurances that non-nuclear-weapon states will not be subject to attack or threat of attack with nuclear weapons, however, subject to various conditions[10]. In April 1995, the UN Security Council adopted a modernised resolution taking note of both existing negative nuclear assurances and the positive assurances where the five nuclear-weapon states undertake to provide "*immediate assistance,. in accordance with the UN Charter, to any non-nuclear-weapon state party to the NPT that is a victim of an act of, or an object of a threat of, aggression in which nuclear weapons are used*" (Op. 7)[11]. But the value of this guarantee is limited as the nuclear-weapon states are also permanent members of the Security Council with a right of veto.

Towards a World of Nuclear-Weapon-Free Zones

These developments now proceed in the general direction of a nuclear-weapon-free world. Considering the scope of existing nuclear-weapon-free zones and the absolute character of the zone concept, further pursuance of the zone concept would be a powerful means for gradually achieving a nuclear-weapon-free world. The part of the world available for deployment of nuclear weapons by the nuclear-weapon states would then gradually diminish.

Such an effort would include

- enhancing remaining entry into force processes of existing nuclear-weapon-free zones,

- strengthening the provisions of existing zonal régimes, and
- establishing new zones,

until the whole world will become such a zone.

The provisions and implementation machinery of the non-proliferation régime would be a dominating basis for negotiating such zones in the future.[12] Whenever a new nuclear-weapon-free zone will be discussed, all or most potential zonal states would thus already by parties to the NPT. Indeed, the NPT itself foresees that role in its Article VII.

A number of zone projects have been proposed in the past which were never established and which could be pursued further. Some were proposed long ago. Among them, several are no longer of interest, others are still relevant. Some were proposed more recently.

When the parties to the NPT in May 1995 extended the duration of the treaty, they also adopted new "*Principles and Objectives for Nuclear Non-Proliferation and Disarmament*", i. e. encouraging the establishment of nuclear-weapon-free zones as a matter of priority.[13] Since then, two such zones were established in South-East Asia and Africa. On the same occasion, the NPT Extension Conference "took note" of "*the declaration by Mongolia of its territory as a nuclear-weapon-free zone*"; and "*the interest of Kyrgyzstan and Uzbekistan in creating a nuclear-weapon-free zone in Central Asia*"[14]. Both proposals are now actively pursued.

On 10 December 1996, the UN General Assembly observing the fact that almost all land south of the equator is part of NWFZs adopted a resolution encouraging a coordinated nuclear-weapon-free Southern Hemisphere.[15]

Following the success of the NPT, the establishment of more nuclear-weapon-free zones is now the dynamic trend in the promotion of the non-proliferation régime.

It should be noted that a codification of the 1991 unilateral declarations on sub-strategic nuclear weapons or a general and global prohibition of such weapons would be a most effective step towards a nuclear-weapon-free world and in one step achieve a major part of what the many proposed nuclear-weapon-free zones are proposed to accomplish.

The NWFZ experience

A discussion of a further implementation of the nuclear-weapon-free zone instrument for nuclear arms control should be based on the experience of the concept so far.

That experience includes firstly four zones established in "densily" populated areas.[16] The Tlatelolco Treaty[17] of 1967, The Rarotonga Treaty[18] of 1985, the Bangkok Treaty[19] of 1995 and the Pelindaba Treaty[20] of 1996 created such zones in Latin America, the South Pacific, South-East Asia, and Africa respectively. Denuclearisation of the Korean peninsula was declared in 1992 by the two Korean states[21], but the agreement remains to be implemented. In addition, a demilitarisation régime and thus by implication also a denuclearisation régime for Antarctica was agreed in 1959[22].

Secondly, there are also authoritative reports on two nuclear-weapon-free zones proposed in the past but which have not been established, i.e. in the Middle East (1990)[23] and in Nordic Europe (1991)[24].

Thirdly, two United Nations expert studies have contributed to establish the scope and the frame of the concept[25] as has a thorough discussion within the United Nations Disarmament Commission 1997–1999 resulting in a set of recommendations for zone-making adopted by consensus on 30 April 1999.[26]

The accumulated knowledge of the zone issues is extensive enough to permit summarising a "theory" of nuclear-weapon-free zones. For the purpose of this analysis, relevant parts will be referred to.[27]

What is a Nuclear-Weapon-Free Zone?

"Nuclear-weapon-free zone" is a summary concept describing régimes for regional security, independently or as a complement to other arrangements, global as well as regional. The concept has been relatively well researched. Geographical, political and other regional specifics related to nuclear-weapon-free zones would make different zones different. No such zone would be an exact copy of another. The term nuclear-weapon-free zone would, however, usually imply the fulfillment of certain objectives and the implementation of certain measures of arms control.

The general objective for establishing a nuclear-weapon-free zone would be to relieve a zonal area from the threat of being involved in mass destruction war. The fulfillment of this objective would usually require cooperation both among prospective zonal states and between them and nuclear weapon states or other extra-zonal states.

But there may be a variety of further objectives for the establishment of such zones in specific cases. Regarding proposed zones in Europe, the objective of geographical separation of the nuclear weapons of the major military blocks during the Cold War has been referred to as such a secondary important objective.

The fulfillment of such zone objectives shall also be considered as a process in time. History shows that so far, establishment of zones is a process over decades rather than years. The time span from the first political proposals to the signing of the guarantee-protocols by all nuclear-weapon powers—a milestone in the establishment process—was 16 years in the Tlatelolco case, 11 years in the Rarotonga case, and 35 years in the Pelindaba case. The Bangkok Treaty was first envisaged in 1971 and 38 years later, there are no guarantee signatures.

In addition, the creation of a nuclear-weapon-free zone would always be considered a temporary step and a contribution to a process eventually leading to general nuclear disarmament.

Geographical Considerations

No precise requirements can be set as regards the suitable size of nuclear-weapon-free zones or other arms control zones, which could range from whole continents to small areas. Sometimes a zone may be initially established in a limited area and later extended as other countries agree to join in. If large parts of the world are to be kept free from nuclear weapons or other weapons of mass destruction, extending such zones to whole continents would provide the natural way to achieve that aim.

The geographical extent of a zone would depend on the specific characteristics of the region and the precise arms control objectives to be realised.

A single state could declare itself a zone. One example is Mongolia presently preparing for establishing itself as nuclear-weapon-free zone. Normally, however, a zone would comprise the national territories of two or several neighbouring states including their territorial waters and airspace and established by agreement among those states. It would also be possible for states separated from each other by high sea areas or otherwise to form a zone.

Sometimes the circumstances are such that only part of a state will be considered for inclusion in a zone. Splitting states in zonal and extra-zonal parts would require special legal reference in zone treaties.

Furthermore, a zone might include geographical areas not under the jurisdiction of any state, for instance sea areas beyond territorial waters, or it may in its entirety be established in such areas, Antarctica being a well known example.

Delimitation of a zone would generally depend on two concepts: the zonal outer perimeter and the territory for which the zone agreement is in force. Both concepts were applied in the Tlatelolco and Rarotonga treaties while no specific perimeter was defined for the Pelindaba and Bangkok zones.

The perimeter is a line encircling the entire area of the zone within which all states would be eligible to become parties to the zone agreement. The perimeter may encompass international areas as well. States outside the perimeter would be considered extra-zonal, but may be invited to assume responsibilities related to the zone and regarding dependencies they may have inside the perimeter.

Where no explicit perimeter is defined, as in the African and the Bangkok cases, the definition of which states would be eligible to be zonal states, would also define an implicit perimeter encircling the combined territories of the cluster of such states. In such cases, the zonal area could, as in the African case, form several sub-areas not in topographic contact.

The area of application of a zone would be the territory, including land areas, internal, territorial, and archipelagic waters, and air space, of parties for which the nuclear-weapon-free zone treaty and related protocols are in force. Zonal commitments could also be applied to international areas such as Antarctica or parts of the high seas upon entry into force of special legal instruments. The area of application would grow in size as more states adhere to the zone agreement but could by definition not extend beyond the perimeter.

The perimeters of the Tlatelolco, Rarotonga and Antarctic zones encompass large areas of international waters.[28] In the Rarotonga Treaty case, the larger part of the area inside the perimeter consists of such waters. As the UN Convention on the Law of the Sea (UNCLOS)[29] prescribes, states parties to a nuclear-weapon-free zone treaty (or any group of states) cannot by agreement among themselves institute generally applicable restrictions in such waters which all states have the right to enter and use. Zonal states can agree to restrict only themselves in international sea areas. In addition, they can invite extra-zonal states to subscribe to special provisions and to respect general zonal restrictions, e.g. by signing special protocols to that effect. This is done in the Rarotonga Treaty and the African zone treaty as regards the prohibition of nuclear testing and dumping of radioactive materials.

Basic Measures and Obligations

There would be three measures of central operative importance for the achievement of the objectives of a nuclear-weapon-free zone in the general case. These are:

- non-possession of nuclear weapons by zonal States,
- non-stationing of nuclear weapons by any State within the geographical area of application of the zone, and
- non-use or non-threat of use of nuclear weapons throughout the zone or against targets within the zone.

The meaning of these measures might seem clear enough. However, their legal representation could be complicated, as shown e.g. by the definition of "nuclear weapon" in the Tlatelolco Treaty (Art. 5).

The non-possession measure would apply to zonal states. Its codification could be much simplified if relying on the concepts of the Non-Proliferation Treaty (Art. II). This measure must be negotiated among the prospective zonal states to become a mutual commitment. If the zone encompasses only territories of states parties to the NPT, most of the non-possession requirement would automatically be in place. If the zone is to encompass states which are not parties to the NPT or states which are former nuclear weapon states, a special régime must be defined. The same would be true in the special case that only a part of a state will be included in the zone. If the whole of a nuclear weapon state is to be included, a procedure for abandonment of all its nuclear weapons must be prescribed.

The non-stationing measure would primarily apply to the territories of zonal states and must be specified in the founding treaty as such stationing is not prohibited by the NPT. This measure must also be negotiated among the prospective zonal states to become a mutual commitment.

There is the exception that zonal states could not by agreement among themselves restrict or prohibit innocent passage (or transit passage) by vessels under foreign flag with nuclear weapons on board in their territorial and archipelagic waters. Non-stationing measures applying to such waters and to international land and sea areas would require special legal arrangements.

Related to the non-stationing measure is "transit" of nuclear weapons through zonal territory. The transit concept would include "innocent" transit over a limited period of time of prohibited weapons by an extra-zonal state, on land, by air or in internal waters including calls at ports by ships or landing of aircraft carrying such weapons.

The transit issue was extensively discussed when the Latin American zone was negotiated in the 1960s. The problem was solved by not being solved. Transit was left to the individual zonal states to permit or deny in each case[30]. The South Pacific zone has a similar transit régime as has the African and the South-East Asian zones.

A zonal treaty should prescribe if transit would be generally prohibited or arranged in a way similar to the Tlatelolco formula. Transit through zonal high sea areas or through territories which are dependencies of nuclear-weapon states could not be permitted without making the zonal régime in such areas an illusion[31].

Recently, the UN Disarmament Commission recommended the 32 year old Tlatelolco formula as a general norm[32]. This norm obviously refers to "innocent transit". "Hostile transit", i.e. passage of delivery vehicles with nuclear weapons across zonal territory towards targets beyond the zone, could probably not be accepted by a community of zonal states. The latter would apply to sea-borne and airborne manned or unmanned vehicles and to ballistic missiles in so far as they penetrate zonal air space, while crossing overhead zonal territory in international space could not be prohibited by agreement among the zonal states[33].

The special transit issue of ships and aircraft which may carry nuclear weapons on board and call at ports or land at airports in zonal states has been particularly sticky because nuclear weapon powers usually "neither confirm nor deny the presence or absence of nuclear weapons on board specific ships or aircraft at specific times"[34]. A political problem of considerable dimension some years ago, the neither confirming nor denying issue has lost most of its former importance following the withdrawal by nuclear-weapon powers of non-strategic nuclear weapons from naval ships.

The non-use measure or the guarantees would be a commitment by nuclear-weapon states. Legally, this provision has been given the form of a separate protocol to existing zone agreements. While it would be the zonal states who will draft such a guarantee protocol and then invite the nuclear-weapon powers to sign, it is obvious that final agreement with them would be facilitated by advance consultations. In several cases, it turned out that guarantee protocols were not fully acceptable to the nuclear-weapon powers resulting in either reservations as in the Latin American case, or in redrafting as in the South Pacific case, or in no signatures as in the South-East Asia case.

Consideration of the non-use measure should be made against the background of existing nuclear security assurances[35] and of ongoing negotiations on general negative security assurances at the Conference on Disarmament in Geneva.

Linked to the non-use measure has been the idea mentioned above that this measure should be complemented by a "thinning-out" arrangement in areas adjacent to the proposed zone where nuclear weapons are deployed. The "thinning-out" idea implies that such weapons should be withdrawn that are targeted against the zone or that have short ranges and are deployed very close to the zone, thus making them usable primarily against the zone. If such weapons are not withdrawn, non-use commitments would be less credible[36]. Unlike drafting of guarantee protocols, negotiations of thinning-out measures could preferably be done through zonal states versus individual states adjacent to the zone.

Special Provisions for Sea Areas

There is a significant difference between applying arms control in sea areas as compared to land areas because of different legal régimes. Almost all land is subject to the jurisdiction of one state, a well-known exception being Antarctica. As a consequence, adversely military forces on land are geographically separated from each other in peacetime. Naval forces of different states, on the other hand, may mix all over the sea, on the surface, in the water, under the ice, and on the sea-bed. Indeed, they frequently do so.

The very elaborate and detailed United Nations Convention on the Law of the Sea (UNCLOS) was agreed in 1982 and entered into force in 1994 and functions as the "constitution" of the sea areas covering more than 70 per cent of the surface of the earth. Many of its sovereignty related provisions are today considered customary law binding for all states whether parties to the convention or not.

According to UNCLOS, coastal states have full jurisdiction over their internal waters only. Their jurisdiction also extends to their territorial seas and archipelagic waters, except that any flag state enjoys the right of innocent passage for its ships in such waters (there is a more liberal régime of transit passage through international straits)[37]. The provisions granting the right of innocent passage to men-of-war make no distinction between ships because of the types of weapon they may carry.

In exclusive economic zones or on the high seas, the coastal states have no jurisdiction related to nuclear weapons.

Zonal states would be obliged not to possess, deploy, or otherwise operate nuclear weapons anywhere including at sea, but they would have no right according to international law to limit by agreement among themselves the general right of flag states to navigate ships in or fly aircraft over waters which all states have the right to enter and use. Denuclearisation of a sea area would require agreement in principle among all states of the world or at least among the nuclear weapon states to make the régime effective. Zonal commitments applying to sea areas should, therefore, preferably be prescribed in a separate legal instrument or protocol linked to the main nuclear-weapon-free zone treaty and expressed in terms referring to the general law of the sea.

Needless to say, negotiating and drafting of a maritime protocol to a nuclear-weapon-free zone treaty would be a delicate matter to undertake. There is no historical precedent so far. In formal principle, it should be done at a special world conference of all states having access to the sea areas concerned. But more practically, it could be done in the same way as guarantee protocols, i.e. by zonal states versus nuclear-weapon states negotiations.

Complaints and Control Procedures

It is traditionally recognised that effective implementation of an agreement on a nuclear-weapon-free zone would require a system of verification to ensure that all states involved, zonal as well as extra-zonal, comply with their obligations.

The precise scope and nature of such a system would vary from zone to zone and depend upon the nature of the obligations prescribed. Generally, a zonal treaty would have to include provisions both for verifying compliance and a complaints procedure for settling issues of suspected non-compliance, should such cases arise. As relevant, the zone verification system could for its implementation be based on verification enforced according to the NPT.

In general, subject to verification under a nuclear-weapon-free zone treaty should be:

a) All activities of zonal states related to nuclear weapons to ensure that peaceful activities are not diverted to the manufacture of weapons;
b) the commitment that no nuclear weapons are present within the zone; special régimes would be required for seas areas;
c) the removal of nuclear weapons that may be present within the zonal area at the time of entry into force of the zone agreement, possibly also requiring an account of the nuclear history of participating zonal states;
d) the implementation of other measures associated with the zone agreement.

The verification system of a zone would have to be worked out and agreed by the zonal states. Most verification related to peaceful nuclear activities of zonal states could be entrusted to the safeguards system of the International Atomic Energy Agency (IAEA). The IAEA is now operating safeguards in very many states, including the non-nuclear weapon states parties to the Non-Proliferation Treaty.

This traditional safeguards system could require extension and reinforcement for the purpose of verifying a specific zone agreement by additional procedures especially defined and described in that agreement. Such an agreement must be negotiated between the zonal states and the IAEA and would have to comply with the provisions of the Agency's statute.

In addition, there would in most cases be necessary to establish standing organs or special bodies for supervising the routine verification by the IAEA, for undertaking verification measures not carried out by the IAEA, and for handling complaints and assessment procedures. In regions where sharp conflicts exist, entrusting the task of verification entirely to an international organisation, perhaps supplemented by bilateral arrangements, might be preferred.

There is also the possibility that an agreement on a zone would provide to any party a right to undertake verification activities in other states parties to the zonal agreement, including on-site inspection. One model for such a system could be the verification system laid down in several arms control agreements adopted within the Conference on Security and Cooperation in Europe (CSCE), i.e. the Stockholm and Vienna Documents on confidence-building measures and the Conventional Armed Forces of Europe Treaty.

The NWFZ in South-East Asia

The idea of establishing a nuclear-weapon-free zone in the south-east Asian region was part of a proposal to establish a zone of peace, freedom, and neutrality in South-East Asia (ZOPFAN) presented in 1971[38]. It lasted to 1992, however, when the US military bases in the Philippines had been abandoned, until concrete work on the idea was started[39] as endorsed by the UN General Assembly[40]. An independent "ASEAN Regional Forum" was set up to discuss security issues including a nuclear-weapon-free zone in the ASEAN area. The Forum held its first meeting in July 1994. At an ASEAN Summit meeting in December 1995, a treaty text was agreed and signed by all ten South-East Asian states[41].

The South-East Asian zone treaty text is generally similar to the Rarotonga and the Pelindaba treaties. Key provisions are the prohibition of development, manufacture or other acquisition, possession or control, testing and use, and stationing or transport of nuclear weapons by member states or in member states (Art. 3:1–2). The treaty also prohibits dumping of radioactive waste at sea, or discharging such material into the atmosphere (Art. 3:3a). Disposal of radioactive material or waste by zonal states is permitted only according to IAEA standards and procedures within its territory or on land in other states (Arts. 3:3b, 4:2e).

The zonal states are requested to accede to the "Convention on Early Notification of a Nuclear Accident" (Art. 6).

The geographical scope of the South-East Asian nuclear-weapon-free zone would encompass the territories of the member states including their internal, territorial, and archipelagic waters, airspace, and in addition their exclusive economic zones and continental shelves (Arts. 1(a), 1(b), and 2:1). No zonal perimeter beyond these territories is defined.

The treaty does not restrict legal rights, according to the international law of the sea, of any state regarding transit and passage (Art. 2:2)[42], and also leaves to the individual zonal party to grant or deny warships or aircrafts of nuclear-weapon powers to call at their ports or land at their airfields (Art. 7). These provisions are very relevant as the zone is very much a maritime area including important sea-routes and international straits. Two zonal states, Indonesia and the Philippines are archipelagic states.

A system for verification and complaints similar to that of the South Pacific zone is also prescribed (Arts. 10–14, Annex).

Use of nuclear energy for peaceful purposes is regulated according to IAEA and NPT standards (Art. 4).

Nuclear-weapon states are invited to adhere to a Protocol prescribing them generally to respect the zone and not to use or threaten to use nuclear weapons against any zonal state or from within the zone. The zone borders one nuclear weapon power, China, both on land and across the sea, but no thinning-out measure is requested in the Protocol.

The ratification processes of the zonal states have been swift and the treaty entered into force in March 1997[43]. There are so far (August 1st, 1999) no signatures by nuclear-weapon states on the guarantee protocol. One reason is that the guarantees are supposed to be granted to all eligible zonal states even those which are not (yet) formal parties to the zonal treaty (Art. 2). Another is perceived contradictions between some zonal provisions and the international law of the sea. A third reason, involving China, concerns disputed small islands in the South China Sea[44]. Negotiations with the nuclear-weapon states will continue, however, on the precise terms of the Protocol in order to pave the way for their signatures.

Some Proposed NWFZs in Asia

The Middle East

The issue of establishing a nuclear-weapon-free zone in the Middle East was raised in the UN General Assembly in 1974 by Iran supported by Egypt[45]. Since that time, the UN General Assembly has every year adopted a resolution recommending the establishment of such a zone. After 1980, it has done so by consensus, i.e. with the support of all Arab states, Iran and Israel.

In 1988, the UN Secretary General initiated "*a study on effective and verifiable measures which would facilitate the establishment of a nuclear-weapon-free zone in the Middle East*"[46]. The report, prepared before Iraq's invasion of Kuwait in August 1990 but issued after the invasion, was adopted by consensus[47].

The report suggested that the zone geographically could include "*all states members of the League of Arab States*[48]*, the Islamic Republic of Iran and Israel*". The report observes that several sea areas would be relevant for a Middle East zone[49]. The UN report was later followed up by a UNIDIR report taking into account the Mubarak plan[50].

The UN report did not propose explicit language for a zone treaty, but did suggest a catalogue of measures to serve as confidence-building measures and as steps to prepare for a régime that would finally become a nuclear-weapon-free zone. Among recommended confidence-building measures were a regional nuclear test ban, the applying of IAEA safeguards on nuclear facilities in the area not covered at present[51], the acceding to the NPT by states currently non-parties, and providing for transparency regarding all major nuclear projects in the area.

The UN report further suggested that nuclear-weapon powers could extend negative nuclear security assurances to prospective zonal states and commit themselves not to station nuclear weapons in the area. Any outside state could declare past, current, and future supply of nuclear material and equipment to recipients in the prospective zonal area in order to put light on projects now creating suspicion that they may have a military role.

The report also recommends that outside support for peaceful nuclear activities in the area would be more appropriate if multilateral or regional in character. The institution of international facilities for nuclear waste disposal would help to ensure against diversion of fissionable material to military purposes. The latter aspect has been analysed by Dr. Mustafa Kibaroglu[52].

The current nuclear programmes of the Middle East countries suggests that only Israel has a nuclear weapon capability[53], or as many experts believe, is already a nuclear weapon power, although the Israeli Government many times declared that Israel will not be the first country to introduce nuclear weapons in the Middle East. This policy of deliberate ambiguity has been said to serve Israel's security interests in three ways: firstly, in times of gloom, it gives hope to the Israelis; secondly, it may caution the enemies of Israel; and thirdly, it relieves other states from the delicate burden of taking an explicit position on the matter.[54] But the mere suspicion that a prospective zonal state is in fact a nuclear-weapon state would require that the zonal treaty includes provisions on the disposal of nuclear weapons present in the zone at its initiation and on verifying that the zone is positively nuclear-weapon-free.

Iraq is a special case. After the Gulf war in 1991, it was revealed that Iraq, despite being a party to the NPT, had for several years pursued a clandestine multi-billion dollar nuclear weapons programme. By UN Security Council decisions, Iraq is now ordered to destroy all facilities related to its programmes for weapons of mass destruction and to their missiles of delivery, a process that will prepare Iraq for membership in a future NWFZ in the Middle East. This process was up to December 1998 supervised and verified by a Special Commission appointed by the UN Security Council[55]. The military nuclear programme of Iraq has now been eliminated. The other prospective zonal states, in the first place Iran cannot at this time support nuclear weapon programmes, but may be able to do so at some future time.

The UN report also observes the view widely held in the Middle East that verification procedures much more far-reaching than those so far implemented under the NPT would be necessary. (The IAEA operations later undertaken in Iraq under a Security Council mandate show what will be necessary). In addition, Israel indicated a need for bilateral verification rights similar to those prescribed in several arms control agreements adopted within the Conference on Security and Cooperation in Europe (CSCE)[56], in order not to be outvoted within international bodies deciding by majority rule.

A further development was the proposal, in 1990, by President Mubarak of Egypt to establish a zone free of weapons of mass destruction in the Middle East (WMDFZME)[57]. The proposal was not intended to replace the earlier idea of a nuclear-weapon-free zone in the area but rather to be pursued in parallel to the earlier proposal.[58]

Today, the expansion of the Middle East zone concept to include all weapons of mass destruction and also their means of delivery seems politically accepted. In May 1995, the 1995 Review and Extension Conference of the NPT parties adopted a resolution of the Middle East recognising that the current peace process contributes to "*a Middle East zone free of nuclear weapons as well as other weapons of mass destruction*" and calling upon all states in the Middle East to take practical steps towards "*the establishment of an effectively verifiable Middle East zone free of* ***weapons of mass destruction, nuclear, chemical and biological, and their delivery systems***" (emphasise added).[59]

A further development took place in April 1996, when the African nuclear-weapon-free zone was established, overlapping about half of the proposed zone in the Middle East. While important geographically, the African zone did not solve the main Middle East problem, as Israel is not covered. For the zone states in north-eastern Africa, the closeness to Israel has been referred to as an intriquate problem.

South Asia

The issue of establishing a nuclear-weapon-free zone in South Asia (NWFZSA) has been an issue on the agenda of the UN General Assembly since 1974[60] following the nuclear explosion test carried out by India on 18 May that same year[61]. A resolution initiated by Pakistan and endorsing the project has been adopted every year up to 1997[62]. But because India, a most important factor in the region, always voted against, the proposed zone has never been precisely defined regarding its geographic scope, operative commitments, verification requirements, etc. A prime argument for India's negative attitude is that nuclear disarmament should be a global rather than a regional issue. Therefore, the proposal is today as underdeveloped and as far from its implementation as in 1974.

But a measure of progress was registered in 1988 when India and Pakistan agreed not to attack nuclear installations and facilities of the other[63].

The NWFZSA project has recently been discussed by S. Yasmeen and D. Banerjee[64]. The nuclear industries of India and Pakistan have been described in a study from the Stockholm International Peace Institute[65].

It has been suggested that the zone geography would include the seven states members of the South Asian Association for Regional Cooperation (SAARC), i.e. Bangladesh, Bhutan, India, the Maldives, Nepal, Pakistan, and Sri Lanka[66]. Later, Afganistan has frequently been mentioned as part of such a zone. Among the states mentioned, all but Bhutan and India favours the establishment of the zone.

A dramatic development occurred in May 1998 when both India and Pakistan carried out several underground nuclear explosions[67]. Both India and Pakistan have now declared themselves nuclear-weapon states and new members of the atomic club, a status that is not recognised by most other states of the world and a concept in non-harmony with the NPT, upsetting both the general non-proliferation norm in the world and the legal régime of the NPT[68]. From a non-proliferation point of view, this development is very unfortunate. But politically, a new situation has emerged. How much nuclear weaponry that lies behind these test is subject to speculation and the two states have not been very forthcoming with information on that point. It could safely be assumed, however, that both countries still have a long way to go before they would become militarily threatening in nuclear terms on level with China and France.

In 1998, after the Indian and Pakistani nuclear tests, the UN

General Assembly did not address the zone issue again. Instead, it passed a different resolution expressing "*grave concern and strongly deplored" the tests*[69]. Both India and Pakistan voted against, however.

For the time being, however, both India and Pakistan are using the new situation for political purposes, such as attempts to upgrade their political status. Various ideas have been mentioned, including that both India and Pakistan could now be recognised as nuclear-weapon powers in exchange for adhering to the NPT as nuclear-weapon states and more importantly by adhering to the Comprehensive Test Ban Treaty. And could be added, by negotiating a nuclear-weapon-free zone in South Asia as a component of a final nuclear-weapon-free world.

North East Asia

The establishment of a nuclear-weapon-free zone in North East Asia with the Korean peninsula as a core element has been discussed for some time[70]. But no thorough investigation into the issue has been done by official authorities and no draft treaty text has been prepared (except for the Korean Joint Denuclearisation Declaration).

An analysis of possible objectives and outlines of an East Asian nuclear-weapon-free zone has been published by professor Andrew Mack suggesting a geographical concept that "*would be one which encompassed the two Koreas, Japan, and Taiwan*", and that an agreement on such a zone should be based on the Japanese "Three Noes" (non-possession, non-stationing and non-use or no-threat-of-use) and in addition a ban on dumping of nuclear wastes within the zone and a ban on reprocessing and uranium enrichment by zonal states.[71] The zone project has recently been discussed also by Bon-Hak Koo and Naoko Sajima[72].

The envisaged geographical scope would touch upon disputed territories, such as some Kurile islands, Taiwan, and groups of tiny islands in the South China Sea[73].

Geography would define such a zone as a "rim" outside two continental nuclear-weapon states—China and Russia[74]—making a discussion of possible "thinning-out" measures on their territories both relevant and significant. The military presence of US armed forces in the region, including at sea, based on positive alliance guarantees, underlines the issue of "thinning-out" possibly also at sea—nuclear weapons earlier deployed by such forces have by now, however, been withdrawn according to unilateral declaration[75].

Geography also invites a discussion of a possible inclusion in the zonal area of adjacent and semi-enclosed sea areas and of possible commitments attached to such sea areas. But that would raise several detail issues of straits and other narrow waters and their relation to both zonal interests and the international law of the sea. China, the Russian Federation, North Korea, and South Korea have established 12 nautical miles wide territorial waters. So has Japan except adjacent to five straits used for international navigation where claims are limited to three nautical miles[76]. The application of law of the sea rules to Taiwan is complicated due to its disputed status.

Consideration of a possible nuclear-weapon-free zone in East Asia would also have to address a number of important area specific matters including the following:

- the fact that all relevant states in the area are parties to the Non-Proliferation Treaty, although Taiwan's status as a sovereign entity is subject to dispute[77];
- the fact that the relevant states in the area but North Korea and all nuclear weapon states but France are parties to the Sea-Bed Treaty;
- the Joint Declaration of South and North Korea on the Denuclearisation of the Korean Peninsula[78];
- the "Three Non-Nuclear Principles" unilaterally declared by Japan in 1967[79], i.e. not to produce nuclear weapons, not to possess nuclear weapons, and not to permit the entry into Japan of nuclear weapons;
- the developed nuclear power industry in the region including the significant plutonium dimension of the energy policies of Japan[80]. The desirability to promote and encourage nuclear energy production for peaceful purposes on one hand and the proliferation prone capabilities of some parts of the nuclear fuel cycle on the other, provides for a difficult and undesirable dilemma. Therefore, it would be worthwhile to explore the possibilities, as recently suggested, to establish an "Asian Atomic Energy Commission" with a wide membership and cooperative verification that would

leave no one in doubt on what is really going on[81]. The structure and almost 40 years of experience of Euratom could then be rather useful as a model. This aspect has been analysed by Dr. Mustafa Kibaroglu for the Middle East case but could be referred to for other regions as well[82].

- the status of Taiwan and its related sea areas, being politically controversial, although hardly impossible to accommodate as the objective of avoiding nuclear confrontation usually takes priority before local sovereignty issues[83];
- the US nuclear test site in the North Pacific, the Amchitka Island (lat. 51° 30' N, long. 179° 0' E) in the Aleuteans, has been abandoned since early 1972.

The participation of Taiwan in an East Asia nuclear-weapon-free zone might facilitate the solution of its controversial status and the territorial dispute with China. A comparison in point is the role of the demilitarisation régime of the Aaland Islands, now part of Finland, in the Baltic Sea. When Finland after the First World War liberated itself from Russian rule, the people of the Aaland Island wanted to become part of nearby Sweden rather than Finland. The demilitarisation of the Aaland archipelago—a consequence of the Crimean War 1853–1856[84]—then became the basis for a compromise solution to the dispute, worked out within the League of Nations[85]. The Aaland Islands stayed as part of the Republic of Finland, was awarded a considerable autonomy, and continued to be demilitarised, now according to a modernised treaty[86]. A similar analysis could be outlined about the possible inclusion of the province of Tibet in a South Asian nuclear-weapon-free zone.

A Maritime Zone

In 1970, a non-aligned summit meeting[87] proposed to establish the Indian Ocean as a zone of peace. A year later, the United Nation's General Assembly declared the Indian Ocean a zone of peace "*within limits to be determined*"[88]. The expressed objective was to remove from the area any manifestation of great power military presence including military bases and nuclear and other weapons of mass destruction, and to establish a collective security system, where "*warships and military aircraft may not use the Indian Ocean for any threat or use of force against any littoral or hinterland states.....*". But there would be no infringement of the right of navigation according to international law.

Geographically and politically, the Indian Ocean zone of peace issue is closely related to the long-time proposal to establish a nuclear-weapon-free zone in South Asia.

The Indian Ocean issue has been on the agenda of the United Nations ever since, the purpose being to deepen the declaration into a more solid and elaborate treaty régime. A resolution on the matter is adopted by the General Assembly each year.

In July 1979, the Littoral and Hinterland states of the Indian Ocean held a meeting resulting in a substantive final document[89]. A preliminary geographical delimitation was agreed as "*the Indian Ocean itself, its natural extensions, the islands thereon, the ocean floor sub-adjacent thereto, the littoral and hinterland states and the air space above*". On denuclearisation, "*the nuclear-weapon states were called upon to undertake not to establish nuclear bases in the Indian Ocean and to refrain from conducting nuclear test activities in the Indian Ocean*", and similarly, "*the littoral and hinterland states of the Indian Ocean should agree not to acquire or introduce nuclear weapons in the Indian Ocean themselves or to allow their introduction by an external power*"

Central Asia and Mongolia

At the NPT Extension and Review Conference in 1995, Kyrgyzstan and Uzbekistan proposed the establishment of a nuclear-weapon-free zone in Central Asia (NWFZCA). Already in 1993, the President of the Republic of Uzbekistan had appealed to the UN General Assembly to support the establishment of such a zone. At the same NPT Extension and Review Conference, Mongolia declared its territory as a single state nuclear-weapon-free zone.[90]

Two years later, the heads of state of five former Soviet republics and prospective zonal states, Kazakhstan, the Kyrgyz Republic, the Republic of Tajakistan, Turkmenistan, and the Republic of Uzbekistan adopted a declaration in Almaty, February 28th, 1997, envisaging such a zonal contribution to the nuclear non-proliferation effort approaching the next NPT Review Conference in the year 2000.[91]

The five Central Asian states continued to pursue their proposal at the 1997 meeting of the UN Disarmament Commission[92] and at an international conference in Tashkent 15–16 September 1997[93]. The Tashkent conference was followed up by several consultative meetings of experts of the Central Asian countries, nuclear-weapon states, and the United Nations for drafting of legal instruments of and preparations for the proposed zone.

The Central Asian zone proposal was unanimously endorsed and encouraged by the UN General Assembly on 4 December 1998[94].

There thus seems to be a realistic prospect that the NWFZCA project will be well under way when the next NPT Review Conference meets in the year 2000.

One area-specific issue to be addressed is the cleaning-up after the nuclear explosions at the Semipalatinsk test site in Kazakhstan. The original proposal was very much based on concerns "*considered through a prism on the ecological condition in the region*"[95].

Another is the zonal delimitation in the Caspian Sea where Russia—a nuclear-weapon power—is also a coastal state[96].

The nuclear industry in the area is modest. Kazakhstan operates one 70 MW(e) nuclear power reactor. There are three research reactors in Kazakhstan and one in Uzbekistan.

When Mongolia in 1997 tabled a document outlining elements of and general guidelines for creating single-state nuclear-weapon-free zones, it also referred to supporting statements from all nuclear-weapon powers regarding itself being established as such a zone[97]. Mongolia's persistent efforts to institutionalise its single state nuclear-weapon-free status has been met with wide support. The UN General Assembly unanimously endorsed and encouraged the project on 4 December 1998[98]. Expert deliberations on the project supported by the United Nations continue.

Mongolia has no nuclear power industry.

The only remaining major step for the establishment of Mongolia as a single state nuclear-weapon-free zone thus seems to be formal nuclear-weapon state signatures on a guarantee protocol.

The five Central Asian states and Mongolia are geographically separated by a narrow land strip, some 40 straight-line kilometers wide (some 55 kilometers along the borderline), where Russia and China share a common border. This separation seems small enough to permit Mongolia to join the other five in one nuclear-weapon-free zone especially if China and Russia agree to include into the zone a small bridging territory between the two areas, possibly under a dependency protocol.

The current proposition is to establish two separate zones, however. Together, the two zones would, when established, encompass a large landmass and could play an important geopolitical role as a denuclearised corridor separating the nuclear powers of China and Russia along most of the former Chinese-Soviet border line. It could just be hoped that this straightforward process would not be "disturbed" by the concrete demonstration of the nuclear-weapon ambitions of nearby India and Pakistan which, however, do not border the proposed zones directly.

Address:

JAN PRAWITZ
The Swedish Institute of International Affairs
S-11182 Stockholm
Sweden

Note:

1. The three states are Cuba in Latin America and Cook Island and Niue in the South Pacific, and the latter two consider themselves bound by the adherence to the NPT by New Zealand administering these two states when NPT entered into force.
2. During the ratification process in the USA, Secretary of State Dean Rusk explained to the US Senate that the NPT "*does not deal with arrangements for deployment of nuclear weapons within Allied territory, as these do not involve any transfer of nuclear weapons or control over them unless and until a decision were made to go to war, at which time the treaty would no longer be controlling*" (Documents on Disarmament 1968, p. (478–495). This statement, indicating an interpretation that the NPT would enter out of force in case of war, reflected a previously agreed position within the NATO alliance. However, in 1985, the third Review Conference of the NPT parties unanimously adopted a final declaration stating inter alia that "*the Conference agreed that the strict observance of the terms of Articles I and II remains central to achieving the shared objectives of preventing* ***under any circumstances*** *(emphasis added) the further proliferation*

of nuclear weapons and preserving the Treaty's vital contribution to peace and security, including the peace and security of non-parties" (Document NPT/CONF. III/64/I, Annex I), thus stating the opposite interpretation. While Mr. Rusk's statement primarily referred to the East-West conflict dominating at the time, the end of the Cold War and the prospects for local wars in the future, now makes the more restrictive 1985 interpretation the only reasonable one. In 1991, the UN Security Council did indeed confirm the 1985 approach in its resolution on Iraq. The opposite interpretation that Iraq's involvement in the Gulf war would have entitled her to acquire nuclear weapons would be beyond reason.

3. Document NPT/CONF. 1995/32/DEC. 3 (11 May 1995), contained in an annex to document NPT/CONF. 1995/32 (Part I).
4. *The Treaty between the USA and the USSR on the Elimination of their Intermediate-Range and Shorter-Range Missiles* (INF Treaty) signed 8 December 1987 and entered into force 1 June 1988. *Treaty between the USA and the USSR on the Reduction and Limitation of Strategic Offensive Arms* (START) signed 31 July 1991 and entered into force 5 December 1994. *Treaty between the USA and the Russian Federation on Further Reduction and Limitation of Strategic Offensive Arms* (START II) signed 3 January 1993 and not yet in force.
5. *Protocol to the Treaty between United States of America and the Russian Federation on Further Reduction and Limitation of Strategic Offensive Arms of January 3, 1993*; for text see Arms Control Today, Vol. 27 No. 6, September 1997, p. 19. In addition, a START III negotiation was scheduled envisaging a combined START II plus START III reduction towards 2000—2500 strategic nuclear weapons on each side by the same date, December 31st, 2007. *Joint Statement on Parameters of Future Reductions in Nuclear Forces*. US-Russian Summit meeting in Helsinki March 21st, 1997; for full text see e.g. Arms Control Today, Vol. 27 No. 3, March 1997, p. 19.
6. Unilateral declarations by Presidents Bush of the USA and Gorbachev of the USSR, on 27 September and 5 October 1991 respectively. For text of the Bush, Gorbachev, and Yeltsin-statements, see e.g. *SIPRI Yearbook 1992*, Oxford University Press, 1992, pp 85–92.

   At the time, the urgency of the situation dictated the form of agreement. The dissolution of the Soviet Union appeared imminent employing a dangerous possibility that 14 Soviet republics seeking their independence could also become de facto nuclear-weapon states by taking over the control of tactical nuclear weapons that happened to be stationed on their soil. The agreement meant that the Russian Federation—the only successor state of the Soviet Union continuing the role of nuclear-weapon power—kept control of all Soviet sub-strategic nuclear weapons and lost no control of such weapons to new states.
7. UN Document A/RES/50/245, 10 September 1996.
8. UN Document Res S 255 (1968).

9. "Negative" guarantees imply that the guarantor abstains from nuclear aggression as different from "positive" guarantees implying that the guarantor actively supports a victim of aggression.
10. The content of these unilaterally declared guarantees are summarised in *The United Nations DISARMAMENT YEARBOOK* VOL. 14:1989 pp 179–180.
11. UN Document Res S 984 (1995), unanimously adopted on 11 April 1995. The basic declarations were made on 5 and 6 April 1995 by the Russian Federation (UN Document S/1995/261), the UK (S/1995/262), the USA (S/1995/263), France (S/1995/264), and China (S/1995/265). On 12 June 1982, The Soviet Union declared a no-first-use policy of nuclear weapons. Neither the "Basic Provisions of the Military Doctrine of the Russian Federation" adopted on 2 November 1993 (Decree No. 1833), nor the "National Security Concept of the Russian Federation" adopted on 17 December 1997 (Decree No. 1300) include the USSR no-first-use declaration, however.
12. Compare J. Simpson, *Inter-Relations Between Regional and Global Approaches to Nuclear Non-Proliferation*, in A. Mack (Ed.), *Nuclear Policies in Northeast Asia*, Research Report UNIDIR/95/16 (UN Sales No. GV.E.95.0.8).
13. Principles and Objectives for Nuclear Non-Proliferation and Disarmament [Document NPT/Conf. 1995/32/DEC. 2 contained in an annex to document NPT/CONF. 1995/32 (Part I)]. The relevant paras (5–7) read:

    "The conviction that the establishment of internationally recognised nuclear-weapon-free zones, on the basis of arrangements freely arrived at among the states of the region concerned, enhances global and regional security is reaffirmed.

    The development of nuclear-weapon-free zones, especially in regions of tension, such as in the Middle East, as well as the establishment of zones free of weapons of mass destruction should be encouraged as a matter of priority, taking into account the specific characteristics of each region. The establishment of additional nuclear-weapon-free zones by the time of the Review Conference in the year 2000 would be welcome.

    The cooperation of all the nuclear-weapon states and their respect and support for the relevant protocols is necessary for the maximum effectiveness of such nuclear-weapon-free zones and the relevant protocols."
14. Documents NPT/CONF. 1995/MC. II/1 as contained in NPT/CONF. 1995/32 (Part II), p. 330, paras 12–14; NPT/CONF. 1995/DC/CRP. 1/Rev. 4; and NPT/CONF. 1995/MC.II/17.
15. UN Document A/RES/51/45 B.
16. The term "densely populated" area is frequently used to distinguish the Latin American and the South Pacific zones from the Antarctica which some states for political reasons prefer to designate as a "populated" area rather than the "unpopulated" place it is otherwise

considered to be.

17. The Treaty for the Prohibition of Nuclear Weapons in Latin America and the Caribbean (*UN Treaty Series*, Vol. 634, No. 9068).
18. The South Pacific Nuclear Free Zone Treaty (*UN Treaty Series* No. 24592).
19. Treaty on the Southeast Asia nuclear-weapon-free Zone signed by member states and potential member states of ASEAN at a summit meeting in Bangkok on 15 December 1995. For text, see *SIPRI Year Book 1996*, Oxford University Press, 1996, pp 601–609.
20. The African Nuclear-Weapon-Free Zone Treaty (Treaty of Pelindaba) was opened for signature at an OAU meeting in Cairo on 11 April 1997. For text, see *SIPRI Yearbook 1996*, Oxford University Press, 1996, pp 593–601.
21. Joint Declaration of South and North Korea on the Denuclearisation of the Korean Peninsula, signed on 31 December 1991 and entered into force on 19 February 1992. For text, see UN Document CD/1147, or J. Goldblat, *Arms Control: A Guide to Negotiations and Agreements*, Sage Publications, London, 1994, pp 643–644.
22. The Antarctic Treaty (*UN Treaty Series*, Vol. 402, No. 5778). For an analysis of the demilitarisation régime of Antarctica, see M. Jacobsson, *The Antarctic Treaty System—Erga Omnes or Inter Partes?*, Kluwer, to be printed.
23. *Towards a Nuclear-Weapon-Free Zone in the Middle East* (UN Document A/45/435). UN Sales No. E.91.IX.3.
24. The studies on the Nordic zone proposal have been summarised in an official five-nation report *Nuclear-Weapon-Free Zone in the Nordic Area. Report from the Nordic Senior Officials Group*. Ministry of Foreign Affairs of Sweden. March 1991.
25. *Comprehensive Study on the Question of Nuclear-Weapon-Free Zones in all its Aspects*. United Nations Document A/10027/Add. 1 (UN Sales No. E.76.1.7); and *Study on the Question of Nuclear-Weapon-Free Zones*. The latter report was not entirely finalised but "exists" as an annex. to a letter of 9 February 1985 from the Chairman of the expert group, Dr. Klaus Törnudd of Finland, to the Secretary General. The formal status of this annex is subject to dispute. It is, however, very informative.
26. Report of the Disarmament Commission's substantial meeting 12–30 April 1999 (UN Document A/54/42), Annex I: *Establishment of nuclear-weapon-free Zones on the basis of arrangements freely arrived at among the States of the region concerned*.
27. For a general reference to the "theory" of nuclear-weapon-free zones, see e.g. J. Prawitz, J.F. Leonard, *A Zone Free of Weapons of Mass Destruction in the Middle East*, Document UNIDIR/96/24 (UN Sales No. GV.E.96.0.19), particularly its Chapter 3: *The Theory of NWFZs and WMDFZs*, pp 35–57.
28. For the purpose of an analysis of nuclear weapons carried by warships and aircraft, the term "international waters" covers the UNCLOS

concepts of "exclusive economic zones" (EEZ) and "high seas". Compare UNCLOS Art. 55–58 and Art. 86–93.

29. *United Nations Convention on the Law of the Sea* (UN Sales No. E. 83. V.5), signed on 10 December 1982 and entered into force on 14 November 1994.
30. The preparatory commission drafting the Tlatelolco Treaty stated that 'transit' in "*the absence of any provision in the Treaty, must be understood to be governed by the principles and rules of international law" and that "it is for the territorial State, in the free exercise of its sovereignty, to grant or deny permission for such transit in each individual case*" (Document COPREDAL/76 p. 8, or UN document A/ 6663). The transit issue was thus considered a bilateral issue between the flag state and the port state in each case and did, in particular, not contradict the traditional US rights to pass the Panama Canal with nuclear weapons.
31. The fact that nuclear-weapon powers are parties to "dependency protocols", i.e. France, the UK, and the USA to Protocol I of the Latin American zone treaty and to Protocol 1 of the South Pacific zone treaty, and France to Protocol III of the African zone treaty poses this problem which has not, however, been raised or referred to politically.
32. The UN Disarmament Commission in April 1999 agreed to recommend that "*States parties to a nuclear-weapon free zone exercising their sovereign rights and without prejudice to the purposes and objectives of such a zone remain free to decide for themselves whether to allow visits by foreign ships and aircraft to their ports and air-fields, transit of their airspace by foreign aircraft and navigation by foreign ships in or over their territorial sea, archipelagic waters or straits that are used for international navigation, while fully honoring the rights of innocent passage, archipelagic sea lane passage or transit passage in straits that are used for international navigation*". (UN Document A/54/42 Annex I, para 31).
33. The problem of drawing a line between the territorial airspace, subject to national jurisdiction of the underlaying state, and the international outer space where the underlying state would have no jurisdiction, has been on the agenda of the United Nation's Committee on the Peaceful Uses of Outer Space for very many years. Many difficult issues must be taken into account when defining such a line. Without a final solution, a reasonable assumption would be to assume that the line would be drawn at approximately 100.000 meters above the sea level.
34. For an account of the consequences of these policies, see i.e. J. Prawitz, The "*Neither Confirming nor Denying" Policy at Sea* in J. Goldblat, (Ed.), *Maritime Security: The Building of Confidence*. Document UNIDIR/92/89 (Sales No. GV.E.92.0.31).
35. Compare footnote No. 11.
36. The "thinning out" idea was first suggested by A. Thunborg in 1975 in relation to the proposed nuclear-weapon-free zone in the Nordic area. A. Thunborg, *Nuclear Weapons and the Nordic Countries*

Today—A Swedish Commentary, A Special Issue of Ulkopolitiikka 1975, pp 34–38.

37. The legal concept of "innocent passage" and "transit passage" are defined in the United Nations Convention of the Law of the Sea (UNCLOS) Articles 17–33, 45, and 52, and Articles 38–44 respectively.
38. UN Document A/C. 1/1019.
39. UN Document A/47/80.
40. UN Document A/Res 47/53 B.
41. Member states of ASEAN are Brunei Darussalam, Indonesia, Malaysia, the Philippines, Singapore, Thailand, and Vietnam. Combodia, Laos, and Myanmar are entitled to join and may do so later.
42. This provision mirrors UNCLOS Art. 6–45 and 52–54.
43. On August 1st, 1999, the only remaining ratification was that of the Philippines.
44. For an account of the complicated negotiations about the delimitation between neighbouring coastal states of the South China Sea and its Paracel and Spratly island groups, see M. J. Valencia, *China and the South China Sea Disputes*, Adelphi Paper 298, IISS, 1995.
45. Compare UN Documents A/9693/Add. 1 and A/RES/3263 (XXIX)
46. UN Document A/RES 43/65.
47. UN Document A/RES 45/52 op. 8.
48. The League of Arab States has 22 member states: Algeria, Bahrain, Comoros, Djibouti, Egypt, Iraq, Jordan, Kuwait, Lebanon, Libya, Mauritania, Morocco, Oman, Palestine, Qatar, Saudi Arabia, Somalia, Sudan, Syria, Tunisia, United Arab Emirates, and Yemen.
49. Both the Red Sea and the Persian Gulf would be enclosed within the area. Prospective parts of the zonal area have coasts in the Meditarranean, the Atlantic, the north western Indian Ocean (and the Caspain Sea) and are adjacent to a few international straits subject to the régime of transit passage, i.e. the straits of Gibraltar, Bab al Mandab, and Hormuz.

    Also important in this respect is the Suez Canal, an iinternational waterway crossing through Egyptian territory open "*in time of war as in time of peace, to every vessel of commerce or of war, without distinction of flag*" according to the Constantinople Convention of 29 October 1888. Only a ship flying the flag of a state at war with Egypt can be prevented from passing the Canal. This convention is also referred to in the Egypt-Israel Peace Treaty of 1978. which provides, i.e. that the Strait of Tiran and the Gulf of Aqaba are "*international waterways open to all nations*".

    As regards the Caspian Sea, compare footnote No. 96.
50. J. Prawitz, J.F. Leonard, *A Zone Free of Weapons of Mass Destruction in the Middle East*, Document UNIDIR/96/24 (UN Sales No. GV.E.96.0.19).

    The issue of establishing a nuclear-weapon-free zone in the Middle East has also been researched by the Egyptian scholar and diplomat Mahmoud Karem (*A Nuclear-Weapon-Free Zone in the Middle East:*

Problems and Prospects. Greenwood Press. New York. 1988. The same author later published "*A Nuclear-Weapon-Free Zone in the Middle East: A Historical Overview of the Patterns of Involvement of the United Nations*" in T. Rauf (Ed.), *Regional Approaches to Curbing Nuclear Proliferation in the Middle East and South Asia,* Aurora Papers 16. Canadian Centre for Global Security. December 1992). Important analytical contributions were made in 1997 by Shai Feldman and Abdullah Toukan (S. Feldman, *Nuclear Weapons and Arms Control in the Middle East,* MIT Press, Cambridge MA, 1997; S. Feldman, A. Toukan, *Bridging the Gap: A Future Security Architecture for the Middle East,* Carnegie, Rowman & Littlefield Publishers, Lanham MD, 1997). The effect of a comprehensive test ban on nuclear proliferation risks in the Middle East, including Iran, Iraq and Israel, has been analysed by Eric Arnett (*Nuclear Weapons After the Comprehensive Test Ban: Implications for Modernization and Proliferation*, SIPRI, Oxford University Press, 1996). Recently, a NWFZ in the Middle East has been discussed by Ibrahim Karawan and Gerald Steinberg (I. A Karawan, The Case for a Nuclear-Weapon-Free Zone in the Middle East, and G. M. Steinberg. *The Obstacles to a Middle East Nuclear-Weapon-Free Zone,* both articles in R. Thakur (Ed.), *Nuclear-Weapon-Free Zones*, MacMillan Press Ltd., 1998, on pp 184–193 and pp 194–209 respectively).

51. International safeguard issues involved was explored at an IAEA workshop in Vienna 4–7 May 1993, See *Proceedings of the Workshop on "Modalities for the Application of Safeguards in a Future Nuclear-Weapon-Free Zone in the Middle East*", Division of External Relations, IAEA.
52. M. Kibaroglu, "EURATOM & ABACC: Safeguard Models for the Middle East?" in J. Prawitz, J.F. Leonard, *A Zone Free of Weapons of Mass Destruction in the Middle East* (Annex), Document UNIDIR/96/24 (UN Sales No. G.V.E.96.0.19), pp 93–123.
53. For recent accounts of Israel's nuclear industry and its history, see A. Cohen, *Israel and the Bomb*, Columbia University Press, New York, 1998; and D. Albright, F. Berkhout, W. Walker, *Plutonium and Highly Enriched Uranium: World Inventories, Capabilities, and Policies*, SIPRI, Oxford University Press, 1997, Part IV "*Material Inventories and Production Capabilities in Threshold States*" pp 257–264.

    On 22 September 1979, a US observation satellite recorded a "suspicious event" in the South Atlantic with a signal signature similar to those produced by nuclear explosions. It was widely suspected at the time that the event might have been an Israeli or a South African test explosion, possibly in cooperation. The case was never hundred percent clarified, despite several thorough investigations including a UN expert study (UN Document A/34/674). The most probable explanation is, however, that the recorded signal was produced by internal equipment malfunction or anomaly and not by an external source.
54. As reported by Dr. S. Freier in C. Attlerling Wedar, S. Hellman, K.

Söder,. (Eds.), *Towards a Nuclear-Weapon-Free World.* Swedish Initiatives. (ISBN 91–972128–0–6) Stockholm 1993, p. 181.

55. UN Document S/RES/687 (1991), 3 April 1991.
56. At the CSCE Summit meeting in Budapest 5–6 December 1994, the CSCE was renamed the Organisation for Security and Cooperation in Europe (OSCE).
57. Document CD/989 (20 April 1990).
58. The Mubarak plan is described by Mohammed Shaker in *Prospects for Establishing a Zone Free of Weapons of Mass Destruction in the Middle East*, Director's Series on Proliferation, No. 6 Oct. 1994, Lawrence Livermore National Laboratory (UCRL-LR-114070-6). Compare also M.M. Zahran, *Towards Establishing a Mass-Destruction-Weapon-Free-Zone in the Middle East*, Institute for Diplomatic Studies, Ministry of Foreign Affairs of Egypt, October 1992. The Mubarak plan concept is further developed in J. Prawitz, J. F. Leonard, *A Zone Free of Weapons of Mass Destruction in the Middle East*, Document UNIDIR/96/24 (UN Sales No. GV.E.96.0.19).
59. Document NPT/CONF.1995/32/RES/1.
60. The first resolutions are UN Documents A/RES/3265 A (XXIX) and A/RES/3265 B (XXIX).
61. India is not a party to the Non-Proliferation Treaty. The (underground) nuclear explosion conducted by India on 18 May 1974 was explained to be "for peaceful purposes", and India was regularly not considered a nuclear-weapon state, neither by itself nor by others.
62. The latest such resolution is included in UN Document A/RES/52/35, 12 December 1997.
63. "Agreement between Pakistan and India on the Prohibition of Attack against Nuclear Installations and Facilities" was signed 31 December 1988 and entered into force on 27 January 1991. For text, see J. Goldblat, *Arms Control: A Guide to Negotiations and Agreements*, PRIO, Sage Publications, London, 1994, pp 537–538.
64. S. Yasmeen, *The Case for a South Asian Nuclear-Weapon-Free Zone, and D. Banerjee, The Obstacles to a South Asian Nuclear-Weapon-Free Zone*, both articles in R. Thakur (Ed.), *Nuclear-Weapon-Free Zones*, MacMillan Press Ltd., 1998, on pp 152–172 and pp 173–193 respectively.
65. D. Albright, F. Berkhout, W. Walker, *Plutonium and Highly Enriched Uranium: World Inventories, Capabilities, and Policies*, SIPRI, Oxford University Press, 1997, Part IV "*Material Inventories and Production Capabilities in Threshold States*", pp 264–281.
66. SAARC was founded in 1985.
67. India undertook three test explosions on May 11th including, according to the press release a thermonuclear test, and two more tests on May 13th. Pakistan undertook five tests explosions on May 28th and one more test on May 30th.
68. NPT Art. IX:3 states that "*.....For the purpose of this treaty, a nuclear-weapon state is one which has manufactured and exploded a nuclear*

weapon or other nuclear explosive device prior to 1 January 1967." The purpose of setting a date was to prohibit a race for acquiring nuclear weapon power status before the entry into force of the NPT. This definition does not, however, cover a new country manufacturing and exploding a nuclear weapon or other nuclear explosive device after the stated date beyond the five established at the time, i.e. China, France, the UK, the USA, and the USSR (in 1991 succeeded by the Russian Federation). Nor does the NPT formula recognise the case that a nuclear-weapon state abondons its nuclear arms to become non-nuclear. South Africa did manufacture 6 nuclear explosive devices but dismantled them before adhering to the NPT. When India conducted a nuclear explosion in May 1974 (and in May 1998) and Pakistan did the same in May 1998, both states have fulfilled the substantial but not the chronological criterion for NPT recognition as nuclear-weapon state.

69. UN Document A/RES/53/77 G.
70. A recent example is the call by the Mayor of Hiroshima, Mr. Takashi Hiraoka, for "*the establishment of a new nuclear-free zone in the Asia-Pacific*" at the Peace Memorial Ceremony on the occasion of the 50th anniversary of the atomic bombing of Hiroshima, 6 August 1995. *The City of Hiroshima Peace Declaration*, August 6, 1995.
71. A. Mack, *A Northeast Asia Nuclear-Free Zone: Problems and Prospects*, in A. Mack (Ed.), *Nuclear Policies in Northeast Asia*. Research Report UNIDIR/95/16 (UN Sales No. GV.E.95.0.8), pp 97–126.
72. Bon-Hak Koo, *A Northeast Asian Nuclear-Weapon-Free Zone: A Korean Perspective*, and N. Sajima, *A Northeast Asian Nuclear-Weapon-Free Zone: A Japanese Perspective*, both articles in R. Thakur (Ed.), *Nuclear-Weapon-Free Zones*, MacMillan Press Ltd., 1998, on pp 123–139 and pp 140–151 respectively.
73. An analysis of the territorial disputes in the South China Sea has been published by M.J. Valencia, *China and the South China Sea Disputes*, Adelphi Paper 298, IISS, 1995.
74. For the geography of nuclear deployment of the nuclear-weapon states, see W.M. Arkin, R.S. Norris, J. Handler, *Taking Stock: World-wide Nuclear Deployments 1998*, NRDC Nuclear Programme, March 1998. Compare also D. Lockwood, *The Status of U.S., Russian, and Chinese Nuclear Forces in Northeast Asia*, Arms Control Today, Vol. 24 No. 9 November 1994, pp 21–24.
75. The US declared on 22 January 1992 that all nuclear weapons had been withdrawn. The US unilateral declaration of 27 September 1991 was declared implemented 2 July 1992.
76. Relates to the Soya Strait, the Sugaru Strait, the channels on both sides of the Tsushima Island, and the Osumi Strait. Compare *The Law of the Sea. National Claims to Maritime Jurisdiction.* (UN Sales No. E.91.V.15). 1992. p. 72.
77. The Republic of Korea, Japan, and the Democratic Peoples Republic of Korea became parties to the Non-Proliferation Treaty in 1975, 1976,

and 1985 respectively. It should be noted that Taiwan (Republic of China) did ratify the NPT in 1970. Among the three depositary governments, only the USA accepted its instrument of ratification. The USSR and the UK did not recognise the government of Taiwan, and the Peoples Republic of China considers "*the signing and ratification of the NPT by Taiwan in the name of China as illegal and null and void*". After the Peoples Republic of China replaced Taiwan (Republic of China) in the United Nations in November 1971, the government of Taiwan has been considered a non-governmental organisation by the UN and the IAEA, and Taiwan could not conclude an NPT related safeguards agreement with the IAEA. Taiwan should, however, be considered a non-nuclear-weapon party to the NPT. Its nuclear activities are subject to IAEA safeguards according to a unilateral submission in October 1969 (IAEA Document INFCIRC/133) and to the transfer of a US/Taiwan agreement in December 1971 (IAEA Document INFCIRC/158).

78. Compare footnote No. 21.
79. For an elaboration of the consequences of the "Three Non-Nuclear Principles, see H. Fujita, *The Three Non-Nuclear Principles of Japan*, Kansai University Review of Law and Politics, No. 5 March 1984, pp 27–66.
80. The nuclear programmes of some of the prospective zonal states are much bigger than those of member states of other zones, nuclear power being an essential part of the electric energy supply of South Korea (41.4% of total), Japan (35.9%), and Taiwan (24.8%). South Korea has currently 15 reactor units in operation producing 12340 MW(e), Japan has 53 units producing 43691 MW(e), and Taiwan has 6 units producing 4884 MW(e). In addition, 6 new units are under construction in these countries. The current nuclear power programme of North Korea is by comparison minor. Besides one research reactor, one small [50 MW(e)] and one medium size reactor [200 MW(e)] were being constructed but are now "freezed". However, according to a recent agreement with the USA, an international consortium would provide North Korea with two light water-moderated nuclear reactors with a total generating capacity of approximately 2000 MW(e) by a target date of 2003. The states in the region operate 25 research reactors. (Power data according to IAEA Reference Data Series No. 1–3, *Energy, Electricity and Nuclear Power Estimates for the Period up to 2020*, July 1999 edition; *Nuclear power reactors in the World*, April 1999 Edition; and *Nuclear Research Reactors in the World*, December 1998 Edition).
81. *Plutonium and Nuclear Power in Japan, in The United states, Japan, and the Future of Nuclear Weapons*, The Carnegie Endowment for International Peace, 1995, p 44.
82. M. Kibaroglu, "EURATOM & ABACC: Safeguard Models for the Middle East?" in J. Prawitz, J.F. Leonard, *A Zone Free of Weapons of Mass Destruction in the Middle East* (Annex.), Document UNIDIR/96/24 (UN

Sales No. GV.E.96.0.19), pp 93–123.

83. There are several such precedents int he history of arms control when ratification instruments have included interpretative declarations referring, explicitely or implicitely, to the status of territories and governments considered controversial. Examples include the German Democratic Republic, the city of Berlin, Israel, Southern Rhodesia, South Africa, the Democratic Republic of Viet-Nam and the Republic of South Viet-Nam, the Democratic Peoples Republic of Korea and the Republic of Korea, and others. A special category are declarations related to sovereignty rights at sea.
84. Treaty on the Demilitarisation of the Aaland Islands, linked to The Treaty of Paris of 30 March 1856.
85. 14th meeting of the Coucil, 24 June 1921; *Official Journal*, September 1921, p 699.
86. The 1921 Convention on the Non-Fortification and Neutralisation of the Aaland Islands, *League of Nations Treaty Series*, Vol. 9, 1922.
87. The Third Conference of the Heads of State or Government of Non-Alligned Countries, held at Lusaka, Zambia, 8–10 September 1970.
88. UN Document A/RES 2832 (XXVI), 19 January 1972.
89. Final Document of the Meeting of the Littoral and Hinterland States of the Indian Ocean, held in New York 2–13 July 1979 (UN Document A/34/45 and Corr. 1). For text, see also e.g. United Nations DISARMAMENT YEARBOOK, Vol. 4:1979 (UN Sales No. E.80.IX.7), pp 325–330.
90. Compare footnote No. on page.
91. Almaty Declaration, 28 February 1997, Op. 2 reads "To call, on the eve of the 50th anniversary of the Semipalatinsk Testing Ground, upon all interested countries to support the idea of declaring Central Asia a nuclear-free zone, open to accession also to other states in the region". Compare also footnote No. 13.
92. UN Document A/CN. 10/1997/WG. I/WP. 12, 6 May, 1997.
93. For report and text of agreed statement of the five Ministers of Foreign Affairs, se A. Karimova, *Possibilities of a Nuclear-Weapon-Free Zone Creation in Central Asia*, in P. Cotta-Ramusino (Ed.), *Nuclear and Conventional Disarmament: Progress or Stalemate*?, Unione Scienziati per il Disarmo (USPID), Milano, 130–134.
94. UN Documents A/RES/53/77 A and A/53/183.
95. Main ecological concerns of a transnational nature in Central Asia referred to are the Semipalatinsk Testing Ground, the Aral Sea basin and the Priaralye region.
96. The Convention on the Law of the Sea does not apply to the Caspian Sea, a lake which used to be divided by demarcation between Iran and the Soviet Union. It has been suggested by some of the coastal states on the Caspian Sea, however, that the Convention by special agreement among the coastal states should be applied also to this lake, thereby introducing new legal concepts such as territorial waters and exclusive economic zones in the area. There could eventually be a

need for a precise delimitation in the Caspian Sea between a Middle East zone including Iran and a Central Asian zone in order to avoid overlaps or grey zones.

97. UN Document A/CN. 10/195, 22 April 1997.
98. UN Document A/RES/53/77 D.

7

Nuclear Weapons' Free Zones: A Comparative Political Analysis

MUSTAFA ELWI SAIF

INTRODUCTION

Since the first nuclear attack on Japan, in August 1945, the world has recognised the unique destructive consequences of nuclear weapons, due to their thermal, explosive and radioactive effects. On the other hand, the world great powers have recognised the political and strategic importance of such weapons, during the post-World War II era, due to their decisive role in the strategy of deterrence that dominated relations between the great powers. Thus, the nuclear arms race emerged, as the world's great victorious powers wanted to balance the United States' superiority in the nuclear field. The Soviet Union was the first to achieve such a goal before the end of the 1940s, followed by United Kingdom in 1952, France in 1960 and China in 1964. The United States presented in the 1940s arms control proposals aiming at monopolising nuclear weapons, while the Soviet Union's counter initiatives aimed at hindering the US efforts to achieve that monopoly.

The Cold War atmosphere made it impossible during the 1950s to achieve any progress in the field of nuclear arms control. Only with the Détente Policy that followed the Cuban missile crisis, the US and the USSR were able to achieve progress in the field of nuclear arms control. In light of the Détente, it was possible to sign the Partial Test Ban Treaty (PTBT) in 1963. The greater achievement, however, occurred five years later by signing the Nuclear Non-Proliferation Treaty (NPT) of 1968.

Nevertheless, these achievements—not denying their importance—were all aiming at controlling or stopping the spread of nuclear weapons rather than abolishing them. This is a reflection of the world great powers' interest in monopolising that type of weapons, in a discriminatory policy that reflects international inequality that has characterised the international community since 1945.

Even the nuclear weapons free zones that have been created up till now can be viewed in two different ways. From the Third World's point of view, nuclear-weapons free zones are meant to regionally abolish nuclear weapons. Meanwhile, and from the great powers' viewpoint, the nuclear-weapons free zones are meant to promote the discriminatory status quo that keeps their monopoly of these weapons. Therefore, these zones are not necessarily an application of a general principle of eliminating nuclear weapons. This is why the big powers may not adhere to clauses, in the treaties of the these zones, that would oblige them to give up their nuclear presence in international waters or their marine visits to these zone's sea ports. Hence big powers have declined to sign or ratify the treaties' protocols that aim to create nuclear-weapons free zones in South East Asia and in Africa.

Despite the great significance of the nuclear-weapons free zones that were regionally created in the aftermath of the 1968 treaty, it is rather difficult to consider them a direct application of Article VI of the NPT. To strictly apply that article of the NPT, states members of the treaty must start a series of negotiations aiming at eliminating nuclear arms by an international treaty that puts into application a time table that achieves such a noble goal. This has not happened up till now. There are no signs that show that this can happen soon, especially after the indefinite renewal of the non-proliferation treaty in May 1995.

THE ESTABLISHMENT OF NUCLEAR WEAPONS FREE ZONES

The Treaty of Antarctica signed in 1959 was not only designed to make the South Pole free of nuclear weapons but also to prevent any military activity on its territory. This enables us to consider that polar continent the oldest and the first amongst nuclear weapons free zones in the world.

The call for creating nuclear weapons free zones in both Africa and Latin America was voiced in the beginning of the

1960s. This resulted in the signing of the Treaty of Tlatilolco in 1967 to create a nuclear weapons free zone in Latin America due to the serious efforts made by Latin American States. It took the world 18 more years, however, to sign another treaty, the Rarotonga Treaty, to create another similar zone in the South Pacific in 1985. Ten years later, in December 1995 the Southern Asian states could sign another treaty to create a nuclear weapons free zone in the Asian region. A similar treaty was signed in Cairo in April 1996 to create a similar zone in Africa. On December 31, 1991, South Korea and North Korea concluded an agreement that prevents any nuclear arms instalment in the Korean peninsula.

Why has the Latin American region surpassed all other regions in creating a nuclear-free zone while most of them started their efforts almost at the same time? The reason why Latin American states moved quickly from the negotiations phase to signing the Treaty of Tlatilolco in 1967, is their adjacency to the biggest nuclear powers, the United States and the Soviet Union, and the alarming nuclear crisis that occurred between them in October 1962. The Cuban missile crisis that resulted from the Soviet attempt to install middle range nuclear missiles in the Cuban land and the consequent American naval siege made these states perceive the possibility of a nuclear confrontation between the two nuclear powers in the area. They felt the nuclear threat much earlier than other regions. They, therefore, moved from the phase of public discussions in the UN general Assembly to the phase of direct negotiations to conclude a treaty that creates a nuclear weapons free zone in Latin America to prevent any possible nuclear confrontation between the Soviets and the Americans in the area. This case is rather unique. None of the other regions was subjected to a similar experience.

Theoretically, Europe should have sought to create a nuclear weapons free zone as it had nuclear weapons belonging to the Soviet Union and the US which may have resulted in a direct nuclear confrontation between the two big powers. Such a threat should have pushed Europe to try to create a nuclear weapons free zone like that of Latin America. However, there are essential differences between the two cases that made Europe reluctant to create a nuclear weapons free zone.

It was impossible for Europe to create such a zone by a unilateral decision taken by European states alone. The continent was divided between the two big powers that had extensive military presence, nuclear and otherwise, in Europe. This was not the case in Latin America. Moreover, Europe was never subjected to a direct threat of nuclear confrontation like that of the Cuban missile crisis. Furthermore, two European powers had already decided to go nuclear and succeeded in doing so. This had terminated—or at least contributed to the termination of any possibility of creating a nuclear weapons free zone in the European continent.

NUCLEAR WEAPONS FREE ZONES: COMPARATIVE ANALYSIS

With the exception of the Middle East and South Asia, the South could be considered a nuclear-weapons free zone, especially after the ratification and entry into force of the treaties creating the above mentioned Nuclear Weapons Free Zones. Despite the attempts to establish nuclear weapons free zones in Scandinavia and in the core land of Europe, the Northern half of the globe failed to establish such a zone. Most probably, it would be rather difficult to imagine a nuclear weapons free zone in the North as long as those states who possess nuclear weapons consider them an essential element in their military doctrine and strategic thinking. Thus, they refuse to seriously negotiate total elimination of nuclear weapons on the international level.

Establishing a nuclear weapons free zone is a political process that necessitates a tremendous effort and a long series of steps and procedures. It starts with the pre-negotiation phase, followed by the negotiations themselves, and finally concluding treaties and their protocols, and putting them into effect.

Till the late 1980s the treaty of Tlatilolco did not include the largest two states in the continent, namely Brazil and Argentina. They only joined in the 1990s after a quarter of a century since the treaty was concluded. Even Cuba did not join the treaty till 1995 after 28 years of its conclusicn. The Cuban accession aroused the US interest in adhering to the treaty and its protocols. This means that it took 30 years after signing the treaty to fully implement it. If we add the years taken to conclude the treaty to the chronic sum, it shows clearly how difficult it is to establish a nuclear weapons free zone and what it takes to get into such a long and sophisticated political process that requires certain psychological, political and intellectual stamina, due to the complication and importance of the goal itself.

Similarly, the nuclear weapons free zone in Africa, which started as an idea of discussion since the independence of African states, took 32 years to be agreed upon in a treaty. First, the African summit called for it in 1964. Second, its goals and procedures were finally clarified in the Pelindaba Treaty that was signed in Cairo on April 11, 1996, after it had taken the African states five years of serious negotiations since the accession of South Africa to the NPT in 1991. Of course, this treaty has not been put into effect yet, as the necessary number of states ratifiers has not been completed. Moreover, the treaty faces a problem concerning the status of the Diego Garcia Island which the treaty considers a part of the area of the Pelindabal Treaty, while it is under the actual control of two nuclear powers, namely the US and the UK. This means that it will take a significant number of years to fully apply the treaty.

The significance of this explanation to the Middle East is that the region should realise that it would not be easy to establish a nuclear weapons free zone in the Middle East and that it would require a long time, unlimited patience and unceasing effort. The call for a Middle East, free of nuclear weapons, was launched long time ago. Both Egypt and Iran co-sponsored the resolution passed by the General Assembly in 1974 that called for a nuclear weapons free zone in the Middle East. Since then, the General Assembly annually passes the same resolution. Similarly, President Mubarak launched an initiative in May 1990 to make the Middle East free of all weapons of mass destruction; chemical, biological or nuclear. Also, the conference held to revise and renew the non-proliferation treaty in May 1995 issued a decision urging Middle Eastern states which did not sign the treaty to do so and consider establishing a nuclear weapons free zone in the Middle East. This issue was discussed also in different track II diplomacy forums.

It is noticeable that all nuclear weapons free zones were established in regions that had been already free of nuclear weapons, except for Africa that could not achieve that goal till the radical change in the nature of the South African political system and it so security strategy occurred. Then, it could dismantle its nuclear weapons after declaring its acquisition. Does this mean that establishing a nuclear weapons free zone in the Middle East will remain idle till Israel changes the nature of its political system and its philosophy and practices?

In fact, establishing nuclear weapons free zones serves certain goals that can be summarised as follows:

- It prevents the horizontal proliferation of nuclear weapons to countries that never possessed them, which serves the big powers' interest.
- It reduces, while not totally eliminates, the possibility of nuclear confrontation, as the nuclear big powers remain the main causes of such possibility. Yet, it prevents nuclear confrontations within the region of the nuclear weapons free zone.
- It promotes security and stability on both regional and international levels, as it creates a new military or strategic balance on lower levels of armament, and it eliminates the fears resulting from possessing the most destructive weaponry systems.
- Any nuclear weapons free zone is considered a confidence-building measure through eliminating some security concerns and reducing some others.
- Every nuclear weapons free zone is a step forward on the way to the full elimination of nuclear weapons on the international level, which is the ultimate objective of nuclear arms control.

SUBSTANTIVE POINTS IN THE NUCLEAR WEAPONS FREE ZONES TREATIES

The main point in all of Nuclear Weapons Free Zones treaties is the strict adherence to eliminating all nuclear weapons and banning the production, acquisition, stockpiling and use of these weapons within the area of the zone. There is a major difference, however, between the treaty of Pelindaba that established a nuclear weapons free zone in Africa and other similar treaties that established similar zones in other regions. This treaty alone prohibits not only the fully assembled nuclear weapons, but also the half assembled ones and the components of such weapons. The prohibition also includes all nuclear explosive devices, even those intended for peaceful use, and any development research for such devices.

The remarkable commitment, all state members of a nuclear weapons free zone have to stick to is the exclusive prohibition of stationing, stockpiling, installing, and/or transferring nuclear weapons within the national territories of these states or in their territorial waters. However, the issue of the prohibition of transferring nuclear weapons is problematic. The big nuclear powers have numerous vital global or international strategic interests. Their nuclear marines are navigating the seas, visiting ports and harbours, and crossing straits that might be within the area of a nuclear weapons free zone. The Treaty of Tlatilolco was not quite clear on this issue. The preparatory committee of this treaty, however, left the members states free to deal with these situations unilaterally and on a case-by case basis. This principles has been crystallised and become an explicit clause in the treaties that followed Tlatilolco.

It is noticeable that all treaties concluded to establish nuclear weapons free zones did not include the abolition of the delivery systems, such as intermediate and long-range ballistic missiles. Perhaps, the regions that established nuclear weapons free zones did not recognise such missiles as a genuine strategic threat in themselves. Therefore, they were not a major or even a minor concern of the states members in these zones. In addition, ballistic missiles could deliver conventional weapons and other types of weapons of mass destruction.

Delivery systems, however, are a major concern in the Middle East because they create a major strategic and security problem. Henceforth, it would be rather important to consider the dismantling or the prohibition of nuclear weapons' delivery systems in any plan for eliminating nuclear weapons in the Middle East. This region is armed to its teeth with advanced systems of medium range and long range missiles that Middle Eastern countries do not really need for their defence. Hence, it would be in the best interest of the region, its stability, and security to consider the nuclear delivery systems as a major element in any future project to eliminate nuclear weapons. The fact that these delivery systems can carry extremely destructive war heads supports the argument that the stability, safety and security for the region cannot be assured only by eliminating the nuclear heads without their delivery systems.

The project intended in the Middle East has developed in the direction of the exclusive elimination of all weapons of mass destruction, including nuclear, chemical, and biological weapons. Finally, the Middle East suffers from a clear imbalance in the area of missile technology in favour of a certain state. Therefore, eliminating nuclear heads alone will not rectify the strategic imbalance which is against the spirit and principle of creating a strategically balanced region, more stable and more secure.

The treaty of Pelindaba is unique in the sense that it stipulates that all member states have to declare all of their nuclear capabilities, including heads and explosive devices, and obliges them to dismantle such capabilities. This clause reflects a characteristic that distinguishes Africa from any other non-nuclear weapon region. This is because Africa used to have a nuclear weapon state, South Africa, which joined the NPT treaty in 1991 and declared the dismantlement of its nuclear capabilities in 1993. Despite the fact that the treaty was signed years after this event, the Pelindaba legislators included that clause in the treaty in order to make sure that South Africa had actually dismantled its nuclear weapons.

Also, this treaty had introduced new concepts which were not known before. It prohibits all types of military assaults on nuclear facilities, such as nuclear reactors, in all member states of the treaty. This concept was developed, as it seems, as a reflection to the impact of the Israeli air raid on the Iraqi nuclear reactor in June 1981, and to the agreement between India and Pakistan which mutually prohibits military assaults on any nuclear facility in both countries.

There are three commitments that the declared five nuclear powers commit themselves to in dealing with any nuclear weapons free zone.

First, the territories that belong to or are under the control of nuclear weapons states within the nuclear weapons free zone including military bases that belong to the five nuclear states, are subject to the obligation of prohibiting nuclear weapons and/or devices stipulated in the concerned nuclear weapons free zones treaty.

Second, the five nuclear weapons states grant negative security assurances to the states members in the nuclear weapons free zone. The five states should not use or threaten the states members in the treaty of using nuclear weapons against them.

Third, the five nuclear powers should stop any type of nuclear testing in the nuclear weapons free zone. This commitment is not mentioned, however, in the case of the Tlatilolco Treaty. That clause, nevertheless has acquired a significant political weight since the international comprehensive test ban treaty was signed.

Sometimes, however, these obligations clash with the interests of the five nuclear powers. For example in 1986, France refused to sign the third protocol of the treaty that eliminates nuclear weapons from the Southern Pacific, because it refused to stop its nuclear tests in the region. Therefore, the US and the UK refused to sign. It is known that France has completed its nuclear tests in January 1996. Then, the three nuclear weapons states; the US, the UK and France signed the protocols of the Rarotonga treaty in March 1996, 10 years after the ratification of the treaty itself.

The problem of the lack of commitment on the part of the nuclear powers exists in the Pelindatba treaty as well. The map annexed to the treaty, which defines the borders of the nuclear weapons free zone, included a critical island, Chagos. The African Unity Organisation considers Chagos an African land. Nevertheless, the island that has a very strategically important location, especially with the Anglo-American base, lies under the control of the US and the UK. The UK refuses to admit that Chagos is part of Africa. Therefore, it refuses to sign the treaty's protocols, and that encourages Russia not to sign the same protocols till the UK signs.

Almost the same problem appeared in South East Asia. Territorial waters were defined in the treaty in a way that clashes with the big five's interests. Thus, they all refrained from signing the protocols of the treaty that created a nuclear weapons free zone in the region. That particular definition of territorial waters in the region is very critical to China, especially because of the conflict over the Sprately Islands in the Chinese Sea and other areas in the region between China and other

Asian states. Enlarging the area of nuclear weapons free zone would limit the Chinese ability to move its nuclear sub-marines.

In the case of the nuclear weapons free zone in Latin America, there had been a problem that persisted over the period of 1967–1995. Because of the American military base in Guantanamo, Cuba, Cuba had not joined the treaty till 1995. The US had not signed the treaty till that year either.

PROBLEMS CONCERNING THE ESTABLISHMENT A NUCLEAR WEAPON FREE ZONE IN THE MIDDLE EAST

One of the basic problems here is the definition of the region. Which definition will be considered: the wide definition from Pakistan to Morocco or the narrow one from Iran to Egypt? Is Turkey a part of the Middle East or not? Turkey is a NATO member, and hence, it hosts nuclear weapons that belong to the NATO. The US refuses including Turkey in any regional security system, especially nuclear free ones in the Middle East. Even the Turkish participation in multilateral talks of Arms Control and Regional Security ACRS, was based on Turkey's capacity as an extra regional power.

There are important states that did not attend the committee of ACRS, such as Syria and Lebanon, other than those who were not invited such as Iran, Iraq and Libya. This arouses doubt about the ability of the efforts made to create a nuclear weapons free zone in the Middle East in absence of such important countries. Waiting for them to join delays the process. One the other hand, starting without them reduces the efficiency of the process. Nonetheless, the effort made by the Arab League's Committee of aiming at creating a region free of mass destruction weapons may solve the problem of inefficiency due to the absence of these Arab states from ACRS talks. These efforts help crystallise a unified Arab stand towards the issue at hand.

There is a problem concerning peripheries (like Sabta and Malila) and the military bases that are under the actual command of extra-regional states, especially if these states are nuclear. The islands that are subject to conflict between states are also considered a problem. There is a general concern about Straits such as Hermuz and canals such as the Suez Canal. If the principle of exercising sovereignty on visits of nuclear marines to ports located in a nuclear weapons free zone has solidified since Tlatilolco, there are rules derived from International Law that organise the passage of military vehicles in canals and straits, that are much more longstanding.

There is an additional problem aroused by the double standard policy and fake rhetoric followed by the big powers. They declare their fully supportive stand for establishing a nuclear weapons free zone in the Middle East while their actual policies hinder achieving that goal.

However, Israel as a nuclear weapon state or at least a threshold state, remains the major problem. Unless Israel changes its nuclear posture or even its entire security strategy and foreign policy, it would be difficult to even talk about a nuclear weapons free zone in the Middle East. Israel should accept the challenge of changing its philosophy to adopt a peaceful approach and give up using its nuclear and non-nuclear power to enforce its political will. Then it can live as a normal Middle East state. Only then, talk about a nuclear weapons free zone in the Middle East could be realistic.

There is another problem concerning the relation between nuclear weapons and other weapons of mass destruction on the one hand and the relation between these weapons and conventional weapons on the other hand, in any attempt to eliminate weapons of mass destruction and build a new more stable regional system. The problem is the nature of the region's prioritisation in eliminating weapons of mass destruction and controlling the arms race. Are we going to eliminate nuclear weapons and other types of weapons of mass destruction and slow down the arms race in a synchronous fashion, or would we go through it in a sequential manner? What type of balance are we seeking in the region? These are all but difficult questions to answer.

For Arabs to deal with this situation with the least needed efficiency, they should adopt a three-dimensional strategy. The first dimension is developing their non-nuclear deterrent capabilities. This will certainly not equate the deterrence of a nuclear weapon but it would reduce the negative effect of the Israeli nuclear supremacy on the balance of power. The second dimension is the development of a nuclear peaceful technology to bridge the technological gap between Israel and the Arabs in the nuclear field. The third is to proceed with the diplomatic efforts on the international and regional levels to persuade the international community that the Middle East should be free of all weapons of mass destruction, or it should be part of a bigger nuclear free zone. That last goal could be a start towards a more stable and more just regional system.

8
The Debate over Nuclear Abolition

STEVEN E. MILLER

INTRODUCTION

Since the dawn of the nuclear age, there has been interest in putting the nuclear genie back into the bottle.[1] In the intervening decades, there have been persistent efforts to promote the elimination of nuclear weapons. The International Pugwash Movement, for example, has struggled for more than 40 years to establish zero as the proper goal of nuclear arms control and to seriously examine the prerequisites and conditions that would permit nuclear elimination to become a plausible policy option. But such voices were always very much in the minority, and their preferred course was always distant from the main lines of debate about nuclear weapons policy and nuclear arms control. Indeed, as is quite evident, during the Cold War the superpower protagonists built vast nuclear arsenals, numbering tens of thousands on each side, and enshrined nuclear weapons at the center of their defense strategies. The notion of eliminating nuclear weapons did not stir wide interest or support.

In the last several years, this has begun to change rather dramatically. There has been a remarkable upsurge of interest in, and support for, the abolition of nuclear weapons. This upsurge has included a series of high-profile studies examining the prospects for nuclear elimination as well as a widening web of prominent supporters of the idea of abolition. The mid- and late- 1990s have witnessed an unprecedented focus on nuclear abolition as a desirable objective worthy of serious policy consideration.

1. A shorter and slightly different version of this essay was prepared for International Institute for Strategic Studies, Strategic Survey, 1998–1999 (Oxford University Press, 1998).

RECENT DEVELOPMENTS IN THE NUCLEAR ABOLITION DEBATE

Several substantial studies in the early 1990s, foreshadowed the more recent attention to nuclear abolition. The Palme Commission Report of 1990 urged serious consideration of the elimination of nuclear weapons. A multi-year project at the Stockholm International Peace Research Institute (SIPRI) produced the volume, *Security Without Nuclear Weapons?*, in 1992. Similarly, a Pugwash project on eliminating nuclear weapons resulted in the edited volume, *Eliminating Nuclear Weapons: Feasible? Desirable?*, published in 1993. But it is really a series of projects, publications, and developments since 1995 that have created an unusual amount of visibility and momentum for the nuclear abolition movement.

• NPT Review and Extension Conference, April-May 1995

In retrospect, it seems plausible that the stage was set for the current wave of high interest in nuclear abolition as a by-product of the process of considering whether and how to extend the Nuclear Non-proliferation Treaty (NPT). Nuclear disarmament figured prominently in the diplomacy associated with the NPT Review and Extension Conference. Indeed, in advance of the conference, the nuclear powers were subjected to several years of mounting pressure from many quarters, including non-nuclear states as well as arms control advocacy groups, to fulfill their obligation to pursue nuclear disarmament, as required by Article VI of the NPT. Many believed that the NPT extension process would put the nuclear powers on the defensive and that their strong desire to achieve the indefinite extension of the NPT would provide leverage to force more progress in nuclear arms control and to compel more serious attention to the process of nuclear disarmament. The community of non-governmental organisations interested in abolition was energised by this perceived leverage and actively pressed the disarmament agenda; on April 25, 1995, at the NPT Review and Extension Conference, an international network of NGOs adopted a statement, labelled Abolition 2000, that called for the elimination of nuclear weapons and urged the adoption by the year 2000 of a convention abolishing nuclear weapons. More importantly, and more influentially, at the NPT Review and

Extension conference, the parties to the NPT accepted a set of "Principles and Objectives for Nuclear Non-proliferation and Disarmament" that explicitly reaffirmed the commitment of the nuclear powers to the objective of the complete elimination of nuclear weapons. By the time the NPT was extended, abolition was in the air.

- **The Stimson Center Project on the Elimination of Weapons of Mass Destruction**

In January 1994, the Henry L. Stimson Center in Washington DC launched a project focused on the elimination of weapons of mass destruction. The project drew together a panel of distinguished scholars, government officials, and retired military officers to reconsider the role of nuclear weapons in US security policy. In December 1995, the project issued a report, *An Evolving US Nuclear Posture*, which called for a fundamental reappraisal of US nuclear policy, asserted the declining utility of nuclear weapons in advancing US interests, urged phased reductions in nuclear forces, and recommended an immediate commitment by the United States to the long-term goal of eliminating nuclear weapons. Because this panel was chaired by General (ret.) Andrew Goodpaster, and included such figures as General (ret.) William Burns, General (ret.) Charles Horner, and Ambassador Paul Nitze, among other prominent figures, its call for abolition was both visible and surprising. In March 1997, the panel of the Stimson Center project published a second report, *An American Legacy: Building a Nuclear-Weapon-Free World*, endorsed by the same collection of individuals, which reiterated even more resoundingly that a nuclear-weapon-free world is in the security interests of the United States and that US policy should undertake the "vigorous pursuit" of such a world. The Project on the Elimination of Weapons of Mass Destruction remains an active standing activity of the Stimson Center and, among other things, it maintains an extremely useful website on this subject at www.stimson.org.

- **Nobel Peace Prize to Pugwash and to Joseph Rotblat**

On December 10, 1995, the Nobel Peace Prize was awarded to the International Pugwash Movement and to its President, Joseph Rotblat. This brought world-wide publicity and acclaim to an institution and an individual strongly associated with nuclear disarmament. Not only did the award put the global spotlight on the abolitionist message of Pugwash's leaders, but it subsequently facilitated extraordinary access to high government officials and the media, thereby amplifying the broadcast of that message. And for many abolitionists associated with Pugwash, this development signified their significant move away from the fringes of the nuclear discourse.

- **International Court of Justice Advisory Opinion on the Legality of the Threat or Use of Nuclear Weapons**

On July 8, 1996, the International Court of Justice (ICJ) ruled that the threat or use of nuclear weapons would be "generally contrary to international law", though permissible in extreme circumstances of self-defense. Though, on balance, this was a somewhat ambiguous finding, the abolitionist movement was buoyed by this result and many concluded that the ICJ had effectively established the illegality of nuclear weapons. There was nothing at all ambiguous about the ICJ's unanimous judgment that "there exists an obligation to pursue in good faith and to bring to a conclusion negotiations leading to nuclear disarmament."

- **Canberra Commission on the Elimination of Nuclear Weapons, November 1995–August 1996**

In November 1995, the Australian Government convened an international commission formed with the specific mandate "to propose practical steps towards a nuclear weapon-free world." This represented the first state-sponsored effort to explore in concrete terms the path to zero. The Commission was comprised of seventeen distinguished, internationally diverse members. It included some long-time stalwarts of the abolitionist community, such Joseph Rotblat and Maj. Britt Theorin, but also included General Lee Butler, former commander of US strategic nuclear forces, Oxford Professor Robert O'Neill, former Director of the IISS, Field Marshall Lord Michael Carver of the United Kingdom, former French Prime Minister Michel Rocard, and former US Secretary of Defense Robert McNamara. This was not a group from whom consensus would be automatically assumed. Nevertheless, in August

1996, the Canberra Commission issued a consensus document (*Report of the Canberra Commission on the Elimination of Nuclear Weapons*) that rejected the utility of nuclear weapons and called for the five declared nuclear powers to make an "unequivocal commitment" to the elimination of nuclear weapons. Announced with fanfare, formally presented at the United Nations, and promoted by a concerted publicity campaign, the Canberra Commission *Report* galvanised interest in nuclear abolition and has remained a focal point in subsequent discussion of eliminating nuclear weapons. Its political significance lay in the demonstration that support for abolition was widening beyond its usual constituency.

- **Statement on Nuclear Weapons by International Generals and Admirals, December 5, 1996**

In a striking development spearheaded by General Andrew Goodpaster and General Lee Butler of the United States, some 60 retired senior military officers from 17 countries issued a statement proclaiming nuclear weapons a threat to humanity and calling for their elimination. As with the Canberra Commission, this was arresting because it suggested that support for abolition was spreading among groups not previously supportive of this notion.

- **National Academy of Sciences Study on US Nuclear Weapons Policy, 1996–1997**

In 1996, the Committee on International Security and Arms Control (CISAC) of the US National Academy of Sciences undertook an assessment of US nuclear weapons policy. This committee of fifteen well-known members is chaired by Harvard Professor John Holdren; John Steinbruner of the Brookings Institution is the Vice-Chair; General William Burns chaired CISAC's nuclear study. In the fall of 1997, CISAC published its report, *The Future of US Nuclear Weapons Policy*. It called for "fundamental changes" in US policy and urged phased reductions in nuclear forces down to much lower levels. It devoted a chapter to the prohibition of nuclear weapons, concluding that "the time has also come to begin to devote serious attention to the prospects for prohibiting" nuclear arsenals. This was another prestigious voice added to the chorus urging that the idea of abolition be taken seriously.

- **Carnegie Commission on the Prevention of Deadly Conflict, 1995–1997**

The Carnegie Corporation of New York, a major American foundation, devoted several years to a high-profile examination of whether and how conflict can be prevented in the modern world, involving another Commission populated with 16 prominent individuals, headed by Carnegie's President, Dr. David Hamburg, and former Secretary of State Cyrus Vance (the Carnegie Commission was also supported by an Advisory Council of several dozen members). The Carnegie Commission issued its report, *Preventing Deadly Conflict*, in December 1997. Though not preoccupied with nuclear issues, the Carnegie report did address nuclear weapons. It applauded and endorsed the Canberra Commission report and commended it to national governments. It advocated the elimination of nuclear weapons as a significant element in building a system of conflict prevention and argued that "The only durably safe course is to work toward the elimination of nuclear weapons within a reasonable time frame....."

- **John McCloy Round Table on the Elimination of Nuclear Weapons, Council on Foreign Relations, 1996–97**

In 1996, the Council on Foreign Relations formed a study group to examine the feasibility and desirability of eliminating nuclear weapons, Chaired by General Larry Welch (USAF, retired, former Air Force Chief of Staff). In January 1998, General Welch issued his *Chairman's Report*, reporting on the results of the study group. Its main finding is that further reductions are both possible and desirable. This group did not endorse abolition nor come close to a consensus that a commitment to zero is presently desirable—though the report does record that there was some support in the group for these propositions. For long-time abolitionists, however, the striking fact is that the Council on Foreign Relations, the single institution most strongly associated with the American foreign policy elite, would have such a study group in the first place. This symbolises the distance that the abolition idea has covered in recent years, from the outskirts of the nuclear debate to the mainstream of the US foreign policy establishment.

- **Statement on Nuclear Weapons by International Civilian Leaders, February 2, 1998**

Seeking to keep the idea of abolition in the spotlight and to preserve the sense of political momentum behind it, a pro-abolition group known as The State of the World Forum, headed by former US Senator Alan Cranston, organised a statement signed by 117 prominent leaders from 46 countries, including 47 past or present prime ministers; notable among these were Jimmy Carter, Helmut Schmidt, James Callaghan, and Mikhail Gorbachev. The statement acknowledges the Canberra Commission Report, invokes a moral imperative to eliminate nuclear weapons, urges the leaders of nuclear states to commit unambiguously to the goal of completely eliminating nuclear weapons, and presses for immediate efforts to marginalise nuclear weapons.

In short, over the past several years, nuclear abolition has received an unprecedented amount of attention and a growing amount of support. Momentum appears to be building, but momentum toward what?

WHAT ARE THE ABOLITIONISTS ADVOCATING?

Each abolitionist project, report, or statement has its distinctive attributes and its singular conclusions or recommendations. But there is a lot of ground that is common in the recent upsurge of attention to the idea of eliminating nuclear weapons, and it is that common ground that constitutes the core of the abolitionist idea.

What is being advocated in this literature? Not, it turns out, the rapid and complete elimination of nuclear weapons. Indeed, it is overwhelmingly accepted by abolitionists that the elimination of nuclear weapons will take decades to accomplish. So the abolitionists case is not that zero is an achievable goal in the near term. Rather, what is being called for in most cases is a genuine commitment on the part of the nuclear powers (particularly the United States) to the *long-term objective* of eliminating nuclear weapons.

There are several paradoxes associated with this position. First, almost none of the studies or statements confidently assert the feasibility of eliminating nuclear weapons. On the contrary, many of the recent studies concede that the feasibility of abolition is uncertain. The National Academy of Sciences study is representative when it says that "It is not clear today how or when" abolition could be achieved. Similarly, the first Stimson report says that "it is not clear whether elimination can ever be achieved." Hence, most calls for a firm commitment to zero are coupled with appeals for serious exploration of the feasibility of eliminating nuclear weapons.

Second, there is not a settled wisdom on what it actually means to eliminate nuclear weapons. Rather, there exist a number of visions of what it means to go to zero. Some see zero as a way to escape and repudiate nuclear deterrence; others think zero may be possible because nuclear deterrence—at least of the existential variety—will continue to function even if there are no nuclear weapons (in this logic, the knowledge of how to make nuclear weapons will cast a shadow over human affairs). Some see ballistic missile defences as an essential and necessary part of a world of zero; others regard such defences as anathema. Some believe that nuclear weapons should be entirely eradicated; others regard a world of zero simply as an extreme state of dealerting or urge the retention of virtual nuclear arsenals that are undeployed but mobilisable if necessary. Some suggest that a fundamental transformation of the international system is a prerequisite or necessary accompaniment of a world of zero; others believe zero can be approached even in the absence of such a transformation. Some emphasise the role of international institutions in managing the transition to and maintaining a regime of zero (perhaps via the international control of nuclear weapons); others see little or no role for such institutions. Those who wish to eradicate nuclear capabilities equate zero with the delegitimisation, and perhaps even the criminalisation, of nuclear weapons; those who envision a deterrence-based world of zero accept the continuing legitimacy of nuclear weapons. In short, there is no agreed end point toward which all abolitionists aim. Rather, visions of zero abound, many of them mutually exclusive.

Third, all of the declared nuclear powers are already formally and legally committed to the long term objective of eliminating nuclear weapons. Most prominently, this was accepted in the NPT and reaffirmed in the NPT Extension and Review Conference. But this commitment has been proclaimed in other contexts as well (for example, at the UN Special Sessions on disarmament) and, in the American case, at least, has been reaffirmed by nearly every President since the end of World War II. The repeated calls for commitment a new to the objective of zero reflect doubt about the sincerity of the commitment heretofore. Hence, the new abolitionist literature is full of calls for "genuine, unequivocal" commitment to abolition, or for "vigorous or determined pursuit" of the objective of eliminating nuclear weapons. But nevertheless, the official position of the US Government is that it is committed to the elimination of nuclear weapons as a long term objective.

When one considers these points together, the abolitionists seem to be in the rather odd position of strenuously advocating a policy objective whose attainment is, at best, decades into the future, whose feasibility is questioned even by its advocates, whose end-points are unspecified, and which has already been accepted as the long term goal by the relevant states. This leads even some who are strongly supportive of nuclear arms control to suggest that the abolitionist upsurge is an unfortunate distraction that deflects attention from the near-term arms control agenda to distant and possibly unattainable horizons.

WHY ABOLITION NOW ?

How, then, does one explain the level and intensity of the present abolitionist upsurge? What is the point of pressing hard now for commitment to abolition decades hence? There is a range of answers to these questions, but undergirding them all is a sense that there exists now a unique, and possibly fleeting, opportunity to advance the abolitionist agenda. The deep and bitter antagonisms of the Cold War rendered abolition irrelevant to the policy realities of the time. The end of the Cold War and the absence of intense great power rivalry has completely altered the international political context in ways that enable the abolitionist agenda to be taken seriously. The drive to capitalise on this opening is lent urgency by the fear that that restoration of great power rivalry—possibly caused by the rise of China or the failure of democracy in Russia—may bring this propitious moment to an end. This sense is exemplified in the Canberra Commission Report, in a sentence that has been picked up and echoed elsewhere in the recent abolitionist literature: "The opportunity now exists, *perhaps without precedent or recurrence*, to make a new and clear choice to enable the world to conduct its affairs without nuclear weapons... ." The push for abolition is happening now because the opportunity is now.

The existence of this opportunity explains why the traditional abolitionists are pressing their cause, but it does not explain why the circle of support for abolition is widening, in the United States and elsewhere. Here a second factor comes into play: the end of the Cold War seems to have dramatically changed the perception of many of the cost-benefit calculus associated with nuclear weapons. Increasingly it has become apparent that for many, including in particular many military officers, nuclear weapons were a repugnant necessity of the Cold War. With the end of the Cold War, with the disappearance of the implacably dangerous foe, the necessity for and the benefits of nuclear weapons have been greatly reduced or (in the view of some) have disappeared, while the risks and dangers of nuclear weapons remain largely unabated. This perception is visible, for example, in the views of US General Lee Butler. In his extraordinarily eloquent remarks at the National Press Club in Washington DC on December 4, 1996, explaining his own conversion to the abolitionist cause after a life-time of service in the nuclear forces of the United States, General Butler said that he felt compelled to speak out because of "a growing alarm that, despite all of the evidence, we have yet to fully grasp the monstrous effects of these weapons, that the consequences of their use defy reason... ." A world in which security is based on nuclear threats, General Butler argues, "was and is intolerable... .The price already paid is too dear, the risks run too great." From his perspective, therefore, the end of the Cold War offers "the promise of reprieve" from the need to live with nuclear weapons. In short, the abolitionist coalition is being swollen in large part due to a growing belief that the costs, dangers, and risks associated with nuclear weapons exceed any conceivable benefits in the post-Cold War era.

Several other considerations reinforce the recent abolitionist upsurge. One is the proposition that abolition should be pressed now precisely because the road is long and full of obstacles. It will take years and decades to build an adequate verification system to support a world of zero, for example, so it is sensible to take the initial steps now. It is hard to quarrel with the logic that it will never be possible to get to the end of this long road if the nuclear powers never start down the road in the first place. A related point is that choices and actions taken today will shape options in the future; for many abolitionists, this reality leads inexorably to the conclusions that it is important to have the long-term objective of eliminating nuclear weapons influencing present policies and priorities. Another immediate concern for abolitionists is sending a credible signal to the world that nuclear weapons are being devalued and (for some) delegitimised; this signal is important, it is argued, in stemming the tide of nuclear proliferation.

Another particularly influential calculation animating the current abolitionist push is the belief that to make their commitment to the long-term objective of zero truly credible (unlike the commitments of the past), the nuclear powers will need to conform their present behaviour to the trajectory of elimination. This, in turn, will contribute to the achievement of various near- and medium-terms objectives advocated by the abolitionist coalition. This is how the commitment to zero in the long run helps reduce the nuclear danger in the here and now.

THE ROAD TO ZERO PASSES THROUGH THE NEAR-TERM

While the broad objective of those who advocate the elimination of nuclear weapons is to secure meaningful commitments to zero from the nuclear powers, nearly every study or statement advocating this result is linked with a roster of immediate or transitional steps that would serve the abolitionist end. Not cvory such list is identical, and there are important differences of selection and priority from one study to the next. But there are a number of recommendations for the immediate nuclear arms control agenda that are common to most such agendas for action. Proponents of zero have an affinity for much of the existing nuclear arms control agenda, including such items as the Comprehensive Test Ban Treaty and the fissile material cutoff. However, several items are especially prominent in their thinking.

First, support for further nuclear reductions, beyond those presently in view in the START process, is nearly universal in the abolitionist literature. Many studies offer their own schemes for phased reductions. Most offer ambitious medium-term targets for the reduction process. The National Academy of Sciences study, for example, recommends seeking, beyond START, a limit of 1000 total warheads for the United States and Russia, including not only strategic nuclear warheads but tactical warheads and reserves as well; this would require dramatic cuts well beyond anything presently on the nuclear arms control agenda. For many, one of the powerful virtues of a focus on zero is that it strongly reinforces the logic of reductions.

Dealerting—that is, reducing the readiness of deployed nuclear forces so that they cannot be used instantly or accidentally—has attracted wide support from proponents of zero. This can involve not only standing down from hair-trigger alert procedures, but can progressively entail removing warheads from delivery systems, storing warheads a sites remote from their delivery vehicles, even partial or full disassembly of warheads. The essential idea is to maximise the amount of time it takes to actually use nuclear weapons. As noted, some regard zero as the inevitable endpoint of the logic of dealerting; the time-to-launch in a world of zero will have been increased from minutes or hours to days, weeks, perhaps months. The logic of dealerting is so compatible with the abolitionist impulse that it figures prominently in much of the recent abolitionist literature. It was at the top of the list of recommendations for immediate steps of the Canberra Commission Report, for example. Similarly, the Carnegie Commission Report advocated moving to zero deployed warheads; that is a long way from zero, but a long way from current force postures as well.

Most, though not all, proponents of zero are strong advocates of embracing no-first-use. This is important to abolitionists because implicit in first-use doctrines is an assumption of the wide utility of nuclear weapons (for example, to deter conventional attacks or chemical and biological threats). The abolitionist goal is to restrict the role and limit or deny the utility of nuclear weapons. Once all nuclear states accept that the only purpose of nuclear weapons is to deter the use of nuclear weapons, there is only one step to the conclusion, deeply compelling in the abolitionist community, that if no one had nuclear weapons, no one would need them. Thus, for example, the National Academy of Sciences study recommends strongly that nuclear deterrence should be confined to its "core function," the deterrence of nuclear attack. And many advocates of nuclear elimination attach high priority to the achievement of an international convention on no-first-use.

Thus, while abolitionists do have one eye firmly fixed on the distant goal of zero, they have an immediate agenda as well. One of their aspirations is to put the long-term goal at the service of this short-term agenda. Nuclear testing, or first use doctrines, or the persistence of large nuclear forces, are simply not compatible, in the abolitionist view, with a serious commitment to zero. If the nuclear powers want to make believable their promises to aim for the elimination of nuclear weapons, they will have to pursue something like the abolitionist agenda in the short run. This, in turn, will move them down the road to abolition, and bring the elusive end a little bit closer. That is why the abolitionists believe that it is worth pressing now for genuine, unequivocal, vigorously pursued commitments to nuclear abolition.

THE GAME'S AFOOT

"We are at the beginning...of a great debate," writes longtime abolitionist Jonathan Schell in his latest passionate article on going to zero in *The Nation* (February 2/9, 1998). If Schell is correct, in the coming years, the advocates of abolition will battle with the still dominant critics of zero to influence and determine the policies and objectives of the nuclear powers. It is a sign of how far the abolitionist cause has come, how much momentum this idea has gained, how much its ranks have swollen, that such a great debate may be in the offing. This contest is only possible because the abolitionists have gained a foothold in the mainstream of policy discussions—a beach head that has been established only in the events and efforts of the last several years.

Nevertheless, there is a substantial distance between this surge in interest and political momentum, on the one hand, and the triumph of the abolitionist policy agenda, on the other. From the abolitionist perspective, some long-coveted break-throughs, such as the signing of the CTBT, have occurred; some very heartening developments, such as the large reductions associated with the START process, are taking place; some significant elements of the abolitionist agenda, such as the fissile material cutoff, are very much on the current policy agenda; and there has emerged considerable overlap between abolitionist recommendations and existing policy interests and priorities, particularly in the area of dealerting. There is much evident in the current scene that gives heart to those interested in the elimination of nuclear weapons.

On the other hand, it is patently clear that the abolitionist influence does not dominate the current realm of policy. Ten years after the end of the Cold War, seven years after the collapse of the Soviet Union, the US-Russia nuclear relationship is marked by the presence not of thousands, but tens of thousands of nuclear weapons. The only legally binding strategic arms control agreement, START I, has its origins not in the conducive post-Cold War period but in the early years of the Reagon Administration—a period known as the new cold war. Subsequent agreement remain mired in political difficulties. And perhaps most disappointing from the abolitionist perspective, neither of the large nuclear powers shows any inclination to disavow nuclear weapons nor do they display any acceptance of the proposition that nuclear weapons are of limited and declining utility. On the contrary, Moscow appears to have reembraced nuclear weapons as the solution to its new and unaccustomed numerical inferiority vis-a-vis its various potential rivals, east and west, and as a necessary compensation for its crumbling conventional military power. Russia's nuclear doctrine, in-so-far as it can be said to have a nuclear doctrine, appears to be the second coming of John Foster Dulles: massive retaliation is the order of the day. Similarly, the United States has reaffirmed its commitment to nuclear deterrence, including the retention of ambiguity about whether or not it might use nuclear weapons in response to the threat of chemical or biological attack. This posture makes no-first-use incompatible with US nuclear policy and demonstrates that Washington continues to see utility in nuclear weapons beyond the "core mission" of deterring the nuclear forces of others. Thus, in some fundamentally important ways, present reality is not conforming to abolitionist preferences.

The goal of the abolitionists in championing a great debate is to change these disappointing realities. For the moment, however, we cannot now know whether this great debate will happen, or whether the abolitionist moment will pass. And if this great debate happens, we cannot know how it will come out, given the mixed evidence so far and the powerful currents that still work against progress toward the elimination of nuclear weapons. But for the short-run, the most important point may be that there is much on the existing arms control agenda that attracts the support of both the abolitionists and their critics; the agendas of these competing coalitions do not necessarily diverge for years or decades to come.

APPENDIX

INVENTORY OF ARGUMENTS FOR AND AGAINST THE ELIMINATION OF NUCLEAR WEAPONS

The Case for Abolition

1. The moral imperative: nuclear weapons are an intolerable, unacceptable, unconscionable, species-threatening technology.
2. The limited, declining, or non-existent utility of nuclear weapons:
   - Nuclear weapons are only useful for deterring nuclear weapons.
   - Therefore, if no one has them, no one would need them.
   - Whatever utility nuclear weapons had in the past, their utility has greatly declined in the post-Cold War era.
   - Nuclear weapons do not provide security to their possessors. Butler: There is no security to be found in nuclear weapons.
   - The net utility of nuclear weapons is not sufficient to warrant their retention: the costs exceed the benefits.
   - The destructiveness of nuclear weapons is so disproportionate to any political purpose that they are unusable.
   - Nuclear weapons diminish security. [Canberra Commission]
3. Escaping the inevitability of nuclear use: it is improbable that nuclear weapons remain indefinitely in the arsenals of

a number of states but are never used. Nuclear abolition greatly reduces this risk.

4. Escaping the inherent dangers of nuclear deployments: the only way to eliminate the risk of accident, unauthorised use, or inadvertent escalation is to eliminate nuclear weapons entirely.
5. Nuclear abolition promotes nuclear non-proliferation:
   - Getting rid of existing nuclear arsenals would eliminate the largest incentive for nuclear proliferation. It is the only way to prevent "nuclear contagion."
   - The NPT regime is unsustainable over the long run if discrimination between the nuclear haves and the nuclear have-nots is preserved.
   - The nuclear powers have made a legally binding commitment, under Article VI of the NPT, to eliminate nuclear weapons.
   - Advocacy of non-proliferation by the nuclear powers is undermined by their own possession of nuclear weapons. Abolition eliminates this inconsistency in their position. Blechman-Fisher: US nuclear policy cripples non-proliferation efforts.
   - The nuclear powers must delegitimise nuclear weapons or risk widespread proliferation.
6. Weapons of mass destruction, including notably nuclear weapons, neutralise the conventional intervention capabilities of the industrial democracies.
7. NWFW may be feasible.
   - Feasibility is linked to timing. Because zero is a long-term objective, probably to be achieved over decades, there will be time to assess and address obstacles to complete elimination.
   - Feasibility depends on what one means by zero. It is possible to create definitions of zero that increase the likelihood of feasibility.
   - The fundamental prerequisite of a NWFW is a political transformation that is not inherently impossible and involves the appropriate configuration of man-made political conditions. In a transition likely to stretch across decades, it is possible that the appropriate political circumstances can be created.

- A new international community is already emerging, in which the use of force is unthinkable. Blechman-Fisher: The modernist community delegitimises force.
- Eliminating nuclear weapons is not more difficult than eliminating chemical and biological weapons, for which international regimes of abolition already exist.
- Nuclear abolition is no more utopian than the end of the Cold War.
- Most of the world is nulcear weapon-free already.

The Case Against Zero

1. Zero is infeasible (A)—transparency:
   - It will be impossible to verify the complete elimination of nuclear weapons. Perle: Not verifiable now, not verifiable ever.
   - Small stockpiles are easy to hide and hard to find.
   - Failure of verification and enforcement could have disastrous consequences.
2. Zero is infeasible (B)—diplomacy:
   - Diplomatically, it will be impossible to reach the multilateral agreement necessary to achieve nuclear abolition. Not all among the nuclear powers have a genuine interest in nuclear disarmament.
   - In the past, it has only been possible to outlaw weapons that were regarded as inconsequential or of secondary importance, whereas nuclear weapons are "the ultimate trump card." [Payne]
   - It will be impossible to achieve the fundamental transformation of international politics that even many abolitionists believe is a prerequisite to zero.
3. A world of zero is dangerous (A) — the endpoint:
   - A world without nuclear weapons will be unstable, perhaps highly so: The problem is the possibility of rapid breakout—the ability of states with nuclear expertise to engage in nuclear remobilisation. Fear of this will keep every state on edge, will produce arms race and crisis instabilities.
4. A World of zero is dangerous (B)—the transition:
   - A commitment to elimination will have uneven effects on the perceptions and credibility of the nuclear forces of nuclear-armed states.

- The process of reducing toward zero may produce undesirable asymmetries in the forces of the nuclear powers.

5. Nuclear weapons are desirable because they promote peace:
   - They induce prudence and sobriety in decision-makers.
   - They enormously raise the risks associated with a large conventional war.
   - Nuclear weapons are, some believe, profoundly defensive in their implications, making states hard and dangerous to conquer, discouraging aggression, and favouring the status quo.
6. The continuing utility of nuclear weapons (A): Nuclear weapons are necessary or useful for a range of WMD deterrent purposes.
   - For the United States, there is the residual deterrence requirement in relation to both Russia and China. Other nuclear-armed states will perceive their own requirements for nuclear deterrence.
   - Nuclear weapons may be necessary or useful in deterring the use of other weapons of mass destruction.
   - Conventional forces will be insufficient for deterring WMD threats.
   - Nuclear weapons may be necessary or useful in deterring rogue states with weapons of mass destruction.
7. The continuing utility of nuclear weapons (B): Nuclear weapons can compensate for conventional deficiencies or disadvantages.
   - States such as Russia, Israel, or Pakistan, who face enormous numerical disadvantages in conventional military forces, continue to rely on nuclear weapons as an equaliser or as a deterrent of last resort.
   - Third World adversaries of the United States may find nuclear weapons to be the only answer to overwhelming US conventional capability.
8. The risks and dangers associated with nuclear weapons are manageable (and acceptable).
   - The nuclear age is more than 50 years old and there have been no major accidents or large-scale failures of nuclear deterrence.

- There is no reason to assume that use or accidents are inevitable.
- It is inconsistent to suggest that mankind can transform international politics and create an effective system of nuclear disarmament but that man is incapable of managing nuclear deterrence effectively.

9. Movement toward zero by the major nuclear powers may stimulate nuclear proliferation:
   - Going to zero would eliminate the nuclear deterrent forces that discourage potential proliferators by negating the value of their nuclear capabilities.
   - Going to zero would deprive US allies of nuclear guarantees that have inhibited their pursuit of independent nuclear capabilities.
   - Going to zero would accentuate the political and military value of small nuclear capabilities within reach of weaker powers.
   - In a world of zero, tremendous advantages would or might accrue to any state that possessed even a few nuclear weapons.
   - US reliance on enormous conventional capabilities for deterrence may invite its potential adversaries to pursue or retain neutralising nuclear capabilities.
   - Even if going to zero does not promote proliferation, there is no reason to believe that it will discourage proliferation because potential proliferators desire nuclear weapons for their own security-related reasons, not because the US or other nuclear powers possess them.
10. The hypocrisy objection: Observers or states outside the East-West domain of the Cold War note that when nuclear weapons were thought necessary for Soviet and American security, those states accumulated vast nuclear arsenals. Now that nuclear weapons are thought more threatening than useful to American security, there is growing interest in nuclear abolition. Why shouldn't those states that see nuclear weapons as necessary for their security have the same right to rely on nuclear weapons that the nuclear superpowers exercised during the Cold War?
11. The dubious motive: From the perspective of some, the growing American interest in nuclear abolition looks like a way to make the world safe for American conventional superiority.

Miller's Crude Summary Comparison of Arguments for and against Zero

	Abolitionists	*Critics of Zero*
moral basis	nukes are intolerable	nukes cause peace
feasibility	may be, possibly, probably	not now, probably not ever, never
utility of nukes	low, waning, nonexistent	high, continuing, essential
proliferation	nukes promote proliferation	zero promotes proliferation
dominant dangers	risks of nuke deployments	instabilities, uncertainties of zero
geostrategic implications	zero preserves conv. advantages	zero unshackles WMD threats
American motivation	zero safer for humanity	self-serving hypocrisy

Address:

Steven E. Miller
Harvard University
Cambridge MA 02138
USA

9
Arms Control Research in China Academy of Engineering Physics

XIANJUE PENG

CHINA ACADEMY OF ENGINEERING PHYSICS

Peace and development has become the trend around the world these days. Arms control and disarmament, as an important element of international security and a crucial way to reduce the war threat, especially the threat of massive destruction, have received more and more attention from the international community. Since 1980's, especially the end of the Cold War, arms control research has far exceed the scope of some traditional area, such as politics, law and diplomacy, to expand deeply into the field of natural science, and get more complicated problems of science and technology involved in it. One can say that the active participation of natural scientists would play a stronger role in the solution of international arms control problem. It is under this background that the scientists of CAEP began to work in the field of arms control research.

Some experts and scholars in CAEP have paid much attention to arms control and some related technology ever since middle 1980's. They have shown strong interest and are capable of conducting research in this field, and they are also intended to push the work forward in China. For the convenience of getting the experts and researchers across the nation associated with each other and promotion of international co-operation and exchange, the Programme for Science and National Security Studies (PSNSS) was proposed and founded in Beijing in 1989. This programme is based on Beijing Institute of Applied Physics and Computational Mathematics (IAPCM) and gets full supports of natural science and technology resources in IAPCM. The member of PSNSS are from various domestic institutes besides IAPCM, including natural scientists and social scientists who work shoulder by shoulder to probe all sorts of problems in international arms control. Of course the programme is intended to pay more attention to the stronger role of science and technology in national and international securities.

The purposes of PSNSS are:

1) To encourage and support the work of natural scientists from IAPCM in the field of international security and arms control.
2) To support the international co-operation and information exchange between Chinese and foreign scientists in terms of arms control and disarmament technology issues.
3) To support the related international meetings and symposiums on arms control research.

So far, PSNSS has conducted efficient works. All sorts of research have been deepening, the academic exchange in the nation and abroad has been increasing, and the increasingly mutual understanding of co-operation and confidence between Chinese and foreign experts has positive effect on international security. In the meantime, PSNSS has gained its reputation in the international arms control research community.

With the development of international arms control and disarmament situation, many scientist of CAEP working in Mianyang district of Sichuan Province have been attracted to join the arms control research, and in Oct. 1995, they set up "Programme for Verification Technology Studies" (PVTS) which involved mostly experimental physicist. Similarly, PVTS is an academic organisation which receives support from CAEP, and the member of it are from CAEP and other domestic research institutes.

The main fields that the scientist of PSNSS and PVTS work in are:

- Comprehensive test ban;
- Nuclear Non-Proliferation;
- Nuclear disarmament;
- Nuclear material safeguard;
- Out-space non-weaponisation and theater anti-missile weaponry;
- Peaceful use of nuclear explosion;
- Global security and environment protection.

The following is a brief description of several key research programmes

1) Verification technology research for comprehensive test ban. Main directions are:
   - The measurement of the gas released from the underground nuclear explosion;
   - Near-site aftershock measurement (near the explosion site);
   - Off-site electronic magnetic pulse measurement.

The radioactive gas is directly indicative of the existence of a nuclear explosion event. The most important gas indicators are ^{133}Xe, ^{133m}Xe, ^{135}Xe, ^{37}Ar, ^{85}Kr, ^{3}H and etc., specifically, the ratio of ^{135}Xe to ^{133}Xe is of great importance to the determination of the nuclear explosion time. The measurement of radioactive gas and its technological feasibility has been proven through on-site measurement of nuclear test. In addition to radioactive gases, other gases, such as ^{4}He, CO_2, H_2S and CH_4 are also important indicators of nuclear explosion. For China is one of the nuclear states which are the last to end atmospheric nuclear explosion, Chinese scientists have done more research of the EMP as one of the effective ways to detect remote atmospheric nuclear events.

2) Studies for nuclear disarmament verification technologies. Here, the main concern is the verification technologies for nuclear warhead.
   Specifically:
   - Passive detection technology;
   - Active detection technology;
   - Tag and signature technology;
   - Monitoring tech for moving objectives.
3) Studies for nuclear material monitoring technologies Here, the main issues are:
   - Accounting technology for nuclear material;
   - Protection and control technology for nuclear facility and nuclear material.
4) Out-space non-weaponisation and theater anti-missile weapon issue
   This is one problem we are concerned about. Our opinions are: the principle of out-space non-weaponisation should be strictlv abided by, and the performance parameter and

deployment of theater anti-missile weapons should be limited, or the process of nuclear disarmament would be affected adversely and another wave of arms race may be triggered.

5) Peaceful use of nuclear explosion.
   This has been a long debate issue, where may people are against it for various reasons. Our opinions are: From the point of science and technology, nuclear explosion technology may contribute to the survival and development of human being. Nuclear explosion could be used to generate electricity and it could be regarded as one practical way of using nuclear fusion for the time being to provide humankind with sufficient, clean and economical energy resource in the future hundreds of years. Some of our Russian colleagues agreed to this point of view. The practical application of the technology, we think, may have to wait for another 20~40 years when the world is in urgent need of energy, due to the concern over nuclear non-proliferation and global security.

In addition, the issues of nuclear non-proliferation and environment have also received wide attention among our experts, who have conducted research accordingly.

Address:

Xianjue Peng
China Academy of Engineering Physics
Chengdu 610003
China

10

The Twin Problem of Warheads and Their Delivery Vehicles. Where to Put the Priority during Future Treaty Negotiations ?

GERT HARIGEL

1. INTRODUCTION

Since the end of World War II there has been a number of treaties dealing with the limitations, reductions and elimination of arms of mass destruction and/or their transport systems (generally called delivery systems). Some of them are bilateral, others multilateral, or in rare case even universal. In the present context only treaties will be discussed, which deal with nuclear weapons and the means to get them to an envisaged target. Weapons are mostly defined by the owner's military establishment or manufacturer as *defensive*. However, to draw a line between defensive or aggressive is close to impossible.

A short list of treaties, not including any of their (present of planned) amendments, follows. Their acronyms, year of signature and/or entry into force (ratification), as well as the main participating parties are indicated. A detailed discussion of these treaties is beyond the scope of this paper; their evaluation as well as their complete text can be found e.g. in ref. [1].

Non-**P**roliferation **T**reaty, **NPT**, (1968 signed, 1970 into force), Nuclear and Non-nuclear Weapon States.

Anti-**B**allastic **M**issile Treaty, **ABM**, (1972 signed and into force), Bilateral: USA and USSR.

Missile **T**echnology **C**ontrol **R**egime, **MTCR**, (1987 signed by Canada, France, FR. Germany, Italy, Japan, UK, USA, but not e.g. by *India, Pakistan and North-Korea*). Not a formal treaty.

Intermediate-range and shorter-range **N**uclear **F**orces, **INF**, (1987 signed, 1988 into force), Bilateral: USA and USSR.

Strategic **A**rms **R**eduction **T**reaty **I**, **START I**, (1991 signed, 1994 into force), Bilateral: USA and former USSR.

Strategic **A**rms **R**eduction **T**reaty **II**, **START II**, (1993 signed, *not yet ratified by Russian Duma*), Bilateral: USA and former USSR.

Indefinite extension of the **N**on-**P**roliferation **T**reaty, **NPT**, (1995, signed by ~180 countries, but not by *India, Pakistan, Israel*), Nuclear and Non-Nuclear Weapon States.

Comprehensive **T**est **B**an **T**reat, **CTBT**, (1996 signed by —131 countries, needed ratification by 44 countries, including India. Not yet ratified e.g. by *USA and Russia*. India and Pakistan are not signatory countries. This treaty may be respected, but never enter into force).

Nuclear and Non-Nuclear Weapon States.
Strategic **A**rms **R**eduction **T**reaty **III**, **START III**,
Bilateral: USA and Russia,
under discussion (Helsinki declaration in March 1997 of Presidents Clinton and Yeltsin).

2. WHAT DO THE TREATIES IMPLY?

The public at large is barely informed about the contents of the above listed treaties. This is mainly the fault of the media, who hesitate to give information which is hard to digest, may not be along the party line of the government, and/or might be somewhat boring to read. A single nuclear arms and their devastating potential is almost completely missing in newspapers. Due to rudimentary information the public is lulled into the belief that nuclear arms are gradually disappearing from our planet due to START I and START II and prohibition of development of new weapons by the CTBT. It must be stressed that these treaties are aiming **mainly** to reduce the carrier of nuclear arms, but only marginally of the weapons themselves. Both processes are at a slow track. Unfortunately, in the declared Nuclear Weapon States (NWS) the *reduction in numbers* is accompanied by improving of weapons' efficiency, transport systems, and targeting accuracy.

Warheads and delivery systems will first be treated separately prior to some suggestions how to make real progress in disarmament.

i) Nuclear Warheads

The bilateral START II agreement between the former USSR (now Russia) and the United States stipulated that by the year 2003 (now revised and postponed to year 2007 both countries could ***deploy*** no more than 3.000 to 3,500 strategic nuclear warheads on inter-continental ballistic missiles (ICBMs), submarine-launched ballistic missiles (SLBMs), and heavy bombers.

Additional limits of the treaty (in a short form) set for by the treaty are:

- No **m**ultiple **i**ndependently targeted **r**e-entry **v**ehicles (MIRVs) on land-based missiles.
- All SS-18 "heavy" Russian missiles must be destroyed.
- No more than 1,700 to 1,750 warheads (about one-half the U.S. Trident warhead loading projected under START I) may be deployed on SLBMs.
- Reductions in strategic nuclear warheads, as well as the deMIRVing of ICBMs, may be achieved by "downloading" (removing) warheads from a missile.
- Once removed, warheads may not be restored to downloaded missiles.

There is no word on the dismantling of the warheads themselves in the treaty.

Progress on START can be summarised as follows: From September 1990 to January 1998 there has been an overall reduction in absolute numbers from 10,563 to 7,986 in the USA, and from 10,271 to 6,680 in the Soviet/Russia Strategic Forces. START I treaty requirements have been obtained by the US, but not yet by Russia. None of the two countries has achieved in any of the different categories of delivery systems the goal set by START II. It is worrying, that the numbers of the sophisticated, modern and more powerful warheads has considerably increased during this period, contrary to the spirit of the treaty. There is clearly no sign that the two powers are acting according to Article VI of the NPT. Other indications for non-compliance are outlined in the Nuclear Stockpile Stewardship Programme [2] and refs. [3, 4]. There is a considerably large grey zone as far as the weapons in hedge are concerned, which consists mainly of tactical weapons.

START III envisages a limit of 3,500 strategic weapons for both partners.

There is no nuclear arms reduction treaty with participation of the other three NWSs.

The costs of producing (and maintaining in operational state) nuclear weapons are high. They peaked in the USA from 1983–1993 at almost $8b/year, and stabilise now at $4b/year for the US Stockpile Stewardship programme [2]. Costs for the former USSR may have been at the height of the Cold War at a similar, if no higher level, but are now drastically reduced.

It can be assumed that most of the industrial countries are now capable to build (rudimentary) nuclear weapons within delays of a few months to a few years, provided fissile material is available or can be procured. The stockpile of such weapon-grade material is enormous and given in Table 1 (on the average only 4 kg Pu and/or 15 kg HEU are needed for building one weapon).

Table 1 : (extracted from [6])

	Pu [to]	HEU [to]
Inside weapons	75	485
Unknown destination	77	553
Declared excess	74	674
Under Safeguards	2	10
Total	**228**	**1,722**

In addition to the above tonnage one has to take into account fuel-grade and reactor-grade plutonium, amounting in the US to 13.2 and 1.3 to, respectively. This material can be reprocessed and then used for weapons. Many of the ~440 operational civil nuclear reactors are able to produce some plutonium. A cut-off treaty for weapon-grade material is therefore highly desirable. A complete stop of nuclear power reactors, however, is not feasible or even desirable in any foreseeable time.

ii) The delivery systems

Delivery systems can be sub-divided in cannons, aircraft, and ballistic missiles (land– or sea-based launchers, **ICMB**s or **SLBM**s, respectively). Generally, systems are categorised in strategic and the tactical—the latter intended to be used on the battlefield. This distinction may make some sense for military planners, but takes not into account that nuclear weapons remain indiscriminate. Explosive power of tactical weapons can be twenty times the one of the Hiroshima bomb and these warheads/bombs can be easily transported by fighter planes, like the F16. This type of weapons employment is becoming almost counter-productive when exploded close to their own troops, mainly due to the spreading of radioactivity.

The total cost for the U.S. Nuclear Weapons programme from 1940–1995 was $3.9 trillion. Strategic nuclear weapon systems need an enormous infrastructure, as can be seen from the following breakdown of cost: Building the bomb ($378 billion), Delivering the bomb ($2 trillion), Command, control and defence against the bomb ($1.1 trillion), Retiring and dismantling the bomb ($15 billion), Legacies of the bomb ($385 billion) and Managing the bomb ($25 billion) [7].

Already prior to the conclusion of the INF treaty a shift away from land-and air-based towards submarine launched missiles became evident in all NWSs. The planned cuts of land based missiles in the former USSR, according to START II, are pointing in the same direction. This does, however, not imply that we should entirely discard the problems with short-range missiles, dealt with in the MTCR (limits range to <300 km, payload <500 kg, the German V2 exceeded the MTCR limit in payload by a factor of two and matched the range limit). Complete blueprints for the Jupiter intermediate range missile of the 1950s (2,400 km range, payload about 1 ton) were contained on more than 500 reels of micro-film, and could be bought for about $ 11.50 per reel [8].

At present, 19 developing countries possess intermediate or shorter-range ballistic missiles, but only a handful have the indigenous capability to produce them or to provide for their maintenance. Of these, only four—India, Israel, North Korea and Pakistan—have produced or flight tested missiles with ranges over 1,000 kilometers [9].

3. MORE TREATIES NEEDED?

Fissile materials have dual use, in weapons (including *peaceful explosions* of nuclear devices) and in civil power reactors. As long as the uranium isotope 235 is not highly enriched, and plutonium isotope 239 is not sufficiently separated from its other isotopes, both materials have limited value for would-be NWSs. The enrichment and separation process for these materials are complicated, expensive and time-consuming.

The conclusion of a **F**issile **M**aterial **C**ut-off **T**reaty (**FMCT**) [10], already proposed long time ago, would be highly desirable, but is not sufficient by itself. We can not close our eyes over the fact, that would-be NWSs can obtain clandestinely sufficient material from the enormous stock of weapon-grade material. Furthermore, most of the industrialised countries have enough knowledge (and raw material from their power reactors) to produce nuclear weapons within a couple of years or even within a few months. Safeguard provisions by the **I**nternational **A**tomic **E**nergy **A**gency (**IAEA**) are not guarantee for correct accountancy.

Missiles have dual-use. They can be used to put communication satellites into orbit, for geological survey, for weather forecast, etc. How to build missiles for military purposes is common knowledge since World War II. The **MTCR** is no longer sufficient as far as the generous lower limit of 500 kg is concerned. Nuclear weapons with explosive power similar to the Hiroshima bomb can be built with less weight.

The **START** treaties aim at reducing operational nuclear weapons, but have a too long lead time and are not cutting the arsenal drastically enough. However, they are a step in the right direction. The ratification of START II by the Duma is long overdue and START III is urgently needed, as well as the inclusion in the disarmament negotiations of the other three declared NWSs, China, France, UK, plus India, Pakistan, and Israel.

Verification of treaties is a vast, problematic and important area, which will not be discussed in the present context due to lack of space. Verification has to go along with trust.

The strict and honest **adherence to existing treaties**, following their spirit (e.g. Article VI of the NPT), could be sufficient to put an end to the nuclear arms race and lead the way to the ultimate elimination of these weapons. It should be complemented by a **Nuclear Weapons Convention (NWC)**, written as a draft version by the Lawyers' Committee on Nuclear Policy [11], and had been distributed to the members of the UN General Assembly in 1997.

The US, the only remaining super-power after the end of the Cold War, could give a positive example by eliminating its arsenal on a time-based frame, then probably/possibly(?) followed by other NWSs, which could lead to a world free of nuclear weapons. However, present indicators of US policy point in the opposite direction [12].

It appears to me that time has come to **aim simultaneously for fissile material cut-off, efficient missile control, nuclear weapons destruction, general disarmament and verification, i.e. to put all these items under one negotiation umbrella**. Ideally, such a comprehensive approach should be accepted by all parties, rather than to continue piecemeal, wasting precious time, money, and effort.

The suggested combined procedure of negotiators may not be sufficient alone for solving the nuclear weapon problem, as long as we have not **changed the mind of people**, have educated them that any war, in particular the one using arms of mass destruction, can not be fought, and is not an acceptable, profitable solution of problems, for any country, continent or the entire planet. Possession of nuclear weapons alone is a threat to survival of mankind, since an accident launch of a single nuclear weapon can cause an all-out war, resulting in Armageddon.

References

1) **Arms Control**, A Guide to Negotiations and Agreements, Jozef Goldblat, International Peace Research Institute, Oslo, SAGE Publications, London (1994).
2) **The U.S. Department of Energy's Stockpile Stewardship and Management Plan (SSMP)** (including Science Based Stockpile Stewardship) a.k.a. "The Green Book", Office of Defence Programmes, February 29, 1996 and **Nuclear Posture Review 1994** (no public report).
3) **End Run**, The U.S. Government's Plan for Designing

Nuclear Weapons and Simulating Nuclear Explosions under the Comprehensive Test Ban Treaty by Christopher E. Paine and Matthew G. McKinzie NRDC, Nuclear Programme, April 1998, (Rev. 1).

4) **Explosive Alliances**, Nuclear Weapons Simulation Research at American Universities, M.G. McKinzie, T.B. Cochran, C.E. Paine, January 1998, NRDC Nuclear Programme, Washington, D.C. 20005.

5) **1998 Military Spending: Behind the Numbers,** The Defense Monitor, Vol. XXVI, Number 3, 1997 Center for Defense Information—Washington, D.C.

6) **Plutonium and Highly-Enriched Uranium 1996, World Inventories, Capabilities and Policies**, D. Albright, F. Berkhout, W. Walker, SIPRI, Oxford University Press, 1997.

7) **U.S. Nuclear Weapons**, Bulletin of Atomic Scientists, Nov./Dec. 1995, p. 39.

8) Missile Proliferation and Dual-Use Technologies, Peter D. Zimmermann in **Missile Proliferation, Missile Defence, and Arms Control**, Eds. G. Neuneck, O. Ischebeck, p. 19 ff, Nomos Verlagsgesellschaft, Baden-Baden (1993).

9) **The Global Proliferation of Theater Ballistic Missiles**, Arms Control Today, April 1994, p 29.

10) **A Treaty on the Cutoff of Fissile Material for Nuclear Weapons—How to Cover? How to Verify?** Annette Schaper, Peace Research Institute Frankfurt, PRIF Reports No. 48, July 1997.

11) **Model Nuclear Weapons Convention**, Convention on the Prohibition of the Developmnet, Testing, Production, Stockpiling, Transfer, Use and Threat of Use of Nuclear Weapons and on Their Elimination, Lawyers' Committee on Nuclear Policy, New York, April, 1997.

12) **The Future of U.S. Nuclear Weapons Policy**, National Academy of Sciences, Washington, D.C., 1997.

Address:

GERT HARIGEL

CERN

69 Rue de Lausanne

CH-1211 Geneva 23

Switzerland

11

Transparency Regime for Nuclear Warheads Elimination: Problems and Solutions

ANATOLI S. DIAKOV

Recent nuclear tests carried out by India and Pakistan demonstrated the fact that without real progress in the nuclear weapons reduction would be impossible to carry out the effective non-proliferation policy. To support the non-proliferation regime effectively to the world community clear evidence must be submitted that the nuclear weapons states are really accomplishing their commitments on the nuclear weapons elimination given to the non-nuclear weapon states during the conference on the NPT extension. First of all this is inevitably linked with assurance that the reductions of nuclear arms carried out by the United States and Russia are irreversible and that other nuclear states will also be involved in this process. A provision of such evidences is required the development of transparency measures with respect to ongoing reductions. It is well known that existing arms control agreements have addressed strategic nuclear delivery systems and only indirectly the nuclear weapons that they carry. No existing arms control agreement requires Russia or the United States to dismantle nuclear warheads. In the context of providing the real evidence with respect to downsizing their nuclear arsenals to the world community the development and implementation of transparency measures to the stockpiles of nuclear warheads and weapons fissile materials is an essential and natural step. In addition, the implementation of this measures will work on strengthening of trust either between the Russia and United States or between other nuclear states and stimulate their involvement in the process of nuclear weapons reductions.

The issues of transparency of nuclear arsenals and irreversibility of their reductions were discussed several times at the Russian-US summit meetings during 1994–95 years and were reflected in the number of the Joint Statements issued by Russian and US Presidents (January 1994 Summit Joint Statement, September 1994 Joint Summit Statement, May 1995 Joint Summit Statement). But no success was achieved during discussions on these issues within joint working groups. Negotiations continued during 1994–95 until November 1995 but when the draft text of the agreement opening the way to the transparency of nuclear warheads stockpiles was practically finished they were cut off by Russia. The transparency and irreversibility issues were raised again at the Helsinki summit in March 1997. In the Joint Statement On Parameters On Future Reductions In Nuclear Forces made by the US and Russian Presidents, both sides agreed that the START III agreement will include, among other things "measures related to the transparency of strategic nuclear warhead inventories and destruction of the strategic nuclear warheads".

The primary obstacle on the way to the transparency of nuclear warheads arsenals is that nuclear weapons continue to exist and play a key role in the security doctrine of all nuclear weapons states. In Russia, due to deteriorating its conventional forces, the role of it's nuclear weapons even increased. Under these circumstances it would be unrealistic to wait that a comprehensive control which will include the data exchange on the history production, storage and utilisation of nuclear weapons and fissile materials, and current nuclear activity is achievable. Taking into account the complexity of this task and experience of recent cooperation of both sides in this area gradual steps must take place to ensure the stability of any measures that are employed. But one should mention that the earlier both sides will reach an agreement on T&I issues their cooperation will be more productive later.

While there is no agreement required, the elimination of nuclear warheads on both sides are carrying out their dismantlement. During the past six years more than ten thousands warheads were dismantled in Russia including about two thousand of strategic warheads. In the same period in the US more than eight thousand warheads were eliminated, from them were more than three thousand of strategic.

Table 1 : Estimated numbers of the US and Russia's (FSU) strategic Warheads

	Beginning of 1990		End of 1997	
	USA	**FSU**	**USA**	**RF**
Deployed	12400	10210	7064	5800
Reserve	600	1300	1836	2170
Scheduled for dismantlement			500	1600
Eliminated			3500	2000

Table 2 : Estimated numbers of the US and Russia's Tactical Nuclear Warheads

	Beginning of 1990		End of 1997	
	CIII A	**Pφ**	**CIIA**	**Pφ**
Deployed + reserve	9500	17100	730	5650
Scheduled for dismantlement			3450	3000
Eliminated			5320	8450

The experience of collaboration on transparency of nuclear warhead stockpiles is demonstrated that the United States are more active in the achievement of transparency than the Russia. Usually to explain this fact a traditional Russian attitude to secrecy is used and, to some extent, this is true. But Russia has other more serious grounds to be restrained on this issue. One of them is the recent experience on verification of already reached agreements. Unfortunately the situation in this area is not sufficient. First of all Russian concerns related with the warhead loading levels of Trident missiles, with the procedures of eliminating MX missiles, and with the final re-orientation of B-1 bombers from nuclear to conventional missions. For example, the matter of problem with Trident missiles is that the number of bus maneuvers during the test flights is more than the number of warheads which has these missiles in accordance with the Treaty. The US also has concerns with regard to the number of warheads deployed at SS-18 and Topol-M missiles. To remove these concerns the US side suggested using the radiation measurements for monitoring of the number of warheads deployed at the Russian missiles. But to the Russian reciprocal proposal to use this technique for monitoring the number of deployed warheads at all Russian and US missiles, the US did not give answer yet. Of course, such monitoring practice is compromising of the efforts on achievement of the nuclear warheads transparency regime because it rises doubts about its real purposes.

An other restrained factor is the cost of verification measures. Their implementation will require essential modification of routine operations at the dismantlement facilities, at the facilities where warheads are maintaining and storage, and transportation operations. Under conditions of lacking financial resources, what Russia spent on the maintaining its nuclear arsenal is several tens of time lesser than the US, the secrecy would be considered as quite natural measure increasing a guarantee against unfavourable development.

The implementation of a START III monitoring regime for nuclear warheads inventories and warheads dismantlement requires to answer several questions. Apparently the key question is related with what categories of nuclear warheads will be covered by the control regime. Here could be expected the main difference between Russia's and the US' approaches.

Table 3 : US Warheads Inventories (Estimated) by the end of 1997

Warheads type	Delivery means	Number
W-62	Minuteman III	610
W-84	GLCM	400
B61–3,4,7,10,11	Tactical and Strategic Bombs	660
W-76	Trident I	3350
W-78	Minuteman III	915
W-80	SLCM/ALCM	2090
B83	Strategic Bombs	600
W-87	Peace-keeper/Minuteman III	525
W-88	Trident II	480
Total		**9630**

Table 4 : Russian Inventories of Nuclear Warheads (estimated) by the end of 1997

Warhead type	Delivery means	Number
	PC-20 (SS-18)	3170
	PC-18 (SS-19)	1100
	PC-22 (SS-24)	105
	PC-22 (SS-24)	380
	PC-12I (Topol)	380
	PCM-50 (I, II, III) (SS-N-18)	660
	PCM-52 (SS-N-20)	1260
	PCM-54 (SS-N-23)	470
	PCM-40 (SS-N-8)	200
	PCM-25 (SS-N-6)	20
	Tu-95, Tu-160 (ALCM)	550
	Bombs, SR ALCM	2060
	Navy complexes	2400
	ABM and Air Defense	1250
Total		**13995**

The Russian interest to monitoring of nuclear warhead dismantlement would be associated with the desire to eliminate the big break-out potential which the US received due to START I and START II Treaties. First of all it related with elimination of all strategic warheads downloaded from ICBM's Minuteman III (W62, W78), MX (W87), and SLBM's Trident (W76, W88). But the US, in accordance with their current national nuclear policy (hedge strategy), are going to support the ability to reconstitute the nuclear weapon stockpile to a level of the START I Treaty. There is no decision yet what types and number of nuclear warheads which the US is going to keep in the stockpile in the context of START III levels. Perhaps, the US have interest to keep their modern warheads W87 and W88 and even there is a plan to deploy W87 warheads at Minutemen III ICBMs.

Another complicated question is related with a nuclear long-range sea launched cruise missiles (SLCM). Russia has regarded a nuclear long-range SLCM as the strategic weapon which role is increasing in the conditions of the depth reductions of the strategic nuclear weapons. The US, based on their own interests, are considering this type of weapon as tactical.

In its own turn, the US would like to establish total numerical limits for both strategic and tactical nuclear warheads stockpiles. The basic motivation for this is the concern about the large Russian tactical nuclear warheads stockpile and it's security. In Russia the suggestion to establish a total limit for tactical and strategic weapons has never been agreed. The US tactical nuclear weapons deployed in Europe are really strategic weapons for Russia, while the Russian tactical weapons can not reach the territory of the United States. Besides this, the tactical nuclear weapon is considered by many military experts in Russia as a "natural" answer to the NATO extension and as a mean to equalise the fighting capabilities of NATO alliance and Russia. Without tactical nuclear weapons the ratio of fighting capabilities for conventional forces is 4:1 in favour of NATO.

Under this condition the optimal solution of the above mentioned problems would be development and implementation of transparency measures only with respect to strategic stockpiles, which would include deployed, reserved, dismantled and awaiting dismantlement strategic nuclear warheads. As a "start" point of time would be 1991. Both sides would be exchanged by data how many warheads were deployed in the strategic reserve that time, as well as how many were assembled new and dismantled old beginning from 1991 up to now. Accompanied by control measures it would create a base for further development of transparency regime.

For decision on long-range SLCM it would be rational to link with the level of START III agreement. If this level is about 1500 warheads, it would desirable to eliminate all nuclear warheads for this category of delivery mean. But if both sides agree on the level of 2000–2500 all long-range SLCM should be counted in this number.

As for tactical nuclear weapons taking the complexity of this problem and its lower negotiability, as well as the desirability to have a progress on the implementation of transparency measures with regard to nuclear warheads, it would be rational to leave the solution of this problem for further stages. At the same time it would be desirable not only to reconfirm the unilateral Bush-Gorbachev obligations but their development too. For example, both sides would discuss and agree to carry out exchange of data on how much tactical warheads were eliminated from 1991.

What organisational measures and technical methods would be used for monitoring nuclear warheads stockpiles and dismantlement of warheads? The basic purpose of this control should be the confirmation that each side is not disturbing the treaty obligations with respect to the agreed number of warheads (deployed and reserve) and that it is carrying out the dismantlement of warheads designated for elimination.

Currently, in accordance with START I Treaty, there is a special type of inspection to check the number of warheads deployed on delivery means. If an agreement could be achieved to use radiation non-intrusive control technology to check the number of deployed warheads during inspections it would provide a high level of confidence. To monitor a warheads reserve it will be necessary to agree on additional measures. They should include the data exchange on the sites of storage and number of warheads stored at each storage, as well as spot-check inspection of the storage facilities with using of non-intrusive measures.

The maintenance of enduring nuclear stockpile would present one .more serious problem for the control. It is known that the service life-time of Russian warheads is about 10–15 years while the life-time of the US warheads is 25–30 years. It means to maintain its nuclear arsenal Russia should produce more new warheads than the US. Also, Russia has four plants for the nuclear warheads production the total capacity of which about 7000 warheads per year. The US have only one plant with the capacity about 2000 warheads per year. Despite there is an opinion that the maintenance activity would not be a subject to monitoring procedure some initial steps are desirable. They would include the establishing of a quota for new warheads production which each side could produce during a year and establishing of a portal monitoring regime at the production facilities.

The implementing of the transparency regime for the nuclear warheads dismantlement will have an impact on operation of whole nuclear weapons complex. The major concerns here are associated with the possibility of releasing of classified information. Today both the Russian and the US weapon designers are considering the information about construction of nuclear warheads as very sensitive. Therefore, the monitoring procedure should be organised in manner that could provide a high level of confidence that elimination of nuclear warheads is really going on and, at the same time, to minimise the possibility for loss of classified information.

By focusing on the control mechanism of the fissile material placed in the containers it would be possible to check their weapons quality. Today the experts of both countries developed methods to determine that the material in the containers has properties consistent with a nuclear-weapon component. But the insurance of what the materials in the containers are really come from eliminated warheads would be achieved by a special arrangement of monitoring measures. The designation of one facility exclusively for dismantlement purposes, portal and perimeter continuous monitoring of this facility, and inspection of disassembly zone before and after dismantlement would provide a high level of confidence that the materials in containers are really come from weapons. One additional measure which will increase confidence would be eliminating under control of body of nuclear warheads. Thus a high insurance of monitoring procedure would be received without release of the classified information.

Today there is not much information available for the public on what options for monitoring warhead dismantlement are considering in Russia. Apparently, on formation of Russia's position on this issue will be effected by all above mentioned factors. It seems the tendency to minimise the volume of information for exchange will dominate, and the monitoring procedure will focus to the control of fissile materials released from warheads. Based on this assumption two extreme options can be assumed.

Minimum level of control:

Data Exchange: exchange of information on the total number of warheads, referred to the agreed moment, the number of warheads produced and eliminated beginning with that time (without clarification of warheads types) and the total amount of fissile materials which were released;

Method of Monitoring: control of the released fissile materials during the loading into the storage facility and monitoring of the components storage facility;

Monitoring Procedure:

1. the use of gamma-ray measurements to determine that the material in the closed containers has properties consistent with weapons-grade plutonium as well as to estimate the minimum mass.
2. monitoring of the components storage facility.

Maximum Level:

Data Exchange: exchange of information on the total number and types of warheads, referred to the agreed moment, the number and types of warheads produced and eliminated beginning with that time and the total amount of fissile materials resulting from the dismantlement process;

Method of Monitoring: control of fissile material during their loading into storage facility, portal perimeter monitoring of storage facility, and portal perimeter continuous monitoring of the facilities especially designated for the dismantlement purposes;

Monitoring Procedure:

1. inspection of every items that passed into and out of the dismantlement facility;
2. limited chain-of-custody within the dismantlement facility;
3. sweeping the disassembly boxes before and after dismantlement.
4. control of nuclear components in the closed containers by use of the gamma-ray measurements.
5. chain-of-custody of nuclear components from dismantlement facilities to the component storage facilities.
6. monitoring of the components storage facility.

Address

ANATOLI S. DIAKOV
Moscow Institute for Physics and Technology
Moscow
Russian Federation

12

Some Thoughts on the Prospect and Technical Challenge of Nuclear Disarmament

JIASHU SONG

There were lots of reports and initiatives on nuclear disarmament in recent years, as the initiative of "nuclear weapon free world (NWFW)" proposed by PUGWASH, the four-steps plan by Henry Stimson Center (1995) the Report of Canberra Commission (1996) and the Report of the CISAC by American National Academy of Science (1996), etc. All these reports indicated the desire of prohibiting nuclear weapons and adopted step-by-step approaches. As we know that the general nuclear disarmament, which finally should reach to total abolition of nuclear weapons, is a time consuming process. It will meet with various difficulties including technical challenges. I believe, however, that the technical obstacles will be overcome by the efforts of scientists, given the positive political will of nuclear powers and through practical approaches. The prospect of nuclear disarmament is bright.

1. Further Steps of Nuclear Disarmament

As a matter of fact, the nuclear disarmament has been carried out in different stages, such as "SALT" in the past and "START" by now. In some of them only delivery vehicles are involved, and in the others, the whole weapon systems.

Up to now, the nuclear disarmament has been limited to the US and Russia, and includes only the reduction of deployed weapons. It does not require the elimination of nuclear warheads.

This kind of reduction could maintain the balance on a lower level of nuclear confrontation, help to remove the bad results of "nuclear arms race" in Cold War and to reduce the probability of risks from accidents. But the retailed warheads will not be really eliminated and it could not do much in reducing the total number of nuclear warheads in nuclear powers. Up to the beginning of next century, the US and Russia will have nearly 10,000 warheads each after the START 2 is finished. In the mean-time, they also lack adequate guarantee of guarding the irreversibility of the reduction. Obviously, the "START" stage could not directly reach the "nuclear weapon free world".

Therefore we believe that further reduction would need a stage of "all nuclear warheads reduction". In this stage the reduction obviously should include all stockpile, that is, not only deployed but also non-deployed (stored) warheads, and include all forms of reduction, including the elimination of warheads, and final disposition of the fissile materials in order to ensure the irreversibility of reduction.

Recently the CISAC study which is entitled "The future of US nuclear weapon policy" calls for a regime of constraint on US and Russian force to the level of 2,000 strategic warheads each first and then to the level of 1,000 total warheads each, with concurrent steps to reduce alert levels and reorient the doctrine away from rapid and massive responses. It is an important recommendation, which would create favourite conditions for multilateral nuclear disarmament. As this stage is started, the reduction to a few hundred warheads will be possible. And then we can talk about the process leading to the final mission—the nuclear weapon free world.

2. The forms of Nuclear Weapon Reduction

From the technical point of view, the nuclear weapon reduction in different stages corresponds to various forms, of which the following three forms of reduction can be listed:

1. Dismantlement of nuclear weapon systems

   As in the INF treaty, the warheads were removed from the missiles, but not required to be further disposed. The meaning of this kind is in the abolition of the delivery vehicles.

2. Dismantlement of warheads.

   There are many possible degrees of dismantlement. It could mean the removal of warheads from re-entry vehicles, the removal of the nuclear explosive from the warheads, or the division of the nuclear explosive into parts (stages separated), and/or complete dismantlement. Under the START only parts of retired warheads are dismantled (but stored as Pits) and it has only symbolic significance, for the Pits can be reused.

3. Disposition of Weapon Fissile Materials.

It should include weapon grade Pu and HEU from retired warheads. Lots of studies about the way of disposition have been done, but so far no treaty has required this disposition.

In addition, total abolition of nuclear weapons, including the elimination of research and production facilities and capability, is a totally new issue which needs further study.

In "START" only the first form of reduction is actually concerned. However we think the real reduction should include all three forms. Despite we have plenty of experience in implementing SALT, START, INF and NPT agreements, the nuclear disarmament in the future very likely will encounter great technical challenge. The important tasks which the technical measures should fulfilled are as follows:

- To ensure the irreversibility of nuclear weapon reduction. The new technology and attitude will be needed to verify the warheads dismantlement and the disposition of fissile materials.
- New verification techniques is disarmament treaties of different steps.
- Controlling the fissile materials in military and civil fuel cycle.
- Ensuring the stability of a world without nuclear weapons. Even more complicated is the verification of a world having less or no nuclear weapons. It will be very important to create a system with high reliability, high accuracy in order to prevent any break-out.

The study on these issues has already been carried out by many scientists. Yet many problems are still to be solved. More new issues could possibly emerge in the future. Anyway, a great uncertainty should be expected in these aspects, so that to predict exactly a time-table of its completion is difficult.

3. Views on the "Nuclear Weapons Free World"

We believe that the final goal of the nuclear disarmament is total abolition of nuclear weapons, which is the fundamental way to reduce nuclear risks and to prevent the proliferation of nuclear weapons. Early in the 1960's when we took part in the research to develop a limited nuclear force, we believed that we need not have developed nuclear weapons if there were no nuclear threats to us. In developing nuclear weapons, our aim was to eventually eliminate nuclear weapons.

The ending of the Cold War provided a good chance to realise this goal. As the consequence, scientists of CAEP have actively taken part in the research of nuclear arms control and related verification techniques. We hope to find possible paths and to develop technology which will promote disarmament. We also hope to strengthen the international exchange with experts working in this area. Of course, we realise that it is a long way to go before the NWFW has come true.

As a matter of fact, the nuclear disarmament is correlated to a series of factors. Technology is not the only key factor. The discussion of nuclear disarmament alone (in isolation) does not make much sense in reality. In order to promote the process of disarmament, the following are needed:

- To further improve the international security environment, to create various confidence-building measures, like mutual promise and accepting the policy of "no first use", nuclear weapon free zones, de-alerting, and so on.
- To prevent new forms of arms race. For example, the US has decided to develop a strong missile defence system in order to build absolute priority, and to carry out Laser Anti-satellite Tests. This policy has surely the risk of inducing new round of nuclear and outer space arms race.
- Preventing the diffusion of weapons of large scale destruction, including the nuclear weapon proliferation.

Without these conditions, the transition to: "all warheads reduction"; would be very difficult and practically impossible; As is well known, the US Government has not yet abolished the extended deterrence policy and continues to keep large number of "targets". In the present circumstances it is difficult to imagine the possibility of reduction to very low level. Therefore the progress in the nuclear weapon reduction should be accompanied by concurrent changes in understanding the role of nuclear weapons and by the decrease in relying on the nuclear deterrence policy. The range of the use of nuclear weapons should be narrowed step by step before the weapons are totally banned. The first step should be to promise not to use nuclear weapons first and not to use nuclear weapons against non-nuclear states. Should the value of nuclear weapons be greatly depreciated, it would be easier to conduct deep reduction, to ensure the irreversibility of the cut and to prevent further proliferation.

Address:

JIASHU SÒNG
Institute of Applied Physics and Computer Mathematics
Beijing 100088
China

PART — 3

ILLEGAL NUCLEAR TRAFFIC: RISKS, SAFEGUARDS AND COUNTERMEASURES

13
The Legacy of the Cold War

M. MARTELLINI

More than fifty-two years have elapsed since the atomic bomb was dropped on Hiroshima and Nagasaki. Over this period, the United States and the Soviet Union have engaged in a constant nuclear arms-race, in an effort to build a growing number of warheads and launching systems of increasingly higher destructive power, reflecting the logic of the balance of terror and the threat of mutually assured destruction. The years of the Cold War have, in fact, been dominated by the incredibly rapid growth of nuclear arsenals. The end of the Cold War, which was marked by the fall of the Berlin Wall in November 1989, coincided with the beginning of the process of nuclear disarmament and a diversion of funds from the industrial military complex.

Nevertheless, 45 years of development and sophistication in nuclear armaments can't be immediately wiped off, for many reasons, the most evident of which being that the nuclear materials that forms the heart of atomic weapons, more precisely known as weapon-grade materials, can't be simply eliminated, as will be explained later in this text. Their dismantlement requires special technical means, which are not yet available to all nuclear powers.

There are also other reasons why nuclear disarmament is such a complex and uneven process.

First, the production of existing arsenals has required huge financial resources. From 1945 to 1995 the United States and the Soviet Union have spent some 4,000 billion dollars each (at 1995 rates), as much as 150 times the cost of the Manhattan project. Consequently, one of the problems arising after the end of the Cold War, and one which belongs to the wider chapter of "military conversion", has to do with turning the huge economic potential of nuclear weapons into readily available resources. Nuclear weapons can't be used, they do not move capitals and they can only serve as a deterrent in a geopolitical situation characterised by the confrontation between a few military powers. Yet, after the end of the Cold War and the collapse of the Soviet Union, the very concept of nuclear deterrence has lost its meaning and a new, more flexible defence pattern has to be developed, along the lines of the UN system.

However, some nuclear powers, like the former Soviet Union, which is facing serious economic difficulties, as it shifts from a centralised to a market economy, cannot afford competing with the West in research and development of new, more sophisticated and accurate war devices to replace mass destruction weapons. For them, nuclear weapons remain the only form of defence available against any aggression. Paradoxically, in the post-Cold War period, nuclear weapons have become the weapons of the poorer countries, unable to afford modern conventional defence, and therefore a last resort in case a conflict that can't be avoided.

The dilemma of nuclear disarmament also involves another difficult problem—environmental safety. Using part of the core of an atomic warhead as nuclear fuel in civilian nuclear power stations, as has been suggested, would be an extremely complex and costly process, because weapon-grade materials are different from the nuclear fuel used in civilian installations. Weapon-grade materials consist of highly enriched Uranium and Plutonium. The former is obtained by raising the concentration of ^{235}Uranium isotope to 90–95 per cent (compared with only 0.4–0.7% in natural Uranium) through electrochemical processes. The latter is man-made element produced in nuclear reactors as a by-product of nuclear fission. Therefore, military nuclear installations always include special plants for enriching or generating weapon-grade plutonium. These inevitably produce a large quantity of highly radioactive sludge and nuclear waste, which are extremely dangerous for human health and the environment which must be removed and temporarily stored. Clearly, there is the need not only to dismantle atomic arsenals, while ensuring nuclear non-proliferation, but also to close and clean all these highly polluted enriching and production sites. As can be seen from these general considerations, nuclear disarmament is a multifaced issue and one that needs to be tackled in all its various aspects.

To better understand the extent of these 50 years of nuclear arms race, let us consider the following figures:

- From 1945 to the present day 130,000 nuclear warheads have been produced: 60,000 in the former Soviet Union and 70,000 in the United States.
- Since 1960 the two super-powers have been confronting each other with some 10,000 operational warheads each.
- There have been 2,050 nuclear tests, 514 of which in the atmosphere.
- The number of countries holding nuclear weapons has gone from 1 in 1948 (the United States) to 8 today: 5 official nuclear powers (the United States, China, Russia, France, and Great Britain) and 3 rogue ones (Israel, India and Pakistan).
- Over 250 tons of weapon-grade Plutonium and 2200 tons of highly enriched Uranium have been produced.
- The United States have produced some 67,500 missiles to carry atomic warheads.
- The estimated overall cost of producing and maintaining the nuclear complex from 1945 until today is about 9,000 billion dollars—that is approximately 9 times Italy's present GDP.
- The production of weapon-grade materials, especially in Russian sites, has lead to the radioactive contamination of dozens of thousands of square kilometers. For example, over 400,000 people were contaminated in the Russian production and storage site of Mayak/Chelyabinsk (Siberia), which spreads over a surface of 26,000 Km^2.
- As much as 80 per cent of all the human and technological resources in the former Soviet Union were taken up by the military complex, which employed over 16 million people. In Russian so called "closed cities", which could not be found on any map, and where weapon-grade materials and nuclear warheads were produced and new mass destruction weapons developed, over 700,000 were employed people as many as 5,000 of whom had all the information needed to build a nuclear bomb.

And yet, the most surprising fact of all is that: 1) despite the wide variety of atomic warheads in terms of range, destructive power and even launching systems, in 50 years of nuclear arms-race, tactical nuclear weapons for limited targets have never been deployed (and would never have been possible); and 2) despite the huge amount of energy and human resources spent to develop and generate nuclear warheads and the natural disasters associated to military nuclear sites, since the atomic bombing of Hiroshima and Nagasaki these weapons have no longer been used.

However cynical it may seem, this last observation epitomises the special character of atomic weapons, so different from conventional weapons, like those based on chemical explosives, for which there exists no international regulation (the first international treaty to ban landmines was signed as recently as 1997).

This special feature has marked the world scenery in the 50 years of the Cold War. The question that arises now is what to do with what international treaties START I and START II refer to as "redundant" nuclear warheads (those which cannot be used for military purposes according to the above treaties).

If these nuclear warheads were dismantled, an estimated total of 100 tons of Plutonium and 700 tons of highly enriched Uranium would be obtained. If you think that, thanks to modern technology, from 2 to 6 Kg of Plutonium or 3 to 16 Kg of highly enriched Uranium are enough to build an atomic bomb, you realise that there is a tremendous risk associated with the smuggling of this "redundant" weapon-grade nuclear material. In addition, from the point of view of nuclear safety, the temporary storage of the weapon-grade portion of the warhead is far from being ideal. Once a nuclear warhead has been dismantled, its supervision is passed over from the armed forces to civilian government bodies, which are responsible, as the MINATOM (Ministry of Atomic Energy) in the Russian Federation and the Energy Department in the United States.

While safety standards in these countries are normally high, they can't be compared with military standards, since the undeclared principle is that once a warhead has been dismantled it is no longer *operationally*, but only *potentially* lethal. Nevertheless, it should be taken into account that, while a non-nuclear country, that is a country that is not included in the official lists of non-proliferation treaties, may have the necessary equipment to produce an atomic warhead, it would be more convenient for a criminal organisation or a terrorist movement to steal weapon-grade nuclear material and assemble a "rudimentary" nuclear warhead, that is non-mobile, on the basis of the civilian technology that is available in all fairly industrialised countries. Consider also that, as has been already mentioned, the critical mass needed to start an uncontrolled chain reaction in an atomic bomb is relatively small compared with the hundreds of tons of weapon-grade materials made "redundant" as a result of dismantling. In addition, Plutonium is chemically highly toxic and a few grams dropped into the water system could poison an entire city, a method that the Japanese sect of Aum Shinriko, responsible for Sarin gas attacks in 1996 in Tokyo's underground, is supposed to have considered.

But even if the smuggling of weapon-grade materials could be reduced to a minimum, there remains the difficult problem of dismantling nuclear weapons while respecting the two principles of environmental safety and non-proliferation. At present, there are only two solutions that seem to meet the above mentioned criteria:

a) the storage of vitrified weapon-grade nuclear materials together with highly radioactive wastes in geologically stable deep underground sites (in such a way that their removal would be extremely dangerous and expensive);
b) the reuse of weapon-grade materials as nuclear fuel in civilian nuclear power installations. While for highly enriched Uranium, enriching plants can be made to reverse the enriching process in order to reduce the concentration of the ^{235}Uranium isotope to generate a low-grade Uranium, there can be no isotope dilution process for Plutonium. The only technological solution elaborated so far consists in MOX (mixed oxides) mixtures, consisting of weapon-grade Plutonium and natural Uranium, in which the Plutonium content should not exceed 30 per cent of the entire mass. This material would be used to make nuclear fuel rods for nuclear power stations based on pressurised water reactors, that is the type of reactor used by most western installations. Unfortunately, MOX production is costly and complicated. Plants equipped to produce it can be found in

Europe and Japan, but not in the United States or in the former Soviet Union. This means that during the fifty years of the Cold War the two nuclear superpowers had never considered the end of the nuclear arm-race! MOX itself is extremely expensive: the official price is about 2,000 dollars per pound, against 8 dollars per pound for natural Uranium. Using MOX instead of naturai Uranium as a source of energy would not be economically advantageous, which means that its use could only be politically imposed. After all, at present the two nuclear superpowers spend some 100 to 200 billion dollars every year only on their military nuclear complexes (according to 1995 data). The cost of producing MOX on a large scale could be included in these costs.

Yet the problem of the disposal of all weapon-grade nuclear materials by "burning" them in nuclear clashes with another aspect: the time factor.

According to START I and STARTII strategic agreements, nuclear warheads available to the USA and Russia for military purposes should be reduced to 3,000–3,500 each by the year 2003 (in March 1997 at the US-Russian Summit in Helsinki the START II treaty implementation time was extended until the end of 2007). If the Plutonium obtained from redundant warheads were immediately converted into MOX, without exceeding 30 per cent of the mixture content as stated above, and using all the available nuclear power stations simultaneously, the process would be over only as late as 2010 and more.

Will the world geo-political situation remain stable until that time?

Those who deal with these problems become soon aware of the global character and inter-dependence of nuclear disarmament issues. Any local solution that does not consider all the different factors involved, such as nuclear and environmental safety, the principle of nuclear non-proliferation, economic and technological sustainability in the management of weapon-grade nuclear materials and timing, is inevitably bound to failure and perhaps even to have detrimental effects. So what needs to be done? The author of the present paper has few certainties and no magic solution to push the bad Genini of History back into his lamp, but a few ideas can be expressed:

i) there will never be any absolute guarantee for nuclear non-proliferation unless and until all nuclear warheads are completely eliminated. What is the point of having 6000 to 7000 operational nuclear warheads after the year 2000? Will the world be safer than in 1997, with about 22,000 operational atomic warheads? What guarantee can be given to non-nuclear States and to developing countries, at a time when the Pacific countries are expected to overtake Europe and the United States in terms of GDP and control over world financial markets in the year 2000, but they will be at risk of being blackmailed by the old nuclear powers, which will preserve their nuclear supremacy? Will these countries, too, want to develop their own nuclear weapons? And finally, the Russian Federation is still very unstable. Will its fragmentation into a number of independent Republics lead to a fragmentation of its atomic arsenals, scattered as they are in about 100 sites all over its territory?

ii) As can be seen, the problem of facing the legacy of the Cold War is a global issue that involves all the countries of the world, and in particular the 5 official nuclear powers. Obviously it can't be tackled on a national scale and perhaps even traditional international organisations such as the United Nations and the International Agency for Atomic Energy have become obsolete, since they reflect the Cold War period, in terms of organisation and internal efficiency.

In the author's opinion, the size and the nature of the urgent issues emerging from nuclear disarmament would require an *ad hoc* international organisation, with independent financial resources to be formed with the contribution of all the nations of the world in different proportions, depending on their relative involvement in the military nuclear complex. This would be, in essence, a large project like the Manhattan project, now aimed not at the creation of weapons of mass destruction, but rather at their total and final destruction. This new international organisation, initially designed to manage the most urgent nuclear issues, may eventually be used to manage other types of global issues and threats, such as food and water resources or renewable and clean energy sources. Theoretically, this organisation could be entrusted with the task of implementing the conversion of the military resources and structures that have been built in 50 years of Cold War to the benefit of the civilian economy and in ways that could be controlled by the free market system. Only a supranational and impartial organisation could achieve the goal of the so-called "dividend for peace", so much called for after the end of the Cold War, by injecting the funds made available from defence budget cuts (after using some to eliminate the "wastes" produced by the Cold War) into wide-ranging peace projects, designed to solve the problems of poverty, hunger, environmental degradation, thus allowing a really sustainable development.

I would like to conclude by quoting the over 50-year-old and yet still valid message of the Einstein-Russel manifesto: "We appeal, as human beings to human beings: remember your humanity and forget the rest. If you can do so, the way lies open for a new paradise; if you cannot, there lies before you the risk of universal death".

Address:

MAURIZO MARTELLINI
Secretary-General
London Network Centro Volta
Como
Italy

14

Identifying the Origin and Intended Use of Seized Samples of Illicit Trafficking of Nuclear Material

L. KOCK

Introduction

Nuclear fissile material of weapon utilisable quality has accumulated on a 100 t scale—often in a form where it can be easily converted into a device of mass destruction. Because of the uncertainties—although in the range of 0.1 to 0.3 per cent—in the analytical determination of fissile material masses it is impossible to establish for bulk handling facilities a mass balance down to the kg level. For this reason it cannot be simply detected from a mass balance mode at a large scale nuclear plant if and how much fissile material in the range of kg has disappeared.

That is why seized nuclear materials from traffickers could not directly be attributed to places "Where it is missing", because such low quantities are within the "material unaccounted for" (MUF). Consequently another approach has to be taken by comparing nuclear and structural characteristics of the seized unidentified item with the material of the plants in question. A pertinent methodology will be described below.

The most effective measures to combat a diversion of special nuclear material are its stringent surveillance and physical protection at the places where it is kept or when moved between facilities. In the former Soviet Union strong emphasis was given mainly to the surveillance of the personnel that had access to the material. That is being shifted now more to material surveillance by highly sophisticated automated instrumentations. Once enriched uranium or plutonium has been diverted it is difficult to detect. A variety of passive and active interrogation techniques exist but—even if they would have been placed at all border crossings—they could not prevent illicit trafficking e.g. across "green border". So far the material seized—as reported in the media—was found haphazardly or by undercover agents, which emphasises the role of the intelligence services in combatting illicit trafficking of nuclear materials.

METHODOLOGY TO IDENTIFY AND TRACE BACK SEIZED NUCLEAR MATERIAL FROM ILLICIT TRAFFICKING

The unauthorised possession of fissile material is a severe criminal offence. In order to comply with the existing legal procedures things like maintaining a chain of evidence, protocols, etc., have to be adopted and extended according to the precautions, which are requested by health physics and for the handling of fissile materials.

The sequence of forensic investigations is illustrated schematically (Table 1).

Table 1 : Sequence of Forensic Investigations

1. *Seizure of material following:*
   - offer for sale
   - detection by passive (n or y) or active interrogation systems (n-n' or n-y)
   - protocols for collection of evidence
   - health physics precautions

   (NDA: high resolution gamma spectrometry)
2. *Blitz analysis*
   - to determine whether the seizure is an illicit contraband
   - categorisation of nuclear material: weapon, weapon utilisable, fuel

   (NDA: high resolution gamma spectrometry)
   (DA: thermal ionisation mass spectrometry)
3. *Characterisation by specialised and qualified laboratories to obtain the required:*
   - endogenous information: histogram, reactor parameter etc.
   - exogenous information (needs historical data base): *nuclear, chemical and material properties: reactor, enrichment plant, fabrication plant, fuel type, fabrication campaign, etc.*

The first steps are the seizure and the classical forensic at the crime scene without unnecessarily touching the nuclear material. Before a seized material is transported and unpacked for investigations, certain additional precautions have to be taken to eliminate any danger. Specialists will determine by a 'Blitz-analysis" by high resolution gamma-spectrometry or thermal ionisation mass spectrometry the composition of the nuclear materials which indicates already in which category it falls (weapons materials weapon utilisable materials nuclear fuel, radioactive sources with no fissile material). Usually this information is sufficient to recognise for the law enforcement the severity of a case.

In a second step the material will be characterised in a specialised laboratory. All features of the material that would reveal the intended use, age, origin and possibly, the smuggling route, have to be measured. A variety of techniques already exist (Table 2) which are being enhanced by adopting known classical forensic techniques or by developing new analytical approaches.

Table 2 : Analytical Techniques to Reveal Origin, Intended use and Smuggling Route

	Analytical Technique	***Results***	***Information***
nucl. materila	Therm. Ionisation MS Induct. Coupied Plasma MS+Ion Chromat.	U-, Flu-isotope Minor actinides	source, process,
bulk	Glow Discharge MS	F.p., nat. elements	type, histogram, intended use in addition via databank:
particles	Electron Nficroprobe Transm. Electron Nficroscopy x-ray diffraction, Sec. Ion MS	chem. composition material properties homogeneity of material	place(s) of production, fabrication
packing			
material	as above plus	nat. elements	via databank: -geolocation
impurities	DNA-analysis	chem. compound pattern, human fingerprint	-smuggling route

Since the types of materials seized have a variety of appearances (Table 3), each investigation has to be tailored to the particular case and follows then the "principles of diagnosis" by guiding the investigator through the results of the previous examinations until he obtains sufficient evidence to answer the question concerning use, age, origin and smuggling route.

Table 3 : Seized Material Analysed at the Institute for Transuranium Elements

- non enriched uranium oxide
- CANDU fuel pellets
- metallic natural uranium fuel
- RBMK and VVER fuel pellets
- 90% ^{235}U oxide granulate
- denaturated weapon plutonium in ionisation detectors
- PuO^2 together with Li-6
- highly enriched plutonium in "red mercury"

To obtain these answers the results to some extent need to be compared to known materials, processes and specific environmental conditions. For the purpose an appropriate data base is being set up to collect such information. However in cases where a material has bene seized in a state where it not originated, it may become difficult to get access to the necessary informations in the state where the material is suspected to have been diverted from.

Nevertheless, it can be concluded that the technical means exist to characterise a seized material to such an extent, that—when the results are available to the last legal owner of the material—he can recognise with certainty his "lost" nuclear material.

To my knowledge the result of the investigations we conducted are accessible, so that the one concerned—without admitting publicly his ownership—can conduct internal investigations, close the leak of nuclear materials and set up appropriate measures to enhance physical protection and surveillance.

NEEDS TO ENHANCE THE COMBATTING OF TRAFFICKING OF NUCLEAR MATERIALS

In its Moscow Kremlin meeting (19–20 April 1996) the P-8 member states agreed on a: Programme for Preventing and Combatting Illicit Trafficking in Nuclear Material.

Illicit trafficking of nuclear material is a public safety and non-proliferation concern. We recognised the importance of this issue at our meetings in Naples and Halifax. As risks continue to exist, we have agreed on, and released, a programme for preventing and combatting illicit trafficking in nuclear material to ensure increased cooperation among our governments in all aspects of prevention, detection, exchange of information, investigation and prosecution in cases of illicit nuclear trafficking."

In an International Working Group (I.T.G.W.) the Terms of References define to:

1. "Identify and prioritise techniques and methods for forensic analyses of non-nuclear materials associated with seized nuclear materials in order to answer questions regarding geo-location and route attribution.
2. Identify and prioritise techniques and methods for forensic analyses of nuclear materials in order to answer questions regarding source attribution and intended use of seized nuclear materials.
3. Formulate and execute inter-laboratory exercises to evaluate and improve the effectiveness of the above mentioned technique and methods for forensic analysis of seized nuclear materials. The suggested exercise include the highly enriched U02 offered by the Czech Republic, and a "known" PuO2 sample originating from a commercial fuel cycle.
4. Supplement and improve technical capabilities:
   Suggest protocols for collection and preservation of evidence in cases involving seized nuclear materials.
   Identify and evaluate available technical equipment for initial hazard evaluation and preliminary assessment of nuclear material composition.
   Identify the impact of specific national laws on international cooperation to execute these terms of reference.
   Facilitate technical assistance to countries in response to specific requests."
   So far the ITWG has given an inventory report on existing and applied techniques: In its third meeting (two days ago at this location) the future tasks have been identified to set up further international cooperation and avoid needless duplication of efforts.

Following the recommendations of the P-8 meeting in Moscow:

"Encourage bilateral and other assistance and cooperation arrangements in the above areas and support their appropriate coordination to ensure that they are complementary and mutually reinforcing and to avoid needless duplication of effort."

The EC is giving technical assistance in the frame of the PHARE and TACIS programmes to several East European states. In particular to cooperate with the Russian Forensic Laboratory at the Bodvar Institute in Moscow by the exchange and support of techniques and instrumentation. To upgrade the technical capabilities in several states of the former Soviet Union to identify seized nuclear materials by non-destructive analytical techniques and to assist in deeper characterisation of seized materials in order to determine its origin and intended use.

Following the recommendations of the P-8 meeting in Moscow: "Support the exchange of scientific information and data to permit the identification of the origin, history, and route of seized illicit nuclear material."

The Institute for Transuranium Elements and the Bodvar Institute are setting up a data base that initially will guide an investigator to select the appropriate analytical techniques in characterising a seized material to the required extent in order to reveal its origin and intended use. So far data on the commercial fuel cycle covering low enriched Uranium fuels are fixed.

Within the existing frame of international cooperation model "Operations Plans" will be worked out, which describe the coordination of services in a state (customs, law enforcement, technical assistance) that then can be adopted to local legislative constraints. Depending on the state's technical capabilities, assistance from outside will be included.

Because of the high concern of the public, many initiatives by individual states are underway. All technical capabilities and scientific information must be joint in a complementary but mutually beneficial way to initiate worldwide actions in preventing and combatting the illicit trafficking of nuclear materials.

Address:

LOTHAR KOCK

European Institute for Transuranium Elements

Joint Research Center

76125 Karlshrue, Germany

15
International Scientific Collaboration in Non-proliferation*

A. TRAVELLI

ABSTRACT

International collaboration is a vital component of every serous non-proliferation effort. Several examples of the experiences that the Argonne Arms Control and Non-proliferation Programme has had in this area are given and, in the process, important components of the programme come to light. Some of the main principles that the programme has learned to follow while pursuing international collaboration projects are shared, as are the pitfalls that the programme has learned to avoid.

1. International Collaboration: the Essential Link in Non-proliferation

The importance of international cooperation in non-proliferation activities should not be underestimated. No matter how powerful or rich in brilliant technologies and scientists, no country can forcibly control nuclear proliferation in other countries. Therefore, the only option for preventing nuclear proliferation is to join forces and coordinate efforts with other nations, creating a fraternity of similarly motivated scientists.

Sharing technical advances developed in other countries can also greatly enhance the ability to stem internal nuclear proliferation. Even the most powerful of countries, when they need assistance to prevent nuclear proliferation within their own borders, can benefit from the exchange of technical information.

---

* The submitted manuscript has been authored by a contractor of the U.S. Government under contract No. W-31-109-ENG-U. Accordingly, the U.S. Government retains a non-exclusive, royalty-free license to publish or reproduce the published form of this contribution, or allow other to do so, for U.S. Government purposes.

As Mr. Redin mentioned yesterday in his talk, the potential for internal proliferation can be especially dangerous in countries which have recently turned from a form of government based on a centrally planned economy to one based on the free market.

2. EXAMPLES OF INTERNATIONAL COLLABORATION

The Argonne Arms Control and Non-proliferation (ACNP) programme is probably the oldest non-proliferation programme represented at this meeting. It started with one programme nineteen years ago, in 1978, when few scientists took non-proliferation seriously, and has thrived on international collaboration since then.

That programme was the Reduced Enrichment Research and Test Reactor (RERTR) Programme, whose goal was and still is to minimise and eventually eliminate the use of highly-enriched uranium (HEU) in research and test reactors throughout the world. Achieving this mission required developing new low-enriched uranium (LEU) fuels, inducing fuel fabricators to produce them on a commercial basis, and, finally, convincing research reactors throughout the world to convert to the use of the new fuels.

The reactors supplied by the West were spread among 30 countries, with active fuel fabricators present in about 10 of those countries. International cooperation was an essential requirement. Within three months from the beginning of the programme, a first meeting was held to gather representatives from most of the HEU reactors all over the world, and from all the HEU fuel fabricators. The goal was to discuss with them our activities and plans, and to receive their comments, suggestions, and—hopefully—their help. It turned out to be a meeting of historic significance for the RERTR programme. Many of the participants had invested heavily in physical protection equipment at their facilities, and were initially strongly opposed to the very purpose of the RERTR programme. At the end of three days of open and intense discussions, however, a plan emerged that allowed the programme to move forward with the support of nearly all the attendees. The meeting has been held every year since, in many different countries, with about 150 attendees from all over the world converging to share their ideas, experiences, and plans. The 20th meeting of the RERTR programme will be held four months from now in Jackson Hole, Wyoming, following meetings which were held in South Korea and France. Meanwhile, fuels that can be used to convert 90 per cent of the Western-designed research reactors have been developed; 24 reactors in 16 countries have been converted; 7 new research reactors have been built in 7 countries using LEU fuels, and 12 fuel fabricators in 11 countries are producing RERTR-developed LEU fuels. Almost none of these results could have been achieved without strong international collaboration.

Collaboration with Russian institutes is a relatively recent experience for the RERTR programme, and is now very intense. The Soviet Union had shared many of the goals of the RERTR programme during the last decade of its existence, and had paralleled many of the programme's activities. It had developed fuels with 36 per cent uranium enrichment, suitable for use in several Russian-designed research and test reactors, and had even converted some of those reactors to the newly-developed fuels. However, the Soviet laboratories that took part in these efforts never made contact with the RERTR programme in that time period, and their efforts were brought to a halt by the economic difficulties of the USSR in the late 1980s.

The political events of the early 1990s, and the resulting transition of Russia from a centrally-planned economy to one based on the free market, brought with them the opportunity for an explicit and intense collaboration between the RERTR programme and several Russian organisations. Today, seven major Russian institutes are cooperating with the RERTR programme in the development of advanced LEU fuels for research and test reactors. These institutes include the Research and Development Institute for Power Engineering (RDIPE); the Yekaterinburg branch of RDIPE (EF-RDIPE); the Bochvar Institute of Inorganic Materials (VNIINM); the Novosibirsk Chemical Concentrates Plant (NZChK); the Russian Research Center "Kurchatov Institute" (RRCKI); the Research Institute of Atomic Reactors (RIAR) in Dimitrovgrad; and the Petersburg Nuclear Physics Institute (PNPI) in Gotchina. Samples of the fuel tubes that could be used to convert many Russian-designed research reactors to LEU fuels suitable for the conversion of all Russian-designed research reactors are being developed.

Additional examples of collaboration with Russian institutes stem from other Argonne non-proliferation projects. For instance, Argonne scientists were among the first to visit the closed cities of Arzamas-16 and Chelyabinsk-70 in 1992, and scientists from those institutes visited Argonne during the same year and on several later occasions, to exchange information and to develop joint proposals.

At present Argonne scientists operating under the umbrella of the ACNP programme are working with scientists from other American national laboratories to improve physical security at the nuclear installations of several technical institutes in various republics that used to be part of the former Soviet Union. These activities include projects related to material protection, control, and accountancy (NWC&A) in the Ukraine, Belarus, Kazakstan, and Uzbekistan. In addition, cooperative activities and training in export control of sensitive nuclear technologies and materials are in progress with those institutes.

Another example of international collaboration that is especially relevant to this meeting is the collaboration between the Argonne Arms Control and Non-proliferation Programme and several Russian institutes aimed at deterring illicit traffic in special nuclear materials. Even without the technical results that are the focus of the collaboration, the development of a common terminology, of common concepts, and of the interfaces defining the areas in which each partner will concentrate its efforts, make a significant contribution to the resolution of the problem.

According to the terminology originally developed by our Russian colleagues and ourselves, and now widely accepted and used, the first line of defence includes all the materials protection, control, and accountancy (MTC&A) activities that ensure the physical protection of special nuclear materials under custody. The second line of defence includes those activities that can lead to detecting the presence, at control points, of special nuclear materials that may have been diverted through some unanticipated weakness of the first line of defence. Finally, the third line of defence includes all the activities that need to be pursued after detection of special nuclear materials by the second line of defence. The most important component of the third line of defence is forensics, which was the topic of the ITWG-3 meeting that preceded this conference. The three lines of defence are sometimes referred to, as in Dr. Koch's presentation, as "Prevention, Detection, and Identification."

The second line of defence to deter illicit traffic of special nuclear materials is an important area in which Russian institutes are cooperating with Argonne. The institutes include the Russian Research Center "Kurchatov Institute" and the Moscow Engineering Physics Institute (MIFI). The cooperation includes developing requirements for the inspections; applying various instruments for detection; and developing a curriculum aimed at training personnel to subsequently implement the technology.

3. LESSONS LEARNED

What are the main lessons that we have learned from these collaboration experiences? The first lesson is one of humility and appreciation for the creativity and inventiveness that many of our partners were able to contribute even though they frequently had a less extensive technological base at their disposal. In almost any cooperation agreement in which we entered we received from our partners, in due time, very significant contributions that frequently exceeded what we had been able to provide.

To quote just a few examples, CERCA (France) demonstrated the feasibility of fabricating fuel with 6.0 g/CM3; M (Germany) developed ways to passivate the fuel powders and to eliminate stray fuel particles in the clad; AECL (Canada) demonstrated the use of U3Si-Al fuel in extruded rods; JAERI (Japan) performed valuable safety tests; KAERI (South Korea) developed a method to produce atomised fuel particles; BATAN (Indonesia) is testing a process to produce 99Mo from low-enriched uranium; and CNEA (Argentina) demonstrated the importance of distributing free uranium in the fuel matrix. In addition, many reactors throughout the world provided free irradiation of fuel elements, and free examination of the irradiated elements.

We have been very successful in our undertakings, but this could not have happened without the cooperation of the international partners involved. We depended on our partners at least as much as they depended on us. Our achievements were their achievements, and vice-versa, and the same was true for the occasional failures. This feeling of inter-dependence, openness and trust are very important to successful cooperation.

Second, we learned that, if we could agree on the ultimate goal, we could disregard less important issues on which we might have differed. People from different countries tend to choose different paths, consistent with their culture, customs, and traditions, to achieve their goals. There is an innate tendency to consider our path, our culture, and our traditions as the best, and to look down on anybody that makes different choices. This is sometimes referred to as the "mirror effect", which inclines us to judge others by how accurately they reflect us. If we share the same fundamental goals, we should be able to work together towards the attainment of those goals and show mutual respect for each others' cultures and traditions.

Finally, we learned that we must strive to be open to discussions and welcome challenges to our ideas or actions. We should encourage debate, if it does not run counter to local customs and traditions, because candid debate will usually result in a better product and will always enhance mutual understanding and trust.

Open discussions in front of diverse and outspoken audiences are especially healthy, as this conference well demonstrates, and I would like to thank the Landau Network-Centro Volta (LNCV) and the Unione degli Scienziati Per Il Disarmo (USPID) for organising this meeting and for giving us one more opportunity to express our differing opinions in a positive and constructive manner.

Address:

ARMANDO TRAVELLI
ENEA, ERG-FISS-FIRE
40129 Bologna
Italy

16

In Search of a Missing Link— Nuclear Terrorism and Nuclear Smuggling

F. GERE

There is a common assumption that because of the development of a nuclear smuggling (NS) generated by the disintegration of the Soviet Union, nuclear terrorism (NT) becomes a clear and present danger. But the analysis of the trends in smuggling as well as the current and foreseeable strategies of terrorist organisations suggest that the two phenomena have little chance to connect. Nevertheless it is necessary to understand why there is so much concern about such a possibility.

NUCLEAR SMUGGLING WHERE ARE WE NOW?

Questions about the market

The alleged existence of an illegal market has been established because of the connection of two kinds of data. On the one hand suppliers in disarray (former Soviet Union), on the other hand the demanding side gathering a bunch of rogue states and terrorists organisation (Iraq). Because it seemed a too simple equation in order to make it more complex or more credible it has been assumed that "Mafias" would spontaneously play the role of dishonest brokers.

Trends

Today we can contemplate two phases.

First, the augmentation of the amount of trafficking activities raised great concern:

1991	41
1992	158
1994	267

This is not the consequence of larger leakages or increasing availability but the results of two phenomenons:

Criminals became more and more aware that it could be a juicy activity regardless of the nature of the materials. Journalists try to identify clandestine networks.

Police increases its efforts (BKA, FBS) to deceive potential smugglers.

In the former Soviet Union the nature of the observed case can easily be explained as a consequence of the old socialist system. Workers used to take out their tools to work outside for their own benefit in the moonlighting system. Others tried more aggressively to steal state-owned machines to sell them outside.

DECREASE BY 50 PER CENT IN 1995 AND 1996 BECAUSE OF THE EFFICIENCY OF COUNTERMEASURES AND THE DIFFICULTIES TO MEET A REAL MARKET

Let's examine briefly the three components of the traffic.

- Serious suppliers remain to be identified if they do exist. So far only four significant traffic have been identified all of them in 1994.
- Mafias remain uninterested with a complex, high-risk-and relatively low profit activity. Drugs, prostitution, illegal gambling, conventional arms traffic, money laundry are much more interesting.
- Serious clients remain serious and will not commit themselves in ridiculous bargains with third level dubious suppliers. It remains to be proved that very well organised smuggling at high level (state level) is underway.

It seems that the whole issue is embodied in nuclear materials alone. Far from that, once you get the material you have to transform it into an efficient weapon adequate to your purpose. Good Aum chemists appeared to be poor in delivering the material. In the case of nuclear device, it is one thousand times more complex. But note that the sect did not try to buy sarin on the "market" or to steal it but choose to produce it.

The only solution is to use radioactive element as a dispersion weapon. In this case the result will be of the magnitude of a chemical product. But a question still remains what would be the goal, the rule of the game and the expected profit?

NUCLEAR TERRORISM AS A STRATEGY: RELATING MEANS TO ENDS

First consider that globally international terrorism is in a phase of reduction.

In 1996, 296 acts versus 665 in 1997, the peak in a generation. Or to put it in a different way is it an easy tool and can it be considered useful?

One question must be addressed. The idea of nuclear terrorism is as old as nuclear weapons themselves. Can we explain why it has never been used during the Cold War?

Many times the Soviet Union has been accused of supporting terrorists activities. But in the cases where it has been possible to identify such connections the support consisted in training, providing explosives, electronic systems, conventional arms. Very classical instruments sometimes sophisticated but the Soviet Union never decided to cross the line and use bio or radioactive substance. To be sure, they did not respect this rule of the game for humanitarian reasons. But it seemed too risky. Distrust towards terrorists organisations, fear that the connection could be unveiled, excess of the magnitude of the effect, impossibility to capitalise from it.

Terrorists entities and political goals.

Terrorism is not an end by itself it is a mean to achieve an end.

Terrorists have goals. They may be willing to offer their lives for the cause. They don't want the cause to die.

Terrorists activities often take place in a very complex and intricated environment. You cannot afford to kill your own people, or at least only at the margin. Even a bomb on a Israelian or Palestinian market appears to be too crude. Imagine a missile with a nuclear warhead against the South of France where areas are crowded with Arab Muslims.

Terrorism is not blind it is good at selecting targets.

A cause supported by terrorist action has sympathisers who are capable to admit a degree of violence, a certain amount of injustice to excuse the death of the victims because of the supreme value of the cause. Therefore the more precise is the target, the larger is the support. No doubt that a massive destruction the killing of thousands of people—even if they were soldiers in a base—would create a massive rejection and condemnation of the cause of the nuclear terrorists.

Terrorists look for a psychological impact through relatively limited activities and instruments. The importance is to get big media coverage. The aim is not to kill everybody but to kill some in order to terrorise the rest and/or to influence leaders through their public opinion.

Now, look at the risk of blackmail by a profit-guided organisation which asks for billions of dollars. Here the goal is to get the money and enjoy it. Can we imagine that such people would get a moment of tranquillity somewhere in the world. Success must not be evaluated in terms of the operation itself but in the perspective of its aftermath. Therefore, the magnitude of the threat has to be carefully tailored to leave some room for real benefits. If risk-taking is part of the job there is no rationale to look for a risk which by itself becomes a challenge against the expected success.

The problem of taking benefits remains the key.

How can you reach your goal using NT without becoming the target of a ferocious retaliation supported by an infuriated international community?

Can you imagine a covert Pearl Harbour?

Covert transaction on the stock exchange may offer a useful comparison.

The question is, can a powerful and complex State achieve its goal through an organisation so remote that it becomes impossible to identify it as the real aggressor? Many argue that the capacity of the Chinese system offers opportunities to play on that key. But the transparency of hypermedia democracy creates such a complexity that it is not as obvious.

In addition why should we, in principle, credit the Chinese political structures of a more complex and efficient capability than the former Soviet Union? It seems that the case presented earlier remains valid in the Chinese case.

So far, we have considered rational terrorists but what about non-rational?

Lunatics and Milienarists

The case has been made by late Defence Minister Aspin as early as 1991 when he was chairman of the House Armed Services Committee.

But after five years of intense debate, it seems that this argument opposed by numerous researchers in many nations has been dropped by the US Administration (Under Secretary Walt Slocombe, Senate, Armed Services Committee, February 1997).

Ultimately, what about Armageddon for the second millennium?

Members of sects are much more at risk than the world outside. They have created their own world. They act inside this paranoia according to their own criteria. It has little to do with the real world. Major massacres have been and still are self inflicted: in Central America, in France (sect du Phenix), even Waco.

Nevertheless let's face a group dedicated to destroy the world. To purify what in order to satisfy whom? How could it acquire such a capability?

No way. But they will try with what they can get a crude nuclear weapon made by whom, delivered in which conditions?

IS NT AN AMERICAN PROBLEM?

You will think that my presentation is focused on the US. Indeed it is worthwhile to ask the question is NT an American fear, phobia, myth? More than 90 per cent of the literature on the topic comes from the US or is inspired by US considerations.

Intensively and efficiently present on the Internet, US universities convey in a repetitive and obsessive way (the characteristics of basic propaganda) the same kind of arguments focusing on technical issues, paying no attention to the linkage between means and goals.

Is it a US problem? In terms of defence planning it is true that the US will be THE target for all in the future? Secretary of Defence Cohen publicly made the case that "American military supremacy... may encourage adversaries to use indirect, what they call asymmetric means to attack our forces and interests abroad and even our people here at home." However, American centric vision could become easily self-deceptive.

The world does not focus on the U.S. Consider the following examples most of terrorist actions in the world are domestic rather than international. American businessmen and officials are targeted all over the world, mainly in Central and South America. Most of the time the goal is to get money from kidnapping. Terrorism is pragmatic and very trivial.

When Tupac Amaru took for the liberation of comrades they went after the Japanese embassy not after the US.

After years of investigations and studies, we can retain with certainty terrorist have serious goals. They do not think at very high strategic levels they are *not* educated and trained in Machiavellian studies.

States using terrorism as a mean can behave in a much more complex way but as mentioned earlier they have to comply by certain rules.

Easy? Not to acquire, not to produce, not to manipulate. Look at the efforts of entire large state-supported organisations!

Useful not much, irrelevant to almost all the cases of terrorism.

NUCLEAR TERRORISM: TWO WAYS TO CONTEMPLATE THE PROBLEM

We must address the problem of the permanent and, today, of the growing concern about nuclear terrorism.

1. NT as a tool to influence (a psychological weapon):

- to enhance the rejection of nuclear weapons. In order to gain greater support to the cause of the abolition of nuclear weapons, one can overestimate the risk of NT. A related activity is to insist on the ability of terrorist organisation to steal plutonium out of the pools. The goal is to increase the fear against the substance itself for many reasons.
- a "Tom Clancy's Approach": to create a political environment supportive of a more robust defence against nuclear terrorism and more broadly nuclear theft by "rogue" actors. Note that the notion of rogue, pariah and ultimately terrorist states create an effective assimilation.

Several states are terrorist in essence or behaviour. Terrorists States look for nuclear weapons therefore, nuclear weapons are terrorist tools.

Nuclear terrorism has proved to be an effective argument during periods of defence budget constraints.

- to promote a more aggressive and efficient policy to eradicate former Soviet nuclear weapons: i.e. in the successors states and in Russia.
- to cast suspicion on a government in order to destabilise it (disinformation by government opponents, Mujahiddin

Qaiq in the case of Iran). Ultimately threats can be used by lunatic politicians to receive attention from the press (is it terrorism?)

Shamyi Basayev, Chechen leader seemed to have threatened Moscow talking about a radiological weapon. It brings us back to that kind of use in a case of political and military desperation.

NT AS A SERIOUS ISSUE FOR DEFENCE PLANNING

As German researcher K H Kamp put it, "Nuclear terrorism falls into a high risk low probability category". Since the consequences could be disastrous, you cannot totally disregard it or just neglect.

What if? What kind of reaction: how to prevent actors or states or covert actors to take benefits. It is important to have a policy and a prudent defence planning.

To conclude, let me say that we are facing a real issue. But it is not at all the present danger of a nuclear threat by terrorists organisations. The issue should be addressed in two ways.

Nuclear smuggling and NT are two different activities whose repression must be addressed differently.

Firstly against NS

It needs continuous process of cooperation, intelligence sharing. To have excellent results through Europol and the cooperation between the EU and the US must continue and deepen.

Secondly against NT—three areas:

- protection of plants and transportation.

To be prepared to react against blackmail or any kind of aventurist threat.

- NEST under President Carter
- DAMOCLES plan in France.

Attacks against nuclear materials, above all, military materials, have been considered and addressed. Simulations have been conducted with significant results, including surprises which led to enhanced measures to properly and fully protect highly explosive materials.

Nuclear terrorism could exist and be practised because of very severe and limited motivations. Therefore, it has to be addressed with caution and efficiency by the relevant institutions and administration working all over the world in close cooperation. Ultimately, prudent defence planning has to contemplate and prepare for adequate responses aimed at dissuading moves in this direction by state-sponsored organisations. It must be made crystal-clear that there will be no reward, no dividends for NT. That it would generate a massive threat of retaliation and that there will be no compromise and no mercy for those who would dare to be tempted to engage in such a way. It is neither in the interest of the international scientific community to address the issue in terms which are purely American nor it is to indirectly support efforts aimed at countering a threat which presently and in the foreseeable future does not exist at the level assumed by its supporters.

Attacks against nuclear plants by whom and to achieve what? In the case of Chechenia was it terrorism or an act of war? Of civil war since it was a "domestic" conflict.

Address:

FRANCOIS GERE
CREST
Ecole Polytechnic
91128 Palaiseau Cedex
France

17
In Field Nuclear Forensics: An Overview of Applicable Instrumentation

S. SCOTT

INTRODUCTION

Recent seizures of nuclear materials [1], and particularly of fissionable nuclear materials [2], have resulted in an increased awareness of the need for fieldable instrumentation for the detection and characterisation of suspect packages and materials.

The ability of perform in-field assessments of materials contained in suspect packages is important for several reasons. First, and most importantly, determination of the type and quantity of material provides information critical to personnel safety. The personnel at potential risk may include both the general population in the area of the package and incident response personnel. Additionally, it would be advantageous to gain information that could result in attribution of the suspect package to a particular facility or pathway as soon as possible after interception. Limited access to a suspect package could also require the use of fieldable characterisation instrumentation as opposed to transporting the package to another location for analysis.

During the initial characterisation process it is also critical to avoid compromising the packaging and internal contents with respect to future investigations; most techniques described in this summary do not require any physical contact with the suspect package and, therefore, address this concern. It is essential to document all characterisation efforts, and any other activities that could potentially affect subsequent investigations, using an appropriate chain of custody procedure.

APPROACH

One approach to identification and characterisation of an intercepted suspect material or package is to use a graded application of instrumentation. The initial phase could be an alarm from either nuclear material pedestrian or vehicle portal monitors, pocket radiation dosimeters/pagers, or visual examination of the package. Pedestrian and vehicle nuclear material monitors are commercially available and are implemented at many nuclear facilities and border crossings. Pocket radiation dosimenters utilise well-established technology and are available from a wide variety of vendors [3]. They can provide a yes/no alarm, require minimal operator training, are robust and have long shelf lives, and are relatively inexpensive. Visual examination includes hand searches of packages/luggage and use of x-ray instrumentation. Personnel training is critical for recognition of suspect characteristics but visual examination can be widely implemented at low cost.

The next phase, alarm response, may include nuclear material verification and characterisation. In addition, awareness of potential critically concerns, detection of denial systems, detection of high explosives, and containment and packaging of the material could be required. Characterisation methods including imaging, mass spectrometry, and x-ray fluorescence may also be employed.

HAND-HELD DETECTORS

There are a range of commercially available hand-held detectors that can be used in an alarm situation to perform passive gamma-ray and neutron measurements. These instruments allow the operator to locate or verify the presence of nuclear material, determine the radiation intensity, and in some cases provide identification of the material. These instruments employ a variety of detector systems including plastic scintillators, NaI(T1), and, in some instruments, 3He tubes. Systems utilising cadmium/zinc/telluride (Cd/Zn/Te) detectors are under development. The CD/Zn/Te detectors have a resolution intermediate to NaI(T1) and HPGe detectors but do not require the liquid nitrogen cooling necessary for high purity Ge (HPGe) systems. The price of he instruments can vary from approximately 1,500 US$ up to 20,000 US$ and is generally a function of the detector type, electronics, software, and degree of ruggedisation (some instruments can be dropped from 1 meter with no resulting damage).

The most simple to operate hand-held detectors, which require less than 2 hours of operator training, allow the operator to locate a radioactive source and measure the gross gamma radiation intensity of the source. Some instruments have added the capability for storage of gamma-ray spectra obtained in the field. These spectra can subsequently be downloaded into a computer and appropriate analysis software (via modem if desired) to perform a full qualitative or quantitative analysis. Most of these systems are based on proven technology and have been used extensively in the field for more than 10 years.

Hand-held instruments that incorporate detectors, electronics, and software for identification of the nuclear material are also available. Some of these instruments allow the operator to choose between a basic or advanced mode of operation. For example, the basic mode of one instrument, which requires minimal operator training, displays an output of "medical", "industrial", or "special nuclear material" and the element that is present with only push-button operation. The advanced mode allows the operator to set spectral ranges and look at these regions of interest while in the field. For an operator that has more extensive training, this important feature could provide additional immediate information regarding the material.

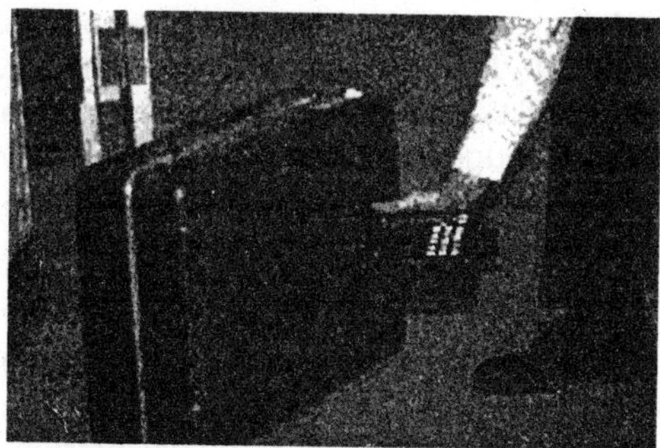

Figure 1 : A Dual Mode Hand-Held Radiation Detector (4).

GAMMA SPECTROSCOPY

Another class of instruments provides the capability to

perform complete gamma spectroscopic analysis in the field. These instruments could be employed after the suspect package has been determined to contain nuclear material, particularly U or Pu. The primary advantage of employing these more complex systems is determination, both qualitative and quantitative, of the isotopic content of the material being analysed; under appropriate conditions, mass measurements can also be performed. Detectors utilised are typically NaI(T1) or liquid-nitrogen-cooled HPGe. Instruments based on Cd/Zn/Te or mechanically cooled HPGe detector systems are currently under development. Multi-channel analysers with from 4096 up to 16,384 channels are typically used in these systems. Battery operation allows for 8–12 hours of field-use time; the instruments weigh from approximately 2–15 pounds.

An experienced operator is required to interpret the spectroscopic data although software that provides simple interpretation of the data for non-technical personnel has been developed and implemented in some of these fieldable instruments. This turnkey type operation is particularly apparent in instruments designed for specific types of analyses, for example, for use in International Atomic Energy Agency verification measurements that utilise this type of equipment routinely [5].

The price of these instruments ranges from approximately 20,000–50,000 US$ and primarily depends on the type and size of the detector. The computer, software, associated electronics, and degree of ruggedisation will also influence the price but may be dictated by the choice of detector. Reviewing clearly defined data quality objectives, the operating environment, and anticipated throughput with technical experts/vendors is essential for optimisation of instrument performance while minimising the cost of the system.

NEUTRON COUNTING

Neutron counting can be used in conjunction with gamma spectroscopy to provide an estimate of mass of identified nuclear material. Extremely user friendly instrumentation, requiring less than one hour of training for a non-technical operator, is available. This type of analysis would primarily be used when a suspect material has been identified as plutonium and a mass estimate is desired, but the analysis can also be used to verify the presence of plutonium oxide, uranium oxide, uranium hexafluoride, uranium tetrafluoride, or metallic plutonium buttons.

Commercially available neutron counting systems weigh up to 40 pounds and include 3He tube detectors and high-density polyethylene moderators. They can be used in either scalar or coincidence modes. Portable neutron probe systems are based on well-established technology and are often used for hold-up and waste measurements. Neutron counting capabilities are also available in hand-held systems; these instruments often include complementary gamma-ray detectors.

HIGH EXPLOSIVES DETECTION

Surveying for high explosives is an important step in assessment of suspect packages or materials. Portable instrumentation for these types of analyses is commercially available and routinely used by law enforcement personnel. Samples are collected by either swiping or vacuuming particulates onto a sample holder. A technically qualified and experienced operator is required for initial set-up of the instrument, calibration before and during the analysis, and for detailed analysis of the data generated. Minimal training is required for personnel that will be obtaining the samples from the package.

IMAGING

Imaging systems facilitate·determination of the interior physical characteristics of a suspect package; this information is critical to law enforcement/incident response personnel for development of a response strategy. Commercially available systems are based on battery powered pulsed x-ray sources and display images on a remote monitor only seconds after the pulse is fired. These systems require minimal training (usually less than one day) and are extensively used by government agencies. Prototype systems that utilise internal energy sources to provide images of material within a suspect package have also been developed.

MASS SPECTROMETRY

With the exception of the high explosives detection systems, all of the instrumental techniques described above employ non-destructive methods of analysis. Mass spectrometry is a destructive analysis technique that requires obtaining ionising,

Figure 2 : Ion Trap Mass Spectrometric Analyzser Assembly and Vacuum Can. The diameter of the upper flange of this assembly is approximately 4 inches. (The associated electronics and computer are not shown) [6]

and then mass analysing ions or fragments of ions generated from the sample material. These instruments can be used to obtain part-per-billion to part-per-trillion sensitivities and, therefore, can potentially enhance inspector, response, or law enforcement personnel safety. Because mass spectrometry provides qualitative and quantitative elemental and isotopic information, these fieldable instruments can also provide immediate information required for attribution of the material or package.

For field use, commercially available ion trap and miniaturised quadropole mass spectrometers have been employed and ruggedised. These instruments require a skilled technical operator to ensure accurate instrument operation and calibration and subsequent interpretation of the mass spectral data generated. Development efforts in this area include enhancement of sample introduction mechanisms and ionisation sources that are coupled to a range of field ruggedised and miniaturised mass spectrometers (e.g., time of flight, quadropole, ion trap). The result of these efforts will be instrument components that can be used to tailor fieldable mass spectrometers for specific field uses while offering improved sensitivities for a wider range of analytes.

PACKAGE CONTAINMENT AND TRANSPORTATION

At any point during the process of examining a suspect package, it may be necessary to contain the material or package for health and safety reasons, to protect the package from contamination, to perform additional destructive analyses that require opening the package in the field, or to prepare the package for transportation.

Figure 3 : Outer Shell Tent of a Fieldable Containment and Packaging System [7]

If transportation of the package will be performed, the packaging system must be designed to meet all current environmental release and shipping regulations. The regulations addressing shipment of nuclear materials are numerous. Systems that address each of these issues have been designed and implemented.

TRAINING

For all personnel that may be involved in the field analysis of a nuclear sample, training is as critical to the process as the instrumentation used for the analysis. Although the radiological analysis equipment itself may require minimal operator training, it is essential that this include hands-on practical training with real nuclear materials. The training should be performed by experienced radiation measurement personnel and should result in the operator learning the instrument response as a function of material type, quantity, physical form and isotopics, the limitations of the instrumentation; and when to defer any further activities to nuclear material measurement and handling experts. Appropriate training will also provide instrument operators with resources and contact information for future technical questions or problems.

CONCLUSION

Numerous types of measurement techniques, both destructive and non-destructive, can be performed in the field. The data generated by these analyses are not only critical to nuclear forensics but to the safety of the general population, first responders, and law enforcement personnel. These instruments can be employed by personnel with a range of technical backgrounds using a graded approach; the results obtained from the simple widely deployable instruments available for field analysis can indicate a progression to more sophisticated specific techniques.

ACKNOWLEDGEMENTS

This information was gathered through discussions with numerous equipment vendors and Lawrence Livermore National Laboratory, Sandia National Laboratories, and Los Alamos National Laboratory personnel. Their time and contributions are greatly appreciated.

References

1) P. Williams and P.N. Woessner "The Real Threat of Nuclear Smuggling", Scientific American 40 (January 1996).
2) J. L. Ford "Nuclear Smuggling: How Serious a Threat?" National Defence University Strategic Forum 59, 1 (1996).
3) Richard Olsher, Los Alamos National Laboratory, personal communication.
4) W.S. Murray and R.W. Blankenship "GN-4, a Lightweight, Battery Operated Gamma-Ray and Neutron Detection System for In-Situ Nuclear Material Surveys," Proc. Institute of Nuclear Materials Management 38th Annual Meeting, Phoenix, Arizona, July 20–24, 1997.
5) Willie Hsue, Los Alamos National Laboratory, personal communication.
6) H.L. Seifert, P.H. Hemberger, M.L. Alexander, and R.E. Morrison "A Highly Miniaturised Mass Spectrometer for Field Monitoring of Volatile Organic Compounds", Proc. 42nd Annual Conference of the American Society for Mass Spectrometry, Chicago, Illionis, May 29–June 3, 1994.
7) J. Phillips, Los Alamos National Laboratory, Personal Communication.

Address:

SARA SCOTT

Los Alamos National Laboratory Safeguards Systems

Los Alamos NH 87545

U.S.A.

18

Polish Prevention System Against Illicit Trafficking of Radioactive Substances and Nuclear Materials

S. STERLINSKI

LEGAL REGULATIONS IN POLAND FOR TURNOVER OF RADIATION SOURCES AND NUCLEAR MATERIALS

1) Act of Parliament Called "Atomic Law" from 1986

Article 4 of this Act says that activities related to turnover of nuclear materials and radiation sources require licence of the appropriate organ of the state administration.

Such an organ is the President of the National Atomic Energy Agency (NAEA) of Poland. Article 62 of this Act reads: "....who, without required licence, produces, processes, introduces to turnover, transports, possesses or applies nuclear materials, sources of radiation or radioactive wastes, is liable to a penalty of 1 to 10 years of depriving of liberty, and a penalty of fine".

2) Act of Parliament from 1988, Dealing with Economic Activities

Under this Act, turnover of some goods with foreign countries requires licence of the Ministry of Economic Co-operation with Abroad.

The list of such goods is as follows:

1. Uranium and thorium ores and their concentrates,
2. Radioactive isotopes—including elements for nuclear fuels—as well as their compounds, mixtures and products containing such substances,
3. Other isotopes, their organic and inorganic compounds, including non-defined ones,

4. Nuclear reactors and not irradiated fuel, machines and devices for separation of isotopes.

3) Act of Parliament from 1993, on Principles of Special Controls of Turnover of Goods and Technologies with Abroad

The Minister of Economic Co-operation with Foreign Countries issued, based on the above Act, the list of goods and technologies undergoing special control.

The list contains two parts:

Part A: primary materials and special fissile materials,

Part B: dual-use devices and materials exhibiting potential for application in nuclear techniques, as well as related technologies.

Primary material: mixture of uranium isotopes as it occurs in the nature, uranium depleted in U-235, metallic thorium, thorium as alloy, chemical compound or concentrate,

Special fissile materials: Pu-239, U-233, U enriched in U-235 or U-233, any material containing one or more of the above.

4) Act of Parliament from 1990 about the Border Guard

By the virtue of this Act the task of preventing of illegal transport is realised by the Border Guard, which was given the authority to control, stop and move back from the State border harmful radioactive materials and wastes.

PREVENTION SYSTEM AGAINST ILLEGAL TRADE AND TRANSPORTATION OF RADIOACTIVE SUBSTANCES AND NUCLEAR MATERIALS

In order to ensure for effective functioning of state responsibility for complying with the above regulations the President of the National Atomic Energy Agency signed agreements on collaboration with the Minister of the Interior.

Under these agreements the Agency and its research institutes provide consultation and help on nuclear–related matters as well as carry out identification of illegal radioactive substances and identification of some nuclear materials.

The fight against smuggling of nuclear and radioactive materials is the task of the Border Guard, Police and the Police and the State Security Office.

The main threat for the Polish territory is resulting from:

— attempts to import radioactive materials without legal permits or with false cargo manifests,
— illegal transport of radioactive and nuclear materials,
— omitting the prohibition of import of radioactive wastes for storage and utilisation.

The largest threat is noticed on the Eastern border. At the Polish-German border such cases are also noticed but most of them are caused by natural radiation of imported materials.

To protect the Polish territory from illegal import of radioactive materials and wastes, beginning from the 10th of October 1990, on all border check points was introduced radiation control of persons and commodity.

The main type of stationary equipment is "radiometric gate". These installations are deployed on entrance direction in front of the check points. They are working constantly and are able to detect all attempts of illegal transportation of materials with raised radiation level.

At present 70 such installations are deployed (36 in road check points, 30 in rail-road check points, 3 at airports and 1 in sea check point). The next 5 installations are mounted (or planned to be installed in 1996).

Portable radiometers and dosimeters are used for surface pollution and dose rate measurements. At present the Border Guard is equipped with 521 such devices.

In June 1996 Border Guard was provided with a special car being equipped with a mobile apparatus, allowing detection of nuclear, other dangerous materials and narcotics. Due to possibility of fast shift this car will be utilised first of all on the Eastern border and at Warsaw-Okêcie International Airport.

Right now the process of securing the Polish border from smuggling of radioactive materials is far advanced. This is very well expressed by number of interventions undertaken in the years 1990–95. In this period were 13,616 interventions and 1405 transports were not permitted (Table 1).

The above mentioned numbers do not obey events recorded on territory of Poland by the Central Laboratory for Radiological Protection (CLRP). In Poland, the CLRP is the institution primary responsible for round-day help in cases of accidents with radioactive substances and nuclear materials.

The following numbers of extraordinary events have been

recorded by the Central Laboratory for Radiological Protection:

1992–69 events, 19 interventions
1993–90 events, 33 interventions
1994–77 events, 22 interventions
1995–95 events, 18 interventions

The CLRP carries out on request the Border Guard, Police, the SSO and Prosecutor's Office identification and analysis of radioactive substances and some nuclear material (LEU). The capabilities of the CLRP in the field of nuclear smuggling forensic analysis are however limited to alfa and gamma spectrometry of radioactive substances and gamma spectrometry of uranium samples. Among others, the non-destructive gamma-ray spectrometric technique for the determination of the U-235 content in LEU-samples is routinely used in the CLRP. So far, the CLRP can not offer the destructive analysis of Pu as well as material origin identification. International co-operation in this field is needed.

Table 1:List of Events Dealing with Nuclear Accidents Recorded by the Central Laboratory for Radiological Protection

Date	Location of event	Description of event	Response Opinion of CLRP
17-10-92	Terespol, Police Station	in a private apartment uranium smuggled from the USSR was detected	Radio-activity measurements at CLRP indicated that uranium exhibited natural proportion of its isotopes
05-11-92	Radom, Police Station	In automobile garage smuggled uranium was detected	As above
08-03-93	Gdynia, Regional Public Prosecutor's Office	Delivery of radio-active material secured by the SSO in a private apartment	The measurements revealed that the materials were: 1. Container with uranium shielding, 2. Small bag containing powdered uranium oxide. The uranium isotopes were in natural proportions. The insertions indicated that it came the former USSR

Table I : Contd.

Date	Location of event	Description of event	Response Opinion of CLRP
02-04-93	Warsaw-Falenica, SSO	Team of SSO detected objects resembling uranium moulder	Team of CLRP established that the material was non-radioactive, probably iron-based alloy
0F-04-93	Nowy Jawornik, roadside near Swidnica	Found 2 big yellow containers with radiation signs and Russian inscriptions	Containers did not have radioactive substances in them
09-04-93	Rzeszów, SSO	In a private apartment SSO secured 2 containers with unknown content	Measurements indicated that it was uranium in its natural proportions
28-04-93	Braniewo, SSO	In a private apartment container was found probably containing Cs-137 source	Measurements confirmed that there was Cs-137 source, activity 1 mCi, of Soviet production
28-06-93	Dorohusk, Customs Office	In a car entering Poland intensely radiating objects of gold platinum were detected. Total weight about 4 kg.	CLRP team established that the objects were contaminated with a substance containing J-131. The objects have been secured in the Customs Office.
21-07-93	Warsaw, Station of Civil Defence	2 containers with radiation sources were brought from Ukraine. Their origin was investigated	It was established that they were brought there under legal contract Poland-USSR in 1985
30-08-93	Szczecin, SSO	Radioactive object found by SSO in a forest near Szczecin	It were Co-60 sources, activity up to 1 mCi each. Soviet province
08-10-93	Œwiecko, border checkpoint	Monitoring detected radioactive contamination of the truck box. Truck entering Poland from Germany	Post-Chernobyl contamination was established. According to the driver's information, the truck was used earlier for transportation of scrap between Ukraine, Belarus and Latvia

Table 1 : Contd.

Date	Location of event	Description of event	Response Opinion of CLRP
18-11-93	Zebrzydowice, border checkpoint	Railway wagons containing contaminated scrap were returned to Poland from Italy	Measurements established presence of radioactivity up to 60 kBq/kg due to Ra-226 in mining deposits
23-11-93	Cmilów (Lublin)	Police detected radioactive material stored illegally on a farm	The material was natural uranium in the form of a cylinder. Smuggled from the former USSR
01-12-93	Cieszyn, border checkpoint	During border control, radiation from a suitcase brought out from Poland was detected	CLRP team found that the radiation came from contaminated US dollar banknotes in the suitcase. The contamination was by Zr-95. The contaminated notes have been secured at the border checkpoint
30-05-94	Dorohusk, border checkpoint	During controls of a bus entering Poland from Ukraine, 7 radiating objects were found	Level gauges (new) containing luminous radium paint were the source of radiation
30-05-94	Przemycel, SSO	SSO team found a container with radioactive substances. It was stored on balcony of a private apartment	Investigation established that there were 2 sources of Sr-90. One was from isotope ice monitor used in aviation. Smuggled from former USSR
04-95	Cieszyn, border checkpoint	A package containing radioactive material found by border guard in a car	Package contained pellets of uranium fuel, enrichment 2 per cent. Pellets were identified as coming from NPP Ignalina, Lithuania

Address:

SLAWOMIR STERLINSKI
Laboratory for Radiological Protection
03194 Warsaw
Poland

19

Report about Prevention of Illegal Transportation of Radioactive Materials and Wastes through Border of the Republic of Poland

T. HADYOE

One of outcomes of political changes in Poland after 1989 was the wider opening of Polish borders. This among other things created threat of smuggling of radioactive materials. To protect the Polish territory from illegal import of radioactive materials and wastes, beginning from the 10th of October 1990, on all border check points was introduced radiation control of persons and commodity.

By the virtue of law (article 1, point 2, subpoint 12 of the Act from October 12th 1990 about the Border Guard) the task of preventing of illegal transport is realised by the Border Guard, which was given the authority to control, stop and move back from the State border harmful radioactive materials and wastes.

The radiation control is realised by controllers trained in the Institute of Nuclear Energy in Œwierk (near Warsaw). At present in the Border Guard is serving 266 controllers. The officers are supervised by Border Movement Control detachments in local Border Guards Divisions. The coordination of all activities in this field is done by The Border Movement Control Bureau of the Border Guards HQ.

The main threat for the Polish territory is resulting from:

— attempts to import radioactive materials without legal permits or with false cargo manifests,
— illegal transport of radioactive and nuclear materials,
— omitting the prohibition of import of radioactive wastes for storage and utilisation.

The largest threat is noticed on the Eastern border. At the Polish-German border such cases are also noticed but most of them are caused by natural radiation of imported materials.

The effectiveness of the Border Guards activities is mostly dependent on the equipment used to control level of radioactivity. The equipment used is either stationary or portable.

The main type of stationary equipment is control installation UK-1 M, known as "radioactive control gate". These installations are deployed on entrance directions in front of check points. They are constantly and are able to detect all attempts of illegal transportation of materials with raised radiation level.

At present 80 such installations are deployed (40 in road check points, 32 in railroad check points and 7 at airports). The next 5 installations are mounted (or planned to be installed in 1996).

We use such portable equipment as:

— surface pollution measurers type RKP 1–2

— radiation level measurers EKO-SK, EKO-C.

At present the Border Guard is equipped with 603 such measurers. This equipment is made in Poland, and is highly rated by the officers although the number of used measurers isn't sufficient.

Right now the process of securing the Polish border from smuggling of radioactive materials is far advanced. This is very well expressed by the number of interventions undertaken in the years 1990–96. In this period were 32,611 interventions and 2045 transports were not permitted. Very important factor is efficient inflow of information to data bank and punctuality of transferred information.

Address:

TADEUSZ HADYOE
Commission Europeenne
2920 Luxembourg

20
Safeguarding Nuclear Weapon-Usable Materials in Russia

T. COCHRAN

INTRODUCTION

A one kilo ton pure fission nuclear weapon can be fabricated with as little as one to three kilograms (kg) of plutonium or three to eight kg of highly-enriched uranium (HEU). There are roughly 300,000 kg of weapon-grade plutonium (WGPU) and some 2 million kilograms of HEU in nuclear weapon programmes globally. Of these amounts over one-half (about 170,000 kg of WGPU and about 1.2 million kg of HEU) is in Russia, and about one-third to one-half of the Russian material is in some 20,000 intact weapons. Most of the remaining Russian materials are presumed to be in the form of warhead components in storage. In addition, Russia has about 30,000 kg of separated reactor-grade plutonium (RGPU) in storage.

Both the United States and Russia are retaining as strategic reserves more plutonium and HEU for potential reuse as weapons, than is legitimately needed. Both have engaged in discussions and have programmes in various stages of development to dispose of excess plutonium and HEU. These fissile material disposition programmes will take decades to complete. In the interim there will be, as there is now, hundreds of tons of separated weapon-usable fissile material stored in tens of thousands of transportable canisters, each containing from a few to several tons of kgs of weapon-usable fissile material.

This material must be secured against theft and unauthorized use. To have high confidence that the material is secure, one must establish criteria against which the adequacy of the protective systems can be judged. For example, one finds such criteria in U.S. Nuclear Regulatory Commission (USNRC) regulations for the protection of special nuclear materials.

REQUIREMENTS FOR THE PROTECTION AGAINST THEFT OR DIVERSION OF WEAPON-USABLE NUCLEAR MATERIALS

In its regulations (10 CFR §73. 1) the USNRC uses the concept of a "design basis threat" to judge the adequacy of the physical protection measures designed to prevent the theft nuclear weapon usable materials. Licensees that possess or transport formula quantities of special nuclear materials are required to protect against:

i) A determined, violent, external assault, attack by stealth, or deceptive actions by a small group with the following attributes:
   a) Well-trained (including military training and skills) and dedicated individuals;
   b) Inside assistance that may include a knowledgeable individual who attempts to participate in a passive role (e.g.) provide information, and active role (e.g. facilitate entrance and exit, disable alarms and communications, participate in violent attack), or both;
   c) Suitable weapons, up to and including hand-held automatic weapons, equipped with silencers and having effective long-range accuracy;
   d) Hand-carried equipment, including incapacitating agents and explosives for use as tools of entry or for otherwise destroying reactor, facility, transporter, or container integrity or features of the safeguards system
   e) Land vehicles used for transporting personnel and their hand-carried equipment; and
   f) the ability to operate as two or more teams.

ii) An individual, including an employee (in any position), and

iii) A conspiracy between individuals in any position who may have:
   a) access to and detailed knowledge of the (facilities), or
   b) items that could facilitate theft of special nuclear materials (e.g. small tools, substitute material, false documents, etc.) or both.

A "formula quantity" means strategic special nuclear material in any combination in a quantity of 5000 grams or more computed by the formula, grams = (grams contained U-235) +2.5

(grams U233 + grams plutonium). In other words, the design basis threats above apply to the protection of more than 2 kilograms (kg) of plutonium, 2 kg of U-233, 5 kg of U-235, or any equivalent combination of these.

The first threat, described in (i) above, is call the "external" threat, while (ii) and (iii) are referred to as "insider" threats. The size of the "small group" in (i) is not defined in USNRC regulations. It used to be taken as six people, and perhaps the USNRC still uses this level as the design basis for the external threat. The conspiracy in (iii) is assumed to involve only two people.

The U.S. Department's of Energy (USDOE) and Defence (USDOD) also use the same "design basis threat" approach to judge the adequacy of security at contractor operated facilities and for the transport of nuclear weapons, weapon components, and weapon usable materials. The USDOD and USDOE requirements are essentially equivalent to or exceed those prescribed in USNRC regulations.

A weakness of any security system, particularly where it concerns a potential theft by knowledgeable insiders, is that the threat can analyse the capability of the physical protection system and bring to bear the necessary people and equipment needed to overcome it an minimise the chances of being apprehended.

Since the dissolution of the Soviet Union in 1991 there have been break-ins at facilities in the former Soviet Union, but no reported incidences of external assaults, of the type described in (i) above, for the purpose of stealing formula quantities of special nuclear materials.

Between 1992 and 1994 there were several cases where single insider—case (ii) above-successfully removes kilogram quantities, but not formula quantities, of weapon-usable highly enriched uranium (HEU). In each of the reported cases the individual was apprehended and the material recovered. At least one of these cases was a sting operation. When I last wrote about these cases in 1995, I offered the following observations which are still valid today:

i) **Kilogram quantities of weapon-usable fissile materials have been stolen from institutes in Russia since the break-up of the Soviet Union.**

ii) **Some fraction of these materials was not intercepted**

before leaving the Russian borders.

iii) **All known cases involved diversions from civil, space, and naval reactor research and fuel manufacturing facilities. No known diversions have occurred that involved nuclear weapons or weapon components.**

iv) **We don't know what we don't know. Given the lack of adequate inventory controls, there may well have been successful diversions that have not been detected.**

One should be cautious in drawing conclusions regarding the absence of reports since 1994 of serious cases—involving kilogram quantities of weapon-usable material—because the security and intelligence bureaucracies may be operating under different reporting guidelines. In any event the limited number of cases during the 1992–94 period suggest that the single insider represents the most probable threat to illicit use of nuclear material.

It is generally agreed that the most cost effective way to protect of weapon-usable materials is to increase security at the source where it is stored. Nevertheless, the fact that insiders have successfully removed kilogram quantities of weapon-usable material from FSU facilities points up the need to improve security across the aborad, including providing additional border security.

While the single insider case (ii) represents the most likely threat, in may view the most serious threat category is case, (iii) involving a collusion of security personnel and/or senior management at facilities storing significant quantities of weapon-usable material. Such a collusion could divert arsenal quantities of weapon-usable materials, and could market it with less chance of being apprehended.

Moreover, we should be more concerned about diversion scenarios where the recipient is a rogue state, rather than an individual or terrorist organisations. As demonstrated by the programme in Iraq, a state can provide the financial and technical support needed to develop sophisticated nuclear weapons. In the case of an individual or terrorist organisation, the consequences are usually described in terms of the damage of a single crude nuclear bomb. On the other hand a rogue state could construct an arsenal of sophisticated weapons. It is generally agreed by all experts within the arms control community that the pacing item in terms of a state's ability to obtain nuclear weapons is the acquisition of the fissile material.

Purchasing the fissile material illicitly would reduce by years the time needed to acquire nuclear weapons. There would be little, if any, time available to bring diplomatic pressure, as in the case of North Korea, or for military intervention, as in the case of Iraq, if a rogue state secretly acquired kilogram quantities of fissile material from Russia by illicit means.

There is compelling evidence for concern over the possibility that a collusion of security and management personnel may become involved in the trafficking of arsenal quantities of plutonium or HEU. First, there was a case in 1993 involving the interception in Vilnius, Lithuania of four tonnes of beryllium that were diverted from a nuclear research institute at Obninsk in Russia.

There was a small incidental quantity of HEU involved but he amount was not significant, and for this reason many analysts have overlooked the significance of this case. The significance of the Vilnius case is that it had all the elements associated with the case (iii) scenario—the most worrisome scenario that potentially could involve the successful diversion of very large quantities of fissile material from Russia to the Middle East. Involved in this single case were a) at least one senior regional Russian government official, b) a senior official of a nuclear institute where tonne quantities of weapon-usable fissile materials are stored under inadequate physical security, c) an organisation believed to be linked to the KGB and organised crime (mob) groups in Russia and Lithuania.

One other aspect of the Vilnius case is noteworthy. This case involved the shipment of four tonnes (2 cubic meters) of beryllium metal in 33 shipping crates. Each shipping crate was about 0.13 cubic meters (about five cubic feet) in volume. It would be very easy to hide a few kilograms of plutonium or HEU in such a shipment. Much larger shipments of metals are routinely exported from Russia legally. The Vilnius case demonstrates how easy it would be to divert enough plutonium or HEU for an arsenal of several nuclear weapons by hiding the fissile material among several tons of metal shipped in the course of a normal commercial transaction. It also demonstrates the importance of good border security.

Also, as noted by David Hoffinan of The Washington Post (May 12, 1997, pp. 1, 15), since the collapse of the Soviet Union, the Russian Federation has failed to establish the rule of law. The judiciary is a weak and ineffective branch of government. People now speak of a fifth power—in addition to the executive, legislative, judiciary and the press—namely, that of organised criminals. Former Soviet bureaucrats, factory directors, aggressive businessmen and criminal organisations have all made a grab for immense riches through inside deals, bribery and brute force. Private businessmen are building their own private armies of security agents, bodyguards and commercial spies. Former elite KGB troops are now operating private security forces to protect Russian businesses. Veterans of the Afghan and Chechen wars have been recruited as bodyguards. There has been a rash of contract killings. Many small businesses must pay for some kind of "krysha", a Russian slang word referring to criminal protection racket. Given these circumstances it is reasonable to conclude that the probability that present or former security personnel will become engaged in nuclear materials trafficking is sufficiently high to be quite worrisome. This threat must be taken most seriously.

To strengthen the nuclear material protection, control and accounting (MPC&A) at some 80 to 100 nuclear facilities at some 40 sites in Russia, in April 1994 the USDOE and its laboratories and nuclear institutes and enterprises of the Russian Federation launched a new programme of cooperation on MPC&A. The programme is called the Laboratory-to-Laboratory Nuclear Material Protection, Control and Accounting Programme (or simply the "Lab-to-Lab MPC&A Programme"). This is one of a series of U.S./Russian bilateral MPC&A cooperative efforts, all designed to reinforce and complement the national MPC&A programmes, particularly Russia's. The programme has provided, and continues and provide, financial, material support and training to Russian facilities, including improved computerised accounting systems, a capability to measure nuclear materials in containers, access control systems, portal detectors, and various other types of monitoring and security equipment, including cameras, motion detectors, cameras and seals.

The Lab-to-Lab MPC&A programme has been a highly successful cooperative effort. Progress during the first two years is summarised in "United States/Former Soviet Union Programme of Cooperation on Nuclear Material Protection, Control, and Accounting, Department of Energy Nuclear Material Security Task Force," December 1996. Kenneth N. Luongo, who until very recently was Director of the Office of Arms Control and Nonproliferation of the USDOE, was in charge of the USDOE effort. Since Mr. Luongo is attending this conference, I assume he will be discussing the successes and current status of the Lab-to-Lab MPC&A Programme.

Without denigrating the remarkable success of the Lab-to-Lab effort, it nevertheless has its limitations. Coverage is not yet universal. Security experts from the U.S. labs do not have access to sensitive military facilities that need to be covered, and must rely solely on their Russian partners for improvements at these facilities. Perhaps most importantly, while the improvements should be most effective in protecting against the single insider threat, they will offer only minimal improved protection against a collusion of security and/or management personnel.

To provide more effective security at Russian nuclear facilities against a conspiracy of insiders, I believe it is now timely for the United States and Russia to complement the Lab-to-Lab effort by creating a more comprehensive bilateral programme, with complete reciprocity for inventory control, including joint maintenance in real-time of a fissile materials database that provides the location and the isotopic, physical and chemical properties of all excess weapon-usable fissile materials. Eventually this could become an IAEA responsibility, but to facilitate access to sensitive facilities and more rapid implementation the programme should be initiated on a bilateral basis through the Lab-to-Lab effort.

Such a programme is also desirable if we are to achieve deep reductions in the nuclear weapon arsenals. Since about 1990, NRDC has been advocating that the United States and Russia (then the Soviet Union) engage in a nuclear weapons and fissile material data exchange as part of a programme to verify the dismantlement of nuclear warheads. The Russian Foreign Ministry proposed a multilateral data exchange in 1992. Prodded by the U.S. Congress, the Clinton Administration agreed to initiate an exchange of nuclear weapons and fissile materials inventory data with the Russians in December 1994. The Yeltsin Administration accepted the concept a year later.

On 10 May 1995, Presidents Clinton and Yeltsin issued a "Joint Statement on the Transparency and Irreversibility of the Process of Reducing Nuclear Weapons". Among the key provisions of this joint statement the U.S. and Russia agreed to establish:

- An exchange on a regular basis of detailed information on aggregate stockpiles of nuclear warheads, on stocks of fissile materials and on their safety and security;
- A cooperative arrangement for reciprocal monitoring at storage facilities of fissile materials removed from nuclear warheads and declared to be excess to national security requirements to help confirm the irreversibility of the process of reducing nuclear weapons, recognising that progress in this area is linked to progress in implementing the joint U.S.-Russian programme for the fissile material storage facility at Mayak: and
- Other cooperative measures, as necessary to enhance confidence in the reciprocal declarations on fissile material stockpiles.

With respect to transparency, the Joint Statement also states that:

The United States of America and the Russian Federation will also examine and seek to define further measures to increase the transparency and irreversibility of the process of reducing nuclear weapons, including inter-governmental arrangements to extends cooperation to further phases of the process of eliminating nuclear weapons declared excess to national security requirements as a result of nuclear arms reduction.

The United States of America and the Russian Federation will seek to conclude in the shortest possible time an agreement for cooperation between their governments enabling the exchange of information as necessary to implement the arrangement called for above, by providing for the protection of that information. No information will be exchanged until the respective arrangements enter into force.

Unfortunately, in late 1995, without explanation, Russia cut off bilateral talks directed toward concluding an Agreement for Cooperation, the legal instrument that would permit the data exchange and transparency measures to go forward. There has been no progress between the United States and Russia on implementation of the agreed upon data exchange, or any warhead dismantlement and fissile material storage transparency and verification measures, except for language in the Helsinki Joint Statement of March 21, 1997 that "the two sides will also consider the issues related to transparency in nuclear materials".

Russia's refusal to move forward with an Agreement for Cooperation has brought to a halt virtually all reciprocal transparency initiatives related to nuclear warhead dismantlement and warhead component storage. It also denies the two sides a real opportunity to significantly improve fissile material security in Russia against the most serious potential threat—a collusion insiders involving security and/or management personnel.

Address:

THOMAS COCHRAN
Director, Nuclear Programme
National Resources Defence Council
Washington DC 20005
U.S.A.

21
Illicit Trafficking of Nuclear Material—A Global Recent Review

S. THORSTENSEN

BACKGROUND

- Presentation aims to cast some light on the problem, especially events which have unfolded over the last few years.
- Illicit trafficking in nuclear materials is a phenomenon of the 1990s.
- The problem of Illicit trafficking from a nuclear non-proliferation perspective began much earlier. The Treaty on the Non-Proliferation of Nuclear Weapons (NPT), came into force in 1970. This early and fundamental "export control" condition came from Article III.2 of the Treaty.
- The Zangger Committee's Trigger List was first published in 1974 by the IAEA as Information Circular 209. In 1978 another set of supplier guidelines and Trigger List created by the 'London Club', later to be known as the Nuclear Suppliers Group, was published by the IAEA as Information Circular 254.
- Countries unable to acquire nuclear equipment and materials because they were unwilling to undertake peaceful use commitments on the special nuclear material created by or used in these imports were forced into acquiring the Trigger List commodities through illicit trafficking and/or by the purchasing of non-controlled commodities and creating the activities through indigenous means. Thus began the nuclear black market.

THE IAEA ROLE

- The direct involvement of the IAEA in the international efforts to address the problems of illicit trafficking began in earnest at the September 1994 General Conference. This led the Conference to adopt a resolution which called upon IAEA Member States "to take all necessary measures to prevent illicit trafficking in nuclear material", in particular with respect to:

1) The establishment and upgrading of national systems of nuclear material accounting and control;
2) The examination of additional options available in the field of collecting, verifying, and analysing data relating to incidents of illicit trafficking: and
3) The field of physical protection

- The Secretariat refined the original mandate and made specific proposals to the March 1995 Board of Governors for activities in the areas of: physical protection; state systems of accounting and control; radiation safety infrastructure related to security of radioactive sources; and, an IAEA data base covering incidents of illicit trafficking of nuclear materials and other radioactive sources.
- In the more than two years since the proposals were accepted there has been susbtantial progress in all of these areas. In the case of the Illicit trafficking data base programme, an initiative new to the Agency, the number of participating Member States has grown to 54. Member States which have not joined the programme are continually invited to do so.
- The number of illicit trafficking incidents in the data base that have been acknowledged and confirmed by Member States is 134 as of May 1997. In addition, there are 52 incidents in the data base still pending resolution and for which no responses have been received from the Agency's requests for confirmation.

THE DATA BASE

- Procedures
- Incidents

THE EMERGING PICTURE

- It would have been impossible to predict in 1994 the scope and the content of the information that was to come from the illicit trafficking data base. Statistics created by the data base of confirmed incidents including the following:

— There are 20 Member States that have confirmed reports of incidents in their countries.

— There were 43 incidents in 1993; 45 incidents in 1994; 27 incidents in 1995; 17 incidents in 1996; and 2 incidents as of May 1997.

— Of the 134 incidents, 34 per cent of the incidents involve radioactive sources and 66 per cent of the incidents involve nuclear materials.

— Of the incidents involving nuclear materials, there are 50 incidents involving source material (natural or depleted uranium and thorium); 34 incidents involving low enriched uranium; 5 incidents involving high enriched uranium; and 4 incidents involving plutonium.

— There are 50 confirmed incidents involving source material, and of these, 20 involved depleted uranium. The sum total of all depleted uranium from the incidents in the data base is slightly more than 200 kilograms. One incident involved only 1 gram of depleted uranium. The number of incidnets involving natural uranium while greater in number only amounted to a cumulative total of just above 60 kilograms.

OBSERVATIONS

- One of the first observations to be made from the confirmed incidents of illicit trafficking in the Agency's data base is that nuclear materials and other radioactive materials have been stolen from facilities in many countries. This underlines the concern that no Member State is immune from the risk of an illicit trafficking incident.
- A second observation is the obvious downturn in the annual number of incidents that occurred after the peak in 1994.
- Regardless of the number of incidents reported, there is the nagging question of how many incidents of illicit trafficking go undiscovered... incidents with far more profound consequences both from a non-proliferation and a health and welfare perspective.
- It could also be claimed with good justification that the number of confirmed incidnets is an irrelevant indicator, and that any incident which indicates the availability or accessibility of weapons usable material to unauthorised persons is one incident too many.

- Thirdly, the two multilateral nuclear supplier arrangements, have concluded that any depleted uranium less than 1000 kilograms, and natural uranium less than 500 kilograms acquired by a given recipient country in a 12 month period is of no consequence for non-proliferation purposes. This means that for a country which has implemented nuclear export controls consistent with the international standards, none of the incidents involving depleted or natural uranium would be a violation of their nuclear export controls. It could, of course, be a crime of theft or unauthorised possession.
- The incidents of greatest concern are those involving the special fissionable materials (enriched uranium and plutonium-239). Here it is clear than even a small quantity is of proliferation concern. Fortunately, only 9 were incidents involving high enriched uranium or plutonium-239.

FURTHER IMPROVEMENTS

- The positive experience gained thus far in the creation and maintenance of the illicit trafficking data base could provide a useful basis for a re-examination of the role of this initiative in the mission of the Agency and to consider how adjustments could be made to the illicit trafficking data base programme to make it more useful and relevant. In that regard, it has been suggested that to use that a good first step would be to consider a redefinition of the categorisation of incidents from "nuclear material (N)" and "radioactive sources (S)" to "non-proliferation (N)" and "safety (S)". This subtle, but important, change would make it clear that the illicit trafficking data base programme is an integral part of the Agency's efforts in these fundamental purposes served by the Agency.
- With such a broadened change in the nomenclature of the categories of incidents, to include equipment and non-nuclear materials identified by Member States in multilateral supplier arrangements as important to control for nuclear non-proliferation reasons.
- For example, the seizure of 100 tons of heavy water under the current criteria for illicit trafficking would not be considered an incident for inclusion in the data base. It would,

however, be an incident of considerable concern for supplier states.

- From a more global perspective, the issue of illicit trafficking, including the programmes to combat the problem (such as physical protection and material accountancy) and the benefits of the illicit trafficking data base as an information sharing mechanism and as a respository of validated information, should be viewed as an important component of the international non-proliferation regime.
- As pointed out at the April 1996 Nuclear Safety and Security Summit, future successes in combating the proliferation of nuclear weapons will depend upon all participants in the non-proliferation regime working together toward the common goal. In that regard the IAEA is prepared to play its role in this effort to address illicit nuclear trafficking.

Address:

SVEIN THORSTENSEN
IAEA
A-1400 Vienna
Austria

22

Creating a Comprehensive Strategy to Prevent Nuclear Smuggling

K. LUONGO

1. Introduction

The end of the Cold War has raised awareness in the international community about the threat posed by large, global stockpiles of weapon usable nuclear material. Particular focus has been directed at the level of protection provided to the fissile materials produced by the Soviet Union and concern has been raised about the growing stockpile of plutonium worldwide. Reported incidents of the diversion of nuclear material have raised the specter of potential nuclear terrorism and of countries of proliferation concern being provided a short-cut to the bomb. In order to address this problem, the international community needs to agree on the rapid implementation of a comprehensive, mutually reinforcing strategy to control existing stockpiles of fissile material, constrain future production and use of these materials, and address the underlying causes of this threat.

2. First Steps

In order to create an effective strategy to prevent illicit nuclear traffic, the international community must aggressively implement those activities that have already been proposed by various governments and organisations. These include:

- Securing existing stockpiles of weapon usable nuclear material.
- Limiting the production and use of fissile materials.
- Eliminating nuclear warheads as par of a new arms control regime.
- Strengthening the international non-proliferation regime.

These policies have a number of components and the implementation of these efforts have met with varying degrees of success.

Secure Existing Fissile Material Stockpiles

Ensuring the highest levels of protection for existing stockpiles of fissile material is an essential first line of defence against potential diversion. The safeguards programmes of the International Atomic Energy Agency (IAEA) are an important barrier but further steps can be taken. These include tightening limits on allowable hold-up during plutonium processing and lowering the parameters of a significant quantity of plutonium and high-enriched uranium to reflect current realities. In addition the approval of the IAEA's programme 93+2 will make an important contribution to the detection of future clandestine nuclear activities and the Agency's recent measures to catalog and evaluate reported incidents of nuclear diversion are welcome additions to their portfolio.

A major new safeguards activity has been the creation of the effort to improve the security of fissile material at all locations in the former Soviet Union where it is known to be stored. This programme is largely funded by the U.S. but is carried out in cooperation with the FSU states. The size of this stockpile is quite large, with some estimates placing the Soviet-produced highly enriched uranium holdings at 1,200 metric tons and the weapon grade plutonium figure at 150–200 metric tons. There is estimated to be an additional 30 tons of reactor-grade plutonium in storage in Russia.

During the period of 1994-1996, the partners in this cooperation to have begun to make improvements at about 40 different locations. This is close to 100 per cent of the locations that are known to possess fissile material. This work is scheduled to expand to the remaining locations in 1997. Significant improvements have been made during this period, with security upgrades instituted for tens of tons of fissile material. In particular, work has been expanded to many sectors of the Russian nuclear complex including cooperation with the Russian nuclear regulatory agency, Gosatomnadzor, the nuclear transportation sector, and facilities focused on fueling nuclear powered ships. This effort needs to be expanded to include nuclear warhead dismantlement facilities in Russia and the pace of improvements needs to be maintained.

Limit Fissile Material Production and Use

While it is essential to secure the existing stockpiles of fissile material, it is also necessary to move forward on a global agenda for limiting the production and use of these materials. One key initiative in this area is the Clinton administration proposal to cut-off the production of fissile material for weapons purposes or outside of international safeguards. Unfortunately, this effort has met with resistance from a number of nations.

In order to take a first step in this area, however, the U.S. and Russia have agreed to cease the production of weapon-grade plutonium. The U.S. has stopped producing this material. Russia continues to produce weapon-grade plutonium in its three remaining plutonium production reactors at Tomsk-7 and Krasno-yarsk-26 because the heat and electricity generated by the reactors is used in the surrounding towns. While the original intend was to shut down these reactors by 2000, the effort has changed its focus to converting the nuclear cores of the reactors by that date so that no additional weapon grade plutonium will be produced. The shut-down of the reactors would come at a later date. Progress in this area is being made but it is slow. For example, the formal agreement to cooperate in this area has been bogged down for over two years. If this effort can be completed, it will serve as an incentive for other nations to join the ban.

An ancillary programme is the decades-long U.S. effort convert the nuclear cores of the research and test reactors that currently utilise highly enriched uranium to non-weapon usable low enriched uranium. This effort has recently been extended to Russia and China, though implementation has been slow. Some nations have resisted the call to convert their reactors, but eliminating HEU fuels is an important step toward a global ban on future HEU production.

Eliminate Nuclear Warheads

The first two policies have focused on preventing the use of fissile materials in nuclear weapons. The flip side of this policy is the dismantlement of nuclear warheads and the disposition of the materials removed from them. A major initiative in the area of cooperative nuclear warhead dismantlement was outlined by Presidents Clinton and Yeltsin at a May 1995 summit. At that time they agreed to work together on a new arms control regime focused on the transparency and irreversibility of the nuclear disarmament process. However, little progress has been made. While both countries dismantle thousands of nuclear warheads a year there have been few steps toward the development of a transparent warhead dismantlement regime. And, and essential part of this regime, the proposal to exchange data on existing stockpiles of weapons and fissile material, is moribund. It is very unlikely that other nuclear weapon states will embrace this type of regime if the U.S. and Russia do not lead the way.

While cooperative warhead dismantlement lags, progress has been made in the fissile material disposition arena. One area that has been working is the U.S.-Russian agreement to blend-down and purchase 500 metric tons of highly enriched uranium derived from dismantled Russian nuclear weapons. To date numerous shipments of blended down HEU have arrived in the U.S. and a five-year agreement has been reached on the amount to be purchased per year and the SWU price.

The future disposition of weapon grade plutonium is also showing some promise, but fewer concrete results than the HEU agreement. In this case, the U.S. and Russia have agreed to dispose of surplus plutonium, though questions continue to remain regarding the exact method by which the plutonium would be eliminated. The U.S. has made a decision to pursue two methods, utilisation in existing reactors, immobilisation or a combination of both methods. Most of the industrialised nations agree that these are the two most feasible options. However, the future path is nuclear. All methods of disposition will be expensive and the reactor-based methods are very controversial. There is considerably more work to be done in this area.

Strengthening the International Non-proliferation Regime

The ability to institute a comprehensive fissile material security strategy rests on the strength of the international nuclear non-proliferation regime. So, efforts to improve this regime and expand international participation in it are very important. The most notable step in this direction was the decision to indefinitely extend the Nuclear Non-Proliferation Treaty. This will ensure that the regime remains on a solid foundation.

However, in order to assure that the regime remains strong into the future, the U.S. and other nuclear weapon states must deliver on their commitments to take steps toward nuclear disarmament. One obvious need is to implement the START II treaty and to move rapidly to the signing of a START III agreement that would further lower deployed weapons. This agreement should also include provisions for transparent warhead dismantlement.

Also, the U.S. and Russia must implement their agreement to allow the international safeguarding of excess fissile materials that were produced for weapons purposes. The U.S. has declared 200 tons of fissile material excess to U.S. national security needs and stated that it will never again be used for weapons. To date 12 tons of this material has been placed under International Atomic Energy Agency safeguards to ensure its non-reuse in weapons. Plans are in place for addressing the safe-guarding of the remaining material. While Russia has agreed to participate in this process under the trilateral initiative, little progress has been made between the three parties to date, although Russia is in the process of determining how much of its fissile material stock-pile is excess to its national security needs.

3. Expanding the Agenda

While the establishment of the existing agenda is a welcome response to the heightened danger posed by fissile materials in the post-Cold War world, new initiatives need to be put in place. Since the security of fissile material in Russia has been highlighted as an acute concern, expansion of the agenda should be focused in this area.

The most important new effort that can be launched is expanded cooperation on the conversion and stabilisation of the Russian nuclear weapons complex. The danger posed by the potential economic collapse of this sprawling complex is serious. These locations are physically isolated, financially strapped and facing a bleak future in the current economic climate. Many of their workers are under- or unemployed as a result of the post-Cold War cutback in nuclear weapons work. These facilities are the custodians of thousands of nuclear weapons and harbour hundreds of tons of weapon-usable nuclear material. Without a fundamental restructuring these locations can serve as incubators for many of the security problems the international community is concerned about, including nuclear smuggling, brain drain and sales of nuclear technology to rogue states.

Despite the useful U.S. and international programmes currently in place to address this situation, the economic decline in the Russian nuclear complex continues. Wages at many of the facilities are paid months late, bank loans with enormous interest charges are taken out to pay for scientists and services, and strikes have occurred to protest the economic conditions. An expansion of international cooperation in this area can make a significant contribution to worldwide security. Ensuring the continuation of positive economic, political social and security trends in Russia is in the interest of all nations. A broader collaboration on the conversion of the nuclear complexes should initially focus on three key areas—non-proliferation, environmental clean-up, and product commercialisation. These would be the first steps toward redirecting the complex away from weapons work and integrating the closed cities that comprise this complex into the evolving Russian market economy.

4. Future Issues and Problems

While there is a clear need to re-invigorate fissile material control efforts, expand their scope and integrate the activities, there are some issues and problems that may complicate this process.

The first and most fundamental is political. On an international scale, nations need to come to agreement that the control and disposal of fissile materials is an important issue. This must go beyond lip service. There are entrenched interests that are seeking to minimise the effectiveness of fissile material control efforts and to slow the implementation of new efforts. These barriers need to be overcome. The extreme danger posed by fissile material diversion must somehow seep into the consciousness of our top leaders. It is they who must direct their governments to fight for an effective regime.

Second is the need to sustain interest in the issue. As the number of reported nuclear smuggling incidents have diminished so has the urgency of the issue in the minds of the media and governments. This attitude cannot be allowed to persist. Effective and vigorous implementation of preventative measure now can save the world from an unfortunate incident in the future.

Finally there is the issue of future funding, this particularly relates to cooperation with Russia. At present the U.S. Nunn-Lugar and nuclear material protection, control and accounting programmes are scheduled to terminate in 2001 and 2002 respectively. This will mean the loss of well over $150 million per year for nuclear security activities in Russia. In the meantime, activities related to a warhead dismantlement regime, plutonium disposition, fissile material limitations and controls, and redirection of Russia's secret nuclear cities certainly will require that tens of millions of dollars be provided for an additional 15–25 years beyond the turn of the century. The current political situation in the U.S. may not support the billions more needed for these efforts, even though their importance is undisputed. Some better, more predictable international financing mechanism must be created to deal with these requirements.

In sum, recent experience has shown that large stockpiles of fissile material pose a serious threat to the world. Some steps have been taken to address the danger, but new initiatives need to be undertaken and all efforts need to integrated into a comprehensive strategy for preventing illicit nuclear traffic. It is not often when the world gets forewarning of a serious security danger. But, threat posed by illicit nuclear smuggling is clear and apparent. There is still time to construct a comprehensive strategy to prevent this threat, but we must act quickly.

Address:

KENNETH LUONGO
Centre for Energy and Environmental Studies
Princeton University
Princeton NJ 08544
USA

23

Nuclear Smuggling in Europe: Real Dangers and Enigmatic Deceptions[1]

A. SCHAPER

1. Introduction

During the last years, alarming news has accumulated on smuggled nuclear material, migrating experts, stolen bombs and new miracle substances that render obsolete all non-proliferation efforts. Together with a look at the economic difficulties, sinking living standards, and rising unemployment in the CIS countries, a pessimistic impression evolves of how to control the huge Russian nuclear complex and how to stem further proliferation of nuclear weapons. This paper tries to assess the dangers of nuclear smuggling and to discuss some possibilities for remedies.

2. What has been smuggled so far?

A small black market of smuggled nuclear material has existed for many years, however, the number of cases and reports has increased during the last years. Cases that have become famous already before the find at the Munich airport was the discovery of strontium-90 and Cesium-137 in a locker of the Frankfurt railway station in 1993, and a case where a smuggler had injured himself with deadly radioactive radiation. Hundreds of cases have become known when vendors offered weapon usable materials but did not deliver them, and there had also been cases of trades which had been observed without arrests in order to catch the agents behind.

Until Spring 1994, none of the findings in Germany were of any use for nuclear weapons, however dangerous for environment and health.[2] Typical examples are low enriched uranium, mainly from light water reactor fuel production, ionising sources containing milligrams of plutonium as are typical for smoke detectors made in the former Soviet Union, and other radioactive substances such as cesium-137, cobalt-60, strontium-90, or californium-252.

The applications of these materials are civilian as well as conventional military. Natural and low enriched uranium is used in nuclear reactors, cesium, strontium, and cobalt sources are used in medicine and science. Some materials are being applied in military decontamination experiments and exercises. The amount of plutonium in smoke detectors is far too low to be of any proliferation significance.

But from 1994 on, some findings have occurred that have a significance for nuclear weapons:

- *Tetigen, May 1994:*

This case was the finding of 5.6 almost pure Pu-239, that was detected in the Garage of Adolf Jäkle, against whom investigations were carried out because of another suspicion. There are hints that this material is one of the samples that had been distributed by the Russian laboratory Arzamas-16 to dozens of East European research institutions: The isotopic composition of this material is astonishing[3] because no nuclear reactor exists that can produce this mixture. Plutonium of this composition can only be produced by enrichment, a technical procedure that is totally uncommon for plutonium and extremely expensive. It is also not necessary for the application in nuclear weapons. For the same reasons it is also likely that not much has been produced, at least for less than kilograms, the minimum amount for nuclear weapons. Probably, the sample has been stolen in an East European institute. The significance of this case is the fact that it had been detected by chance and that apparently no police sting operations have been involved.

- *Landshut, June 1994*

This was the finding of 800 mg highly enriched uranium (HEU) that probably was destined to be used for nuclear submarines or for research reactors. Although this quantity is far below of what is necessary for one nuclear warhead, the case gives rise for concern because such HEU is being produced and used in large quantities.

•Prague, December 1994

Indeed, material with an identical isotopic and chemical composition has been found in Prague in the much larger quantity of 2.72 kg, apparently stemming from the same source.

•Munich, Airport 1994

On a Lufthansa flight from Moscow to Munich, 560 g mixed oxide (MOX) have been found, containing 363 g Pu-239, so far the largest quantity of plutonium whose smuggling has become known. About ten times as much would be sufficient for one warhead. Also plutonium is being produced and used in large quantities. The composition was also unusual: firstly, the Pu content is astonishingly high, secondly the isotopic composition is between what is generally known as civilian reactor Pu and military weapon Pu 4.[4] The Munich case has caused a scandal which is described further down.

More cases of diversion of weapon usable materials have become known in Russia[5]

•Podolsk (Russia), May–September 1992

1.5 kg HEU (90% enriched). This material stems from the "Luch Scientific Production Association" in Podolsk. The theft has been discovered and the material been secured at the same location. The case is significant because of the large quantity of the material and because it demonstrates that thefts are possible.

• Adrejewa Guba (Russia), August 1993

1.8 kg HEU (30% enriched), that had been stolen at a submarine storage site. Legally, uranium of this enrichment ocunts as HEU, however it is more difficult to use it for weapon purposes (see next section). The significance of this case is the demonstration that the security of the Russian submarine fleet must be imporved.[6]

•Poljarni (Russia), November 1994

Theft of 4.5 kg uranium with an enrichment of 20 per cent. The use for nuclear warheads is limited.

So far, these cases are the only well-known ones that involved weapon grade materials. The fourth section will describe several materials that can or can not be used for nuclear weapons.

As a summary it can be stated that so far, no smuggling case has become known in which it has been proven that the origin of the material was the Russian military complex. Also thefts of weapon grade material within Russia's military complex so far have not been unambigously confirmed. But this does not necessarily imply that such cases have not taken place or that they would not take place in future.

3. The German Nuclear Smuggling Summer

In Summer 1994, the problem of nuclear smuggling caused big headlines in Germany, although many cases have occurred in East Europe.[7] The reasons are the involvement of weapon grade materials for the first time, and the timing: because of the summer holidays there was a shortage of other interesting news. Another factor was the German public that was already very sensitised, mainly because of environmental concerns. Especially the case of the MOX-finding at the Munich airport caused a tremendous press echo[8] and a substantial scandal. Other cases also found much attention, but they belong into the category of non-proliferation relevant incidents.

Already at the end of 1993, undercover agents of the German intelligence service (BND) and the Federal Police Office (BKA) who operated in Spain had contacts to traders claiming to be able to sell plutonium. These talks had been intensified in May and June 1994. They resulted in the transport of a plutonium sample by three Spanish smugglers on board of a Lufthansa aircraft from Moscow to Munich. The smugglers who had lived in Moscow for a long time were later convicted. Their trading partners were BND and BKA agents. The trials of the smugglers have revealed that the undercover agents have persuaded the smugglers to bring the plutonium to Munich, and that they knew in advance that the transport would take place.

This was justified by some officials citing perceived problems in the cooperation between and German authorities: there was uncertainty whether the risk of leakage would not have been too high, and in order not to spoil the undertaking, a cooperation attempt did not take place. Meanwhile, a Federal and a Bavarian Parliamentarian Investigation Committee are examining to which extent which authorities were involved in the case, especially how much has been known in advance, approved, or even coordinated by the Secret Service Coordinator Bernd Schmidbauer on whose role a lot of doubt has been cast.[9] So far, both committees are still at work and a final report is not yet ready. Earlier this year, it was revealed in a closed session of the investigation committee, that the successor agency of the KGB has spread some misinformation, that fueled the suspicion against Schmidbauer.[10] A severe diplomatic mistake was the early alert of the press on the findings by Shcmidbauer himself. This has deeply offended the Russians who had note been consulted. Many officials and observers deplored the diplomatic damage.

An attempt for remedy was a cooperation agreement between Kohl and Yeltsin. Meanwhile, it seems very likely that the plutonium originated in Obninsk, where the security is known to be not sufficient.[11]

4. What Does a Proliferator Need?

A proliferator needs either plutonium or HEU, each in quantities of several kilograms. Natural uranium contains only 0.7 per cent U-235, but HEU in nuclear warheads must be enriched, preferably to more than about 800 per cent. Indigenous enrichment technology is only available to industrialised states, others would also require technology transfers.[12] Low enriched uranium contains only 3 per cent U-235 and is not weapon usable. Plutonium is being produced in nuclear reactors together with radioactive fission products. It can be separated by chemical means together with radiation protection technologies, called reprocessing. It consists of several isotopes, especially Pu-238 to Pu-242, whose ratios vary depending on the reactor and the burn-up. In established nuclear weapon states, the portion of Pu-239 is maximised. However, in principle, almost any isotopic composition can be used to create a nuclear explosion.[13]

Apart from the application in nuclear weapons, there are some more applications of Pu and HEU that might pose diversion risks and could constitute a proliferation danger. HEU is being used by some states as submarine fuel, among them Russia. A civilian application is research reactor fuel, whose use has been reduced in the last years because of proliferation concern.[14] Plutonium is being used as mixed oxide fuel for light water reactors in some industrialised countries, among them Russia. It is also being used as fuel for breeders. All these applications can be sources for nuclear thefts and smuggling.

5. The Fabulous "Red Mercury"

In Spring 1994, the First German TV channel broadcasted a film dealing with a mysterious substance called "red mercury".[15] The quintessence of this film was the claim that with this substance, the Russians have managed to manufacture nuclear warheads of the size of a beer glass. Since it is already being smuggled everywhere, terrorists will soon be capable to blow up whole cities. It culminated in the claim made by the American former nuclear weapon scientist Samuel Cohen that this Russian wonder weapon is a so-called "clean bomb", rendering obsolete any non-proliferation efforts that had taken place so far.

The term "clean bomb" needs some technical-historic explanation: so far, all nuclear weapons need a fission trigger for ignition which is the major source of the fall-out produced by a nuclear explosion. An important motivation for the first negotiations of a Comprehensive Test Ban (CTB) that were started at the end of the fifties was the environmental concern because of the fall-out. The nuclear weapons labs at that time had a strong interest to prevent a CTB. A strong argument was made by Edward Teller who claimed that it could be possible to construct a clean bomb, e.g. a warhead whose energy comes only from fusion but not from fission and which accordingly would hardly produce any radioactive fallout. But meanwhile it is well known that such a device is principally not possible because there is not chemical composition that could produce the energy densities necessary for the ignition of a secondary. The revival of this old claim coincided with the resume of CTBT talks that had started in January 1994. S. Cohen is a nuclear weapon scientist similarly as E. Teller, and well known as test ban opponent. In case the claim would have been believed that red mercury would enable a clean bomb, this would have revived interest in nuclear testing and would have undermined the interest in a CTB. It seem reasonable to assume that the motivation for this claim comes from CTBT opposition.

The film made several technical assertions which cannot be maintained under scientific scrutiny. One allegation was that red mercury is an explosive of sensational high energy density, another was that it is a neutron reflector, similarly with extreme properties. This is hard to believe because the one quality would be due to special chemical properties but the other due to special nuclear properties, two completely different things whose coincidence is very unlikely.[16] Even if red mercury had these properties, the construction of warheads would not have become easier. Plutonium or HEU would still be necessary. Some of the findings involved indeed mercury compounds. But similar can be bought everywhere at low cost, e.g. for anti-rust protection paint.

The film showed several interviews with "Russian nuclear weapon scientists", however they were not identified. The film also failed to explain how it was checked whether the interview partners were indeed weapon scientists or somebody else who was happy to earn some Western money.

6. Who Drives the Market?

The case "red mercury" could easily be shrugged of as a typical case of yellow press, but such cases have a fatal side effect: the authors of the film went to Moscow several times in order to find interview partners pretending to be potential buyers searching for weapon grade materials. The effects on criminal communities in Moscow that rapidly communicate the appearance of sound customers can be imagined, also the temptation of people who indeed could be able to offer relevant goods such as HEU or Pu. A similar case had already happened in 1993 when journalists instigated Russian soldiers to smuggle ostensible weapon grade material from Moscow to Berlin where at that time they were still stationed. The soldiers indeed brought the material but later claimed that they knew it was a fake.

A similar stimulating effect can be caused by undercovers of secret services or the police. The degree of intrusiveness must be carefully weighed when gangs are penetrated, similarly as in the investigation of drug smuggling. Apart from journalists and undercovers, another group has been active as buyers, namely other criminals. Similarly to drug markets, also nuclear materials are first passed on from one dealer to the next. Each hopes for the big business by meeting the rich end users from Iraq or another proliferating country.

The question arises whether these end users exist. Apart from the five recognised nuclear weapon states, also India,

Pakistan, and Israel are assumed to possess nuclear weapons. They have their production capacities and have produced already larger quantities of weapon grade materials. They are therefore not dependent on smuggling. The other well known proliferation problems are Iraq whose advanced nuclear programme has been discovered after the Gulf War, North Korea, and Iran, who is said to have nuclear programme although this claim has not yet been proven.

North Korea was building up a clandestine Pu production and had provoked the international community by non-compliance with safeguards requirement. But at the same time, it also has interests, especially the acquisition of modern civil nuclear technology as an exchange for its renunciation of plutonium and military ambitions. Indeed, its prospects are good, however any possibility of a successful deal would be destroyed when it would be caught in a nuclear smuggling affair. It is very unlikely that North Korea would take such a risk, especially in Germany where the danger of being detected is high. However this does not imply that it might have some direct contracts, but it is more likely that they take place in some remote areas such as in Central Asia.

Similar considerations also apply to Iraq and Iran. Both countries have specific interests and goals that they would not endanger for the doubtful benefit of snapping up some kilograms Pu or HEU on a black market in a Western industrialised country. Iraq has the clear and urgent interest to get rid of the UN sanctions. Iran at the time being is embargoed by the U.S. and does not obtain any nuclear technology from any Western country. A nuclear reactor project that was begun by Siemens still during the time of the Shah is only completed by the Russians without any German assistance. Iran therefore has a strong interest to demonstrate its reliability in non-proliferation matters. An example is the invitations of IAEA inspectors to visit any location at any time.

The question remains whether there is an unknown proliferator: who uses this method to acquire materials and technology. Principally, this possibility cannot be excluded, but so far no concrete indications have become known. It is also doubtful, whether such a proliferator would send his buyers just to Germany which is dense of specialised police undercovers; instead of sending them to the CIS. Nevertheless, this possibility is the major reason for concern. Although the events in Germany have been greatly exaggerated, the possibility must be taken into account that a proliferator will try to acquire materials at the source, which is in Russia.

There are some more theories why so spectacular findings have been made just in Germany. They all start more or less from complicated conspiracy theories or even manipulations of the Federal Government. An example of the several variants is the claim that the found samples are only substituted and in fact originate in Germany.[17] Another is the suggestion that the results of the analyses have been manipulated.[18] The reasons given by proponent of such theories include election campaign tricks,[19] a plot to exert pressure by the German Government in order to participate in the disposition of Russian disarmament Pu,[20] and the attempt to enhance Germany's international role, e.g. to acquire a permanent seat in the Security Council of the UN.[21] As proof it is claimed that at that time indeed there had been some German initiatives towards nuclear disarmament, that categorically had been rejected by the nuclear weapon states. A prominent example is the proposal of the Foreign Minister to implement a nuclear weapons register at the UN to create an overview on the number of existing warheads.[22] The mistakes such as the unnecessary provocation of the Russians by the precipitate claim that the material originates from the Russian nuclear complex or such as the dangerous transport of the plutonium in a civilian passenger flight can be explained much more by lacking professionality than by conspiracies.

7. The Problems in Russia

The origins of the smuggling problem cannot be found in Germany but in Russia, where the security of the nuclear installations has rapidly deteriorated.[23] However, it is still better in the military complex than in civilian sites. The security is mainly based on human staff, not on technology. This system has worked in the former Soviet Union, when incentives and social pressure were strong enough to guarantee the compliance of the staff. At that time, there would have been no possibility for private businesses with nuclear materials. Meanwhile, the social situation has tremendously deteriorated, so that the likelihood has risen that nuclear materials are diverted. An indicator of the crisis is the recent suicide of the director of Chelyabinsk-70. His motive was desperation that he was not able to secure the future of his institute and to pay the salaries of the staff that were outstanding since half a year. The desperation is also voices by employees of the nuclear complex, who also emphasise that "decency and honour" still guarantee that no assistance towards proliferation and criminal activities would take place.[24]

Another potential proliferation danger is the brain drain of nuclear weapon specialists[25] estimated that about 10,000 to 15,000 persons have such knowledge. This includes specialists on aspects such as enrichment as they also exist in several industrialised non-nuclear weapon states like Germany. Indeed, there also had been proliferation cases, e.g. German experts have advised Iraq in the design of enrichment technology. About 2000 employees have special knowledge that concerns directly the construction and functioning of nuclear warheads. An example is the art of spherical compression of nuclear material with special arrangements of high explosives and ignition mechanisms. A specialist on this technology might be able to accelerate a nuclear programme of several years.

The traditional social system of the comparatively privileged life in the closed cities hardly exists any more and in only maintained with tremendous force and Western aid. The living standard has deteriorated enormously.[26] Russian declarations maintain that so far no nuclear weapon specialist has emigrated, in contrast to highly qualified staff from nuclear basic or civilian research.[27] The nuclear weapon scientists themselves are searching alternatives that allow them non-criminal professional work which means Western financing of civilian projects.[28]

Without Western aid, Russia would be unable to implement a reliable material control and accountancy system. This applies especially to civilian facilities in Russia and in other CIS states, e.g. the storage sites of weapon grade HEU submarine fuel in Kasakhstan, e.g. the HEU storage of the Kasakh Atomic Energy Institute near Almaty. An American college who has visited this institute in Spring 1996, reported about a guard, dressed civilian, who was on duty only four days per week. He is not paid by the state but by the institute itself that is sport of money. Since months, his salary had not been paid.[29]

The situation in Russia is somewhat better, and partially first successes of Western aid can be observed.[30] But much more has to be done. In Russia, the own unability to solve the problems quickly is also perceived as a humiliating defeat. It is therefore necessary also to develop diplomatic skills and intuition in the collaboration with Russia.

8. Potential Remedies

Some potential remedies have already started, an example is the International Technology Center (ITC) in Moscow, that has been founded at the end of 1992 after a long negotiation period. The goal of the ITC is to offer incentives to Russian scientists and engineers who have sensitive knowledge on the construction of nuclear weapons, to use their skills for the benefit of their own country for peaceful purposes. The center mainly organises projects. Applied projects on subjects from environmental protection, energy, and nuclear safety prevail. It is financed by the European Union, the United States, and Japan with about 80 million U.S. dollars. Also several smaller Western countries have contributed. In the beginning, the bureaucratic process has been very slow, so that there was a long time delay between approval and arrival of money. Fortunately, meanwhile the Center has been well established and can report on successful projects.[31]

A major measure is the "Cooperative Threat Reduction" (CTR).[32] It aims at assisting in nuclear disarmament and in enhancing security and material accountancy of fissile material. Also the CTR had starting problems. They had several origins: one was that the Americans asked more transparency from the Russians of the disarmament process than they were willing to offer in return. Also the bureaucratic processes until the aid arrived had been very slow. But meanwhile, lessons have been learned, and the quality of the aid and cooperation has changed. In the beginning, only goods were offered, but no financial aid. All money was supposed to be spent in the U.S. Therefore, suspicion was growing in Russia, it was feared that only the precious Russian knowledge was being exploited, but that only Western industry benefited from the money. Also, all cooperation was organised from top, e.g. by the governments, so that the many buraucractic actors on both sides petered out all progress. Lessons were learned, for these experiences, and a positive result was the so-called "Lab-to-lab programme", that enabled direct interaction between the scientists on both sides, some of it also financed by the ITC. The goal of this cooperation is the destruction of weapons of mass destruction and the disposition of materials, the improvement of the security of these processes, the prevention of brain drain, conversions and also military cooperation. Both sides emphasise that this approach has been the most successful so far.

A similar approach had been started even earlier by Euratom, but with a lower budget. Its goal is to assist the Russians to develop material accountancy of standards comparable to those of Euratom.[33] Great care is being taken that Russian existing skills and technologies are integrated. The project aims at implementing a training center for inspectors and maintenance staff in Russia. In addition, seminars and visits at Euratom and the Joint Research Center are being organised. Meanwhile, there are a number of further measures by the EU[34] Member States, with additional governments, especially the U.S. and Japan, and the IAEA and NATO.[35] They include activities by the Safeguards Directorate (DG XVII-E), that aims at a modest system for material accountancy and control and that trains experts, by the Joint Research Center, especially its Transuranium Institute, that analyses materials and builds up a database for analyses, and the technical aid programmes TACIS and PHARE. The latter, however, are predominantly occupied with environmental and safety projects. The proportion, that has been devoted to physical protection so far, is only about 95–99 per cent.[36] An example for a TACIS project are radiation detectors for Kasakhstan and Ukraine. The cooperation partners on the Russian side are the Atomic Ministry ("Minatom") and the agency Gosatomnadzor. The German Government has concentrated its aid on the destruction of chemical weapons. In Spring 1995, the proposal has been discussed to make use of the MOX fabrication facility in Hanau for the disposition of Russian disarmament plutonium.[37] The advantage would have been that the whole process would have been subject to international control according to the most modern standards of physical protection and safeguards. But because of lacking domestic acceptance, the idea has been abandoned.

All these measures point into the right direction, but nevertheless they are still far from solving the problem soon. It must be emphasised that all these activities must be maintained and intensified. In the long term, it will be necessary to implement international safeguards also in the nuclear weapon states, similarly as they are taking place already in the non-nuclear states since a long time. Although safeguards would not detect diversions of such small quantities as they have occurred in the smuggling cases, they would nevertheless have the effect of creating discipline in the material accountancy and control, whose implementation in Russia is necessary anyway. The proposal of introducing safeguards also in the nuclear weapon states—is on the agenda also in several other contexts, e.g. it surfaced in the 93+2 negotiations, in the negotiations on an international plutonium management, and in preliminary discussions on negotiations of a fissile material cut-off treaty.

On the short time agenda are the following goals must be achieved:

— Implementing Western standards of material accountancy and control
— similarly of physical protection of civilian and military nuclear facilities
— similarly of nuclear transport
— centralisation of nuclear material accountancy
— reform of the export controls
— conversion of jobs in the military nuclear establishment
— construction of safe and secure storage sites
— safe and secure dismantlement of nuclear warheads.

Notes

1. An earlier version of this paper has been presented at the Investigation Committee of the Bundestag on Plutonium as: "Vagabundierendes Nuklear-potential der ehemaligen Sowjetunion", Papier für die Anhörung vor dem 1. Untersuchungsausschu β "Plutonium", 5 Dezember 1996 [WD0]11.
2. Cf. Harald Müller, Annette Schaper, Nuklearer Schwarzhandel?, Friedensforschung Aktuell, Ausgabe 33, Februar 1993 [WD0]12.
3. Pu composition: 0.067% Pu-238, 99.75% Pu-239, 0.18% Pu-240, 0.003% Pu-241, 0.0002 Pu-242
4. Pu composition: 0.16% Pu-238, 87.23% Pu-239, 10.99% Pu-240, 0.91% Pu-241, 0.7% Pu-242.
5. William C. Potter, Before the Deluge? Assessing the Threat of

Nuclear Leakage From the Post-Soviet States, Arms Control Today, October 1995, p. 9–16.

6. This case and its investigation is described in: Mikhail Kulik, Guba adreeva: Another Nuclear Theft has been detected, Yaderny Kontroi, English Digest, Spring 1996, No. 1, p. 16–21. Mikhail Kulik was the chief investigator of the Russian prosecutor.
7. Draft of proposed remarks by R. James Woolsey, Director of Central Intelligence before the House Foreign Affairs Committee Int. Organisations, Int. Security Human Rights Sub-committee, on Russian Organised Crime, 27 June 1994.
8. See for example the summarising documentation of the Deutsche Welle: Der Atomschmuggel und seine Hintergründe (The atomic smuggling and its backgrounds), October 7, 1994.
9. A critical view has been presented by Hermann Bachmeier who is a Social Democrat and member of the Federal Parliamentarian Investigation Committee: Hermann Bachmeier, Die Spur f'ührt ins Kanzleramt, Documentation, Frankfurter Rundschau, May 17, 1996.
10. M. Hibbs, 1994 plutonium smuggling was deed of KGB Bonn now claims, Nucleonics Week, February 6, 1997.
11. Gennadi Pshakin, Security of nuclear materials and installations in Russia, presentation at the Conference on Non-proliferation, Carnegie Endowment, Washington, D.C., June 10–11, 1997.
12. In its clandestine nuclear programme, Iraq had tried to obtain enrichment technology. Thereby it had especially exploited loopholes of the former German export control system. See H. Müller, M. Dembinski, A. Schaper, "From Black Sheep to White Angel? — The New German Export Control Policy", PRIF Reports No. 32, January 1994.
13. During the 80s, a debate on the explosive properties of plutonium took place in Germany. It was maintained, that civilian Pu from light water reactors is not weapons grade and therefore no proliferation danger. Meanwhile, this claim is refuted. See E. Kankeleit, C. Küppers, and U. Imkellcr, Bericht zur Waffentauglichkeit von Reaktorplutonium Arbeitsbericht IANUS-1/1989, Darmstadt 1989 (this report has been translated by the Livermore Laboratory); and J.C. Mark, Explosive Properties of Reactor-Grade Plutoinum, Science and Global Security, Vol. 4, pp. 111–128, 1993.
14. A debate on this application is currently taking place because there are plans to construct a new HEU fueled research in Garching near Munich. See A. Schaper/D. Richter, J. Kalus, Soll der in Garching geplante Forschungsraktor FRM-II mit hochangereichertem Uran betrieben werden? Physikalische Blätter, Jahrg. SO, Nr. 7/8, S. 621 (1994).

15. ARD-Exlusiv: 'Die Terrorbombe—Auf den Spuren einer neuen Waffe', 15 April 1994, 21.50–22.30 h, Author: Gwynne Roberts, Editor: Peter Winterberg, Scientific Advisor: Frank Barnaby
16. F. Barnaby, who was scientific advisor or the film has published a article on this subject which also does not bear close scrutiny: F. Barnaby: Red Mercury—Is there a pure-fusion bomb for sale?. International Defence Review, June 1994. As an example, be claims that mercury and antimon form a covalent chemcial bond which is wrong.
17. This is the opinion of several Russian commentators.
18. See Mark Hibbs, Plutonium, politics, and panic, Bulletin of the Atomic Scientists, November/December 1994, p. 29: "In response to questions regarding the accuracy of the technical analyses of seized materials. Euratom officials refused to discuss whether Bonn had manipulated its findings." This sentence does not directly claim a manipulation but it suggests a supposition.
19. The political Günter Verheugen during the election compaign
20. M. Hibbs, Bonn pushing for weapons Pu role in wake of smuggling escapades, Nuclear Fuel, October 24, 1994. p. 9.
21. M. Hibbs, A second agenda, Bulletin of the Atomic Scientists, November/December 1994, p. 30.
22. K. Kinkel, "German 10-point initiative for nuclear non-proliferation", Bonn, 15 December 1993. For the significance of this proposal and the reaction of the nuclear weapon states see: H. Müller, Transparency in Nuclear Arms: Toward a Nuclear Weapon Register, October 1994, p. 3.
23. Vladimir A. Orlov, Accounting, control, and physical protection of fissile materials and nuclear weapons in the Russian Federation: Current situation and main concerns, Paper presented at the International Seminar on MPC&A in Russia and NIS, Bonn, April 7–8, 1997.
24. E.g. at the "3rd PugWash Workshop on The Future of the Nuclear-Weapon Complexes of Russia and tbc USA" in Moscow, 24–26 March 1996.
25. R. Adam Moody, Re-examining Brain Drain from the Former Soviet Union, The Non-proliferation Review, Spring-Summer 1996, Vol. 3, No. 3, p. 92.
26. Office of Technology Assessment, Proliferation and the Former Soviet Union, Washington, September 1994.
27. K. Hart, Russian weapons scientists said to be working on foreign nuclear projects, Nuclear Fuel, November 21, 1994, p. 2.

28. An example is a project proposal from Arzamas-16: Project Summary for Open Publication: Technical Problems of International Stability (Safety) Under Conditions of Nuclear-free World. The project would finance 18 scientists for 24 months. The budget proposal is 300,000 U.S.-$.
29. William C. Sutcliffe, Draft Foreign Trip Report, Russia, Uzbekistan, Kazakstan, Ukaraine—May 1996.
30. An example is the Kurchatov Institute in Moscow that has implemented physical protection with cameras and additional sensors. See Sutcliffe and Alexander Rumyantsev, The Accounting and Control of Nuclear Material and Radioactive Substances in Russia, Yaderny Kontroi, English Digest, No. 1, Spring 1996.
31. The International Science and Technology Center, January-December 1995, Second Annual Report, Moscow 1996.
32. Department of Defence, 1996, Annual Defence Report, Chapter 8: Cooperative Threat Reduction, See also: The White House Fact Sheet: US Nunn-Lugar Safety, Security, Dismantlement Programme, March 21, 1994.
33. Commission of the European Communities, Reprot of the Implementation of Euratom Safeguards 1991–1992, KOM (94) 282 endg, Brussels, 6 June 1994.
34. European Commission: Communication from the Commission to the Parliament and the Council. Illicit Trafficking in Nuclear Materials and Radioactive Substances—Implementation of the guidelines laid down int he communication from Commission of 7 September 1994 (COM(94)383) and in the conclusions of the Essen European Council, COM (96) 171, Brussels, 19 April 1996.
35. The activities of the IAEA are described in: Sven Thorstensen, Nuclear Material Accounting and Control: Coordinating assistance to newly independent states—An overview of IAEA-supported activitics to help former Soviet Republics establish State systems of accounting and control, IAEA Bulletin, p. 29, January 1995.
36. Contract Information. Nuclear safety programmes—Budget 1995, published by the European Commission, Directorate General External Relations: Europe and the New Independent States, Common Foreign and Security, Policy and External Missions (DG 1 A)
37. German American Academic Council and National Academy of Sciences, German-American Cooperation in the disposition of excess weapons plutonium, Bonn 1995; A. Schaper, Using

Existing European MOX Fabrication Plants for the Disposal of Plutonium from Dismantled Warheads, in: W.G. Sutcliffe (Ed.), Selected Papers from Global '95, UCRL-ID- 1 24105, Livermore, June 1996, p. 197.

Address:

ANNETTE SCHAPER
HSKF (PRIEF)
60322 Frankfurt
Germany

24
Action to Counter Illegal Trafficking in Nuclear Materials in the Northern Region

O. REISTAD *and* M. MAERLI

Norwegian Radiation Protection Authority, Nuclear Safety Department

The break-up of the Soviet Union and economic conditions in Russia have probably increased the risk of proliferation of nuclear and radioactive material. There is, reason to believe that previous lines of command have been weakened and that military control over fissile material is therefore less firm. Much of the uranium fuel is highly enriched and represents a proliferation threat. The existing storage capacity for spent fuel and other radioactive material is said to be fully used, and some of these facilities are also in poor condition. The dangers arising in this connection are primarily radio-ecological, but there is also the possibility of sabotage or terrorism. To round off the picture, one must also mention the very difficult living conditions in Russia. Russian society is marked by political instability, economic hardship and poor control to such an extent that there is a risk of fissile and other radioactive material going astray.

The primary aim of action to counter smuggling of radioactive material is to prevent weapons-grade material from reaching terrorist organisations and countries which have renounced nuclear weapons under the terms of non-proliferation agreements. Weapons-grade material means plutonium with a Pu-240 isotope content of less than 7 per cent and high enriched uranium, i.e., enriched to over 93 per cent U-235. Both new and spent reactor fuel from the Kola power station, the icebreaker fleet and the Northern Fleet contain some U-235 and plutonium in varying quantities. Spent fuel also contains fission products such as Cs-137 and Sr-90. Cs-137, Co-60 and Ir-192 are also examples of easily detectable gamma-emitting industrial sources, which are another concern in anti-smuggling arrangements.

In 1994, the Norwegian Government submitted a report to Parliament, Stortingsmelding 34, "Nuclear activities and chemical weapons in areas close to our northern borders". The report formed the basis for the "Nuclear Action Plan", which provides a framework and financial basis for a number of nuclear safety projects. The action plan was submitted in 1995 and between 100 and 130 million Norwegian Kroner per year have since been allocated to this work. In geographical terms, the focus is on north-west-Russia and the Kola peninsula, where Norway has a common border with Russia and major economic interests dependent on maritime resources.

Lack of International Supervision

As a nuclear weapons state, Russia is not subject to the fissile material accounting rules of the International Atomic Energy Agency (IAEA). At the same time, much of the material is in military hands and thus not accessible to observation and supervision by the Russian civilian authorities or foreign supervisory bodies. However, Russia claims that the Russian authorities are aware of the danger of proliferation and are working hard to reduce it.

However, Norway's main concern is the specific situation on Kola and the planned disarmament measures to be carried out around the turn of the century. Several thousand atomic warheads are stored on the Kola peninsula an board submarines and other naval vessels and probably also on aircraft and in army installations. Every such warhead contains fissile material. A large proportion of the nuclear weapons belong to the navy and are kept in weapons stores or on ships. The largest Russian submarines, for example, each carry 20 missiles, each missile carrying 10 warheads and each earhead a charge of 5-10 kg of plutonium. The smallest, tactical warheads may have as little as a few kilograms of plutonium or 4–6 kg or uranium. Larger, strategic warheads can contain several tons of kilograms of plutonium or uranium. Fissile material is also present as fuel in reactors. Most submarines and icebreakers operate with uranium fuel enriched to 20–40 per cent U-235, but some classes of submarines probably have fuel enriched to as much as 90 per cent. The START II Agreement Russia and the USA, which has not yet been ratified lays down that in the year 2003 Russia is not to have more than 3000 warheads and imposes severe limits on the number of submarines in the Northern Fleet.

The Kola peninsula also has the greatest density of nuclear reactors on earth: almost one fifth of the world's reactors are located in this area. In addition to military submarines, three nuclear-powered naval surface vessels are in operation and the Murmansk Shipping Company has eight nuclear-powered civilian ships. On land, there are research reactors and research establishments, especially in the area around St Petersburg, together with the Kola power station with its four reactors, two VVER 400/230, and two VVER 440/213. Over a 10-year period, there Russians plan to build three new VVER 640 reactors, i.e. the same as is also planned at St Petersburg.

Currently, more than 70 Russian nuclear submarines have been taken out of service and are lying along the coast of the Kola peninsula awaiting decommissioning. Most of the submarines have two reactors. The missiles have been removed but there is still fuel on board about 50 of these mothballed vessels. Russia plans to cut up 125 submarines by the year 2010. As a result of disarmament and the scrapping of nuclear submarines, large quantities of plutonium and uranium, both weapons-grade and less highly-enriched, will be stored ashore. The treatment, transport and storage of this material is both an environmental and a safety problem.

Action to Prevent Proliferation of Nuclear and Other Radioactive Material

There are two complementary approaches to preventing the proliferation of nuclear and radioactive material: security and materials accounting on site and surveillance and anti-smuggling measures at the frontier. The Norwegian authorities are today following both lines of approach: upgrading and training the Norwegian customs service, and undertaking accountancy and physical protection projects.

Trilateral Cooperation with Gosatomnadzor

No one is better placed to supervise and map radioactive sources in Russia than the Federal Inspectorate for Nuclear and

Radiation Safety, Gosatomnadzor (GAN). This organisation is responsible for supervising nuclear material and radioactive substances in the country and knows the situation from the inside. A decree issued by Yeltsin in September 1995 unfortunately removed GAN's right of inspection and supervision of military installations. Today, GAN basically supervises all civilian use of nuclear and other radioactive material. In the north-west region of Russia alone, there are more than 100,000 radioactive sources in over 3000 undertakings and establishments.

GAN today suffers from poor funding and equipment. Much of the record-keeping and supervision is carried out manually and there is no comprehensive overview of the sources. Together with the Swedish nuclear inspectorate (SKI), the Norwegian Radiation Protection Authority has begun cooperation with GAN, providing office and computing equipment so that inspections and supervisions can be made more effective.

Consistent systems for supervision and recording of radioactive material can be built up by developing a uniform database structure for the region and equipping inspectors at all its local offices with PCs. So far, two stationary units have been delivered to develop the database structure centrally, together with standardised recording schemata. As a pilot project for the local offices, the Murmansk office received two lap-top computers and work on registering sources is well under way. Experience from this project, which so far has been positive, will indicate how the work should proceed.

In parallel with the Nordic cooperation project, GAN has begun to collaborate with Euratom on developing its capability for nuclear materials accountancy systems.

Murmansk Shipping Company

The Murmansk Shipping Company (MSCO) has five seagoing icebreakers, two river ice-breakers and one freighter. The reactors in the ice-breakers are based on the same design as is used in submarines. Present standards of security for new and spent fuel are not satisfactory. Given the lack of permanent onshore storage facilities, new and spent fuel is stored in two ships (the Imandra and the Lotta respectively) lying at the quayside at Atomflot. The harbour area is fenced and there is a measure of access control. MSCO cooperates to some extent with the

Russian Northern Fleet, e.g. in refuelling of surface vessels and processing of low-level liquid waste.

The Americans are particularly concerned with the security of new uranium fuel. Under its "Lab-to-Lab programme", the United States DOE has installed physical protection systems at a number of Russian research reactors. The work with the Murmansk Shipping Company represents the first time the Americans have carried out security work on board vessels and in harbour areas. The body which works most closely with the Americans is the Kurchatov Institute.

For trafficking purposes, spent fuel would be less attractive than new because of its high radioactivity. However, the fuel poses other threats: the danger of accidents, unintended criticality and possibly sabotage means that the possibility of radioecological consequences cannot be excluded. The effects can be serious locally and any atmospheric emissions can endanger economic interests in the north, while marine emissions can be a threat to Norwegian fishing interests in the Barents Sea.

The Norwegian Radiation Protection Authority, together with the SKI, has begun to cooperate with the Murmansk Shipping Company in the areas of accountancy and physical protection. The joint Nordic approach seeks to forestall accidents, sabotage or pollution in connection with the use, processing or storage of fuel, and is thus 'broader" than the American approach. Cooperation between all interested parties is, however, of vital importance and the work is closely coordinated with the American operation. The players draw on each other's experience and the security and detector systems and radio communications equipment on the ships and in the port area must be compatible and capable of integration. The project is at an early stage, but the ideal would be for the same camera/surveillance equipment and locking and alarm systems to be used on all ships. The design and selection of equipment packages is planned as a joint American–Russian-Nordic undertaking, but as yet no specific equipment lists have been produced. Equipment deliveries and installation are scheduled to begin in May 1997.

MSCO is also interested in an accountancy systems for its fuel and the Swedish consultants AMC are already working on producing a suitable system. It is anticipated that the software will have been fully tested by the end of April 1997. Training, hardware and support are included in the package.

Norwegian Action to Prevent Smuggling of Fissile and other Radioactive Material

Norway has not traditionally maintained any defence against he smuggling of nuclear material. Until very few years ago, the border was very quiet and little stress was laid on civil and military nuclear installations and the associated nuclear material. The 1993/94 report to Parliament on nuclear activities in areas adjacent to Norway did, however, mention the risk of such smuggling in the context of non-proliferation. The following measures were taken in the light of this perceived threat:

- *Establishment of a Contact Group*

In the spring of 1995, the "contact group on smuggling of fissile and other radioactive material" was set up with representatives from the Norwegian customs administration, police, armed forces and Radiation Protection Authority. The Radiation Protection Authority provides the secretariat and directs the project of which this group forms part, inter alia with respect to the Norwegian action plan for nuclear issues relating to northwest Russia. The group today has responsibility for the introduction of countermeasures against nuclear material smuggling in Norway; its members work well together and have established good inter-departmental communication. There is thus a clear picture of what expertise is available in each department.

- *Establishment of Norwegian measurement facilities to combat smuggling of nuclear material.*

These arrangements are based on three lines of defence. The first will typically be the individual officer, the second, the specialised staff deployed following warnings or other indications of smuggling, and the third would be any laboratory examination of specimens carried out in Norway or abroad.

First Line

For some years, the customs administration has had Geiger-Müller counters known as "Automess equipment", which is easy to use and indicates quickly and in a simple form whether radiation in excess of the background level is present. Plutonium and high enriched uranium will be ducted if unsealed, but it is unlikely that smuggling of such materials would be attempted in the unsealed condition, passing openly through the controls. Simple metal shielding disguises low-level gamma radiation, while water or paraffin screens neutrons. Strong gamma emitters such as Co-60 and Cs-137 will always be sealed and in many cases it could be easier to detect lead shielding because of its high density than a signal on the Automess equipment.

Other instrumentation is now available besides the Automess equipment currently in use and the Radiation Protection Authority is in the process of checking out these options, which are recommended e.g. by the IAEA. These include various types of scintillation detector without energy resolution in addition to Geiger-Müller tubes.

Second Line

The Radiation Protection Authority considers the obvious solution at this level to be to station one measurement and verification unit at Osterås and another at its Svanhovd field office. Both units today have the necessary expertise to undertake such anti-smuggling duties. Two 35 per cent coaxial p-type Ge detectors and two NaI detectors with 4 × 4 crystal and associated electronics have been purchased. The software has in part been developed at the Radiation Protection Authority. The equipment is portable, can be operated by a single person and will improve the capability for detecting plutonium and high enriched uranium.

Third Line

The possible origins of the material can be determined by means of isotope analysis and the IAEA's collection of "fingerprints" from various producers. This third line could be included in the Radiation Protection Authority's general laboratory procedures for analysis of specimens of radioactive material and would encompass any analysis of specimens in foreign laboratories.

Seminar at Svanhovd

In June of this year, the Radiation Protection Authority arranged a seminar at its Svanhovd field office near Kirkenes, one of the subjects being smuggling of nuclear material. It was attended by representatives from the Norwegian contact group, the local customs authorities and police in Kirkenes, the Swedish and Russian radiation protection authorities and the IAEA. It was thus possible to compare notes on national and international efforts to prevent smuggling of nuclear material, to detect attempts at smuggling by means of measurements and to assess the risks of such trafficking in the northern region. One of the conclusions was also that contact should be established with the Russian customs authorities, which have been given responsibility for controlling the frontier, and this will be arranged through the Norwegian contact group.

Manual

A manual has been written for distribution to members of the defence forces (primarily the coastguard service), the police and customs to allow an effective network of personnel to be built up who are capable of detecting any attempted smuggling. The manual is intended as an aid to coordinating action to counter what is internationally recognised as a steadily growing threat: smuggling and illicit use of fissile material. The main aim is to prevent proliferation of nuclear weapons and nuclear explosives.

An abridged version of the manual is split into two parts. Part A contains specific instructions on the procedure to be followed on receipt of tip-offs or discovery or seizure of the materials concerned. Part B contains background information to make it easier for other user to come to grips with the difficult problems which the manual seeks to address.

Training

Customs and police officers are being given training as an effective way of sharpening awareness and building up expertise as regards radioactivity in general and smuggling of nuclear material in particular. This is done by means of in-service and induction courses and is specially geared to persons working on the Norwegian/Russian frontier.

25

Summary Report of the International Technical Working Group (ITWG) on Nuclear Smuggling 10–11 June 1997

L. KOCK

The third meeting of the International Technical Working Group Nuclear Smuggling (ITWG) was held at Villa Olmo in Como, Italy, on 10–11 June 1997. Arrangements for the meeting were kindly provided by Paolo Cotta-Ramusino and Maurizio Martellini of the Landau Network-Centro Volta and USPID and by Carlo Bui (Polizia, Rome). The meeting agenda was organised by Lothar Koch and Sidney Niemeyer.

Opening remarks were made by Lothar Koch (European Institute for Transuranium Elements), Carlo Bui (Polizia, Rome), Maurizio Martellini (University of Milan), and Joaquin Gonzalez-Ducay (European Commission, External Relations Directorate). Mr. Martellini made the point that nuclear smuggling is an important issue for a nculear-free zone country, and he belived that the work of the ITWG could help to bring attention to methods for combating the problem. Mr. Gonzalez-Ducay encouraged the ITWG to continue its work as it is an important part of helping to deter nuclear smugglers. He also advised that the ITWG continue to develop nuclear forensics and avoid overlapping work with other technical groups. Responsibility for chairing the meeting was then given to Sidney Niemeyer (Lawrence Livermore National Laboratory). He first gave an update on ITWG activities since the last meeting, e.g. distribution of the synopsis of the previous ITWG meeting; a summary paper given at the International Conference on Materials Protection Control and Accounting held in Obninsk, Russia; and a presentation by Nikita Nilciforov at the same International Conference on a broader programme for countering nuclear smuggling. The steps taken to prepare a new draft of the ITWG Terms of Reference were also described, and it was further noted that a planned task for this meeting was to discuss the possibility of expanding the ITWG tasks to include subjects that are not covered under the strictest definition of nuclear forensics. Another important topic would be to idnetify means for helping the ITWG to become more task-oriented and to work together between meetings.

The upcoming International Nuclear Forensics Round Robin was presented by Gordon Dudder (Pacific Northwest National Laboratory) and Georges Herbillon (EURATOM). They relayed that the formal invitations for participation had been sent out, and they provided information concerning the time-line for the exercise and steps that needed to be taken in order to participate. A major problem identified as slowing down the progress with the exercise was all of the considerations related to shipping SNM. Both U and Pu oxide samples will be shipped for the exercise; only one set of samples will be shipped to each participating country. Euratom will ship the samples but the receiver will have to pay any transportation costs. The results are intended to remain anonymous by using code names. The point was emphasised that the key goal for the exercise is to provide a learning experience, *and* it is *not* a performance evaluation. The objectives are to prioritise forensic techniques, evaluate attribution capabilities, and examine the utility of existing data-bases. In the discussion it was pointed out that the exercise did not address the field assessment aspects fo nuclear forensics, rather it was a laboratory exercise that is focused solely on source attribution. It was also noted that some participants will have access to databases that others do not, so in each participant's evaluatiion of their results, they can identify what types of information would help them to go further with their interpretation.

Two, law-enforcement talks presented their approach to forensics and its relation to the special case of nuclear forensics. Graham Mason and Sandy Mackenzie (New Scotland Yard) described the rather extraordinary lengths they have taken in order to respond to court challenges regarding possible cross contamination when trace evidence is used. Trace evidence is important for forensics, but its use requires that one can give a credible account that contamination is not a problem. Effective use of cordons is a key element for protecting people and the forensic evidence. They use sterile kits that are prepared ahead of time for collecting evidence. An example of the measures taken to prevent contamination is that potential suspects are transported in vehicles from a non-police fleet, the interior surfaces are covered with plastic, and the suspect is also placed in protective clothing to prevent contamination.

Tom Jourdan (Federal Bureau of Investigation) next discussed the step-by-step process they envision in a nuclear smuggling case in which classical forensics would be performed along with the special measurements on the nuclear materials themselves. He described how in the U.S. nuclear forensics will be performed using a formal working relationship between the FBI and the Department of Energy. For an intact suspect package, the FBI will send an Evidence Response Team and DOE will also send a team of experts. Current plans then call for the package to be sent to a DOE lab. The FBI and DOE will then jointly develop the plan for exploring the package and collecting evidence. A critical feature is to proceed step-by-step, collecting evidence at each step, and doing so in a manner that does not interfere with later steps.

Sara Scott (Los Alamos National Laboratory) presented an overview of in-field instrumentation that could be used for the on-site portion of nuclear forensics. The talk provided information regarding the types of equipment that are available differences between the commercially available equipment, and trends in development of new fieldable equipment that oculd be applied to the nuclear smuggling problem. Many fo the detectors are based on NaI, but the packaging and accompanying analysis codes vary greatly in their sophistication and cost. For the most cost-effective solution, users need to carefully evaluate their needs to avoid buying more capability than they really require. Then more sophisticated (and more expensive) detectors were reviewed, including high purity Ge detectors, that enable much more reliable identificatiion of radionuclides. Other technologies were also touched upon, e.g. ion mobility measurements in the field.

Vladmir Minkov (Argonne National Laboratory) gave an informal presentation on a Russian/DOE collaboration on "Second Line of Defence Technologies". The second line of defence focuses on finding or detecting nuclear materials once they are illicitly taken from a nuclear facility. The four major tasks are:

1) defining the requirements.
2) conceptual design for technologies,
3) training the experts, and
4) preparing a model system.

This programme is interested in exploring coordination links with the ITWG.

Slawomir Sterlinski (Laboratory for Radiological Protection) discussed the "Polish Prevention System Against Illicit Trafficking of Radioactive Substances and Nuclear Materials". This system consists of three elements: Registration systems (legal regulations), gamma detection at the borders, and laboratory analyses. They now have 80 "radioactive gates" that consist of two large gamma spectrometers; they also use hand-held detectors for assessments. A notable feature of their experience is the high number of incidents where alarms were set off at borders and the corresponding number of "turnbacks" resulting from these alarms (19,000 interventions with 640 turnbacks in 1996). The number of incidnets has increased dramatically during recent years, but it was also noted that the detection capabilities have improved over time.

Sidney Niemeyer presented a talk on "Technologies for Detecting Illicit Materials and Identifying Their History Using Geolocation Clues". He emphasised the need to develop these detection technologies in the context of a systematic approach. A first step is to develop strategies for deterring smuggling through detection at critical locations; a pathway analysis can provide the basis for evaluating detector performance trade-offs and for identifying deployment strategies that provide the greatest deterrent value. A multitiered approach is recommended for each checkpoint in which target objects are analysed using various approaches (passive radiation detector, radiography, density measurements, etc.) in order to assess which objects should be actively interrogated with either neutrons or gammas. An outline was given of various research issues that need to be addressed to determine the feasibility of these two active interrogration modes. Turning then to the route attribution aspect of nuclear forensics, a case was presented for analysing materials adhering to or incorporated into the surface of nuclear materials to provide clues to post production handling. One example is the possibility of exploiting the variations in $^{18}O/^{16}O$ ratios by geographical location in order to constrain where the surface oxide layers formed.

Nikolai Redin (Minatom) gave an overview of the "Russian Nuclear Material Safeguards Development." This included a review of Minatom's formation and structure, and the various government declarations and laws that have provided for an improved NTC and A.

He emphasized the Russian interest in deterring nuclear smuggling and the resulting implementation of their own programme of safeguards for nuclear materials. Based on a Novemeber 21, 1995 Federal law, "on use of atomic energy," all facilities developed much documentation on how to protect nuclear materials. An October 14, 1996, declaration described the concept of a state system for nuclear materials accounting and control (NMAC), and that on the federal level, NMAC is to be carried out by Minatom. He also noted the importance of assistance by western natioins for implementing their improved MPC and A.

Diana Kowitz (Department of Energy) presented an overview of the U.S. Nuclear Assessments Programme. The first element of the programme is the "communicated threat credibility assessment". Three types of information are provided in the final report technical, behavioural and operational threat assessments. The talk described the structure by which these tasks are accomplished, the primary customers and a description of the assessment produce and how it is communicated to the customer. A second element of the programme is a database on black market in nuclear materials. Customers include a variety of U.S. agencies, as well as the international community. A third element is "Technical Assistance" by which precise and accurate responses can be given to a wide range of questions.

The second day of the ITWG meeting was devoted to round-table discussions that focused on continued international cooperation to reveal and prevent illicit trafficking of nuclear materials, the new draft of the Terms of Reference for the ITWG, specific future tasks, and corresponding action items. Lothar Koch started the day by revieweing the justification for this group, and he suggested the need to discuss increasing the number of states participating in the effort (in fact, Finland, Sweden, and Norway were invited to meeting, and Elina Martikka from Finland was able to attend despite the last-minute invitation). He also noted that he is gradually learning about various support programmes that are relevant to the ITWG work, and suggested that the ITWG should take steps to learn more about these programmes. He presented a three-step process to accomplish this task: inventorying ongoing technical support programmes, making suggestions for harmonisation, and identifying gaps that supplementary proposals could address. These suggestions prompted a debate on the feasibility of the ITWG taking on such potentially wide-ranging tasks. It also prompted a variety of new suggestions for the ITWG, e.g. prepare a brochure showing how materials might appear in nuclear smuggling cases by using pictures provided by a variety of countries.

Another suggestion that was prompted by the previous day's talks was to develop principles that guide the interplay between classical and nuclear forensics. This work should also address specific issues for law enforcement, e.g. can we raise a fingerprint on nuclear material itself, is DNA evidence destroyed by long exposure to radiation from nculear material? A related suggestion was to develop a model for a coordinated "action plan" that includes detection, seizure material (i.e. weapons grade, reactor fuel, etc.) and external specialised technical support to identify the origin of the seized material. Most of these steps should and can be done at the site of the seizure. The most reliable categorisation may require a quick analysis at a laboratory, although future developments may enable this task to be done reasonably well in the field. In any case, the categorisation should be accomplished quickly (24 hours) so the appropriate government agencies can be notified. In practice, the third, much more complicated set of laboratory analyses, can be conducted over a much longer time scale.

During these discussions Elina Martikka (Finnish Centre for Radiation and Nuclear Safety) described their cooperative approach with Russia, Ukraine, and the Baltic states to protect their common borders. The hardware includes automatic gate monitors for trains and vehicles at crossing points on the border, monitors in airports and harbours, radiation search monitors, and radiation dose monitors. Training of both the FBI and Russian border personnel is an integral part of the programme and includes basic radiatioin training, as well as training in the measurement of nuclear materials.

In the next segment of the round-table, the draft of the revised ITWG Terms of Reference was discussed. The comments clearly showed that there is fairly broadspread concern with expanding the scope of tasks of the ITWG at this time. It was agreed to remove the listing of other nuclear smuggling programme elements in the first paragraph. An added difficulty is that the definition of the term "nuclear forensics" has not clearly been made and, therefore, difficulties were encountered with respect to the identifying the scope of the tasks that the ITWG should undertake in the future. A number of participants also spoke to the importance of having a relationship with the P-8, or its NPEC, to provide an umbrella for the ITWG. We need a realtionship that provides support and endorsement for the ITWG, but it should not be a controlling relationship. The issue of whether we are a governmental or non-governmental working group was also raised. While this issue was not resolved, the group agreed that our meetings should not be public meetings, and that the participatioin by the attendees should be supported by their respective governments or organisations. The group also wanted a stronger statement regarding the need to avoid duplication with other technical groups. It was also noted, though, that the ITWG has provided a forum for people with different experiences and backgrounds who share information, and that this feature is one of the distinguishing and valuable characteristics of the group. The communities involved include scientists and several user communities. Numerous other specific revisions to the draft were made and recorded by the meeting chair; in addition, two written inputs were given to the chair after the open discussion.

Based on all this input, the meeting chair agreed to generate a new draft that incorporates the requested changes. This new draft will then be sent to all participants for their review, which hopefully will bring the ITWG to the point that it can seek the endorsement of the governments and organisations of the ITWG participants.

The meeting concluded by summarising the various suggestions that were made during the day for future tasks for the ITWG. Volunteers for coordinating some of these efforts were also identified; it is hoped that other participants will also be able to help with these various tasks during the next year. The tasks identified at the meeting were:

1) Pursuing an ITWG presentation and participatioin at the upcoming Vienna meeting sponsored by the P-8 (Niemeyer).
2) Prepare an inventory of plans for nuclear forensics in various countries (Koch).
3) Inventory the broader technical issues in nuclear smuggling efforts and identify technical areas not being addressed by other groups.
4) Develop the first version of a model (or preferred) actio plan for nuclear forensics (Jourdan, Dudder).
5) Investigate means for assisting recognition of possible illicit radioactive material trafficking (for example instrumentation, handbook, training video, workship for border personnel, etc.) (Scott; Thompson).

Finally, perliminary plans for the next meeting were made. Possible locations for the next meeting are the UK and Finland. The preferred time frame is May-June 1998. Possible agenda items for this meeting are:

1) Discussion of the Internatioinal Round Robin Analysis (likely to be the primary agenda item).
2) Initial reports on tasks undertaken since this meeting (definition of tasks, objectives).
3) Update on experiences in actual nuclear smuggling incidents.
4) Portable detector demonstrations.
5) Progress in the science of nuclear forensics.

The meeting chair agreed to provide a written summary of the meeting. Copies of some of the formal presentations are also available upon request.

Address:

LOTHAR KOCK
Nuclear Unit, EITEJRC
76125 Karlshrue
Germany

PART — 4

PRESENT RESEARCH IN THE FIELD OF NUCLEAR WEAPONS, THE FUTURE OF NATIONAL LABORATORIES AND LAB-TO-LAB COLLABORATION

26
The Stockpile Stewardship Programme

PAUL S. BROWN

INTRODUCTION

In the 1990s, the U.S. conducted its last nuclear explosive test and halted the production of new nuclear warheads. Thus ended an era in which the U.S. modernised its nuclear weapons stockpile by the steady replacement of aging systems with new systems and in which nuclear testing served as the ultimate arbiter of the safety, reliability, and performance of the nation's nuclear weapons stockpile.

With the decision to cease production of new nuclear warheads and to end nuclear testing, the U.S. now faces the challenge of maintaining its existing nuclear weapons stockpile with other tools and different kinds of tests. To meet this challenge, the Department of Energy has developed the Stockpile Stewardship Programme.

Significant progress has been made in the implementation of the Stockpile Stewardship Programme. A Record of Decision on the Programmatic Environmental Impact Statement was issued in December 1996, establishing the architecture for the future U.S. weapons complex. An implementation plan for the Stockpile Stewardship Programme has been developed and is already in its second annual revision. The first two Annual Certifications of the stockpile have been completed. Many new capabilities and facilities for the Stockpile Stewardship Programme are well under way. For example, the Dual Axis Radiographic Hydrodynamic Test Facility is being constructed at Los Alamos, and construction of the National Ignition Facility has begun at Livermore. Industry has begun delivering the advanced computers required for the Accelerated Strategic Computing

Initiative, and record-breaking teraflops operation has been demonstrated on the first of them. Life extension programmes for the enduring stockpile are being developed, and the dismantlement of U.S. nuclear warheads retired from the stockpile is continuing. We are already applying the principles of stockpile stewardship to a number of stockpile systems, and we are on track to restoring the nation's capability to produce tritium.

In this paper, which has been extracted and edited from a Department of Energy publication[(1)] that LLNL helped write, we present an overview of the current programme and highlight some of the accomplishments and progress made to date.

PROGRAMME GOALS

"As part of our national security strategy, the United States must and will retain strategic nuclear forces sufficient to deter any future hostile foreign leadership with access to strategic nuclear forces. In this regard, I consider the maintenance of a safe and reliable nuclear stockpile to be a supreme national interest of the United States."

"I am assured by the Secretary of Energy and the Directors of our nuclear weapons labs that we can meet the challenge of maintaining our nuclear deterrent under a Comprehensive Test Ban Treaty through a science-based stockpile stewardship programme without nuclear testing."

President Clinton, August 11, 1995.

The Department of Energy (DOE) programme for stockpile stewardship has the fundamental objective of maintaining high confidence in the safety, performance, and reliability of the U.S. nuclear weapons stockpile. However, the conditions under which we must accomplish this objective have changed greatly since the early 1990s. DOE is meeting and will continue to meet that challenge by using non-nuclear experiments and computer simulations in lieu of nuclear testing. The changed conditions include:

- ***No nuclear testing :*** Without nuclear testing, we no longer have a way to do integrated, direct tests of warhead operation. Without such testing, we have also lost the ultimate measure and demonstration of warhead performance and weapons scientists' competence. Under the CTBT, nuclear testing would be used only as a last resort, requiring invocation of the "supreme national interest" clause, if all other means prove insufficient to resolve a serious stockpile issue.

- ***Tritium production*:** The U.S. has not produced tritium for warheads since 1988 and will require a new tritium production source possibly as early as 2005. A dual-track approach using commercial light water reactor and accelerator production of tritium is being vigorously pursued.
- ***An aging stockpile* :** Nuclear warheads are not static objects. Materials change over time (e.g., radioactive decay, embritlement, corrosion). Some changes do not adversely affect warhead reliability or performance, but others have done so in the past, and some changes and their impacts on reliability, safety, and performance are yet unknown. With the average age of the stockpile older now than it has ever been in the past (including a number of warheads approaching the end of their originally anticipated deployment), it is expected that previously unencountered aging-related problems will arise. To meet this challenge, each stockpiled warhead is undergoing a thorough assessment to determine vulnerabilities and to establish refurbishment schedules that will ensure stockpile life extension.
- ***An aging cadre of stockpile stewards* :** Many of the scientists and engineers with actual weapons design, production, and test experience have already retired, and most of those remaining are within ten years of retirement. A new generation of weapons scientists and engineers must be trained and their competence validated before the current generation leaves the work force. Knowledge preservation programmes are under way that include video-recording the experiences of senior designers as well as mentoring of new stockpile stewards. The senior designers are also reviewing achieved nuclear weapon test data and showing the future stewards how to interpret and extract useful information previously not needed when nuclear tests were conducted.
- ***A smaller, less diverse stockpile:*** The nation's stockpile now has fewer warheads and fewer warhead types than any time since the 1960s. Further reductions are expected as we proceed to START II and the next strategic arms

reduction treaty. Thus, the stockpile is more susceptible to common-mode failures—for example, defects arising from a refurbishment-fabrication process or material used in several warhead types or failures that affect warheads with similar design features.

- ***No requirements for new-design nuclear warhead production***: Without new production programmes, warheads will remain in the stockpile well beyond their anticipated lifetimes and beyond the base of experience. Without requirements for new warheads, existing warheads will be refurbished and modified to extend their stockpile lifetimes. To address this challenge, the capability is being maintained to design and fabricate replacement warhead parts, as well as to design (but not produce) replacement warheads for existing stockpiled weapons.
- ***A reconfigured production complex:*** The production complex of the Cold War years is being downsized and consolidated. The future complex, with its reduced capacity and capability, will not be configured for high-rate production programmes. Thus, improved manufacturing processes, including the integration of system design, component design, and process development, will be needed to achieve timely production at a reduced cost. To address this challenge, a Programmatic Environmental Impact Statement was developed that specifies significant reductions in the size of the DOE production complex and the development of an agile, capability-based manufacturing enterprise that will use advanced design and production methods to respond to both normal and contingency requirements.

The focus of the Stockpile Stewardship Programme is the U.S. nuclear stockpile. All elements of the programme are directed at ensuring the safety, reliability, and performance of this stockpile. We will develop the fundamental understanding needed to ensure our ability to anticipate and fix problems, to deal with future unknowns that affect stockpile safety or reliability, and to be prepared to demonstrate that our assessments and evaluations are accurate and credible.

At the heart of the Stockpile Stewardship Programme is the issue of confidence. The programme must provide three types of confidence—in the weapons themselves, in the systems and infrastructure that maintain the weapons, and in the judgment of the people who assess the weapons. These three elements of confidence are tightly linked and must be effectively integrated to provide overall confidence are tightly linked and must be effectively integrated to provide overall confidence in the nation's nuclear deterrent. The ultimate measure of success of the programme is to certify that the stockpile remains safe and reliable without nuclear testing. The first two annual certifications of the stockpile have been completed, and both the DOE and the Department of Defence (DoD) have concluded that the stockpile is safe and reliable.

PROGRAMME STRATEGIES

The goals of the Stockpile Stewardship Programme will be achieved through an integrated process of surveillance, assessment, certification, and manufacturing. Today's programme is characterised by three integrated strategies, or phases, of stockpile stewardship:

- ***Surveillance:*** *predicting and detecting problems.* We must identify defects and aging-related changes before they degrade warhead safety, performance, or reliability. To the extent possible, we must predict—relying on experiments coupled with computer modeling and simulation—the occurrence and impact of changes—both the kinds of changes that have been dealt with previously and changes that have not been encountered before.
- ***Assessment:*** *analysing and evaluating effects of changes on warhead safety and performance.* We must assess the effects of identified and predicted age—and environment-related changes in stockpile warheads to determine whether or not the changes adversely affect safety, performance, or reliability. We must determine if the degradation is severe enough to require the replacement or rebuilding of warhead components or even entire weapons. We must also evaluate new materials, new fabrication techniques, and new manufacturing processes to make sure they are functionally equivalent to the originals.
- ***Design and Manufacturing:*** *refurbishing stockpile weapons and certifying new parts, materials, and processes.* We must periodically replace limited-lifetime components (e.g.,

tritium reservoirs, neutron generators) and rebuild or manufacture parts to replace those that have or will soon experience detrimental aging-related changes. We must also certify that the new components do not introduce defects that degrade warhead safety, performance, or reliability.

A distinguishing feature of the Stockpile Stewardship Programme is its integration. Not only are the laboratories and plants working closely together, particularly for surveillance and manufacturing, but the activities themselves are tightly interconnected. In addition, the Department of Energy's laboratories and plants work closely with the Department of Defence and the military services to make sure that the enduring U.S. nuclear stockpile meets its mission requirements.

The stewardship process is continual, with no clear ending of one phase before the beginning of another. Assessment and certification pervade all activities, from surveillance through manufacturing. Likewise, computational modeling and prediction are integral to every activity, from the assessments of aging-related changes, to the design and certification of replacement components, to projections of stockpile life extension.

PREDICTING AND DETECTING PROBLEMS: THE SURVEILLANCE PROGRAMME

Stockpile surveillance has been a major component of the U.S. nuclear weapons programme ever since the first weapons were put into the stockpile. Since 1958, approximately 14,000 weapons have been examined and subjected to a variety of non-nuclear laboratory experiments and flight tests. In cases where laboratory tests could not provide conclusive answers, nuclear tests of stockpile warheads or warhead components were conducted.

Problems requiring corrective action have arisen in nuclear and non-nuclear warhead components. All warhead types in the enduring U.S. stockpile have had repairs or retrofits, and several have required repairs to the nuclear package. Without replacing older warheads with new ones, the stockpile will age beyond our experience base. There have never before been large numbers of 30-, 40-, or 50-year-old warheads in the U.S. stockpile. (The typical age of a stockpile warhead has been less than 13 years). As a result, we expect that previously unencountered changes and problems will arise in the older warheads.

To succeed in this new reality, we need new surveillance methods and enhanced predictive capabilities so that we can detect the full range of problems that may arise in the enduring stockpile. We also need to be able to anticipate and identify aging-related changes and understand the significance of these changes and their effect on warhead safety and performance. Some changes have little or no effect, whereas others can make a major difference.

Defects occur throughout the life-time of a warhead. Typically in complex manufactured systems, initial defects associated with design or fabrication form a large fraction of all defects found. With a high sampling rate during the early years, these defects can be detected and corrected. During middle age, the defect rate typically declines to a lower but non-zero level. As a system ages and components deteriorate, the defect rate climbs. The U.S. nuclear stockpile has followed the pattern of these first two stages. However, we have essentially no experience with the third stage and must develop the capability to predict when it will be reached.

The goal of enhanced surveillance is to anticipate or detect the precursors and onset of aging-related defects before they jeopardise warhead safety or reliability. Predictive modelling and simulation are central to our ability to assess warhead aging and the effect of potential defects on warhead performance. With sufficient lead time, we can make the necessary redesigns, refurbishments, and recertifications efficiently and cost effectively as we work within the capabilities and capacity of the downsized production complex. The alternative to this approach is a large costly manufacturing complex that can fix problems quickly when they occur.

An Enhanced Surveillance Programme has been established to develop the technologies and methods as well as the fundamental understanding of material properties and physical principles to significantly improve our detection and predictive capabilities. The major activities to be pursued are:

- Testing and researching the age-related behaviour of existing stockpile materials, components, and systems, including those from retired warheads.
- Developing improved computational models of materials aging and materials performance.

- Developing and conducting high-fidelity non-nuclear tests to examine the behaviour of nearly all actual warhead components in realistic flight environments. (Historically, most flight tests have been performed using only a limited number of warhead components from the stockpile).
- Developing techniques for advanced analysis of existing surveillance data, including complex numerical models and simulations as well as improved access to and analysis of archieved data.
- Developing sensors to detect material failure in stockpile warheads for those mechanisms whose failure times cannot be adequately predicted.

As these enhanced surveillance technologies and methods are prototyped and validated, they will be integrated into the core stockpile surveillance programme. Improved predictions of component lifetimes, made possible through the Enhanced Surveillance Programme, are key to the strategy for extending indefinitely the life of stockpile warheads.

ANALYZING AND EVALUATING: THE ASSESSMENT AND CERTIFICATION PROGRAMME

Data and test results from many sources must be analysed, assessed, and evaluated before conclusions can be drawn as to the safety, performance, or reliability of stockpile warheads. The Assessment and Certification Processes are designed to:

- Develop an understanding of and an ability to predict the lifetime and aging-related changes that occur in every warhead component.
- Identify and understand those variables, processes, and changes that really matter in terms of warhead safety, performance, and reliability.
- Validate the new experimental and computational tools.
- Validate new manufacturing processes and materials to ensure that rebuilt parts and warheads are functionally equivalent to the originals.
- Develop and demonstrate the judgment of the next generation of weapons scientists and engineers.

The science and engineering of nuclear explosives are extremely complex. There are many parameters and unknowns that greatly influence the performance of nuclear warheads.

Some of these have, in the past, been identified only in nuclear test failures. Even when nuclear testing was permitted, we were never able to test nuclear warheads to a statistical certainty. In addition, various testing constraints (e.g., the Threshold Test Ban Treaty) required extrapolations to evaluate full-warhead performance and safety characteristics. The key to accurate extrapolations, then as now, is the expert judgment of our weapons scientists.

Confidence in the accurate judgement of weapons scientists and confidence in the safety and reliability of the U.S. stockpile are closely linked. In the past, a weapon scientist's judgement was developed and validated through nuclear testing and new warhead development. The Stockpile Stewardship Programme is developing other means for honing and demonstrating the expert judgement of the next generation of stockpile stewards.

We are accomplishing this through a balanced and integrated programme of computational simulation, fundamental scientific research, and non-nuclear experiments. In particular, we are designing experiments that test and expand the boundaries of our understanding. There are many areas of warhead operation that cannot be adequately addressed with the tools and knowledge we currently possess. To close these gaps, the Stockpile Stewardship Programme is making significant investments in enhanced computational capabilities and advanced facilities for above ground experiments.

Of particular concern is the assessment challenge posed by the unrecognised problem—the "unknown unknown". We must have rigorous experimental and computational processes that not only confirm and extend what we know and expect, but also fill in gaps in our current understanding. This ability to fill in the gaps is especially important in those areas where previously we would have used nuclear testing, and where a successful tests (that is, one in which the nuclear explosive device worked) would have clearly defined the margins of our concerns. Therefore, we need an aggressive verification and validation process for both our tools and our results.

The Stockpile Stewardship Programme provides for demonstration-based assessment and certification of warhead safety and reliability. In the absence of nuclear testing, we must rely on different experiments and tools to obtain data relevant to nuclear warhead performance. We have identified a suite of enhanced capabilities that we believe are needed to fill in the gaps of our knowledge and provide data relevant to various stockpile concerns. Advanced experimental facilities will provide high-fidelity data on the stages of the nuclear explosion—primary implosion, boost, primary-to-secondary coupling, weapon effects, etc. Wherever possible, our goal is to obtain data experimentally by at least two different methods.

Under the Stockpile Stewardship Programme, computational modelling and numerical simulation provide the critical integration of theory, existing data and knowledge, new experimental results, and predictions into results that can be verified and validated. Advanced computational capabilities (application codes, computers, and various tools and techniques) are being developed under the Accelerated Strategic Computing Initiative and incorporated into ongoing stockpile computational activities. Numerical simulations will both drive the experimental programme and be driven by it. Our goal is to combine past nuclear test data, computational modelling, and new data from advanced experimental facilities to fill in gaps in our current knowledge and extend our scientific understanding in other vital areas.

To assess aging-related changes that occur in nuclear and non-nuclear warhead components, we need complex three-dimensional computational simulations that are beyond our current computational capabilities. Through the Accelerated Strategic Computing Initiative, we are developing enhanced capabilities. For example, increases of more than ten-thousand fold in computational speed, network capacity, and data storage are planned to provide simulations of weapon safety and performance of increased detail and complexity. These efforts are closely linked to experiments to validate new and evolving computer models and provide improved physics data.

These new capabilities will be used in addition to the experimental and computational capabilities developed during the nuclear testing years. However, because these older tools were designed to complement nuclear testing, they are not, in and of themselves, sufficient in the absence of nuclear testing. As new facilities and capabilities come on line and are validated, we will incorporate their data into our assessments. Our goal is to accomplish this transition in less than ten years.

Peer review is a key component of stockpile stewardship. Because assessment and certifications of stockpile safety, reliability, and performance rely to so heavily on expert judgement, it is essential that the judgement of our weapons scientists, and the assessment process itself, be vetted and validated. Formal and informal peer interactions take place among the three weapons laboratories. In addition, periodic independent reviews by outside experts help provide confidence in the credibility of the laboratories' assessments and in the process by which the assessments are made.

REFURBISHING AND RECERTIFYING: THE DESIGN AND MANUFACTURING PROGRAMME

Nuclear warheads are not static objects. They contain radioactive materials that decay and organic materials that decompose with time. Some materials, like tritium, decay rapidly and must be replaced every few years throughout the warhead's lifetime. In addition, radioactive decay produces heat and decay products that cause changes in the radioactive materials themselves and in adjacent materials. For example, plastics and other organic materials change with age and exposure to heat and radiation. Many of the metals used in nuclear warheads are chemically reactive and are damaged by long-term exposure to radiation. As a result, all warhead parts must be considered limited-lifetime components, and all warheads in the enduring stockpile will require periodic refurbishment and remanufacturing.

With an improved understanding of the effects of aging on warhead safety, performance, and reliability, developed through our enhanced surveillance and assessment efforts, we will be able to take a proactive approach to refurbishment. Our goal is to replace or fix components systematically, before aging-related changes jeopardise warhead safety or performance.

We have instituted the Stockpile Life Extension Programme (SLEP) to extend the life of a number of stockpile warheads. The SLEP provides the framework for our research and development activities and our production planning. A number of specific life extension programmes are being defined for each warhead type, allowing the laboratories, plants, and the DoD to anticipate and plan for future maintenance and refurbishment requirements.

The schedule guides stockpile-related research and development–at the laboratories to design and certify replacement components and validate new materials, and at the plants to develop and certify new manufacturing processes.

ENSURING CONFIDENCE IN THE SYSTEM: INTEGRATED PROGRAMME MANAGEMENT

The Stockpile Stewardship Programme is an integrated set of activities performed by an integrated complex of laboratories and plants. The technical challenges involved, combined with the downsizing of the production complex and the consolidation of activities at the laboratories, create the need for seamless, effective, and efficient program management. Indeed, programme management lies at the heart of the new paradigm for stockpile stewardship.

The laboratories and plants are collaborating with each other and with U.S. industry to develop enhanced surveillance, advanced manufacturing and computer simulation and mode lling capabilities, and tritium production technologies. Once the technologies and capabilities are developed and validated, integration and collaboration will continue as surveillance results are evaluated, replacement parts and manufacturing processes are designed concurrently, and refurbished warheads and components are certified.

An essential element of maintaining confidence in the stockpile—and in the system that maintains the stockpile—is the informed, vigorous, and skeptical set of interactions that take place among the laboratories, the DOE, and the DoD and its advisory groups. These include:

- ***Peer Review***. In scientific research, peer interaction and review are essential for maintaining excellence and providing confidence in the quality of the work. In the absence of nuclear testing, the need for peer review in stockpile stewardship and management is greater than ever before. Vital peer interactions take place through integrated and collaborative activities among the laboratories and through formal reviews of independent activities.
- ***Dual revalidation.*** This formal review process was developed in consultation with the Department of Defence. Teams from Los Alamos and Sandia-Albuquerque and

from Livermore and Sandia-California independently evaluate the safety and reliability of each warhead. The teams perform the evaluation, one team with personnel from the laboratories that originally designed thewarhead and the second team with personnel from the other laboratories. The teams independently review existing calculations and experiments pertaining to the warhead, evaluate relevant stockpile surveillance results and predictive analyses, and conduct separate programmes of experimental and calculational work to investigate issues of concern and improve the baseline of understanding.

- ***Stockpile life extension (SLEP).*** The Stockpile Life Extension Programme addresses the need to extend indefinitely the life-time of existing warheads. It is designed to balance the concern that aging-related changes will degrade warhead safety or reliability against the concern that stockpile modifications may introduce new uncertainties. Life extension options have been defined for each warhead type in the stockpile. These scheduled refurbishments also provide the opportunity, while the warheads are disassembled, to make modifications to improve safety, reliability, or longevity. All stockpile life extension activities are closely coordinated with the Department of Defence before they are initiated.
- ***Annual certification.*** The Secretary of Defence and Secretary of Energy formally certify to the President that the U.S. nuclear stockpile is safe and reliable and that no nuclear testing is required. This certification is based on rigorous technical analyses that lead to formal concurrence by the Nuclear Weapons Council, the Directors of the three nuclear weapons laboratories, and the Commander in Chief of the U.S. Strategic Command.

Other important interactions take place between the Department of Defence and the Department of Energy on issues related to stockpile safety and security. In addition, the Nuclear Weapons Council (NWC) carries out executive decisions on stockpile actions. The NWC, which is supported by a highly knowledgeable staff of military officers, members of the Office of the Secretary of Defence, and the DOE's Assistant Secretary for Defence Programmes, also provides valuable review of DOE's stockpile plans.

STOCKPILE STEWARDSHIP PROGRAMME ACCOMPLISHMENTS

Almost six years have passed since the last U.S. nuclear test and almost seven years since the last new warhead entered the U.S. stockpile. The decision to end nuclear testing and new warhead production significantly changed the way in which the U.S. maintains the safety and reliability of its nuclear weapons stockpile. As the accomplishments highlighted below illustrate, much progress has been made in the development and successful implementation of the Stockpile Stewardship Programme.

Programme Architecture: the PEIS

Beginning in May 1995, Department of Energy held a series of public meetings as part of the process for preparing the Programatic Environmental Impact Statement (PEIS) for the Stockpile Stewardship Programme. Meetings were held at each laboratory and plant site and in Washington D.C. The comments, questions, and discussions arising from these meetings provided extremely useful input for refining the programme.

The Record of Decision for the PEIS was signed by the Secretary of Energy in December 1996. This document formally defines the architecture of the weapons complex for the Stockpile Stewardship Programme. It covers the future capabilities required of the three weapons laboratories, the four plants, and the Nevada Test Site. It calls for construction of several advanced experimental facilities at the laboratories, for downsizing production capabilities in place, and for reestablishing some manufacturing capabilities at the laboratories.

The weapons complex outlined in the PEIS preserves the critical capabilities and assets of the laboratories, plants, and the Nevada Test Site. It is consistent with the reduced U.S. nuclear weapons stockpile under current and projected START options. It also supports the U.S. nuclear weapons policy of "lead plus hedge", as set forth in the most recent Nuclear Posture Review (conducted by the Department of Defence and approved by the President in September 1994). With this complex, the U.S. will be able to maintain a reduced nuclear arsenal while sustaining the capabilities needed to reverse course (in terms of stockpile size, nuclear testing, and new warhead production), should future circumstances so dictate.

Implementation Plan: the Green Book

The laboratories and plants have worked with each other and the Department of Energy to lay out a detailed implementation plan for the Stockpile Stewardship Programme. This strategy is discussed in *The Stockpile Stewardship Plan*, often referred to as the Green Book. Specific roles and responsibilities have been defined and unique facilities and capabilities identified. As the programme has evolved, the strategy has been revised and modified. With an up-to-date and detailed implementation plan, jointly prepared and agreed to by the plants and laboratories, we will be able to execute the Stockpile Stewardship Programme efficiently and cost effectively, without duplication of effort, overlooked responsibilities, or gaps in necessary capabilities.

Administration and Congressional Support: the Budget

As the President has stated,

"In order for this programme to succeed, both the Administration and the Congress must provide sustained bipartisan support for the stockpile stewardship programme over the next decade and beyond. I am committed to working with the Congress to ensure this support".

The President's 1999 budget request includes a five-year plan that meets this commitment. With such budgetary support, we are proceeding with the design and construction of vital new experimental facilities and with the acquisition of the next-generation supercomputers required for the Stockpile Stewardship Programme.

Confidence in the Stockpile: the Annual Certification

A primary responsibility of DOE is to certify the safety and reliability of the nation's nuclear weapons stockpile. Stockpile safety and reliability issues are continually assessed by the Department and the Los Alamos, Livermore, and Sandia Laboratories.

As part of the Stockpile Stewardship Programme, a formal annual certification process has been established. This process incorporates technical evaluations from DOE and DoD, the Directors of the laboratories, and advice from the Commander in Chief of the Strategic Command and the Nuclear Weapons

Council. The certification that the stockpile is safe and reliable rests on the expert judgement of these senior officials.

The first two Annual Certifications have been completed. In their letters to the President, the Secretaries of Defence and Energy stated that there was no need to conduct and underground nuclear test to address safety or reliability problems.

New Capabilities

We have made major strides in the development and application of new capabilities for stockpile stewardship. Data and information from these new capabilities will be integrated with data from experiments in existing facilities and from past experiments and nuclear tests. Restoration of tritium production is essential and the development of all the new experimental facilities and capabilities are needed to provide confidence in the safety, reliability, and performance of the U.S. nuclear stockpile, now and in the future. The sum total of all these developments is needed for success.

DARHT, NIF, Subcritical Experiments, and Tritium Production

Important decisions were reached regarding two vital new experimental facilities, subcritical experiments, and tritium production.

- ***DARHT:*** The Environmental Impact Statement for the Dual Axis Radiographic Hydrodynamic Test (DARHT) Facility has been completed, and construction is under way at Los Alamos. When completed, DARHT will be the nation's most advanced facility for hydrodynamic experiments. These experiments are essential for validating the implosion performance of nuclear primaries.
- ***NIF:*** Lawrence Livermore has been selected as the site for the National Ignition Facility (NIF), engineering design work is under way, and ground was broken for construction in May 1997. This immense laser facility will provide a means for weapon physics and weapon effects experiments, and for improving the validating new physics models and computer codes.
- ***Subcritical Experiments:*** In April 1997, the Secretary of Energy announced a schedule for experiments, an essential component of the Stockpile Stewardship Programme. The first two of these experiments were conducted in 1997.

These experiments will provide valuable scientific information about the behaviour of nuclear materials during the implosion phase of warhead operation. This information is needed to predict accurately the performance of stockpile warheads as they age.

- *Tritium.* DOE is currently pursuing a dual-track approach for the most promising tritium supply alternatives: 1) to initiate purchase of an existing commercial reactor or irradiation services with an option to purchase the reactor for conversion to a defence facility, and 2) to design, build, and test critical components of an accelerator for tritium production. By late this year, the DOE plans to select one of the tracks as the primary source of tritium. The other alternative, if feasible, would be developed as a backup source. Substantial progress has been made for both alternatives.

Strategic Computing and Simulation Advances: Teraflops Operation

Advanced computational tools to model nuclear weapons performance in three dimensions and with complete physics (that is, with minimal approximations) are being developed through the Accelerated Strategic Computing Initiative (ASCI). In partnership with industry and universities, we are developing the high-performance computing platforms, network connectivity, integrated codes, and problem-solving environment that will enable numerical simulation to provide as much of the integrated understanding of stockpile health as possible in the absence of nuclear testing.

The first instalments of the advanced new ASCI computers have been delivered. Our ultimate goal is to develop 100-teraflops computers that can perform one hundred trillion mathematical operations per second. These computers will use thousands of processors, linked together using a technique known as massively parallel processing.

- ***ASCI Red:*** Intel Corp. and Sandia-Albuquerque are developing the ASCI Red machine. In December 1996, this machine demonstrated record-breaking teraflops operation—one trillion operations per second.
- ***ASCI Blue Pacific:*** IBM and LLNL are developing the ASCI Blue Pacific machine. The first instalment of this machine

was delivered in September 1996, and stockpile-related problems were run on it two weeks later.

- ***ASCI Blue Mountain:*** Silicon Graphics and LANL are developing the ASCI Blue Mountain machine. The initial delivery for this machine was made in late December 1996. In January 1997, a 9-hour test run of a hydrodynamics code was successfully completed.
- ***ASCI White:*** In February 1998, DOE announced that IBM and LLNL will develop the ASCI White machine. It is scheduled to be demonstrated in March 2000 and delivered to LLNL in June 2000. Its speed will be 10 teraflops, with a memory of 4 terabytes and a disk capacity of 150 terabytes.

Significant progress has been made by all three laboratory-industry teams in developing the problem-solving environment (the operating system, data storage and retrieval, utility and applications codes) needed to support terascale computing. Even at this early stage in their development, advanced weapons simulation codes are providing unprecedented capabilities to our weapons scientists. For example, coupled three-dimensional thermal-chemical-dydrodynamic calculations of warhead safety (for example, a warhead in a fire) are now possible. As a practical matter, some standard weapons-related calculations have been performed 100 times faster than before. Through this initiative, the time it took to run one simulation was reduced from 74 days to 7 hours.

The Academic Strategic Alliances Program was launched in July 1997. Five major U.S. universities—Stanford University, the California Institute of Technology, the University of Chicago, the University of Utah at Salt Lake, and the University of Illinois at Urbana/Champaign—will collaborate with LLNL, LANL, and Sandia on extremely challenging projects that will drive the advancement of large-scale computational modeling. The challenges posed by ASCI—and the potential civilian as well as stewardship applications of the computational advances to be gained—demand the broad collaboration of the national laboratories, academia, and industry. There will be benefits in scientific areas such as turbulence shock physics, astrophysical thermonuclear flashes, atmospheric modeling, numerical simulation of accidents involving explosives and propellants, and other basic science and civilian applications, in addition to stockpile stewardship.

Meeting the Day-to-Day Needs of the Stockpile

We have continued to maintain the U.S. nuclear weapons stockpile. Problems in the stockpile have arisen in the last five years. Some of these problems are similar to those for which, in the past, we conducted nuclear tests to investigate or resolve. However, using the stockpile stewardship approach, we are drawing on test-related expertise and on emerging new capabilities to evaluate and resolve these problems. We have made major progress in dual revalidation, stockpile life extensions, in developing new processes and production capabilities, and in transferring production from plants that are shut down to those that remain.

Warhead Dismantlement

Dismantlement of the Cold War arsenal continues, even as we are developing the enhanced capabilities and facilities needed to support to enduring U.S. stockpile. As a result of START I and separate Presidential decisions, thousands of warheads have been removed from the U.S. stockpile and are slated for dismantlement. More than 1000 warheads were dismantled in 1996, and over 500 warheads in 1997. START II will result in further reductions. In March 1997 at Helsinki, Presidents Clinton and Yeltsin agreed as part of the next START agreement to reduce strategic stockpiles even further to levels of 2000–2500 warheads, as well as to explore measures that would involve transparency of warhead inventories and their destruction.

Conclusion

Under the Stockpile Stewardship Programme, our goal is unchanged from previous years—namely, to provide confidence in the safety and reliability of the U.S. nuclear stockpile. Absent nuclear testing, the tools have changed significantly—stockpile confidence will now rely on non-nuclear demonstration-based assessments of warhead safety, reliability, and performance. New experimental and computational capabilities are being developed. With these tools, we will be able to offset the loss of critical expertise that will result from the retirement of our remaining nuclear test-and design-experienced weapon scientists and engineers. In the coming years, we will validate these new tools, integrate the information they provide with the suite of weapons-related data and models, and train the next generation of stockpile stewards. On the basis of progress made to date, we are confident that we will be successful and will be able to maintain the safety, security, and reliability of the enduring U.S. nuclear stockpile, now and in the future.

References

1. "Stockpiel Stewardship Programme—Overview and Progress", Department of Energy, Office of Defence Programmes, October 1997.

Address:

PAUL BROWN
National Security Directorate
Lawrence Livermore National Laboratory
Livermore CA 94550
USA

27

The Persistence of the Missile Defence Illusion

JOSEPH CIRINCIONE

Sir Martin Rees, the Astronomer Royal of England, notes that there have been just as many remarkable astronomical discoveries in the past two years as in any earlier period. Evidence of planets around other stars, new evidence of very distant galaxies, news that the expansion of the universe might be accelerating, even the possible discovery of a new basic force in the universe that is the reverse of gravity. The same might be said for the field of non-proliferation. Every week brings new stories of deadly chemical agents discovered in Iraqi warheads; reports of new, deadly strains useful for biological weapons terrorism; shipments of special steel from Russia halted at the Azerbaijan border before it can be used in Iranian missiles; emergency airlifts of poorly-guarded highly enriched uranium from Tbilisi, Georgia; or reports of the deterioration of Russia's command and control system for its nuclear forces. There is a proliferation of proliferation concerns.

The response should be to redouble efforts to stop the spread of these deadly weapons, including the ratification of treaties and agreements to prevent and reduce the threats. In fact, the reverse is occurring. The reaction from influential members of the US political establishment has been to call for increases in military spending, to stiffen opposition to arms control treaties and to demand the rapid deployment of new weapons systems, particularly a national missile defence system.

For example, the chairman of the Republican Party's National Committee, Jim Nicholson, this June launched a new initiative to make national missile defence an issue in the November 1998 congressional elections in the United States.

"The Republican Party is prepared to have this become a political issue," he wrote in an editorial published in the conservative newspaper, *The Washington Times*, June 21. "We are prepared to ask the American people if they agree the United States should be defenseless against weapons of mass destruction, relying instead on outdated treaties and the good intentions of our adversaries."

His new campaign builds on Republican efforts in the Senate to pass a bill mandating the deployment of a national missile defence system as soon as it is technological feasible. Although the legislation was blocked by Democrats, Senate Majority Leader Trent Lott said in support of the bill, "Only effective missile defence, not unenforceable arms control treaties, will break to offensive arms race in Asia and provide incentives to address security concerns without a nuclear response." Numerous other Senators took to the Senate floor in the days after the India tests, citing the "India threat" as justification of a crash programme to field a missile defence system.

Dozens of articles and speeches by conservatives have used the South Asian tests as proof that future threats are inherently unpredictable, our intelligence estimates are consistently unreliable, the proliferation of weapons of mass destruction fundamentally unstopable and, thus, the only truly effective response is reliance on American defence technology. Gary Bauer, president of the fundamentalist Christian group, Family Research Council, warned with Edwin Feulner, president of the Heritage Foundation, in another *Washington Times* opinion piece June 14, "The nuclear club is getting larger, not smaller. The world is getting more threatening, not less. American needs to make its house secure again." They are appalled that the U.S. remains defenseless against missiles. "This is not because the military couldn't field such a system, but because the Clinton administration opposes anti-missile weapons and believes the best defence against missiles is the threat of a U.S. counter-attack."

The South Asian tests, the continuing reports of efforts by "rogue" states to acquire weapons of mass destruction, and the highly-politicised controversy over alleged American corporate assistance to China space-launch vehicles and ballistic missiles have increased public and political awareness of proliferation risks. "All this has reinforced the public's impression that the world is quite a dangerous place," says Frank Gaffney, a staunch conservative proponent of national missile defence.

In short, there is a concerted effort by the right-wing of the Republican Party to use these events and this new unease to promote, yet again, a vision that has assumed almost religious stature in conservating circles despite years of failed tests, bloated budgets and dashed expectations. Will it work this time?

It might.

The growing drumbeat on the right could converge with three other developments:

- The selection, for the first time, of a prime contractor whose job will be to design, develop, and prepare for deployment of a national missile defence system.
- Changes in the national intelligence assessments of the possibility of additional countries fielding long-range ballistic missile or of China significantly expanding its limited ICBM force.
- The lack of a clear, persuasive plan from the Clinton administration on counting the adverse proliferation trends and the lack of an administration leader on these issues.

THE POLITICS OF MISSILE DEFENCE

To understand the current development, it is important to briefly review previous efforts to force the deployment of a national missile defence system (a brief history of ballistic missile defence efforts is provided as an appendix.) Republican congressional leaders have tried to make President Clinton's defence strategy a political issue, first in their Contract with America in 1994 and 1995, and then as the leading defence issue in the 1996 presidential campaign. It didn't work. The public did not feel threatened by enemy missiles; military leaders did not see a technology capable of meeting the challenge; and deficit-conscious members of both parties in Congress balked at the $60 to $116 billion estimates for deploying the proposed national systems.

Still, the administration felt it necessary to protect its right flank. The Administration developed a "3+3" plan for developing a rudimentary missile defence system over three years and if conditions then warrant-deploying the system over the next three years. Meanwhile, under congressional pressure, missile defence funding has steadily climbed. The budget for fiscal year 1999 now before Congress calls for spending $950 million of the total $3.6 billion missile defence budget on national missile defence. Clinton has also pumped an extra $2.3 billion for the 3+3 programme into the budgets for the next five years, almost double what was originally allocated for the programme. Suddenly the programme is flush with funds, and it now has a lead contractor to direct and pump up the effort.

This May, the Department of Defence chose the Boeing Co. as its "lead system integrator." Jacques Gansler, the assistant secretary of defence for acquisition, told Congress in the late February that he was absolutely committed to deploying the system. The "only question," he said, "is when." And General Lester Lyles, director of the Ballistic Missile Defence Organisation (BMDO), defended the programme schedule before the Senate in late March, promising "We are on the verge of fielding a system."

Almost all the billions spent on national missile defence up to now have been for hundreds of paper studies and long-shot experiments. But once factories start "bending metal," weapons systems acquire a serious constituency of sub-contractors, chambers of commerce, labour unions, and workers' families, not to mention congressional hearts and minds. The first production funds, for long-lead items such as rocket boosters, guidance systems, and radar, will likely appear next year in the Pentagon's

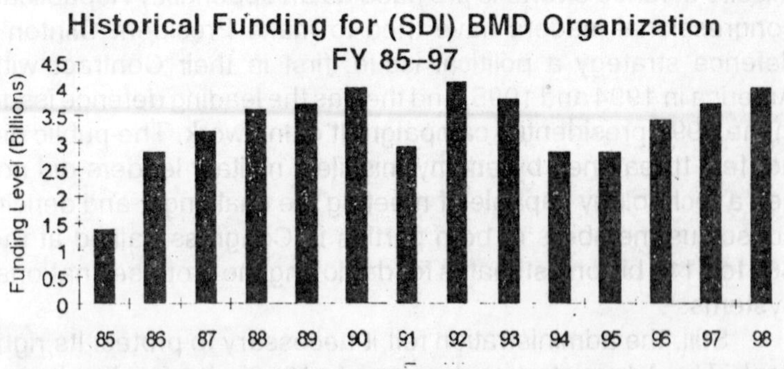

"Does include additional funds appropriated for Department of Energy or Military Services BMD Programmes

budget for fiscal year 2000. General Lyles noted during his Senate testimony, "We're trying to keep the contractor focussed on successful testing and not buying hardware. This is very difficult."

TEST FAILURES FAIL TO DETER PROPONENTS

Missile defence advocates accept as a matter of faith that inexpensive technology to defend against missile attacks is at hand. The only obstacles to deployment, they believe, are slavish adherence to the Anti-Ballistic Missile Treaty and the lack of political will in the Clinton Administration. The solution is both "available" and "doable", says House Speaker Newt Gingrich, but he US "governmental refuses to protect Americans" by immediately deploying a missile shield.

Three recent independent reports, however, have highlighted the formidable technical failings of both the national and theater missile defence efforts thus far. The most recent, and devastating, is a review by former defence officials chaired by retired General Larry Welch warning that the programmes were in a "rush to failure." In particular, the Welch panel said, the national missile defence programme was "highly unlikely" to succeed, lacked coherence and a realistic plan and should be fundamentally restructured. Congressional proponents and Administration and industry officials have ignored these warnings.

The Welch panel detailed 17 tests conducted by the Department of Defence since 1982 of "hit to kill" interceptors (now 18 with the recent failed THAAD test). Only four actually hit their targets. Worse, of the 14 tests (now 15) directed against high-altitude targets, only two high, for a 13 per cent success rate.

The most advanced system now under development, the Army's Theater High-Altitude Area Defence system (THAAD), has failed in all five of its intercept attempts, most recently on May 12. These tests were against specially designed targets with known trajectories and characteristics that fell well within the expected performance range of the system. Further, the maximum ranges of these targets were 750 to 1,100 kilometers, or about one-tenth the range of an ICBM.

The Navy's Upper Tier (or Theater-Wide) programme has failed no better. Based on an upgrade to the Standard missiles used by AEGIS cruisers and destroyers, this system as been trumpeted by conservative advocates as a quick, cheap solution to both theater and national missile defence. But, here too, the Welch panel documents, the proposed LEAP interceptor has four failures in four intercept attempt. The panel noted that "the general planning and execution of the THAAD and LEAP programmes are inconsistent with the difficulty of the task. These programmes are pursuing very aggressive schedules, but these schedules are not supported by the state of planning and testing."

As General Lyles explained last year, "The THAAD systems have been very successful in every aspect except the very critical end-game." That is, they hadn't actually hit anything. The low number of tests and the weak kill rates warrant deep skepticism regarding future success.

Finally, there have not yet been any intercept tests of the proposed national missile defence system itself and few are scheduled before a deployment decision is to be made Noting that the national missile defence schedule is shorter than most other major system acquisition programmes, the General Accounting Office recently warned:

- "Because of the compressed development schedule, only a limited amount of flight tests data will be available for the system deployment decision in fiscal year 2000. By that time, BMDO will have conducted only one system-level flight tests, and that test may not include all system elements or involve stressing conditions such as targets that employ sophisticated countermeasure or multiple warheads. As a result, not all technical issues, such as discrimination, will be resolved by the time of the deployment review. Also the current schedule will permit only a single test of the integrated ground-based interceptor before production of the interceptor's booster element must begin. If subsequent tests reveal problems, costly redesign or modification of already produced hardware may be required."

In February, the Pentagon's director of operational test and evaluation warned Congress that the programme was filled with technical and scheduling risks. "If deployment is required by 2003," he said in his annual report, "the NMD programme will have to compress the work of 10 to 12 years into six years." As a result, he says, the programme will have to do basic testing after production has already started, and it will have to rely heavily on modeling and computer simulations instead of actual tests.

The director also said that there are several factors that will "significantly limit the ability of the NMD Test and Evaluation programme to test, analyse and evaluate system performance" including: planning for only one system-level flight test before the Pentagon holds its deployment readiness review; not testing the actual capability of the ground-based interceptor boosters before the readiness review; not testing the interface between the battle management command, control, and communications system and U.S. Space Command before the review; not planning for any tests against multiple targets before the review; and having low confidence in the models and simulations used instead of real tests since they will have "minimal validation by real flight data."

THE WELCH WARNING

The Welch panel's critique of the missile defence programmes has been the most biting, perhaps because it comes from missile defence advocates familiar with the programmes and frustrated by political pressures to rush into production systems they know are nowhere near ready to be tested, let alone fielded. The panel points out that "we are still on 'step one' in demonstrating and validating HTK [hit-to-kill] systems." The test failures, they say, actually had little to do with the kill vehicle performance (the warhead that will intercept the incoming missile), but from failures in supposedly mature components in the missiles carrying the warhead to the target "basket". "And even when this first step is achieved, these programmes will still have to go through steps two and three: demonstrating reliable HTK at a weapon system level and demonstrating reliable HTK against likely real-world targets." We are still years away from these kinds of tests.

The most basic lessons highlighted by the Welch panel is that Congress cannot legislate physics. Forcing programmes into unrealistic schedules has been counter-productive. "The study group was not surprised to find that accepting higher risk is not accelerating fielded capability," their report said. "The virtually universal experience of the study group members has been that high technical risk is not likely to accelerate fielded capability. It is far more likely to cause programme slips, increase costs, and even programme failures." These experts should know. The panel included Gen. Eugene Fox (ret.), former programme manager of the Army's BMD programme; Gen. Don Lionetti (ret.) former commander of the Army Space and Strategic Defence Command, Rear Admiral Wayne Meyer (ret.), often cited as the "father" of the AEGIS air defence system, former Under Secretary of Defence for Acquisition Paul Kaminksi and other veterans of the Star Wars programme and defence testing programmes.

The Welch report document that the national missile defence is shaping up as a classic example of a programme rushing to failure. Historically, weapon systems that have been fast-tracked through concurrent production and testing, such as the B-1 and B-2 bombers, have experienced severe production problems, cost growth, and performance degradation.

The experts on the Welch panel are saying as loudly and as clearly as they can that the Administration and Congress should slow down and rethink these programmes. "In the judgement of the study group," they conclude, "successful execution of the 3+3 formulation on the planned schedule is highly unlikely. The programme will benefit from the earliest possible restructuring to contain the risk."

And even if everything in a national missile defence system works as planned, a system of 100 ground-based interceptors with space-based sensor satellites might be able to intercept only a few warheads. Deputy Secretary of Defence John White said this to Congress in June 1996:

"If the number of threats increases of the complexity of the threats increases then this basic system is likely to provide poor protection of the [United States]. This poor protection is due partly to a lack of sufficient discrimination capability against complex threats, which will cause the interceptor inventory to be depleted by shooting at warhead decoys, allowing some real warheads to penetrate the defence... The systems is not designed to protect against an unauthorised launch which might contain a large number of warheads (e.g., a full load of warheads from a Russian [submarine])."

BUDGET BUSTING

The deficit-busing potential of the entire programme ultimately ended previous Republican efforts to pass legislation mandating deployment of a national system. Proponents had anticipated House passage of the Defend America Act in May 1996, when a required cost estimate from the Congressional Budget Office (CBO) landed like an incoming Scud. The CBO estimated that the bill's requirement for a "highly effective defence of all 50 states....augmented over time to provide a layered defence against larger and more sophisticated ballistic missile threats" would cost up to $60 billion for the first stages of its deployment of ground-based interceptors, tracking stations, satellite sensors and hundreds of yet-to-be-invented space-based interceptors and space lasers. A July 1996 CBO report showed that the total cost of the proposed systems could reach $116 billion over 20 years, including operation and support costs.

Freshman Republican deficit hawks in the House revolted. They refused to support a major new government programme, even for national defence. After a raucous Republican Caucus session, the leadership was forced to pull the legislation from floor consideration. Despite repeated efforts to reintroduce the bill, they could never muster sufficient support. Many members felt betrayed; they had been led to believe that the programme was almost cost-free.

National missile defence is expensive. The types of systems proposed by Congressional advocates have always been estimated to cost tens to hundred of billions of dollars. For example, the G-PALS system proposed by President George Bush was similar to the Defend America proposal and would have cost an estimated $63 billion. The simple anti-ballistic missile system the United States built and deployed at Grand Forks, North Dakota in the early 1970s cost $22 billion in current dollars before President Gerald Ford shut it down.

THREAT INFLATION

The weak public support for national missile defence, the technological setbacks, the absence of a credible and immediate threat and the general fiscal conservatism of the Congress accounted for the failure of missile defence to advance as a political debate in 1997. Undeterred, conservative leaders now seem to believe that the time is right to try again.

A key focus of their strategy has been the national intelligence estimates of the threat from the proliferation of ballistic missile technology.

The missile threat is not as severe or immediate as frequently suggested by the proponents. Only Russia and China have nuclear-armed ICBMs that can reach the United States, and China has only a dozen or so. Building an ICBM is a difficult and expensive process. National intelligence experts agree that it is very unlikely that additional countries will build ICBMs within the next fifteen years. While there is ample reason for concern about a black market in technologies relevant to ballistic missiles, the idea that a huge ICBM, with all its support equipment could be stolen by or sold to a rogue state seems very farfetched.

Indeed, in its review of the November 1995 National Intelligence Estimate (NIE) of emerging missile threats to the United States, an independent panel of experts chaired by former-CIA Director Robert Gates states that...

"[F]or sound technical reasons, the United States is unlikely to face an indigenously developed and tested intercontinental ballistic missile threat from the Third World before 2010, even taking into account the acquisition of foreign hardware and technical assistance. That case is even stronger than presented in the NIE."

These assessments are under renewed attack. A congressional-appointed panel will soon release its independent review of the ballistic missile threat to the United States. It is likely that this panel will criticise the official findings as overly optimistic and warn that missile threats can develop more quickly than previously estimated. Once case in point is the 1000-kilometer *Nodong* missile, which has now reportedly been fielded by North Korea even thought it has only been tested once in 1993. (This is quite unusual, since new missiles are usually tested several times before military commanders feel confident enough to field them.) Another case is the surprise test April 6 of the Pakistan 1500-kilometer *Ghauri* missile, Pakistan's announced plans to test at least two new previously unknown missiles in the near future and India's announced intentions to renew testing of its 2000 kilometer-range *Agni* missile.

These intelligence uncertainties bolster claims by deployment proponents that systems such as the mysterious North Korean Taepodong-1 and Taepodong-2 could materialise much more quickly than experts anticipate. The latter missile is said to have sufficient range to hit the outer islands of the Alaskan archipelago, while testing for the former "could begin at any time", according to the Pentagon's annual report on weapons proliferation released last November.

MISSING MATTER

The South Asian nuclear blasts, the renewed push for missile defence, and charges of compromising vital national security interests for the sake of Chinese business deals and campaign contributions have rattled the arms control strategy of an administration already reeling under the constant attacks and investigation from its Republican congressional opponents.

Even prior to the recent events, the Administration lacked a clearly articulated strategy for dealing with the problems of global weapons proliferation. Its nuclear arms reduction efforts are on perpetual hold pending Russian Duma ratification of the START II treaty, the implementation legislation for the Chemical Weapons Convention contains killer provisions that will seriously compromise challenge inspections, and the only arms control treaty that President Clinton has negotiated and signed while in office, the Comprehensive Test Ban Treaty, is blocked by a determined Chairman of the Foreign Relations Committee, Senator Jesse Helms.

The Administration throughout has lacked a single, determined champion for non-proliferation and threat reduction efforts. The most logical spokesperson, Arms Control and Disarmament Agency Director John Holum, has been constantly distracted by efforts to abolish his agency. Even with his new appointment as acting Under Secretary of State, his status, portfolio and authority remain uncertain. Nor has the administration organised itself as it did during the 1996 presidential campaign. Then, the president delivered a series of speeches, coordinated with cabinet secretaries who unveiled the "3 + 3" programme and its integration in a vision of "three lines of defence" against the spread of weapons of mass destruction: treaties to prevent and reduce the threat, conventional forces to respond to attacks, and lastly, missile defence programmes.

Now the pressures are all pushing the administration towards further compromise. Republican efforts to push their bill mandating deployment through the Senate failed on May 13 by only one vote. History has shown that lack of proven technical capability has never prevented congressional approval of military systems. If the projected budgets for a national system can be kept low enough (even if they rise later), international events and political pressure could very likely combine to either pass missile defence legislation or force an administration compromise commitment to deployment. As presidential politics 2000 come into play, the odds increase that the Democratic candidate will seek a way to remove missile defence as a campaign issue.

Despite the low threat, dismal test record, and lack of adequate technology, it appears more likely than anytime in the past decade that the United States will eventually commit to the deployment of a rudimentary national ballistic missile defence system.

APPENDIX I

A BRIEF HISTORY OF BALLISTIC MISSILE DEFENCE

The Early History of Ballistic Missile Defence

The earliest recorded use of powered missiles in warfare was in 1232 at the military siege of Kaifeng, former capital of the Chinese province of Henan, in which rockets were used to set fire to tents and wicker-work fortifications. European technology developed these rockets into larger and longer-range weapons. In 1807, for example, Copenhagen and a large French fleet in its harbour were almost totally destroyed by a British naval attack using thousands of iron rockets. The national anthem of the United States reflects the common use of these weapons in naval battles in the 17th and 18th centuries, when Francis Scott Key saw the American flag "by the rocket's red glare."

The first true ballistic missile—one that has a brief period of powered flight, continues on a ballistic trajectory outside the atmosphere, then curves back to an impact point on earth—was developed at the end of World War II. Serious efforts to find a defence against ballistic missiles began shortly after the first German V-2 slammed into London, without success. Overall, the United States has spent more than $100 billion (in current dollars) in the pursuit of missile defence since the mid-1950s (plus $17 billion on the Patriot system, developed separately by the Army as an anti-aircraft system.) The United States remains the only nation devoting a significant portion of its national defence budget to missile defence.

President Eisenhower began the search for a defence to these missiles when he authorised the operational development of a nuclear-tipped interceptor missile, Nike-Zeus, and commission Project Defender to develop components for a nation-wide ballistic missile defence system. In the late 1960s, President Richard Nixon approved the deployment of the Safeguard Anti-Ballistic Missile (ABM) system, in response to the Soviet development of an ABM system around Moscow. Although many in Congress were concerned that the system would be ineffective, vulnerable to attack, and easily overwhelmed, it was approved in order not to undermine America's negotiating position in the Strategic Arms Limitation Talks.

In 1972, the Soviet Union and the United States announced the first Strategic Arms Limitation Treaty (SALT 1) as well as an agreement limiting defensive systems—the ABM Treaty. Both nations agreed "that effective measures to limit anti-ballistic missile systems would lead to a decrease in the risk of outbreak of war involving nuclear weapons." In attaining both of these agreements at the same time, the negotiators intended to insure strategic stability by stopping large scale deployment of strategic defensive systems while attempting to limit offensive forces.

The broad purpose of the ABM Treaty is to prevent either party from fielding a nationwide ballistic missile defence of its territory. The Treaty prohibits the development, testing or deployment of sea-based, air-based, space-based, or mobile land-based ABM systems, as well as components based on advanced physical principles (Article V and Agreed Statement D). The U.S. Arms Control and Disarmament Agency notes the ABM Treaty is designed to "decrease the pressures of technological change and its unsettling impact on the strategic balance."

The proven logic behind the prohibition against a nationwide defence is that an arms race in strategic defence systems fosters the proliferation of offensive missiles and the development of countermeasures to defeat the defence.

The ABM Treaty permitted a limited deployment of defences. Russia for years maintained today a site of 100 nuclear-tipped interceptors around Moscow. Administration officials have always been confident that United States missiles could penetrate and overwhelm this defence. If the Soviets had deployed more advanced or proliferated defences, the United States would surely have deployed more advanced devices to ensure the continued capability to penetrate. Russian officials recently indicated that they have recently taken all nuclear war-heads off the Moscow ABM interceptors.[1]

In the Administration of President Gerald Ford, officials and military advisors determined that defences permitted us under the Treaty were not worth maintaining since they could easily be penetrated by Soviet ballistic missiles. As a panel of the George C. Marshall Institute (proponents of deploying a space-based defensive system) noted during the Star Wars debates of the late 1980s, the problems with the 1970s defensive systems were that "a 'ground-based' defence is readily overwhelmed" and that the fixed, ground-based radars on which the system depends are 'easily targeted by the Soviets and vulnerable to destruction in a surprise attack." Ultimately, although the Safeguard system was deployed, it was operational for only a few months in the mid 1970s, and then shut down as obsolete.

THE MODERN ERA OF BMD RESEARCH

Under President Jimmy Carter, the United States continued an active research programme into strategic defences, averaging just under $1 billion per year. At the beginning of the Reagan Administration, the consensus in the defence community was that ballistic missile defences could not be militarily effective. Some, however, disagreed and promoted two systems—High Frontier and space-based lasers—each considered by the Reagon Administration and rejected before the President's surprise "Star Wars" speech of March 23, 1983.

High Frontier

Retired General Daniel Graham promised that by using "off-the-shelf" technology the United States could build a network of several hundred satellites carrying rocket interceptors that would defeat any Soviet attack. He claimed the U.S. could launch this system for some tens of billions of dollars. This plan met with nearly universal rejection by defence analysts. In November 1982, Secretary of Defence Caspar Weinberger wrote Graham:

"Although we appreciate your optimism that 'technicians will find the way and quickly' we are unwilling to commit this nation to a course which calls for growing into a capability that does not currently exist....While there are many instances in history where technology has developed more quickly than the experts predicted, there are equally as many cases where technology developed more slowly."

Weinberger based his view, in part, on an Air Force Space Division analysis that had concluded High Frontier "has no technical merit and should be rejected... No alternate configuration supported a favourable conclusion." Another Defence Department analysis stated, "It is the unanimous opinion of the Air Force technical community that the High Frontier proposals are unrealistic regarding state of technology, cost and schedule."

Lasers in Space

Another possible defensive system, space-based lasers, was also found to be unpromising. In 1981, the Department of Defence's Science Board concluded unanimously:

"It is too soon to attempt to accelerate space-based laser development towards integrated space demonstration for any mission, particularly ballistic missile defence."

On March 23, 1983, the day of President Reagon's speech, Air Force officials were on Capitol Hill testifying about the space laser programmes they ran. General Donald L. Lamberson told the Senate that he could not recommend an acceleration of the space-based laser programme on technical grounds "at this point in time." General Bernard Randolph told the House of Representatives that a laser weapon system would require many megawatts of power, would need a precision mirror much larger than any yet manufactured, would weigh 150,000 pounds, and would cost "many, many billions of dollars." He explained that "to point the system at a target would be like pointing from the Washington Monument to a baseball on the top of the Empire State Building and hold it there while both of you are moving.... I view the whole thing with a fair amount of trepidation."

The President's Vision

That night, President Regan made his speech, asking:

"What if free people could live secure in the knowledge that their security did not rest upon the threat of instant retaliation to deter a Soviet attack, that we could intercept and destroy strategic ballistic missiles before they reached our own soil or that of our allies?"

President Reagan called on the scientific community "to turn their great talents now to the cause of mankind and world peace, to give us the means of rendering nuclear weapons impotent and obsolete." The President said he was directing "a long term research and development programme to begin to achieve our ultimate goal of eliminating the threat posed by strategic nuclear missiles." He said, "It will take years, probably decades, of effort on many fronts."

Congress responded to the President's vision—continued by President Bush—appropriating over $44 billion between 1983 and 1993 for the Strategic Defence Initiative (SDI). The annual budget for ballistic missile defence research quadrupled, from $991 million in 1983 to a peak of $4 billion in FY 1988.

Throughout Reagan years, the President's personal sponsorship and persistence propelled the SDI programme through Congress far beyond its technical and strategic merits. The President kept his vision alive with repeated references to his hope that SDI could, as he told a 1986 high school graduating class, "put in space a shield that missiles could not penetrate—a shield that could protect us from nuclear missiles just as a roof protects a family from rain."

During those years, the SDI programme produced technological advances in some areas, such as sensors, miniaturisation, and lasers. In the summer of 1987, the SDI Organisation presented to the Department of Defence's Acquisition Board a plan to move from research to the "phased" deployment of strategic defences. SDI officials advocated the development of a "Strategic Defence System." The Department of Defence approved the new plan, despite the sharp warnings of its own Defence Science Board, which said "there is presently no way of confidently assessing" the system's price or its effectiveness.

The primary mission of Phase I of this system, which was to be deployed in the 1990s, was protecting only nuclear forces any key command centers. The lowered goal now was to "reduce Soviet Confidence in the military utility of its ballistic missile force" and to "complicate Soviet attack plans." As SDI officials explained in a 1988 report to Congress:

"These first phases could severely restrict Soviet attack timing by denying them cross-targeting flexibility, imposing launching window constraints, and confounding weapon-to-target assignments, particularly of their hard-target kill capable weapons."

The centerpiece of Phase I was the space-based interceptor, much like BAMBI and High Frontier. Phase I was to have—also include ground-based interceptors; ground-and space-based sensors to detect, track, and target Soviet missiles; and a command, control, and communications network to manage the battle.

Technical Analysis

The major technical problems that remain unresolved and eventually forced the cancellation of these ambitious plans are the same obstacles that have ruled out an effective ballistic missile defence for forty years. The basic problems are:

- the ability of the enemy to overwhelm a system with offensive missiles;
- the questionable survivability of space-based weapons;
- the inability to discriminate among real warheads and hundreds of thousands of decoys;
- the problem of designing battle management, command, control and communications that could function in a nuclear war; and,
- low confidence in the ability of the system to work perfectly the first and, perhaps, only time it is ever used.

This problems have been detailed at length in many independent expert studies, including two that played a major role in the Star Wars debates—the 1987 American Physical Society Directed Energy Weapon study and the 1988 Office of Technology Assessment Ballistic Missile Defence study.

But there are literally hundreds of other major technical problems that would have to be resolved before an effective defence can be deployed. In the long term, new technologies, particularly directed energy weapons, hold out the prospect that some of the problems might be solved in the next century. In the short term, there is little reasons for technological optimism.

Protecting Against Accidental Launches

The technological problems, the exorbitant cost estimates for deploying a system such as SDS (over $250 billion), and the collapsing Soviet threat led to a restructuring of the programme. In January 1991, President Bush abandoned plans to protect against a massive Soviet first strike and redirected the programme to a "Global Protection Against Accidental Launch System" (GPALS) to protect the United States, its forward deployed forces, and its allies and friends from limited ballistic missile attack. This policy shift (though not this particular still-expansive system architecture) had been long urged by some influential members of Congress, including then–Chairman of the Senate Armed Services Committee Sam Nunn. He had said as early as 1988 that:

"If carefully redirected, our research efforts could produce options for limited deployments to deal with the frightening possibility of an accidental or unauthorised missile launch.... This should be coupled with a rigorous unilateral review by both sides of their respective fail-safe procedures and safeguards."

However, analysis of proposed ABM Treaty-complaint accidental launch protection options have raised serious technical strategic, cost and arms control implications. Dr. Theodore Postol, former scientific advisor to the Chief of Naval Operations, has done extensive work in this area, including an analysis for the House Democratic Caucus in 1988 of the first Accidental Launch Protection system.

Dr. Postol concluded that a limited deployment of 100 missiles, such as the original ERIS system proposed by Lockheed Corporation, might be effective against a small accidental launching, if no penetration aids were used, but the system would have trouble dealing with a launch of more than five missiles. Dr. Postol noted that such a system could not defend the major industrial and population centers of the entire North-east against a missile launched from a submarine off the coast of the United States. Just what kind of coverage is possible is not clear.

Many analysts believe that an accidental launch protection system would require interceptor sites on both coasts. Such a multiple-site system might require several new large radars and would be much more expensive than a single site system. It would require an abrogation or amendment of the ABM treaty.

Most disturbing, this type of system deployed by either country would have a very large capability to target satellites. Dr. Postol noted in 1988,

"Such a system would introduce a qualitatively new scale of ASAT threat to the satellites of other nations. For example, if the U.S. has an ERIS-like defence system with geosynchronous altitude capabilities in place today, it could be-used to launch a simultaneous attack on two thirds of the Soviet Union's satellites. A similar Soviet ERIS-type system, in turn, could simultaneously attach more than ninety per cent of U.S. satellites."

If the Russians viewed the limited nation-wide defence as a foot-in-the-door effort to break out of the ABM Treaty or as a partial defence of ICBM fields (a capability noted by several contractors), they could develop methods to overcome it with relative ease. If the U.S. deploys such a system and the Russians do too, then both sides would certainly deploy penetration aids on their missiles. In that case, our ability to even deal effectively with an accidental launch of a single ICBM would be undermined. (The U.S. now has penetration aids on some of its missiles in response to the limited Russian defence around Moscow).

The Budget Decline and the Rise of Patriot

From a peak of $4 billion in 1988, funding dropped to an average of $3 billion per year as technology and interested waned. President Bush's down-sizing of the programme's mission did not convince Congress to provide funding above that level until 1991. Then, the perceived success of the Patriot missile in the Gulf War was used to boost funding back up to $4 billion.

For example, Secretary of Defence Dick Cheney told the Senate Armed Services Committee on February 21, 1991:

"Patriot missiles have demonstrated the technical efficacy and strategic importance of missile defences. This underscores the future importance of developing and deploying a system for GPALS, to defend against limited missile attacks, whatever their source...Defences against tactical ballistic missiles work and save lives. The effectiveness of the Patriot system was proved under combat conditions."

SDI Director Henry Cooper told the House Committee on Appropriations on April, 17, 1991:

"It is difficult to imagine a better validation of President's Bush's redirection of SDI and continued approach to our negotiations with the Soviet Union than the clear lessons of the recent Gulf War."

The SDI Director also told the Government Operations Committee that year:

"Another observation of the Gulf War is that missile defence can 'work' well enough to be extraordinarily useful. In the Gulf War, Patriot intercepted 51 to 52 Scuds engaged. This level of effectiveness against a very limited threat would be extremely useful whether the offensive missiles were armed with conventional warheads or weapons of mass destruction."

Israeli officials and experts agree that the Patriot failed in its military mission. The only debate in Israel is whether the

Patriot hit one or none of the Scuds it attempted to intercept. Israeli officials tracked each Scud to the ground and thus had the craters to prove that the initial claims of intercept success were false.

In the United States, confusion over the Patriot's performance still fuels overly optimistic estimates of the effectiveness of new, proposed defensive systems. Many officials, journalists and experts rely on the Army report on Patriot. The Army evaluation was performed by a small team of nine officials from the Patriot Programme Office and related Army offices and others from the prime contractor on the programme, the Raytheon Company. On average, between three and nine Raytheon personnel supported the Army in the post-war performance analysis and approximately 12 Raytheon personnel provided support to the Army in Saudi Arabia and Israel in analysing Patriot performance and operations. The Army paid Raytheon $520,000 to provide analysis of Patriot performance in the war. Still, this report remains the only official government assessment ever conducted on the Patriot's performance.

As a result of Congressional investigations into the performance of the Patriot, the Army revised its claims in 1992. The Army now reports that during Desert Storm, 88 Scuds were launched by Iraq. The first 12 were launched at Israel prior to the deployment of Patriot units in that country. Of the remaining 76 Scuds, somewhat less than 45 were actually engaged by Patriots.

A total of 158 Patriot missiles were fired during the war:

- 86 Patriots were launched at Scud targets in Saudi Arabia and Israel.
- 30 per cent of the Patriots were launched as Scud debris mistaken for targets.
- 15 per cent of the Patriots were launched against false targets caused by radar backlobe and sidelobe interference (including one launched by accident in Turkey).

The Army claims, with varying degrees of confidence, that the Patriot Missile system destroyed 52 per cent of the Scuds.

The General Accounting Office does not share that confidence. Independent review of the evidence in support of the Army claims reveals that, using the Army's own methodology and evidence, a strong case can be made that Patriot hit only

9 per cent of the Scud warheads engaged, and there are serious questions about these few hits. *(GAO Report: "Operation Desert Storm: Data does not exist to conclusively say how well patriot performed," September 1992, NSIAD 920340)* The speed of the Scuds, the limitations of the Patriot missile system, and the confusion and targeting difficulties caused by the break-up of the Scud missile as it re-entered the atmosphere seem to have contributed to the high failure rate.

APPENDIX II

BALLISTIC MISSILE DEFENCE PROGRAMME DESCRIPTIONS

Current Budgets and Plans

President Clinton in 1993 reduced the missile defence budgets to under $3 billion dollars and renamed the Strategic Defence Initiative Organisation, the Ballistic Missile Defence Organisation (BMDO), while leaving the personnel directing the programme largely unchanged. President Clinton also reversed the programme's funding priorities to favour spending on theater ballistic missile defences over national missile defence system,

In so doing, he followed previous Congressional preferences. Congress had begun the shift by enacted the Missile Defence Act of 1991, which stated that it is:

"...a goal of the United States to provide highly effective theater missile defences to forward deployed and expeditionary elements of the armed forces of the United States and to friends and allies of the United states."

By 1994, funding for national missile defence was relegated to research only programmes with total funding measured in the low hundreds of millions of dollars. Since the Republican Party assumed control of Congress in 1995, however, the Congress has consistently added funds to the Administration's request, increasing funding once again to $4 billion per year.

This year, as part of the Fiscal Year $271 billion defence budget, Congress will likely appropriate $3.4 to $3.8 billion for the programmes of BMDO. The Administrations budget request was $3.6 billion. The Senate Appropriations Committee approved a budget of $3.43 billion on June 12, for example, while the House approved a budget of $3.8 billion earlier. These budgets will be reconciled in conferences later this year. These funds include:

- $950 million for national missile defence (matching the president's request).
- $401 million for the Army's Patriot PAC-3 programme ($100 million below request)
- $310 million for the Navy's Upper Tier (theater-wide) programme ($120 million above request)
- $497 million for the Army's THAAD ($150 million below request, cutting the procurement funds)
- $289 million for the Navy's Lower Tier (area-wide) programme (at request)
- $255 million for the Airborne Laser programme ($57 below requests
- $43 million for the Army's Medium Extended Air Defence System (MEADS) (at request)
- $ 38 million for Israel's Arrow programme (at request)
- $ 73 million for the Space-Based Laser programme ($20 million below request)
- $ 47 million for Atmospheric Interceptor Technology

Programme Descriptions

The so-called lower-tier systems are designed to intercept missiles within the atmosphere and to try to protect relatively small areas, with dimensions of tens of kilometers, and are intended to counter short-range missiles with ranges under 1,000 kilometers. They include:

The Patriot PAC-3 is a major upgrade to the PAC-2 Patriot missile used in Israel and Saudi Arabia during the Gulf War with little, if any, success. The actual Patriot missile will be replaced by an entirely new system, the ERINT, which will be much more maneuverable than the original Patriot interceptor and have to hit-to-kill warhead instead of an explosive warhead and proximity fuse. Four ERINT will fit in each of the Patriot launcher canister and be deployed along with PAC-2 Patriots as part of the overall system. An improved "multi-mode" Patriot failed in two out of three intercept tests conducted after the war. The Army found it "operationally unacceptable," choosing the new ERINT instead. The ERINT will not be fully tested and deployed until 1999.

The Navy Tower Tier (Area-Wide) system is an upgrade to the US Navy's AEGIS radar system and Standard missile to give them anti-TBM capabilities. The Navy would like to deploy this system on over 50 AEGIS cruisers and destroyers.

The Medium Extended Air Defence System (MEADS) is an entirely new missile defence system. Formerly known as Corps SAM, it is being developed jointly with Germany and Italy. France, an original partner, has dropped out, citing the high cost of the programme. MEADS is intended to be a highly mobile defence to protect forces on the battlefield.

The higher-tier systems are designed to protest larger areas (hundreds of kilometers) and counter missiles with ranges up to 3,500 kilometers. No nation hostile to the United States currently has such missiles, but this is the threat represented by systems reportedly under development in North Korea. Both the Administration and the Congress favour developing systems to intercept these missiles, with Congress trying to force a faster development and deployment schedule. The systems would attempt to intercept targets above the atmosphere.

THAAD has been the primary system proposed by the Administration for this mission. The current design has an interceptor with a peak speed to about 2.6 kilometers per second, with intercept altitudes above 40 kilometers.

The Navy Upper Tier (Theater-Wide) system would use the AEGIS phased-array radar and vertical launcher systems deployed on the Navy's AEGIS cruisers and Arleigh Burke class destroyers. The interceptor would probably be a LEAP kill vehicle deployed on a variant of the Navy's Standard anti-aircraft missile, with a peak speed of about 4.5 kilometers per second and a minimum intercept altitude of 80 kilometers.

There also are three laser systems of interest. The most immediate is the **Tactical High Energy Laser system (THEL)** that would be deployed against very short-range targets. The MIRACL laser which was used to test anti-satellite capabilities recently, was also used successfully to damage a Katyusha rocket on February 9, 1996 at White Sands Missile Range, New Mexico. The MIRACL is a huge facility, however, and it will be technically challenging to shrink this capability into a rugged system for the battlefield. The current plan is to deploy the system in six or so large trailers.

The Airborne Laser (ABL) would attempt to fit a laser into a Boeing 747 for a demonstrated shoot down in 2002. The Air Force proposes to build a fleet of such planes for $1 billion a copy to patrol friendly airspace. The plane would cruise about 13 kilometers above the surface (and most cloud layers). It would attempt to shoot down tactical ballistic missiles in the boost-phase, the brief period of powered launch, as the missile rises above the clouds. While the beam travels at the speed of light, it will have to dwell on the target for several seconds in order to weaken the 1-3 mm of steel skin of the missiles, triggering either a rupture from the internal pressure of the fuel tanks, or a collapse of the missile along the weakened area.

In order for the laser to bring its laser beam on target and experts estimate it would have to be within 320 to 470 kilometers of an Al-Husayn Scud (the 650 km-range missile encountered in the Gulf War) or within 185 to 320 km for a North Korean Nodong-1 missile (with a reported 1000 km range). This would mean, for example, that it could not be used to counter a Nodong launched from Iran, as the closest friendly territory (Turkey, Israel or the Gulf) would put it out of range.

Finally, the **Space-Based Laser** would attempt to make real the original cartoon images of the SDI programme. At a cost of $1.5 billion, the BMDO hopes to orbit a half-size demonstration satellite in 2005. Based on the Zenith Star programme SDI programme the so-called "Star Lite" project hopes to build a 17,500-kg laser weapon that can be launched into space atop a Titan IV rocket. It would target missiles in their more vulnerable boost-phase as they rise from the atmosphere. While there have been some advances in mirror technology, miniaturisation and beam compensation since cost and technical failures caused the cancellation of the original Zenith Star programme in 1993, space lasers may still be decades away. Even if the formidable technical problems can be solved, officials estimate it could cost as much as $100 billion to build, launch and maintain a constellation of laser weapons in space. And, of course, there are countermeasures that could be deployed by determined opponents to thwart such lasers.

APPENDIX III

COUNTRIES POSSESSING BALLISTIC MISSILES[2]

This chart is adapted from the forthcoming *Tracking Nuclear Proliferation 1998* (July 1998, Carnegie Endowment for International Peace). It lists the countries, other than the five nuclear powers, that have operational ballistic missiles with range capabilities over 100 kilometers. Although some countries have demonstrated the ability to use surface-to-air missiles in a surface-to-surface role, these systems are not listed unless they are deployed as dedicated ballistic missiles such as China's CSS-8. Range is given in kilometers and payload in kilograms.

Country	*System*	*Status*	*Range/ Payload*	*Country of origin*	*Notes*
Afghanistan	Scud B	O	300/1000	USSR	
Algeria	Scud-B	O	300/1000	USSR	
Armenia[3]	Scud-B	O	300/1000	Russia	
Azerbaijan	Scud B	O	300/1000	USSR	
Belarus	SS-21	O	70-120/480	USSR	
	Scud-B	O	300/1000	USSR	
Bulgaria[4]	Scud B	O	300/1000	USSR	
	SS-23	O	500/450	USSR	Banned by INF
Treaty					
Czech Republic[5]	SS-21	O	70-120/480	USSR	
Egypt	Scud-B	O/U	300/1000	USSR	
	Project T	O	450/1000	I/DPRK	
	Scud Mod C[6]	O	500/700	DPRK	
Georgia	Scud B	O	300/1000	USSR	
India	Prithvi-150	O	150/1000	I/USSR	From Russian SA-2
	Prithvi-250	D/T	250/500	I/USSR	From Russian SA-2
	Prithvi 350	D	350/500	I/USSR	From Russian SA-2
	Agni	D/T	2000/1,000+	I/US/ France	From Scout
	Sagarika	D	300/500	I/Russia?	From Prithvi/SA-2
	Surya	D	12,000/?	I	From PSLV

(Contd)...

Ballistic Missiles (Contd...)

Country	***System***	***Status***	***Range/ Payload***	***Country of origin***	***Notes***
Iran[7]	Mushak-120	O/U?	120/500	I/PRC?	
	Mushak-160	O/U?	160/190	I/PRC?	
	Mushak-200	O/U?	200/500	I/PRC?	
	CSS-8	O	150/190	PRC	Mod SA-2
	Scud-B[8]	O/U	300/1000	Libya	
	Scud Mod B	O/P	300/1000	DPRK	
	Scud Mod C	O	500/700	DPRK	
	Zelzal-3	D	1000-1500/ 1000		I/?
	Tondar 68	D	1000/500	I/PRC?	Chinese M-18?
	Shahab-3	D	1300/700	I/DPRK	from Nodong?
	Shahab-4	D	2000/1000	I/Russia	from Russian SS-4
Iraq	Ababil-100	P	100-150/300	I	
	Al-Samoud	P	150/?	I	From Scud[9]
	Scud B	Hidden?	300/1000	USSR	
	Al Hussein	Hidden?	600-650/500	I	From Scud
	Al Hijarah	Hidden?	600-650/250?	I	From Scud
Israel	Lance	O/S	130/450	US	MOU
	Jericho I	O	500/500	France	
	Jericho II	O	1500/1000	France/I	
	Jericho III	D	2500/1000	I	
Kazakhs-tan	SS-21	O	70-120/480	USSR	
	Scud B	O	300/1000	USSR	
Libya[10]	Scud B	O/U	300/1000	USSR	
	Al Fatah (Ittisalt)[11]	D/T	950/500	I/?	
North Korea[12]	Scud Mod B	O/P	300/1000	USSR	
	Scud Mod C	O/P	500/700	I	
	Nodong[13]	D/T	1000/700-1000	I	
	Taepo Dong 1[14]	D	1500+/1000	I	Nodong+
Scud	Taepo Dong2	D	4000-6000/ 1000	I	
Pakistan	M-11	S	280/800	PRC	
	Hatf 1	O	80/500	I/France?	
	Hatf 1A	O	100/500	I/France?	
	Hatf 2[15]	D	280-300/500	I/PRC?	M-11 derivative?
	Hatf 3	D?	600/500	I/PRC?	M-9 derivative?
	Ghauri (MK-III)[16]	D	1500/500-750	I	

(Contd)...

Ballistic Missiles (Contd...)

Country	*System*	*Status*	*Range/ Payload*	*Country of origin*	*Notes*
Poland	SS-21	O	70-120/480	USSR	
	Scud B	O	300/1000	USSR	
Romania[17]	Scud B	O	300/1000	USSR	Unilateral?
Saudi Arabia	CSS-2/DF-3	O	2650/2150	PRC	Non-nuclear
Serbia	K-15 Kraijina	D	150/?	I	
	Scud Mod[18]	D	400/700	???	
South Korea	Nike-Hercules-1	O	180/300	U.S./I	Mod SAM
	Nike-Hercules-2	D	250/300	U.S./I	Mod SAM
Slovakia	SS-21	O	70-120/480	USSR	
	Scud B	O	300/1000	USSR	
	SS-23[19]	O	500/450	USSR	Banned by INF Treaty
Syria	SS-21	O	70-120/480	USSR	
	Scud B	O	300/1000	USSR	
	Scud Mod C[20]	O	500/700	DPRK	
	M-9	D?	600/950	PRC?	
Taiwan	Ching Feng	O	130/400	I/Israel?	Green Bee
	Tien Ma	D?	950/500	I/?	Sky Horse
	Sky Spear[21]	D	300/?	I	Mod SAM
UAE	Scud B	O	300/1000	Russia?	
Ukraine	SS-21	O	70-120/480	USSR	MOU
	Scud B	O	300/1000	USSR	
Vietnam	Scud B	O	300/1000	USSR	
Yemen	SS-21	O	70-120/480	USSR	
	Scud B	O/U	300/1000	USSR	
Zaire	Scud Mod B	O	300/1000	DPRK22	

ABBREVIATIONS:

Status	*Country of Origin*	*Notes*
D: Development	I: Indigenous	MTCR:Member of Missile Technology Control Regime
O: Operational		Unilateral: Unilateral Commitment to MTCR
P:Production		MOU: Memorandum of Understanding on adherence to MTCR

(Contd)...

Status	***Country of Origin***	***Notes***
S: Storage		SAM: Surface-to-air missile
T: Tested		Mod SAM: SAM modified for use as a ballistic missile
U: Used		From SAM: ballistic missile based on SAM technology

Notes

1. At a panel discussion April 21, 1998 at the Carnegie Endowment in Washington, D.C., General Alexander Yesin, Deputy Secretary of the Security Council of the Russian Federation (and former chief of staff of the Strategic Rocket Forces) confirmed the Russian de-nuclearisation of its ABM system. For decades, Russia has maintained the world's only operatioinal anti-ballistic missile system, with up to 100 *Galosh* and *Gazelle* nuclear-tipped interceptors. General Yesin explained that the interceptors remained with conventional warheads and could now provide a defence against a limited missile attack. This is unlikely, however, since the *Galosh* and *Gazelle* interceptors were configured with nuclear warheads to compensate for their poor accuracy. Russia has reportedly deployed new versions of the S-300 air defence missiles around Moscow, the same missile involved in a controversial arms sale to Cyprus. Similar to the short-range *Patriot* system, this system, too, would have no capability against long-range missiles.
2. Principle sources for this table include Humphry Crum Ewing, Robin Ranger, and David Bosdet, "Ballistic Missiles: The Approaching Threat," Bailrigg Memorandum 9, 1994, Center for Defence and International Security Studies; Robert Shuey with Craig Cerniello, "Ballistic and Cruise Missile Forces of Foreign Countries," CRS, June 5, 1995; Defence Intelligence Agency, "Global Missile Proliferation Threat," Presented to the Missile Technology Control Regime Transshipment Seminar, July 15, 1996; Department of Defence, *Proliferation: Threat and Response*, November 1997; "Missile Proliferation" in *The Military Balance 1995–1996* (London: International Institute of Strategic Studies, 1995) pp. 281-284; *The Military Balance 1997/98* (London: International Institute of Strategic Studies, 1997); Duncan Lennox, "Ballistic Missiles," *Jane's Defence Weekly*, April 17, 1996, p. 40; National Intelligence Estimate, November 1995, printed in Bill Gertz, "Intelligence Report Warns of Missile Launches Against U.S.," *Washington Times,* May 14, 1996, p.

A3; "Artillery Rocket, Ballistic Missile, Sounding Rocket, and Space Launch Capabilities of Selected Countries," *Non-proliferation Review*, Spring-Summer 1996, pp. 162–165; and *The Proliferation Primer* (Washington, D.C.: U.S. Senate Committee on Governmental Affairs, January 1998).

3. Russia shipped 8 Scud launchers and 24 missiles to Armenia between 1992–95. Nikolai Novichkov, "Russia details illegal deliveries to Armenia," *Jane's Defence Weekly*, April 16, 1997, p. 15.
4. Vaseil Lyutskanov, "Existence of Eight SS-23 Missile Complexes Viewed," *TRUD*, September 13, 1996 in *FBIS-FEU-96-179*, September 16, 1996.
5. The Czech Republic dismantled its Scud-B inventory between 1988 and 1991. The last SS-23 and associated laucher and support equipment in the Czech Republic was destroyed by mid-1996. "Czechs Destroy Last Soviet Missiles," *OMRI Daily Digest*, July 26, 1996, p. 4.
6. According to Admiral Studeman, "Pyongyang has provided Scud missiles and production equipment to Egypt." See Senate Committee on Armed Services, *Worldwide Threat to the United States*, January 17, 1995, p. 39. The DIA lists Egypt as a recipient of missile-related transfers from North Korea. Egypt reportedly received seven shipments of Scud Mod C-related material, possibly including production equipment, in 1996. Bill Gertz, "Cairo's Misile Buy Violates U.S. Laws," *Washington Times*, June 21, 1996, p. 1.
7. The DOD lists a 200 km Zelzal missile and a 150 km Nazeat missile, which may be variations of the Mushak series. Iran has also tried to acquire a complete North Korean Nodon system and the Chinese M-9 and M-11 missiles.
8. During the Iran-Iraq War, Libya and Syria shipped Soviet-built Scud-Bs to Iran DOD lists the Libyan-supplied Scuds as still in Iran's inventory.
9. A recent intelligence report called the Al-Amoud a "scaled down Scud." See "Iraq's Weapons of Mass Destruction Programmes," U.S. Government White Paper No. 3050, released February 17, 1998.
10. The DOD and DIA say Libya's only operational missiles are Scud Bs acquired from the USSR. However, then Director of Central Intelligence Deutch listed Libya as one of the recipients of North Korean Scud missiles (possibly the Scud Mod B or C). See Senate Select Committee on Intelligence, *Current and Projected National Security Threats to the United States and Its Interests Abroad*, February 22, 1996, p. 9. The DIA also lists Libya as a

recipient of missile-related technology from North Korea, but according to a March 1995 CIA report, *The Weapons Proliferation Threat*, Libya possesses only Scud-Bs. Libya has also sought to acquire the North Korean Nodong missile, but is not reported to have made the purchase yet.

11. According to the DOD, Libya's indigenous missile programme has only succeeded in producing missiles with ranges of about 200 km. Libya hopes that the Al Fatah will reach ranges of up to 950 km, but so far it has only been successfully tested to 200 km. A Serbian firm, JPL Systems, is reportedly aiding Libya's Al Fatah missile programme. Bill Gertz, "Serbia Is Helping Libya With Ballistic Missiles, CIA Says," *Washington Times*, November 12, 1996, p. 3.
12. "North Korea builds and is likely to offer for export earlier Scud-based, short-range ballistic missile systems in the 300, 500, and probably 800 kilometer range. We are talking here about what is known as Scud B, C and D systems." Admiral Studeman, Senate Committee on Armed Services, *Worldwide Threat to the United States*, January 17, 1995, p. 18. An 800 km Scud Mod D has not been listed by any other sources.
13. A December 1995 DIA report estimates that the Nodong can carry a 1,000 kg payload to 1,000 kilometers. Defence Intelligence Agency, *North Korea: The Foundations for Military Strength Update 1995*, December 1995, p. 6.

    "We assess the Nodong is capable of delivering a 700 kg payload to 1,000 kilometers." Admiral Studeman, Senate Committee on Armed Services, *Worldwide Threat to the United States*, January 17, 1995, p. 39. Reportedly, Syria, Libya and Iran are interested in purchasing the Nodong when it is operational.
14. The Department of Defence *Proliferation: Threat and Response*, November 1997, notes that the Taepo Dong 1 has a range of at least 1,500 km. Other sources including Shuey and Cerniello, *op. cit.*, and Ewing *et al.*, suggest a range of 2000 km and a payload of 1,000 kg. The CIA believes "it is unlikely Pyongyang could deploy Taepo Dong I or Taepo Dong II missiles before three to five years." Responses to Questions for the Record, dated April 3, 1995, from Admiral William Studeman, Acting Director of Central Intelligence in Senate Select Committee on intelligence, *Worldwide Intelligence Review*, January 10, 1995, p. 105.
15. One analysis suggests that Pakistan developed the Hatf 2 based on French sounding rocket engines that it had obtained. See S. Chandrashekhar, "An Assessment of Pakistan's Missile Capability," *Jane's Strategic Weapon Systems*, March 1990, p. 4.

16. Several reports raised speculation about the development of the new IRBM Ghauri missile in Pakistan prior to its surprise test April 6, 1998. The missile has an apparent range of 1500–2000 km. See Ben Sheppard, "Too close for comfort: ballistic ambitions in South Asia," *Jane's Intelligence Review*, January 1998, pp. 32–5; "Pakistan: Pakistani Daily Reports Ghauri Missile Development," *Rawalpindi Jang*, January 3, 1998, p. 4, translated in *FBIS-TAC-98-005*, January 5, 1998.
17. The United States is assisting Romania in the dismantlement of its Scud missile and launchers. Prepared Statement of Thomas McNamara, Assistant Secretary of State for Political-Military Affairs, Senate Sub-committee on International Economic Policy, Export and Trade Promotion, March 12, 1997.
18. See the CDISS website at http://www.cdiss.org/btablea2.htm#SAUDI.
19. Slovakia's possession of SS-23 missiles has been confirmed by Prime Minister Meciar. Nora Sliskova, "Meciar Comments on NATO Referendum, SS-23 Missiles," *PRAVDA*, November 30, 1996, in *FBIS-EEU-96-232*, December 4, 1996.
20. Admiral Studeman lists a Scud Mod C transfer from North Korea to Syria. Senate Committee on Armed Services, *Worldwide Threat to the United States*, January 17, 1995, p. 39. The Nonproliferation Center's *Weapons Proliferation Threat*, March 1995 states that Syria has both the Soviet-supplied Scud-B and North Korean-supplied Scud Mod-C.
21. This programme was reportedly begun in the fall of 1995 and is based on the Sky Bow II SAM. Lu Chao-lung. "Taipei to Test Surface-to-Surface Missiles," *CHUNG-KUO SHIH-PAO*, September 11, 1996 in *FBIS-CHI-96-180*, September 17, 1996.
22. Duncan Lennox, *op. cit.*, lists Zaire as a possessor of Scud B variants. North Korea is the only known supplier of such missiles, but this transfer has not been otherwise confirmed.

Address:

JOSEPH CIRINCIONE

Carnegie Endowment for International Peace
Washington DC 20005
USA

28
U.S. Nuclear Weapons Research and Development under the Comprehensive Test Ban Treaty

THOMAS B. COCHRAN

INTRODUCTION

U.S. nuclear weapons policy, as currently implemented by the Department of Energy (DOE) Stockpile Stewardship and Management Programme (SSMP), seeks to retain Cold War capabilities and develop new ones. In this paper we focus on the emerging technical capabilities of the United States to design and certify nuclear weapons absent underground testing with super-critical assemblies of fissile material. Recently-available government documents depict a strategy for acquiring advanced research facilities which moves in this direction. Details are also now emerging about the Submarine Launched Ballistic Missile Warhead Protection Programme, a joint DOE/Department of Defence (DoD) initiative to design and certify "replacement" warheads for the currently deployed Trident strategic weapon systems. A discussion of x-ray laser research in the SSMP is given, both as an example of the scientific work supporting warhead certification under the CTBT and as an example of continuing advanced nuclear weapon design activities.

SSMP CERTIFICATION OF WEAPONS

The DOE defines certification as to "render judgment on the ability of weapons or their components to perform their functions as defined by military requirements."[1] Two reports play a significant role in the certification process: the *Major Assembly Release* and the *Final Warhead Development Report*. "(Certification) is initially reported in the weapons *Major Assembly Release* (MAR). The MAR is updated as necessary when there are changes in the military requirements or changes in the weapon's ability to meet military requirements."[2] The *Final Warhead Development Report* is signed by the design laboratory director-either Los Alamos National Laboratory (LANL) or Lawrence Livermore National Laboratory (LLNL) when a weapon is introduced into the stockpile, and Department of Defence (DoD) uses this document in its own certification process, which is carried out by the Design Review and Analysis Group (DRAAG).

DOE performs both "Non-recurring Certifications" and "Recurring Certifications". While technical issues are presumably common to both activities, Non-recurring Certifications are of interest with respect to changes to the U.S. stockpile and new nuclear weapon design work. Five planned Non-recurring Certifications are shown in Table 1. Three "W87, W88, and SWPP) represent a certification following modifications affecting the nuclear explosive package.[3] Since the nuclear explosive package, unlike most components of a modern. U.S. nuclear weapon, cannot be fully tested under the CTBT, modifications to it represent a more significant departure from current experience than changing out other components. NRDC has publicly argued that modifications to the nuclear explosive package should serve as a demarcation between weapons activities consistent and inconsistent with the intent of a test ban. The controversial B61-11 modification is being performed to enable an earth-penetrating mission (i.e., penetration of 3–6 meters of concrete or hard rock before detonation in order to destroy targets buried hundreds of meters deep). While this is not a modification to the nuclear explosive package, the B61-11 does represent a change in the warhead which provides the U.S. with a new military capability.[4]

There are two principal Recurring Certifications: "Annual Certification" and "Yield Certification." Annual Stockpile Certification was initiated by President Clinton in 1995 and is scheduled for completion in August of every year:

> The goal of this process is to certify the safety and reliability of each type of weapon in the active stockpile, or identify significant issues that may need to be addressed by a nuclear test. The certifications are provided to the Secretaries of Defence and Energy, who are advised by the NWC (i.e., Nuclear Weapons Council), the Directors of the DOE nuclear weapons laboratories, and the Commander in Chief of the U.S. Strategic Command (CINCSTRAT).[5]

The Senate has just passed an amendment to pending legislation setting up what is in effect a parallel annual certification process.[6] The future of the other Recurring Certification-Yield Certification-is currently undergoing review:

In the past, DOE has provided weapons on a more or less periodic basis. The most recent certification was transmitted in March 1995. That certification was based on a re-examination, by the laboratory technical staff that designed and tested the stockpiled weapon, of all the new data that might impact the weapon's yield. With the cessation of underground nuclear testing, DOE is currently reviewing the methodology that has been used in the past to revalidate all aspects of stockpile nuclear weapon performance, including the yield. The results if this review, in part, will determine if future certifications will be issued periodically or only when circumstances warrant.[7]

Table 1 : Non-recurring Certificaitons of U.S. Nuclear Weapons. From the "Stockpile Stewardship and Management Plan: First Annual Update", pp. 4–10 to 4–11.

Non-recurring Certification	***Schedule***	***Description***
B61–3,4,10 Surety Upgrades	To be determined	Weapon modificaiton includes improvements to the traiectory sensing signal generator, an encryption modification of the permissive action link (PAL), use-control life extension, and other possible safety and security modifications.
B61–11	FY 1997	Modification of the B61–7 to meet the mission of the B53 (destruction of deeply buried structures).
W87 Life Extension programme	FY 1998	Structural modications for both retention of the Peacekeeper and to support single reentry vehicle deployment on Minuteman III. Modifications will affect the nuclear explosive package.
W88 Surveillance Rebuilds require	Not given	Pits manufactured at LANL to replace pits destroyed in the surveillance programme re- certification.
Submarine Launched Ballistic Missile Warhead Protection Programme (SWPP)	To be determined	"An assessment of each design and a proposed process for their certification will be conducted..."[8]

Given the above discussion of the form that warhead certification processes now take, it is of interest to also consider the technical content.

> Certification is guided by extensive calculations with validated three-dimensional codes, and independent peer review.[9]

Calculations with validated three-dimensional codes are the focus of the long-term (i.e., to 2010) SSMP certification strategy.[10] Here the computer simulation of a nuclear explosive is being advanced in pieces which are then to be integrated. For example, the Dual-Axis Radiographic Hydrodynamic Test Facility (DAHRT), under construction at LANL, represents an advance in the experimental capability to collect data on some aspects of primary performance for comparison to new calculations. While some stages of a nuclear explosion or some bomb components can be experimentally studies and modelled in isolation, key processes such as boosting cannot:

> ...the entire implosion and explosion history affects the boost process... Non-nuclear experimental capabilities do not adequately address the boosted explosion phase of a primary. The complexity of the coupled physical processes and the catastrophic consequences to warhead performance would result from primary boost failure and require significant advances in theoretical and calculational capabilities.[11]

Clearly the integration of sub-models into a comprehensive model of weapon behaviour also cannot receive experimental validation under as testing moratorium, thus "...the existing surveillance strategy of testing at the highest attainable level of systems integration remains a cornerstone of the certification strategy."[12] It is possible, however, to compare new calculations with past test data (a process referred to as "post-diction"). This work is being performed under the SSMP, for which test data is being reviewed and achieved.

MODIFICATIONS TO NUCLEAR WEAPONS IN THE U.S. STOCKPILE: "REFURBISHMENTS"

Currently there is no requirement for DOE to design and build new nuclear explosive packages to replace those in today's stockpile weapons. On the other hand, a broad range of weapon design activities are ongoing and are referred to as "refurbishments" of existing warheads. "Refurbishment" of the nuclear weapon stockpile includes period maintenance activities, as well as modifications to nuclear weapon components through design, development and production. In 1997 DOE stated: refurbishment will be derived from a need to:

- perform modifications in response to a change in military requirements;
- redress performance or surety[13] concerns developed through the surveillance and certification functions; and
- replace components whose performance can no longer certified."[14]

The on-going B61-11 conversion is characterised as "a change in military requirements." Furthermore, DOE states that to "(p)rovide and demonstrate the capability to design and developed new nuclear weapons and associated components" is a stockpile refurbishment programme need.[15] And, of course, certain components of nuclear weapons such as boost-gas generators, boost-gas transfer systems, boost-gas reservoirs, neutron generators, and power supplies are replaced on a periodic schedule, sometimes with components of improved design.[16]

All components replaced in refurbishment will be certified, individually and in the sub-system or system, to meet Military characteristics (MCs)[17] and Stockpile-to-Target-Sequence (STS)[18] normal, abnormal, and hostile environment criteria. DOE states that if this cannot be accomplished, a joint re-examination of the Military Characteristics may be initiated and/or a request for a nuclear test may be initiated.[19]

SLBM WARHEAD PROTECTION PROGRAMME

In 1995 the U.S. Navy requested DOE to begin a joint design and certification process for potential replacement warheads for the two Trident strategic weapon systems. The subsequent development of this programme, called the Submarine Launched Ballistic Missile Warhead Protection Programme (SWPP) is described in detail in the Appendix. In the SWPP, two warhead designs were selected for Phase III[20] development: one that employs the technique of pit reuse and one that incorporates a newly designed pit. Unlike for the B61-11, these designs include changes to a tested nuclear explosive package and the development of a new one, although no decision to produce and deploy these warheads has yet been made.

Both replacement warhead design were intended to improve upon the W76 and W88 in several areas-ease of surveillance, resistance to aging, safety (i.e. use of intensivite High

Explosives (IHE) and Fire Resistant Pits (FRP), and possibly use-control improvements-while possessing nearly identical Military Characteristics and a Stockpile-to-Target-Sequence as the W88. Both design are intended to be compatible with the Mark 5 (Mk5) reentry vehicle.

Currently available documents do not describe the SWPP certification strategy in detail. LLNL has made clear that the certification process itself will not occur during the SWPP for the reused pit design, but is relegated to a follow-on programme. It should be noted that budgetary, not technical issues, are cited in this decision. An indication of the certification activities for the reused pit design was made in 1997:

> Calculation modeling of pit reuse primary performance will incorporate all improvements in models of IHE behaviour, plutonium equation of state, mix, and other parameters discussed in Chapter 7 (i.e., Chapter; 7 of the first Annual Update to the Green Book), as they become validated and in the application codes capabilities are anticipated during this time frame and will be a critical part of the final establishment of certification performance.[21]

This quotation reaffirms that new weapons models supported by new data will serve to certify and modified nuclear explosive package without underground testing. Note the specific reference to IHE behaviour: IHE was not used for the W76 or W88 but is envisioned for the SWPP designs.

The new design pit option for the SWPP is being pursed at LANL. Here the goal is explicitly to design a primary which could be certified without underground testing:

> The replacement warhead for the Mk5 RB [i.e., reentry body] is currently in a conceptual design stage. A major certification issue being addressed includes a review of the design envelop for robust or high margin primary designs based on past nuclear test data. The design envelop for non-nuclear and nuclear weapons components has been historically determined by constraints of the delivery system and its STS and MCs. The replacement warhead option for the Mk5 will be further constrained by certification issues without additional nuclear testing and by future manufacturing and production capabilities.[22]

One goal of the SWPP is to pursue design and certification to the point where complete full-scale engineering development may be completed three years from the time that the replacement option is authorised. Currently, warhead flight tests for the pit reuse design are scheduled for fiscal years 1999 and 2000, and flight tests for the new pit design for 2000 and 2002.[23]

X-RAY LASERS IN THE SSMP

The National Ignition Facility (NIF) is intended to provide

experimental data for the SSMP at conditions similar to those created in a nuclear explosion. The principle categories of weapons-physics experiments planned for the NIF are: opacity experiments; radiation-flow experiments; equation-ofostate experiments; non-LTE (Local Thermodynamic Equilibrium) and x-ray experiments; hydrodynamic experiments; and capsule implosion experiments.[24] Research in x-ray lasers is currently pursued in the U.S. nuclear weapons programme because of their demonstrated potential as plasma diagnostic instruments (i.e., to produce calibration data for nuclear weapons and inertial confinement fusion codes) and in dual-use applications, as well as for their relevance to advanced nuclear weapons concepts.

Recall that laser is an acronym for "Light Amplification from Stimulated Emission of Radiation." In the Lasing process, atoms or molecules are stimulated by ambient photons to make a quantum transition to a lower energy state and emit photons of similar frequency. For this to occur, more atoms or molecules must be in a higher-energy state than would normally be the case in thermodynamic equilibrium (e.g., as given by a Blotzman distribution). This condition necessary for lasing is called a "population inversion," and is an example of non-LTE physics.

A 1995 LLNL report discusses a three-year experimental plan for non-LTE x-ray laser experiments on the NIF, totalling 200 shots.

> These are needed to maintain our core competency in non-LTE design and to develop short-wavelength x-ray lasers for dual-benefit applications. ...Our plan includes developing a test-bed for non-LTE physics, exploring new regimes and schemes in x-ray laser research, and using x-ray lasers for plasma-imaging diagnostics. These lasers are used for plasma imaging because they provide absolute plasma-density measurements. This information is needed to validate codes and benchmark data for both the ICF (i.e., Inertial Confinement Fusion) and weapons-physics communities.[25]

LLNL posits three categories of non-LTE and x-ray laser experiments: plasma imaging experiments; ionisation balance experiments; and "new regimes and schemes for x-ray lasers."[26] Plasma imaging is the application of short-wavelength x-ray lasers as a tool to diagnose rapidly evolving (at time scales less than one nanosecond) high electron density plasmas (10^{21} cm^{-3} plasma electron density 10^{24} cm^{-3}) in two dimensions.[27] X-ray lasers have yet to be fully developed as reliable diagnostic instruments. Ionisation balance experiments involve the study of photo-ionisation and (electron-ion) recombination kinetics and radiation transfer in non-LTE plasmas. Nuclear weapon code calibration is emphasised for these kinds of experiments ("a test-bed for non-LTE physics").

Research is planned for the NIF both on the physics of x-ray lasers and on new x-ray laser architectures. "The large energy available on NIF will use to extend existing x-ray laser schemes into new regimes and to test new schemes for producing short-wavelength x-ray lasers."[28] Furthermore:

> To maintain core competency in advanced nuclear design while developing laboratory x-ray lasers for dual-benefit applications. Defence Science is planning non-LTE and x-ray laser physics experiments for the new National Ignition Facility (NIF). In the first two sections [i.e., of this LLNL Report] we review some of the important physics experiments that were done underground at the Nevada Test Site (NTS) and describe NIF experiments that can help improve our physics understanding of these issues. ...[deleted lines) The goal (presumably the goal of the nuclear tests] was to assess the technical feasibility of an x-ray laser weapon to avoid technological surprises from an enemy. [deleted lines] The basic idea was to create a bright x-ray beam that could destroy a target, such as an ICBM [i.e., and Intercontinental Ballistic Missile] or satellite, at a great distance.[29]

Such work continued on the NIF is referred to as "enhancement of the technology infrastructure."[30] Finally, prefacing a discussion of the "Fast Ignitor"[31] concept for ICF, LNNL notes: "...we describe a number of advanced concepts that could help maintain the design skills of weapons physicists in areas ranging from secondary physics to nuclear directed-energy weapons concepts."[32]

NRDC has learned anecdotally that in the 1980's a capability like ignition with gain on the NIF was viewed by some at the National Laboratories as important for progress in the Strategic Defence Initiative. It is clear from recently-declassified documents that there is a technical connection between the NIF and such weapon concepts, and that development work will continue in the SSMP.

CONCLUSIONS

The U.S. certification process for nuclear weapons historically involved nuclear explosion. Under the CTBT, the SSMP strategy for certification stresses advances in computer modeling validated by data from new research facilities. Five Non-Recurring certifications, performed for example when a modified or new warhead is introduced into the stockpile, are now planned.

Three of the warhead certifications are required due to modifications of the nuclear explosive packages. Two certifications are planned because of significant modifications to components outside the nuclear package. One of these, the B61-11 retrofit, provides the United States with a new nuclear military capability.

Certification issues for new-design warhead under a test ban are being addressed in the SWPP. While entailing modifications to the nuclear explosive package, the SWPP designs do not appear to be qualitatively far removed from tested designs now deployed in the U.S. stockpile. By contrast, x-ray laser research is planned for the NIF which is technically linked to nuclear directed-energy weapon concepts that were pursued unsuccessfully as part of the Strategic Defence Initiative.

DOE is putting in place the necessary tools to permit certification of changes to non-nuclear components as well as to the nuclear explosive package under the CTBT. A significant future threshold will be the design and certification without testing of an entirely new weapon in order to increase U.S. military capabilities. Achieving this capability will ultimately depend on the robustness of the Stockpile Stewardship and Management Programme, now funded at about $4.5 billion annually. Two international security impacts of this course of development are plausible. First, when explosive testing which typically produces remotely-detectable seismic waves is a requirement for militarising nuclear capability, a well-monitored test ban serves as an effective means of arms control and non-proliferation. However, if the United States pursues the capability to design and certify nuclear weapons without such testing, the perceived security benefits of the CTBT for rival nuclear powers diminishes. Second, the capabilities to be generated by the U.S. SSMP enable continued effort towards advanced weapons concepts which were not realised during the Cold War.

APPENDIX

Detailed Chronology of the U.S. Submarine Launched Ballistic Missile Warhead Protection Programme April 1995 to January 1998

The following chronology of the Submarine Launched Ballistic Missile (SLBM) Warhead Protection Programme (SWPP) is derived from 41 documents released to NRDC pursuant to the Freedom of Information Act (FOIA), spanning the time period April 1995 to January 1998.

In a 14 April 1995 memo to Harold P. Smith, Chairman of the Nuclear Weapons Council Standing and Safety Committee (NWCSSC), the Director of Navy Strategic Systems Programmes stated:

> As a result of the continuing need for a reliable strategic deterrent and the results of the recent Phase II Study, it is requested that the DOE join with Navy Strategic Systems Programmes in a cooperative effort to ensure the availability of alternate warhead designs for the MK4 and MK5 SLBM re-entry systems. It is envisioned that this effort, coordinated through the existing Project Officers Group (POG), will address the analysis and above ground testing of a limited number of design options.[33]

Here the Navy requested (jointly with the DOE) permission from the NWV to draw up a "detailed plan and schedule" for completing design work on a few warheads for the submarine leg of the U.S. strategic triad, potentially "to a point where (they) could be productionised."[34] In the past, nuclear weapons design, production and retirement was structured into seven phases as shown in Table 2. The Phase II or "Joint Feasibility" study referred to in this memo implies a prior Phase I or "Concept Definition" study, possibly derivative of the original W76 and W88 Phase I studies. As Phase III—"Development Engineering"—in the past culminated in a nuclear warhead design and Phase IV—"Production Engineering"—adapted the design into a manufacturing system, the Navy in April 1995 effectively solicited preliminary approval for both Phase III and Phase IV work, while not explicitly using these terms. Prior to its termination, the Military Liaison Committee (MLC) had responsibility to approve or deny a Phase III request, which is now the responsibility of the NWC.

Table 2 : Summary of the seven phases of nuclear weapon research, development, production and retirement practised by the United States during the Cold War.[35]

Phase I	Concept Definition Studies	Definition of a new nuclear warhead or component concept, including performance parameters, transportability, employment concepts, delivery techniques, yield and/or effect selection, fusing options, typical targets, safety considerations and control requirements.
Phase II	Joint Feasibility Studies	Determination of the technical feasibility of developing a nuclear warhead to meet the Phase 1 requirement.
Phase IIA	Joint Design Definition and Cost Studies	Estimate of costs and production schedules. Analysis of tradeoffs involving the safety, security, survivability and control features of the weapon.
Phase III	Development Engineering Project	Start of warhead development at a DOE weapon laboratory, culminating in a proposed warhead design.
Phase IV	Production Engineering	Adaption of the Phase 3 design into a manufacturing system that can mass produce the warhead.
Phase V	Initial Production	Delivery of the first warhead and limited production
Phase VI	Quantity Production	Quantity production
Phase VII	Retirement	Retirement

While the Cold War roles of DOD as nuclear weapons "customer" to DOE is reflected in the SWPP, programme responsibility and organisation within DOE required definition. A 28 June 1995 memo from the Deputy Assistant Secretary for Military Applications and Stockpile Support (DP-20) to the Deputy Assistant Secretary for Research and Development (DP-10) states:

To ensure success in this endeavour (i.e., SWPP), DOE should clearly delineate areas of responsibility with Defence Programmes. I envision a process in which DP-10 is responsible for funding and leading this effort during the programme's development phase. While DP-10 has the lead, my office, DP-20, will serve as overall coordinator/observer. If production of one of the alternative warheads is required by the Navy, the funding and lead for production and stockpile support will transition to DP-20. ...This approach to the division of responsibility provides the necessary framework for us to accomplish our required programme function while minimising concerns over questions of "who is in charge." This effort will be an excellent foundation for delineation of responsibilities in any future stockpile warhead modification or replacement activity.[36]

This division of responsibility appears to have been implemented at about the same time.[37]

On 25 May 1995 Smith wrote to Vic Reis, DOE Assistant Secretary of Energy for Defence Programmes, advising him that the SWPP "must be consistent with the findings of the Nuclear Posture Review and not preclude in any way similar efforts by other Services."

Subsequently on 22 June 1995, a joint planning meeting which involved Los Alamos National Laboratory (LANL), Lawrence Livermore National Laboratory (LLNL), and Sandia National Laboratory (SNL) was held at the DOE complex in German town, Maryland. Acting Director of DOE Defence Programme's Office of Development and Testing (DP-12) Kenneth J. Adney called the meeting so that DP-12 could be briefed on the results of the Phase II study, be provided suggestions by the National Laboratories for follow-on activities, and generate debate on the budgetary competition between the SWPP and other DP-12 "new programme activities". "The results of this meeting will be either a DOE coordinated and integrated development plan supported by each design laboratory or a path to achieve such a DOE plan."[38] Since all three labs briefed DP-12 on the Phase II study it can be assumed that it was a collaborative effort, as the SWPP continues to be. Both the W76 and W88, however, are LANL designs.[39]

The DOE planning session at German town was certainly in preparation for the SWPP "kickoff meeting" held one week later at Kaman Sciences Corporation in Arlington, Virginia. The Kaman agenda lists four potential candidate nuclear warhead types to be discussed: Safety Pod, Pit Reuse, New Pit, and Replacement. A memo accompanying the agenda notes: "The candidates that will be used for this initial planning effort will be those briefed at the recent Stratcom sponsored Stockpile Confidence Symposium. Additional candidates may be proposed by the principals as appropriate."[40] This statement is noteworthy, since the Phase I and II studies should logically have served to provide the set of options for Phase III. The Safety Pod is nowhere else mentioned in the documents released to NRDC, and the Replacement may refer to the W94 warhead (a new warhead type listed in one of Cuneo's viewgraphs in preparation for the 29 June 1995 SWPP "kickoff" meeting, discussed below).

PIT REUSE and NEW PIT were subsequently developed in the SWPP, as discussed below.

Also on the agenda for the SWPP "kickoff meeting" were there topics to be presented by the National Laboratories:

- Identify aging mechanisms for nuclear and non-nuclear components.
- Yield certification process
- Expanded tests and analysis techniques to eliminate need for UGT [UnderGround Tests]

These three issues are central the SWPP, and to a general discussion of the Stockpile Stewardship Programme with US development of new nuclear designs under a CTBT.

Predicated on the continuing US adherence to the test moratorium, the SWPP entails development of alternatives to underground nuclear testing in order to certify the yield and other military characteristics of future designs. One day prior to the kick-off meeting, Rear Admiral Beers, DOE Deputy Assistant Secretary for Military Application and Stockpile Support, directed the DOE Albuquerque Operations Office to: "allow for the integration of new science-based stockpile stewardship facilities [into the SWPP plan and schedule] as they become available. These new facilities should become integral elements for maintaining the reliability and safety of the MK4 and MK5 re-entry systems and developing alternate warhead designs".

Language in Navy's earlier memo to the NWC was reiterated to participants at the SWPP "kickoff meeting".

> A working group of the W76 and W88 POG is being formed to develop a detailed plan and schedule to support the SLBM Warhead Protection Programme recently requested by Navy Strategic Systems Programmes. The plan and schedule should focus the technical efforts on warhead candidates which protect the reliability and safety margins of the SLBM strategic deterrent while maintaining the design expertise needed to respond to operational system needs. This group will address the analysis and above ground testing of a limited number of design options. These options would address candidates of increasing safety margins. The candidates would be time phased to provide alternatives in the near-term, mid-term, and far-term, should they be needed. If a candidate design is brought to a point where it could be productionised, but is not yet needed, due to the continuing longevity of the exi[s]ting W76 and W88 Warheads, it can be shelved and work continued on the longer time frame candidates.

This basic concept for the SWPP was subsequently developed and approved by the NWCSSC. A viewgraph from Lt. Col. Jeffrey Cuneo (DOE Office of Development and Testing, DP-12) distributed for comment prior to the kickoff meeting is shown in

Figure 1. Importantly, the SWPP is characterised as an opportunity to apply Stockpile Stewardship concepts to a potential future weapons system, and is required to maintain weapon design expertise and capabilities.

SLBM WARHEAD PROTECTION PROGRAMME

GROUND RULES

- Must be manageable within budget/manpower
- Not a directed programme—opportunity to realistically apply science-based stockpile stewardship concepts
- Stated requirement is to maintain weapon design expertise and capabilities
- Must consider effects to stockpile management plans

Fig. 1: DP-12 vu-graph distributed for comment prior to the 29 June 1995 SWPP "kickoff" meeting held at Kaman Sciences Corporation, Arlington, Virginia.

On 3 July 1995, Victor Reis wrote to Harold Smith, indicating that DOE "is pleased to join the Department of Defence (DoD) in an effort to propose a plan and schedule for ensuring the availability of alternate warheads for Submarine Launched Ballistic Missile (SLBM) reentry systems". Reis noted: "... as science-based stockpile stewardship facilities become available, DOE can direct their use to support this activity".41 Thus prior to defining the SWPP plan and schedule for the NWC, Reis has indicated to the NWC that new Stewardship Facilities, such as the LANL's Dual-Axis Radiographic Hydrodynamic Test Facility (DARHT) and LLNL's National Ignition Facility (NIF), can play a role in alternate warhead development.

During the following six months, DOE and DoD organised the SWPP and briefed the NWCSCC. Meetings involving all three National Laboratories and DoD occurred at LANL on August 1. DOE management's concerns were that the SWPP should not be seen as distinct from the Stockpile Stewardship Programme:

> The plan and schedule should emphasise the strategies depicted in our Stockpile Stewardship and Management Programme. It should also allow for the

integration of new science-based stockpile stewardship facilities as they become available. Finally, the plan should ensure that the alternative warhead designs considered will maintain design expertise and capabilities that are necessary to respond to the Navy's operational system needs.[42]

Was DOE management concerned that weapons scientists and engineers would approach the SWPP with strategies that would differ from Stewardship, and perhaps not include facilities such as the NIF? The SLBM warhead protection programme, although described as a warhead design effort, is not here characterised as a response to Navy "operational system needs", as no decision to proceed to production is envisioned in the near term.

By 18 September 1995, LLNL had generated a "SWPP Pit Reuse Tier-0 Draft Plan", developed by James Tyler of LLNL and Doug Henson of Sandia National Laboratory's California site (SNL/CA). In a cover memo dated 15 September, Tyler describes the plan:

It proposes a schedule, three different phases with deliverables and reviews at the end of each phase, and a final deliverable at the end of the project. It also describes generic activities and objectives within each phase. It is a "Tier-0" project plan, which is intended to provide an overall structure within in which to plan major project activities, such as ground and flight tests. This letter plan of major, specific activities will constitute the Tier-1 plan for the project.[43]

Moving cautiously, Tyler remarks: "Once the DOE community has arrived at a mutually agreeable Tier-0 plan, we would propose to forward it to the navy SP (Strategic Programmes] Office. We solicit comments from DOE/HQ [DOE Headquarters] regarding how to do that."[44]

The Pit Reuse Project within the SWPP aims to achieve a warhead design consistent with the Trident II/MK5 re-entry body (RB) delivery system with a reused pit. The final product of this Phase III effort will be a Data package, which Tyler defines as:

...a collection of information that defines the SLBM WPP pit reuse warhead design option and its integration to the MK5 RB and that provides the rationale, from both the experiments and calculations, that this design option would meet requirements and could be fabricated rapidly and entered into the stockpile with high confidence.[45]

It is unclear whether a Data Package was generated for past warhead designs, for which underground test results would have contributed to confidence assessments. Tyler and Henson anticipate a tight budget for the Pit Reuse Project:

The project will be constrained by limited funding, significantly less than was traditionally expended on Phase 3 projects, and some expensive tests will be deferred to the hypothetical, future period when the project might be restored in response to a future stockpile need. This could especially be true for large demonstration tests. The strategy will be to best utilise our limited funds to address the maximum number of vital (or "stowstopper") issues.

A few months later, Tyler would liken the SWPP to the comic strip character "Fealess Fosdick" who gets shot full of holes but nevertheless pushes on.[46] In their Tier-0 plan, a five year schedule for he Pit Reuse project was suggested, and is shown in Table 3.

In November 1995 the Nuclear Weapons Council Standing and Safety Committee was briefed on the SWPP, and approved it on 15 November. A 6 December meeting at Kaman Sciences Corporation included a discussion of what Navy told the Committee:

Mr. Maglich, SSP [Strategic Systems Programme], who will be the Navy's principal Manager for this programme then went over the ground rules for the programme and outlined the proposed sub-groups of the POG/Executive Working Group. This basically reiterated the principals outlined in earlier discussions. His description of the purposes of the different subgroups (Design certification, Design Integration, Safety and Joint Test) elicited some general discussion on the role and philosophy of certification in particular. *All parties view this programme as a forcing function for the DOE to define certification for a new weapon and apply it to the products in SWPP as a trial run* (italics added).[47]

This description of the SWPP as a "forcing function" for the DOE is significant. It is a recognition that sizable facilities in the Stockpile Stewardship Programme will not become operational for years, yet their purpose is predicated in part on an as yet untested strategy for designing and certifying weapons without underground testing.

Table 3 : LLNL/SNL proposed schedule for the SWPP Pit Reuse Project. Each phase concludes with a design report and a review of the project and design (review panel similar to a DRAAG [Design Review and Analysis Group]). The Final Phase design report is intended to be a major component of th eData Package. The Tier-0 plan suggests: "The LLNL and Sandia Directors formally certify the design, as appropriate for this level of development (policy still to be developed)."

Project Phase (duration	**Preliminary** (15 months)	**Interim** (24 months)	**Final** (21 months)
Fiscal Year(s)	1996–97	1997–1999	1999–2000
Activities	Develop "documents similar to MCs, STS, and ICDs" Develop the preliminary	Finalise design and programme requirements	

(Contd...)

Project Phase (duration*	**Preliminary** (15 months)	**Interim** (24 months)	**Final** (21 months)
Fiscal Year(s)	1996–97	1997–1999	1999–2000
	Design concept Conduct early preliminary	Develop complete set of detailed drawings	Produce final set of design drawings (CER-like)
	and tests to examine crucial issues, especially potential "showstoppers" Develop the methodology	Conduct majority of calculations and tests	Complete tests and calculations, emphasize use of test units
	Certification criteria and strategy Develop the methodology		Conduct certification study
	for robustness analysis Develop the surety theme	Begin the robustness analysis	Complete robustness analysis
	Determine the future		
	Fabrication complex and technologies to assume for the project	Developed plans to validate fabricability; make components for test units	Demonstrate fabricability
		Establish methodologies to quantitatively estimate longevity	Provide quantitative estimates of longevity

In late December 1995 a Memorandum of Understanding (MOU) was drafted delineating DOE and DoD responsibilities for the SWPP. Both Departments are tasked without ground and flight tests of the replacement warheads, and separate responsibilities are listed in Table 4. Here the Data Package was identified as the "Deliverable" form this effort, which in addition called for the "Identification of any additional activities required for Full Scale Engineering and Development (FSED) and certification...". The MOU derived its authority on the 21 March, 1953 Agreement between the Atomic Energy Commission (now DOE) and the Department of Defence for the development, production, and standardisation of weapons, AT (29-2)-230, as amended.

On 20 February 1996 the SWPP Design Integration Sub-group (DISG) met at Lockheed Martin Missiles and Space in Sunnyvale, CA. The DISG was tasked with the following responsibilities:

The Design Integration Sub-group (DISG) is responsible to the SWPP Study Group for performing Mechanical and Electrical integration of DOE Hardware into DoD produced MK4 and MK5 hardware to meet SWPP Requirements. The Sub-group will conduct design integration studies and activities, and document findings as required. In conjunction with these activities the Sub-group will recommend appropriate ground and flight tests.[48]

At this meeting, the flight test requirements were discussed, along with issues relating to the development of replacement components. In a second meeting of the DISG at Kaman Sciences Corporation, Colorado Springs, CO, the agenda included a presentation on quantifying the advantages of "using proposed advanced technologies for advanced AF and F [Arming, Fuzing, and Firing] systems by Sandia, and "Shipboard Vibration Data Analysis" by Kaman.[49]

Table 4 : Delineation of DOE and DoD responsibilities for the SWPP, as described in the December 1995 draft MOU. Note the explicit requirement for the DOE to develop alternatives to underground testing in the certification of new warhead designs.

DoD SWPP Responsibility	DOE SWPP Responsibility
Provide reentry body (RB)/missile physical and functional interfaces definitions and DOD requirements (such as the MCs)	Provide component, subsystems, and systems alternative design options including the Advance Fuzing and Firing
Assess the reentry body with the replacement options	Provide a warhead certification methodology which does not rely on future Underground Nuclear Testing (UGT) and is consistent with the forecasted capabilities for both above and underground non-nuclear testing and computational techniques of Science Based Stockpile Stewardship (SBSS)
Assess the acceptability of changes required to the DoD hardware and facilities	Fabricate and evaluate components and subsystem hardware to demonstrate the feasibility of the designs as replacement options

Other SWPP subgroups formed and met in the first half of 1996: the Producibility Subgroup on 13 March (at DOE Albuquerque, Albuquerque, NM) and the Design Certification Sub-group on 17 April (at Kaman Sciences, Colorado Springs, CO). The

Producibility Subgroup (or Process Realisation Team) was tasked to "facilitate coordination between the laboratories and production facilities to ensure that warheads are designed that can be produced within the reconfigured nuclear weapons complex."

The agenda for the Design Certification Subgroup meeting was held "to review the issues involved for certification within this programme and in the larger context of DoD and DOE policy; and to discuss joint testing requirements planned to support the certification portion of the design data report at the end of the programme."[50]

Jim Tyler and Doug Henson, authors of the Pit Reuse Project Tier-0 plan, reacted to the March 13 Producibility Subgroup meeting with a "white paper" discussing the apparent lack of DOE/Albuquerque support for the goal of demonstrating that this design option can be fabricated:" ...it became clear that there is not a common understanding between the laboratories and DOE/AL regarding the level of support the project should expect from DOE/AL."[51] Tyler and Henson stated that they expect the SWPP to include a demonstration that a design option can be fabricated (and should achieve a Product Realisation Plan stating how that could be done in the future). But at the 13 March Sub-group meeting" ... it appeared that at present DOE/AL has no significant funding to apply to the SWPP, it has no planning to supply funding to apply to the SWPP, it has no planning to supply such funding is questionable". They go on to warn: "If this is indeed the DOE/AL position regarding the SWPP, then the programme will have to be drastically downscoped from that we have understood up until now. In particular, the demonstration of fabricability will have to be deleted."[52]

With respect to the Pit Reuse Project, Tyler and Henson reiterate the major programme technical of design and fabricability:

> There are two overall technical objectives for the SWPP/PRP. The first is to certify a replacement design option without further nuclear testing. We plan to use the existing and developing techniques and capabilities of Science Based Stockpile Stewardship (SBSS) to accomplish this end. Indeed, the SWPP/PRP is intended to provide a demonstration of the practical utility of the SBSS approach. The laboratories will take the lead in this part of the effort.
>
> The second objective is to demonstrate the capability to fabricate the design option within the context of the changing production complex. The organisation and locations of fabrication capability are presently undergoing change, and various fabrication technologies may change. As design agencies, the laboratories expect to work closely with DOE/AL and the production agencies with concurrent engineering throughout the design and demonstration cycle.[53]

Tyler and Henson proposed that the first year of effort on Pit Reuse fabrication should consist of study and planning, particularly to identify *expected future* nuclear weapon fabrication technologies. Subsequently, "a substantial amount of funding through DOE/AL (presumable of the SM (Stockpile Maintenance) variety) for the production agencies will also be required, if fabricability is to demonstrated."[54]

In what was sure to be an important issue for the Navy, Tyler and Henson state that if the SWPP Pit Reuse Project was truncated such that demonstration of fabricability was not attempted: "There would be no Product Realisation Plan and no reliable estimate of a future, hypothetical schedule to FPU (First Production Unit), should the programme ever be restarted for entry into the stockpile."[55] DOE funding for fabrication demonstration of Pit Reuse was estimated as modest until fiscal year 1998. As shown below, the SWPP schedule placed the Pit Reuse option several years ahead of New Pit, therefore fabrication issues arose earlier. However the fabrication issues with respect to Pit Reuse and New Pit are clearly very different, and LANL has only recently fabricated a W88 pit (but not a War Reserve pit).[56]

In parallel with the Spring 1996 efforts of the Producibility Subgroup, the Certification Subgroup was also defining its tasks. Jeffrey Cuneo prepared a draft "SWPP Design Certification Subgroup Charter," shown in Figure 2. The Charter calls for the development of a set of criteria for design certification. Notably, Cuneo referred to "emerging DOE certification guidelines." One inference from this statement is that DOE was developing broad or generic guidelines for warhead certification under a test moratorium.

A second general meeting for the SWPP was held on 5 June 1996 at Kaman Sciences Corporation in Arlington, Virginia. One day later a DOE meeting was held in German town, Maryland, to "discuss the DOE perspective of the programme."[57] The meeting minutes contain the following passage:

> Karen Lombardo DoE HQ (DP-17) explained for the benefit of the Labs and AL the political climate that is evolving in Washington concerning the development of new nuclear weapons. Questions of whether WPP constituted a new weapon have been raised, in the wake of the comments on WPP in the Inside the Pentagon" and the Secretaries request for information. The Office of Development has responded to several inquiries whether the DOE was developing any new weapons. Karen further explained the importance of the WPP as not only the means to exercise our expertise, but also to exercise the tools being

developed in Science-Based Stockpile Stewardship. The WPP will set a precedent[t] for future design activities as well as influence the future relationship with the Navy. Therefore, it is paramount that the programme develops and maintains a direction that emphasises Stockpile Stewardship and de-emphasises new warhead development.

While emphasizing the political dangers DOE HQ saw in a perception that SWPP represents the development of a new nuclear weapon, the message to the National Laboratories was to prepare the way for future design work by incorporating Stewardship Program capabilities within the SWPP

DRAFT

SWPP Design Certification Subgroup Charter

The SLBM Warhead Protection Programme Study Group (SWPP) has delegated to the design Certification Subgroup the following responsibilities:

1. Develop a DoD/DOE consensus on the basic requirements for the SWPP as articulated in the Military Characteristics (MCs) and the Stockpile to Target Sequence (STS).
2. develop a set of criteria for design certification to guide design requirements.
   a) Criteria should be consistent with and help clarify the emerging DOE certification guidelines.
   b) Criteria shall be consistent with national policy guidance—e.g. so dependence upon underground nuclear tests.
   c) Criteria will satisfy customer's confidence requirements in the end product.
3. Interface Control Document (ICD) details shall be addressed and resolved through this subgroup.

The subgroup will consist of a Navy SSP and DOE/AL co-chairs and representatives from participating national laboratories, contractors and the DOE Albuquerque field office.

Reports on progress and issues will be provided to the SWPP Executive Committee. Issues which cannot be resolved or which have a significant minority opinion (to be decided by the co-chairs) will be referred to the Executive Committee for further consideration.

This subgroup will be active throughout the SWPP or until dissolved by the Executive Committee.

Fig. 2 : Fax from Jeffrey Cuneo (DOE DP-12) to personnel at LANL, SNL/NM, SNL/CA, LLNL, Kaman Sciences Corporation, LMMS, and DOE/AL Cuneo wrote on the cover. "Attached is a DRAFT Charter for our subgroup. It is my no means carved or even written in stone or anything firmer than pencil + paper. I hope it\ provides us with a starting point to discuss what we are going to address in the meeting next week. Call me if you see a burning issue that won't wait till the 17th. See you in C-Springs".

In addition the issue of nuclear weapon Use Control was raised at the DOE meeting:

> Another important topic that was discussed was Use Control. It was discussed that the policy may not be completely clear (see Attachments) as was the case for the HPRF [High Power Radio Frequency Warhead] programme. The DOE and DoD interpretation were slightly different. Likewise, the NPR [Nuclear Posture Review] indicates that the policy is for improved Use Control and to improve it. It was discussed that the WPP was an excellent opportunity to establish a better working relationship with the Navy on the appl[ication of] Use Control, recognising that the Navy's first response is no. This discussion lead to a plan that called for the labs to brief the Use Control Effectiveness Committee (UCEC) on possible Use Control options. It was expressed that the minimum Use Control option should not be zero. The UCEC will evaluate the effectiveness of each option. The UCEC positions will be evaluated against safety and cost benefit analysis by a subgroup within the DOE team of WPP. This group has not been organised and the details about the group were not discussed. This group with the UCEC would then make recommendation to DP-20 for a DOE position. DP-10 will then present and negotiate the DOE Use Control position for WPP with the Navy. These options would of course recognis the Navy original requirements. The question was raised as to LLNL drop dead date for implementing Use Control and response was soon of not already past. LLNL was therefore to present their option at the next UCEC the week of LLNL was therefore to present their option at the next UCEC the week of June 10th.[58]

An attachment to the meeting minutes describes DOE Use Control Policy, revised guidelines for which were apparently issued 16 April 1990. This attachment is reproduced in Figure 3. Apparently the DOE meeting and the prior SWPP general meeting on 5 June 1996 clarified the issue of fabricability raised by Tyler and Henson: "Dr. Hannah at the programme review stated that demonstrating the production process was not a requirement. The Navy did expect the programme to work on real hardware, but this could be test hardware."[59] Furthermore: "A plan that defined how the components would be produced was required. Therefore, the only unfunded assistance from the plants and AL would be to provide support in the design phase as part of concurrent engineering and in the development of the production plan."[60] This is in some contrast to Tyler and Henson's "White Paper," where they concluded that without substantial plan funding, production uncertainties would be significant. A 13 December 1996 Sandia memorandum notes that Jim Tyler was replaced as LLNL Programme Manager, however this occurred prior to mid June, 1996.

"Department of Energy (DOE) USE Control Policy—Protection of Nuclear Weapons and Nuclear Explosive Against Deliberate Unauthorized Use", issued April 16, 1990

- An integrated system of positive measures shall be developed, implemented, and maintained to protect all nuclear weapons and NEDs in the custody of the DOE against deliberate unauthorized use.
- To the maximum extent possible, UC measures shall include positive design features.
- DOE will assist and coordinate with the DoD and other federal agencies to provide protection against deliberate unauthorized use for all U.S. nuclear weapons throughout their life cycle.
- Applies to all DP nuclear weapons development or modifications unless there are compelling reasons to the contrary which are approved by the Assistant Secretary for Defence Programmes.

Fig. 3 : Attachment to a DOE memorandum, describing recent policy on Use Control for nuclear weapons.

As noted above, DOE management indicated that the SWPP should incorporate Stewardship tools, however in the 1996 the vulnerability expensive components of the Stewardship Programme-notably the Advanced Design and Production Technology (ADaPT) programme and the Accelerated Strategic Computing Initiative (ASCI)—where in question:

This issue of connections and dependence on other programmes was briefly discussed. The labs did not see a concern over the inter-connections. As part of new business practice, the use of tools being developed in other programmes was seen as an effective and necessary means to achieving the objective of the programme. The concern from HQ was that the heavy dependency on programme such as ADaPT, Enhanced Surveillance, and ASCI in the out years could cause some difficulties or delays if those programmes were cut back or redirected. It was therefore crucial that the Labs Keep HQ informed on the dependencies on these programmes in the out years, so HQ could help work to resolve any issues. The support from other programmes will vary within the Labs. LLNL stated that they were not very dependent on other programmes as this time. The concern from HQ was more directed at the elements of the programme that are much farther out in time and not as well defined a this time. For instance, the Certification Plan may plan to use facilities or codes that are being developed under other programmes and what happen(s) if the programmes are cut or redirected.[61]

The Certification Plan for the SWPP necessarily involves a substitution for nuclear testing in the process whereby nuclear weapon designs are determined to satisfy military (user) requirements for performance, reliability and safety. To the extent that the SWPP designs deviate from tested designs, the Certification Plan will likely entail Stewardship experimental data and computer calculations intended to extrapolate from or supplement past underground testing results. It is these programmatic ties between SWPP and elements of Stewardship that are of concern with respect to DOE obligations to the Navy. However, the meeting minutes note that both the Certification Process and "links to other projects (ASCI, ADaPT, Dual-Revalidation, etc." were not discussed.

On 26 June 1996, the Design Certification Sub Group met at Sandia National Laboratory to discuss:

- Draft SWPP Military Characteristics (Comparison with Phase 2 Military Characteristics);
- Vulnerability Certification without underground tests; and
- Yield Certification without underground test;

Also at Sandia, the Design Integration Subgroup met on 18 September, 1996. "Action Items" included:

- Prepare qualitative description of advantages for advanced Arming, Fuzing, and Firing systems; and
- Review Lockheed Martin Missile System proposal for change to Military Characteristics regarding the reentry body/warhead allowable dynamics.

On 27 November, 1996, a draft paper entitled "SLBM Warhead Protection Programme Requirements and Constraints Document" was issued. It was noted at the 6 June, 1996 meeting that SWPP requirements seemed loosely defined orally and in viewgraphs. The Requirements and Constraints document notes: "This document provides requirements in addition to the Military Characteristics (MC's) and Stockpile-to-Target Sequence (STS) documents to define the deliverables, goals, and constraints of the Submarine Launched Ballistic Missile (SLBM) Warhead Protection Programme."[62] In particular it spells out the constraints that may arise from future Stewardship Programme activities:

6.1.1 SWPP will be relying on the enhanced Surveillance Programme to provide tools and information needed to determine the longevity of the replacement design. Delays in the Enhanced Surveillance Programme may impact the ability to accurately predict the longevity of the design.

6.1.2 The ability to provide evidence for both the confidence in the design of some components and the certification of the design in the future will require computational tools that will be provided by (the) Accelerated Strategic Computing Initiative (ASCI). These tools will have to be developed and validated in a timely in order to support certain aspects of SWPP. Delays in ASCI may adversely affect confidence in the SWPP designs as well as the future certification process as applied to the deployed stockpile.[63]

In addition to DOE's effort to design potential replacement warheads for Trident which possess well defined, even extended service lifetimes, the SWPP will incorporate changes to the W76 and W88 warheads intended to increase safety:

4.5 The replacement design option will provide increased safety margin.

Deliverable: The replacement options will include safety features defined in Section I of the Design Data Package and the supporting evidence will be documented in Section II that the design features will increase the present W76/Mk4 and W88/Mk5 safety margins.

As will be shown below, DOE plans to incorporate Fire Resistant Pit (FRP) technology, as well as Insensitive High Explosives (IHE), into both Trident replacement warhead designs.

Production, or fabricability issues and the extent to which the SWPP served to break ground both in the new Stockpile Stewardship certification processes, and in the Stockpile Management Production processes, was apparent again in a 6 December, 1996 memo from Thomas P. Seitz (Deputy Assistant Secretary for Military Application and Stockpile Management, DOE DP). Seitz set up meeting at DOE HQ to address the "zero defects issue" and SWPP:

... I recently hosted a meeting where we discussed approaches to achieving zero defects in the remanufacture process. Among the methods discussed were concurrent engineering, product realisation teams, integrated product and process design (IPPD), "virtual prototyping and model based manufacturing. In the past 10 years industry has successfully applied these techniques to improve quality and reliability, reduce time to market and even reduce cost. During that period the nuclear weapons complex has spent significant R&D effort in developing technologies supporting these methods but, in general to practice has not followed. It is time to change that trend.

...Over the last year the SLBM Warhead Protection Programme (SWPP) has become a prominent activity in the complex where new ways of meeting our product realisation mission are being examined. I think it is very important that the zero defects issue be addressed in the SWPP programme.[64]

On 16 January, 1997 the SWPP Design Certification Subgroup met for the fourth time. The agenda for the meeting included:

- Current Mass Properties for LLNL Warhead Designs
- Participants' MCs [Military Characteristics] Changes Recommendations
- Laboratory End-to-End Certification Plan
- Core Punch Results for SWPP.

Action Items at the meeting included: "provide current mass properties for LLNL warhead designs", and "Promulgate 4 per cent weight limit due to vulnerability considerations to other concerned SWPP subgroups, e.g, Design Integration Subgroup".[65] On 15 January 1997 the Design Integration Subgroup also met for the fourth time, at Lockheed Martin Missiles and Space in Sunnyvale, CA.

A document dated 4 December, 1997 and bearing the Kaman Sciences Corporation Letterhead is entitled: "SLBM Warhead Protection Programme (SWPP) Programme Plan." In the first (Background) section of the Plan, it states:

> ... the SLBM Warhead Protection Programme (SWPP) was conceived as a multi-year effort that concentrates on replacement point designs rather than development of a new or increased capability. The programme focuses on devising replacement alternatives to non-reproducible technologies in the stockpiled W76/MK4 and W88/MK5 weapons. It also focuses on development of certification methods which are not based on future Nuclear Underground Tests (UGTs), but rather on non-nuclear Above Ground Tests (AGTs) and evolving science-based computer modelling and simulation technologies. This is a cooperative Department of Defence (DoD) and Department of Energy (DOE) programme, being managed by a Study Group of the W76/MK4 and W88/MK5 Project Officer Groups (POGs). The programme does not include fabrication of hardware for the stockpile.[66]

Furthermore, the Plan states that the SWPP warheads will be designed so that they can be integrated into the existing aeroshells without modification. The designs will be time-phased to provide a mid-term option in the next few years (i.e., the Pit Reuse), followed by a longer term option (i.e., New Pit). Note that this a change from earlier planning, in which three alternate designs (also time-phased) were envisioned for the SWPP.

The SWPP Plan states two objectives of the programme:

A. To maintain and exercise the DOE expertise, facilities and capability to design, fabricate and certify replacement weapon components, subsystems and systems.

B. To provide for a continuum of replacement options for the W76/MK4 and

W88/MK5 reentry systems. These replacement options should have decreased sensitivity to aging, enhanced safety features and maintain current capability to the maximum extent possible.[67]

The plan also enumerates Departmental requirements for the SWPP, and Programme goals, listed in Table 5.

Table 5 : Departmental Requirements for the SWPP, and Programme Goals

DoD and DOE SWPP	*SLBM Warhead Protection Programme Goals*
1. MCs for the Warheads for theTRIDENT II (D-5) MK4A and MK5A Reentry Bodies dated 3 2. Trident II (D-t)-WYY 0/MK5A Reentry Body Assembly (RBA) Preliminary Stockpile to Target Sequence (STS) dated 15 December, 1993. 3. WXX-0/M4A Reentyr Body Assembly (RBA) Preliminary Stockpile to Target Sequence (STS) dated 15 December 1993 4. DOE Order 0452.4 - Security and Control Nuclear Explosives and Nuclear Weapons issued 4 June, 1997. 5. Affordability shall be primary consideration in all programme aspects. 6. Be consistent with the strategies depicted in the Stockpile Stewardship and Management Plan and apply the principles of Science-Based Stockpile Stewardship, as applicable. 7. Maintain current safety margin. 8. No new Underground Tests (UGTs) are permitted.	1. No changes to the TRIDENT II (D-5) missile. 2. No changes to the MK4 or MK5 reentry body aeroshells 3. Maintain the size, shape and to the extent possible the mass properties of the present reentry system. Mass properties changes to be within Navy-provided limits. 4. No changes to the missile physical and functional interfaces for the MK5 RBA 5. Minimise changes to the functional interface of the MK4 to allow for MK5 fuzing functionality and MK5 nuclear safety implementations. Changes will be coordinated with the Navy. 6. No changes to the MK4 physical interfaces. 7. Minimise dependence on new flight tests. Flight test configurations must not invalidate the MK4 and MK5 flight test data. 8. Proposed replacement design unit cost should be 1/4 of the W88/MK5 flight test data. 9. Proposed replacement design should achieve first Productioin Unit (FPU) within 3 years of start of E&MD. 10. Use control features will be incorporated in such a manner as to minimise changes to DoD hardware and minimise impacts while in DoD custody 11. Minimise changes to operational capability. 12. Minimise changes to the missile fire control software. 13. Maintain the existing V-Gamma map. 14. Provide increased safety margin, where practical

Related to the issue of maintaining design capabilities, the SWPP Plan states: "At appropriate milestones of each project on the SWPP DDP review will be modelled after the design Review and Acceptance Group (DRAAG) review process defined in DOD Instruction 5030.55 of 21 January, 1974".[68]

The SWPP products will not be warheads for the U.S. stockpile, but a set of documents called the Design Data Package, which includes the Certification Plan and Concurrent Engineering Report. A description of each document is presented in Table 6.

Table 6 : Descrption of SWPP Products.

SWPP Product	*Design Data Package*	*Certification Plan*	*Concurrent Engineering Report*
Description	design information; evidence to support feasibility; description of required follow-on developments, certification and production activities; "The DDP also provides the rationale, from both experiments and calculations, that the design options can meet the requirements, reduce development time, be fabricated, be certified and enter the stockpile with high confidence".	"provides the experiments and calculations conducted during SWPP, the experiments and calculations need to be performed during E&MD in order to formally certify the designs, and the rationale for these tests. It also defines the risks associated with not performing certain experiments until E&MD and not performing certain experiments due to restrictions".	"summary of the processes needed to produce the components. It describes the issues and risks that need to be addressed during E&MD. It also provides information on the process used to manufacture the test hardware".

From the SWPP Master Schedule given in the Plan, the Technical Strategy for Certification was intended to be complete by 1 November,1997, however it is not marked as complete as of December. The two documents which NRDC received under FOIA are from early and mid-January, 1998. The "PRP Response to [SNL/NM] Comments on Technical Strategy for Certification"[69] dated 2 February, 1998 probably indicate why. (Note that SNL/CA—not SNL/NM—are involved in PRP). A viewgraph entitled "PRP response to SNL/NM comments on Technical Strategy for Certification" states: "The SWPP/PRP will not certify the design:

- Building a technical base for certification
- Basis will be incomplete in the data package
- Certification plan will provide a guide to complete the technical basis
- Budget a priority decisions will determinate how far we get".

It is probably a correct supposition that the PRP team was responding to criticisms about the sufficiency of their proposed certification strategy. Despite Navy's original description of the project, the LLNL-SNL/CA team states now that the end product will not be a certified design. However, the PRP team cities management/budgetary huddles, not ones of a technical nature.

To emphasise their point, the PRP team presented a further viewgraph entitled: "Certification will not be done in the Pit Reuse Project (PRP)", shown in Figure 4. Here the design validation process is depicted as less than half complete by the end of PRP, and certification would occur only afterwards.

White the PRP team's final viewgraph is entitled "Peer review is expected before certification", paradoxically it goes on to note:

- [peer review] Probably required for certification
- Peer review is a fiscal issue
- Independent analysis peer review could increase confidence in data package
- Peer review is not necessary for PRP data package.

This position must be contrasted with the fundamental rationale for maintaining two nuclear weapon design laboratories—providing for adequate peer review for classified national defence research and engineering.

Finally on the manufacturing issues for SWPP, Daniel Rose of DOE/AL presented a viewgraph entitled: "SWPP Issues". In this he reiterates the issues stressed by the former LLNL Programme Manager for Pit Reuse (James V. Tyler).

Minimal Production Complex funding/involvement

- No concurrent engineering
- No process development
- No Test Hardware

Manufacturing is not being addressed

Fig. 4 : Viewgraph from "PRP Response to Comments on Technical Strategy for Certification"; a presentation by LLNL and SNL/CA on the SWPP.

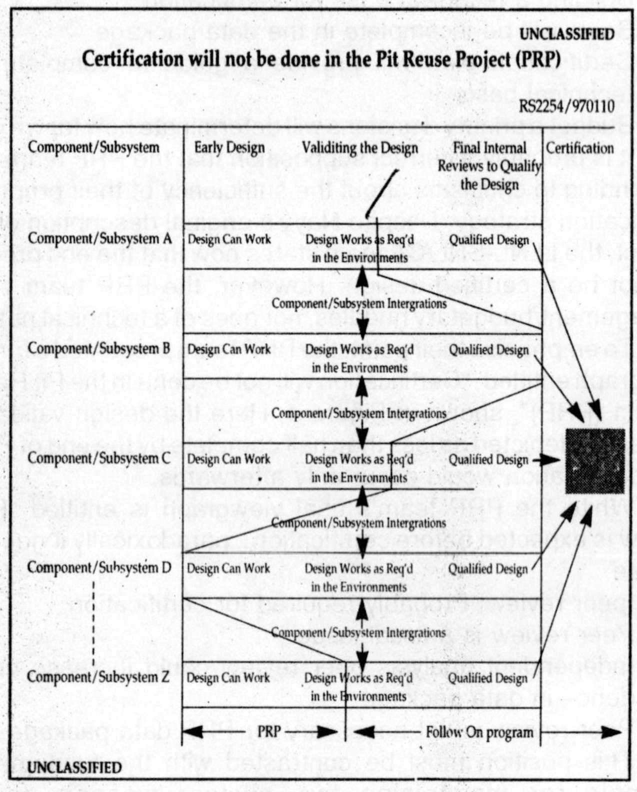

Rose's recommendations are to "Fully Fund Production Plant Support of SWPP". Cost estimates—shown in Figure 5—are given, but is nuclear whether they refer to SWPP as a whole, PRP, or New Pit. Note that LLNL and SNL/CA are not listed.

The indications from the available documentation is that SWPP in an uncertain programme—in both the tasks of certifying the designs and of establishing the ability of the downsized DOE production complex to fabricate the designs. Perhaps in response, a Programmre Realisation Team (PRT) was established for the SWPP, and initially met on January 22 and 23,

Fig. 5 : SWPP cost estimates, from a presentation by Daniel Rose (DOE/AL). It is unclear whether these cost estimates refer to the SWPP as a whole, to the Pit Reuse option, or to the New Pit option.

(Dollars in '000)	*FY97*	*FY98*	*FY99*
Kansas City Plant			
Required	$ 788	$ 2,220	$ 434
Funded	$ 788	$ 1,578	
LANL			
Required		$ 300	$ 300
Funded			
Pantex Plant			
Required		$ 300	$ 300
Funded			
Savannah River Plant			
Required		$ 200	$ 200
Funded			
SNL/NM			
Required	$ 1,600	$ 1,980	$ 325
Funded	$ 1,600	$ 1,775	
Y-12			
Required		$ 1,860	$ 1,535
Funded		$ 3,507	$ 825
Total Needed		**$ 3,507**	**$ 825**

1998. Daniel Rose (Programme Manager for Navy Strategic Programs, Weapons Divisions Programme, Albuquerque Operations Office, DOE) wrote that the purpose of the meeting was to:

1. Discuss the SWPP path forward based on the decisions from the DP-10 and DP-20 Meeting of December 5, 1998: **No additional funding was provided** (emphasis in original).
2. Develop and Integrated Laboratory and Production Plant Programme Plan.
3. Determine resource requirements and budget planning for SWPP.
4. Identification of SubPRTs, taskings and identify team members. Specifically proposed Sub-PRTs are for the advanced AF&F, Pit Reuse Option and New Pit Option.[70]

To summarise, the Navy conceived of the SWPP as a means to force DOE to apply Stewardship strategies/capabilities in the development of designs for existing strategic delivery systems (Navy called it a "forcing function"), in part to probe the DOE Stewardship Programme and in part to keep SLBM warhead skills in practice. The Navy wanted DOE to complete Phase III and what would have been part of Phase IV for two designs : a reused pit and a new pit. DoD emphasized that the MCs, STS, etc. would remain the same as for the Trident systems, and characteristics like the warhead mass distribution would be constrained to very close to the W88.

However, National Laboratory personnel indicate that they do not have sufficient funds either to a) get very far in the certification—without underground testing process or b) demonstrate that the Trident replacement warhead designs can be manufactured in the downsized DOE complex. LLNL claims that peer review is "probably required for certification", but "not necessary" for this project. That is an incredible statement, as LLNL lobbied for their continued existence largely on the issue of classified peer review.

One can postulate several reasons for the SWPP programmatic uncertainty:

1) NRD's report "End Run" and other press reports diminished upper management or administration support for SWPP;
2) DOE is reluctant to fund SWPP because certification without underground testing is politically risky: if the National Laboratories can achieve this capability now, why invest billions more in the SSMP; if the National Laboratories run up against general "showstoppers", this may cast doubt on the over-arching Stewardship strategy even with new future planned capabilities;
3) DOE simply views the money as better spent on other parts of Stewardship, so the Navy will get something, but additions to the stockpile in the near term are not planned as part of the SWPP.

Notes :

1. "Stockpile Stewardship and Management Plan: First Annual Update (DELETED VERSION, CLASSIFICATION MARKINGS REMOVED, Derivate Classifier: Dr. Ruben Sanchez, DP-22)",

DOE Office of Defence Programmes, Approved for Release to Congress October, 1997, p. 4–3.

2. Ibid. And further on p. 4–10: "The document, prepared by SNL, approved by the two applicable design laboratories and then DOE, states that specific war-reserve material is satisfactory for release to the DoD on a designed effective date for specified uses qualified by exceptions and limitations".
3. The "nuclear explosive package", or alternately "physics package" or "nuclear assembly", is a collective term for the primary, secondary, and radiation case.
4. See for example Kristensen, Hans, "Targets of Opportunity: How Nuclear Planners Found New Targets for Old Weapons", *Bulletin of the Atomic Scientists*, No. 5, Vol. 53 (September 1997), p. 22.
5. "Stockpile Stewardship and Management Plan: First Annual Update", p. 4–12.
6. An amendment offered by Senator Jon Kyl (Republican-Arizona) was adopted by voice vote in the Senate as an amendment to the Fiscal Year 1999 Defence Authorisation Act (S2057). "(a) ESTABLISHMENT. There is hereby established a commission to be known as the "Commission for Assessment of the Reliability, Safety, and Security of the United States Nuclear Deterrent". (b) COMPOSITION—(1) The Commission shall be composed of six members who shall be appointed from among private citizens of the United States with knowledge and expertise in the technical aspects of design, maintenance, and deployment of nuclear weapons...".
7. "Stockpile Stewardship and Management Plan: First Annual Update", p. 4–11.
8. Ibid., p. 4–11.
9. Ibid., p. 7–29.
10. This overching Stockpile Stewardship strategy is analysed in the NRDC reports: *End Run* (August 1997) and *Explosive Alliance* (January 1998).
11. "Stockpile Stewardship and Management Plan: First Annual Update", p. 7–13.
12. Ibid., p. 7–28.
13. The DOE defines "surety" as an umbrella term for safety, security, and use control of nuclear explosives.
14. "Stockpile Stewardship and Management Plan: First Annual Update", p. 1–8.
15. Ibid., p. 5–3.
16. Ibid., p. 1–3. This process is referred to as Limited Life Component Exchange (LLCE).
17. An example of a Military Characteristic of a weapon is its yield.
18. DOE defines the Stockpile-to-Target Sequence as the range of

environmental conditions, including temperature, moisture, acceleration, and vibration, which a weapon must be to withstand and still function properly.

19. "Stockpile Stewardship and Management Plan: First Annual Update", p. 7–27.
20. "Phase III development of a nuclear weapon is the start of warhead development at a Natioinal Laboratory, culminating in a proposed warhead design. Phase I is the definition of a new warhead or component concept and Phase II addresses technical feasibility, cost, and production schedule. See Appendix. p. 13.
21. "Stockpile Stewardship and Management Plan: First Annual Update", p. 10–19.
22. Ibid., p. 10–20.
23. Ibid., p. 5–9.
24. "NIF System-Design Requirements for Nuclear-Weapons Physics Experiments", Lawrence Livermore National Laboratory, April 1995, UCRL-ID-120738. "The NIF will be an ideal facility for studying much of the phsycis involved in nuclear weapons both as isolated processes and as compound events, (pg.iii)".
25. Ibid., p. 15.
26. Ibid., p. 15–17.
27. A soft-x-ray laser interferometer operating at 155 Angstroms was developed at Nova around 1995. Ibid., p. 15.
28. Ibid., p. 16.
29. "Nuclear Weapons Physics on the NIF: Experimental Opportunities to Improve Predictive Modeling Capabilities of Nuclear Weapons Phenomena (Stephen B. Libby, Conference Chair; Mark J. Eckart, Co-Chair; John E. Heidrich, Scientific Editor; Shirley R. Jennings and Jane C. Olivera, Technical Editors, January 1995, UCRL-MI-119994, p. 57.
30. Ibid., p. 65.
31. The technique now most actively pursued towards achieving fusion ignition in the laboratory is to symmetrically implode a small but complex capsule target containing a mixture of tritium and deuterium. The capsule implosion is performed either by direct illumination by—for example—a laser, or x-ray illumination of the capsule by laser-heating a high-Z (usually gold) shell or hohlraum which contains it. In the "Fast Ignitor" concept a capsule implosion is initiated, a channel is then bored through the surrounding plasma to the capsule center by a laser, and a second laser beam travels down that channel to ignite the fusion fuel.
32. "Nuclear Weapons Physics on the NIF: Experimental Opportunities to Improve Predictive Modelling Capabilities of Nuclear Weapons Phenomena", p. 69.

33. G.P. Nanos, Jr., Memo, Subj: SLBM WARHEAD PROTECTION PROGRAMME, 14 April, 1995.
34. Some background on the Navy's strategic forces may be useful in understanding the SLBM Warhead Protection Programme. Under the second Strategic Arms Reduction Treaty (START II), the submarine leg of the U.S. strategic triad will account for about half of all accountable, deployed warheads, Most of the Navy's strategic warheads are of the W76 design, about 3500 of which were produced between 1978 and 1987. About 400 W88 warheads were produced from 1988 to 1991; W88 pit productiion was terminated prematurely due to the closure of the Rocky Flats plan near Denver, Colorado. Both the W88 and the W76 warheads were designed at Los Alamos National Laboratory. The W88 is heavier and has a much higher yield (475 kilotons) than the W76 (100 kilotons).
35. Cochran, Thomas B., Arkin, William M., Norris, Robert S., Hoenig, Milton M., *Nuclear Weapons Databook Volume II: U.S. Nuclear Warhead Production*, Ballinger, 1987, p. 104–106.
36. Charles J. Beers, Jr., Rear Admiral, U.S. Navy, Deputy Assistant for Military Application and Stockpile Support (DP-20), DOE Defence Programmes, 28 June, 1995, to the Deputy Assistant Secretary for Research and Devleopment DOE DP-10, SUBJECT: SUBMARINE LAUNCHED BALLISTIC MISSILE WARHEAD PROTECTION PROGRAMME.
37. A memo written on the same day to DOE's Albuquerque Operations (DOE/AL) Office described the above organisational structure as in effect. "The Deputy Assistant Secretary for Research and Development, DP-10, is responsible for funding and leading this effort during the programme's development. If production of one of the alternative warheads is required by the Navy, the funding and lead for production and stockpile support will transition to by office, DP-20". Charles J. Beers, Jr., to Manager, Albuquerque Operations Office, 28 June, 1995, SUBJECT: SUBMARINE LAUNCHED BALLISTIC MISSILE WARHEAD PROTECTION PROGRAMME.
38. Kenneth J. Adney, Acting Director Office of Development and Testing, Defence Programmes, Department of Energy, Memo, Subject: SEA-LAUNCHED BALLISTIC MISSILE (SLBM) WARHEAD PROTECTION PROGRAMME PLANNING MEETING, May 22, 1995.
39. A 28 June, 1995 memo from the Deputy Assistant Secretary for Military Applications and Stockpile Support, DP-20 (Rear Admiral Charles J. Beers, Jr.) to the Deputy Assistant Secretary for Research and Development, DP-10, stated: "The SSPO [Stra-

tegic Systems Programme Office] wants the alternative warhead effort handled via the MK4/W76 and MK5/W88 Project Officers Groups (POG). It is anticipated the POGs will estbalish a working group of POG members, a Lawrence Livermore National Laboratory member, and selected observers. This is similar to the approach used for the Navy Phase 2 Study".

40. Cliff Dejong, Kaman Sciences Corporation, Memo, SUBJECT: SLBM WARHEAD PROTECTION PROGRAMME, 15 June, 1995.
41. Victor H. Reis, Assistant Secretary for Defence Programmes, DOE, letter to Harold P. Smith, Jr. Chairman, Nuclear Weapons Council Standing and Safety Committee, Department of Defence, July 3, 1995.
42. Roger E. Fisher, Deputy Assistant Secretary for Research and Development, DOE, memorandum to Siegfried S. Hecker, Director, LLNL; Bruce C. Tarter, Director, Sandia; and Al Narath, President, Sandia: July 27, 1995.
43. James V. Tyler, memorandum to "Distribution", Subject: Proposed Project Structure for the SLBM WPP Pit Reuse Project", September 15, 1995.
44. Ibid.
45. Jime Tyler (LLNL) and Doug Henson (Sandia), "SLBM Warhead Protection Programme Pit Reuse Project Proposed Tier-0 Project Overview", September 15, 1995, p. 1.
46. Jeffrey A. Cuneo, Lt Col., USAF, Programme Manager, Office of Development and Testing, Defence Programmes, DOE, memorandum to "Distribution", Subject: Meeting Minutes, SLBM Warhead Protection Executive Working Group, Dec. 6, 1995, December 18, 1995.
47. Jeffrey A. Cuneo, Lt Col., USAF, Programme Manager, Office of Development and Testing, Defence Programmes, DOE, memorandum to "Distribution", Subject: Meeting Minutes, SLBM Warhead Protection Executive Working Group, Dec. 6, 1995, December 18, 1995.
48. David Gibson, Kaman Sciences Corporation, memorandum to SWPP DISG Members, Subject: Action Items, 5 March, 1996.
49. Ron Brunsvold, Kaman Sciences Corporation, memorandum to "Distribution", Subject 2nd Meeting of the SWPP design integration subgroup, 25 March, 1996.
50. Jeffrey A. Cuneo, memorandum to "Distribution", Subject: SLBM Warhead Protection Programme (WPP) Design Certification Working Group, March, 29, 1996.
51. Jim Tyler, memorandum to "Distribution", Subject: LLNL/SNL Perspectives on DOE/AL Support Needed for the SWPP, April 2, 1996.

52. Jim Tyler (LLNL) and Doug Henson (SNL/CA), "Support Needed from DOE/AL and the DOE Production Agencies for the SLBM Warhead Protection Programme Pit Reuse Project", April 2, 1996, p. 1.
53. Ibid., p. 2.
54. Ibid., p. 3.
55. Ibid.
56. Reference to be provided.
57. "Meeting Minutes"; DATE: June 12, 1996; RECORDED BY: DP-17, R. Glenn Bell (SNL); MEETING: DOE Team meeting for the SLBM WPP; LOCATION: Germantown, June 6, 1996, p. 1.
58. "Meeting Minutes"; DATE: June 12, 1996; RECORDED BY: DP-17, R. Glenn Bell (SNL); MEETING: DOE Team meeting for the SLBM WPP; LOCATION: Germantown, June 6, 1996, p. 2.
59. Ibid., p. 2.
60. Ibid.
61. Ibid.
62. Draft SLBM Warhead Protection Programme Requirements and Constraints Document", Dated November 27, 1996, p. 1.
63. Ibid. pp. 5–6.
64. Thomas P. Seitz, Deputy Assistant Secretary for Military Applications and Stockpile Management, DOE DP, memorandum to "Distribution", 6 December, 1996, Subject: "SLBM Warhead Protection Programme (SWPP) Focus on Zero Defects".
65. Bob Nuttleman, Kaman Sciences Corporation, Memorandum to "Distribution", 12/16/96, Subject: 4th SWPP Design Certification Subgroup Meeting.
66. SLBM WARHEAD PROTECTION PROGRAMME (SWPP): PROGRAM PLAN, K97-84U(R), 4 December 1997, Kaman Sciences Corporation, p. 1.
67. Ibid., p. 2.
68. Ibid., p. 7.
69. John Heidrich and Dana Rowley (LLNL), Larry Weingarten, Steve Vasey, and Doug Henson (SNL/CA), "PRP Response to Comments on Technical Strategy for Certification", January 14, 1998. "Presented to Design Certification Sub-group".
70. Daniel Rose, Memorandum, Subject: "SLBM Warhead Protection Programme (SWPP) Programme Reazation Team (PRT) Meeting, December 22, 1997.

Address:

THOMAS COCHRAN

Director, Nuclear Programme
Natural Resources Defence Council
Washington DC 20005
USA

29

AÏDA: The French-Russian Aid Programme for Dismantling of Russian Nuclear Weapons

HERVE BERNARD

INTRODUCTION

The French-Russian Programme AÏDS—French acronym of Aide au Dèmantélement des Armes Nuclèaaires russes—contributes to the application of the START agreements signed in 1991 and 1993 by the United States and the Russian Federation. Nuclear strategic warheads must be reduced to 6000 (START 1) then 3500 (START 2). So, the end of the Cold War has created a legacy of surplus weapon-grade fissile materials both in these two countries and the need to establish policy for peaceful and safe disposition of such materials.

France with USA contributes to the Russian warhead dismantling programme, in which also take part United Kingdom, Italy, Germany, Japan.

In February 1992, a general cooperation agreement is signed between the President of the French Republic and the President of the Russian Federation. Technical discussions between French and Russian experts allow the selection of several items of cooperation. As soon as September 1992, the French Government commits a financial aid of 400 MF (65 MUS $) to the Russian Federation.

The Military Applications Division of the French Atomic Energy Commission (CEA/DAM) and the MINATOM (the Ministry of Atomic Energy of the Russian Federation) are the two operators of this agreement. The AÏDA programme is born.

Shared in two sub-programs, AÏDA covers the following points:

- Sub-programme A:
  - A1 : supplying radiological protection and assessment equipment
  - A2 : supplying high integrity container overpacks
  - A3 : supplying machine tools
  - A4 : constructing an interim storage building in Novosibirsk, Russia
- Sub-programme B: AÏDA-MOX 1

  evaluating the possible use of weapon-grade Plutonium (W-Pu) in MOX fuel for loading in the Russian nuclear power plants.

  The following milestones were:
  - November 12, 1992, signature of the general agreement and signature of two specific agreements on the use of military nuclear materials for peaceful nuclear power generation in Russia (sub-programme B) and on the supply of equipment (sub-programme A-1)
  - October 20, 1993, signature of the specific agreements on sub-programmes A2 and A3
  - November 17, 1994, signature of the specific agreement on sub-programme A4.

  In addition, two administrative arrangements were signed:
  - March 25, 1994 between MINATOM and CEA/DAM setting the practical terms of the cooperation : industrial property, right of use, confidentiality and publication rules, liability, etc..., and the practical commitments on various issues : insurance, transport, customs, personnel at disposal, training of operators, etc...
  - November 4, 1994 between MINATOM and the French Ministry of Defence setting terms of the French inspections in Russia on delivered equipment in order to verify their right use.

THE AÏDA PROGRAMME

If the sub-programme A is mainly devoted to the technological aspects of the dismantling of nuclear weapons and the interim storage of components, the sub-programme B, more ambitious, contributes to the definite use of W-Pu- with no possible return to nuclear weapon purpose—while producing electrical energy for the benefit of Russian population.

The sub-programme A1

The objective is to ensure safety and surveillance during transport, interim storage and dismantling operations of the nuclear weapons.

Numerous equipment was supplied to Russia since 1995:

- equipment for radiological assessment;
  - 2 calculation systems for the preliminary mapping of the consequences of an accidental contamination
  - 3 "Helinuc" gamma mapping systems to be installed under Mi-8 helicopters with a "Trident" radiolocation system for mapping of the real contamination of the ground
  - 3 Renault "Inspector" vehicles equipped with in situ gamma spectrometry equipment
  - 3 Renault "Visa" vehicles equipped with environment monitoring equipment
- equipment for personnel dosimetry, facility monitoring and irradiation/contamination checks : dosimeters, masks, alpha check systems, ...

To use all this equipment, Russian personnel have been trained by French operators.

THE SUB-PROGRAMME A2

The 100th and last high integrity container overpack has been delivered to Russia last April. These containers will ensure the safe transport of nuclear munitions from their current storage sites to the dismantling facilities and the transport of sensitive components to their final storage site.

These armored super-containers (cf. figure 1) are characterised by their large dimensions (4.3 m x 1.9m x 1.7 m), a very good resistance to severe attacks (gunshots, water immersion, 9 m fall, fire ...) and to climatic changes

Well appreciated by our Russian Colleagues for their remarkable performances, these containers were designed, developed and tested by the CEA/DAM.

THE SUB-PROGRAMME A3

The CEA/DAM supplied in January 1995 one Berthiez vertical milling machine for opening up the re-entry vehicles and in March 1996, 3 Missler cutting machines for cutting the metal sub-structures.

Fig. 1 : Super container

THE SUB-PROGRAMME A4

Hydrogen and lithium bearing materials resulting from the dismantling of nuclear weapons have to be safely stored. For that purpose, the CEA/DAM began in 1995 the construction of a storage building in Novosibirsk on a basis of a "turnkey delivery".

This 5000 m^2 floor area building (cf. figure 2) is designed to store 3240 containers and ensure the safety and security of the stored materials even at the extreme Siberian external temperature. The French Company, SPIE Batignolles realised the construction work with the Russian subcontractor, SIBENERGOSTROI between September 1995 and December 1997. The :key" of the building was officially delivered to Russian operator on December 16,1997.

THE SUB-PROGRAMME B : AÏDA-MOX 1

AÏDA-MOX 1 is intended to specify a complete industrial system to use plutonium recovered from Russian nuclear weapons by fabricating MOX fuel in Russia. This MOX fuel loaded in currently operating or planned commercial Russian nuclear

Fig. 2 : Storage Building

power plants will produce electricity while diminishing the W-Pu inventory.

The CEA, operator of the French-Russian inter-governmental agreement coordinates the French contributions by COGEMA, FRAMATOME and EDF. The Russian Ministry of Nuclear Power, MINATOM, has the same role with respect to the research organisations (VNIINM, RIAR, IPPE, Radium Institute, Kiurchatov Institute,...), and project institutes (GSPI, GYDOPRESS, VNIPIET,...).

The final report of the French-Russian studies conducted from 1993 to 1996 was internationally distributed in mid-1997. The report provides the following recommendations:

- **For the short term:** the use of Russian W-Pu as MOX fuel inexistent Russian BN 600 reactor and VVER 1000 units with the construction of a facility known as TOMOX 1300 in Russia for converting the plutonium into MOX fuel. The proposed MOX fabrication capacity of 1300 kg of plutonium per year would be sufficient to supply the BN 600 reactor and four VVER 1000 reactors (about 30 tons of MOX per year)
- **For the medium and long term:** various scenarios involving the construction of new reactors (e.g. BN 800, VVER 640) loaded with 100 per cent MOX fuel, designed to enhance the utilisation rate of plutonium.

MAIN CONCLUSIONS OF THE AÏDA-MOX 1 PROGRAMME

The French-Russian studies established the advantages

and technical feasibility of the W-Pu MOX option in certain existing nuclear reactors within the Russian Federation.

Broadly speaking, the PWR and Fast Reactor—MOX options for using excess dismantled weapons plutonium for peaceful commercial nuclear power generating purposes—like the CANDU-MOX and HTR-MOX options—offer several advantages over the remaining option (interim storage, disposal in vitrified form,...) notably from a non-proliferation standpoint. More-over, the LWR-MOX and FR-MOX options are the only ones implementing proven technology: over 500 metric tons of PWR and BWR-MOX fuel have already been fabricated in Europe, over 22 European reactors have been using MOX for a number of years, over 100 metric tons of FR-MOX fuel have been fabricated to date in Europe. The PWR-MOX and FR-MOX options are also the only ones compatible with existing Russian reactors capable of using large quantities of W-Pu as known to be economically viable.

Concerning the implementation of these options by the Russian Federation, the joint French-Russian studies have led to the following conclusions:

1) Using 30 per cent MOX fuel is feasible in certain VVER 1000 reactors, after implementation of design changes similar to those made in France from 1985 to 1990 when EDF began using 30 per cent MOX fuel in its 900 MWe PWR. Each VVER 1000 should load about 270 kg of W-Pu per year.
2) Using 100 per cent MOX fuel in the BN 600 fast reactor is certainly the most promising option in a reasonable time frame. In the short term, an hybrid core with about 18 per cent of MOX fuel assemblies can be implemented without substantial modifications and would enable dispositioning of 240 kg of W-Pu per year.
3) A reference process for converting W-Pu metallic alloy into MOX fuel has been defined after 4-year bilateral studies. The reference process chosen by the French and Russian specialists comprises two steps:
   - the conversion of the metallic alloy into oxide powder suitable for pellet fabrication by a Pu alloy dissolution in a nitric-fluorhydric solution, followed by a purification by solvent extraction of plutonium nitrate and finally a

conversion of Pu nitrate solution and PuO_2 powder by an oxalate precipitation. This process has the best mutual industrial experience and technologies.

In fact, several processes are available for this conversion step: aqueous or dry ones[(2)] Some of them have now reached industrial maturity : dissolution in an acid mixture (Russia), calcination/ electrolytic dissolution (France), purification and oxalate precipitation (France and Russia). Others are still in the development stage: hydridation-dehydridation/ oxidation (USA), pyrometallurgical process (Russia)

- the fabrication of the MOX fuel pellets by the well known MIMAS process for VVER fuel and COCA process for BN fuel.

Two process variants have been kept as back-up solutions: the first one uses the direct coprecipitation of mixed powder $(U,Pu)O_2$—instead of pure PuO_2—followed by powder blending without grinding—instead of MIMAS and COCA processes; the second one uses a calcination of the plutonium alloy followed by a PuO_2 dissolution by electro-generated Ag(II). Neither variant has been adopted to date for lack of sufficient industrial experience; further research and development work will be required in both cases.

4) A preliminary design work on the TOMOX 1300 facility implementing glove-box handling technology has been realised in Russia with French participation by the CEA, COGEMA and SGN. A detailed proposal is planned for the AÏDA MOX phase 2 programme to be carried on next two years.

NEXT STEP: THE AÏDA-MOX 2 PROGRAMME

AÏDA-MOX 1 programme was undertaken between France and Russian Federation from 1993 to 1996. A similar programme was implemented between Germany and the Russian Federation regarding MOX fuel fabrication and safety aspects of loading MOX fuel in reactors.

A trilateral initiative between the Russian Federation, Germany and France was announced in October 1996 during the G8 experts meeting in Paris.

The corresponding trilateral agreement has been signed 2 June 1998 in Moscow. Open to other partners, this joint action can constitute the groundwork for a broader cooperation mobilising international financing for the elimination of surplus W- Pu. This agreement covers the following areas:

1) TOMOX 1300 plant basic design, consisting in:
   - the design of the CHEMOX plant for converting W-Pu alloy into suitable a plutonium oxide powder.
   - the design of the DEMOX plant for fuel fabrication using the previous powder.

   The TOMOX plant will include the basic facilities already in place at Russian nuclear sites, particularly Cheliabinsk 65 and Krasnoyarsk 26.
2) MOX utilisation in VVER and BN reactors with specification and cost assessment of required reactor modifications, safety reports and licensing for loading MOX fuel.
3) "MOX Demonstration" for demonstrating at an experimental scale the feasibility of the MOX-VVER 1000 option in Russia—the MOX-BN 600 option has already been demonstrated—To date, the objective is to fabricate 3 (full or partial) MOX subassemblies in Russia to be loaded in one of the four Balakovo VVER 1000 reactor.

The duration of this AÏDA-MOX Phase 2 programme is expected for two years: June 1998-June 2000 and the French funds will be of 60 MF shared between public and corporate financing.

CONCLUSION

Weapons plutonium is a valuable material: 50 tons of W-Pu represent the energy equivalent of 50 to 100 million tons of oil. The use of W-Pu with mature commercial fuel cycle technologies provides a satisfactory answer to strategic concerns, such as the irreversibility of W-Pu disposition, and international control of non-proliferation. For the future of such W-Pu use in Russia, the major issue will be to mobilise international financing to enter the AÏDA-MOX phase 3 with the construction of MOX facilities.

References

1. N.N. EGOROV, E.G. KOUDRIAVTSEV, V. POPLAVSKY, A. POLYAKOV, X. OUIN, N. CAMARCAT, B. SICARD, H. BERNARD, : AÏDA-MOX 1 Programme: Results of the

French-Russian Study for Peaceful Use of W-Pu" Global 97, Yokohama, Japan, October 6-8, 1997

2. H. BERNARD, J.C. THIEBLEMONT, P. PILUSO, A.S. POLYAKOV, B.S. ZAKHARKINE, O.S. SKIBA, R.I. LIOUBSEV, " AÏDA-MOX 1 : 1993-1996, Evaluation of Processes for the Fabrication of MOX fuel for Weapon-Grade Plutonium", Global 97. Yokohama, Japan, October 6-8, 1997.

Address:

HERVE BERNARD
French Atomic Energy Commission
91680 Bruyere Le Chatel
France

30

Reducing the Global Threat from Civilian Plutonium through International Development of New Technology Integrating Unique Russian Know-how

CURT MILEIKOWSKY

1. THE GLOBAL ENERGY PROBLEM FOR THE COMING HALF-CENTURY

The Energy Issue

On the eve of the 3rd millennium, humanity is increasingly faced with the formidable problem not only of achieving an ever-higher standard of living for a continually-increasing world population but also of doing so without endangering our planet's fragile environment, without exhausting its resources, some of which are limited, and without leading to social unrest, political turmoil or war.

One of the major issues that will dominate the first half of the coming century is the energy problem. At present more than 80 per cent of humanity's energy consumption is covered by burning fossil fuels: coal, oil, natural gas and wood. The rest is mainly hydro and nuclear power, while very little power is derived from renewable sources such as wind, solar power and heat and earth heat for example in the form of hot springs.

Fossil Fuels

Within a few decades from now, supplies of some of the main fossil fuels will be dwindling: oil companies predict that by about the middle of the century oil and natural gas will be scare and far more expensive than today. Coal will last longer, even through it will take over the markets currently occupied by oil and natural gas. But the issue—still not completely clarified—of global warming being used by the greenhouse gases, mainly CO_2, in that are emitted during the burning of fossil fuels, could lead to a requirement of mitigation and consequently result in higher energy costs. For example, existing technology allows the CO_2, in the exhaust of a fossil-fuel-fired turbine to be captured and deposited into an aquifer 1000m below the Earth's surface, where it dissolves due to the high pressure at that depth. Implementing this method increases power cost by about 50 per cent compared to just letting the CO_2 into the atmosphere.

Hydropower

As to hydropower, most of the available sites are already exploited: perhaps 50 per cent more could be built, which is totally insufficient compared to increasing global demand. Furthermore, the ecological problems are far from negligible and the investment structure is unfavourable to developing countries: low operating costs but very high capital outlay.

Nuclear fission

Nuclear power has for the time being lost momentum in the U.S. and Western Europe (except in France where new plants are being built). One of the reasons for this is deregulation: faster payback is achieved from fossil-fuelled turbines because of their lower capital cost. Another reason is non-fixed interest rates on construction loans. A third is the complex bureaucratic procedure to be gone through in order to obtain a construction licence, A fourth is scepticism among a certain portion of the population with regard to safety and waste disposal aspects of nuclear reactors. Nevertheless, in East and South-East Asia, on the other hand, a considerable nuclear power programme is planned, with projects in China, the Koreas, Japan, etc.

For the above reasons, particularly the unsolved problem of civilian waste and its plutonium content, nuclear power in its present form seems not to be a promising candidate for covering humanity's future energy needs unless a reliable solution is developed, tested and implemented.

Nuclear fusion

Rather large budgets have been devoted to international research on fusion reactors, but the problem has proved to be more complex than expected, so we do not know when a real breakthrough may come.

Solar power

How about solar energy, photo-electricity or simply solar heat? Space projects have been needing solar electric panels for several decades, and obtained them. Important developments have been accomplished in this field and financed thanks to space needs. The resulting technology however is not competitive in ordinary commercial applications on Earth, except at some distant spots where demand is so very small that power lines or microwave transmission are uneconomical and diesel generated power is impractical—for example at desert vacation cottages in California and certain parts of Africa. In order for solar energy to supply basic power, very large areas are needed for the photoelectric panels. That land is expensive, as are the panels themselves and also their supports, which must be strong enough to resist violent storms. One way to decrease the cost of the panels is to integrate them into the buildings' walls and roofs, but this surface area is insufficient. Furthermore, the panels must be regularly cleaned of dust and sand and sometimes, in the higher latitudes, snow and ice. It is not known when solar power will become commercially competitive.

Solar power produced outside Earth and transmitted by microwave or laser beams to Earth's surface may become competitive enough to be implemented but probably not until rather late in the next century.

As regards energy cost and worldwide availability, one can therefore predict, for the late first half of the coming century, an energy crisis on a scale previously unknown, entailing serious risks for world economy, environment and peace, *unless some technical breakthrough* brings a solution.

2. TOWARDS SOLVING THE ENERGY PROBLEM: ACCELERATOR-DRIVE, SUBCRITICAL NUCLEAR POWER

It has been seen above that the problem of energy for the next half-century is characterised by the problems and insufficiencies of existing sources of power. What then can be envisaged for that period and a long time thereafter? A very promising solution is offered by a machine type which combines a high-power accelerator with a subcritical nuclear reactor. Why is it so promising? This is because subcriticality enhances safety and the extra neutrons supplied by the accelerator mean superior flexibility.

Protons, accelerated to 1-2,5 GeV, hit a metal target in the centre of the subcritical reactori causing so-called nuclear spallation in the target nuclei. Each beam proton liberates about two dozen neutrons, which diffuse into the reactor blanket, replacing thosė neutrons which are missing due to subcriticality. The proton current has to be several 10:s of mA.

The idea is far from new, having been considered as early as the 1950's, but sufficiently powerful accelerators could not be built until the 1980's. Linacs of the required strength are today's technology. They can be either normal-conducting or superconducting. Cyclotrons are also being considered for the same purpose (by Rubbia and his group at CERN). A development project is planned.

The choice of the subcritical reactor blanket has its own history. It was shown early on that light water was not feasible: its properties for neutron diffusion required many proton beams. Heavy water was studied by LANL with the Russians. D_2O was abandoned and the work was concentrated on molten salt. Now both the American and European development groups have chosen liquid metal, namely liquid lead / bismuth. This material is the beam target; it serves as an ideal moderator and also as a coolant. The Russians have the most know-how in the world about using these metals as coolant thanks to their long experience from a separate class of their submarines.

Now, what is it that this system is expected to achieve in the energy sector in the not-too-distant future?

1. It can efficiently incinerate the bomb material plutonium and the other transuranium elements in spent fuel, eliminating them almost completely. This it can do better than any other method.
2. It can transmute the most dangerous fission products in the spent fuel—technetium, iodine and cosium—to stable isotopes. These are the most dangerous fission products because they cause specific biological damage to the human body and are moreover soluble in water, so that if a leak occurs in the copper canning of the fuel assemblies in terminal repository, they may end up in the ground water. This dual capability eliminates two of the greatest disad-

vantages of nuclear power as currently implemented. Furthermore, subcriticality makes it impossible for the reactor to "run amok".

3. WHY IS INCINERATION OF THE CIVILIAN PLUTONIUM IN SPEND FUEL IMPORTANT?

Because any rouge country, mafia, terrorist organisation or even wealthy individual can make nuclear bombs from civilian plutonium relatively easily and cheaply. Such bombs present one disadvantage in that it cannot be known before they are actually ignited what explosive power they will provide, but it will be somewhere between 10 per cent and 70 per cent of the power of the Hiroshima bomb for a bomb containing 10kg of plutonium. And terrorism on that scale can have devastating effects and lead to war.

The scale of the risk of nuclear bombs made from civilian plutonium is certainly not small. The amount of spent fuel in the world is ever 100,000 tons, spread in cooling ponds in over 30 countries, and not always well guarded against theft or coup. These 100,000 tons will have become 200,000 tons within a decade or so. Most of this is uranium-238 but 1 per cent or 2,000 tons, is plutonium. And since less than 10kg of civilian plutonium is needed to make a bomb, this material is enough to produce 200,000 bombs. This is the scale of the problem.

This risk must of course be kept under best possible control. Thus the United Nations International Atomic Energy Agency (IAEA) in Vienna has been entrusted with checking that this widely disseminated plutonium is not misused to make bombs. Because the half-life of plutonium is 24,000 years, this IAEA control must be maintained for about 100,000 years. If this international agreement concerned any less serious problem, it would have to be regarded as totally absurd: no human institution lasts that long, and particularly no international agreement. No wonder some people have doubts about nuclear power!

4. THE SOLUTION TO THE PLUTONIUM THREAT

If a nuclear waste treatment system of accelerator-supported subcritical reactors is set up, however, the situation becomes totally different. For every 10,000 MWE (30,000 MWTh) of first-generation nuclear power in a given country, a subcritical capacity of 25 per cent i.e. of (7,500 MWTh) is sufficient to clean up both the waste of the first-generation reactors and the waste of the subcritical clean-up reactors themselves. And the need for IAEA safeguard is restricted to the actual life-span of the clean-up reactors, i.e. 35 to 40 years (instead of 10,000 to 100,000 years under the current configuration).

Before implementation, normal technological development must be carried out, mainly with regard to materials properties and chemistry, and some pilot plants must be financed, designed, constructed and put into operation. This phase takes about seven years. If these pilot operations work as expected, one or two full-scale plants must then be built and tested before widespread implementation can begin. This second phase would take some time, perhaps 12 to 15 years.

For the time being two pilot plant projects seem to have a chance of receiving funding in the relatively near future, one in the U.S. and one in Europe. In a month or two we will know more about how close to begin financed they are.

What are the economic aspects of a full-scale clean-up plant? The thermodynamical efficiency of a plant of say 3,00 MWTh is similar to that of a light water power reactor, if not a little higher, so it produces the same electric power. But the accelerator "steals" almost 10 per cent of the power produced and also adds somewhat to the total capital cost of the plant. This increases the cost of the power to be sold to the net. However, cost-reducing factors are also present. While the reactor tank and other tanks, pipes, valves, heat exchanges and so on are all built to resist high pressure in a light water power reactor, no high-pressure designs are needed in a liquid-metal-called plant, the regulation system is simpler and the final repository for fission products not be treated is simpler and cheaper.

Even if the power available for the net from the clean-up plant turned out to be slightly more expensive than power from critical LWR:s, the cost differential would be further reduced by the fact that only one LWR- sized clean-up reactor is needed to handle the waste from four LWR:s (and itself)

5. THE NON-PROLIFERATION ISSUE

One of the important properties of an accelerator- assisted subcritical reactor system is its capacity to burn the bomb material civilian plutonium almost completely. But it is a fact that high energy accelerators can be used to produce the bomb materials plutonium or uranium 233 if the beam is (clandestinely) diverted to a different target. Can this threat be avoided? The answer is yes. First of all, IAEA personnel and instruments would discover the vacuum line to the clandestine target, on condition that the personnel were allowed to make routine visits and that their measuring instruments were not manipulated. Manipulation of the instruments would immediately be apparent to IAEA, and refusal to allow access to a country or plant would clearly signal unauthorised activity. IAEA could do nothing about it except alert the UN. But the UN could decide to stop the system by bombing the accelerator, an action that would not release any significant amount of radioactivity. In contrast, the UN cannot bomb a reactor which has already been or is in operation because of the huge amounts of radioactivity it contains.

6. THE COLLABORATION BETWEEN THE UN, RUSSIA AND WESTERN EUROPE

The starting point : the Saltsjöbaden Conference

In early 1991, the Russian suggested a meeting in Moscow with American experts about the newly interesting possibility of accelerator- supported subcritical reactor systems for energy production and treatment of nuclear waste in spent fuel. However, at that time the cold war was not yet over and it was too early for American experts to go to Moscow, so Los Alamos-with whom I had a long history of cooperation—proposed that I organise a conference on the subject in neutral Sweden. As there were no difficulties for either Russian or American scientists to come to Sweden, after checking with both the Swedish government and the opposition leaders, I accepted, and organised the conference in Slatsjobaden, near Stockholm, around midsummer 1991. It attracted about a hundred participants, scientists from the U.S., the former Soviet Union, England, France, Germany, Italy, Japan and elsewhere.

7. THE INTERNATIONAL SCIENCE AND TECHNOLOGY CENTRE, MOSCOW (ISTN)

ISTC was formed in 1995 by the U.S., the European Union and Japan, who contributed 25 M USD 22 M ECU and -12 M USD respectively. The purpose was to finance peaceful R&D

projects in Russian laboratories in order to give meaningful work to the Russian weapons experts who, following the reduction in the Russian weapons industry, would otherwise have been to a large extent unemployed, and could have been lured by irresistibly lucrative offers from other nations trying to develop nuclear arms.

Two more countries jointed the ISTC: Sweden and Finland. As at that time they were not yet members of the European Union, they made independent contributions: in the case of Sweden, 4 M USD to ISTC Moscow and 1,5 M USD to ISTC Keiv, and in the case of Finland 1.75 M USD.

After the Saltsjöbaden Conference we organiSed a small scientific/ technological research team, the ATW group, with members from Uppsala University, Royal Institute of Technology, Stockholm, and Chalmers institute of Technology, Gothenberg. We suggested to the Swedish Ministry of Foreign Affairs and the Ministry of the Environment that 1 M SUD of the Swedish contribution to ISTC Moscow should be used for the study of the accelerator- driven subcritical technology, with its interesting possibilities for plutonium and treatment of spent fuels. This proposal they agreed to.

We in the Swedish ATW group felt that this Swedish money should best be used for having the Russians at the Obninsk Institute design and manufacture a lead/ bismuth beam target several negotiations in Stockholm with the Russian experts, and with representatives from LANL and the Paul? Scherrer Institute in Switzerland present as consultants, we arrived at an agreement on the specification and the price, 1 MSUD, for the target.

The Russian Institute then applied to the ISTC board for financing of the project, knowing that Sweden would pay for it. The project was accepted by the ISTC, and then it happened that the US offered to pay half of it and the European Union 25 per cent so in the end only 25 per cent of the financing comes from Sweden.

After a few bureaucratical delays, the work is now proceeding smoothly. Russian experts will participate in mounting the target in Los Alamos, something that would have been unthinkable seven years ago. The research around the target in planned to be international.

8. CONCLUSION

Looking forward to the coming century and humanity's growing energy needs, accelerator- assisted subcritical reactors seem to hold good potential. Nuclear fusion and solar power, in principle promising, appear still to need long research and development times and may not be commercially available until later into the century, by which time fossil fuels will already have become scare and very expensive. Nuclear power. In the form of the currently-exploited light water reactors, leaves plutonium in the spent fuel, plutonium which can be used to make bombs and which, if stored (as it is today), needs to be guarded for about 100,000 years.

The technology to construct accelerator-assisted subcritical reactors exists today and the pilot-plant phase could get under way as soon as financing is supplied. Accelerator-assisted subcritical reactors present certain advantages compared to light water reactors: they cannot run amok, do not produce plutonium that has to be stored, and can clean up the existing waste from the first generation of nuclear reactors (as well as their own). And in terms of cost, any slight differential in the cost of the electricity supplied to the net compared to current light water reactors would be many times compensated for by the benefit in terms of their superior safety, their plutonium clean-up and their transmutation capability.

With accelerator-assisted subcritical reactors, nations and utilities which, under the pressure of increasing fossil fuel costs during the coming century, want to invest in nuclear power may have the opportunity to do so with a significantly improved nuclear system. And all current nuclear-power nations could have the means to clean up waste which has hitherto been stored with its plutonium and other biologically damaging isotopes intact.

Address:

CURT MILEIKOWSKY

International Security Programme
Harvard University
Cambridge MA 02138
USA

31

Disposition of Arctic Rad Waste and Decommissioning of Propulsion Nuclear Power Facilities

ANTONINO LANTIERI

SUMMARY

This report points out the difficulties of the Russian Federation about the management of the nuclear waste and of the submarine decommissioning located in the Arctic Region.

All the industrial and naval enterprises involved with the nuclear submarine utilisations are unable to solve these problems and the national programmes are currently limited by a number of reasons, the main of which is that the facilities and more in general the support complex is not sufficient to effect in the time required the necessary operation to reduce the risk of the nuclear and radiation accident.

More resources should be devoted to decommissioning and clean-up problems.

Considering the economical situation of the Russian Federation and the urgency of the problems, an international emergency plan, coming from a Key Study of the real situation, is required to support the Russian intervention.

GENERAL STATUS OF THE PROBLEM

A host of environmental problems of the Arctic region related to the nuclear waste have come to the attention of the world in the last years. In this context, the issue of nuclear powered submarine utilisation occupies a particular place.

The operation of nuclear powered submarine generates considerable amounts of nuclear waste; liquid and solid radioactive wastes need to be removal from submarine and stored; periodically the submarine needs to be refuelled, thus spent fuel needs to be removed and also stored. In addition decommissioning a nuclear submarine generates a big quantity of nuclear wastes.

Analysing the data comprised in the free reports, in the Arctic Region every step of the process concerning the submarine activities is facing problems.

The situation is aggravated by the fact that many of these submarines are loaded with irradiated highly radioactive fuel and exceeded by designed service time limits and therefore their quickly deteriorating integrity may create, in the case of emergency, significant probability of radiation accident on international scale.

The support complex which was already in poor shape and accident-prone during the Soviet time has been particularly burdened in the last few years. Shore-side waste sites are full of low-level radioactive waste and spent fuel. Shipments of the spent fuel to Mayak for reprocessing have problems for funds and equipment. The service ships which unload the spent fuel from submarines are also full and in poor shape. The shipyard where the work is done has several problems. There are no final land-based storage sites for decommissioned reactor compartments removed from submarine, so they are being stored afloat in bays near naval bases.

The number of the decommissioned NPPS in the Russian Federation, in accordance to the Strategic Weapons Reduction Treaties and owing to the resource exhaustion, is at the moment over 150, and the number will continue growing in future, because constructed submarines have been more than 250.

Actually the decommissioned submarines have been anchored in the special shipyards that are unprepared to the long term secure floating storage.

The lifetime of the major part of the decommissioned submarines ranges from 25 to 30 years, about 35 per cent of them were launched more than 30 years ago. Up to 40 per cent of the decommissioned nuclear vessels have ;not been under any maintenance or reparation for more than 10 years.

By the beginning of 1996, 18 nuclear submarines have been utilised according to the scheme of the three compartment block cutting out (this scheme consists: the reactor compartment is cut out together with two adjacent ones and this three compartment unit it tugged to special shipyard allocated for the temporal reactor storage afloat), 8 vessels have been prepared for the long term storage afloat and 13 vessels were under the procedure of utilisation and preparation for the long term storage.

RADIOLOGICAL RISK

Actually all the industrial and naval enterprises involved with the nuclear submarine utilisation are unable to dismantle more then 5-6 submarines a year.

At this utilisation performance the submarines that are currently destined to the dismantling will remain waterborne for a last 10-15 years; this situation presents a serious potential nuclear and radiation environmental hazard, and in the case of emergency, significant probability of radiation accident on international scale.

The risk of nuclear and radiation is increased by the poor technical condition of the anchored nuclear submarine and by the fact that more than 60 per cent of these submarines still have highly radioactive spent nuclear fuel on board.

However irradiated fuel of naval reactors has specific features, such as a long cooling time and the retention of the high enrichment (and this is the difference between this fuel and the irradiated fuel of ordinary nuclear power plants). Due to these features in combination with a large amount of the fuel, the irradiated nuclear fuel of the Navy poses a serious threat to the international security since it is attractive from the point of view of the proliferation of unauthorised use and pose the radiation danger far environmental.

It should be emphasised that the amount of the highly enriched uranium irradiated fuel is much larger than its amount in the fresh fuel and sufficient international attention should be devoted to minimise their environmental and proliferation risks.

Such risks may be essentially diminished by speedy unloading; of irradiated fuel from decommissioned submarines and by reducing to safety standards disposed reactors and other radioactive components of such submarines.

The real magnitude of the potential and radiation hazard is conditioned by the spent fuel accumulated; this is more than 300 cores and more than 70,000 fuel assemblies.

The transportation of the fuel from North Fleet to Mayak processing plant, under normal condition, is 18-20 cores for years while the expected annual increase is 30-32 cores.

Lack of effective financial and organisational/technical mechanism of nuclear submarine disposition poses both environmental and nuclear proliferation risks not only for national but also for international security and therefore must resolved by international cooperation.

So, to take the situation in normal condition and to provide the nuclear and radiation safety, extraordinary measures are to be taken.

We consider relevant to promote in this Country an extraordinary international plan to give solution to these problems, in particular the European Union would have a primary rule considering, also, that same installation are near to European Countries.

PROPOSAL FOR THE PROBLEM SOLUTION

To provide the nuclear and radiation safety, considering the state of the decommissioned submarines, the first priority is to defuel the reactors, to empty the cooling circuits and to isolate the compartment containing the primary circuit including the reactor vessel.

This problem provided the focus for serious studies in Russia which give different approaches, but the main obstacle to the realisation of the programmes is the condition of inadequate and irregular finances and the absence of a head organisation for its implementation.

We consider that the head organisation would be the Russian Navy with the support of the interdepartmental scientific council; we consider this solution necessary to guarantee together the safety and the security.

The first action would be the management of the emergency, consisting in programmes for the implementation, inside the bases, of the fuel storage capacity in line with the capacity of the Country processing plants and more in general for the implementation of the safety waste storage.

In parallel an international Key Study on problems of decommissioning and disposition of propulsion nuclear power facilities of the Russian navy with the aim to originate the condition for an international programme to solve definitively the problem could start.

The Russian organisation involved in this study should be the Russian Navy and the main scientific organisation involved in this problem.

THE INTERNATIONAL KEY STUDY

The whole cycle of decommissioned submarine disposition and industrial utilisation of equipment and material from disposed compartments include several operations:

1. Temporary storage of decommissioned submarine till reactor fuel unloading.
2. Procedure of irradiated fuel unloading from submarine reactor.
3. Temporary storage of submarine with unloaded reactor fuel.
4. Transportation of submarine to dismantling plant.
5. Cutting out reactor compartment and preparing it for storage.
6. Transportation of reactor compartment to the place of "otstoi".
7. Cutting other compartments in parts at dismantling plant.
8. Industrial utilisation of equipment and materials from disposed compartments.

"The Key Study" would comprise the complex investigation concerning the problems of decommissioning and disposition of propulsion nuclear power facilities and include the following main objectives.

- generalisation of the investigation results and gained technological and industrial experience related to the problem:
- technical and economical comparison of various technologies to one another and with the method of long-term storage of the radioactive constructive elements of the nuclear facilities without preliminary full-scale processing;
- assessment of the ecological risk related to various technologies of the radioactive material processing. Comparison of the ecological parameters of the long-term storage and various reprocessing methods;
- investigations and development of the universal methodology for minimisation of the amount of the dangerous waste

produced in decommissioning of the propulsion nuclear power facilities.

The first stage of the investigation assumes search and development of the universal methodology of the waste minimisation separately for stationary and propulsion facilities.

On our opinion, the key problems of the NPSS decommissioning and utilisation are:

- providing safe long-term (at least, 15-20 years) on float storage of the decommissioned NPSS;
- unloading and transportation of the spent nuclear fuel accumulated in the ships and at the fleet bases for the storing at the specially equipped sites;
- cutting the reactor compartment and development of the infrastructure for their long-term storing at the specially equipped sites; in the framework of this task, a special place is occupied by the problem of managing the reactor compartments with emergent steam-generating facilities;
- processing and safe storage of the significant amount of liquid radioactive waste;
- dismantling of the vessel tails of the utilised NPSS using efficient and ecological technologies that provide minimization of the harmful impact on the environment;
- providing conditions for nuclear non-proliferation on all stages of decommissioning and utilisation of the NPSS;
- development of the required legislative background, in particular, in the ecological sphere;
- improving the managerial structure of the NPSS utilisation process as a unified complex programme.

So "the Key Study" must investigate the aspects of these problems and will be based on published materials and documents and on other open sources of information.

Existing in Russia at present methods for disposition and utilisation of nuclear submarines are not environmentally clean, they require large human and financial resources and are not cost-effective. Therefore it is necessary to explore the possibilities of creating new ecologically clean and cost-effective technology for nuclear submarines disposition.

"The Key Study" would analyse technical, organisational and other aspects of these operations and would made recommendations for further possible activities of cooperation.

EXPECTED RESULTS

Analysis of organisational and technical aspect

The analysis will be devoted to collection and processing of available open source information, to analysis of general organisation and technical aspects of the problem of nuclear submarines disposition.

The expected results will include:

1. Realistic estimate of the scope of the problem;
2. Description of present organisational scheme of nuclear submarine decommissioning and disposition;
3. Comparable evaluation of different options for preparing nuclear submarines to disposition;
4. Comparable evaluation of complex industrial nuclear submarine disposition options;
5. Cost estimates of possible disposition options;

Fuel Problems Investigation

Major amounts of such fuel are located:

- In reactors of decommissioned nuclear submarines of the Russian Navy;
- in land based fuel storage facilities;
- in storage facilities of floating service shops and fuel storage (floating platforms).

The expected results will include:

1. Analysis of real situation of the temporary storage;
2. Analysis of the reprocessing plants problems;
3. Suggestion for the short and long term strategy.

Waste Problems Investigation

It will pay special attention to the problem of managing the radioactive construction materials of the nuclear power facilities, so it will be effected:

1. Inventory of the wastes;
2. Analysis and generalisation of the world scientific, technological, and industrial experience gained by the world nuclear power industry in the problem of the radioactive waste compaction;
3. Investigation of the advantages and shortcoming of the existing technologies and approaches to the compaction of the radioactive waste of the nuclear power industry; tech-

nical and ;economical comparison of these methods with the method of the long-term storage of the radioactive construction elements of the nuclear facilities without preliminary scale processing;

4. Investigation of the environment impact factors and assessment of the ecological risk related to various technologies of the radioactive material processing. Comparison of parameters of the ecological load of the environmental from the long-term storage and various reprocessing methods;
5. Investigation of the approaches and development of the universal methodology for minimisation of the amount of the dangerous waste; produces in decommissioning of the propulsion nuclear power facilities.

BRIEF FINAL CONSIDERATIONS

The aim of the Key Study to originate the condition for an international programme complementary to the Russian programmes and financed by the international community to solve definitively the problem.

The Russian organisation involved in this study should be the Russian Navy and the main scientific organisation involved in the problem.

We think that the Key Study should be financed by the E.U. programmes.

Italian and Russian organisations, already, cooperate to optimise the proposal of the Key Study.

Address:

ANTONINO LANTIERI
ENEA, ERG-FISS-FIRE
40129 Bologna
Italy

32

Cooperative Threat Reduction Programme

SARAH LENNON

COOPERATIVE THREAT REDUCTION PROGRAMME

Nunn-Lugar Legislation

It is in the National Security Interests on the United States:

A) To facilitate on a priority basis the transportation, storage safeguarding, and destruction of nuclear and other weapons in the Soviet Union, its republics, and any successor states; and

B) To assist in the prevention of weapons proliferation.

(Soviet Nuclear Threat Reduction Act of 1991, PL 102-228)

CTR MISSION

Cooperative with FSU Republics to get rid of WMD threat and to provide top notch physical security during the process.

PROGRAMME OBJECTIVES

1. Assist Ukraine, Belarus, and Kazakhstan in becoming non-nuclear weapons states and in meeting their START requirements
2. Assist Russia in accelerating strategic arms reduction to START levels
3. Enhance safety, security, control, accounting, and centralisation of nuclear weapons and fissile material in the former Soviet Union to prevent their proliferation and encourage their reduction:
4. Assist states of the former Soviet Union to eliminate and prevent proliferation of biological and chemical weapons capabilities
5. Encourage military reductions and reform and reduce proliferation threats in former Soviet Union recipient states.

CTR Funding*

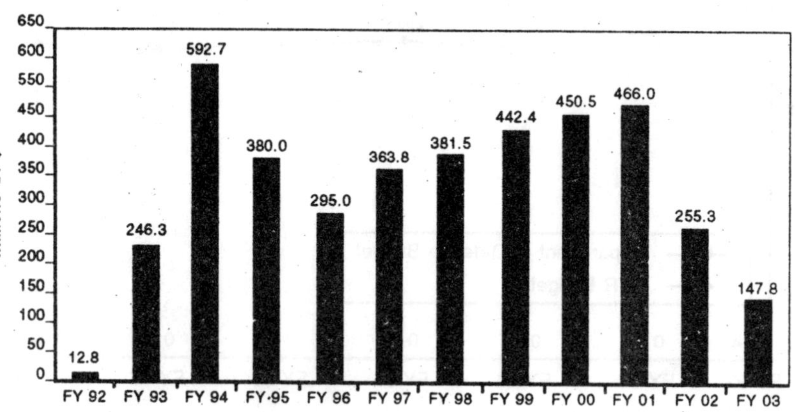

*Based on current CTR budget authority.
FY 9–03 authority is contingent upon Congressional approval

Planned Cumulative CTR Funding 1992 2001 by Country

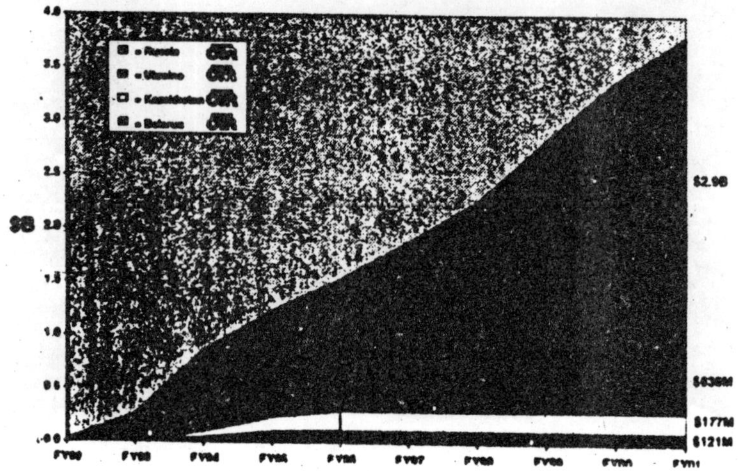

CTR Portion of the Department of Defence Budget

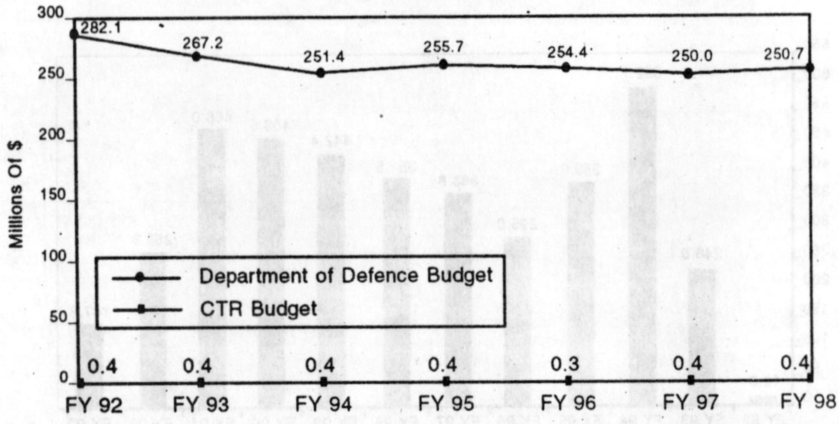

The Nuclear Chain of Custody

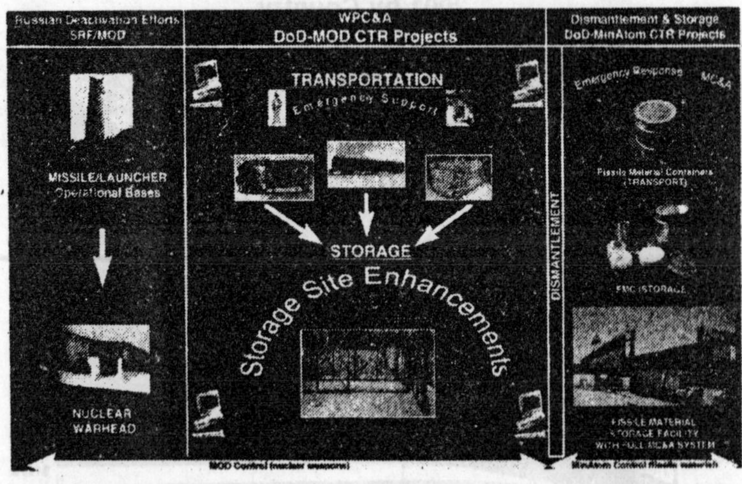

U.S. – Russia
Overview of Implementation

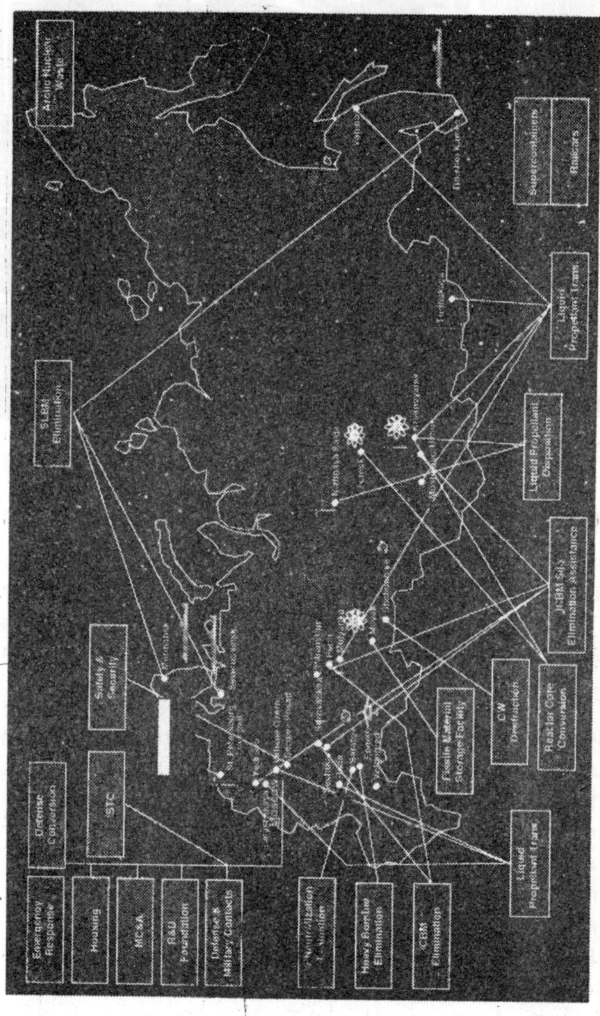

The Chain of Custody

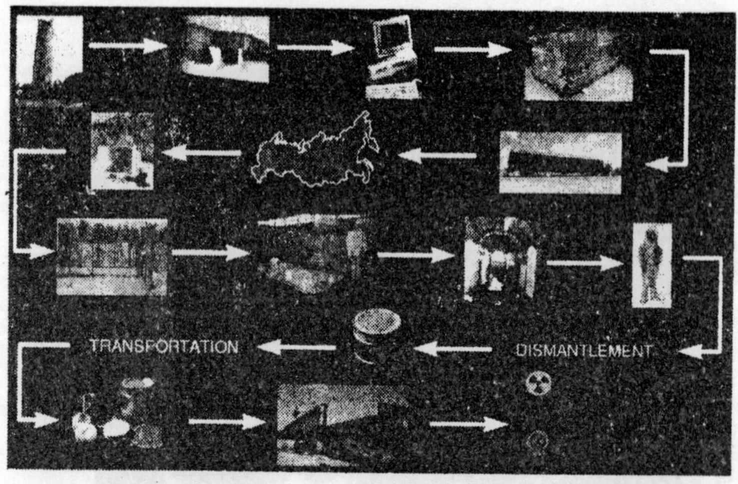

WEAPONS PROTECTION, CONTROL AND ACCOUNTING

Umbrella Agreement

Agreement Between the United States of America and the Russian Federation concerning the Safe and Secure Transportation, Storage and Destruction of Weapons and the Prevention of Weapons Proliferation, June 17, 1992 (duration 7 years).

Projects

- Automated Inventory Control and Management System
- Armoured Blankets
- Supercontainers
- Security Enhancements for Railcars
- Emergency Support Equipment
- Information Analysis System
- Training Centre
- ASSESS
- Quick Fix
- Security Assessment and Training Centre
- Site Enhancements
- Personnel Reliability Programme
- Dosimeters

AUTOMATED INVENTORY CONTROL AND MANAGEMENT SYSTEM (AICMS)

- *Objective:* To computerise inventory control of nuclear weapons at all levels in the chain of command.
- *Providing:* Modern computers and data management software installed in a modern computer network connecting headquarters, regional, and storage sites
- *Use:* Enhance MOD's capability to account for and track nuclear weapons
- *Total project value:* $ 22.6M

ARMOURED BLANKETS

- *Objective:* To enhance security during transportation
- *Provided:* 4,020 armoured blankets
- *Use:* Protects nuclear warheads from small arms fire
- *Total project value:* *3.2M

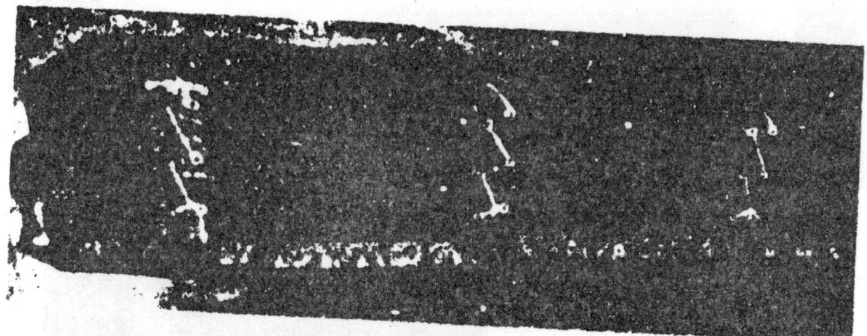

SUPERCONTAINERS

- *Objective:* To enhance security during transportation
- *Provided:* 150 supercontainers
- *use:* Protects nuclear warheads from small arms fire and enhance physical security during handling by precluding direct access to weapons
- *Total Project Value:* $ 23.6M

RAILCAR ENHANCEMENTS

- *Objective:* To enhance security during transportation
- *Provided:* Upgrade kits for 100 cargo and 15 guard railcars
- *Use:* Improved physical security through tamper and intrusion detection sensors; thermal protection enhanced with fire detectors and thermal insulation
- *Total project value :* $ 21.5M

EMERGENCY SUPPORT

- *Objective:* To enhance ability to respond to accidents/incidents
- *Provided:* 5 mobile sets of equipment with containers
- *Use:* Recover weapons and mitigate radiation effects
- *Total projects value:* $ 8.5M

STORAGE SECURITY ENHANCEMENTS—PHASE I

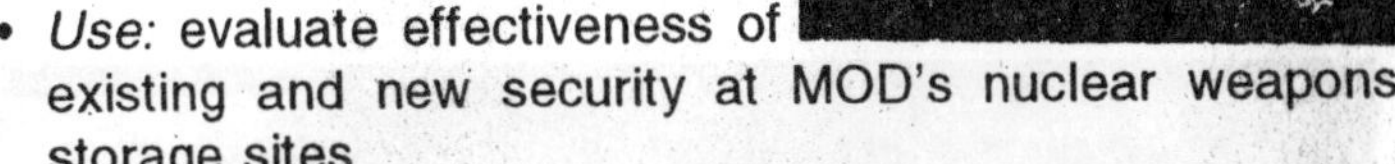

- *Objective:* To assess security weaknesses at nuclear weapons storage sites
- *Provided:* Analytical System and Software for Evaluating Safeguards and Security (Assess) computer model, computers, and training
- *Use:* evaluate effectiveness of existing and new security at MOD's nuclear weapons storage sites
- *Total project value* (all phases + personnel safety): $ 167M

CTR Scorecar

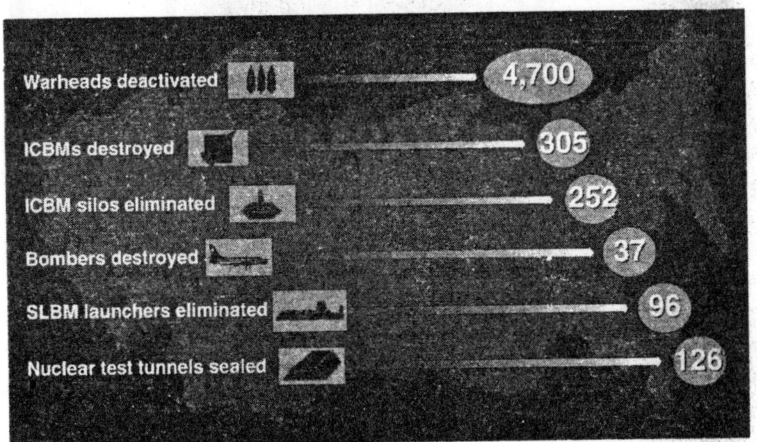

STORAGE SECURITY ENHANCEMENTS—PHASE II

- *Objective:* To address urgent MOD requirements for security upgrades at nuclear weapon storage sites
- *Provided:* 50 one km sets of security fencing and microwave security systems and 2 additional levels of engineering fencing per set
- *Use:* To provide immediate upgrade of MOD nuclear weapons storage sites

STORAGE SECURITY ENHANCEMENTS—PHASE III

- *Objective:* To assess security weaknesses and to upgrade physical security at MOD's nuclear weapons storage sites
- *Providing:* Establishing and outfitting Security Assessment and Training Center (SATC)
- *Use :* Evaluate new physical security equipment, technology, materials, and to conduct training

STORAGE SECURITY ENHANCEMENTS—PHASE IV

- *Objective:* To upgrade physical security at MOD'S nuclear weapons storage sites
- *Providing:* Physical security upgrade kits for up to 50 nuclear weapons storage sites
- *Use:* Upgrade nuclear weapons sites with new enhanced security systems

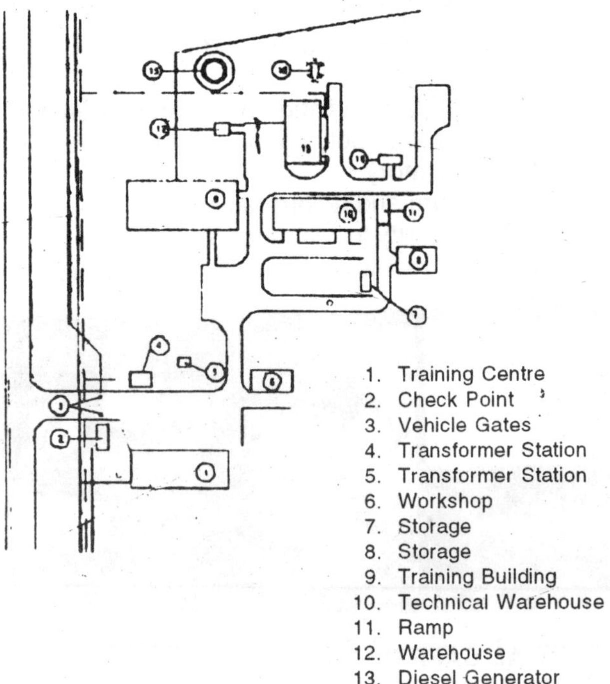

1. Training Centre
2. Check Point
3. Vehicle Gates
4. Transformer Station
5. Transformer Station
6. Workshop
7. Storage
8. Storage
9. Training Building
10. Technical Warehouse
11. Ramp
12. Warehouse
13. Diesel Generator
14. Warehouse
15. Fire Protection Tank
16. Diesel Fuel Storage Tank

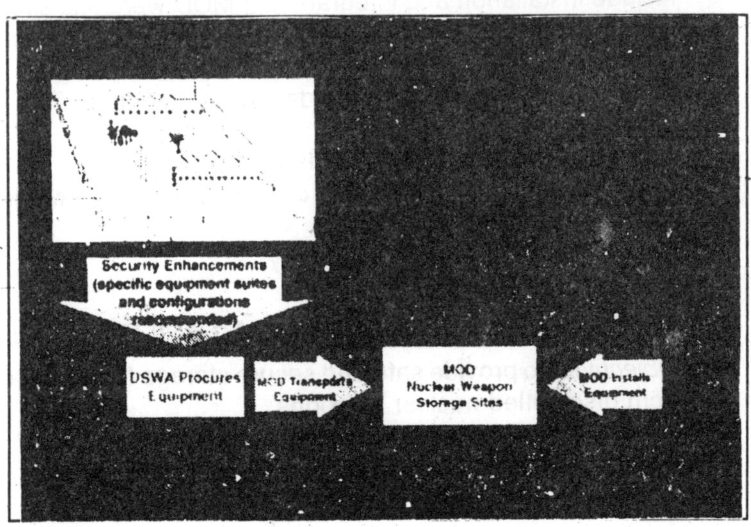

WEAPONS PERSONNEL

- *Objective:* Enhance safety and reliability of nuclear technicians
- *Providing:* 3.500 radiation dosimeters, portable drug and alcohol test kits, polygraphs, and fixed-site analysis and testing laboratory
- *Use:* Measure and record radiation dosages and enhance MOD's capability to accurately evaluate and test personnel who have direct and indirect access to nuclear weapons

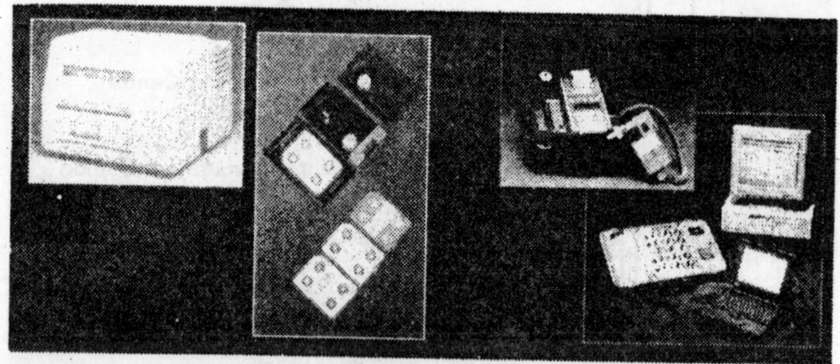

POTENTIAL SAFETY AND SECURITY COOPERATION WITH MOD

- Extend Storage Site Security Enhancements project to include installation and integration at MOD weapons sites
- Extend the AICMS project to include all sites for the interim capability
- Extend AICMS project to include site preparations
- MILES 2000 or commercial alternative
- Additional railcar enhancements
- Armoured Trucks for transport of nuclear weapons within bases and to/from rail lines
- Efforts to improve weapons safety
- Projects with the SRF

MAYAK FISSILE—MATERIAL STORAGE FACILITY

- *Objective:* To provide safe and secure storage for material from dismantled nuclear weapons
- *Provided:* design and construction assistance
- *Total project value:* $ 165 M

AT-400R FISSILE MATERIAL CONTAINER

Address:

SARAH LENNON
Defence for Strategy and Threat Reduction
Washington DC
USA

33

Research and Development in Nuclear Material Accounting, Control and Physical Protection at the Kurchatov Institute Russian Research Centre

VLADIMIR SHEMELEV

ABSTRACT

Research and development in nuclear material accounting, control and physical protection at "Kurchatov Institute" Russian Research Centre is briefly described. Main part of this activity is performed in cooperation with the US National Laboratories under the US Department of Energy Laboratory-to-Laboratory cooperative programmes.

INTRODUCTION

Due to significant political, social and economical changes occurred due to the disintegration of the Soviet Union, since the end of 1992—beginning of 1993 the "Kurchatov Institute" Russian Research Centre (KI RRC) has intensified its activities in the area of nuclear materials accounting, control and physical protection. The importance of these activities for the strengthening of nuclear non-proliferation was timely evaluated by KI RRC management who strongly supports them.

After disintegration of the former Soviet Union and transition of Russia's economy from the strictly centralised type to the market economy type, the problems of nuclear material control and its security assurance became very acute. Non-proliferation efforts now are focused on the weak points of MCP & A system at the State level, and at research centres, materials production and processing facilities, nuclear power plants.

Kurchatov Institute is an example of large research centre where vast amounts of various nuclear materials are accumulated.

"KI" RRC HISTORY AND PRESENT STATUS

Established in 1943 to solve the problem of nuclear weapon development, Kurchatov Institute became eventually the largest nuclear centre aimed towards resolving of a very wide range of military, civil, applied and fundamental problems. This range was expanded constantly to cover many additional areas of natural science.

In November 1991 Kurchatov Institute was transformed in the State Research Establishment named "Kurchatov Institute" Russian Research Centre, subordinated to and reporting directly to the Government of Russia.

At present KI RRC is multi-disciplinary scientific and technical establishment with 11 research institutes and with several so called scientific-technical complexes.

Due to the multi-discipline character of KI RRC, it is equipped with many nuclear facilities and laboratories, which are different from each other by physical and technical parameters and by types of nuclear materials and radioactive substances used for research and development.

KI RRC has also considerable infrastructure of auxiliary facilities such as specialised storages of irradiated fuel and other radioactive substances.

The largest amount of nuclear materials is located in the Central Storage, in irradiated fuel storages, in research reactors and in critical and subcritical assemblies(2)

The work on nuclear materials safeguards is not new for KI RRC.

R&D for international safeguards were initiated there as early as in 1970's. One of two facilities submitted for IAEA safeguards under the USSR voluntary offer was research reactor "IR-8" in Kurchatov Institute. Several projects were implemented in Kurchatov Institute under USSR support programme for IAEA safeguards.

However really extensive work on MCP&A began in KI RRC in 1993 with the development and implementation of NUMACS Project (Nuclear Materials Accounting, Control and Safeguards).

The purpose of NUMACS Project was development, implementation and deployment in Kurchatov Institute of modern automated system of accounting, control and safeguards of nuclear materials and other radioactive materials and radiation sources. This model system would allow to use it at any nuclear facility, complexes of facilities, installations and plants or may be used as scientific and technical basis for Regional or State systems. (2)

At present main part of the work on MCP&A in KI RRC is being performed in cooperation with several US DOE laboratories under Laboratory-to-Laboratory programme.

Agreements for cooperation and direct contracts, providing the basis for technical works are concluded by KI RRC with LANL, LLNL, ORNL, BNL, SNL AND PNWL.

The work is performed by KI RRC is cooperation with the US National Laboratories in the following directions:

1. Upgrading MPC&A at the nuclear facilities at the territory of KI RRC
2. Upgrading MPC&A at the nuclear facilities of the Russian Navy.
3. Upgrading MPC&A at the nuclear facilities of Murmansk Shipping Company (MSC).
4. Remote Monitoring Technology.
5. Nuclear export analysis and control.
6. Regulatory documents development and training.

NUCLEAR MATERIAL CONTROL AND ACCOUNTANCY AT KI RRC

Under this heading fall the following projects:

- Vulnerability assessments;
- Physical protection;
- Computerised material accounting;
- Physical inventory taking;
- Accounting procedures for transfers between MBA's;
- Bar-code technology;
- Scales and weight measurements;
- Tamper indicating devices;
- Radiation measurements;
- Tamper indicating devices;
- Radiation measurements;

- Criticality measurements;
- Integrated demonstration.

In less than three years the MPC&A Programme has led to important advances for the protection of nuclear materials at the KI RRC.

UPGRADING MPC&A AT RUSSIAN NAVY

In March 1995 Commander-in-Chief of the Russian Navy wrote a letter to KI RRC Vice President suggesting that KI and the Russian Navy could work together on MPC&A upgrades for Navy nuclear fuel storage and handling and that possible US cooperation with this effort should be explored A series of communications and exchanges followed that culminated in a joint statement issued by US Secretary of Energy, representative of the Russian Navy and President of KI RRC in July 1996. This statement indicated that "the Russian Ministry of Defence and United States Department of Energy decided to jointly cooperate to ensure the highest possible standards of control, accounting and physical protection for all storage locations of the Navy of Russian Federation containing fresh highly-enriched uranium fuel for naval nuclear reactors" and that this effort would be conducted through the MPC&A Programme. As a result of this joint statement, a Russian Navy MPC&A project was formally initiated.

The objective of this project is to place all Russian Navy fresh fuel into a high-level domestic safeguards system during the next three years. Fresh fuel, upon delivery at the Northern and pacific fleet sites, will enter into an integrated MPC&A system that addresses physical protection and item control and accountancy through all stages of consolidated hand-based storage, intransite land transportation, water-based storage, and intrasite water transportation. The local commands of the Northern and Pacific fleets and the Navy Central command in Moscow will have the capability to monitor the local or overall system status and material control remotely.

A MPC&A project to increase domestic safeguards for Russia Navy nuclear materials is being executed. This cooperation includes enhanced protection and ;control features for storage facilities and fuel storage ships, computerised accounting systems for naval fuel, methods and equipment for rapid inventories, status monitoring capabilities, plans for improved security of fresh fuel during truck transportation, and training. Over the last 12 months, emphasis has been placed on the design and implementation of integrated MPC&A upgrades for Northern Fleet Storage Facility (Site 49). Also, preliminary work has begun on related tasks for the fuel storage ships, intransite truck transportation, regulatory document development, and an operations training programme.

UPGRADING MPC&A AT FACILITIES OF MSC

Kurchatov Institute participates in cooperative multinational efforts aimed to upgrade MPC&A at the facilities of MSC. Under this multinational cooperative efforts, weapon usable nuclear materials at the MSC will be brought under high level safeguards during next two and a half years.

KI RRC and US Laboratories focus areas include upgrading MPC&A of highly enriched uranium fresh fuel at the service ship "Imandra" and "Atomflot Port" while Norwegian and Swedish focus areas include MPC&A of nuclear fuel for the ships "Lotta", "Sevmorput" and the Icebreaker fleet.

REMOTE MONITORING TECHNOLOGY

In the context of US and Russian Lab-to-Lab cooperation SNL contracted with KI RRC to demonstrate the feasibility of remotely monitoring the storage of nuclear material.

Kurchatov Institute selected the storage at "Gas Plant" facility containing about 70 kilogram of HEU and USA selected the storage at Argonne national Laboratory-West containing about 130 kg of HEU. At both sites the system is provided by various containment and surveillance means including: door monitors, infra-red and micro-wave volumetric sensors, item motion sensors and fiber loop seals, break beam monitors and computerised data acquisition system (DAS). The DAS perform data acquisition, storage and long term archiving of event and ;image data to an optical storage disk. Data transfer is accomplished via commercial dial telephone access.

Installation was performed in similar manner at both ANLW and KI RRC facilities. At both sites installation and trial operations went very smoothly and expeditiously. The successful installation and operation of the system is the evidence for the maturity of the RMS technology and design. RMS description was given in the INMM meeting paper (3).

We want to emphasis that we see enormous potential of RMS for applications in Physical Protection, in international Safeguards and in various disarmament measures.

NUCLEAR EXPORT CONTROL

Increasing volume of trade involving nuclear materials, nuclear technology and nuclear related dual use equipment leads to ever greater diffusion of nuclear technology and to increasing danger of nuclear weapon proliferation. Nuclear export control became indispensable element of prevention or deterrence to illicit diversion of materials and technology useful for nuclear weapon development.

Several political, organisational and regulatory measures of national or multinational character were undertaken in the last few years. However, scientific and technical basis for export control support is still rather weak. Not many scientific and technical organisations are performing systematic research and development work in this area. Noting this shortcoming KI RRC together with LANL has initiated several projects related to improvements of scientific and technical basis for nuclear export control.

One of these project is aimed to developing the criteria and methodology for technical export licensing review and defining the technical expertise and Data Bases needed to support this methodology.

Another project is devoted to evaluation of New Independent State export control Authorities with respect to scientific and technical guidance and to identify technical and scientific resources required and available to assist export control authorities in their activities.

Separate project is devoted to evaluation and development of methods and software for computerised system for nuclear export analysis and information support of decision making procedures.

In the framework of these projects the Export Control Data Base (ECDB) was established and is constantly updated. The ECDB contains the information on legal and regulatory documents on export control, control authorities, on potential exporters and export goods etc.

As a result of this work the prototype computerised nuclear export analysis system NEAS was designed, developed, tested and evaluated.

TRAINING

Training Centre was established at RRC "KI" in order to implement to the most modern technique and methods of detection, as well as to train personnel. Vast protected territory and existence of laboratory facilities gives the good possibilities to arrange and conduct training classes and to demonstrate samples of the equipment. Main goals of the training centre are the following: training of nuclear dangerous facilities (NDF) personnel and specialists on physical protection, MC&A, demonstration of samples of equipment (produced in both Russia and USA), comparative analysis of their efficiency, selection of optimal combinations of elements, spreading of advanced methods onto other Russian nuclear facilities.

CONCLUSION

Lab-to-Lab cooperation programme being implemented in KI RRC demonstrates that Russian and US specialists can work together to implement real systems that protect, control and account for nuclear materials. During the programme implementation many of the organisational, coordination and technical problems have been discovered, solved and noted for future work. The Kurchatov systems have been demonstrated to numerous organisations as an example of the outreach activity.

REFERENCES

1) "Russian Research Centre Kurchatov Institute". Booklet published by INTERSVIAZ, Moscow, 1994.
2) "Model automated system of nuclear materials accounting, control and safeguards for complex nuclear facilities (NUMACS) system)". Technical Specification, Copyright 1994 by KI RRC, Moscow, Russian Federation and American Technical Institute, Brunswick, Tennessee, USA.
3) "Demonstration of bilateral US and Russian remote monitoring system for special nuclear materials". Paper presented at INMM 36th Annual Meeting, July 9-12, 1995, Palm Desert, CA, USA.
4) "US-Russian Lab-to-lab Programme of MPC&A at the KI RRC". INMM 37th Annual Meeting Proceedings, July 1996.

5) "Partnership for Nuclear Security". DOE Nuclear Material Security Task Force, December 1997.

Address:

VLADIMIR SHMELEV
Deputy Director
Division of Non-Proliferation and Central
RRC Kurchatov Institute
123182 Moscow
Russian Federation

34

Sharing the Challenges of Non-Proliferation

PAUL HERMAN

The changes brought about by the end of the cold War have created a surprising turn of events. Once unthinkable collaborations and partnerships to reduce the threat of proliferation are now happening with increasing frequency.

In these post-Cold War days, the secret cities that contain Russia's weapons complex remain closed, still surrounded by fences patrolled by armed guards. But changes are going on within them. Scientists and engineers from Lawrence Livermore can now be found inside, engaged in meetings with their Russian counterparts. This change has occurred largely because of the convergence of two events: the shift from an arms race to arms reduction, and the dissolution of the Soviet Union, with its attendant economic upheaval.

One of the many risks introduced by the first event is that of increased nuclear proliferation if the disposition of nuclear weapons technology and materials is not managed carefully. Russia has, for example, large amounts of surplus weapons grade nuclear materials in various forms. These materials are highly desirable to potential proliferators and terrorists. They have become more vulnerable to theft or diversion because Russia now has fewer resources to apply to safeguarding its nuclear materials. U.S. and Russian scientists and engineers are working together to reduce such risks.

U.S. policy makers recognise that Russian nuclear scientists have essential roles to play in global arms reduction and non-proliferation causes. Alleviating the scientist' economic hardships and uncertainty would greatly aid the stabilisation of Russian nuclear weapons complex. To these ends, the U.S.

Department of Defence has formulated a policy to aid Russian scientists through stimulating commercial economic development in the closed cities. One large component of the policy is the Nunn-Lugar Cooperative Threat Reduction bill, passed in 1991, which initiated collaborations between the U.S. and the newly independent states (NIS), principally Russia. The effectiveness and positive reception of Nunn-Lugar initiatives led to similar and complementary initiatives by the Energy and State departments.

Dubbed "defence by other means" by former Secretary of Defence William Perry, this policy depends as much on scientific capabilities as on political expertise. Thus, Lawrence Livermore staff have found themselves travelling thousands of miles between Livermore and various parts of the NIS to collaborate with NIS scientists on worthwhile, non-weapons-related projects as well as to monitor and assist the progress of arms reduction.

PROGRESS IN ARMS REDUCTION

The arms reduction taking place in the U.S. and Russia is an important step for global nuclear security. Because verification activities for the strategic arms reduction treaties (START) are concerned with the destruction of weapons launchers and do not deal with the warheads, the Biden Condition was appended to START I during the ratification process to ensure that warheads would be verifiably dismantled in future arms reduction.

Developing transparency measures to deal with the fissile materials derived from dismantled weapons is the task of the Safeguards, Transparency, and Irreversibility Working Group, a joint effort between the U.S. and Russia. Formed as a result of agreements made between Presidents Clinton and Yeltsin over several summit meetings, the group is chartered with developing mutually acceptable ways to keep fissile materials derived from dismantled nuclear weapons secure, account for and control their quantities, and prevent them from ever being used again in nuclear weapons.

Jim, Morgan is one of the Livermore scientists working with this group to implement its complex task. He has been involved in discussions about sharing information on fissile materials. The most difficult negotiations involve key proposals brought to the table by the U.S.:

- Regular exchanges of detailed information about weapons and fissile materials stockpiles;
- Reciprocal inspections at storage facilities to confirm the amounts of plutonium and highly enriched uranium removed from weapons;
- Various arrangements to monitor fissile material stockpiles.

These have been difficult proposals from the beginning, starting with fundamentally differing views on information sensitivity. Russia classifies its information differently than the U.S. In addition, because of former Energy Secretary Hazel O'Leary's openness initiatives, the U.S. has already published some general information about U.S. fissile materials stockpiles, which goes well beyond the type of information the Russians are willing to share.

The progress of the negotiations has been slow. The U.S. delegation has been trying to maintain some momentum in the talks by suggesting negotiating patterns to keep negotiations moving.

Whatever the course of action, these negotiations will not end when agreements on information exchanges and monitoring procedures have been made. There must also be U.S.-Russian agreements on what measuring devices and instrumentation are allowable for deriving specific information during reciprocal inspections at nuclear facilities.

In parallel to Morgan's work in negotiations, scientists and engineers at Livermore are designing special measuring technologies for use inside U.S. and Russian facilities. One candidate device that has been demonstrated to Russian scientists is a portable, battery-operated, germanium gamma-ray spectrometer. This instrument can determine whether plutonium stored inside containers is consistent with material that may have been removed from dismantled nuclear weapons. The spectrometer measures the plutonium's gamma-ray intensities in a narrow band of energy (630 to 670 thousand electron-volts) to reveal whether its ratio of plutonium-240 to plutonium-239 is consistent with weapons-grade material; it also estimates what minimum mass of plutonium is necessary to produce the observed intensities.

The narrow band of energy measured by the spectrometer

intentionally leaves some details of the material being measured unknown to satisfy Russian security concerns and make the spectrometer acceptable to the Russians. Tools used for transparency measurements must observe a careful balance between yielding enough to confirm crucial verification requirements but not revealing so much as to threaten the security interests of either side.

REDUCING HEU HOLDINGS

Even as the negotiations for safeguards, transparency, and irreversibility continue, the U.S. has found another way to safeguard some Russian weapons uranium—by buying it. In 1994, the U.S. signed a 20-year, $12-billion deal to purchase 500 metric tons of highly enriched uranium (HEU) recovered from Russian weapons. The contract calls for this uranium to be blended down to low-enriched uranium (LEU) and then shipped to the U.S. to be used for making commercial reactor fuel.

The transparency protocols for the HEU purchase are those that strive, on the one hand, to confirm for the U.S. that the shipped material has indeed been derived from Russian weapons material and, on the other hand, to confirm to Russian satisfaction that the LEU is not going to end up in the U.S. weapons programme. These confirmations require access to the uranium processing facilities of both sides. The negotiations for such access, normally complex and difficult, were made even more so when they became subsumed by a host of other issues surrounding the deal, including pricing and LEU market competition.

The final agreement allows Russian monitors access to the U.S. Enrichment Corporation's Portsmouth Gaseous Diffusion Plant in Piketon, Ohio, and to the five U.S. fuel fabrication facilities receiving the Russian Uranium. In turn, U.S. monitors are allowed access to the three principal Russian plants involved in the conversion of HEU to LEU. Lawrence Livermore is taking a lead role in support of DOE programme activities related to monitoring activities at those three plants.

What the agreement has meant for Livermore's Doug Leich, HEU transparency technical leader and a member of the U.S. monitoring team, is several long trips to Russia each year, to the cities of Seversk, Zelenogorsk, and Novouralsk. At the plant in Seversk, HEU metal is processed into an HEU oxide before being shipped to the electrochemical plants in Novouralsk or Zelenogorsk. In these facilities, the oxide is fluorinated and combined with a slightly enriched blending material to turn it into LEU suitable for making civilian power reactor fuel.

MONITORING ACTIVITIES

Describing the monitoring tasks at Seversk, Leich says that monitors can observe the whole oxidation procedure, from the beginning when the uranium metal is analysed by portable gamma-ray spectrometry to confirm its weapons-grade status, through its feed into and withdrawal from oxidation process equipment, to the final analysis of the withdrawn oxides. Leich and the other monitors apply U.S. tags and seals to some containers of the oxides before their shipment to Novouralsk or Zelenogorsk.

When the containers of oxide arrive at those sites, monitors first check the tags and seals on them. Then, says Leich, "We can request non-destructive assay of containers of HEU oxide, observe the feeding of oxide into a process that chemically converts the HEU to a hexafluoride form, and perform an assay of the HEU hexafluoride withdrawn from the conversion process. During the blending-down process, we can request random samples of the HEU hexafluorides, the blending materials, and the resulting LEU right out of the process piping and put them through an analysis procedure." U.S. and Russian monitors also have the right to measure the total flow of uranium at the blending point. Before the LEU is put on railcars to start its journey to the U.S., the monitors observe the application of Russian and U.S. tags and seals.

Monitoring at Seversk and Zelenogorsk is confined to periodic visits, but monitors have continuous access to the Novouralsk plant through the U.S. Permanent Presence Office there, which Lawrence Livermore manages for DOE. At all three plants, U.S. monitors have access to relevant documentation and accountability records.

TOWARD PEACEFUL ENTERPRISES

Lawrence Livermore is currently active in several programmes that provide collaborative project opportunities for scientists from the newly independent states, principally Russia,

Ukraine, Belarus, and Kazakhstan. The goal of these programmes is to direct the scientists toward work that will help develop free-market economies in their home states.

The first of these programmes is the laboratory-to-laboratory programme, which began in 1992 shortly after the directors of the Russian and American nuclear weapons design laboratories exchanged visits. Supported and monitored but not directly funded by DOE, the lab-to-lab programme involves interactions between NIS institutes and DOE laboratories for the purpose of "encouraging exchanges of information between U.S. and NIS scientists, thereby building confidence and openness between the two sides," according to Janet Hauber, Group Leader for Cooperative R&D and facilitator of Livermore's laboratory-to-laboratory efforts. Funding for projects that result from these collaborations comes from the sponsoring DOE laboratory with the stipulation that the work is neither related to weapons development nor enhances weapons capability.

Hauber reviews the work between the NIS and U.S. scientists to access the benefits derived by the participants. Although DOE is kept informed about lab-to-lab projects, the technical contacts are made directly by the scientists and involve only the laboratories and institutes. Thus, scientific collaborations are both informal and easy to initiate.

The laboratory-to-laboratory model for doing business has been so successful that it has been adopted by the Initiatives for Proliferation Prevention (IPP) Programme, another source of cooperative work for NIS scientists. Hauber is a member of the Inter-laboratory Advisory Board of the IPP, its primary project responsibility. Sponsored and directly funded by DOE, the IPP programme supports collaborations between NIS and DOE national laboratory scientists. The objectives of the IPP, like those of the lab-to-lab programme, are to strengthen non-proliferation and keep NIS scientists employed in their current institutions, but unlike the lab-to-lab programme, the focus of IPP-sponsored projects is clearly on their commercial potential.

Although projects must be mutually beneficial and not related to weapons, the major emphasis of IPP projects is on promoting economic recovery in the NIS. To that end, a large effort is expended on developing NIS know-how in the areas of intellectual property rights, entrepreneurship, and commercialisation. To facilitate these collaborations, DOE has simplified the project review and approval process and promoted uniform administrative procedures, such as uniform contracts and general patents, which make it easier to protect intellectual property.

Projects done under the IPP programme are carried out in three stages. In the first stage, the collaborating laboratories and institutes perform a feasibility study. Since the beginning of the programme in 1994, some 200 projects in technical areas such as materials manufacturing, biotechnology, energy, and waste management have been initiated. Projects considered to be feasible move into a second stage, one in which private industry can participate through cost-sharing (by matching government funding) and by assisting in prototype development. A number of projects are currently in this second stage.

A typical project—an analysis of the use of superplastic deformation technology to make automobile wheels—is being performed by a consortium made up of Lawrence Livermore, the All Russian Institute of Technical Physics, the (Russian) Institute of Metals Superplasticity Problems, Kaiser Aluminum, and Rockwell International. Lowrence Livermore's specific role at this juncture is to characterise wheel design and material for compliance with U.S. Department of Transportation standards and to determine whether the wheel will be able to meet U.S. requirements. Once the superplastic technology has been fully developed, it has potentially many more uses than for making car wheels. Because it uses nearly all of its starting materials to form the final product, it is a beneficial technology that produces few industrial waste byproducts. Also, because it is a technology previously used to make weapons components, it will be a true swords-to-plowshares project.

The third stage of IPP projects involves production of the developed products in the context of a purely commercial agreement between the Russian entity and a U.S. industrial firm.

While the progress of IPP projects is sometimes slow. Hauber is enthusiastic about the programme, believing that it will be an important factor in developing ;strong economies for the NIS. She says that "we just need to continue this work a little longer. The Russians are determined, and that determination will go a long way toward a successful outcome."

INTERNATIONAL SUPPORT

A third programme provides project opportunities to NIS scientists through the International Science and Technology Centre (ISTC) in Moscow and the Science and Technology Centre of the Ukraine in Kiev. Established by agreements among participating governments, the centres develop and fund non-proliferation projects whose primary objective is to provide peaceful, non-weapons-related opportunities to weapons scientists and engineers from the NIS, particularly those with knowledge and skill in the development of weapons of mass destruction (nuclear, chemical, and biological).

Although headquartered in Moscow, the ISTC is available to other states of the former Soviet Union—so far, scientists from Russia, Armenia, Belarus, Georgia, Kazakhtan and Kirgizia have submitted proposals. The ISTC is supported by the U.S., the European Union (EU),[1] Norway, and Japan. The EU, Japan and U.S. each place a deputy director at the Centre and provide staff support for Centre operations such as finance and programme management. The parties rotate the Centre directorship as well as the chair of its governing board. The current chairperson of the governing board is Ron Lehman, Director of the Centre for Global Security Research at Lawrence Livermore.

The ISTC sponsors projects focused on developing scientific and technical solutions for national and international problems, reinforcing the transition to a market economy, developing basic science and technology, and promoting the further integration of NIS scientists into the international scientific community. Project proposals submitted to the ISTC are evaluated for scientific merit by the funding parties. Eileen Vergino, Lawrence Livermore's Programme Manager for the ISTC, sits on the U.S. scientific advisory committee and provides technical support to the U.S. State Department, both by finding scientific reviewers for submitted technical proposals and advising ;them on funding decisions. Proposals are evaluated for technical merit as well as for conformance to ISTC policy. Overall approval is provided by the ISTC governing board, and final funding decisions are made by the funding party.

1) The member nations of the European Union are: Austria, Belgium, Denmark, Finland, France, Germany, Greece, Ireland, Italy, Luxembourg, Netherlands, Portugal, Spain, Sweden, and the United Kingdom.

U.S. scientists, including those at DOE laboratories, are encouraged to express support for or, better yet, collaborate on proposals they find interesting and significant to their area of expertise. A firm commitment by U.S. collaborators to a project will improve its chances for funding. While the U.S. collaborators will not receive any funding, they will play a key role in project development and review. U.S. collaborators often see the ISTC as a means for leveraging funds and enabling collaborations between themselves and NIS scientists on projects ranging from reactor safety to treaty verification to environmental assessment and cleanup.

The Science and Technology Centre of Ukraine is modelled after the ISTC. Its main difference is its sponsors, currently composed of the U.S., Canada, and Sweden and soon to include the EU and Japan.

SECURITY AND ACCOUNTABILITY

Russia's transition toward democracy has changed its state mechanisms for controlling and securing nuclear materials. Because of the economic and social changes in Russia, the borders around weapons complexes are now more permeable; gaining access to weapons materials has become easier. These factors increase the potential for the theft of nuclear materials. Therefore, one of the larger U.S. efforts in Russia is to provide assistance for improving the physical protection, control, and accounting of Russia's nuclear materials.

The DOE Materials Protection, Control, and Accounting (MPC&A) programme is a cooperative effort with Russian institutes and enterprises that process or store nuclear materials usable in weapons. Lawrence Livermore is one of seven DOE national laboratories involved in the programme and is working directly with Russia's nuclear institutes to provide them with technical support, training, funding, and equipment. The goal is twofold: enhance Russian physical protection and nuclear material accounting capabilities and encourage an overall change of philosophy about physical and material protection. The programme is intended to foster support from institutes and scientists for enhanced security concepts and methodologies that will be the foundation for enhanced national standards throughout the newly independent states.

Begun 1994 with pilot projects at three Russian institutes and modelled, like the IPP programme, after the laboratory-to-laboratory programme discussed earlier, the MPC&A programme has expanded to more than 44 institutes and enterprises. One or more project teams have formed at each institute or enterprise. One of the several project teams led by Lawrence Livermore has responsibility for Chelyabinsk-70.

T.R. Koncher, leader of Lawrence Livermore's MPC&A work, says, "We think of Chelyabinsk-70 as Russia's equivalent to lawrence Livermore because it is their second oldest weapons complex, just as we think of Arzamas—16 as their Los Alamos." Chelyabinsk-70, now called Snezhinsk, is east of the Ural Mountains, approximately 1,900 kilometers east of Moscow and about 80 kilometers south of Ekaterinburg. Several other nuclear facilities located nearby have close relationships with it, so it is expected that any security improvement techniques developed at Chelyabinsk-70 will ultimately be beneficial to these other institutes as well.

SECURITY UPGRADES

Lawrence Livermore's approach to upgrading safeguards and security at Russian weapons complexes is to work with Russian colleagues to identify areas where upgrades are required and then rapidly install those upgrades. The MPC&A programme first installed safeguards such as barriers, alarms, communications systems, and portal monitoring systems. Subsequently, pedestrian and vehicle portals were installed to improve entry and exit systems. Older Russian manual systems are being replaced with automated control systems that will incorporate nuclear materials monitors, metal detectors, and ballistically hardened booths for the guards. The new systems can detect nuclear materials being smuggled out, improve the capability to discover anyone trying to sneak inside, and offer better protection for guards in the event of an attack.

Lawrence Livermore is also working to enhance Russian transportation systems for nuclear materials. The Automatic Transportation Security System (ATSS) is an ongoing project to use readily available technologies to make rail systems more secure. The three-phase project, scheduled to be completed in the year 2000, covers some 375 development tasks. The first phase, now under way, includes installing intrusion and environmental sensors, security seals, on-train data communications and display, voice communications, physical barriers, locks, active delays such as high-intensity explosive sound generators and smoke generators, and off-train data communications and tracking. In parallel to the physical controls, U.S. and Russian scientists are developing safety methodologies—for example, procedures for coordinating emergency response from a central command post.

Improvements for the ATSS were designed at Moscow's Eleron Institute, which is devoted to the development, manufacture, and implementation of security equipment and systems. Actual implementation of the improvements will be done in conjunction with seven other Russian institutions, which will assure that the system has been incorporated into the Russian transportation infrastructure.

Efforts are also under way to obtain an accurate measure of nuclear material inventories and to establish procedures for checking and evaluating material balances regularly throughout all operations. This work requires the use of non-destructive assay methods to measure or verify nuclear inventories efficiently. U.S. scientists, for instance, are providing a gamma-ray spectrometer that can measure plutonium, isotopes or uranium enrichment and thus determine and verify nuclear inventories. Lawrence Livermore scientists developed the codes required to interpret the gamma-ray measurements. The codes analyze the complex gamma-ray spectra of plutonium or uranium to determine the actinide isotopic distribution for samples of any physical form, size, shape, or chemical formula. The system is easy to use: it does not require calibration of the instrumentation, and its measurement and analysis times are short.

LONG-TERM ASSESSMENTS

In addition to upgrading security weaknesses, U.S. scientists are helping Russian scientists assess long-term security infrastructure needs and establish priorities for implementation. Lawrence Livermore, in conjunction with Sandia National Laboratories, has been working with Russian institutes to conduct vulnerability analyses. This work, which generally begins with a training workshop, teaches quantitative probabilistic risk analysis, the technique that DOE uses to evaluate protection systems for special nuclear materials. The focus of these workshops is on using a computer-based analysis tool called ASSESS (Analytical System and Software for Evaluating Safeguards and Security) to quantify the detection, delay, and neutralisation probabilities of various protection systems. The quantitative values depend on the objectives of the protection system. These objectives, in turn, are defined through an analysis that asks: What needs protection? What are the consequences of losing the material? What possible types of threat does it face? What is the maximum level of acceptable risk for it? The objectives of the protection system must be identified and understood before an evaluation can be made of its effectiveness.

In addition to the workshops, subsequent vulnerability analyses, performed solely by Russians or jointly with U.S. scientists, are used to evaluate and prioritise physical and procedural security upgrades. The approach of these analyses differs from the present Russian approaches, so the rationale of the analysis tools must be communicated. The work also has to do with inculcating an MPC&A culture throughout the Russian institutes, so that both physical protection to fight off outsider threats and resistance to insider threats will be improved.

ADDITIONAL BENEFITS

Lawrence Livermore's work in the area of non-proliferation and arms control is one of critical importance to national and global security. This work draws on the expertise of personnel from directorates throughout Lawrence Livermore: Non-proliferations, Arms Control, and International Security as well as Engineering, Physics and Space Technology, Environmental Programmes, Energy Programmes, Chemistry and Materials Science, Computation, and Plant Operations. These staff are involved in the non-proliferation efforts because their technical expertise allows them to work directly with scientists from Russia and other newly independent states in ways that diplomats and politicians could not. Their face-to-face interactions are yielding benefits beyond the goals of their various collaborative efforts. As Bill Dunlop, Programme Leader, Proliferation Prevention and Arms Control, notes, "The access that U.S. and Russian scientists now have to each other's secure facilities is remarkable. It would have been unimaginable not too long ago. This level of trust results from common technical expertise, our similar background in national security issues, and our mutual respect."

About the Team

(*) Lawrence Livermore personnel who contributed to this article are: PAUL HERMAN, JIM MORGANM, and SCOTT MCALLISTER; BILL DUNLOP, EILEEN VERGINO, DOUG LEICH, and T.R. KONCHER, DEBBIEBALL, and JANET HAUBER.) All, except Leich, are members of the Proliferation Prevention and Arms Control Programme, which is part of the Non-proliferation, Arms Control, and International Security Directorate. Leich is part of the Fusion Energy and Systems Safety Programme in the Energy Directorate.

The work of these scientists and engineers is performed under the auspices of the U.S. Departments of Energy and State and focuses on reducing the risks of nuclear proliferation through collaboration and partnership with scientists and engineers in the newly independent states of the former Soviet Union. Projects range from negotiating mutually acceptable ways to monitor arms reduction and the disposition of excess nuclear materials to developing technologies to safeguard nuclear materials from theft or diversion to promoting commercial, non-weapons applications of nuclear weapons know-how and technology.

Address:

PAUL HERMAN
Lawrence Livermore National Laboratory
Livermore CA 94550
USA

35

How Can We Detect Clandestine Nuclear Activities

LOTHAR KOCK

INTRODUCTION

The recent bomb shots in India and Pakistan prove once again that the know-how of producing a nuclear weapon is accessible. Form the components needed to construct a nuclear device the bomb-grade fissile material is still the most difficult to obtain. If therefore one wants to timely detect a nuclear warhead development one has to monitor the acquisition of fissile materials. In the past there were several incidents where states attempted or succeeded to obtain bomb-grade fissile materials:

- by diversion to another state
- by clandestine production in an isolated area
- by clandestine production collocated to a civil fuel cycle facility.

There have been cases of illicit trafficking of fissile material reported [1]. Some have been detected by chance or when offered to undercover agents. The diversion in a state with a larger nuclear programme will be unnoticed if it is within the uncertainties of the balance in the accounting of the nuclear material, unless a strict, trustworthy physical protection system exists.

Under the second scenario the breeding of plutonium in secret nuclear reactors is easily detected by satellite surveillance due to the discharge of heat from the reactor—however, the enrichment of uranium not. Especially if a laser enrichment process is employed not needing a cascade of numerous units—as it is for the enrichment process by diffusion or centrifuges.

As for third possibility in a country with a fully developed nuclear energy generating industry:

- A clandestine high enrichment of U-235 in a existing enrichment facility takes place.
- Or in a state having a declared non-closed fuel cycle a clandestine reprocessing of low burn-up commercial fuel yields bomb grade Pu. Since this operation requires radiation shielded facilities—easy to locate by satellite surveillance—the reprocessing has to be made in a declared facility.
- For a state having a closed fuel cycle the illicit processing of spent fuel in a declared facility is easy to accomplish but also easily detected under a full scale international safeguards for nuclear materials.

DETECTION OF CLANDESTINE ACQUISITION OF BOMB-GRADE NUCLEAR MATERIALS

Passive interrogation of nuclear materials in bulk quantities can be performed by gamma radiation detectors and—in addition for plutonium—by neutron detection, if the material is not shielded. In such a case the induced fission by photons or neutrons could help, but any kind of **active interrogation** can only be applied under strict health physics regulations. Once nuclear material is detected sampling with subsequent destructive analysis will give detailed information. This type of detection is applicable in conjunction with physical protection measures of nuclear installation for border controls in combating illicit trafficking. Several stationary (fig. 1) and mobile units are installed e.g. the Yantar IV (Scientific Production Centre "Aspect", P.O. BOX 62, Dubna, 141980 Russia). It consists of two metal pillars installed opposite to each other on both sides of the road at a distance of (3-8) m between them. The pillars house gamma and neutron detectors. The gamma detectors are made of organic plastic scintillates. The neutron detectors are proportional He-3 counters. The minimum amount of nuclear material that can be detected at a speed of 10 km/h is 10 g weapon grade Pu and 1 kg pure U-235.

For the other scenarious—as discussed above—a suspected nuclear activity could be detected by monitoring the operational releases of materials used in the processes, which

Fig. 1 : Stationary detector for vehicles carrying gamma— and neutron emitting nuclear materials.

can be measured with present ultra sensitive nuclear analytical techniques even under stringent precautions of the operator. The **sampling** of the released material takes place in the **environment** of the facility or—if accessible—inside.

Sampling by **swipes** is common where (under controlled circumstances to avoid any cross contamination e.g. carried with the inspector) precipitated material is wiped off with a small cotton cloth from various places inside and around a suspected facility.

Cotton swipe kits (fig. 2) are used for sampling in enrichment facilities and outside of irradiated fuel handling hot cells. These standardised kits contain cotton swipes individually packed in mini-grip bags and also a larger mini-grip bag for double bagging, labels, two pairs of clean-room gloves, a pen, and a sample collection sheet with sampling instructions. Each hot cell kit consists of two PVC sampling tools with covers and swipes attached, plastic material for packaging in the provided lead-shielded aluminium can, clean-room gloves, and a working paper. The aluminium can is pre-labelled with an identification number. Based on field experience, the sampling term generally consists of three people: the sampler, an assistant, and a note taker.[2]

Chemicals involved in the nuclear processes such as fluorine in enrichment or tributylphosphate in reprocessing would

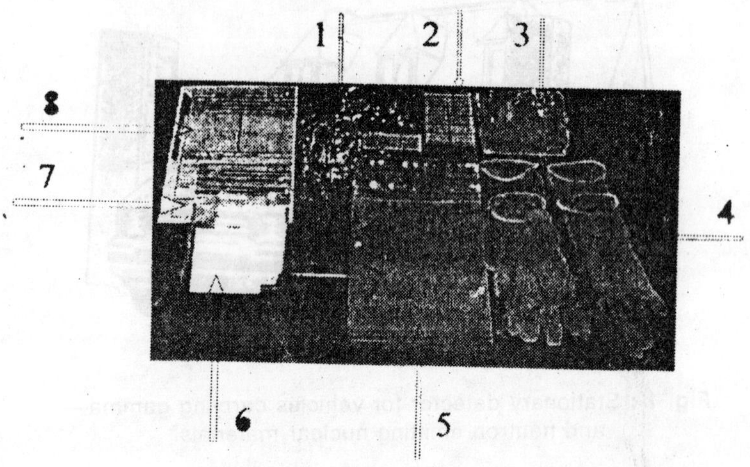

Fig. 2 : Kit for swipe sample collection

1 = Containment kit bag; 2 = Labels; 3 = Aluminium foil;
4 = 2 pairs of gloves; 5 = double bgs; 6 = Bag with control cotton swipe;
7 = 6 swipes; 8 = pen and working paper

also point to clandestine activities but are not as confirmative as the radionuclides. If—under the scenario two—uranium is detected—depleted or enriched in U-235 compared to its natural occurrence—we would have strong evidence for an enrichment process. Under the scenario three of a collocation of a declared and clandestine activity of course only an enrichment higher than the declared one is relevant. In the reprocessing of spent fuel a larger number of radioactive nuclides are involved which would allow for a dating of the different process steps because of the differences in their radioactive half-lives. If the measurement is focused on short lived nuclides a real time detection seems possible (3). In the off-gas of the facility the fission gases Xe and Kr as well as J can be found and in the aqueous effluents among a variety of fission products as well as activated nuclides of construction materials and transuranium nuclides can be observed.

A clandestine weapon Pu production by irradiation of natural uranium e.g. in an existing PWR is detected by the differences in their radionuclide abundances. Table 1 shows isotope pairs for various elements. The effluents after a short cooling (which is normal in weapon Pu production) show still short lived nuclides. Even a long cooling of short irradiated U will be detected by the distinct differences in nuclide abundancies e.g. for Pu-239/Pu-240.

Table 1 : Differences in the build up masses (g) of Pu-isotopes and selected fission products in low burn-up, short cooled nat. U fuel and standard PWR fuel after 50d (for nat.U) and 7a cooling (for PWR fuel).

Nuclide (T 1/2)	*50 d*	*7 a*
Sr-89 (50d)	9.4	—
Sr-90 (28a)	15	421
Zr-93 (10^6a)	21	720
Zr-95 (64d)	16	—
Ru-103 (39d)	11	—
Ru-106 (373d)	5.4	0.16
1–129 (10^7a)	4.3	177
1–131 (8d)	1	—
Cs-134 (2a)	0.2	4
Cs–137 (30a)	30	943
Pu-239 (24000a)	1253	5248
Pu-240 (6563a)	72	2121

Particulate matter released from installation precipitate on plants, soil or in sediments of nearby lakes or downstream of rivers. Suspension of indicator nuclides can be filtered out or absorbed in aquatic biota. Filter feeders such as oysters, mussels, and clams are particularly good bioconcentrators and may accumulate radionuclides or other trace constituents by 4 to 5 orders of magnitude. Their collection and subsequent laboratory analysis provides a good indication of the operation of nuclear processes. Specialised aircraft flights through areas downwind from suspected nuclear activities will filter out airborne particulate [4].

Some of the released nuclear material will dissolve in the environment, e.g. fission gases, iodine, caesium etc. Such environmental samples require an enrichment by established analytical chemical separation procedures before an analysis becomes possible. Sometimes the enrichment occurs in nature, where e.g. caesium is enriched in the food chain and iodine radionuclides can be collected from bovine thyroids(4).

In order to eliminate any cross contamination of the samples they are unpacked in clean laboratories of class 10 and—in the case of our institute—prepared for measurement inside special clean hoods of class 1. To begin with a non destructive high resolution gama spectrum in a low background counter is taken. Particles are washed off from the cloth-swipes and collected on adhesive conducting substrates for Secondary Ionisation Mass Spectrometry (SIMS). By this technique the isotope measured to about 1 per cent relatively. A distribution of the nuclide in question inside any one particle is described below in fig.4 and the distribution of the U-234 vs. U-235 enrichment in particles stemming from an enrichment plant is displayed in fig 3 (5). Trace elements homogeneously diluted in a bulk sample can be analysed with high accuracy of about 1 per cent by separation in Thermal Ionisation Mass spectrometers (TIMS) with special detectors down to a sample size of e.g. 10^{-15} g for Pu. A direct analysis of a solid sample by glow discharge mass spectrometry will detect radionuclide down to the ppt-level (10^{-12}) provided the mass is not covered by a natural occurring isobar. In such a case the dissolution of the sample and separation and enrichment of the nuclide in question by ionchromatography will provide a reliable answer.

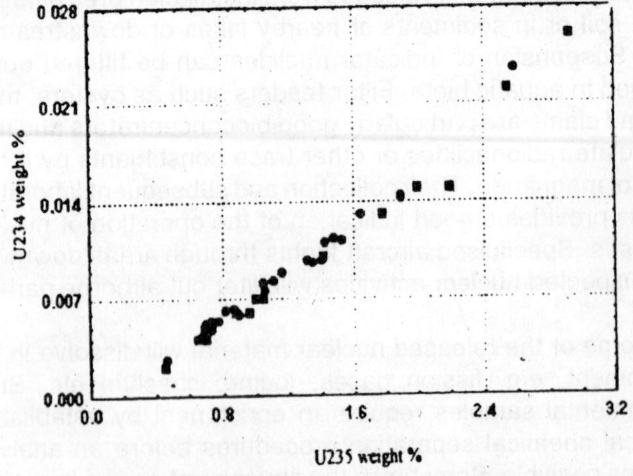

Fig. 3 : Distribution of the U-234 vs. U-235 enrichment in particles

Fig. 4 : Distribution of U-235 inside a particle (upper picture) and electron microscopy image.

FALSE ALARM PROBLEM

The above mentioned techniques to non-destructively detect the illicit trafficking of nuclear material or its clandestine production by analysis of traces stemming from the environment are only a hint of a suspected acquisition and consequently not false alarm proof. A definitive confirmation of this charge needs a sample from bulk material. Because a suspected case under investigation—possibly turned by the media into an accusation—may become a political sensitive issue utmost care and strict confidentiality should be observed during the investigations. Especially in trace analysis where the signature of the normal (base line) nuclear activity. Several incidents of this kind have been observed (6) and the following example describes the case where subsequent analysis of the particle material properties allowed the case to be proven as a false alarm.

Distant from nuclear facilities inside an industrial compound slightly contaminated material was found. The Secondary Ionisation Mass spectrometry (SIMS) analysis showed UO_2 particles enriched up to 90 per cent in U-235. Transmission electron microscopy revealed, however, solid sintered UO_2 and not the amorphous structure of hydrolysed UF_6, which definitely proved that this was material not released from an unknown enrichment facility but residue on scrap from a nuclear test reactor.

CONCLUSION

Although the technical means exists to detect clandestine acquisitions of material wide spread application is still to be institutionalised. Regardless whether investigations are done in a regional or international legal frame they have to meet the criteria of preservation of evidence, quality control etc. to withstand any foreseeable attempts to challenge the results.

REFERENCES

1) L.Koch, Science for Peace, Unesco Venice Office, p. 51 (1998)
2) J.N. Cooley, E. Kuhn, D.L. Donohue, IAEA-SM-351/182 (1997)
3) R. Hooper et al, IAEA-SM-333/69 (1994)
4) N.A.Wogman, R.W.Perkins, G.R. Holdren, IAEA-SM-333-98 (1994)
5) TUAR 97, EUR 17746 (1997)
6) L. Koch, M. Betti, J. Dolgov, K. Mayer, I. Ray, A. Schubert, A. D. Stalios, M. Wallenius, EUR 17665, p. 43 (Esarda 28) (1997).

Address:

LOTHAR KOCK
Nuclear Chemistry Unit, EITEJRC
D-76125 Karlsrue
Germany

36

The U.S.-Russian Nuclear Cities Initiative: Origins and Status

KENNETH LUONGO

1. ORIGINS

The United States and Russia have entered into numerous cooperative non-proliferation efforts since 1992. Most of these have been the result of activities initiated under the cooperative Threat Reduction programme and the Department of Energy's (DOE) laboratory-to-laboratory efforts. All of these efforts are aimed at increasing U.S. - Russian partnership in this important area and decreasing the threats posed by the collapse of the Soviet Union. But, despite the success of many programmes in this area, important problems still exist.

In particular, western nations are concerned about the lingering instability of the Russian nuclear weapon complex. In the mid and early 1990s, a few programmes were created to help promote the conversion of Russia's weapon scientists to non-weapon activities. The two most prominent were the International Science and Technology Centre (ISTC) and the Industrial Partnering Programme (now named the Initiatives for Proliferation Prevention Programme) (IPP). The ISTC focused on supporting weapon scientists by providing funds for them to pursue research projects. The IPP was aimed at creating U.S.-Russian business partnerships revolving around Russian-designed technologies. While the ISTC has essentially met its objectives, the IPP programme has had more limited success, thought its mission remains relevant and important.

In part, the difficulties for IPP, and other programmes, were linked to the uniqueness of asking former Cold War enemies in the U.S. and Russian nuclear weapon production complexes to cooperate. This concept took some time to take hold. It was not until the U.S. and Russia began broad scale collaboration on the protection and control of fissile material that the defence related complex began to open up. And, at present, U.S. and Russian specialists are working at virtually all locations in the Russian nuclear complex (one major exception being the warhead dismantlement plants where access is severely limited).

The MPC&A programme was designed to address what many considered to be the most dangerous proliferation danger in Russia—the loss or theft of highly-enriched uranium and plutonium from the vast Russian stockpile. But it quickly became apparent that while this effort was essential and could open the Russian complex to cooperation, it and other similar cooperative programmes, could not solve the severe crisis fueling the proliferation concerns about the Russian nuclear complex—that it was still Cold War sized, while its budget had been slashed significantly. This situation had created an economic vacuum and many were concerned that weapons experts could be tempted to sell materials, technology, or knowledge to proliferant nations.

Recognising that this underlying economic problem needed to be addressed, the U.S. Department of Energy raised the issue of working more closely with Russia inside the U.S. Government in the winter of 1997. Informal discussions were held on the subject, and then—Minister of Atomic Energy, Viktor Mikhailov, indicated that he was positively inclined to the suggestion. However, he stated that he wanted such cooperation to focus on concrete suggestions and proposals for action. DoE then suggested that a U.S.-Russian working group be created to explore concrete projects.

At the February 1997 commission meeting between U.S. Vice President Gore and Russia's then-Prime Minister Viktor Chernomyrdin (called the Gore-Chernomyrdin Commission or GCC) the issue of working more collaboratively on stabilising the Russian nuclear complex was raised.

However, Minister Mikhailov at that meeting suggested that the issue be addressed under a new nuclear issues committee that the Russian government wanted to be formed under GCC. This was a controversial suggestion that had been raised before and pushed off by the U.S. government because of a variety of concerns that they had. Unfortunately, the debate over the establishment of the nuclear issues committee dominated the subject of cooperation on the problems faced by the Russian complex and no action was taken to further the dialogue.

In March 1997, the Russian-American Nuclear Security Advisory Council (RANSAC) was formed and many of its members were former U.S. Officials who had worked extensively on improving U.S.-Russian nuclear relations. RANSAC decided to take on the issue of the deterioration of the Russian nuclear weapons complex as a top priority. A meeting was convened in May 1997 in Moscow. The attendance from the Russian side included a number of Russian laboratory directors as well as the First Deputy Minister Lev Ryabev. The attendance was a good indicator of the high level of Russian interest in moving forward with the conversion of its nuclear complex.

RANSAC had proposed to use the meeting as a vehicle for creating a proposal for action for consideration by the GCC. This proposal was written during the summer of 1997 and produced in time for consideration at the September GCC meeting. The RANSAC proposal had three recommendations: expand non-proliferation cooperation, expand cooperation on the development of environmental technologies and begin to seriously cooperate on business development.

An advance copy of the RANSAC recommendations was given to the Russian Ministry of Atomic Energy (Minatom) and it approved them. The recommendations were then sent to key U.S. officials. The goal was to get the White House to accept the broader collaboration in the recommended areas. U.S. officials, particularly in the Office of the Vice President (OVP) and at the Department of Energy, responded that they were interested in furthering the dialogue with Russia in the area of the conversion of the nuclear complex. A statement to work on the issue on an expedited basis was issued by both governments at the September GCC meeting and this formed the beginning of the breakthrough to new cooperation in this area.

Now that high level cover to pursue the issue vigorously had been provided, the two questions that remained were (1) how to put in place a comprehensive and effective programme of work and (2) how to pay for it. RANSAC members believed that they had outlined a path forward on the first issue and went to work on the second. Since the debate on the budget for FY98 was in its terminal phase, It proved impossible to get the Congress to earmark new funds for effort in 1997. However, sympathetic members of congress did agree to instruct the executive branch to focus the efforts and funds of the Initiatives for Proliferation Prevention on the danger posed by the economic conditions in the Russian nuclear weapons complex.

From October of 1997 to the end of that year, RANSAC briefed its recommendations to many government officials and met and communicated with many different Russian officials on the subject. The goal was to move beyond the vague, but ground breaking language in the September GCC statement and get the governments in both countries to launch a new initiative on this matter at the March 1998 meeting of the GCC. As a first step, RANSAC received a formal letter from Minatom agreeing that a new initiative should be launched and suggesting that an experts meeting on the subject be convened. In tandem with the acceptance of the Russians, officials in OVP and DOE were also seeking ways to formally launch this new effort.

To facilitate the process, RANSAC offered to host the experts meeting proposed by the Russians. The U.S. Government responded by preparing a written initiative for presentation at March 1998 GCC. The U.S. proposal, titled the "Nuclear Cities Initiative" included: training in commercialisation and business affairs; a refocussing of the IPP and ISTC programmes; and concentrating larger U.S. and multilateral programmes on the conversion of the closed cities. The proposal was approved by Gore and Chernomydrdin. This was progress, but the focus of the U.S. - proposed NCI was more narrow than the scope of the recommendations proposed by RANSAC, which had promoted greater cooperation in the non-proliferation and environmental clean-up areas as well as business development.

II. STATUS

The Gore-Chernomyrdin agreement did not spell out in great detail what actions would occur under the NCI. A meeting between DoE Secretary Pena and new Minatom Minister Adamov provided a structure for how to proceed—they agreed to the formulation of a Joint Working Group, but both governments were searching for ways to create NCI content and progress. To assist in facilitating progress on the issue. Both on the structural and substantive issues, RANSAC convened another meeting in Moscow in April 1998. Again, high level Minatom and Russian laboratory officials were invited to participate. The discussion focused on three issues: business development and investment; new proposals for non-proliferation cooperation; and new work on environmental remediation technologies. Interesting and productive proposals were put forth in all three areas.

Non-Proliferation Cooperation

The major non-proliferation cooperation recommendation was to create non-proliferation centres similar to those that exist U.S. labs at 4 Russian nuclear institutes—two in closed cities (Arzamas-16, Chelyabinsk-70) and two outside of closed cities (the Institute of Physics and Power Engineering (IPPE) at Obnisk, and the Kurchatov Institute in Moscow). The proposal for the development of non-proliferation centres was raised at a May 1997 meeting in Moscow convened by RANSAC. The original proposal for these centres contained three components:(1) the development of dedicated centres that could produce high quality technical and policy analysis of non-proliferation issues for the use by the Russian Government and in U.S.-Russian security cooperation; (2) the identification of experts from these four institutes that would come to the U.S. for training at U.S. universities and national laboratories (3-6 months at each location); and the expansion of existing U.S.-Russian non-proliferation cooperation to new areas of mutual interest.

At the April 1998 RANSAC meeting it was clear that the Russian institutes had given serious consideration to this idea over the past year. The two most advanced concepts were put forth by Chelyabinsk-70 and the Kurchatov Institute. Both institutes offered to send two of their people to the U.S. for training and each expressed a commitment to form a non-proliferation centre if funds could be identified. The Kurchatov Institute had previously submitted to RANSAC a written proposal for the structure of its centre. Funding for this centre is now under consideration by a U.S.-based philanthropic foundation and it is likely to be approved.

A proposed visit of Chelyabinsk-70 experts to Princeton University and Sandia National Laboratory also was made at the RANSAC meeting. If final approval is provided by the Russian

Ministry of Atomic Energy, these people will arrive in the U.S. in September 1998. IPPE provided a non-page overview of their proposal for an Analytical Centre. The representatives from Arzamas-16 suggested that they were content with the current way in which their non-proliferation work was conduced.

Surprisingly, the Russian labs indicated that they were rarely, if ever, asked by the ministries of the Russian Government or the President or Prime Minister's offices for analysis and evaluation of non-proliferation policy and technology issues confronting Russia and that no equivalent to the Lawrence Livermore Z Division existed. Still, all the institutes believed that they could play a role in assisting the Russian Government with its non-proliferation responsibilities and identified areas of existing work that could be expanded and new areas of cooperation with the U.S. and its laboratories. Key areas where little or no work was occurring at Russian labs but where they would like to begin work included: verification of past fissile material production (nuclear archaeology), verification of a fissile material cutoff treaty, alternatives to the use of HEU fuels, identification and analysis of the declassification of information in Russia, assessments of foreign nuclear programmes, non-proliferation technology detection/verification, arms control transparency, training of new generation of non-proliferation specialists. Collaboration with the U.S. could be expanded to many of these areas.

Environmental Activities

Both the United States and Russia face an enormous challenge in cleaning up their nuclear weapons complexes. This will require development of many new technologies. At the RANSAC meeting it was recommended that further U.S.-Russian cooperation in this area be pursued as it is in the interest of both countries and relevant to the mission of converting the Russian nuclear complex. One example provided at the RANSAC meeting was radionucleide migration. Here the Russian complex has considerable experience that the U.S. could use. In particular, the Hanford reservation in the U.S. is facing this migration problem and little work has been done to characterise and correct it.

The United States spends roughly $300 million per year developing new technologies for nuclear cleanup but the results have been disappointing. In fact, according to a 1997 report by the U.S. Congressional Research Service, since 1991 DOE has spent $2 billion funding 1,370 cleanup R&D projects. Fifth of these technologies have been deployed at a projected lifetime cost savings of $309 million. By comparison, a small U.S.-Russian cooperative programme in this area has been financed at $10 million since 1992 and 30 Russian-developed technologies have been identified as being potentially applicable to the U.S. cleanup problem. Five of these technologies are now available for use in the U.S. and the most expensive of them cost $800,000 to produce. However, none of the Russian technologies have been deployed because of insufficient U.S. funding.

Contracting a small portion of future U.S. research and development activities to Russian rather than U.S. experts could provide multiple benefits: the United States would get technology developed more cheaply than developing its own; contaminated Russian facilities that can serve as demonstration sites for new technologies; nuclear weapons experts would be redirected to peaceful tasks; and Russia could utilise these technologies at its numerous contaminated nuclear sites.

Russian nuclear officials have responded positively to the opportunity that environmental clean-up cooperation with the U.S. presents, and the U.S. national laboratories have also been enthusiastic. However, U.S. officials do not yet seem convinced of the benefit.

Business Development and Investment

It is clear that in recent months Minatom and the Russian complex have given considerable thought to the conversion task and some priorities have been established. It seems that the top three cities in need of conversion assistance are Chalyabinsk-70, Krasnoyarsk-26 and Arzamas-16, ranked in that order. If successful, the next targets may be warhead dismantlement plants followed by other production facilities. It was the view of some prominent officials in Minatom that if one or more commercial successes could be established at the top three priorities, the job would become much easier for the other cities. However, it was clearly emphasised that even if conversion was successful and military missions were downsized, it is unlikely that any of the cities will be closed. Most of the Russian believed that expanding the Nuclear Cities Initiative to include more cooperation on non-proliferation and environmental clean-up would be worthwhile complements to the business development activity.

The view in Minatom is that a central decision is how to structure the joint Working Group established for this initiative by Pena and Adamov. They would like to see it include a broad spectrum of people including government officials, private sector investors and other specialists. An advisor to the Ministry has already been chosen by Minatom and they are awaiting a similar decision by the U.S. Government. All the Russians asserted that they want help meeting experts in marketing, product development and evaluation and investors. In general the impression that was given is that the time is now ripe to move forward in the commercialisation area and that some success needed to be registered in the next 12-18 months.

Despite this basic agreement on what needs to be done, there are some differences between the way the individual cities would like to proceed with business development and the way some in Minatom believe this process should be carried out. For example, some in Minatom want to limit the interaction between the nuclear cities and potential investors, while some of the cities want to make direct contact with these people without having Minatom in the middle. Also, there are differences over whether existing programmes are capable of making commercialisation of products a reality. While all parties support the funds provided by ISTC and IPP for basic research, the more skeptical individuals wanted to know why more hasn't been done up to this point. There are also questions of whether the initiative should concentrate on large or small projects.

Among the cities represented at the RANSAC meeting, there are varying degrees of progress on the conversion front and no one common approach. Krasnoyarsk-26, for example, is pursuing the building of a large silicon purification plant that could employ several thousand workers. This project has been funded by IPP and the Defence Enterprise Fund and so Krasnoyarsk will remain dependent on these programmes until private investors can be fund. Arzamas-16, on the other hand, has worked with the local government of Nizny Novgorod to create a fund to support the development of new enterprises in the city and has also developed an off-shore investment zone where business are able to register and receive tax breaks. The representatives of this city are less inclined to believe that ISTC and IPP can produce commercialisation for them and they very much want direct contact with investors. Chelyabinsk-70 has the offshore registration zone but non-enterprise fund. So they are still willing to see if existing programmes can deliver results while awaiting opportunities to meet with investors. All the facilities and Minatom would be interested in utilising funding from international financial organisations.

At the RANSAC meeting, many of these issues were discussed and the key areas identified for further work were: structuring the joint committee; assuring access to investors; improving project evaluation/selection; assisting with business training; initiating analysis of Russian legal and infrastructure issues; and expanding non-proliferation and environmental co-operation.

CONCLUSION

The creation of the Nuclear Cities initiative is a major step forward in creating intensified cooperation between the U.S. and Russia on the stabilisation, conversion and downsizing of the Russian nuclear complex. The activity has been needed for some time. However, the project currently is faced with numerous challenges, including: lack of funding; disagreements over the scope of cooperation; minimal organisation; and political questions in both countries. These issues must be addressed in the coming months if the programme is to succeed. However, it will take more than the U.S. and Russian Governments to overcome the NCI's difficulties and make the programme work. It will require a coalition of international governments, businesses, academics, and non-governmental organisations to assist. The Russian-American Nuclear Security Advisory Council has made a commitment to keep watch over this issue and use its members to help continue to move the cooperation in a productive direction. The addition of other interested parties to this process would be a major step forward in assuring a successful NCI process.

Address:

KENNETH LUONGO
Princeton University
Princeton NJ 08544
USA

37

A Model for the Implementation of the Nuclear Cities Initiative

STEPHEN V. MLADINEO

Cooperation between the U.S. Department of Energy and its National Laboratories and the Closed Cities of the Russian nuclear weapons complex has evolved from emphasis on strictly scientific cooperation to issues of nuclear non-proliferation and arms control. The most notable success has been the extensive cooperation on nuclear material protection, control, and accounting (MPC&A). Other successes include the sale of blended-down Russian HEU to the United States for use in commercial nuclear power plants, and cooperative programmes dealing with nuclear weapons dismantlement transparency and irreversibility.

An important next step in expanding cooperation on non-proliferation of weapons-usable nuclear materials is the Nuclear Cities Initiative. This programme developed out of discussions in the tenth Gore-Chernomyrdin Commission meeting in Washington in March 1998, and has begun implementation through an agreement between Secretary of Energy Peja and Minister of Atomic Energy Adamov. The purpose of the Nuclear Cities Initiative is to help provide alternative employment for the scientists, engineers, and technicians of the ten closed Cities of the Russian Ministry of Atomic Energy (Minatom) nuclear weapons complex. The first organising meeting of the U.S. and Russian Government officials responsible for implementing the initiative took place in Moscow on 29 May 1998.

Shortly thereafter, a previously planned workshop became the first cooperative event of the Nuclear Cities Initiative. An economic diversification workshop was held between officials from Arzamas-16 (Sarov) and local officials from Richland, Washington and Pacific Northwest National Laboratory (PNNL).

The workshop took place at Pacific Northwest National Laboratory on 1-8 June 1998. Richland was one of the first three Nuclear Cities in the United States, where the first plutonium used in production of nuclear weapons was produced. It has also been the first of the U.S. Nuclear Cities to undertake significant economic diversification as the plutonium production mission disappeared. There is an extensive experience base in Richland and surrounding cities on implementing economic diversification programmes.

During a December 1997 visit to PNNL by technical specialists from Russian nuclear weapons institutes, for discussions related to nuclear weapons dismantlement transparency and irreversibility, the extent of economic diversification efforts in Richland became apparent to specialists from the All Russian Scientific Research Institute for Experimental Physics (VNIIEF). VNIIEF is located in the Closed City of Arzamas-16, now known as Sarov, and is one of the two nuclear weapons design institutes in Russia. As a result of their visit, Sarov specialists asked to return to PNNL at their own expense to learn more about the economic diversification efforts in Richland and the surrounding area. The resulting workshop became the first step in the process of cooperation under the Closed Cities Initiative.

The agenda for the workshop included a series of presentations by American experts on economic diversification in a community undergoing reductions in its government sponsored workforce. Presentations were organised around the three methods of building wealth in a community. These are expanding existing businesses, starting new businesses, and attracting businesses to the community. The discussions were supplemented by daily visits to a total of over a dozen new businesses in Richland and the surrounding area. Details of the PNNL entrepreneurial leave of absence programme, local business incubators, work force retraining programmes, and other related programmes were discussed and analysed for their applicability to the Russian closed cities.

In parallel with these presentations and visits, Russian participants related their experiences with economic diversification. Many of the experiences were remarkably similar, down to the number of new jobs that were expected to be required in the two cities, about 10,000 each. Russian presentations included a description of the Investment Zone of Sarov, a concept similar to an enterprise zone in the United States. Through national and local legislation, tax preferences have been created which favour the development of new businesses in the closed city of sarov.

The Russians also described 42 projects in various stages of development. Some of these were enterprises that have begun operation, while others were still ideas looking for funding. Progress in business plan development also varies among the projects. Many of the new start-up businesses have received assistance from a subordinate organisation to VNIIEF known as VNIIEF-Conversion, and some have received financial assistance form the Sarov Fund for Development of Conversion Enterprises.

A key outcome of the workshop was the description of a possible model for investment in the closed cities. This model developed out of conversations regarding the Sarov Fund for Development of Conversion Enterprises. This fund received its initial infusion of money from four sources: The Ministry of Atomic Energy through its agent Rosenergoatom, VNIIEF, the City of Sarov, and the Nizhny Novgorod Region. Local tax legislation has provided a mechanism to allocate additional funding to the Fund from a percentage of the tax revenue received by the city.

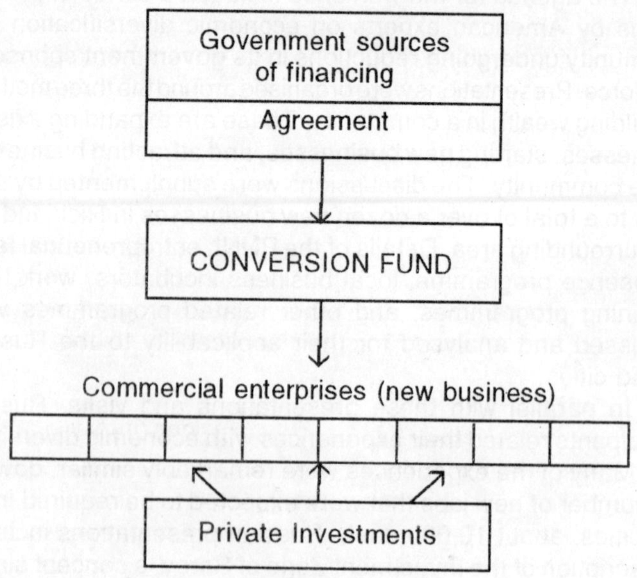

The purpose of the Fund for Development of Conversion Enterprises is to provide financial assistance to start-up business projects to create jobs in the city of Sarov. The fund provides zero interest loans to help defray start-up costs, and anticipates repayment over a period of about two to three years. The money is allocated by decisions of a board of directors of the fund based on evaluation of business plans.

The existence of this fund creates a potential opportunity for targeted financial support to the economic diversification of the closed cities. For example, the directors of the fund suggest that through execution of an agreement between the U.S. Department of Energy and the Fund, money could be provided to the fund on a tax-free basis allocated in accordance with provisions of the agreement. The agreement could include criteria for choosing business recipients of assistance, and should include an outside auditing arrangement to ensure that the funds are used only for the purposes for which they are intended.

A major aspect of the model that makes it an attractive vehicle is the potential for the donated funds to serve as a catalyst to attract private investment. The western analogy is gap financing, which is sometimes provided to projects by local, regional, or state governments. This kind of gap financing could help private businesses successfully complete their due diligence process by improving risk prospects for investment in projects supported by the fund.

If the model were to find acceptance and serve as a vehicle for U.S. Government investment, it might also persuade other potential donors to contribute to the fund. For example, foundations might be willing to make grants to such a fund in support of non-proliferation goals, or for humanitarian purposes. Other western governments might accept the model as a basis for their contributions.

Organisers of the fund have stated a willingness to use their fund to promote economic diversification in other closed cities. However, a more likely scenario would be imitation of the fund mechanism in other closed cities if the fund were to demonstrate success in developing new businesses and attracting private capital. The Ministry of Atomic Energy is also considering a slightly different model, where a conversion fund might be centrally managed to provide economic diversification assistance to the entire Minatom complex. Another possible variant could be a national level not for profit fund set up outside the direct control of Minatom.

Successful implementation of the Sarov model or some variant now depends on the actions of the Joint Department of Energy/Minatom Steering Committee. If they agree on the merits of the model, execution of an agreement and provision of funds could happen quickly. The next few months will be the test of the success of this innovative approach.

Address:

STEPHEN V. MLADINEO

Pacific Northwest National Laboratory
Washington DC 20014–2115
USA

38

Collaboration and Conversion Programmes at RFNC–ITP

VADIM A. SIMONENKO

TRAITS OF HISTORY

Russian Federal Nuclear Centre—All Russian Research Institute of Technical Physics (RFNC-VNIITF, ITP for short) was founded in 1955 to develop explosive devices (NED) and nuclear weapons (NW), to research physical processes of nuclear explosion and its effect, to run author's control over production and exploitation NED and NW, to provide basic scientific and technological support for all stages of their development and life cycle. The Institute was established as the second nuclear weapon centre of the Soviet Union in the very period of Cold War when nuclear confrontation of two systems—capitalist and socialist—was only gaining force.

There is a deep parallel in establishing, development and activity of American and Russian (for that time Soviet) nuclear weapon centres. The first American centre—Los Alamos National Laboratory (LANL) was founded in March-April 1943 when the Second World War was in full swing. Its goal was to develop nuclear bomb energy released during fission of nuclei of heavy chemical elements—uranium-235 and plutonium-239. There was a danger that such weapon can be made in Hiiler's Germany. The pressing target was to prevent such possibility. This effort resulted in successful test of American nuclear bomb on 16 July 1945 already after Germany's defeat. Soon, on August 6 and 9, nuclear bombs were applied against Japan. After that, nuclear weapon was begun to be used in its new quality, i.e. for political pressure on the Soviet Union.

The first Soviet weapon nuclear centre was founded three years later, in April 1946 in Sarov of Nizhegorodskaya (Gorky) region when it became clear that force confrontation of former allies on anti-Hitler coalition, i.e. the USA and the USSR, is inevitable. Efforts of Soviet specialists also resulted in successful nuclear test on August 1949, but this did not lead to force balance.

With desire to retain the US overwhelming leadership in NW, on 31 January 1950, President Truman issued a directive to Atomic Energy Commission to continue its work on "so-called hydrogen or superbomb". In 1951, the first thermonuclear ideas were verified in experiments of operation GREENHOUSE GEORGE (equilibrium thermal X-radiation was used with the help of tube with opaque walls to burn thermonuclear mixture of tritium and deuterium placed separately from fission device in special ampoule) and ITEM (fission reactions boosting with the thermonuclear T-LD burning inside the fission system). The next essential step was done in experiment MIKE of operation IVY, in 1952. It was checked experimentally the basic principle of cascade arrangement of thermonuclear charges using X-ray radiation energy transfer from the primary charge to provide subsequent implosion the secondary charge. Large amounts of energy (in comparison with HE energy of primary systems) allows to compress large mass of fission materials and provide good conditions for burning of various thermonuclear materials. The MIKE experimental facility was rather complicated with liquid deuterium as a main "thermonuclear fuel". But even after modernisation, it could not be considered as a thermonuclear weapon.

To provide more intensive advance in the development of new type of super powerful nuclear weapon, in July 1952, US Government established the second weapon nuclear centre-Livermore National Laboratory.

The Soviet Union was forced to accept the thermonuclear challenge. On 12 August 1953, it responded by thermonuclear device test which used a new thermonuclear fuel—Li6D, very perspective for military systems, but radiation implosion principle was not used yet. It was the beginning of nuclear race escalation. The next step was made by the US in the tests of operation CASTLE. The most famous of them is the first explosion BRAVO (1 March 1954, with 15 Mt yield) with radiation implosion and Li6D as thermonuclear fuel. These experiments opened wide opportunities for thermonuclear weapon development. Soviet specialists used both principles in 1955.

ITP was established as the second nuclear weapon centre of the Soviet Union in 1955, three years later after foundation of LLNL. Besides providing competition with the first Soviet NW centre, there were some separation of activity areas. It was also a duplication function, taking into account possible danger of nuclear was existing in those years.

INSTITUTE AND CITY

The Institute's professional basis comprised the specialists of the first nuclear centre (KB-1 at that time, RFNC-VNIIEF now). Besides that, intensive selection took place for engineers, designers, researchers, workers from other institutions of nuclear industry, other ministries, universities and educational institutes, technical schools around the country.

Completely non-settled area was chosen for the new Institute location, in the depth of the country, at the eastern foothills of the Middle Urals, approximately between Ekaterinburg (Sverdlovsk) and Chelyabinsk. New city was founded on the southern shore of Sinara lake. The city was later known as Chelyabinsk-70 or Snezhinsk. It is located in about 15 km from a highway connecting these largest cities of the Ural. The city of Snezhinsk was built exclusively to support activity of the Institute. Its population now comprise about 55,000 people.

The choice of the place for ITP was successful in many other respects. The Ural region appeared to be the concentration of several important enterprises of nuclear industry of the country. IT facilitated successful collaboration of the Institute with them in solving the problems of the main programme and helps now in organising conversion programmes. The neighbourhood of important industrial and cultural centres, Ekaterinburg and Chelyabinsk, helped also to successful development of the Institute and the city, allows to use their potential to solve a numeral complicated problems. There was also successful collaboration of the Institute and city with other scientific and cultural centres of the country.

Young scientific and technical personnel of the Institute gained force quickly. The first thermonuclear device put into the service in the Soviet Union was developed and tested by the specialists of new Institute in 1957. The first basic large scale physical experiment was successfully performed in the same 1957 in Novaya Zemlya its scientific leader was E.N. Avrorin. It opened one more important direction for the Institute activity—unique basic study in area of extreme states of matter and high intensive dynamic processes. The early years of successes of the Institute looks to be a miracle today. Just to mention the comparison of the first full-scale tests of Livermore Laboratory in 1954 were unsuccessful.

From the very beginning, at first under scientific leadership of K.I. Schyelkin and then (since 1960 through 1984)—under the leadership of the second Scientific Director, E.I. Zababakhin, the Institute began to follow its own independent technical policy. In particular, the orientation was taken on the development of miniature nuclear systems to provide high efficiency, improved technical and operating characteristics critical attitude was formed to the development of systems of anti ballistic missile defence. It is worthy to underline that the Institute displayed restraint attitude to the programme of supper bombs development with high energy release. However, in this class of nuclear devices, the Institute developed a highly efficient system.

Many projects were performed in competition with VNIIEF. However, in some areas, e.g., strategic submarine complexes, cruise missiles, air bombs, artillery shells, the work was performed mainly by our Institute.

EARLIER PERIOD CONVERSION

We would like to pay special attention to the development of nuclear charges for peaceful purposes. This programme was attractive for many specialists of the Institute. Even when the Cold War was in full swing, in the end of 1960's and in the beginning of 1970's. E.I. Zababakhin considered it as out conversion direction, though at that time, this term was not yet used.

Special nuclear explosive devices (NED) were developed with high purity on fission products and low level of induced activity. It is clear now, that there is no analogues in the world. They were designated for excavating application to build channels, harbours, dams, to remove dead rocks from ore deposits. The wide research programme of effects for similar applications was performed also. These programmes were terminated on the reasons of large impact on environment as well as strong restrictions on radioactive products spreading according to Peaceful Nuclear Explosions Treaty.

For the purpose of contained nuclear explosions at a large depth, a series of special nuclear devices was developed. They characterised by regulated energy yield, small diameter of the devices, high temperature resistance, low yield of residual tritium. It allowed to use such systems in regular oil and gas wells, in particular. The main part of the national programme of peaceful explosions—several applications for extinguishing of emergency oil and gas gushers, construction of underground cavities and reservoirs, seismic monitoring for deep geophysical survey—was performed with the help of PNE devices of our Institute. The majority of mentioned systems has no analogues in the world.

Even at the initial stage of creation of the Institute, it was clear that complex arrangement of all efforts was necessary including the selection and substantiation of physical scheme of systems, theoretical and experimental substantiation of mock-ups and experimental patterns, laboratory and full-scale tests of nuclear charges, warheads and their components, investigations of the actions and effects of nuclear explosion. This was the very fact that determined the structure of the Institute which included the complex of theoretical divisions, two design bureaus, which included design, experimental and testing divisions complex testing and proving facility, internal test site for non-nuclear explosive experiments, two experimental plants, technological division. The Institute structure also contained several supporting divisions. Currently, some of them are transferred to municipal.

It is well known that to develop NW and NED, study their effects and applications, of great importance are theoretical selection and mathematical substantiation of performance ability of selected physical schemes. Analysis of explosion processes, prediction of their results in various operational conditions are also very important. Many data on matter properties and characteristics of corresponding high-intensive processes running can not be obtained experimentally. This predetermined the decision to organise in the Institute a strong divisions of theoretical physicists, mathematicians, specialists on mathematical simulations, numerical methods and computer support.

Such structure gives favourable conditions for theoretical investigations, development of physical models of complicated phenomena and processes. Unique results were obtained by our investigators in basic studies with the help of nuclear explosion, in particular. Unique experimental data were obtained on thermonuclear detonation in cylindrical system on shock compressibility of several important substances in gigabar pressure range (by two-three orders higher than the values obtained in laboratory explosive tests) and studies the influence of shell electron structure of atoms on thermodynamic properties of matters, their "optical" opacity under high temperatures. Unique data were obtained on turbulent mixing of substances in dynamic processes. These results comprise a part basement of modern high energy density physics.

The most important results were also obtained in the field of initial confinement fusion, action of high intensive laser beams on a matter, laboratory X-ray lasers, high velocity collisions of bodies, high velocity passage of bodies through gas atmospheres, description of high intensive electromagnetic processes. Investigations of cumulative phenomena like implosion, convergence of shock waves and hollow shells, which provide essential increase of density of internal and/or kinetic energy, takes a special place in the Institute activity. Currently, such processes have found an extending applications, e.g., in inertial confinement fusion, studies of matter properties in extreme states, while using cumulative jets in engineering. The knowledge on occurring such phenomena in nature becomes wider, e.g., formation of snow avalanches, tsunami wave impact on ocean shore, star formation and supernova explosions. Great part of such studies was made earlier and is performed in theoretical divisions of the Institute. Academician E.I. Zababakhin was great enthusiast of cumulative phenomena studies.

It is characteristic for the Institute to have high level of design work. In several cases, the development of new up-to-date designs required parallel experimental studies at internal and external test sites. It was the reason why experimental and testing divisions were included in design bureaus. Industrial and technological basis of the Institute allows to solve complicated scientific and technological problems within the shortest time.

It was also created in the Institute a good basis for experimental investigations. There are series of pulsed reactors, large current electron accelerator, unique facilities to study gravitational instability and turbulent mixing development, impact of high intensive laser beams, etc. There is also a complex to perform radio chemical studies with tritium, fissile and other radioactive materials. Several facilities and techniques were developed for unique explosive experiments. In particular, with the help of them, jets of dense matter were obtained with the velocity up to 35km/s; investigations are made on matter properties study by recovering of samples after dynamic loading in multimegabar pressure range; fine structures of wave processes in condensed media are studied. Original high informing diagnostic techniques were developed.

PROGRAMMES OF COLLABORATION

Under the present-day conditions of nuclear arms reduction, the functional tasks of the Institute have changed essentially. In accordance with political decisions of the country's leadership, the course for reduction of nuclear arsenals has been taken, nuclear tests have been stopped, the task to develop new nuclear explosive devices has been in fact withdrawn. The tasks to maintain operational ability of the arsenal without full-scale nuclear tests, to ensure its reliability and safety have been brought to the forefront. New important technical and organisation tasks have arisen, such as: providing the author's supervision for maintaining and reproduction of arsenal in the absence of full-scale tests, protection and disposal of released nuclear weapon materials, ensuring of parity and balance during the process of nuclear arms reduction, keeping transparency which is necessary and sufficient for mutual confidence, non-proliferation of nuclear technologies, control over observance of the nuclear test-ban treaty.

Some of these programmes can be implemented with partial international collaboration. The Institute has a broad programme of collaboration with the U.S. nuclear laboratories. The main directions of this collaboration are following:

- accounting, protection and control of surplus nuclear weapon materials;
- transparency and mutual confidence measures in the area of nuclear arms reduction;
- development of methods and means for control of the Comprehensive Test Ban Treaty;
- ensuring the regime of non-proliferation of nuclear weapon technologies, in particular, by organisation of export control.

Such collaboration is carried out in the framework of purpose-oriented programmes on the "lab-to-lab" basis. In particular, large programmes of work are performed on materials protection and transparency. Some problems on safety, transparency, materials protection, non-proliferation of nuclear technologies, international treaties control are solved by the Institute in collaboration with specialists for the nuclear centres of France and China.

Introduction of our fundamental results into developing areas of scientific research is also of great importance. For example, physical models and mathematical programmes developed at the institute are used to investigate inertial confinement fusion alternatives to study luminescence of gas bubbles in a liquid which cavitate under action of high-frequency external sound fields; to research of intensive magnetic fields, dense plasma, etc.

MODERN DIRECTIONS OF CONVERSION

Owing to the reduction of defence programmes, the development of conversion directions to ensure professional employment of considerable part of the Institute's highly qualified specialists is of fundamental importance for the Institute. This process, which is labour-consumptive in itself, is complicated also by the fact that it takes place under acute crisis of the country's economy. In conditions of our town, an additional complication is caused by the fact that the town is a closed one. It causes sometimes certain difficulties for interaction with partners. Nevertheless, good results have been obtained for a number of high technology conversion directions.

In particular, highly efficient explosive devices to perforate walls of oil and gas wells have been developed. They significantly increase production and have won a high appraisal of domestic oil and gas specialists. Currently the Institute organises their expanded production together with some industrial enterprises of the region. Equipment and some original technologies to produce optical fibers have been developed and cooperation to introduce them has been established.

There are also a number of efficient technologies and developments (completed and close to completion) which can be used in economy. The difficult state of the country's economy hinders their usage at the present time. Nevertheless, the demand analysis allows to expect good prospects in the future. Examples are: the technology for hydrocutting of high-strength materials and articles made of them, the technology for explosive cutting of large-sized metal structures, explosive production of ultra disperse diamond power and technologies of its usage. Some samples of medical equipment, equipment for treatment of agricultural products have been built, the possibilities of producing a number of unique articles have been demonstrated.

Ecological investigations, development of technologies and systems for environment clean-up from domestic and industrial pollution are one of essential directions of conversion for the Institute. In particular, wide investigations were performed on interpretation of the accident of 1957 at the Mayak Production Association, on development of the radioactive release which occurred then and evolution of its trace the work is underway to describe and predict the underground transfer of radioactive contamination of Lake Karachai at Mayak; on evolution of undergound radioactive contamination caused by the influence of the tailing storage of the Kara-Balta Ore-Enriching Plant in Kyrghyz Republic. Efficient methods for purification of industrial effluents are under development.

Solving some urgent problems of nuclear power engineering is another important direction of conversion. In particular, the Institute builds containers of the new generation to transport and store spent fuel cells of nuclear power plants, investigates some possible emergency processes. Investigations on development and future usage of the technology of molten-salt reactors are of special interest. Its introduction can solve the most urgent problems of nuclear power engineering and thus improve essentially its acceptance by the broad public. The technology of molten-salt reactors provides essential increase in internal safety of nuclear power engineering, creates favourable conditions for non-proliferation of nuclear weapon technologies, opens up the possibilities of burning of long-lived radioactive waste from modern nuclear reactors, provides radically new solution of ecological problems of nuclear power engineering as a whole. It should be emphasised that the major financing of conversion programmes is carried out at the expense of Russian sources.

Partner collaboration with other scientific and production centres of the region and country and with foreign centres is a very important element of the science and technology policy of the Institute. Thus, essential part of ecological investigations has been carried out and is now underway under projects of the International Science and Technology Centre (ISTC) in Moscow, in collaboration with foreign laboratories, with the largest scientific and production centres of the country, such as the Mayak Production Association, the Kurchatov Institute, the NPO Khlopin Radium Institute, with the CIS partners, e.g. with the Kyrghyz

Academy of Sciences, with the centres of the Ukraine and Belorussia. Involvement of scientific potential of the Ural region in such work is especially valuable. In particular, a number of investigations on dynamic loadings, radiation effects, chemical processes, etc. are carried out in collaboration with the institutes of the Ural Branch of Russian Academy of Sciences. Broad collaboration is underway with other scientific and production centres of the country as well.

A number of investigations were performed in the past decade and are currently underway at the Institute in collaboration with laboratories of the USA, France, China, Japan, with the International Atomic Energy Agency, the CERN.

The Institute has a strong staff of scientists: 257 Candidates and 30 Doctors of Science, 12 Processors, two members of the Russian Academy of Sciences. A number of scientists are members of other academies and international scientific societies. There are two academic doctoral boards on defending doctoral theses. To ensure continuity of generations is one of the most acute present-day problems at the Institute. Systematic work is carried out to create favorable conditions for the progress of young scientists. The Institute interacts continuously with the Snezhinsk Physical and Technical Institute (branch of the Moscow Engineering Physics Institute) in the area of student's education. Many scientists of the Institute deliver lectures, play an important role in the work of key departments. The Institute also interacts with other higher educational institutions of the region and country to take on new specialists. The Institute and its specialists pay great attention to creation of a high educational and cultural level in the town. In particular, the project on computerisation of the town schools was implemented with their participation in 1997, as well as the All-Russian conference of the school youth "The Step into the Future" was held. The Additional Education Centre and the School Academy have been successfully working for many years. The town is famous for a high level of knowledge of school students.

PULSE OF LIFE

The work of the ITP's specialists is winning ever-increasing recognition. The Institutes specialists present regularly their work at international scientific forums. A number of international workshops were held at the Institute on the problems of ecology, mathematical modelling, technology of molten-salt nuclear reactors, radiation effect on materials, transparency, reduction of nuclear arms (the Pugwash Workshop), etc. The Zababakhin Scientific Talks, which will be held for the fifth time in 1998, received wide recognition as an international conference on high energy densities physics.

Of great interest is the specialised international conference on the problems of space protection of the Earth against collisions with dangerous space objects—asteroids and comets. The third conference on this subject is planned to be held in 1999. Some time ago even most scientists believed the problem of the danger of colliding space objects to be factitious. Nowadays it is generally recognised: collisions of cosmic bodies can result in great destruction, numerous victims, up to global elimination of mankind. Simultaneously, the necessity of developing the technologies to protect the Earth against such collisions, among which explosive nuclear impact is of first importance, is the subject of general discussions. It becomes of special value the PNE experience of the Institute.

The scientists of the Institute firmly hold the position of continuity of technological development. A half century experience in development of complicated nuclear arms systems is their basis. Even more complicated systems than those already developed for military purposes will be necessary to develop and use in the Space new technologies for efficient action on threaten objects. However, such systems cannot be developed immediately when the real danger is detected. Their development will take at least decades and huge resources which can be accumulated at an acceptably low level of annual expenses for many years only. Development of new systems will be efficient and they will ensure sufficiently reliable protection of humankind, if such work is purpose-oriented, is carried out by stages and consistently. Naturally, such programmes should be carried out on an international basis. Even the development and implementation of the principles of international approach to their realisation need time and great efforts.

The joint experiment on control of the threshold treaty on nuclear tests which was carried out at the Nevada and Semipalatinsk testing sites in 1988 was of fundamental importance for beginning of the era of nuclear arms reduction. Our Institute was a leading one for carrying out this experiment from the Soviet side and for providing hydrodynamic control in subsequent underground U.S. explosions. This provided a unique experience in joint solution of major scientific and technical problems which was used many times later on, in particular to carry out the projects under bilateral agreements with U.S. nuclear laboratories, e.g., on analysis of emergency situations with the Sandia National Laboratories, on investigation into turbulent mixing with the Livermore National Laboratory, on nuclear materials protection, etc. A great number of projects are carried out under the ISTC programme. There are separate projects under programmes of other foundation.

CONCLUSION

To summarise this brief description of the RFNC-VNIITF history and present-day state, it can be said that the Institute solved the country's most important defence problems on development and ensuring reliable performance of nuclear weapons. It was formed as a highly equipped complex science and technology centre. Its unique scientific and technical potential enabled in the past and allows now-a-days to carry out the most complicated investigations not only in the field of nuclear weapons development. This facilitates the process of integration of the Institute work into national and international economic relations and technological processes.

In the present-day conditions the functional tasks of the Institute have essentially changed. In particular, the task to develop new types of nuclear weapons, which was among the main ones at the time of the Institute formation, has been withdrawn. On the contrary, more emphasis has been put on the tasks to maintain the arsenal in new, more difficult conditions of absence of nuclear tests, to ensure safety and reliability, to prevent proliferation of nuclear technologies, to develop methods and means for control over observance of the Comprehensive Test Ban Treaty, to provide the processes of nuclear arms reduction, etc. An acute need for development of conversion directions has arisen. Some conversion developments have been completed or are close to completion. However, their introduction encounters difficulties caused by the crisis state of the country's economy. Hard search for efficient mechanisms of their introduction domestic and international economy is underway.

RFNC-VNIITF is one of the most important science and technology centres of Russia. Its high positions have won recognition in the world.

Address:

VADIM SIMONENKO
Deputy Scientific Director
Russian Federal Nuclear Centre
456770 Snezhinsk
Russian Federation

PART—5

DISPOSAL OF RADIOACTIVE WASTE/FISSILE MATERIALS AND THE ENVIRONMENTAL CLEAN-UP

39

Current Progress Toward a High-Level Radioactive-Waste Repository at Yucca Mountain, USA

ROBERT J. BUDNITZ

I. INTRODUCTION

This paper will provide a short summary of the current status of the proposed repository that the United States Government is developing at Yucca Mountain, Nevada, in the south-western part of the U.S. This project has been underway for many years, but is just now (finally) reaching a critical stage where a political decision, based in part but not entirely on technical criteria, must be made as to whether to proceed.

The basic facts about the proposed Yucca Mountain repository are quite well-known, and I will not go into much detail here. The repository itself would be located deep within an arid but low-lying mountain in the desert region of south-western Nevada, in an area with very low rainfall and essentially no other current (or future) productive uses. The mountain rock is comprised of unsaturated tuff, and the repository would sit 200 to 300 meters below the mountain top and also about 350 meters above the water table, which flows north-to-south in that region.

The repository's capacity is planned for the equivalent of 70,000 metric tons of spent commercial nuclear-power-plant (LWR) fuel; but of that capacity, 10 per cent has been set aside for radioactive waste accumulated over the years by the U.S. Government's defence activities—mostly, these are wastes left over from the manufacturing of nuclear weapons and related waste streams. Thus 90 per cent (or 63,000 metric tons) of commercial LWR spent fuel is destined for the repository; the defence-waste allocation, equivalent to 7,000 metric tons of spent fuel, is mostly vitrified fission-product wastes and defence-production-reactor spent-fuel wastes. [The 63,000 tons represents about 3/4 of the total spent LWR fuel anticipated to be generated in their lifetime by the currently operating 106 nuclear power reactors; the defence-waste allocation will accommodate most of the total current defence-waste inventory, depending on how the allocation/equivalence algorithm is worked out.]

The current schedule calls for permission to proceed, in the form of a licence from the U.S. Nuclear Regulatory Commission (NRC), in 2002. This would be followed by a several-year construction phase, and emplacement of the first waste about the year 2010. The repository would continue to accept waste for 50 to 75 years, after which it could be closed, but current plans are to keep it open for monitoring purposes for many years thereafter before finally closing it permanently.

The costs of developing the Yucca Mountain repository are being financed by a charge of $0.001 for every kilowatt-hour of electricity generated by nuclear power in the U.S. This charge, in place for many years, has accumulated so that it now represents a multi-billion-dollar fund, out of which money for the repository is allocated each year by the Congress. I estimate that the total cost of the repository will be in the range of $25 to $30 to perhaps $40 billion dollars—nobody quite knows because plans for the repository design itself, and plans for its operating mode, are constantly changing (see below).

II. THE CURRENT LEGAL/REGULATORY STATUS

There is no authorisation, as of today, actually to build the Yucca Mountain repository. That authorisation is still for the future. The U.S. Department of Energy (DOE), which is developing the repository, has been authorised only to *explore the feasibility* of the Yucca Mountain site, in order to determine whether a repository there would be viable—here "viable" means that it would meet all applicable regulations, and could be built and operated at an acceptable cost.

The problem is that as of today there are no regulations in place that would govern the repository. The U.S. Environmental Protection Agency (EPA) is charged to develop a public-health standard specifically tailored to the Yucca Mountain repository, after which the NRC is charged with developing regulations to enforce the EPA standard, and then with determining whether or not to issue a licence to build and operate the repository. But neither the EPA standard nor the NRC regulations exist: EPA has been working on developing a public-health standard for many years, but after several different versions, a court decision that threw out certain earlier provisions, various Congressional authorisations urging EPA to proceed, and political pressure, nevertheless EPA has not yet issued its standard.

The status is that EPA has finally developed a draft of its site-specific standard, and has had it "ready to go" since perhaps one year ago. However, EPA has not received permission to issue it because of objections to certain of its provisions by the NRC and by DOE. This impasse, which probably can only be overcome by either a White House (Presidential) decision or a new Congressional law, is very unlikely to be overcome until after the federal elections of November 1998. As I understand the situation (although the facts here are not officially in the public domain), the impasse is mainly over whether EPA (charged with issued a public-health standard) should issue separate standards to protect ground water-as-a-resource near Yucca Mountain (EPA's desire), or whether the proposed public-health-protection standards are adequate to protect ground-water without the need for separate standards (NRC's and DOE's position).

The other basic elements of the regulatory framework of the proposed EPA public-health standard are known, although informally, and they are as follows:

1) The standard would protect nearby populations by requiring that no member of the public receive an annual dose exceeding 15 millirem/year (perhaps it will be 25 millirem/year; nobody yet knows) from the Yucca Mountain radioactivity.
2) This would apply to all persons living in the vicinity for the next 10,000 years.
3) The assumption in the draft standard is that today's current population, in terms of its size, living habits, and locations, would persist for the entire 10,000-year period (of course, nobody can actually predict who will live where, or what their habits might be, in the far future).
4) The DOE repository-development team is charged with performing an analysis of the proposed design to demonstrate with high confidence that this standard can be met. This analysis, intrinsically probabilistic in nature and called

a "performance assessment" (PA), must account for all known natural processes that might occur in the next 10,000 years.

5) The possibility of an inadvertent human intrusion into the repository is examined by asking for an analysis of a postulated single exploratory borehole into the repository, which would pass through it all the way to the underlying groundwater; the additional doses arising because of this borehole penetration are to be evaluated also to assure that they are not unacceptable.
6) The period beyond 10,000 years must also be studied by DOE, although in its current form the draft EPA standard does not place a regulatory limit on doses more than 10,000 years in the future. The purpose of this examination is to assure that major public-health problems do not arise in the far future, and also to determine whether simple engineering changes now could reduce those far-future doses, in which case they should be considered.

The licensing process will be carried out by the NRC, under NRC's regulations, which must conform with and enforce EPA's standard. The licensing process will be a public process in which members of the interested public can participate either as parties or as commenters. The current schedule calls for DOE to submit its licence application to NRC in 2001, with a one-year licensing process leading to a licence (or a denial of the application!) in 2002. But of course that schedule pre-supposes both an EPA standard and NRC regulations, neither of which now exist!

Whether the political impasse over the EPA standard can be solved soon is not clear. There is a fundamental clash of authorities, because EPA's office that regulates water quality, and has authority over groundwater quality throughout the U.S., apparently desires to protect groundwater-as-a-resource under its separate authority, quite apart from the regulatory authority under which NRC, enforcing EPA's standards, will regulate the public-health impacts of the proposed repository. NRC, however, believes that it has overriding authority to regulate all long-term aspects of the repository at Yucca Mountain. DOE, another (very large) government department, is the "applicant" in this funny game, but it is an influential party and not merely standing-by waiting for the dust to settle—it wants to get on with the repository development programme, specifically to get on with determining whether the Yucca Mountain site will ultimately be acceptable.

III. THE CURRENT STATUS OF THE ENGINEERING DEVELOPMENT

Good progress has been made on the design of both the underground repository itself, and of the surface facilities needed during the century-long operating phase. A multi-kilometer-long exploratory shaft tunnel has been dug into the mountain, to understand the underground environment, to learn about the "rocks" and the "water", and to enable various underground experiments to take place. A preliminary design of the repository has been completed, calling for a series of about 250 parallel tunnels, each about 5.5 meters in diameter and 1000 meters long, and spaced about 25 meters apart. The plan is to dispose of the radioactive waste (spent reactor fuel; waste in vitrified glass; etc.) in large metal canisters, each about 1.8 meters in diameter and 5.7 meters long; there would ultimately be about 7,000 to 8,000 of them in the entire repository.

The century-long operational phase is complex because the wastes are all highly radioactive: lethal doses would accrue to anyone standing in their vicinity for more than a few minutes. Thus all handling must be by remote operations. Much effort has gone into designing these facilities and operations. But as difficult as that job is (and I will not dwell on it here at all), the crux of problem is to design a system to assure acceptable performance over the very long (10,000 year) future time period.

To accomplish that, it is necessary to carry out a probabilistic "performance assessment" (PA), which is a very complex task because the analysts must attempt to model, for very long times into the future, a set of interacting conditions. The PA will be discussed next.

IV. THE PERFORMANCE ASSESSMENT

The first PA for Yucca Mountain was performed several years ago, long before either the specific design features of the repository or the detailed hydrological geological and mechanical conditions underground within the mountain were understood very well. Several more PA analyses have been performed since, and each one has included more realism as both the mountain has become better understood and the engineering design has matured.

The PA lies at the crux of the Yucca Mountain Project: its interim findings have guided the specific decisions among options for the repository design, and have also indicated which aspects of the overall performance are not well enough understood, such that more information must be gathered (in the form of experiments in the laboratory, investigations within the mountain, development of analytical tools, and improved analyses of specific processes). The PA also has given some clues as to whether the overall performance of the repository is likely to fall well below, or well above, the regulatory limits being considered by EPA and NRC.

On this last point, I won't keep the reader in suspense for long: *As best I can tell*, from my vantage point as an independent scientist currently being paid by DOE to review the current PA, *the current bottom-line on whether the Yucca Mountain repository will meet the suggested EPA/NRC regulatory limits is that we cannot say for sure, one way or the other.* I will elaborate on this conclusion below.

The proposed regulatory limits apply to members of the general public who will live in the distant future in the Amargosa Valley, some 20 kilometers to the South of Yucca Mountain, down-gradient of the water now flowing southward from beneath the repository. Today, people in the Amargosa Valley are the closest population group that is making use of this aquifer, and the regulatory concept is to protect future populations (in Amargosa Valley) as if today's population patterns were to persist. Today, this population drinks from the local groundwater, and a certain fraction of them eat some of their food from crops and livestock using that same groundwater. It is postulated that future groundwater, containing concentrations of radionuclides from the repository, will be used by this population, and will give them radiation doses. The regulatory issue is to compare those doses to the limits, in the range of 15 to 25 millirem/year (the exact limit has not yet been decided).

The problem is that to analyse for the aquifer's groundwater concentrations requires a set of (linked) analyses that, in detail, are very difficult to carry out. The analyses require data on a wide variety of issues, and analytical models covering an equally wide variety of processes, that must all be linked together. The analysis task is staggering in its complexity and its diversity. Even though the PA analysis team is very talented, has drawn on a wide variety of experts beyond its own inner circle for advice, and has a multi-million dollar, multi-year budget, the uncertainties in the bottom-line conclusions remain large, and likely will remain large even at the very end (three or four years hence when the NRC will compare the PA's results with its regulatory limits).

The key question is whether, despite these large uncertainties, there will nevertheless be an adequately-supported case that the repository will meet the limits, so that NRC can make a positive finding—can issue a licence. So far, the situation does not look bleak, but by no means does it look entirely favourable either. The best current estimates for doses in the future 10,000–year regulatory period are in the same range as the likely dose limits: not orders-of-magnitude above them, but not way below them either. Most importantly, the uncertainties are large, and likely to remain so.

These uncertainties arise from the various processes that must be modelled. First, a small fraction of the rain falling on the mountain seeps into it (the vast majority does not), and seeps downward into the mountain by various flow paths: fracture flow, fast and slow, and matrix flow. Ultimately, some of this water, flowing through this unsaturated zone for a minimum of 200 meters (the over-burden varies across the repository), reaches the repository horizon, where it makes its way into the tunnels (drifts) and onto the waste canisters. But in the initial period (the first many centuries), the waste is very hot from radioactive decay, and it heats up the surrounding rock for many hundreds of meters, keeping liquid water from flowing through until finally the heat dissipates. Modelling the complex thermal, mechanical, and chemical behaviour of this system during and after this hot thermal period requires hydrological-geological-geochemical-thermo-mechanical models, and lots of supporting data, that are at the cutting edge of today's state-of-the-art. Approximations must be made: no fully linked model of these processes has been developed, nor could it be developed today. Judging the adequacy of these approximations is a job for experts, who find the judgments difficult.

Next comes the issue of behaviour inside the tunnels (drifts). The current design has open drifts (although a design with backfill is being contemplated). The water would seep into the drifts, drip onto the canisters, and ultimately cause corrosion of the metal. Corrosion-resistant alloys are being selected, which (if they can perform as well as some hope) may last well beyond the 10,000-year regulatory period. If this can be demonstrated, then the demonstration of 10,000-year compliance will be trivial. But if not, then the modelling must capture the process whereby corrosion damage allows water into the canisters. The water would then interact chemically with the radioactive wastes inside (spent LWR or other reactor fuel; vitrified defence wastes; etc.), releasing a variety of chemical forms of waste to seep out and into the ground below.

Modelling these canister-corrosion processes and these in-canister chemical-leaching processes is extremely difficult. Understanding the behaviour of the large number of different chemical species, and the dependence of this behaviour on the physical/chemical environment (temperature, oxidation potential, pH, etc.) is difficult in many cases. Much uncertainty exists, and even with numerous efforts to pin down the parameters through laboratory experiments, many of the uncertainties will likely remain high—unless the canisters can be shown to survive intact, corrosion-resistant, for well beyond 10,000 years. (Showing this conclusively is also very difficult even for the best-understood corrosion-resistant alloys).

The next modelling task is the flow and transport of radionuclide-laden water down from the repository horizon about 350 meters to the water table. Flow and transport here, in fractured tuff, are governed by fracture-flow and matrix-flow phenomena, influenced by chemical effects. Understanding these again requires coupled-effects modelling. The data to support these models are sparse and hard to gather, and the current models are not fully coupled, so approximations are needed. The travel time from the repository to the water table is currently thought to be in the range of a thousand years or so, but getting this part right is very tricky.

Finally, the radionuclide-laden water would reach the water table, where saturated-zone processes carry it down-gradient, southward, for 20 kilometers to where it is under the people who would ultimately put down wells to draw it up and use it. The saturated-zone processes are another area where the models are difficult to develop and support with good data. Currently, not enough data exist for this 20-km region: hardly any wells exist to illuminate the actual geological-hydrological-chemical conditions. This must be remedied soon to enable the modelers to get the needed information. Flow is thought to occur in flow-tubes, and perhaps in sheets with little vertical mixing; but the phenomena here are still being explored. Flow in this saturated zone to the Amargosa Valley is estimated to take a few thousand years.

Then the radionuclides are available for uptake by humans. The postulated scenario is that wells are dug, and the contaminated water is used for drinking, for irrigating crops, and for livestock. Dose pathways to humans come from food and from drinking water, and the individual annual doses must be analysed using data (and assumptions) on food habits and so on.

This base-line analysis is not the entire story, however. There are so-called "upset" conditions to study, such as potential volcanism (thought to be too unlikely to affect things), earthquakes (currently thought mainly to cause rockfalls in the tunnels that could compromise the canisters earlier than would otherwise be the case), human intrusion (a stylized single-bore-hole scenario must be evaluated), and nuclear criticality (currently thought not to produce much additional impact on the overall doses, even if it were to occur).

Finally, there is climate change: the current inter-glacial period will likely end in another few thousand years, and although glaciers have not historically reached as far south as Yucca Mountain, a doubling of rainfall is predicted, which would vastly increase infiltration into the mountain and would influence all of the other processes, as well as changing the amount and speed of saturated-zone flow. The current draft EPA standard, for the purposes of regulatory compliance, allows glaciation to be ignored for the first 10,000 years (even though climatologists predict that its onset is likely sometime during that period, although more likely toward the end than nearer the beginning of that period). However, the onset of glaciation is thought to be almost certain before 20,000 years hence, and base[illegible] historical record the glaciers are predicted to last for [illegible]t 50,000 years or more thereafter. Modelling these phen[illegible]a is also a challenging task.

What I have presented here is a very brief and abbreviated overview of the modelling/analysis task that the Yucca-Mountain performance-assessment (PA) team must accomplish. The modelling and data needs are vast, complex, and inter-disciplinary, and there are genuine technical controversies in many of the areas, because much of the PA work is literally on the forefronts of the many current scientific capabilities.

Thus it is not surprising that, when the whole PA analysis is assembled from the many sub-parts, the overall uncertainties are great. The challenge for DOE is to develop an overall PA model that, despite these limitations, can provide a confident finding that doses in the Amargosa Valley will be below the regulatory limits, for 10,000 years into the future.

Is this analysis task a feasible one? The answer depends on the following:

i) If the canisters can be shown to last for well beyond 10,000 years, in the complex environment in which they are predicted to lie, then the rest of the problem is trivial. Work on canister performance is a high-priority area of engineering evaluation that is underway now.

ii) If the bottom-line individual annual doses are small enough, then even with their uncertainties, there may be enough confidence to allow NRC to make a finding that the regulatory limits will be met.

There are several areas within the detailed modelling where work is now underway that could substantially affect the ultimate PA results. The outcome of this work may help provide higher confidence that the repository will perform adequately vis-a-vis the regulatory limits:

iii) Even if the canister-failure model is not as robust as above it is possible (perhaps likely) that the leaching of radionuclides from within the canisters and its transport out of the canisters may be at very low rates, or very low concentrations. Work on this area is now proceeding vigorously.

iv) If there is enough retardation (holdup) in the unsaturated [illegible]he below the repository, then the concentrations reach[illegible] the water table (and ultimately reaching the Amargosa [illegible]ey) may be low enough to fall well within the regulatory [illegible]s.

v) The model for flow in the saturated zone is now being substantially upgraded based on new insights. It is possible that the flow in the unsaturated zone will provide more dilution than currently thought, which would decrease concentrations and hence would lower the estimated doses.

vi) Some of the models for biological uptake are currently in need of reevaluation: the ones used until recently are believed to be too pessimistic based on current knowledge. This could decrease currently-predicted doses substantially.

For all of the above reasons, it is possible that the analysed doses will be much lower when the PA analysis is redone two years hence. Much new knowledge, as well as some design improvements (changes in the canister design, possible backfilling of the drifts, possible ventilation of the drifts for the first century to decease the thermal effects, possible changes in understanding of the spent-fuel chemistry, etc. etc.) will come into play.

But the improved PA analysis that will emerge perhaps two years from now may not produce such a result. Some of the phenomena may turn out more pessimistically than now thought, including saturated-zone flow, the modelling of the coupled thermal-chemical-mechanical situation in the unsaturated zone above the repository during the hot-thermal initial period, and the in-canister modelling. Perhaps the effects of earthquakes may be much worse than now thought. And so on—the situation is so complex, and the challenge of modelling it over 10,000 years is so difficult, that in the end perhaps it simply will not be feasible to show, with high confidence, that the annual dose limits can be met.

Time will tell.

And that time will be soon: the licence application is due in 2001, only about three years from now. A full year or more before that, the engineering design must be frozen, at least so that it can form the basis for the licence application.

V. SUMMARY

The effort to design a high-level radioactive-waste repository at Yucca Mountain, Nevada USA is well underway. The challenging task of performing an analysis of how such a repository would perform for the next 10,000 years is also underway. It involves a complex set of analytical studies that must be linked together to provide an assessment of annual doses to postulated nearby residents for the full 10,000 years.

Whether the analysis can demonstrate adequate performance in terms of public-health risk for 10,000 years (in the form of annual dose limits to the nearby population) is not yet clear. Today, our best analysis seem to show that the repository is not clearly "unsuitable" in the sense that the projected doses are much higher than could meet a reasonable standard; but neither are the projected doses so small that meeting the regulatory limits will be quite easy. Compounding this is the fact that, as of today, no legal (regulatory) framework yet exists, because neither the EPA nor the NRC has promulgated the actual standards (EPA) or regulations (NRC) to which the proposed repository will be held.

Will the repository ultimately meet the (not-yet-promulgated) regulations? I certainly hope so, and I believe that the chances are good. But today nobody can be certain.

The more important question is: *Will the repository ultimately be built?* That depends not only on technical considerations, but also on political conditions at both the federal and the Nevada-state level that are totally unknown—"unknowable" would be a better word, because the relevant political conditions will arise only a few years from now, when different politicians, responding to different political forces, national and local, will be in place.

Address:

ROBERT J. BUDNITZ
President
Future Resources Associates, Inc
Berkely CA 94704
USA

40

Spent Fuel and Nuclear Proliferation

HAROLD A. FEIVESON

INTRODUCTION

I would like briefly to make the following six points:

- There is a lot of plutonium in spent fuel from civilian reactors.
- This spent fuel must be safeguarded, since the contained reactor-grade plutonium can be used for weapons.
- There are limited alternatives on what to do with spent fuel.
- The most important near-term goal is to find interim storage sites for the spent fuel.
- In the long term, mined geological repositories for spent fuel do not present unacceptable proliferation risks.
- This is especially so in comparison with the principal alternative to geologic disposal, the destruction of plutonium through partition and transmutation.

1. THERE IS A LOT OF PLUTONIUM IN SPENT FUEL

From the beginning of the nuclear era, through the end of 1996, about 1300 tonnes of plutonium had been produced worldwide. See Figure. Of this total, approximately 250 tonnes were produced for weapons, and about 1050 tonnes were produced in civilian power reactors. Essentially all of the weapons plutonium has been separated from reactor spent fuel. Of the civilian plutonium, 200 tonnes were separated, and 850 tonnes remained unseparated in spent fuel. Approximately 50 tonnes of separated civilian plutonium had been burned in reactors. So at the end of 1996, the civilian stockpile of plutonium included 150 tonnes of separated plutonium (almost all of this in Britain, France, and Russia) and 850 tonnes of plutonium still locked in spend fuel. (My source here is David Albright, Frans

Berkhout, and Willian Walker, *Plutonium and Highly Enriched Uranium 1996: World inventories, Capabilities, and Policies* (Oxford, 1997) and an occasional paper by Frans Berkhout, "Plutonium: Storage, Disposal, or Utilisation", Science Policy Research Unit, University of Susses, 1998.)

A typical one gigawatt light water reactor generates annually about 20 tonnes of spent fuel, containing roughly 200 kilograms of plutonium—that is, about 1 per cent. Today, worldwide, civilian reactors generate annually about 8000 tonnes of spent fuel, containing roughly 800 kilograms of plutonium. About four-fifth of this spent fuel is being stored and will in all likelihood be disposed as waste directly without any separation of the contained plutonium.

Looking to the future, the reactors currently operating will generate on the order of 160,000 tonnes of spent fuel containing 1600 tonnes of plutonium over the next 20 years or so. (1 use 20 years as a average remaining lifetime of current reactors—it might actually be longer).

2. THE CIVILIAN SPENT FUEL MUST BE SAFEGUARDED

Reactor-grade plutonium (that is, plutonium with the isotopic mixture found in civilian reactor spent fuel) presents difficulties to a bomb designer: compared to weapon grade plutonium, it contains a higher fraction of the isotope Pu-240, which has a high rate of spontaneous fission, and therefore has the potential to start a chain reaction prematurely it has a higher fraction of the isotope Pu-238 which decays rapidly thereby significantly increasing the rate of heat generation in the material and it has a higher fraction of the isotope Pu-241, which decays to Am-241, an emitter of a highly penetrating gamma ray, making handling of the material more difficult. See Figure.

This has led, over the years and still today, to many statements that reactor-grade plutonium is not suitable for weapons. For example, in a September 1994 Financial Times conference in London, Jean-Pierre Rougeau, Vice President of the French reprocessing combine which separates more than one half of world's plutonium, stated that: "reactor-grade plutonium is not realistically a potential weapons material...it is _ practically speaking _ virtually impossible to convert reactor-grade plutonium to military use."

Unfortunately, this is far forrn the case. According to U.S. weapons experts, in a Department of Energy release, dated January 1997:[1]

"a potential proliferating state or subnational group using designs and technologies no more sophisticated than those used in first-generation nuclear weapons could build a nuclear weapon from reactor-grade plutonium that would have an assured, reliable yield of one or a few kilotons....Advanced nuclear weapon states such as the United States and Russia, using modern designs, could produce weapons from reactor-grade plutonium having reliable explosive yields, weight, and other characteristics generally comparable to those of weapons made from weapons-grade plutonium."

The reasons for this conclusion given by the expert group are several: First, while pre-initiation of the chain reaction due to the production of neutrons by the spontaneous fission of Pu-240 might be difficult for an unsophisticated sub-national group to overcome, even the "fizzle yield" of a relatively simple first-generation nuclear device would be on the order of 1-2 kilotons. Proliferating states using designs of intermediate sophistication (that would allow faster assembly of the weapon) could produce weapons with assured yields substantially higher than the kiloton range.

Although dealing with Pu-238 and Pu-240 requires careful management of the extra heat generated in the device, there are well-developed means for addressing these problems, and they are not considered a significant hurdle to the production of nuclear weapons, even for developing states or sub-national groups and although the greater radioactivity due to Am-241 and Pu-238 means increased radiation doses to personnel working on the weapons, this too is not considered a significant drawback to a determined proliferator.

3. WHAT THEN TO DO WITH THE SPENT FUEL?

In the short term, the basic decision a country faces in dealing with its civilian spent fuel is whether or not to reprocess the spent fuel to separate plutonium. If the spent fuel is not reprocessed, it could be stored either at the reactor site or at one or more centralised away-from-reactor sites. see Figure.

In the long run, the spent fuel could either be disposed permanently unreprocessed in a mined geologic repository, or conceivably by some other means such a sub-seabed disposal, or disposed through a process of "partition and transmutation" which would seek to burn-up all the plutonium in a reactor or accelerator. Note that even if the plutonium is initially separated and burned in a reactor, the spend fuel generated by this process will eventually have to be disposed.

4. FINDING INTERIM STORAGE FOR SPENT FUEL IS THE MOST IMPORTANT SHORT-TERM GOAL

Much has already been written about the diversion risks and poor economics of reprocessing and recycling of plutonium in light water reactors. I won't repeat these analyses here but I do wish to point out that one of the principal goals to commercial reprocessing is the fact that no country has yet satisfactorily figured out what to do with its spent fuel. This has most notably been the case in Japan, Germany, and a few other countries. Because of concerns raised by environmentalists and other political factions, these countries have not been able to find politically-acceptable places to store their spent fuel out utilities have bot been blocked from sending the spent fuel but of the country to be reprocessed in France or Britain. Eventually, the separated high-level wastes and plutonium will have to come back, but in the meantime, utilities put off for several years having to deal with the problem of spent fuel disposal.

So, even for countries that are now reprocessing much of their spent fuel, it is important to explore ways to store spent fuel for an indefinite period, without reprocessing. I don't have a global programme to suggest but let me mention a couple of possible elements of such a programme.

First, the reprocessing complexes of Great Britain and Russia, which are separating plutonium from civilian spent fuel on a significant scale, could over time be used mainly to store spent fuel, without reprocessing. Both countries are now accumulating large stockpiles of separated plutonium with no realistic plans in sight to use the plutonium. Britain, for example, already has a stock of about 60 tonnes of separated plutonium, produced by the reprocessing of domestic spent fuel. It is also reprocessing foreign spent fuel to recover separated plutonium. As alternative to this reprocessing, my colleague, Frank von Hippel, has suggested that Britain could undertake a kind of" virtual reprocessing" that is, it could take in the spent fuel from a country and then send back to the country already-separated plutonium from the British stockpile, without running the reprocessing plant at all! This would both reduce the large British stockpile of separated plutonium and avoid the costs and environmental problems of reprocessing. See Figure.

Russia also is reprocessing civilian spent fuel, hoping eventually to use separated plutonium in breeder reactors but this use is very far off and it does not make economic sense to reprocess now for that remote contingency. It also seems especially paradoxical for Russia to continue to separate plutonium from spent fuel while at same time exploring complicated and expensive ways to turn its separated weapons plutonium back into spent fuel.

The United States, of course, unlike Britain and Russia, or France, does not have any plans to reprocess civilian spent fuel. But it also does not yet have a politically acceptable strategy to deal with the spent fuel. The problem, as seen by nuclear utilities in the U.S., is that reactor-site pools are filling up, and reactors will have to shut down unless the utilities are allowed to put spent fuel into dry cask storage on site or to send the spent fuel to some centralised store. Both the public and the utilities are nervous about the former since they fear that dry stores on site could entail indefinite obligations to monitor and secure the spent fuel even after the reactors are shutdown. As a consequence, the utilities have strongly supported a bill that would allow them to send spent fuel to a temporary store in Nevada, near Yucca Mountain, the proposed permanent disposal site for spent fuel. But, so far, neither Nevada nor any other state or community has been willing to accept a centralised away-from-reactor spent fuel store. See Figure.

In the midst of all this confusion, there is a growing realisation that Yucca Mountain, if and when it is certified to receive spend fuel, will in effect be a monitored, retrievable store for 200-300 years, before the repository is closed. Although this realisation may not help immediately to break the political deadlock, it could do so eventually by convincing the public that there is a relatively safe place to store the fuel without absolutely committing to its final disposal.

But, although there need not be an irreversible commitment to final disposal at Yucca Mountain or other geologic repository, the public would have to be convinced that such final disposal probably makes sense. And I turn in the remainder of this talk to this long-term issue, focusing on the proliferation implications of disposal of spent fuel in a mined repository.

5. THE PROLIFERATION IMPLICATIONS OF GEOLOGICAL DISPOSAL OF SPENT FUEL

Several observers have raised concern that a spent-fuel repository could be "mined" for plutonium. For example, Glenn Seaborg at a 1995 meeting of the American Nuclear Society (ANS) in San Francisco, attacked "the widespread assumption that by leaving spent fuel intact, proliferation risks are avoided." On the contrary, seaborg argued: "those who advocate the disposal of spend fuel... do not necessarily occupy the high ground in the non-proliferation debate.'

Seaborg's and other concern stems principally from two factors. First, the time and effort necessary to recover spent fuel from a repository, although significant, may compare favourably with other ways that a nation may acquire spent fuel, the "ore" from which plutonium can be extracted. Second, as a result of the relatively short (30-year) half-life of cesium-137, the spent fuel radiation barrier will decay to a low level after a few centuries of cooling, so that older spent fuel can be handled and reprocessed with lower risk of injury than spent fuel more recently discharged from a nuclear reactor.

But how serious is this risk of plutonium mining really? In a repository, spent fuel will be maintained in a retrievable state for about a century or two after emplacement. After this period, access tunnels and ventilation shafts will be backfilled and sealed, and support facilities will be dismantled. After closure, recovery of the material, although still possible, would be more difficult, time-consuming and expensive.

Nevertheless, even after closure, the spent fuel could in principle be mined by drilling new access tunnels into the repository. Recently it has been suggested that due to advances in the technology of tunnel boring machines (TBMs), a substantial quantity of plutonium could be recovered from a repository in under a few months for a cost of less than $10 million, simply by boring a small-diameter tunnel to the repository level.[2] For example, for the proposed design of the U.S. repository at Yucca

Mountain, Nevada, recovery of single disposal package would net about 10 tonnes of spent fuel, containing 100 kilograms of plutonium.

However, this estimate does not account for the fact that the driller would, in all likelihood, have to dig several tunnels to intercept a spent fuel canister. For instance, in the Yucca Mountain repository, about 10,000 spent fuel packages would be distributed in a single layer over an area of 500 hectares. For this configuration, the probability of intercepting a package with a single 3.6-metre diameter tunnel perpendicular to the repository layer is only 2 per cent, so that on average 50 tunnels would have to be drilled to locate a single canister.

For a typical tunneling cost of $500 per metre, a single 1-kilometer long, 3.6 metre diameter tunnel would cost $5 million fifty such tunnels would cost $250 million. At an average advance rate of 30 metres per day, each tunnel would take about a month to complete given that tunnelling operations usually involve a 'learning curve" of 5-10 weeks to reach the average rate, and require two to twelve weeks for preparation, an estimate of three to six months is more realistic. If the tunnels were drilled in series, it could take several years before the first canister were located. If they were drilled in parallel, the time could be reduced, but this would require simultaneous acquisition and use of many TBMs, substantially increasing the cost, personnel requirements and observability of the operation.

On balance, the range of conditions under which repository mining will look attractive compared to other routes to plutonium acquisition is extremely narrow. In circumstances where a country has a working nuclear fuel cycle, mining a repository is not likely to look attractive compared to more direct ways to obtain weapons-usable material—for example, from spent fuel in above-ground retrievable stores. In a nuclear-free world, repository mining might be quicker and cheaper way to obtain plutonium than by building reactors and reprocessing plants, although given uncertainties in technological development in the next several hundred years, it is hard to make any definitive comparisons. At a minimum, it seems likely that mining to obtain significant quantities of plutonium will take at least several months and will be readily detectable if there are reasonable safeguards applied at the repository sites.[3]

In any case, if the spent fuel accumulated in a nation's civilian power programme is not put into a repository, and is instead left in retrievable storage and eventually processed in a P-T programme, that course will itself generate significant risks of plutonium diversion or theft, as is described below.

6. SHOULD PLUTONIUM BE DESTROYED RATHER THAN PUT UNDERGROUND?

If one does not put spent fuel into a repository, then what? If one disregards exotic alternatives, such as shooting plutonium into the sun, the main alternative to geologic burial is partitioning the plutonium, as well as higher actinides that will eventually decay into plutonium isotopes, from other constituents in the spent fuel and then fissioning them in a nuclear reactor or an accelerator-driven spallation source. This alternative is often referred to as "partition and transmutation" (P-T), an ambitious procedure which aims to eliminate a large fraction of long-lived actinides and fission products that are present in spent fuel.

Implementation of P-T would involve reprocessing and recycling on a truly grand scale. For example, consider what would be involved in treating the U.S. spent fuel inventory, which is now about 35,000 tonnes and is increasing by approximately 2000 tonnes per year. Reprocessing this amount would require four plants each the size of the THORP reprocessing plant in Great Britain, assuming a 40-year plant lifetime.

Since there are about 10 kilograms per tonne of transuranics (plutonium, neptunium and higher actinides) in U.S. light-water reactor (LWR) spent fuel, the P-T system would have to be able to absorb 28 tonnes of transuranics per year. Assuming for specificity a fast neutron reactor of 1400 megawatts-electric capacity with conversion ratio of 0.76 and lifetime of 40 years (characteristics of the PRISM fast reactor concept developed by General Electric), such a reactor could fission in its lifetime about 37 tonnes of transuranics. Assuming that U.S. LWR capacity will remain constant over forty years, the LWRs will generate about 1150 tonnes of transuranics, and the transmutation effort will involve thirty fast reactors. Because the transuranic inventories in the fast reactor spent fuel will still be substantial, this fuel also would have to be reprocessed and the recovered transuranics fed back into process. Depending on how completely one wishes to destroy the transuranics, the transmutation effort could take from hundreds to thousands of years.[4]

In addition to reprocessing plants and reactors, the P-T system would include an array of additional facilities, such as plutonium fuel fabrication plants, scrap recovery plants, waste handling plants and storage areas for in-process materials, as well as numerous transport links between the facilities. This unprecedented scale of activity would be extremely difficult and expensive to safeguard and physically secure. Even if separations technologies can be devised so that the plutonium nominally stays mixed with fission products throughout the entire fuel cycle it would always be possible for a country to re-configure the separations process to allow the purification of plutonium.

Therefore, safeguards would have to be maintained in perpetuity on a P-T system, just as they would on spent fuel repositories. Moreover, the burden would be much greater for P-T, both with respect to the number of facilities and the type of safeguards involved. Safeguards on bulk-handling facilities, such as reprocessing plants, are based on material accountancy methods that are intrinsically less reliable and more difficult to apply than the containment and surveillance procedures needed for repository safeguards. Even if the long-term risks of plutonium mining in a repository were significantly reduced through P-T, this advantage would be overwhelmed by the near-and medium-term proliferation and environmental risks of reprocessing and recycling.

7. CONCLUSIONS

For these reasons, recent suggestions to scale back efforts to develop a geologic repository in the U.S. are regretable. Such a scale-back could delay for decades the movement of spent fuel out of retrievable storage in the U.S. and abroad. Geologic disposal may not be the ideal solution of the nuclear waste problem, but it is the least risky option available now for the backlog of spent fuel and plutonium the world has already produced.

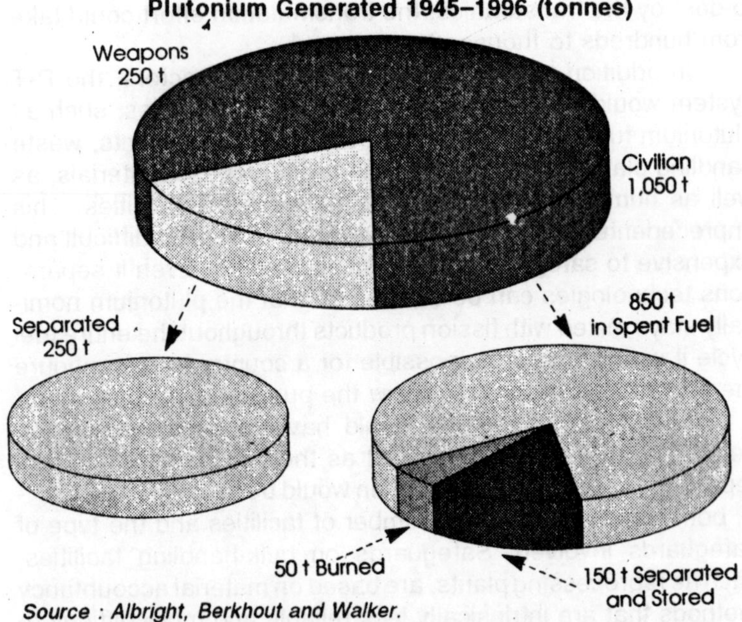

Source : Albright, Berkhout and Walker.

Table 1 : Approximate Isotopic Composition of Various Grades of Plutonium

Grade Isotape	***Pu-238***	***Pu-239***	***Pu-240***	***Pu-241***	***Pu-242***
Super-grade	---	.98	.02	--	--
Weapons-grade	.00012	.938	.058	.0035	.00022
Reactor-grade	.013	.603	.243	.091	.050
MOX-grade	.019	.404	.321	.178	.078
FBR blanket	---	.96	.04	--	--

Department of Energy release, dated January 1997:

"a potential proliferating state or subnational group using designs and technologies no more sophisticated than those used in first-generation nuclear weapons could build a nuclear weapon from reactor-grade plutonium that would have an assured, reliable yield of one or a few kilotons.... Advanced nuclear weapon states such as the United States and Russia, using modern designs, could produce weapons from reactor-grade plutonium having reliable explosive yields, weight, and other characteristics generally comparable to those of weapons made from weapons-grade plutonium."

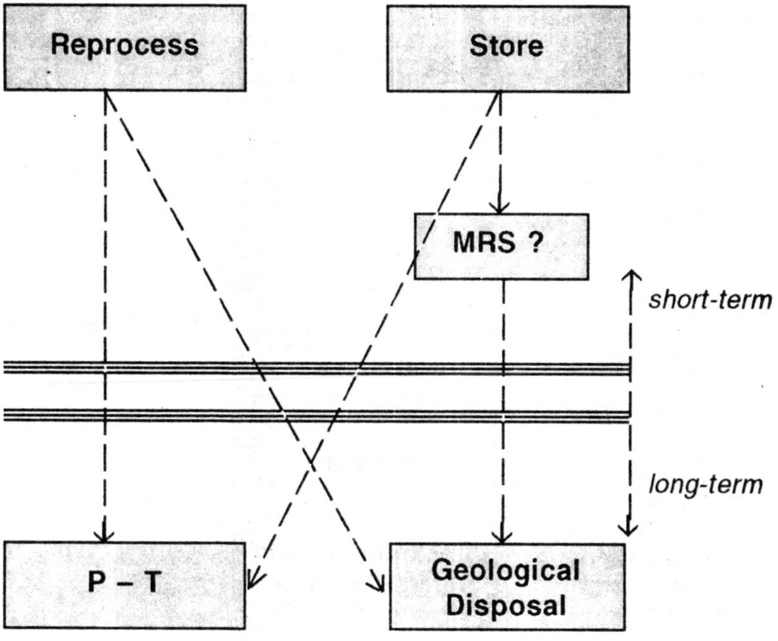
Alternatives for Disposal of Spent Fuel
Reprocess
Store
MRS ?
short-term
long-term
P – T
Geological
Disposal

Virtual Reprocessing in Great Britain

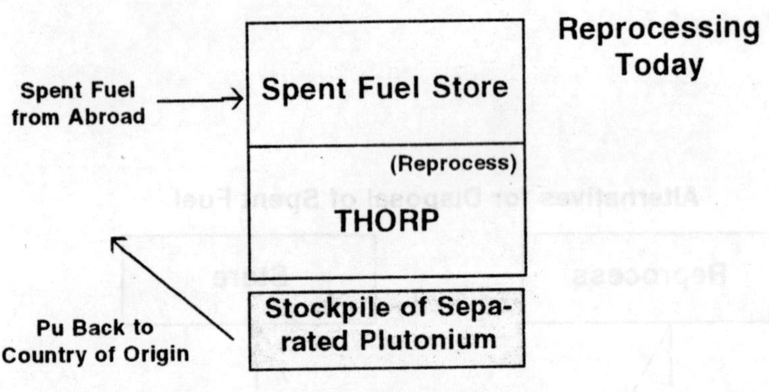

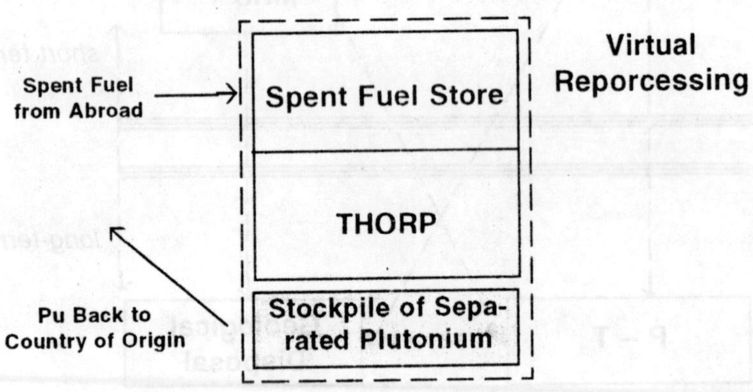

U.S. Spent Fuel

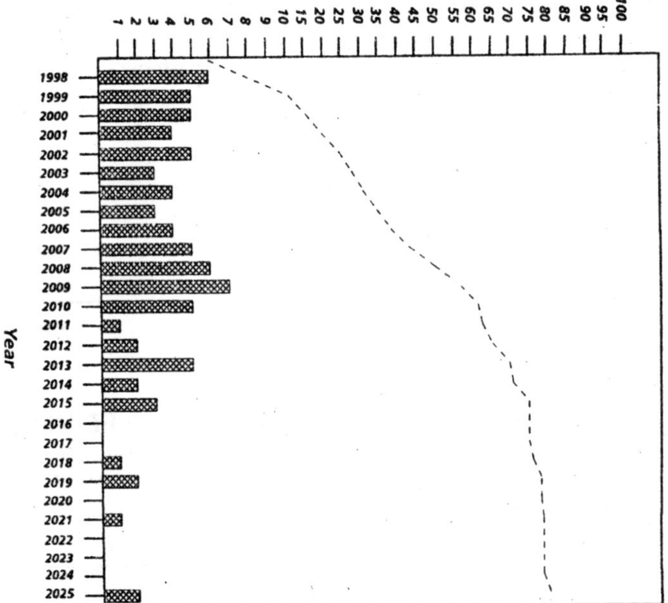

Notes:

1. *Non-Proliferation and Arms Control Assessement of Weapons-Usable Fissile Material Storage and Excess Plutonium Disposition Alternatives* (U.S. Department of Energy, 1997), pp. 38–39. The classified reference for this conclusion is, *Extraction and Utility of Reactor-Grade Plutonium for Weapons*, W.G. Sutcliffe and T.J. Trapp, eds. (Lawrence Livermore National Laboratory, UCRL-LR-115542, 1995 {S/RD}).
2. Pers Peterson, "Long-term Safeguards for Plutonium in Geological Repositories, *Science & Global Security*, (1996).
3. A more complete discussion is Edwin Lyman and Harold Feiveson, "The Proliferation Risks of Plutonium Mines". *Science & Global Security* (1998).
4. Partition and transmutation systems have been described and analysed recently by the Lawrence Livermore National Laboratory (Lawrence Ramspott, et al, *Impacts of New Developments*

in Partitioning and Transmutation on the Disposal of High-level Nuclear Waste in a Mined Geologic Repository, UCRL ID-109203, Lawrence Livermore National Laboratory, Livermore, California, March, 1992.] and by the Panel on Separations Technology and Transmutation Systems (STATS panel) of the National Research Council [*Nuclear Wastes: Technologies for Separations and Transmutation*, National Academy Press, 1996.] Both reports drew on earlier work by Thomas Pigford and collaborators.

Address:

HAROLD FEIVESON
Princeton University
Princeton NJ 0844
USA

41

Immobilisation of Excess Weapons Plutonium: A Rapidly Achievable Disposition Strategy for Russia

ALLISON MACFARLANE

INTRODUCTION

In remarks made a few weeks ago, Madeline Albright, U.S. Secretary of State, declared that she was "directing U.S. negotiators to conclude agreements by the year 2000 to make 'excess' U.S. and Russian plutonium permanently unusable for weapons" (Albright, 1998). In contrast, a recent report from the centre for Strategic and International Studies (1998) criticised the slow pace of excess weapons plutonium disposition in the United States and Russia. Making plutonium "permanently unusable for weapons" is a daunting, if not impossible, task that will require a swift and focused effort on the part of policy-makers in both the United States and Russia. The process of plutonium disposition is complex and the roles of both countries are intricately intertwined.

The lack of progress in plutonium disposition in part derives from the desire of the United States to move in parallel with Russia on this matter. The United States has already begun to implement policy decided upon by the Department of Energy, known as the Dual-Track Approach, which includes both the burning of plutonium in nuclear reactors and the immobilisation of plutonium in a ceramic matrix encased in radioactive glass. (Department of Energy, 1997) Russia has yet to declare an official policy on plutonium disposition, although it has tentatively agreed to follow the U.S. plan. (U.S. - Russian Independent Scientific Commission, 1997) Russia's Ministry of Atomic Energy (Minatom), however, has openly declared its strong preference for the use of reactors to dispose of all separated plutonium in Russia, including 30 metric tons of reactor-grade plutonium separated from civilian spent fuel. (e.g., Mikerin and Kudryavtsev, 1994; Egorov et al., 1995; Karelin, 1996; Mikhailov et al., 1996)

Plutonium disposition in Russia is complicated by the fact that Russia does not have the funding to cover the costs of disposing 50-100 metric tons of plutonium declared excess to military needs, let alone its 30 metric tons of civilian plutonium (U.S. - Russian Independent Scientific Commission, 1997). The United States and the other G8 countries will therefore have to cover the majority of the costs of plutonium disposition in Russia. The incentive to do so is to prevent Russian excess weapons plutonium from being reused by either terrorists or the host country, by first securing it against terrorist threats through physical protection programmes1 and then by providing a clear signature of some type if the host country attempts to reuse the plutonium in weapons. Another incentive to cover the costs of Russian disposition is to expedite the disposition of excess military plutonium from both the United States and Russia. As stated above, the United States will not go forward with their disposition programme until Russia makes a concerted effort with its own programme. As a result, a source of funding will be able to move the Russian plutonium disposition programme forward, and accordingly, the U.S. plutonium disposition programme will also be expedited.

Consequently, the United States and the G8 should have a strong interest in funding the most efficient and cost effective programme for plutonium disposition in Russia. Unfortunately, no robust technical and economic comparisons of the disposition options in Russia exist. Although Minatom has pressed the use of reactors to disposition plutonium, it is not at all clear that this is either the most efficient or cost effective route. In fact, as I will show in this paper, a short analysis reveals that Russia is actually better positioned to immobilise its plutonium than to burn it in reactors.

BACKGROUND

Minatom's view of plutonium disposition originates from its plan to develop a fully functional closed nuclear fuel cycle (e.g., Egorov et al., 1995). The recycling of plutonium is an important part of this scheme, with the hope that it will one day make the country less dependent on other energy resources. Consequently, Minatom takes a dim view of the Immobilisation of plutonium.

Minatom's main objections to disposition of plutonium through immobilisation are threefold. First of all, plutonium, is seen as a resource that has been created at great expense, not to be disposed of lightly. Secondly, use of plutonium in mixed oxide (MOX) fuel in reactors will provide long-term employment, revenue, and energy for the country. Jobs are an especially important part of the picture in the currently economically strapped country.

The third objection has to do with the end result of immobilisation. As it was first presented to the Russians by the United States, immobilisation of plutonium was synonymous with vitrification, the incorporation of plutonium in glass. Minatom was concerned that[1] immobilisation would not change the isotopic composition of plutonium, thus it would remain weapons grade (in contrast to the MOX option, e.g., Karelin, 1996) and[2] it would be relatively easy to extract plutonium from the vitrified end-product (Hodren et al., 1996). Their concerns about extraction of plutonium were well justified, it turns out. In the U.S. case, the composition of glass that the Department of Energy considered was originally developed as a temporary storage form for the actinides americium and curium (Rudisill, 1997; Macfarlane, in press). Plutonium could be extracted rapidly with high efficiency. Fortunately, the Department of Energy decided instead to immobilise their plutonium in ceramic, a material for which there is no standardised industrial-scale method to extract radionuclides (Department of Energy, 1997).

The use of ceramic instead of glass should somewhat alleviate the earlier concerns of the Russians. For the case of

1. Presently no mandatory international standards for the physical protection of fissile materials in Russia and the United States exist, but a few efforts are underway to create such a set of standards including, work at the Internatioinal Atomic Energy Agency, the U.S. Arms Control and Disarmament Agency, and the Centre for International Security and Arms Control at Stanford University.

2. Costs for these options are not well established and those cited in this paper likely contain substantial error.

the U.S., extraction of plutonium from ceramic would require the construction of new facilities, or at least the substantial refitting of old facilities. Such activity would provide a "signature" to other countries and the International Atomic Energy Agency that the U.S. was beginning to rebuild its nuclear arsenal. Recovery of plutonium from spent MOX fuel, on the other hand, would not provide as obvious a "signature," because the U.S. could simply restart its shutdown PUREX plants for the extraction of plutonium. This, after all, was the method employed to recover plutonium during the cold war. Although the plutonium recovered from spent MOX fuel would not be weapons grade, it is still usable in nuclear weapons (Mark, 1993). Furthermore, Minatom officials have even allowed that it is acceptable to use ceramic for the long-term storage of actinides, such as plutonium, at low concentrations (Mikerin and Kudryavtsev, 1994).

THE DE FACTO SOLUTION IN RUSSIA: REACTORS AND PLUTONIUM

The remainder of the paper will focus on a simplified comparison of the reactor and immobilisation options for plutonium disposition. Such a comparison starts with an examination of the facilities, timing, and costs[2] for the reactor option. The burning of plutonium in nuclear reactors would begin with the conversion of plutonium pits and other metal forms to plutonium oxide. Because this process will be necessary for both burning of plutonium in reactors and its immobilisation, it is not necessary to include it in the analysis. The plutonium oxide will then be combined with uranium oxide to form MOX fuel in a MOX fuel fabrication facility. MOX fuel will be burned in one of two types of reactors" breeder or light water reactors. This paper will consider both reactor options. Finally, the spent MOX fuel would need storage and perhaps eventually a geological repository. Again, because both the immobilisation and MOX options would need the same kind of long-term storage, this facility will not be considered in this analysis.

Mox Fuel Fabrication Facility

To date, Russia has no full-scale MOX fabrication plant. Currently, they have pilot scale plants at Dmitrovgrad and the Mayak facility at Oxersk (formerly Chelyabinsk-65), which can process between 70-80 kg (Mayak) and 350 kg (Dmitrovgrad) of plutonium per year (Antipov et al., 1995; Egorov et al., 1995; Mikahilov et al., 1996). The much larger complex-300 MOX fabrication facility at Ozersk is only half-finished and would be capable of processing 5-6 MT of plutonium per year (Antipov et al., 1995; Egorov et al., 1995; Mikahilov et al., 1996). Complex-300 was originally designed to make MOX fuel for breeder reactors, not light water reactors, so it would have to be substantially modified to fabricate light water reactor fuel (Antipov et al., 1995). The cost of completing this plant is not well established, but would be on the order of $35 million to greater than $250 million, depending on which type of fuel was made (Diakov, 1997).

Breeder Reactor Option

Minatom's preference is to dispose of plutonium with breeder reactors, although their breeder programme is not presently able to accommodate the volume of available plutonium. Only one breeder reactor, the BN-600 at Beloyarskaya, remains on Russian soil the other breeder, the BN-350, is in Kazakhstan at the Aktau site. Both reactors were designed to use uranium-based fuel. Because neither of the two operating breeder reactors that Russia built were designed to use MOX fuel, Minatom envisions constructing BN-800 models, which can operate with MOX fuel (Egorov et al., 1995; Mikahilov et al., 1996). Each BN-800 reactor would be able to use approximately 1.6 MT of plutonium per year, so that 3 of these breeders could use 100 MT plutonium in little over twenty years (Mikahilov et al., 1996; diakov, 1997). This estimate does not include construction time (on the order of a decade or more, most likely) and down time due to problems with the reactors (Diakov, 1997).

The cost of the BN-800 reactors will likely be prohibitively expensive. Estimates from 1991 suggest that construction costs for the completion of the two partially constructed BN-800 reactors at the South Urals plant would be on the order of $3 billion (Joint U.S.- Russian Plutonium Disposition Steering committee, 1996). Construction at these plants was stopped due to lack of funding. A new plant would cost upwards of $1.5 - 2 billion (Joint U.S. - Russian Plutonium Disposition Steering Committee, 1996). These estimates do not include the costs of operations, although such costs may be somewhat offset by electricity sales.

In terms of the viewpoint of some members of the G8, the choice of timing for Minatom to pursue the use of breeder reactors is rather unfortunate. Many countries that were once enthusiastic about breeder technology have recently moved to abandon it. France has just announced the closing of the Superphenix reactor, which spent most of its lifetime shutdown due to accidents or technological difficulties. Japan is reconsidering its breeder programme due to recent accidents at the Monju site and subsequent local protests against breeder technology. Britain, Germany, and the United States never fully entered into the breeder programme, and have now abandoned any future plans to do so. Consequently, if Russia plans to burn MOX fuel this way, it is likely that they will receive little or no support from the United States to do so.

Light Water Reactor Option

The Russian reactors most likely to be supported by the United States and the G8 for plutonium disposition are pressurised light water reactors. In particular, the best Russian models to be used are the VVER-1000 series, which are the only Russian light water reactors closest to satisfying international safety standards. Russia currently has seven operating VVER-1000 reactors at 3 sites, Balakovskaya (4), Kalinin (2), and Novovoronezh (1). Minatom officials have declared that only the 4 VVER-1000 reactors by the National Academy of Science (National Research Council, 1995) suggests that all VVER-1000 reactors (including the 4 at Balakovskaya) would require safety retrofits such as the installation of additional control rods and the reracking of existing fuel rods for safe and efficient burning of MOX fuel (Levina et al., 1995).

Although some Minatom estimates suggest that 15-20 VVER-1000 reactors would be needed to complete the job of plutonium disposition (Mikerin and Kudryavtsev, 1994; Mikahilov et al., 1996), my calculations suggest that fewer would be required. Based on current operating methods of the existing VVER-1000 reactors, including an average fuel burn-up of 42.5 MWd/kg, plutonium fuel loading of 2.5-5 per cent, a capacity factor of between 70-92 per cent, and loading of a third of a core of MOX fuel, each VVER-1000 reactor would use between 200-300 kg of plutonium a year. The current reactors will complete their 30-year lifetime within the next 25 years. Consequently, even if all seven VVER-1000 reactors were employed (assuming all modifications were made for MOX fuel) at best they could process about 25 MT in their lifetimes. If it were possible to extend the lifetimes of these reactors another 10 years, then approximately 50 MT of plutonium would be processed, which is half of Minatom's projections of excess weapons plutonium and about one third of the total separated military plutonium. As a result, Russia will need to build new VVER-1000 reactors if they plan follow this route to disposition.

Minatom does have plans to complete three or more new VVER-1000 reactors (FNS Transcript, 1998). As is the case for the breeder reactors, Minatom has repeatedly halted construction of the new reactors due to lack of funding. Three additional VVER-1000 reactors would process between 0.6-0.9 MT plutonium per year. If three new VVER-1000 reactors are operational by 2015, then by 2035, at best approximately 65 MT of plutonium would be dispositioned. Present estimates indicate that the cost of a new VVER-1000 is on the order of $1 billion (Diakov, 1997). For three new reactors and modifications to existing ones, the VVER-1000 option would cost over $3 billion, not including the costs of retrofitting the existing reactors for MOX fuel use. Again, these estimates do not include operating costs.

The Ukraine could aid Russia through the use of its VVER-1000 power reactors. Ukraine has 10 at 4 sites, Khmelnitski (1), Rovno (1), South Ukraine (3), and Zaporozhe (5). Although this plan would certainly speed the process, it is unlikely due to the fluctuating political situations in the Ukraine and Russia.

Two other factors have not yet been accounted for: costs due to transportation of materials and increased proliferation risk from transportation of proliferants. If MOX fuel is manufactured at the Mayak plant, then it will need to be transported hundreds of kilometres to at least three sites of the VVER-1000 reactors are used. Transportation costs will be much higher than those for immobilisation, which will presumably occur at one facility. Security risks will also increase from transportation over such long distances in isolated areas. Again, security will not be as large a factor for immobilisation due to the lack of need for transportation.

AN ALTERNATIVE SOLUTION: IMMOBILISATION

Like the MOX fuel approach, immobilisation of excess weapons plutonium will require the construction and operation of new facilities. Unlike the MOX fuel approach, the number and cost of new facilities for immobilisation should be lower and consequently, the disposition schedule should be shorter. Below I present an analysis of the immobilisation of plutonium in Russia based on published work by Russian authors. The basis for this analysis is the approach adopted by the United States to first incorporate plutonium in a ceramic matrix and then surround the ceramic with high-level waste glass, which provides a lethal barrier to reuse, otherwise known as the can-in-canister approach. In particular, the U.S. approach involves placing the plutonium-containing ceramic into small cans and placing a number of them inside a much larger canister (2 m high and 0.6 m in diameter), into which high-level waste glass will be poured (Department of Energy, 1997). The final outcome is a two-ton canister that will deliver a lethal dose of radiation (200-500 R/hr) to a person standing 1 m from its surface (Department of Energy, 1997). This method involves two steps, first creating the ceramic with a cold-press and sinter process and second, adding the high-level waste glass in a vitrification facility. I will treat each step separately.

Ceramic Immobilisation

In terms of ceramic immobilisation, Russian scientists have vast experience with crystalline materials. Recently, a number of Russion scientists have been working on their own or in collaboration with researchers from the United States and other countries on waste forms for excess weapons plutonium. Researchers in Russia involved in projects on immobilisation are located at a number of institutions, including the V.G. Khlopin Radium Institute in St. Petersburg (e.g., Anderson et al., 1995), the A.A. Bochvar Institute in Moscow, Arzamas-16 (Trutnev et al., 1996), the All Russian Scientific and Research Institute of Inorganic Materials in Moscow (Matyunin et al., 1996), All Russian Scientific and Research Institute of Chemical Technologies (Nardova et al., 1995), and the Obninsk facility (Starkov and Konovalov, 1995). In addition, a small programme at the Lawrence Livermore National Laboratory in the United States is funding research on waste form production and characterisation for both glass and ceramic waste forms at both the Khlopin and Bochvar institutes in Russia (Jardine, 1997; Jardine and Borisov, 1998).

Immobilisation, if carried out similar to the U.S. plan, would require that a glove-box scale production facility be constructed for immobilising plutonium directly in the waste form. This facility would most likely contain a calcining step, in which all reactant oxides are calcined, a milling step, which would reduce the grain size of reactants to <5 μm, and a pressing step to form the final product. The United States plans to use a cold press and sinter method to create its ceramic hot pressing, hot isostatic pressing, or a sol-gel process could also be adopted. These processes can be modified to fit the glove-box scale required for working with fissile material like plutonium.

Vitrification Facility

Russia is certainly not without experience in the immobilisation of radioactive materials. Minatom continues to operate a programme to immobilise its high-level radioactive waste from the reprocessing of spent fuel for plutonium production. This process is carried out at the Vitrification Facility associated with the RT-1 reprocessing facility at Ozersk (Chelyabinsk-65). Minatom has used two different glass compositions, phosphate and borosilicates to vitrify their wastes (Aloy et al., 1996). The Vitrification Facility is still operating and has successfully immobilised over 9100 m^3 of high-level waste (Kushnikov, 1995). This facility could easily be adapted to use for the can-in-canister method of immobilisation.

There are no adequate estimates for the total cost for the immobilisation option in Russia. The Russian situation is similar to the American one in that they have an already-operating vitrification facility and would need only to build a glove-box ceramic facility adjoining it. The cost estimates for immobilisation of plutonium in the United States are on the order of $2 billion, including both construction and operating costs for 50 MT of plutonium (Department of Energy, 1997). Consequently, the immobilisation option in Russia would most likely cost under $2 billion for the disposition of 50 MT of plutonium.

Policy Suggestions

Although Minatom has made a strong case for the use of reactors to disposition plutonium, there are a few persuasive incentives which may encourage immobilisation for at least a portion of the excess weapons plutonium. First of all, as stated earlier, Russia is awash is separated plutonium. It is unlikely that all 130 metric tons will be burned in reactors within a reasonable time frame that will ensure the safety and security of the material. Consequently, it will pose a security threat whil it remains in separated form.

Perhaps more significant is the situation of the 30 metric tons of plutonium separated from commercial reactors. This material contains relatively high levels of the isotope Pu-241, much of which has already decayed to Am-241, a gamma-ray emitter. Consequently, the material has become more dangerous and will require costly radioactive shielding methods to handle it. Moreover, plutonium with high Am-241 levels is not really suited to MOX fabrication or use, hence it is best immobilised.

From both a technical and an economic point of view, immobilisation appears to be the best option for Russia. Russia already has half of the necessary facilities in place to carry out the can-in canister approach to immobilisation of plutonium. It only requires the construction of a ceramic production facility. The burning of plutonium in reactors, on the other hand, requires a MOX fabrication facility, modifications to existing reactors and the construction of new reactors, the latter of which is a time-consuming project.

Although it is difficult to make absolute statements about costs this early in the game, it appears that it would be more cost-effective for the United States and the G8 to fund the immobilization process than the reactor process. Most likely immobilisation will cost at least $1 billion less than using light water reactors and most certainly at least $1 billion less than using breeder reactors.

My suggestion is to fund a small-scale immobilisation facility and then scale up to a large facility, if it is found to be agreeable by Minatom. This policy would include funding of current waste form research and development, funding of research on geologic repositories, and funding of a pilot immobilisation facility. Actually, all of the above are already being funded by Lawrence Livermore National Laboratory perhaps it is time to fund such programmes at a larger scale. It will also be important to fund Russian prepared comparative studies of the two options in terms of technical and economic aspects. Finally, the United States should fund a full-scale waste form production facility adjacent to the already-existing vitrification facility in Ozersk. Such an approach would not preclude Minatom's use of MOX fuel in the future, but would ensure that the United States and G8 countries pay for the most cost effective method of disposition.

Such a programme will allow the United States to move away from the MOX approach to plutonium disposition, a source of controversy and potential delay. The United States currently maintains its non-proliferation policy, first established in 1976, against the use of reprocessing of spent nuclear fuel to recover plutonium and uranium for reuse in reactors. Although the use of MOX fuel fabricated from excess military plutonium does not legally violate this policy, a number of anti-nuclear groups such as the Nuclear Control Institute claim that use of MOX fuel would violate the spirit of the non-proliferation policy. If the United States attempts to use MOX fuel for plutonium disposition these groups will sue the government for breach of policy. Potentially causing major delays in the plutonium disposition programme. Consequently, if the United States immobilised all of its plutonium, along with Russia, it would be much more likely that these countries would rapidly disposition their excess weapons plutonium.

Finally, the plan outlined above in no way prevents Minatom from pursuing a closed fuel policy. It just ensures that the United States is consistent in its pursuit of international security,

REFERENCES

Albright, Madeline, 1998, Secretary's Remarks to the Stimson Centre, June 10,1998, U.S. Department of State.

Aloy, A.S., Bel'tyukov, V.A., Demin, A.V. and Revenko, Y.A., 1996, Experiences with vitrification HLW and development of new approaches in Russia, *in* National Research Council, ed., Glass as a Waste Form and Vitrification Technology: Summary of an International Workshop: Washington, National Academy Press, v.p. E.50-E.51.

Anderson, E.B., Bukarov, E.E., Galkin, B.Y., Starchenko, V.A. and Vasiliev, V.G., 1995, Crystalline matrices for the immobilisation of plutonium and actinides, in Plutonium Stabilisation and Immobilisation Workshop: Washington, Department of Energy, v.p. 269-277.

Antipov, S., Astafiev, V., Borisov, L., Menshikova, T., Reshetnikov, F., Revjakin, V. and Zakharkin, B., 1995, Manufacturing experience on pelletised mixed oxide (MOX) fuel for fast reactors, *in* Merz, E.R., Walter, C. E. and Pshakin, G. M. ed.s, Mixed Oxide Fuel (MoX) Exploitation and Destruction in Power Reactors: Dordrecht, Luwer Academic Publishers, NATO ASI Series: Disarmament Technologies, Vol. 2, p. 211-224.

Centre for Strategic and International Studies, 1998, Disposing of Weapons-Grade Plutonium.

Department of Energy, 1997, Final non-proliferation and arms control assessment of weapons-usable fissile material storage and excess plutonium disposition alternatives: Department of Energy.

Department of Energy, 1997, Record of decision for the storage and disposition of weapons-usable fissile materials final programmatic environmental impact statement: Department of Energy.

Department of Energy, 1997, Technical recommendation for the immobilisation of plutonium: Lawrence Livermore National Laboratory.

Diakov, A., 1997, Disposition of weapons-grade plutonium in Russia: Evaluation of different options, *in* Schulte, N.T ed., Dismantlement and Destruction of Chemical, Nuclear, and Conventional Weapons: Dordrecht, Kluwer Academic Publishers, NATO ASI Series: Disarmament Technologies, Vol. 10, p. 171-180.

Egorov, N.N., 1996, Separated plutonium utilisation in Russia: International Uranium Fuel Seminar, St. Petersburg, Florida.

Egorov, N.N., Murgorov, V.M., Kagramanian, V.S., Troyanov, M.F., Poplavski, V.M. and Mateev, V.I., 1995, Management of Plutonium in Russia, *in* Merz, E.R., Walter, C.E. and Pshakin, G. M. ed.s, Mixed Oxide Fuel (Mox) Exploitation and Destruction in Power Reactors: Dordrecht, Kluwere Academic Publishers, NATO ASI Series, Disarmament Technologies, Vol. 2, p. 1-15.

FNS Transcript, 1998, Press Conference with Nuclear Energy Minister Viktor Mikhailov:

Holdren, J.P., Ahearne, J.F., Garwin, R. L., Panofsky, W. K. H. Taylor, J.J. and Bunn, M., 1996, Excess weapon plutonium: How to reduce a clear and present danger: Arms Control Today, v. November/December, p. 3-9.

Jardine, L., 1997, Disposition of Excess Weapons Plutonium from Dismantled Weapons: Lawrence Livermore National Laboratory.

Jardine, L. J. and Borisov, G. B., 1998, Russian-American Strategy for Stabilisation and Immobilisation of Excess Russian Weapons Origin Plutonium: Lawrence Livermore National Laboratory.

Joint U.S. - Russian Plutonium Disposition Steering Committee, 1996, Joint United States/ Russian Plutonium Disposition Study.

Karelin, A., 1996, Weapons plutonium management: The present situation, the methods and the prospects, in Merz, E. and Walter, C. ed.s, Disposal of Weapon Plutonium: Dordrecht, Kluwer Academic Publishers, v.p. 1-7.

Kushnikov, V., 1995, Radioactive waste management and plutonium recovery within the context of the development of nuclear energy in Russia: Plutonium Stabilisation and Immobilisation Workshop, Washington, DC, Department of Energy, p. 25-34.

Levina, I. K., Sparykin, V. V. and Morozov, A.G., 1995, The safety criteria and VVER core modification for weapon plutonium utilisation, *in* Merz, E., Walter, C.E. and Pshakin, G. M. ed.s, Mexed Oxide Fuel (MoX) Exploitation and Destruction in Power Reactors: Dordrecht, Kluwer Academic publishers, NATO ASI Series: Disarmament Technologies, Vol. 2, p. 83-91.

Macfarlane, A., in press, Immobilisation of Excess Weapons Plutonium: Reasonable Alternatives to Vitrification: Science and Global Security.

Mark, J. C., 1993, Explosive properties of reactor-grade plutonium: Science and Global Security, v. 4, p. 111-128.

Matyunin, Y.I., Kushnikov, V. V. and Demin, A.V., 1996, Possible ways of the weapon plutonium processing to produce the steady glass materials suitable for further safe disposal, in Merz, E. and Walter, C. ed. s, Disposal of Weapon Plutonium: Dordrencht, Kluwer Academic Publishers, p. 123-128.

Mikhailov, V.N., Bogdan, V. V., Murgorov, V.M. and Kagramanian, V.S., 1996, Prospects for using plutonium as a fuel in Russia, *in* Merz, E. R. and Walter, C. E. ed.s, Disposal of Weapons Plutonium: Dordrecht, Kluwer Academic Publishers, NATO ASI Series, Disarmament Technologies, Vol. 4, p. 37-49.

Mikerin, E.I. and Kudryavtsev, E.G., 1994, Problems of Russian Plutonium Utilisation, *in* Garwin, R., Grubb, M. and Matanle, E. ed.s, Managing the Plutonium Surplus: Applications and Technical Options: Dordrecht, Luwer Academic Publishers, p. 37-41.

Nardova, A. K., Filippov, E. A., Glagolenko, Y. B., Dzekun, E. G., Korchenkin, K. K., Mashkin, A. AN., Rovnij, S. I. and Kudryavstev, E. G., 1995, Immobilisation of plutonium from solutions on porous matrices by the method of high temperature sorption, *in* plutonium Stabilisation and Immobilization Workshop: Washington, Department of Energy, v. p. 189-197.

National Research Council, 1995, Management and Disposition of Excess Weapons Plutonium: Reactor-Related Options: Washington, DC, National Academy Press.

Rudisill, T., 1997, Recovery of americium/ curium isotopes from strontium aluminoborosilicate glasses: 21st Annual Actinides Separations Conference Abstracts, p. 84.

Starkov, O. and Konovalov, E., 1995, Immobilisation of actinides in stable mineral type and ceramic materials (high temperature synthesis), *in* Plutonium Stabilisation and Immobilisation Workshop: Washington, Department of Energy, p. 263-268.

Trutnev, J.A., Nadykto, B.A. and Prudova, N.S., 1996, On the long-term storage of weapons plutonium in form of criticality-free ceramics containing lithium, *in* Merz, E. and Walter, C. ed. s, Disposal of Weapon Plutonium: Dordrecht, Kluwer Academic Publishers, p. 91-92.

U.S. - Russian Independent Scientific Commission, 1997, Final Report of the U.S. - Russian Independent Scientific Commission on Disposition of Excess Weapons Plutonium.

Address:

ALLISON MACFARLANE
Stanford University
Stanford CA 94305–6165
USA

42

A French-European Kazakhstanese Comprehensive Programme of the Ecobiological Consequences of Nuclear Tests in Kazakhstan (1949–1989): a Five Years Follow-up

CHRISTIAN CHENAL *and* JACQUES MONGNET

The north-east and central part of Kazakhstan is a well known region contaminated by radioactivity due to the nuclear tests performed on the semi-palatinsk Soviet Union test site. From 1949 to 1962, 113 atmospheric tests were done and from 1963 to 1989, 343 underground or pseudo underground (in small hills) tests were performed.

During the atmospheric test period, 10,000 people were heavily irradiated by external irradiation (several Gys) and half million people were irradiated with lower doses. (severl cGy).

From 1949 until now, several thousands of people were irradiated by small doses of internal contamination due to the fallout of atmospheric tests and the release in the environment of radionuclides coming from the pseudo-underground tests.

The actual radioecological situation is characterised by a residual contamination on the polygon and to a lesser degree in the next vicinity around it, where people are still living.

THE PESK PROGRAMME

French physicians were invited in Kazakhstan to participate to a meeting about the nuclear tests for the first time in 1991. The meeting was held in Semipalatinsk. Several colleagues, physicians and biologists, asked us to establish a research programme about the possible consequences on environment and health of the 40 years Semi-palatinsk nuclear polygon activity The "Kazakhstanese environment and health programme" was born. (PESK in French).

During two years our objectives were to define precisely the project and to convince French and Kazakhstanese authorities. We were supported mainly by the French foreign office, and the Rennes 1 (F), Almaty and Karaganda (KZ) Universities. Several meetings in France and Kazakhstan were organised. we decided in 1993 to focus the research programme on the following subjects:

1) to gather all the available dosimetric data, on and around the polygon;
2) to establish a cartography of the polygon vegetation;
3) to study the potential mechanisms of adaptation of radioactive low dose irradiation stress on selected species of plants and on animals (blood and testis);
4) to assess the possible long term effects of external low dose irradiation on the female fertility and on children;
5) to develop the links between French and Kazakhstan researchers through the training in biology of Kazakshtan colleagues;

The main difficulties to implement the PESK programme were:

1) the subject by itself: indeed the assessment of the possible effects of nuclear tests on environment and health is a matter of controversies in western countries;
2) the objectives of the programme: the effects of low dose irradiation consequences is in a same way a matter of controversies in nuclearised for a civil purpose western countries;
3) the socio-economical problems in Kazakhstan: the research activity doesn't seem to be the actual priority;
4) the difficulties to find a substantial international support: in matter of radioactive contamination, this international support has been mainly devoted to Tchernobyl. Therefore the situation in Semipalatinsk has not been considered as a priority until a recent period.

Despite all these difficulties, the PESK programme, initially supported by the French Embassy in Kazakhstan, effectively started in 1994.

At this time, two studies are on going

- the first one is sponsored by the EC through an INTAS programme: it involves 8 teams including the Rennes 1 University (F), the National French Institute for Medical Research (INSERM), the London School of Hygiene and Tropical Medicine (UK), the National Environmental Research Council (UK), the National Nuclear Centre of Kazakhstan, two teams of the National Academy of Sciences of Kazakhstan, the Research Institute of Epidemiology of Almaty.
- the second one is sponsored by the Rennes 1 University through a grant of EDF (French Electrical Company).

A third grant coming from the French Ministry of the Research allowed the training in France of a Kazakhstan researcher in radiobiology.

THE PESK PROGRAMME: STATE OF THE ART

We are now at the end of the four years duration of the programme:

1) **We have gathered a lot of dosimetric information in several areas**
   i) *the characteristics of nuclear tests:*
   This work has been done in collaboration with the Institute for Protection and Nuclear Safety, Fontenay aux Roses, France). The information comes from the official publications of the Minatom in Moscow. Some data on atmospheric, ground, underground and excavation tests are presented.
   ii) *the spatial distribution of external doses* for the most contaminating tests.
   iii) for some settlements we know *the dynamics of irradiation of people after the tests*
   iv) *the level of external residual actual contamination on and around the polygon*
   v) *the level of contamination of biological samples* on and around the polygon: ground, plants, animals (wild and domestic species)
   vi) *the past and actual level of human beings contamination* mainly by Cs 137 and Sr 90.

All these informations can be implemented on a database to access the doses reconstructions in environment.

2. The cartography of the Vegetation

During the last two years we have collected the data in two areas:

- three small scale maps are now available covering the vegetation of the three most contaminated areas: the experimental field, the Balapan lake and the mounts Degelen
- we have got three satellites scenes covering the same spaces. The scenes are actually informatically studied in Rennes I University (F).

We plan to superimpose the ground observations, the satellite digitalised informations and the dosimetric available maps. The aim of this part of the PESK programme is to try to find some vegetal biomarkers of external irradiation and/or of radionuclides contamination. The work is still in progress.

3. The Mechanisms of Adaptation of Plants and Animals to Low Doses and Low Dose Rate Irradiation

- For the studies of the consequences of the radioactive pollution on the metabolisms and reproduction of plants, the chosen model was Stipa sareptana which is usually grazed by cattle. Mature caryopsis of Stipa were collected in 1993 and 1995 and 1996, in six different sites of the polygon and its nearly contaminated area along a radioactive gradient, measured at the ground level, ranging from 0.23 to 25 micro Sv/h. The cytogenetic, biochemical and biometric parameters were accessed.

Experiences conducted during 1994 on material collected in 1993 lead to the conclusion that chronicle radionuclides contamination at low doses in polygon region (5,4 micro Sv/h) does not significantly modify the detoxification metabolism. Curiously we observed a stimulation in growth and development but the reproductive function is perturbated with a numerous abnormal seeds in the collected material. These results contrast with those obtained on the same specy collected in Temirtow region (Karaganda oblast) where there is an important atmospheric and heavy metals pollution: on this material we observe (i) genetic modifications leading to quantitative and qualitative variations at the detoxifying enzyme level and (ii) growth and developmental perturbations (low biomass and fertility).

From experiences conducted in 1995 on material collected in 1994 (wider range of dose-rates, 0,23–25 micro Sv/h), we have drawn the following conclusions:

- — no correlation between exposition and micromorphological changes (scanning electron microscopy).
- — increase of caryopsis size at low dose rate (LDR) and decrease at high dose rate (HDR).
- — stimulating effect of LDR on growth, inverse effect for HDR.
- — inverse correlation between exposition and germination/ survival ability.
- — no influence on DNA content
- — the isoenzyme expression is highly influenced by irradiation mainly for phospho glucosisomerase, esterases (tendency to monomorphism).
- — some detoxification enzymes are over expressed in irradiated populations (superoxide dismutase).

- The observations on animals started one year ago: two kinds of experiences were undertaken, one on wild species of small mammals living on the contaminated areas and one on domestic species (sheep), grazing on the same areas. The results are compared to those taken from a control group.

Wild species: animals of two different species have been captured. The radionuclides have been measured in soil and in animals. External irradiation dose rates have been measured. The following parameters have been accessed: weight of animals, sex ratio, abnormalities of spermatozoid heads, histological studies of testis. The preliminary results show the presence of Cs and Sr in the body of these animals in small quantities; the other parameters are under investigation. Some laboratory investigations done in France seem to show some reversible abnormalities in lymphocytes DNA content.

Domestic species: the same parameters have been studied on sheep grazing on contaminated areas and usually eaten by the local population. The work is one progress.

4. The Effects of Low Doses Irradiation on Female Fertility and Children

This epidemiological protocol started one year ago and has two objectives: malformations and infecondity/sterility potentially related to nuclear tests. The statistical and demographic data available in Kazakhstan have been accessed. A questionnaire for congenital malformations has been elaborated. This questionnaire has been pre-tested in Almaty and translated in Russian: several hundred questionnaires were distributed in three selected areas including a control group.

At this time two important results must be mentioned:

— a functional epidemiological network involving Kazakhstan and European countries has been implemented.

— we receive regularly the questionnaires distributed in the areas of interest and gather by our Kazakhstanese colleagues. A first rough analysis made in Kazakhstan would show an increase of malformations in the most exposed areas. We must gather all the questionnaires and especially those distributed in the control areas to validate these very preliminary results.

5. Links and Exchanges between Kazakhstanese an occidental Researchers

This is a very important aspect of the PESK programme; our experience shows that this scientific programme gives a marvellous opportunity to have very fruitful exchanges with our Kazakhstanese colleagues;

In conclusion, we can say that the first period of implementation of the PESK programme was very difficult. We can notice some important improvements in the communication facilities. Therefore it seems easier now to propose some new competitive projects. Despite all these difficulties, the initial scientific objectives have been assessed. This data will be published in a short time. In particular they concern the adaptative mechanisms of plants to chronic radioactive stress. Similar experiments are done in France (Rennes 1 University) in laboratory. We hope to draw some conclusions on human beings (fertility) in less than one year. These on field observations and laboratory experiments should contribute to bring new information on the effects of low doses irradiation on environment and health. Moreover the descriptive cartographic part of the PESK programme should contribute to given new recommendations about the rehabilitation of the contaminated areas of the Semipalatinsk nuclear test site.

We take the opportunity of this talk of claim the importance to have a global overview on the consequences of nuclear tests on environment. Physics and field dosimetry are of course of great importance and are the first step of such a study. But the final end point is biology, including the effects of irradiation on the whole ecosystem and especially human beings. This is our responsibility as physicians.

Address:

CHRISTIAN CHENAL
Rennel 1 University
Rennes
France

JACQUES MONGNET
Secretary, AMF PGN
77130 Monterean-Fault-Yionne
France

43

Accelerator Driven System for Energy Production and Radioactive Waste Disposal

GIUSEPPE GHERARDI

WHY A R&D PROGRAMME FOR ACCELERATOR DRIVEN SYSTEM ?

In the next decade the R&D will give a demonstration of feasibility of the ADS as a waste disposal system; then it will be possible to upgrade to a competitive power system.

1. transmutation is a very appealing system for waste disposal,
2. many technical problems have to be solved to make it convenient,
3. many technical solutions are of interest in other fields of application,
   i.e. the accelerator is a challenge that some Lab is facing for fundamental or applied research (medical, materials, biotechnology, etc.)
4. the research for ADS maintain a bridge between the present time (low cost fossil fuel) and the future (need of economic, abundant and sustainable energy sources, in alternative to fossil fuel)

WHY IS THIS OLD IDEA NOW STUDIED BY SO MANY LABS IN THE WORLD ?

- the results of the studies and experiments conducted in CERN by Prof. Carlo Rubbia;
- innovative technologies in the accelerator field;
- interest of the scientific community to the realisation of reliable high current accelerator;

- interest of industries to the development of the new technologies (materials, elector-chemical process...).

WHAT DOES TRANSMUTATION REALLY MEAN ?

Physics is well known if you have neutrons enough, with a suitable energy spectrum, you can transmute every long lived radioactive element in another element, which is stable or short lived. There is no R&D need on this argument. The problem is how to produce all those neutrons.

A fission reactor is not flexible enough to do the job; neutron flux, energy spectrum, core materials are strictly connected to the safety issues.

An ADS system has more flexibility, and the safety is not affected when traditional fuel is replaced by innovative fuel and by the waste to be burnt.

The reduction of the long lived radioactive inventory is limited by technical and economic considerations: today it seems possible an abatement of 100 times.

This doesn't mean that a repository is not necessary, but it can be realised with less demanding requirements.

WHERE ARE WE NOW ?

A consensus on the general features of a ADS based transmutation system has been reached in Europe.

The accelerator required performance are not far from those of the existing facilities (6 mA vs. 1 mA at 1 GeV), except reliability.

The target is the same kind that the Russian laboratories are building for LA, with the ISTC funding.

The design of the plant requires some year of engineering work and supports testing.

A demonstrator can be realised in the next ten years, with the existing technologies, with a supporting R&D that has been defined.

The European labs are presenting proposals for the ADS R&D in the 5° framework programme of CE.

Complementary to the national programmes (France, Germany, Italy, Spain: 40–50 researchers for each country; strong interest of Belgium, Holland, Sweden and UK).

An international WG headed by Prof. Rubbia, with three members for each country (now France, Italy and Spain, but they will be more next autumn) has been established last month by the Ministries of research; members represent R/D organisations and industries.

The task is to agree in 1998 on a detailed planning for the realisation of the demonstrator.

WHAT IS A DEMONSTRATOR ?

A facility to experiment the coupling of the accelerator with the subcritical system.

Materials and control systems as for the industrial plant, power is about 100 MWth, no electricity produced.

A conventional fuel (SPX like) is foreseen for the first core, the time for qualification of an innovative fuel is longer than ten years.

The fuel treatment technology in order to separate the elements to be transmuted is studied in a parallel R&D line.

PROLIFERATION CONCERNS

In no phase there is separation of pure fissile material, more attractive for proliferation purposes than the irradiated fuel coming from the NPP.

At the end the fissile materials will be completely burnt (except U233, mixed with enough U238 to be useless for the bomb).

The burner can be built near a power plant park, or near a treatment plant, or near the final repository, so there is no additional transportation of irradiated fuel.

The only proliferation concern, to be compared with the reduction of long term fissile material disposition, is the potential of this technology to be converted in a proliferant technology.

Address:

GIUSEPPE GHERARDI
ENEA
40120 Bologna
Italy

44

Rethinking Geologic Disposal of High-Level Nuclear Waste and Spent Nuclear Fuel

WARNER D. NORTH

The disposal of spent nuclear fuel and high-level nuclear waste has the deserved reputation as one of the most intractable policy problems facing the United States. This policy problem also confronts the other nations using nuclear reactors for electric power generation. This paper advocates reformulation of this policy problem. "Disposition" should replace "disposal". What to do about spent nuclear fuel and high-level waste should be regarded as a choice among ongoing management strategies, rather than a choice of a geological site, a repository design, and a schedule for acceptance of waste and repository closure. It is uncertain whether the proposed US repository at Yucca Mountain can overcome substantial public opposition and the need to satisfy US regulatory criteria. A rethinking of the US high-level waste programme may be needed, with more focus on the assurance of safety during the next thousand years and more flexibility in considering spent nuclear fuel as a potentially valuable energy resource.

1. INTRODUCTION: THE CHALLENGE OF NUCLEAR WASTE

The legacy of both civilian and military nuclear technology is nuclear waste, a mixture of radionuclides created by nuclear fission and neutron capture. Reactors in the United States are projected by 2020 to produce a quantity of nuclear waste of the order of 80,000 metric tons of heavy metal (uranium, plutonium, or thorium equivalent) of spent nuclear fuel and other nuclear waste[1]. The comparable worldwide estimate is about 450,000 metric tons[2]. Most of the radioactivity is found in high-level waste

(HLW), although the volumes of intermediate and low-level waste are much larger.

Facilities for low and intermediate level nuclear waste are in operation in many countries, but forty years after the National Academy of Sciences recommended geological disposal[3], neither the US nor any other country has succeeded in developing a repository for HLW. Most HLW in the United States is stored where it was produced, at civilian power generation reactors and in major sites of the Department of Energy has failed to meet its obligation under the 1982 Nuclear Waste Policy Act to begin accepting civilian spent nuclear fuel beginning in 1998. The proposed high-level waste repository at Yucca Mountain, Nevada is far behind its original schedule for site characterisation, licensing, and construction, with initial operation postponed until at least 2010. The State of Nevada and many of its citizens have vigorously opposed the development of HLW repository at Yucca Mountain. Eventual operation of this proposed repository is by no means assured[4, 5, 6, 7]. The US regulatory criteria for Yucca Mountain are still being developed, and it is uncertain whether the proposed repository can meet these criteria[8].

In the mid-1970's this author presented a paper at a conference convened by the U.S. Environmental Protection Agency, offering the opinion that sitting and licensing repositories for HLW should not pose insurmountable difficulties[9]. Like others with a background in physics, the author failed to anticipate that managing nuclear waste would become such a difficult issue. This article and several other recent publications[10, 11] reflect the author's experience from five years as a member of the Nuclear Waste Technical Review Board (1989-94) and subsequent service on the National Research Council Board on Radioactive Waste Management (1995-1999). The opinions expressed are the author's alone and not reflective of any organisation with which he is, or has been, associated.

2. A DIFFERENCE IN PERCEPTIONS

It is tempting to assume as self-evident that national policy should be based on a good understanding of science. It is important for scientists to understand that, in a democracy, public policy is determined through electoral and legislative processes that reflect public sentiment, rather than science. If national policy is to reflect a sound understanding of science, the public must either understand the science or trust those who do.

While radiation and radioactive materials merit great care and respect, there is nothing unusually mysterious to a trained scientist about nuclear energy, ionizing radiation, or radioactive isotopes. Radioactivity seems much like high voltage electricity, toxic materials, exothermic chemical reactions, and disease causing microorganisms, all of which have been studied by scientists in laboratories with the knowledge that these phenomena can be dangerous, and mistakes in dealing with them can be fatal. Members of the public generally accept that scientific research can be dangerous, and that technology emerging from research can be used for good or for evil.

Nuclear waste is a case where there is a large divergence between the views of trained scientists, who view radioactive isotopes as dangerous but well-understood and readily managed, and many members of the public who view nuclear waste as a long-lasting, malevolent legacy from nuclear weapons and reactors. The problem of managing nuclear waste is not a problem that is likely to be solved by a "technical fix" emerging from further research. The essence of the nuclear waste issue is a difference in perceptions. This difference in perceptions has led to the erosion of trust in scientists and in the institutions that have been given responsibility for dealing with nuclear waste[12]. The problem of managing nuclear waste has become a serious challenge to the nations of the world using nuclear technology. The failure to date to establish repositories for HLW is generally regarded as a major impediment to further deployment of nuclear power generation in the US and in other nations.

Progress on managing nuclear waste involves more than deployment of technical systems for storing the waste, transporting the waste from reactors to repository sites, and emplacement of the waste so as to isolate the radionuclides. To achieve a public perception that these systems will provide adequate protection will require more effective communication between scientists and the public and development of trustworthy institutions for managing nuclear waste.

3. PERFORMANCE ASSESSMENT

Physical scientists and engineers examine safety of materials by asking how much might be released over a period of time and how the materials (or their transformation products) might reach receptors that could be damaged. Such analysis for chemicals is often called fate and transport modelling. In the nuclear waste community it is called performance assessment[13]. The objective of such analysis is to examine how well dangerous materials can be isolated so that they will not cause damage.

Performance assessment models are used to calculate the concentrations of radionuclides that could result from releases from a nuclear waste repository. These models address time periods of thousands to millions of years. The decay processes for some radionuclides involve these long time scales, which are comparable to the time scales for many geological processes.

Performance assessment is intended to summarise information about migration of radionuclides from nuclear waste repositories due to natural phenomena such as water intrusion, material dissolution and transport in ground water, and migration and subsequent release into the biosphere. Based on past geological records and our understanding of geological processes, we can assess the probabilities of geological events, such as earthquakes, volcanic eruptions, or renewed glaciation, that could possibly compromise the integrity of a repository and lead to the release of radionuclides.

Performance assessment has a much harder time dealing with inadvertent or deliberate human intrusion into waste repositories as a mechanism for nuclear waste release. Such analysis is highly dependent on assumptions about future human behaviour. The National Research Council report commissioned by Congress to advise on appropriate standard for high-level waste site at Yucca Mountain cautioned that risks from future human intrusion are not susceptible to analysis in the same manner as risks from geological processes[14]. Risk analysis of nuclear waste sites and of chemical sites has tended to focus on transport processes, degradation of engineered barriers, and naturally occurring changes in the site, rather than scenarios for human intrusion.

The Waste Isolation Pilot Plant (WIPP) for transuranic waste, located near Carlsbad, New Mexico, has now been provisionally approved by EPA for an operating licence to accept small quantities of waste. The bedded salt at WIPP provides a geological setting that is extremely stable against intrusion by groundwater, so that there is no significant risk of release of radioactivity to the biosphere unless WIPP is disturbed by human activity. The human intrusion event assumed in the EPA regulation is the penetration of the waste emplacement area by drilling equipment as the result of future exploration for oil and gas. (Extensive oil and gas deposits are located in this area of southern New Mexico.) The probability of release of radioactivity under such a "human intrusion' scenario has been judged to be acceptably low, so as to meet the regulatory criteria. However, some critics have recommended further characterisation and containerisation of the waste drums to reduce the probability of release still further. A recent report by the National Research Council concluded, "It is disconcerting that a repository such as WIPP, which appears to be an excellent choice based on geological considerations, could be judged 'non-compliant' solely on the basis of a criterion (the assumed human intrusion scenario) that has a poor scientific basis."[15], (p. 81).

A simple bounding calculation was used in the National Research Council's review of issues on the proposed Ward Valley low-level radioactive waste site in California[16]. This site, in the Mojave desert, lies about ten miles from the Colorado River, and some geologists asked about the possibility that radionuclides in the waste could be transported through ground water to the river. The National Research Council committee suggested that the consequences of this very unlikely scenario be examined. Suppose the repository did release the small amount (10 curies, Ci) of plutonium estimated to be in the repository, evenly into the Colorado River over a period of 30 years.

The committee concludes that the addition of 0.33 Ci per year would be insignificant compared to the natural alpha load of 44 Ci per year carried by the Colorado River. The addition of 0.07 pCi/1 (picocurie per litre) from the waste plutonium to the existing 4.4 pCi/1 would result in a total load to 4.47 pCi/1, well below the health-based standard of 15 pCi/1 for gross alpha-emitting radionuclides[16], (p.7).

Compared to the radioactivity from uranium naturally present in the Colorado River, the additional radioactivity from the plutonium would be small and well below the health standard.

Despite this performance assessment argument that even

if a release of plutonium to the Colorado River were to occur, the amount would be a small fraction of natural background, the Ward Valley site remains mired in controversy. Opponents of nuclear waste sites assert that any releases of radioactivity from a repository are unacceptable. Many scientists find it surprising that this argument receives significant public support. Investigation of why this public support occurs lead to the topic of risk perception.

4. RISK PERCEPTION

Nuclear energy, radioactive materials, and nuclear waste have received a great deal of attention in risk perception studies, because research has clearly shown that much of the public is quite fearful about these 'things nuclear." It is useful to inquire how these attitudes were formed. One useful background reference is a book by a historian of science[16], which explores the portrayal in the press and in popular culture of nuclear energy and radioactive materials, beginning in the late 19th century when radioactivity was first discovered. Although lacking in data measuring public attitude that social scientists now collect, the book provides persuasive arguments that, well before the first controlled releases of nuclear energy, strong public perceptions had already been formed:

By the 1930s most people vaguely associated radioactivity with uncanny rays that brought hideous death or miraculous new life with mad scientists and their ambiguous monsters with cosmic secrets of death and life; with a future Golden Age, perhaps reached only through an apocalypse, and with weapons great enough to destroy the world, except perhaps for a few survivors[17] (p.421).

Scientifically inaccurate portrayals of radioactivity in US popular culture continue into the present time. Spiderman is a comic strip superhero whose abnormal strength derives from being bitten by a radioactive spider. Godzilla is a giant radioactive lizard, who emerged from the depths of the Pacific following the 1954 nuclear weapons tests, to star in a highly successful series of Japanese movies. (Godzilla reappeared in a new Hollywood "summer blockbuster" movie in 1998.) Many other examples are discussed in[17].

In developing a perspective on nuclear waste, we must consider the difference in perceptions between the scientists and members of the public who lack background about radioactive waste. Engineers and physical scientists tend to regard radioactive waste as a mixture of radionuclides whose properties and behaviour are very well understood. Naturally occurring radionuclides and cosmic rays from space create a natural background of ionising radiation that is orders of magnitude higher than permissible levels of human exposure from nuclear materials. Biologists and health scientists regard the biological of facts of the ionising radiation released by nuclear waste as reasonably well understood, in so far as we have knowledge that ionising radiation within cells can lead to cancer and to heritable alterations of genetic material. Health physicists have excellent instruments for detecting ionising radiation from radionuclides, so that exposure can be determined even when the radionuclides are present only in very low concentrations.

Many members of the public fear radioactive waste as a poorly understood threat imposed on them by scientists who could not foresee the consequences of developing nuclear technology. From the social science perspective, many members of the public are not convinced that nuclear waste is a manageable problem, The available data indicate that nuclear waste evokes fear and concern from these people. The conclusion that this author draws from experiences over the past decade is that the policy issues posed by nuclear waste are not going to be solved by advances in knowledge form engineering or physics, or form more detailed performance assessment models. We need better understanding of the public's concerns, and we need better ways to assure the public that nuclear waste can be managed to assure safety.

5. POSSIBLE CONCLUSIONS ON YUCCA MOUNTAIN EMERGING FROM ANALYSIS

Since the 1987 Nuclear Waste Policy Amendments Act the US programme for high-level nuclear waste has focused on the characterization of one site, Yucca Mountain, Nevada, as a potential repository. Much has been learned about this site in the past decade[7,8]. While, to date, information has not emerged that provides a clear rationale that the site should be abandoned as suitable, the unsuitability of the Yucca Mountain site remains a subject of vigorous discussion involving many complex technical issues. The Department of Energy now has underway a Viability Assessment, scheduled for completion in 1998, which should provide a good technical basis for a national decision on whether to continue the effort to develop a repository at Yucca Mountain. This Viability Assessment does not address many policy and risk perception issues that may be crucial for national decision making on HLW, but only performance assessment at the Yucca Mountain site. Whether the performance assessment will show that a proposed repository design will satisfy the regulatory criteria is as yet uncertain[8].

If Yucca Mountain is judged a viable site for the first US HLW repository, many technical issues of repository design and timing still remain to be resolved. Among the design issues still to be resolved are the spacing of the spent nuclear fuel, which determines the increase in temperature that will occur in the rock adjacent to the waste as the result of radioactive decay heat. A temperature above the boiling point of water may reduce the effect of water in corroding the waste container. Over thousands of years as the heat generated from radioactive decay diminishes, liquid water will again come in contact with the waste containers. Recent findings from tunneling through Yucca Mountain indicate that there are fast paths by which water flows from the surface to the layer of rock where waste would be emplaced. So although the repository is above the water table in unsaturated rock, this rock is not naturally dry, and aqueous corrosion processes may occur. The performance of the engineered barriers in retarding the migration of radionuclides will be an important determinant of the extent and timing of radionuclide releases. Appropriately designed engineered barriers could last at least ten thousand years at present levels of infiltration, so the combination of engineered barriers and geological barriers would fail more rapidly only under greatly altered climatic conditions. Under conditions of increased precipitation, tens to hundreds of thousand years in the future, failure of the containers might lead to transport of radionuclides from the waste through the unsaturated zone to groundwater. Long-lived radionuclides such as neptunium-237 in groundwater used for drinking or for food crops could lead to elevated human exposure to radioactivity[18].

An underground repository can provide assurance that, even in the event that civilisation should collapse, radionuclides will remain contained for a period of the order of at least ten thousand years. Any subsequent release should occur slowly, at a rate determined by corrosion processes for the engineered barriers and migration of the radionuclides through hundreds of metres of unsaturated zone rock to reach the water table. Inadvertent or deliberate human intrusion could cause a larger and faster release, but placing the waste at least 200 metres below the earth's surface makes human intrusion much more difficult than for waste in surface storage. We should consider it a success rather than a potential failure if we can predict with confidence what a repository will keep radioactivity from reaching the biosphere for at least ten thousand years, in the absence of human intrusion. Performance assessment may show that exposures to radioactivity exceeding regulatory standards are predicted only for climate change scenarios such as new ice ages. Such climate change scenarios will imply much more extreme consequences to human civilisation than the radioactivity releases from HLW repositories.

In this author's opinion, enough has been learned from research in the US and other countries characterising sites from high-level waste repositories, to motivate a paradigm shift[11]. The geological characteristics of a repository site are not going to assure containment of the nuclear waste, especially when human intrusion is considered. For most sites under investigation for national HLW repositories, the geology is complex, isolation is not assured by geology alone, and engineered barriers are judged necessary to improve HLW containment. Human intrusion is a distinct possibility, especially if materials in the HLW such as plutonium provide motivation for deliberate intrusion.

The US should make a shift from its present policy of seeking a geological site for HLW where we can "set it and forget it", and instead implement a policy of establishing facilities for ongoing management of HLW. Such facilities might be placed underground and might be called repositories. Even under the current policy, repository design involves establishing engineered barriers as well as geological barriers. Surveillance systems, designed to last many centuries, perhaps millennia, might provide ongoing verification for concerned members of the public that radionuclides are not being released and will therefore not cause adverse impacts. From a public perception standpoint it may be far better to maintain ongoing research and inspection so that the public can be assured that the waste is behaving as predicted and not migrating out of containment via some unexpected process. Since the planned repository at Yucca Mountain lies about 350 meters above the water table, water reaching the HLW can be readily drained away rather than remaining in prolonged contact with HLW containers. Placement of the HLW waste in the unsaturated zone makes it possible, at low cost, to postpone the closure of this repository for at least 100 years and perhaps indefinitely. The decision to close the repository can be left to future generations, who can evaluate the trade-off between further opportunity for monitoring, maintenance, and upgrading of engineered barriers, versus sealing the repository so that further human access would require penetrating the order of 200 meters of volcanic tuff rock.

6. THE ALTERNATIVE TO A GEOLOGICAL REPOSITORY: ONGOING SURFACE STORAGE

An alternative to a geological repository is ongoing storage of HLW in containers on the earth's surface, rather than underground. Surface storage is easier and cheaper. The large cost of constructing an underground repository is avoided or deferred. If the HLW is left at reactor sites the cost of transportation is also avoided or deferred. Surface storage containers will need ongoing programme of monitoring, maintenance, and replacement, but this alternative to geologic repositories may be considered inexpensive if costs more than a century in the future are discounted at the interest rates typically used for government investments.

Placing the waste underground can provide assurance that HLW will be in a carefully engineered protective environment, such that weathering and corrosion processes and human activities will not lead to breaches in waste isolation. On the surface, accidents, extreme weather events, and a range of inadvertent and malevolent human activities may result in higher probabilities that isolation of the HLW will be breached and radioactivity released without the benefit of a geological barrier to retard migration into the biosphere.

Both houses of the US congress have passed versions of a bill that would require that, after a short time for additional study, spent nuclear reactor fuel and other high level waste must be shipped to the area of the nuclear test site in Nevada adjacent to Yucca Mountain. (The margins of passage were not sufficient to overcome a threatened veto by President Clinton, and the legislation now appears dead for the current session of Congress. This legislation is likely to be reintroduced after the US Congressional election in November of 1998.) The interim storage provisions of this proposed legislation imply the need to transport a large quantity of spent reactor fuel to Nevada. The nuclear electric utilities are highly supportive of this effort to transfer management responsibility for spent reactor fuel from the reactor owners to the federal government, as mandated under US law.

The movement of a large quantity of nuclear waste into Nevada will be a controversial undertaking. The technology for transporting spent nuclear fuel and high-level waste is very well developed, and the large transportation containers have been carefully designed and tested to withstand extreme accidents without releasing radionuclides. But public fears and concerns may be substantial.

Consider the following scenario: A truck carrying nuclear waste is involved in a collision with another vehicle. The heavy nuclear waste container comes off the truck bed and rolls down a hill, coming to rest in a river. This scene is photographed and shown on national television news, with the observation that if the radioactive contents of the container were to get into the river, the dose of radioactivity could be sufficient to kill anyone drinking water from the river, or even living alongside this river. Even if there is no release of radioactivity (which is overwhelmingly likely because of the design of the container) many members of the public may assume that radioactive contamination of the river did occur. They may then refuse to drink the water , to eat food that might have been grown with water from this river, to visit the area, or to buy goods that could have been produced in this area. Even with no adverse impacts from actual radionuclide release, the economic loss resulting from the stigma of perceived radioactivity could be very large[19, 20].

7. INTERNATIONALISING HLW: ALTERNATIVE TECHNOLOGIES

Many nations are now using nuclear reactors for commercial electric power generation. Each of these nations has the responsibility to deal with its HLW. At present cooperation among the national HLW programmes is limited to collaborative research relating to geologic disposal. While a few experts in the field have called for international cooperation in developing multinational repositories[21], essentially no progress has been made to date in developing international organisations that could assume responsibility for developing repositories that would accept HLW from multiple countries.

Disposal of HLW in the deep seabed is an alternative that has been proposed by US oceanographic experts[22]. Many areas of the seabed under deep ocean waters are far more stable geologically than in Yucca Mountain. The thick clay-rich mud sediments under the seabed should provide an excellent barrier to migration of radionuclides into the ocean waters. Even if some migration should occur, adverse effects should be minimal, because the deep seabed has little if any life, and any radioactivity released above would be diluted into extremely large volumes of sea water. Human intrusion to retrieve waste placed under several miles of ocean would be possible only at great effort and expense, and such efforts could be readily observed with satellite imagery. Despite its potentially attractive technical characteristics, most experts have dismissed seabed disposal as infeasible politically. Existing international law prohibits disposal of radioactive waste at sea, and it could be extremely difficult to obtain an international consensus that seabed disposal should be regarded as a serious alternative to geological repositories on land. These obstacles are viewed as so formidable that no significant research on seabed disposal is currently being carried out.

Another alternative to geological repositories is separation and transmutation (S&T) of radionuclides, especially the actinides produced by neutron capture and long-lived fission products such as technetium-99 and iodine-129. While "burning" these radionuclides in specially designed reactors is possible, this approach would involve the construction of many new reactors, require more advanced reprocessing of nuclear fuel and take several centuries to accomplish. The National Research Council recently reviewed these technologies and concluded the following: "The committee found no evidence that applications of advanced S&T have sufficient benefit for the US HLW programme to delay the development of the first permanent repository for commercial spent fuel.[23]

8. A PERSPECTIVE FROM ONE THOUSAND YEARS HENCE

Public policy discussions of HLW have tended to focus unduly on the length of time that nuclear waste remains radioactive[11]. Performance assessment for Yucca Mountain indicates that an isotope of neptunium (Np^{237}) may dominate the releases of radioactivity in a time period many tens of thousands of years into the future. Is this release likely to be a threat to humans? Assuming people inhabit the area surrounding Yucca Mountain, the release could be a threat to people who drink the local groundwater or use it to grow their food. But release of neptunium and other long-lived radionuclides after tens of thousands of years assumes that spent nuclear fuel will remain in the repository for this period. Assuming that human civilisation endures and progresses, the assumption that radioactive releases at that time will damage human health seems questionable. Whatever humans are living in southern Nevada thousands of years from now are likely to have access to adequate means to detect radioactive contamination in their groundwater and either to remove this contamination or to refrain from using the groundwater.

Let us consider instead that in the less distant future, within one thousand years, there is likely to be a strong economic incentive to recover the spent nuclear fuel and recycle it to reclaim the fissionable materials. An alternative of placing HLW underground but not sealing the repository may provide protection from both human activities and natural processes that could compromise waste isolation. This alternative also would permit our descendants to make the choice to use the fertile and fissionable isotopes of uranium and plutonium in what we are calling HLW to meet their needs for nuclear energy.

Most of world energy needs today are now being met by fossil fuel: natural gas, oil, and coal. Nuclear reactors for electricity generation are barely competitive with generation from coal and hydroelectric generation in much of the world. Nations such as Japan and France that generate much of their electricity from nuclear have chosen to do so because these countries lack domestic coal and other fossil fuel, and so must depend on imported fuel supplies. In the United States, Canada, and portions of Europe, natural gas has become available at prices that allow this fuel to be competitive for base-load power generation, especially in situations where high efficiencies can be obtained by producing electricity plus process heat or steam. Reactors built 20 years ago in the US are now regarded as "stranded" investments that are no longer competitive with new generation that uses natural gas as fuel.

Let us consider the situation as it might be viewed from the transition to the next millennium, 1000 years into the future. Long before ten centuries have past, geologists now tell us that the low-cost oil and natural gas resources of our planet will be depleted. This depletion may occur at an accelerating rate as the nations of the former soviet bloc and the "Third World" industrialise and their citizens gain access to energy consuming technologies similar to those enjoyed in North America and Europe during much of the 20th century. Such energy consumption includes space heating, air conditioning, cooking, and personal transportation, and high energy content in many manufactured goods. Coal resources might possibly be sufficient to meet world energy demand for another five hundred years, but combustion of this quantity of carbon will almost surely lead to unacceptable changes in the earth's climate. It seems highly probable that beginning in the 21st century, strong economic and environmental forces will increase the cost and decrease the availability of fossil fuels as an energy source. Some combination of the use of solar and renewable energy (geothermal, biomass hydroelectric) and nuclear energy (fusion or fission) must replace fossil sources in providing for the world's energy needs. It is also likely that the rich, low-cost deposits of uranium ore will be depleted during this time period.

While energy from the sun may contribute substantially, the low energy density for either biomass or solar photovoltaics implies that very large land areas must be devoted to these technologies for large-scale energy production. Nuclear energy, either from fission or from fusion, has extremely high energy density: very large amounts of energy can be released from the fission of the fissile isotopes of uranium or plutonium, or from the fusion of light nuclei such as deuterium, tritium, or lithium. So on a planet where land is valuable for other purposes, there will be an economic incentive for these technologies that allow large amounts of energy to be generated from a small area of land, without involving large quantities of fuel that must be transported long distances. Fusion could produce extensive amounts of nuclear waste through neutron activation of the materials used in a fusion reactor. The inherent engineering limitations implied by obtaining energy from highly energetic neutrons obtained in most fusion reactions imply that fusion reactors may have grave difficulties competing with conventional light-water and breeder reactors. It is even possible that the commercialisation of fusion energy will be through a hybrid system using fusion-produced neutrons to transform U^{238} into fissionable isotopes of plutonium, which are then burned in light-water reactors.

The reasoning above suggests that our descendants may view a repository such as that planned for Yucca Mountain as a rich lode of energy materials to be recovered and used when these materials represent that lowest-cost sources of available energy. It may be very difficult for people at the end of the 30th century to understand why we at the end of the 20th century did not foresee that the era of cheap fossil fuels was rapidly coming to an end that within a few hundred years, recovery of fissionable material from spent nuclear fuel would be far more economical than either pure fusion, once-though use of natural or slightly enriched uranium, or the competing energy technologies.

If the likely long-term use of spent nuclear fuel is for energy, then we should not focus performance assessment efforts on the releases of neptunium and other actinides from the repository after tens of thousands of years. We should be assuring safe storage over the next thousand years with some attention to assuring that if civilisation does undergo a precipitous decline, the waste is sequestered well enough that it should provide no significant threat to humanity during the time needed for human civilisation to rebuild itself. In this context the threat that neptunium may cause contamination of deep ground water many tens of thousands of years in the future may not appear significant.

Will future generations continue to be as concerned about diversion of plutonium from spent fuel as a means for nations or subnational groups to acquire a weapon of mass destruction? We might consider scenarios such as a civilisation recovering from nuclear war that might seek fissionable material for more nuclear weapons so as to resume hostilities. Would a nation under such circumstances be motivated to dig up its spent nuclear fuel to obtain plutonium for weapons manufacture? At the end of the twentieth century such motivation might seem highly unlikely. A variety of such scenarios should be carefully examined. In considering policies for disposition of HLW we should extrapolate from current proliferation concerns to questions of when, under what conditions, and at what level of expense should future civilizations be able to retrieve spent nuclear fuel. We will manage the potential for human intrusion by the choices we make for managing the spent nuclear fuel portion of HLW. This aspect should be at least as important as analysis of geological scenarios for release of radionuclides.

9. RETHINKING THE HLW DISPOSAL PARADIGM

Given the state of technology and the political realities of the late 20th century, this author believes that a geological repository remains the best alternative for the disposition of the US inventory of HLW. However, the US programme to develop a repository at Yucca Mountain needs a significant shift toward a revised paradigm of ongoing safe management to replace the current paradigm of disposal. Maintaining retrievability for at least 100 years has been recommended by the National Research Council[23]. The US Department of Energy has now taken steps toward adopting this viewpoint with a policy to defer repository closure for at least 100 years following initiation of waste emplacement, with the decision to close the repository left to future generations. This view is not yet incorporated into US national policy as established in legislation. Many members of the public still view the US national HLW policy as disposal (i.e., "set it and forget it"). These members of the public have serious concerns as to whether today's repository technology will be adequate to contain HLW for the time period (up to a million years) that the HLW will remain radioactive.

As the discussion on Yucca Mountain proceeds, it may be appropriate to place more emphasis on scenarios involving human activities over the next thousand years, rather than scenarios for migration of radionuclides into the biosphere tens of thousands to a million years in the future. Management of uranium, plutonium and other actinides in commercial spent fuel may be accomplished by continuing the current US policy of a once-through fuel cycle (i.e., no reprocessing) and placing spent fuel in repository storage for an indefinite period. Future generations can make the decision whether to retrieve this waste for its energy value, which could become much higher as fossil fuels and high-grade uranium ore deposits are depleted. The future availability of low-cost energy technologies will determine whether spent commercial nuclear fuel, which contains uranium and plutonium, becomes a valuable energy resource or remains a waste. The obligation of the current generation is that this spent fuel and other HLW be safely managed, and that we pass on to succeeding generations the opportunity to continued the safe management of these materials.

REFERENCES

1. John Ahearne, "Radioactive Waste: The Size of the Problem," *Physics Today*, 50(6), 24–31 (1997).
2. Charles McCombie, "Nuclear Waste Management World-wide," *Physics Today*, 50(6), 56–62 (1997).
3. National Research Council, *The Disposal of Radioactive Waste on Land*, National Academy Press, Washington, D.C., 1957.
4. Task Force on Radioactive Waste Management, Secretary of Energy Advisory Board, *Earning the Public Trust and Confidence: Requisites for Managing Radioactive Wastes,* final report, US Department of Energy (1993).
5. Douglas Easterling and Howard Kunreuther, *The Dilemma of Siting a High-Level Nuclear Waste Repository*, Kluwer Academic Publishers, Boston (1995).
6. J. Flynn, R. Kasperson, H. Kunreuther, and P. Slovic, *Environment* 39(3), 6(1997).
7. Chris G. Whipple, "Can Nuclear Waste Be Stored Safely at Yucca Mountain?" *Scientific American*, 274, 72–79, 1996.
8. Robert J. Budnitz, "Current Progress Toward a High-Level Radioactive-Waste Repository at Yucca Mountain, USA," this conference, 1998.
9. Stephen M. Barrager and D. Warner North, "Analysing the Decision on Radioactive Waste Management," paper presented at the EPA Workshop, "Issues Pertinent to the Development of Environmental Protection Criteria for Radioactive Wastes," 1977.
10. D. Warner North, "Unresolved Problems of Nuclear Waste: Motivation for a New Paradigm," *Physics Today*, 50(6), 48–55 (1997).

11. D. Warner North, "Nuclear waste management: shifting the paradigm," *Reliability Engineering and Systems Safety*, 59, 123–128 (1998).
12. Paul Slovic, "Perceived Risk, Trust, and Democracy," *Risk Analysis*, 13(6) 675–682, 1993.
13. William E. Kastenberg and Luca J. Gratton, "Hazards of Managing and Disposing of Nuclear Waste," *Physics Today*, 50(6), 41–47 (1997).
14. National Research Council, *Technical Basis for Yucca Mountain Standards*, Washington, D.C., National Academy Press, 1995.
15. National Research Council, *The Waste Isolation Pilot Plant: A Potential Solution for the Disposal of Transuranic Waste*, Washington, D.C.: National Academy Press, 1996.
16. National Research Council, *Ward Valley: An Examination of Seven Issues in Earth Sciences and Ecology*, National Academy Press, 1995.
17. Spencer R. Weart, *Nuclear Fear*, Cambridge, MA: Harvard University Press, 1988.
18. Robin McGuire et al, *Demonstration of a Risk-Based Approach to High-Level Waste Isolation*: *Phase 2*, report prepared for the Electric Power Research Institute, EPRI-TR-100384, May 1992.
19. John S. Petterson, "Perception vs. Reality of Radiological Impact: the Goiania Model," *Nuclear News*, 31(14), 84–91, 1988.
20. Joseph D. Rosen, "Much Ado About Alar," *Issues in Science and Technology*, pp. 85–90, (Fall) 1990.
21. Klaus Kuhn, "Technical and Political Aspects for Site Characterisation, Evaluation, and Acceptance," pp. 299–311, Vol. 1, *Proceedings of the International Conference, Safewaste '93: Safe Management and Disposal of Nuclear Waste,* Avignon, France, 1993.
22. Charles D. Hollister and Steven Nadis, "Burial of Radioactive Waste under the Seabed," *Scientific American*, 278(1), 60–65, 1998.
23. National Research Council, *Nuclear Wastes: Technologies for Separations and Transmutation*, Washington, D.C., National Academy Press, 1996.

Address:

D. WARNER NORTH
Vice President
Decision Focus Inc
Mountain View, CA
USA

45

On Utilisation of a Nuclear Warware in Russia: Ecological and Legal Aspects

YURY ROGOZHIN

INTRODUCTION

During the first years of the nuclear era both in America, and in USSR neither experts, nor the state figures discussed what would be necessary to do with nuclear installations after they would fulfil the resource. With the little knowledge that was known they could not foresee the impressive effects in the application of atomic energy. Only after one decade the problem has risen to utilise objects of nuclear industry, nuclear power unit and nuclear powered submarines (NPS) with the whole acuteness.

In the former Soviet Union the nuclear submarines have begun to be removed from battle structure of tis Navy since 1986. In 1988, the first special decision by the Central Committee of Party, Presidium of the Supreme Body and Ministerial Council of USSR devoted questions of an utilisation removed NPS was accepted. After that some documents signed by management already of Russian Federation, called were let out to regulate the decision of the appropriate problems. Unfortunately, they are executed only partially, proportionally is real (instead of on a paper) financing, allocated for these purpose.

Since 1993 in articles of the Russian budget appearing under the common name "the state defence order", the means to utilise NPS were provided. However, first, in the budge the financing for these purposes was provided in volume, making only half from necessary. Secondly, even these means have reached the Navy again only in half to a share.

The problem of financing of maintenance and utilisation of removed NPS in Russia remains unsolved so far. The situation continues to become aggravated, because to NPS, earlier removed from operation, newly removed NPS are added.

On some rather reliable data, in Russia now-a-days there are approximately a hundred and fifty nuclear submarines which have been removed from battle structure of the Navy. These served its resource NPS of the first generations repose basically at piers of ship docks and factories, coastal technical bases and naval bases of "otstoi".

Removed NPS stand at piers by float, many of them for several years, from which a part of the nuclear fuel is not unloaded. It is not only radiating but also nuclear failures.

As it was marked in the documents of Russian competent bodies, removed NPS are actually transformed into floating poorly protected and out of time under repair. Storehoused highly active spent nuclear fuel creating not only threat of their floating and huge subsequent material and financial expenses, but also constant radioecological loading on seaside territories.

Not a less serious danger than NPS, is represented by the significant amounts of nuclear materials demounted from a nuclear ammunition. In Russia the plutonium and the high-enriched uranium is stored at conditions, which do not fully satisfy the demands of safety. There is an unsolved question about further destiny of nuclear materials, demounted from a nuclear ammunition, not only in Russia.

1. UTILISATION OF ATOMIC SUBMARINES

There are three main causes complicating utilisation of atomic submarines in Russia. First of all, designers did not reflect on inevitability of utilisation, when atomic submarines had been projected. Secondly, the rates of nuclear submarines withdrawal from military use have appeared higher than it was possible to expect. Thirdly, the wide-scale expansion of atomic submarines utilisation is limited by low rates of the unloading of nuclear fuel from reactors of atomic submarines.

In spite of the fact that the above mentioned problems are represented as purely technical, for their decision it is necessary first of all to create legal and normative base. Russia has not yet neither that, nor other, though some motions recently were planned. The administration bill of the utilisation of all military objects recently was developed in entrails of Russian parliament and is planned for the firs reading to the nearest autumn. As one of the developers of this administration bill, I can testify, that it contains some articles specially devoted questions of utilisation of nuclear submarines.

The group of experts, led by Prof. Sarkisov, puts forward the initiative to develop a concept of nuclear submarines utilisation. This document should plan ways to decide the appropriate social, economic and technical problems.

2. UTILISATION OF WEAPON NUCLEAR MATERIALS

At the end of the last year the Russian Government has organised a commission for development concept on weapon nuclear materials. This document should plan ways to decide basic problems connected with the weapon nuclear materials, released from Russian nuclear ammunition, in particular:

- The status of released (from weapon) nuclear materials;
- Principles of utilisation of released nuclear materials;
- Features of the handling with released nuclear materials.

Unfortunately, the work above the concept till now is not completed. It is possible, that the law on utilisation of military objects, which, in particular, concerns questions connected to nuclear weapon materials, will be accepted before the concept.

3. LEGAL SUPPORT OF THE NUCLEAR WARWARE UTILISATION

Despite of the half-century experience of atomic energy use in military purposes, Russia till now has not the laws, regulating not only utilisation, but development and operation of military nuclear objects. Only recently attempts made to correct this situation, seeming now as abnormal, but quite explainable in conditions "of Cold War".

I already mentioned the bill, prepared by Duma, of utilisation of military objects. Although this bill has a very wide scope of implementation from tanks to airdromes, it is concerned also with the problems of utilisation of the atomic submarines and the demounted nuclear weapon materials. With acceptance of this law, the basis of legal regulation of general questions of warware utilisation will be incorporated in Russia.

Most likely it will appear, that one such law to decide all the specific questions of mentioned above on nuclear objects utilization. Then the development of additional bills can be required.

Already today it seems necessary to develop the following bills:

- "On utilisation of atomic ships and submarines";
- On utilsation on nuclear materials demounted from atomic weapons".

There are legislative accumulations in Russia to quickly make a draft of these laws.

CONCLUSION

I am convinced, that it would be rather useful and productive to connect the experts from western countries to joint work for the decision not only technical, but also legal questions of nuclear warware utilisation. It could become one of the perspective opportunities of International lab-to-lab cooperation for the decision of problems of nuclear warware utilisation.

Address:

YURI ROGOZHIN
Head of Laboratory
Gosatomnadzor
Moscow
Russian Federation

46

Radio-ecological Legacy of the Past and its Overcoming on the Territory of Kazakhstan

UMMIRZAK M. SULTANGAZIN

INTRODUCTION

After the disintegration of the Soviet Union, new independent states on its territory have been formed. Among them, Kazakhstan experienced negative consequences of war politics of the former Soviet Union during the period of Cold War. The huge part of the territory of the Republic has been allotted under disposition of war-industrial complexes and test sites for bacteriological, nuclear and chemical weapon.

As a result, Kazakhstan received weak economy, oriented on extractive industry and unfavourable ecological situation. Kazakhstan became the owner of a great nuclear heritage including uranium industry, enterprises on uranium processing and fuel producing, experimental and industrial reactors, powerful scientific-technical potential and nuclear weapon, in particular 104 SS-18 rockets of stationary basing with 1040 nuclear warheads.

Besides that on the Repúblic territory there have been disposed 40 strategic bombers E-95 with 240 winged rockets. "Baikonur" cosmodrome, including its entire infrastructure, became the property of Kazakhstan. This fact demanded from the country leadership to work out the national policy concerning military complexes, in particular atomic weapons. During the USSR existence according to the will of the public movement 'Nevada-Semey' and the Parliament of the Republic the President N.A. Nazarbaev closed the Semipalatinsk nuclear test site by his decree.

The Republic has signed the Lisbon Protocol to the agreement between USSR and USA on reduction and limiting of strategic arms pledging in shortest period to adhere to the non-proliferation treaty as the state did not own nuclear weapon. On 13 December 1993 the Republic had subscribed this agreement. After, in accordance with its articles the covenant on guarantees between the Government of Kazakhstan and International Atomic Energy Agency (IAEA), came into force after its ratification, had been signed. At present time, all nuclear objects are under guarantee IAEA. The Republic voted for termless prolongation of NPT, joined its voice to the treaty of nuclear test stopping. On 26 May the problem of nuclear weapons on the territory of Kazakhstan had been solved. That very day the President had made the statement. The last charge underground Semipalatinsk test site had been destroyed with all means of safety.

There are developed atomic industry and science in the country, and we are holding the principles of non-proliferation. All nuclear weapon is moved away from the Republic territory, all activities out for military field are stopped. On 5 Dec. 1994, during a Budapest meeting in accordance with the CSCE principles the memorandum on security on the behalf of Russian Federation, USA, United Kingdom confirming their respect to independence and sovereignty of Kazakhstan was signed. The memorandum contains important obligations on non-forcing against territorial integrity and political independence of Kazakhstan, refusing on economic compulsion. Later China and France gave their guarantees.

Wide programmes on former military-research infrastructure conversion has been carried out under support of foreign countries.

For instance, about 600 kgs. of high-enriched uranium had been stored for 20 years on Ulba metallurgical plant. This was dangerous for Kazakhstan and the whole world, we had signed NPT and it was non-profit to cover the expenses on its guarding. For this reason, the decision on tis selling had been taken. "Sapphire" operation on its removal to USA had been carried out. Thus, nuclear materials were made inaccessible for potential terrorists and new atomic regimes. Programme fulfilling is the bright examples of joint efforts of Kazakhstan, USA and Russia in the field of non-proliferation.

To provide nuclear safety, the number of legal acts has been established in the Republic. The Law "Atomic Energy Using", being based in atomic laws, came into force, the project of the Law on "Radiation Safety" has been developed.

All this demands the creation of information and scientific base for liquidation of consequences of radioactive pollution on the territory of Kazakhstan. The National Nuclear Centre has been established on the base of Semipalatinsk nuclear test site and a number of scientific-research institutes of the Academy of Sciences. The main task is conversion of Semipalatinsk nuclear test site and integration of efforts of scientists and specialists in the various fields of atomic technology. The main directions of research are radioecology, atomic energetics and safety, geophysical monitoring and study of the effect of nuclear explosions on the state of terrestrial crust.

In the paper the development of geo-information system for evaluation of radiation situation on the base of space monitoring is discussed. As well space monitoring allows to trace tradition pollution dynamics and discover accidents on atomic stations and other objects connected with radioactive sources.

1. THE MAIN SOURCES OF RADIOACTIVE POLLUTION ON THE TERRITORY OF KAZAKHSTAN

Radioactive pollution of the territory of Kazakhstan is defined by several main factors:

- former Semipalatinsk nuclear test site activity;
- nuclear explosions carried out with economic purposes;
- activity of atomic industry, as well as enterprises using radioactive materials;
- radioactive waste on the Republic territory;
- natural radioactivity.

Brief characteristics of the main pollutant sources are given below. Semipalatinsk nuclear test site was active from 1949 to 1989. During this period there had been conducted 470 nuclear explosions, among them 90 in atmosphere, 26 on ground and 354 underground. Analysis of demographic characteristics, data on morbidity and mortality rates of population showed negative effect of test site on population health. At present time investigation of radioactive pollution in the districts contiguous to the site is continuing. According to preliminary data pollution levels of Cs-137 and Sr-90 are at background values, except some parts equal to 2,160 sq. km., total square is 18,000 sq. km. Situation on pollution by plutonium-239 and other transuranium elements is investigated weakly. The waste from atomic blasts totals about 12 million tons with 12.9 million Curie out of which 6.5 million tons are due to underground explosions and will be long deposited in the molten rock[1].

On the territory of Kazakhstan 32 underground tests outside military sites have been conducted, for example, for seismic sensing of the earth's crust, as well as for creation of underground reservoirs in salt cupolas of the pre-Caspian depression.

On the theme of deep seismic sensing of the earth's crust 6 explosions had been conducted.

For creation of underground reservoirs 23 explosions had been made.

The volume of every reservoir is 60,000 cubic m. Controllable removal of radioactive gases from the cavities to environment had been carried out on test grounds. To the present time potential pollution of underground waters on these grounds had not been studied and monitoring had not been conducted. All this causes anxiety of population and public opinion in the Republic. At present time 6 underground cavities are flooded with brine, but the causes are not revealed. Four of these cavities are in exploitation as reserve barns of Karachaganak deposit's gas-condensate. On Fig. 1 the map of explosions on the territory of Kazakhstan (based on the materials of MAEP and paper of Deriglazov V.I. et al.) is given.

One of the most dangerous sources of radioecological pollution is radioactive waste, the danger is multiplied by non-accounted wastes.

In our country there have been accumulated 232.9 million tons of radioactive waste, their activity is 13.2 million Curie, among them:

- waste of mine and processing industry—252.9 ths Curie, 220.6 million ton;
- atomic expositions waste—12882.4 ths Curie, 12.3 million ton;
- waste of energetic and research nuclear reactors—27.0 ths Curie;
- worked-off ampule sources of isotopes—32.5 ths Curie.

Radioactive pollution of the territory is also connected with activities or uranium—ore mining, nuclear research and energetic reactors, extraction and processing of polymetallic ores, oil and gas accompanying by extraction of rocks with increased radioactivity—elements of uranium—radium and thorium rows. Radioactive pollution is of local character (up to several sq. km) and has exposition dose from dozen of microroentgen per hour to 250 roentgen per hour. In the Republic there are 776 enterprises using 79,650 source of ionising radiation, total activity is 250 ths Curie. 19,700 of them have to be buried.

2. EVALUATION OF RADIOACTIVE SITUATION IN KAZAKHSTAN

Principal methods of investigation are:

- aerogamma-spectrometric and radio-hydrolytochemica survey;
- compiling of passports on underground tests carried out in peaceful purposes;
- investigation of ore mining enterprises;
- agroradiation and veterinary-radiation monitoring;
- monitoring of food radioactive pollution;
- monitoring of population morbidity caused by radiation.

In 1993–94 the area of 914.9 ths sq. km. had been covered by aerogamma-spectrometric survey and 480 ths sq. km. by radio-hydrolytochemical survey with scale 1:1000000[2].

First results of regional aerogamma-spectrometric and radio-hydrolytochemcal works showed their keen necessity at present time. The districts of pollution within the borders of Semipalatinsk zone of environmental disaster and adjoint territories are identified, now the measures on population irradiation limiting could be taken and, therefore, it is possible to change the territory's status. Besides that, earlier unknown parts polluted by both radionuclides and other chemical toxic elements, where more investigations need to be conducted and measures on ecological situation improving should be taken, are discovered.

On adjacent territories (Semipalatinsk, Karaganda, Pavlodar, East-Kazakhstan districts) there have been revealed a lot of caesium anomalies (0.2–0.27 Curie/sq. km) from 5 to 20 km., unifying into anomalous zone, mainly in north-east and south-west directions, oriented on areas of elevated relief.

Based on the data of aerogamma–spectrometric survey of the territory of Atyrau region there have been discovered several anomalous spots of high (30 microroentgen per hour and more) radioactivity caused by increased concentration of radium and thorium. Visually the connection between spots placing and ground oil pollution due to oil deposits is observed. On the south of West Kazakhstan region two anomalous tails have been revealed, with the length of 100–140 km, containing Caesium-137 equal to 0.1–0.18 Curie/sq. km: southward the Zhaltyr lake and between the Aralsor lake to the boundary with Russia along the north-east direction. One more caesium anomaly (0.13–0.17 Curie/sq. km) occupying the area of 80*120 km is founded out near the Azgir ground.

According to the results of radiohydrolytochemical topographic survey data in scale of 1: 1000000, nine conditionally-anomalous zones with concentration of Caesium exceeding 0.07 Curie/sq. Km, at average values over Kazakhstan being equal to 0.065 Ku/sq. km, were revealed on the territories of East-Kazakhstan, Kokshetau, Pavlodar and North-Kazakhstan regions.

In the 1993–94 on the 25 objects of peaceful underground tests (Region, Meridian, Batolyt, Lira, Azgir, Mangyshlak) there have been compiled the passports of test conditions involving the schemes and data on test purposes, power, depth of charge, geologo-hydrogeological conditions of site, mechanical and seismic effects of explosion. Analysis of this information gives the possibility to make conclusion about the negative influence of some exposition cavities on the environment.

While studying radioactive waste of uranium mining on the territory of Kazakhstan there have been determined high values of doses, from 120 to 3000 micro Roentgen/h, practically at all dumps. Obtained estimates are not interconnected as components of a unified data base. For this reason, surface radioactive waste in all regions of Kazakhstan needs development of GIS and programme of radioactive waste localising. The main aim of the programme is to provide present and coming (future) generations and human environment with reliable and economically effective protection from radioactive pollution both in normal and extreme situations.

3. INFORMATION SYSTEM ON RADIATION SITUATION IN KAZAKHSTAN

The structure of any information system is defined by aims of its application. The main objective of the system is to provide information basis for concrete investigations and further make decisions in order to minimise risk and possible negative consequences of activities connected with radioactive environmental pollution. For example, the question "Whether economic exploitation of the territories adjacent to Semipalatinsk nuclear test site is allowable?" is very acute one.

Such issues of explicit economic character stimulated and continue to stimulate investigations and collection of information at the different regions of Kazakhstan. As it was mentioned above, bulk of information has been collected, but it is necessary to provide free access to this information and give the opportunity to experts to carry out complex analysis. Moreover, it should be taken into account, that the information and give the opportunity to experts to carry out complex analysis. Moreover, it should be taken into account, that the information, as a rule, is territorially distributed, i.e. the principal bulk of data needs additional characteristics—geographical coordinates administrative, topographic and other connections.

Thematic processing suggests the usage of special models adequately reflecting spatial-temporal laws of radionuclides migration in different natural media and along trophical chains. Therefore, development of mathematical and imitation models for radionuclides store and migration as well as their effect on vegetation, animals and human beings is of great importance at the creation of the system. To create and implement such comprehensive and expensive system through traditional scheme (draft, technical project, working project) is impossible at present time. Therefore the strategy is aimed to be carried out with the usage of two following principles[3].

The essence of first one consists in the following: different parts of the system are developed independently forming individual "islets" further merging into united system. This concept is expedient for applying in territorial (regional and local) subsystems. To begin with, it seems reasonable to use the principle of maximal utility; i.e. to prefer the regions where expenses are minimal and effects is maximal. In particular, the more territory is known, the more it is preferable.

Second principle. While developing individual territorial sub-systems of SĖMK, at first the minimal kernel of the system for limited circle of users and tasks is constructed. After that worked out and well-approved decisions are distributed among other users and territories.

The bulk of collected data in different organisations provides evidence, that Semipalatinsk nuclear test site should be chosen as standard territory for devęlopment of base complex of the system and testing of principal project decisions (Fig. 2). On the first step of the system development, the tasks of radionuclides migration in natural environment and along trophical chains are the most actual. These tasks at the present time have sufficient scientific, personnel and technical back-up.

Structure and loading of database for any programme complex are defined by directions of its application. The main objective of data collection is to provide basis for preliminary investigations and further decision making to minimise risk and possible negative consequences of activities connected with radioactive environmental pollution.

Suggested data base consists of set of functional components describing physical objects containing or producing radioactive materials which are potential environmental pollutants. Each component of database contains description of such characteristics as physical, chemical, geographical socio-economical features of different objects containing radioactive substances of being sources of radioactive pollution.

Set of data for each component is divided on the following main groups:

- general information;
- physical, chemical and technical characteristics;
- geographical and geological data;
- socio-economical data.

Whole information about the object includes general description and references on possibly occurred incidents connected with radicactive pollution.

Data on physical, chemical and technical characteristics are divided into two groups:

– data on radioactive materials accumulated in the places of their compact storing or using and being potentially dangerous due to any unforeseen or insuperable circumstances;

– data on radioactive materials distributed in biosphere owing to accidents, tests or other artificial reasons.

Geographical and geological data describe location of the object, soil features and demographic situation of the district.

Socio-economical data are used together with other public information while developing measures connected with the cases of occurred or possible radioactive pollution.

At the present time the principal blocks of the system are realised in application to the Semipalatinsk test site.

Let us briefly describe this local GIS.

4. LOCAL GEO-INFORMATION SYSTEM OF SEMIPALATINSK NUCLEAR TEST SITE (STS)

At present time this system consists of three principal blocks:

- information data base;
- block of satellite monitoring;
- block for calculation of inhabitant's dose load.

Block of radionuclides migration under the influence of surface water is being developed.

Available Information

The available information for STS is divided on following groups:

- general geographical information, i.e. maps;
- geological information on maps;
- maps of chemical composition of ground waters.
- aerogamma-spectrometric maps;
- data of aerogamma-spectrometric survey along the two main traces of radioactive fall-out;
- data on nuclear explosions;
- geological soil maps;
- hydrogeological maps;
- maps of underground waters chemical composition;
- maps of underground waters levels;
- maps of clay soils depth;
- analysis of different samples;

Analysis of samples includes:

- records of analysis of soils on gamma-radiation level;

- records of water chemical analysis;
- analysis of soil samples during the 1990s;
- analysis of tritium levels.

In addition, there is the bulk of data on radioactive pollution of animal food and data on radioactive pollution of nature.

Information Base Structure

Carried out investigation reveals the necessity for development of database and GIS on the base of standard software: Arc/Info (UNIX version), Arc/View (PC and UNIX realisation)

Geographic and geological data may be stored in the database as set of layers (surfaces) and their attributes.

Layers of data base may include:

a) Geographic and geological data:
   - — Hydrological net with the following attributes: type (tributary, river, lake etc.), length, area and so on.
   - — Administrative borders of districts and regions.
   - — Elevations.
   - — Elevation isolines.
   - — Populated area with attributes: type, area, population etc.

b) Geological and hydrogeological data:
   - — Stratigraphic division.
   - — Magmatic complexes.
   - — Regions of tektonic anomalies.
   - — Emission regions.
   - — Underground water level.
   - — Clay soil depth.

c) Social geography:
   - — Population density.
   - — Season population migration.

d) Aerogamma survey data:
   - — Contours of radioactive pollution spot.

e) Analysis of samples.
   - — Analysis of water samples.
   - — Analysis of surface soil samples.
   - — Analysis of soil samples on the level of gamma-radionuclides concentrations.

f) Data on conducted explosions.

Information of general type, physical, chemical and technical characteristics as well as the data of public interest may be stored in data base of common purpose. Records of this database should be connected with GIS layer's elements.

Block of Satellite Monitoring

For the space monitoring of Kazakhstan territory, NOAA polar orbiting operational satellite data have been used. These satellites revolve at a height of about 850 km with circulation period of 102 minutes. The Advanced Very High Resolution Radiometer 2 (AVHR/2) placed on these satellites provides the earth remote sensing data in the visible (0.58–0.68 ìm·channel 1), near-infrared (0.73–1.10 ìm channel 2), three-infrared (3.55–3.93 ìm channel 3, 10.3–11.3 ìm channel 4, 11.5–12.5 ìm channel 5) wavelength. The width of the survey zone is 2700 km with 1.1 km resolution in nadir. At present there are three satellites simultaneously in space, which allow for receiving space images of the Kazakhstan territory 6 times per day. Reception and processing of the information are prompted by workstation from the Telonics firm. The visible-range data (ch1) are used for forming survey maps of the republic's territory. In winter these maps allow for controlling the state and dynamics of the snow cover. For vegetation monitoring, maps of Normalised Difference Vegetation Index (NDVI) are made with NDVI=(ch2-ch1)/(ch2+ch1). The thermal channel data (ch4 and ch5) are used for the definition of the ground surface temperature.

By producing ten-day survey maps of the Kazakhstan territory for the end of the 1995–1996 and the 1996–1997 winter[4], we concluded that the territory of the Semipalatinsk test site was practically snowless during the entire winter. Structure analysis of the thermal field reconstructed by the ch4 channel data determines clearly contoured zones, the temperature of which is 10 degrees higher than that of the surrounding snow cover. (Figure 3).

The location of anomalous zones and field configuration practically did not change in the February-March images, and the amplitude of the anomalies was also preserved in spite of the total temperature increase. Zones of higher temperature on the STS only disappear when the ground is heated to 35–40°C. At those temperatures it becomes difficult to detect the difference.

The above facts do not enable one to give an unambiguous explanation of the phenomenon observed. Particularly, the phenomenon could be caused by local features of the site (relief, type of underlying surface etc.) resulting in snow cover denudation and consequently in more intensive of snowless ground spots by sunrays. The phenomena could also be stimulated by numerous nuclear explosions. To test these ideas we carried out additional investigations.

Block for Calculation of Current Dose Status

This block was realised on the base of RESRAD model[5,6] developed in Argonne National Laboratory (USA) and is intended for calculation of current doses and risk for health of critical group of population permanently residing at the polluted zone. At the calculation of equivalent dose received by the population there is used the conception of dose coefficient describing relative importance of different radionuclides, ways of their transfer and transport in the environment.

After calculations preliminary digital maps of integral doses received by population permanently residing at the STS (Cesium–137 + Potassium-40 + Thorium) have been designed. Uranium doses occurred to be less than 1.2 mbar/year and were not accounted in sum dose. Values of sum dose in accordance with data used change within the range from 11 to 195 mbar/year (maximal values refer to Shagan artificial lake district). Some results of calculation are given on Fig. 4, 5.

REFERENCES

1. U.M. Sultangazin, T. Tsukatani. Modelling of the Kazakhstan Economy and Environment. Kyoto University. Kyoto, 1995, 38p.
2. Radioecological situation in Kazakhstan (1954–94). Ministry of Ecology and Bio-resources RK, Almaty, 1997.
3. Hanter R. Methods for Software Engineering Control. M., 1981, 390 p.
4. Sultangazin U.M., Zakarin E.A., Spivak L.F., Arkhipkin O.P., Muratova N.R., Terehov A.G. Monitoring of temperature anomalies in the former Semipalatinsk nuclear test site. C.R. Acad. Sci. Paris, t.326, S. 2b, p.135–140, 1998. Metrology, instrumentation.
5. Manual for Implementing Residual Radioactive Material Guidelines Using RESRAD, Version 5.0 Environmental Assessment Division, Argon National Laboratory, 1993.

6. Nuclear Regulatory Commission Regulatory Guide 1.109. Calculation of Annual Doses to Man from Routine Releases of Reactor Effluents for the Purpose of Evaluating Compliance with IOCFR, Part 50, Appendix 1, 1977.

Address:

UMMIRZAK M. SULTANGAZIN
First Deputy Minister of Science
KS 480021 Almaty
Kazakstan

47

Innovative Uranium-Free Fuels for Weapons Plutonium Disposition in Current LWRs

FORTUNATO VETTRAINO

About 100 MT of excess plutonium are going to be originated from warheads dismantling under the START II Agreements (50 MT by each side US and former Soviet Union), and another about 200 MT are already stockpiled from commercial spent fuel reprocessing. The civilian or reactor grade plutonium (RGPu), amounted at about 1000 MT total world inventory at the end of 1995, and is forecasted to attain some 1500 MT by the year 2000.

Therefore, there is a pressing need in finding novel and ever more safe methods to deal with all kinds of excess plutonium in the aim of rendering it ultimately unusable for proliferation purposes.

As to weapons Pu, one leading option (US and Russia) is to burn it in LWRs after having converted it to MOX fuel. However, among the possible types of fuel which can be envisaged to burn plutonium in LWRs, innovative U-free fuels such as *inert matrix and thoria fuel* are novel concepted that find justification since, especially for weapons Pu transmutation, the precisely required fuel is not immediately available and even for the LWR reapon-MOX, processes and facilities have to be set-up before reaching the operational stage. The dimension of such a delicate task and the perspective of finding even more effective and ultimate solutions from both security and safety standpoint, we think are sufficient reasons for putting attention on a new fuel concept such as IMF. Inert matrix fuel is an non-fertile oxide fuel consisting for PuO_2, either weapon-grade or reactor-grade, diluted in inert oxides such as stabilised ZrO_2, $Al2O_3$, MgO or $MgAl_2O_4$, its primary advantage consisting in no-production of new plutonium during irradiation, because it does not contain uranium (*U-free* fuel) whose U-238 isotope is the departure nuclide for breeding Pu-239. Some thoria addition in the matrix (*thoria-adoped fuel*) may be required for coping with reactivity feedback needs. The full *thoria-plutonia fuel* though still a U*-free* variant cannot be defined non-fertile any more because the U-233 generation. The advantage of such a fuel option consisting basically on a remarkable already existing technological background and a potential acceleration in getting rid of the Pu stocks.

Al U-*free* fuels are envisaged to be operated under a *once through cycle scheme* (the spent fuel is outlooked to be sent directly to the final disposal in deep geological formations without requiring any further reprocessing treatment).

From calculations related to inert matrix U-*free* fuel in a commercial PWR, in a once-through cycle mode, emerges that IMF has the highest Pu burning capability:> 93 per cent of fissile plutonium and 74/85 per cent of total Wg-Pu and RG-Pu, is burnt at end of fuel life. The fabrication tests on simulate fuel demonstrate that the nuclear-grade specifications can be met and a very limited solubility under the current fuel reprocessing techniques, coupled to the quality-poor residual Pu in the spent fuel, will make proposed fuel a potentially strong anti-proliferation product.

Fabrication and irradiation of HEU and Pu oxide bearing fuel is expected to follow under cooperation with the OECD-Halden Project and possibly within an ENEA-IPPE (Obninsk), Russia cooperation expected to be supported by the Italian Ministry for the Foreign Affairs.

BACKGROUND

Safe plutonium control and disposal has become more and more a major issue at world level in the last few years. The stockpiles have reached elevated dimensions either for excess-weapons Pu or for the reactor-grade (civilian) plutonium.

Weapon-grade Pu (*Wg-Pu*) amounts to about 200 MT (~ 94% Pu-239) expected to be recovered from nuclear weapons dismantlement (100 MT already declared available equally by US and Russia under START II agreements).

Large amounts of *civilian or reactor-grade Pu (RG-Pu)*, about 71% Pu-239+Pu-241, coming from nuclear power reactors spent fuel have been already stockpiled (1995):

~1000 MT generated in 180.000 MT of spent fuel; ~185 MT separated; ~50 MT already recycled; ~135 MT stored (estimated to become ~175 MT by the end of 1999).

Proliferation risks, radiotoxicity, environmental safety and energy conservation reasons suggest to burn as fast as possible excess Pu in existing world-wide available LWRs.

INTRODUCTION

There exists a pressing need in finding novel and ever more safe methodologies to deal with all kind of excess plutonium in the aim of rendering it ultimately unusable for proliferation purposes.

Three main technology alternatives in weapons Pu disposition have been addressed in so far: reactor (LWRs); direct immobilisation (vitrification or ceramics); deep porehole entombment.

Most recently attention has been concentrated only on the first two options (US) and the reactor alternative is referred basically to MOX fuel utilisation in LWRs

Three generic fuel types could be used to burn plutonium in LWRs:

a) uranium-based mixed oxides (U, Pu) O_2 called also MOX fuel;
b) thorium-based mixed oxides (Th, Pu) O_2;
c) inert matrix fuels (IMF), where PuO_2 is dispersal in a neutron transparent ceramic inert carrier.

The mixed-oxide $(U,Pu)O_2$ fuels have been thoroughly studied and experienced and a rather wide operational experience is now available on a commercial basis. In fact this option has been indicated by US-DOE as the reference one for disposing its WG-Pu. The main drawback for this solution is that new plutonium is continuously generated during the fuel irradiation so that a significant fraction (70%) of the loaded plutonium is still present in the spent fuel, although denatured.

The $(Th, Pu)O_2$ fuel option is interesting because of its potentiality of increasing the Pu consumption with respect to $(U,Pu)O_2$ due to the lack of U-238. From calculation emerges that the total plutonium consumption, compared to MOX, can be higher by a factor>2 and the fissile plutonium burnt fraction increases of a factor 1.6. Moreover, there exists experience, although limited on (Th,U)O2 fuels, that indicates a good behaviour under irradiation and a rather high stability for the resulting exhausted fuel, so that it is likely to be appropriate for direct disposal. A major disadvantage, however, for Th-fuel is the breeding of U-233, even though the high gamma activity of the separated uranium mixture, coming mainly from U-232, should highly discourage this proliferation pathway.

R&D activity on innovative U-free fuels such as inert matrix not-proliferant fuel, appears widely justifiable since, for weapons Pu transmutation in LWRs, the precisely required fuel is not readily available and even for MOX, processes and facilities have to be set-up before reaching the operational stage. The dimension of such a delicate task and the perspective of finding even more effective and ultimate solutions from both security and safety standpoint, we think are sufficient reasons for supporting the development of new fuel concept such as IMF-thoria fuels.

OBJECTIVES

- Prove Pu definitive elimination be feasible via U-*free* inert matrix-thoria fuel aimed at commercial PWRs.
- Potential contribution in the mid-long term to ultimately solve the pressing Weapons Plutonium disposal problem as arised from the international disarmament agreements.

THE INERT MATRIX AND THORIA FUEL CONCEPT

It consists basically in a dispersion of Pu oxide in inert ceramic oxides and/or Thoria, where the Pu content is set in order to give approximately the same reactivity performances as in the standard U-235 enriched fuel for LWRs. The constituent materials are then:

- Pu oxide, under powder or sol-gel microsphere form (Metal W-Pu assumed converted to oxide);
- Matrix oxides consisting in: stabilised ZrO_2-CaO; $MgA1 2O_4$ (spinel); mixture of a1203, ZrO_2-Y_2O, MgO; ThO_2
- Burnable poisons such as ZrB2-IFBA, $G2O_3$, $Er2O_3$, whose presence is foreseen for controlling peaking factors and burn-up reactivity swings.

Annular pellets are foreseen for zirconia and spinel use, and introduction of ThO_2 is considered mainly, for:

- improve reactivity feedbacks (Doppler) in pure inert matrix fuel;
- strengthen chemical stability;
- accelerate Pu stock disposition;
- profit of some good already existing in-reactor experience;
- look at applicability to Pu-Th cycle as in hybrid systems (ADS).

The possible fabrication processes are the *GSP (Gel Supported Precipitation) sol-gel* process or the *standard dry-blended powders technology* process.

A priority is being give to microspheres produced via the GSP "wet route" process, which is less contaminant during fabrication phase (no fines). Pellets are pressed and sintered following the current LWR fuel technology. Fuel pin and assembly will respect the standard LWR geometry.

SPECIFIC ADVANTAGES OF INERT MATRIX FUEL

There are a number of favourable justifications for envisaging plutonium transmutation in commercial LWRs via innovative inert matrix uranium-free fuel:

- Commercial LWRs, with their consolidated technology, are the only world-wide available facilities candidate in the near-mid term to destroy Pu and recover its huge energetic content.
- Inert Matrix U-free Fuel shows a very high anti-proliferation potential compared to MOX fuels (no generation of new Pu and nearly total destruction of fissile Pu).
- Elimination of either RG-Pu or WG-Pu and radiotoxicity load not negatively affected by inert oxides.
- Residual Pu in spent IMF is "poor-quality" and highly difficult to be recovered due to characteristics of high dissolution resistance (*inherently not-proliferant fuel*).
- *Once-through fuel cycle* viability enhanced (Burn & Bury).
- Inert matrix ceramics suitable carrier for HEU and MAs.
- Applicability to other alternative Pu and MAs disposition options (CAPRA, Fusion reactors and ADS-hybrid subcritical systems).
- Potential for HEU inert matrix fuels for special applications.
- Moreover, the IMF concept (sol-gel microsphere version) is suitable also as direct ceramic immobilisation option in Pu disposition strategy.

THORIA BASED FUEL

There are some drawbacks associated with pure Inert Matrix Fuel such as the limited available data base of under irradiation behaviour; low thermal conductivity; lower melting point for some mixed compound (avoid eutectics); hollow pellet solution needed for some matrices like zirconia; burnable absorbers needed for reducing peaking factors, linear power and burn-up reactivity swing. Most relevant is the reactivity feedbacks (Doppler) decrease due to lack of U-238. To cope with all this drawbacks a possible way is to induce ThO_2 in the matrix, partially (thoria-doped fuel) or totally (thoria-plutonia fuel). U-238 also is good for Doppler recovering, but it would produce new Pu and it would be also source of further radiotoxicity because of the MAs production during the irradiation.

A possible remedy to cope with the problems arising from proliferation standpoint (U-233 breeding from Th-232) is to add however some limited amount of depleted U in order to denature the U-233 which is produced during irradiation.

Amongst the all-outs there is also the applicability to the ADS "Rubbia concept", which is envisaged to act as a burner machine either for plutonium or Mas and long-lived radwastes. In this latter context Th utilisation could constitute an additional option of the solution here proposed, which permits the U-233 recoverability under suitable anti-proliferation countermeasures, this possibility is undergoing a renewed interest though it will introduce more complexity in the fuel cycle management with respect to the standard UOX-MOX fuel cycle.

NEUTRONIC FEATURES AND PLUTONIUM BURNING PABILITY

The neutron physics investigations are aimed at enlighting the best core layouts which are possible to draw when using U-free IMF and Thoria fuels in commercial PWRs operating in a once-through cycle scheme. Compared evaluations on different Pu fuels in PWR-AP-600 (pin cell, assembly and core level) have been performed.

The considered fuels were (Table 1): Inert Matrix Fuel (IMF); Mixed Oxide fuel (MOX); Thoria-doped IMF; ThO_2-PuO_2.

A possible arrangement for IMF bundle is reported in Fig.1.

Calculation Method

- The Pu content (WG-Pu and RG-Pu) is set to ensure a

Table—1 : Considered Fuel Composition in Neutronics Calculations

Fuel	*Description*	PuO_2	UO_2	*Matrix*	ThO_2
PUR	$(U,Pu)O_2$, R-Pu	6.01	93.99	--	--
PUW	$(U, Pu)O_2$, W-Pu	3.58	96.42	--	--
IMR	Inert Matrix Fuel, R-Pu	4.82	--	95.18	--
IMW	Inert Matrix Fuel, W-Pu	4.05	--	95.95	--
TDR	Thoria-doped IMF, R-Pu	4.91	--	72.97	22.12
TDW	Thoria-doped IMF, W-Pu	3.84	--	65.43	30.73
PTR	$(Th, Pu)O_2$, R-Pu	7.26	--	--	92.74
PTW	$(Th, Pu)O_2$, W-Pu	4.78	--	--	95.22

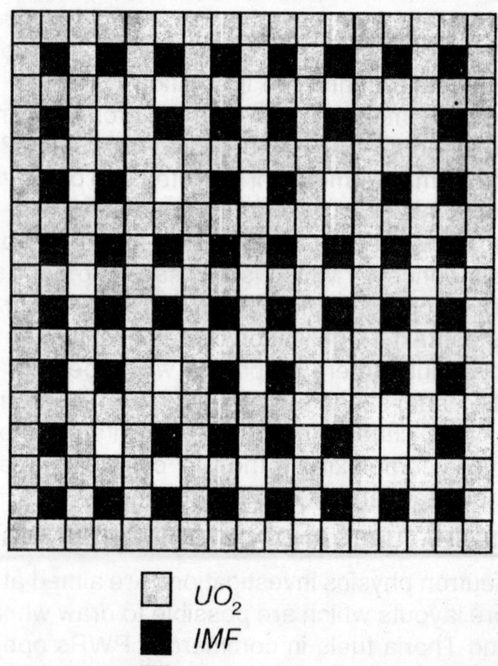

Fig. 1 : Possible arrangement for a PWR (17×17) IMF Assembly

number of EFPDs for an equivalent brunup of 33 Mwk/kg.

- Depletion analysis is made via SAS2H of SCALE-4.3 integrated code system (44-group cross section library).

Inert Matrix Fuel

The Inert Fuel shows a high initial reactivity (lack of absorptions in fertile materials), rapidly decreasing vs EFPD (Fig. 2). Highest reactivity swing among all fuel alternatives.

Steep and large variation in power generated along fuel lifetime (burnable absorbers needed) are evidentiated.

The reactivity coefficients are considerably lower than for the UO_2 reference fuel, Table II).

Full core loading with IMF appears not viable for control reasons.

Table—II : Dynamic Coefficients (BOL)

	Doppler Coefficient pcm/°C	*Void Coefficient pcm/% void*	*Boron Worth pcm/ppm*
UO_2	−2.28	−130.5	−12.9
PUR	−2.95	−186.5	−4.6
PUW	−2.81	−182.7	−6.1
IMR	−1.08	−85.0	−6.1
IMW	−0.81	−48.3	−6.4
TDR	−2.38	−135.9	−6.0
TDW	−2.39	−112.3	−6.5
PTR	−3.35	−173.1	−4.5
PTW	−3.24	−161.3	−5.6

IMF shows very high plutonium elimination potential:> 93% of fissile plutonium and 74/85% of total RG-WG-Pu is burnt at end of fuel life (Tables III and IV).

Table III : Plutonium Consumption

	$Pu_{fiss(kg/TWh)}$ *BOL*	Pu_{fiss} *burnt%*	$Pu_{tot(kg/Twh)}$ *BOL*	Pu_{tot} *burnt%*
PUR	55.0	35.2	76.4	23.3
PUW	42.7	49.5	45.4	28.4
IMR	44.4	92.6	61.7	74.4
IMW	48.8	96.6	51.9	84.8
PTR	61.7	63.2	85.7	46.7
PTW	52.8	78.6	56.2	63.0
TDR	44.4	86.0	61.6	67.1
TDW	44.9	92.0	47.8	78.3

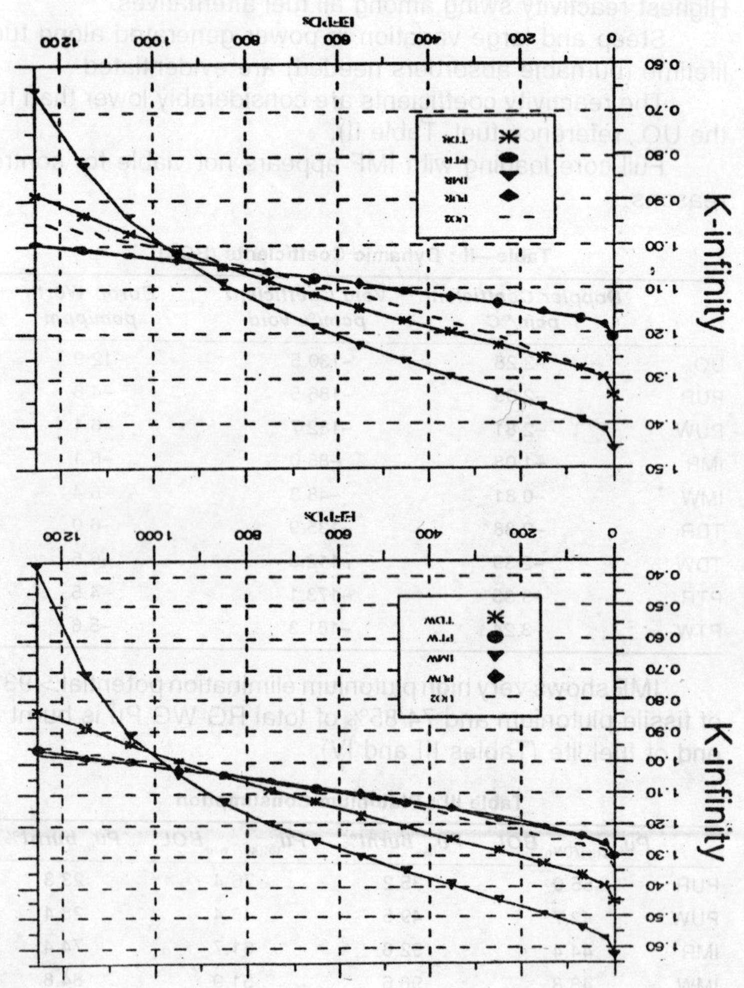

Fig. 2 : Reactivity curves (K∞) for W-Pu and RG-Pu

Table—IV : WG-Pu Disposition Capability Inert Matrix-Thoria Fuel vs. MOX

Reactor Type	: 1000 MWe (3000 MWth) commercial PWR			
Operational period	: 15 years			
Capacity factor	: 0.8			
WG-Pu Mass	: 100 MT			
Fuel Type	**MOX**	**IMF**	**IMF-ThD**	**Pu-Th**
Loaded Putot (kg/TWh)	45.4	51.9	47.8	56.2
Loaded Pufiss (Kg/TWh)	42.7	48.8	44.9	52.8
No. of Reactors (full core loading)	7	7	7	6
No. of Reactors (30% core loading)	23	20	22	19
Pu in spent fuel (% Loaded Pu)	71.6	15.2	21.7	37
Burnt Fissile Pu (% Loaded Fissile)	49.5	96.6	92.0	78.6

On the basis of the above results can be concluded that best arrangements for increasing-controlling Pu loading in the core are: *a*) mixing IMF and UO_2 pins/assemblies (~1/3core); *b*) inserting Pu within a thoria matrix. The advantage of solution *a*), though limiting the loaded core fraction, would couple IMF high plutonium elimination capabilities with the adequate reactivity coefficients of UO_2, while being a truly non-fertile option.

Thoria Fuel

The Pu burning efficiency (Tables III and IV) for the IMF option with a 30 per cent core partial loading, leads to about 44 kg WG-Pu per TWhe generated by the whole core. This means that a large number of *burner* reactors would be needed to eliminate the excess Pu if using pure IMF for that core fraction. Trying to increase the above figure by loading higher inventories of Pu or by inserting in the fuel assembly larger percentage of IMF rods, would worsen the values of the reactivity feedbacks or induce higher power peaking. The fertile Th-232 addition is in order to enhance the resonance absorptions and then improve the values of reactivity coefficients.

A parametric study has been carried out in order to investigate the possibility of increasing Pu inventory in the core or possibly to feed the whole core with Th-based fuel.

Fuels containing ThO_2 in various percentage (9-95 % mol) have been compared by focusing on reactivity coefficients and Pu elimination efficiency. The higher the Th-232 concentration, the less sharp the reactivity swing (Fig. 2), thus implying low power peaks in the fresh fuel at BOL and limited variations in power generation along the fuel lifetime. The Doppler coefficient for IMF is rather low and close to 0 at EOL; this is one of the main reasons which would render the implementation of a core completely fed with IMF not viable. The Doppler coefficient for Th-based fuels increases with Th content and can reach easily values similar to those of UO_2. A similar behaviour is obtained for the void coefficient. However, the improvement of reactivity coefficients is followed by a reduced relative Pu elimination capability for the higher Th contents, due to the higher initial Pu inventory and to the contribution of U-233 to the total fissions. Nevertheless, a larger Pu inventory with respect to pure IMF is required to be loaded in the core (Pu concentration has to be increased for coping with Th high absorption rate) and the Pu disposition rate for a whole core loaded with Th-based duels can be as high as 112 kg WG-Pu/TWhe (thoria-doped fuel) or 106kg WG-Pu/TWhe (thoria-plutonia fuel).

Table—V : Inert Matrix and Thoria Fuel Specifications

Assembly: 17 × 17 PWR fuel bundle geometry		
a) Al_2O_3-ZrO_2-MgO (solid pellets)		
outer diameter	:	8.20 mm
height	:	10.16 mm
density	:	95% T.D.
fuel content	:	5 mol%
matrix content	:	95% mol (47 mol% Al_2O_3, 38 mol% ZrO_2, 10 mol% MgO)
b) Zirconia and Spinel (annular pellets)		
outer diameter	:	8.20 mm
Inner diameter	:	4.10 mm
height	:	10.16 mm
density	:	95% T.D.
fuel content	:	7 mol%
matrix content	:	93% mol (Stab. ZrO_2, or $MgAl_2O_4$)
c) Thoria Matrix		
– Thoria-doped fuel	:	22 mol%; 73 mol% matrix, 5% mol% PuO_2
– PuO_2-ThO_2 fuel	:	93 mol% ThO_2; 7 mol% PuO_2

Proliferation Concerns in Case of Thoria Fuel

During irradiation, thoria-bearing fuels generate the highly fissile ^{233}U which via (n, g) reactions on Th-232, may arise a major concern from a proliferation standpoint. A potential deterrent to diversion is, however, the strong g activity of daughter products of ^{232}U (Tl-208) which emit highly penetrating 2.6 MeVg photons. Another countermeasure to curb the proliferation risk, could be the addition of some small amount of ^{238}U in order to denature the uranium mixture. The substitution in the fresh fuel of a small amount (4-7%) of thorium by depleted uranium is sufficient for setting a limit of 20 per cent to the fissile content in the uranium at EOL. The simultaneous presence of the two fertile isotopes ^{232}Th and ^{238}U bring the reactivity to decrease (Fig.6 at p. 537), by leading in turn to a burn-up reduction of some 10-15 per cent (Fig. 7 at p. 537).

Synthetising IMF vs. MOX and Thoria Fuels

$(U,Pu)O_2$: MOX fuel is ready on a commercial basis but more than 70 per cent of the loaded Pu is still present at EOL. Full core loading possible.

$(Th,Pu)O_2$: Pu consumption is higher than in MOX due to lack of U-238. A main drawback from proliferation standpoint is U-233 production, but gamma activity deriving from U-232 should be a strong barrier against diversion. Very good reactivity feedbacks (full core loading viable). Fuel sub-assembly same as standard, but the fuel cycle is more complex.

Inert Matrix Fuel (IMF): The lower value of the dynamic coefficients (reactivity feedbacks) implies that only partial core loadings can be conceivable.

A suitable IMF variant, *thoria-doped* inert matrix fuel, will permit to achieve larger Pu loadings (full core is likely) because the reactivity feedbacks do increase as close as to std. UO_2 fuel, and should maintain stability and proliferation characteristics nearly same as truly IMF which characterises for the highest Pu destruction potential (about 85% of total loaded Pu and about 97% of loaded fissile Pu).

FABRICATION-CHARACTERISATION AND IRRADIATION

An R&D activity aimed at assessing the experimental response has been undertaken, starting from fabrication-characterisation tests on simulate cold material, in the aim of fabricating a small lot of Pu bearing Inert Matrix Fuel and Thoria to be irradiated in LWR.

The main steps of the programme are listed below:

1) Fabrication of pure matrix and sim-fuel sample pellets (under-way)
2) Thermo-physical characterisation, IID (ion irradiation damage) testing and solubility testing on "cold" material (under-way)
3) Fabrication of HEU bearing pellets of IMF, including ThO_2 with characterisation and irradiation in Halden reactor
4) Design, fabrication and irradiation of a representative Pu inert matrix and thoria fuel test experiment, expected to take place in Halden experimental HWBR and/or in a Russian PWR.

Simulate-fuel fabrication tests are progressing on $ZrO_2(CaO)$, $MgAl_2O_4$, Al_2O_3-$ZrO_2(Y_2O_3)$-MgO and ThO_2 matrices. The PuO_2 is simulated by CeO_2. The sol-gel GSP is assumed as the reference process however (Fig. 3 at p. 535).

Fabrication test results: Fabrication tests carried out so far, have demonstrated the feasibility of the IMF in respect of the most relevant nuclear grade specifications such as for elemental content, pellet density, geometry and microstructure (Fig. 4 at p. 536 and Fig. 5 at p. 536). The pre-irradiation picture of the materials, is going to be completed through characterisation tests as reported below.

Characterisation Testing on Simulate Fuel

- Measurement of thermo-physical properties such as: thermal conductivity, specific heat, thermal expansion, melting point and molten material behaviour.
- Ion Irradiation Damage (IID) testing: First exploration on targets by using accelerated 180 ke V ions (He, Xe), for different doses ranging from 10e 14 to 10e 17 ions/cm_2.
- Solubility tests (corrosion and dissolution). Different conditions are simulated:

long term geological disposal (about room T, pH ~ 8); fresh-spent fuel reprocessing (HNO3-8M, 120 °C); PWR operation (300 °C, 155 bar, pH 7.2).

Corrosion and dissolution test results: from a set of preliminary test turns out that the investigated inert matrix materials are

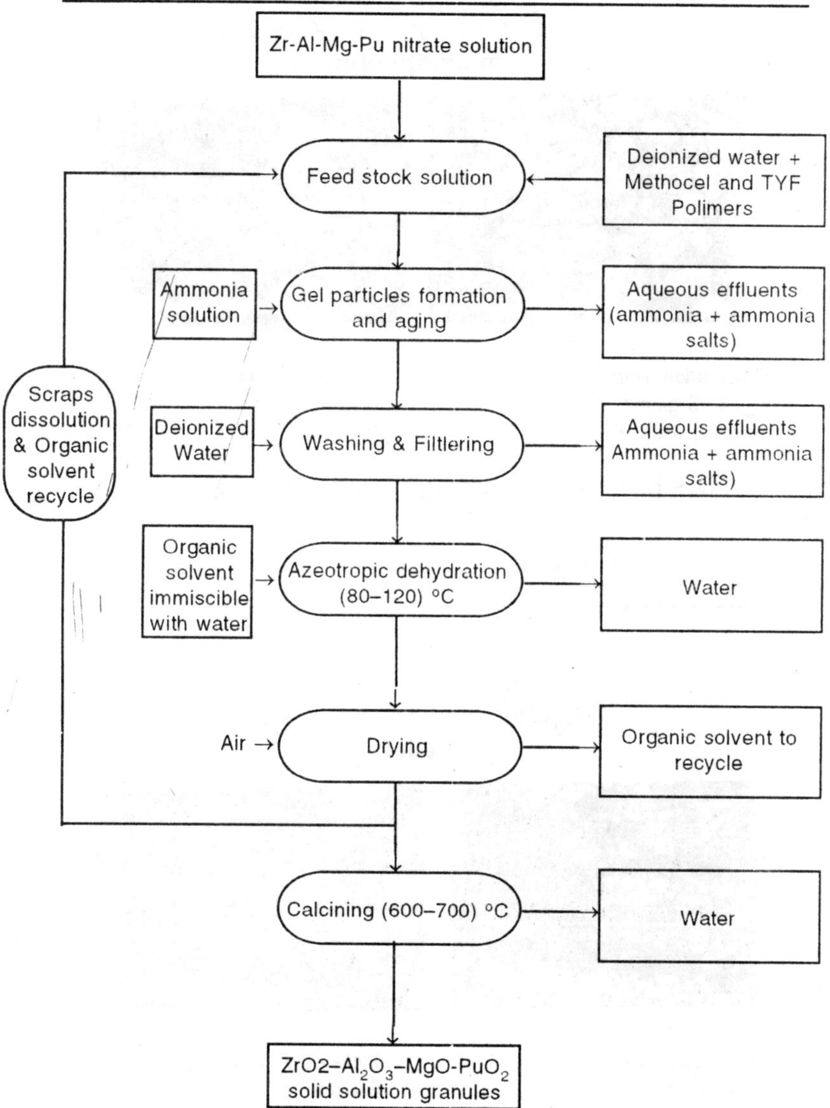

Fig. 3 : The GSP Process for Inert Matrix Fuel Microsphere Production

particularly dissolution resistant and that is especially evident for the simulate fuel material containing Ce.

SOL-GEL Process

ZrO_2+CaO and ZrO_2+CaO+CeO_2

microspheres

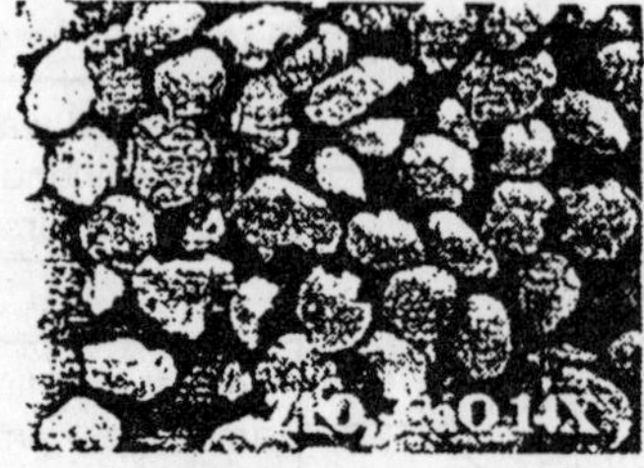

Calcination Temperature=800°C	Calcination Temperature=800°C
r_{He}=4.95 g/cm^3	r_{He}=5.37 g/cm^3
Specific Area (B.E.T.)=38.6 m^2/g	Specific Area (B.E.T.)=41.43 m^2/g

FN S.p.A. New Technologies and Advanced Services

Fig. 4 : Microsphere Micro-structures

SOL-GEL Process

ZrO_2+CaO and ZrO_2+CaO+CeO_2

pellets

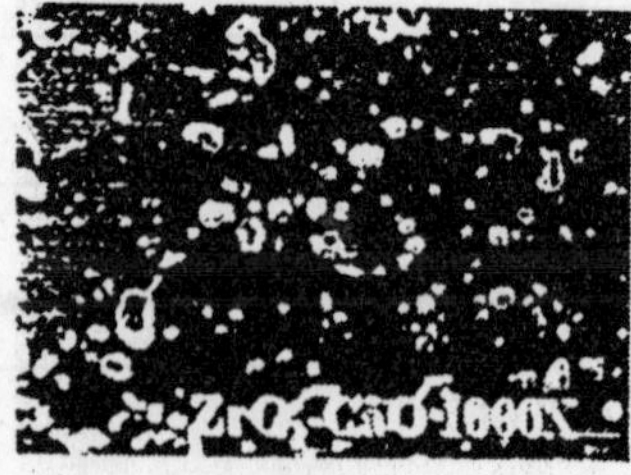

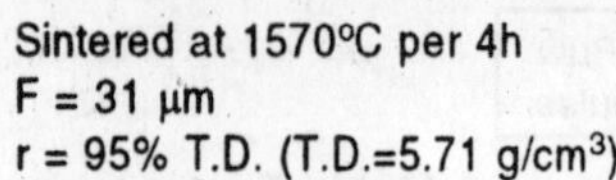

Sintered at 1570°C per 4h	Sintered at 1600°C per 2h
F = 31 μm	F = μm
r = 95% T.D. (T.D.=5.71 g/cm^3)	r = 96% T.D. (T.D. = 5.84 g/cm^3)

Fig. 5 : Pellet Microstructures

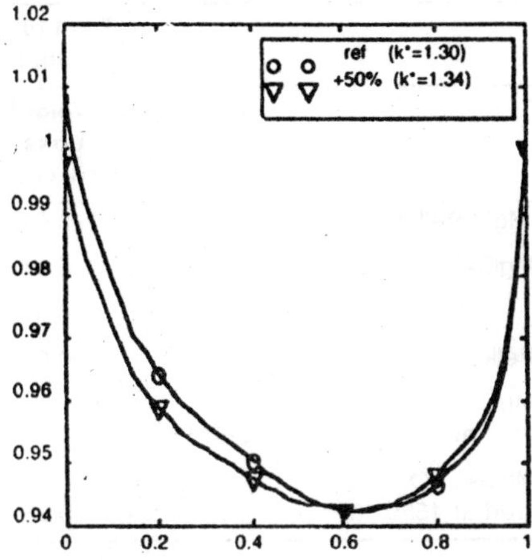

Fig. 6 : Relative Reactivity vs. ^{232}Th fraction

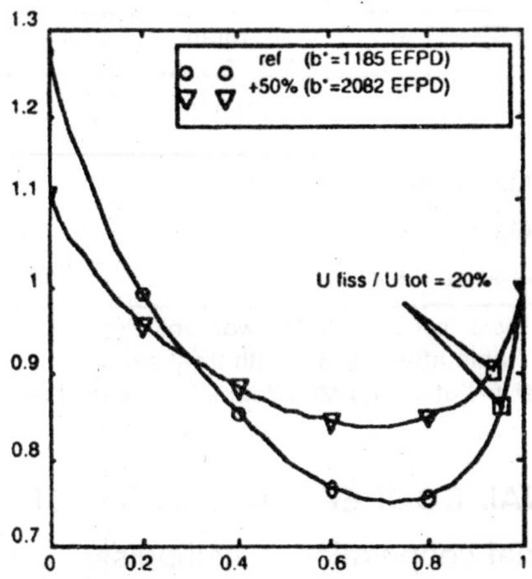

Fig. 7 : Relative burn-up vs. ^{232}Th fraction

Only in very aggressive conditions such as in the combination HNO^3 + HF solubilisation is partially attained. The extrapolated corrosion velocities are extremely low (Tables VI and VII).

Table—VI : Solubility Tests: Dissolution in HNO3-8M, 1 hour

		T = 110°C room P	220°C 16.4 bar
Sample Type	**Sample Weight (g)**	**Weight Loss (% wt.)**	**Weight Loss (% wt.)**
$ZrO_2(Y_2O_3)$–Al_2O_3-MgO Pellet	2.5124	<0.01	0.01
$ZrO_2(Y_2O_3)$-Al_2O_3-MgO+CeO_2 Pellet	2.5633	<0.01	0.03
ZrO_2(CaO) Pellet	4.9872	<0.06	0.2
ZrO_2(CaO)+CeO_2 Pellet	5.2836	<0.01	0.1
$ZrO_2(Y_2O_3)$-Al_2O_3-MgO Microgranules sintered at 1550°C	2.5217	<0.11	0.6
$ZrO_2(Y_2O_3)$-Al_2O_3-MgO+CeO_2 Microgranules sintered at 1550 °C	2.5066	<0.06	0.4

Table VII : Solubility Tests: Dissolution and Leaching Rates

	HNO3-8M (1 hour)		Leaching (room P and T, 8 months)	
Sample Type	room P; T-110° (g/g/hr)	C16.4 bar; 220°C (g/g/hr)	pH-3.65 (g/g/day)	pH=8[^^] (g/g/day)
$ZrO_2(Y_2O_3)$-Al_2O_3-MgO Pellet	10–5	2x10-4	4x10-8	1x10-8[^]
$ZrO_2(Y_2O_3)$-Al_2O_3-MgO Microgranules sintered at 1550°C	5x10-5	5x10-3	8x10-8	1x10-8[^]

^ In the leaching tests (pH=8) only Mg was solubilized (Al, Y, Zr and Ce was never detectable after the 8 month treatment)

^^ The groundwater pH at Yucca Mountain US projected repository is about 8

INTERNATIONAL COOPERATION CONTEXT

International cooperation is compulsive condition for the programme to achieve final objectives. EU in the context of the 4-th Framework Programme 1994-98 (Nuclear Fission Safety), provides a useful but minor support at moment. Stronger cooperation efforts will be needed for the programme to progress and an enlarged-harmonised action at European level would be desirable. In this perspective the ENEA irradiation experiment on IMF, to be carried out in the Halden HWBR, has the potential to become a common experiment.

The experimentation in Halden will start from HEU based IMF as a first step and is outlooked to of follow with Pu bearing IMF in a second step. In either case the presence of Th is envisaged also, not only for having the IMF reactivity feedbacks increased but also because the U-Th and Pu-Th oxide fuel is foreseen in the ADS-Rubbiatron devoted to trans-uranium nuclides (TRU) and long lived fission products transmutation, whose R&D programme is presently supported by ENEA too.

Moreover an ENEA-IPPE Obninsk (Russia) joint proposal on Excess Weapon-Pu Disposition, has been presented to the Italian Ministry for the Foreign Affairs under the framework of the "International Non-proliferation Agreements" and is on wait to be supported by it. The main objective of this project is a demonstration experiment based on design, fabrication and irradiation in a Russian LWR (possibly commercial VVER), of a representative IMF test assembly bearing Russian excess-weapons Pu. There is the possibility that this cooperation will be accessed by some US participants also.

CONCLUSION

The Pu stockpiles, either excess-weapons or civilian, do represent a tru challenge for which the present peaceful nuclear technology can find a rational solution. The urgent risk put particularly by weapons-plutonium coming from dismantled warheads, can be offset by setting appropriate and innovative solutions like inert matrix-thoria fuel aimed at current commercial PWRs, for converting a not negligible energetic potential into useful electricity in these already existing nuclear power plants. In that way the huge economic investments made during the last decades on nuclear weapons, could at least partially be recovered and anyway converted to a peaceful goal. The once through U-*free* fuel concept is moreover an highly promising anti-proliferation option for the safe-ultimate Pu disposition.

The drawbacks arising with respect to standard MOX fuel can be overcomed by using possible remedies like durable poisons and Thoria addition. The fabrication technology is available. Extensive irradiation testing is required, however. The effort and time needed for the fuel qualification can be compatible with resources and time outlooked in the excess-weapons Pu disposition programmes (US-DOE Surplus weapons-Usable

Plutonium Disposition Programme envisages a 25 years period starting from 1995, with first 10 years for solutions set-up).

ACKNOWLEDGEMENTS

Acknowledgements are due to Drs. Magnani and La Torretta of ENEA-Research Centre on New Materials, Faenza-BO, Drs. Marmo, Bruno and Coelli (stageaire) of Fabbricazioni Nucleari, Bosco Marengo-AL, Drs. Luzzi, Ossi and Pastorelli of Polytechnic of Milano, Dr. Zappa of ENEA-Casaccia Research Centre, for their fundamental contribution to the on-going experimental activities on inert matrix and simulate fuel pellets.

REFERENCES

1. E. CERRAI and C. LOMBARDI," Burning Weapon-Grade Pu in Ad Hoc Designed Reactors?", *Proc. International Symposium on Conversion of Nuclear Warheads for Peaceful Purposes, II (1992).*
2. H. AKIE et al., " A New Fuel Material for Once-Through Weapons Pu Burning", *Nuclear Technology,* 107 (1994)
3. C. LOMBARDI, A MAZZOLA, " A Non-Fertile Fuel Concept for Pu Burning in Thermal Reactors", *HPR-345, II, Halden Reactor Project (1994).*
4. J.M. PARATTE and R. CHAWLA, "On the Physics Feasibility of LWR Pu Fuels Without Uranium", *Ann. Nucl. Energy, 22,7,471 (1995)*
5. C. LOMBARDI, A. MAZZOLA, F. VETTRAINO, "Pu Annihilation in PWRs Via Non-Fertile Matrices", *Proc. Global 1995-International Conference on Evaluation of Emerging Nuclear Fuel Cycle Systems, September 11-14,1995, Versailles, France, 2,1374*
6. C. LOMBARDI, A. MAZZOLA, "Pu Burning in PWRs Via Non-Fertile Matrices", *Nuclear Science and Engineering, 122,229(1996)*
7. C. LOMBARDI, A. MAZZOLA, M. RICOTTI, "Neutronic Analysis of a Plutonium Burner PWR Partially Fed With Inert Matrix Fuel", *Proc. 4th International Conference on Nuclear Engineering 1996 (ICONE-4), New Orleans, March 10-14, 1996,4,221*
8. C. LOMBARDI and A. MAZZOLA, "Exploiting The Plutonium Stockpiles in PWRs By Using Inert Matrix Fuel", *Ann. Nucl. Energy, 23,14,1117*

9. C. LOMBARDI, A. MAZZOLA, F. VETTRAINO, "Inert Matrix Fuels for Burning Civilian and Weapons Plutonium in Thermal Reactors", *EHPG-Meeting, Loen, Norway, May 19-24, 1996*
10. C. LOMBARDI, A. MAZZOLA, F. VETTRAINO, "Non-Fertile Fuels for Burning Weapons Plutonium in Thermal Fission Reactors", *Int. Conference on Military Conversion and Science "Utilisation/ Disposal of the Excess Fissile Weapons Materials: Scientific, Technological an Socio-Economic Aspects", Como, Villa Olmo, Italy March 18-20, 1996, SCIENCE FOR PEACE series, Unesco Venice Office-Landau Network Coord. Centre Publisher*
11. C. LOMBARDI, A. MAZZOLA, "Use of Plutonium Stockpiles in Uranium-free fuel for Energy Production", *Proc. of the Embedded Topical Meeting on DOE Spent Nuclear Fuel & Fissile Material Management, Reno, NV, ANS Publisher, 1996*
12. F. VETTRAINO, C. LOMBARDI, A. MAZZOLA, "Inert Matrix Fuel for Plutonium Burning in LWRs", *IAEA Symposium on Nuclear Fuel Cycle and Reactor Strategies: Adjusting to New Realities", Vienna, Austria, June 3-6, 1997 (Poster IAEA-SM-346/17P)*
13. C. LOMBARDI, A. MAZZOLA, M. RICOTTI, "Analysis of a Pu-Th Fuel Cycle for the Reduction of the Plutonium Stockpiles", *Proceeding of ICONE-5: 5th International conference on Nuclear Engineering, Nice, France, May 26-30, 1997*
14. P. JELINEK-FINK, "Summary of the Symposium", *IAEA Symposium on Nuclear fuel Cycle and Reactor Strategies: Adjusting to New Realities", Vienna, Austria, June 3-6, 1997*
15. F. VETRAINO et al., "Inert Matrix Fuels for Plutonium Disposition in PWRs", *ICONE-6 Sixth International Conference on Nuclear Engineering, San Diego, Calif., USA, May 10-15, 1998*

Address:

FORTUNATO VETTRAINO
Nuclear Fission Division, ENEA
40129 Bologna
Italy

NUCLEAR MATERIALS
Issues and Concerns

NUCLEAR MATERIALS
Issues and Concerns

Vol. II

Editor

Dr. Digumarti Bhaskara Rao
M.Sc., M.A., M.A., M.Ed., Ph.D.

Secretary
"Academy of Communication Culture, Education Science and Service"
1-22-10 Srinivasa Nagar
Guntur 522006
Andhra Pradesh
India

2001

DISCOVERY PUBLISHING HOUSE
NEW DELHI—110 002

Published by :
Discovery Publishing House
4831/24, Ansari Road, Prahlad Street
Darya Ganj, New Delhi—110 002 (INDIA)
Phone : 327 92 45
Fax : 91–11–3253475

First Published—2001

ISBN 81–7141–611–X(Set)

Laser Typeset by :
Allied Computers,
Karnal (Haryana)

Printed at:
Arora Offset Press
Laxmi Nagar, Delhi 110 092.

Preface

Nuclear materials have civilian utilisation too along with military utilisation. The dual nature of the utilisation of nuclear materials has originated many problems but has also created several opportunities. If the nuclear materials are used for the production of civilian nuclear energy even to a limited extent, then there will be no energy crisis atleast in sectors like electricity. But, many world countries are developing and or developed sophisticated technologies for arms development at the cost of civilian safety. It is the time now for elimination of the danger of nuclear conflicts from human life and for developing nuclear technologies that work for the welfare of mankind.

This book covers a wide range of vital issues and concerns related to nuclear materials such as nuclear research and development, disarmament and arms control, illegal nuclear traffic and its associated aspects, disposal of radio-active waste/fissile materials, decommissioning and management of surplus nuclear equipments and materials, etc. A special attention to Indian and Pakistan nuclear tests has also been given.

The articles included in this book are selected from the (1) International Conference on Military Conversion and Science" Utilisation/Disposal of the Excess Fissile Weapons Materials: Scientific, Technological and Socio-Economic Aspects", 18–20 March 1996; (2) International Forum on "Illegal Nuclear Traffic: Risks, Safeguards and Countermeasures", 12–13 June 1997; (3) International Forum on "Nuclear and Biological Decommissioning: Management of Global Security Threats", 27–28 June 1997; and (4) Forum of the UNESCO International School of Science for Peace on "Nuclear Disarmament, Safe Disposal of Nuclear Material for new weapons Developments ? Where are the National Laboratories Going ?", 2–4 July 1998 organised by the UNESCO Venice Office and others. The articles have been written by the experts in nuclear issues and concerns.

I am grateful to Prof. Pierre Lasserre, Director, UNESCO Venice Office, for giving me permission to reproduce articles from volumes one, four, five and eight of Science For Peace series. I am thankful to Dr. Rossanna Santesso, Asst. Information Officer, UNESCO Venice Office for readily responding to all of my queries. Special thanks are extended to UNESCO Venice Office, Landau Network-Centro Volta, Italian Ministry of Foreign Affairs, Italian Ministry of Universities and Scientific and Technological Research, ENEA, and ACCESS.

I hope that this book will have an impact on politicians and policy makers, scientists and technologists, economists and educationists and advocates and activists with regard to nuclear issues and concerns.

Dr. Digumarti Bhaskara Rao

Contents

PART–7
UTILISATION/DISPOSAL OF THE EXCESS FISSILE WEAPON MATERIALS: SCIENTIFIC, TECHNOLOGICAL AND SOCIO-ECONOMIC ASPECTS

PART–8
INDIAN AND PAKISTAN NUCLEAR TESTS

PART—6

DECOMMISSIONING AND MANAGEMENT OF SURPLUS NUCLEAR EQUIPMENTS AND MATERIALS

48
Decommissioning Strategy in the Russian Federation

A. ABAGYAN AND V. ZIMIN

DECOMMISSIONING STRATEGY

In the period up to 2020, in connection with expiration of the specified lifetime, 28 units with total capacity of 20,242 MWe are expected to be decommissioned in Russia (Table 1).

There are two main types of reactors, which have been designed and operate in Russia:

- the RMBK- 1000 reactor—a heterogeneous channel-type nuclear thermal neutron power reactor which uses graphite as moderator. This type of reactor has two generations which include 11 units with total capacity –11,000 MW;
- the WWER (PWR) type reactor—a pressurised water reactor with water as moderator and coolant. This type of reactor has three generations—WWER—440 and VVWER-1000 which include 12 units with total capacity—8640 MW. At present, 4 units with the total capacity of 875 MWe have already been shut down in Russia:
- Beloyarsk NPP Units 1,2 (RBMKs prototype);
- Novovoronezh NPP units 1,2 (VVERs prototype).

The shutdown units are the first ones to be decommissioned and their decommissioning experience will be used to master the methods and principles to be employed for decommissioning subsequent units. In view of that, for the above units the final phase and the timing of decommissioning have not been defined.

Final shutdown of an NPP unit is conducted in the following cases:

- on expiration of the specified lifetime;
- after an accident resulting in the impossibility or impracti-

Table 1

NPP Name	*Reactor Type*	*Unit No.*	*Power MW(el)*	*Unit start-up year*	*Shutdown year*
1. Novovoronezh-skaya	WWER	1	210	1964	1984 *
	WWER	2	365	1969	1990 *
	WWER	3	440	1971	2001
	WWER	4	440	1972	2002
	WWER	5	1000	1980	2010
2. Beloyarskaya	AMB	1	100	1964	1981 *
	AMB	2	200	1967	1990 *
	BN	3	600	1980	2010
3. Bilibinskaya	EGP	1	12	1974	2004
	EGP	2	12	1974	2004
	EGP	3	12	1975	2005
	EGP	4	12	1976	2006
4. Balakovskaya	WWER	1	1000	1985	2015
	WWER	2	1000	1987	2017
	WWER	3	1000	1988	2018
5. Kolskaya	WWER	1	440	1973	2003
	WWER	2	440	1974	2004
	WWER	3	440	1981	2011
	WWER	4	440	1984	2014
6. Kaliniskaya	RBMK	1	1000	1984	2014
	RBMK	2	1000	1986	2016
7. Leningradskaya	RBMK	1	1000	1973	2003
	RBMK	2	1000	1975	2005
	RBMK	3	1000	1983	2013
	RBMK	4	1000	1985	2015
8. Kurskaya	RBMK	1	1000	1976	2006
	RBMK	2	1000	1978	2008
	RBMK	3	1000	1983	2013
	RBMK	4	1000	1985	2015
9. Smolenskaya	RBMK	1	1000	1982	2012
	RBMK	2	1000	1985	2015
	RBMK	3	1000	1990	2020

* — the units have already shut down and their capacity does not count up in total capacity

cability of unit operation;

- if safety and reliability requirements have changed and they cannot be met using the existing design criteria;
- if further operation is economically impractical.

Main principles of NPP decommissioning conception in

Russia are:

- while planning NPP unit decommissioning it is necessary to proceed from the principle of maximum restoring or replacing units shutdown by new more safe NPP units;
- to use sites of decommissioning NPP as much as possible;
- maximum possible usage of NPP sites to be decommissioned for nuclear power needs;
- over profiling conversion of NPP units to be decommissioned.

At present the following conditions exist in Russia which significantly influence and complicate resolution of the decommissioning problem:

- there is a lack of legislation and normative information regulating the issues of NPP decommissioning and radwastes management;
- there are no national and regional storage and disposals for storage of various types of radwastes, arising during operation and decommissioning of NPP unit;
- decommissioning of NPP unit will be carried out at NPP site where other units continue in operation, so the effect of the decommissioning activities on the safety and operating conditions of these units must be considered;
- complexity and lack of experience with technical procedures, high cost of liquidating units or entire NPPs which become prohibitively expensive when the sites are required to be brought into a "green field" condition;
- decommissioning has to be carried out on financing from a single decommissioning fund created from contribution of basic activities of NNP which at present are very limited amounts.

Decommissioning is a long and complex process and preparation for it is made at all stages of the NPP unit life cycle.

In accordance with regulatory body requirements the operator have to fulfill a feasibility study of unit discontinuance options before five years of its schedule shutdown and have it approved by supervisory and regulatory bodies. The goal of the feasibility study is an analysis of technical and economic feasibility of options for extending the operational life time or decommissioning the unit and defining the main option. (Fig. 1).

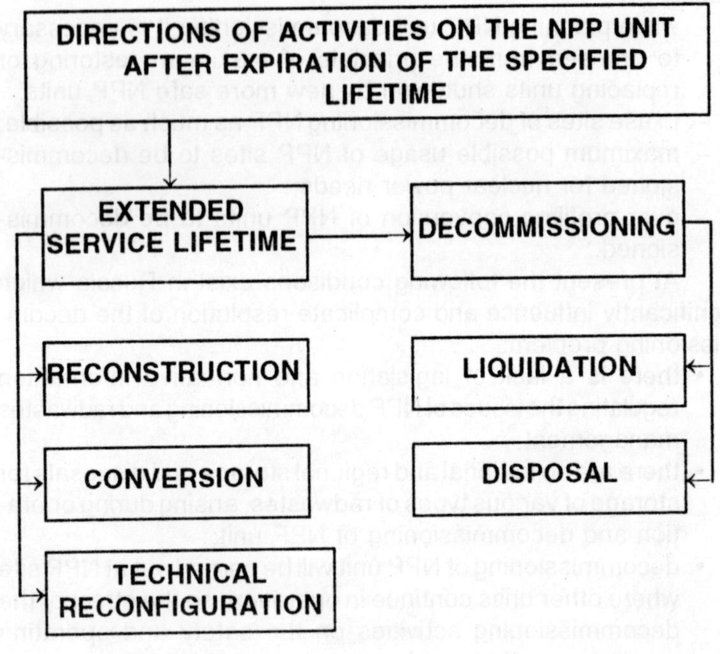

Fig. 1

If it is decided to decommission the unit a programme for unit preparing for decommissioning and decommissioning programme are prepared.

After final shutdown of the unit the period of preparation the NPP unit for decommissioning begins. In accordance with regulatory body requirements the unit undergoing preparation for decommissioning is considered to still be in operation in regards to safety regulations and specially prepared and approved set of procedures is established for this purpose.

The period of preparation includes removal of all fuel from the unit, standard decontamination of equipment, processing of operational radwastes, development all necessary documentation for decommissioning activities.

The period of preparation the unit for decommissioning has length from three to five years.

Decommissioning is the final stage in the life cycle of a NPP unit and of a whole set of activities after shut down and removal all nuclear fuel from the unit which ensure the unit will never again be used as a source of energy and provide for the radiation safety of personnel, population and environment.

To implement NPP decommissioning technical policy a "Concept of NPP Unit Decommissioning After Lifetime Expiration" was developed, which in 1991 was adopted by the Russian Ministry of Atomic Energy in the form of a regulatory document.

This concept specifies the implementation of the following stages for an NPP unit having expired its lifetime or its extend lifetime:

- unit preparation for safe storage under surveillance:
- facility moth-balling;
- localisation of highly-active equipment (reactor as an assembly) in reactor building rooms;
- the simultaneous dismantling of non-active and low contaminated equipment, buildings and structures; processing of radioactive waste for safe storage at NPP site;

  The length of this stage is from three to five years.

2) Safe storage under surveillance of localised (isolated) equipment during long time. The total activity reduction during toe 20-30 year period of hold-up is estimated to be 60-100 times.

The length of this stage varies according to the type of reactor, service period of the construction elements in which the localised equipment is located, necessity of using the site for new construction, etc.

The length of this stage can vary from 30 to 100 years.

3) Unit complete dismantling with preparing of the unit site for new construction. During this stage the following activities are carried out:

- complete dismantling of localised equipment;
- dismantling of those buildings and construction elements of the unit not to be used further;
- processing and removal all radwastes to a regional storage or for disposal;
- relandscaping of the site for construction of a new NPP unit or its use for some other industrial use.

  The length of this stage is about five years.

RADIOACTIVE WASTES

During NPP decommissioning the quantity of radioactive

waste being a subject for processing develops from:

- Operational liquid (solidified) and solid radioactive waste accumulated during period of NPP operation (Table 2);
- Operational liquid and solid radioactive waste arising during NPP decommissioning and using of installations on processing (Table 3);
- Solid waste from dismantle of NPP equipment, buildings and structures (Table 4).

Management of radioactive waste is one of the most important stages of NPPs' decommissioning, since just engineering provisions for transportation, storage and disposal of waste define the more preferable way for plant decommissioning.

Table 2 : Operational radioactive waste arising on one NPP unit during 30 years operation

Reactor Type	**Solidified Waste**		**Solid Waste,**
	Bituminised, cubic m.	***Cementing, cubic m.***	***cubmic m.***
WWER-440	3900	5200	6000
WWER-1000	7500	10000	9000
RBMK-1000	22500	30000	12000

Table 3 : Operational liquid and solid radioactive waste arising during NPP unit decommissioning and using of radwaste processing installation

Reactor Type	**Solidified Waste**		**Solid Waste,**
	Bituminised, cubic m.	***Cementing, cubic m.***	***cubmic m.***
WWER-440	390	520	600
WWER-1000	750	1000	900
RBMK-1000	2250	3000	1200

Table 4 : Solid waste from dismantle of NPP equipment, buildings and structures

Reactor Type	***Concrete, ton***	***Metal, ton***	***Equipment, ton***
WWER-440	390	520	600
WWER-1000	750	1000	900
RBMK-1000	2250	.3000	1200

Implementation of multi-barrier conception for environmental protection while managing liquid and solid radwaste is planned to be carried out in some stages.

At the first stage, collection and grading of liquid radwaste on activity level, salinity, content of surface-active agents are being carried out. Solid radwaste are being graded on activity level as well as on combustible, pressible, decontaminatible and other groups depending of the decision taken for their following reprocessing or storage.

The second stage, namely temporary storage of waste at the plants where it was generated without reprocessing, may be caused by the absence of reprocessing facilities or necessity to decrease radioactivity of waste with decay of short-living radionuclides.

The third stage, namely waste conditioning, foresees to transform liquid and solid waste into the suitable condition for storage, transportation and disposal. The criteria to choose this condition are: chemical, thermal and radioactive stability explosion-proof and fire safety, mechanical strength etc. To the methods for liquid waste conditioning belong solidification, packing of waste solidified into drums or different containers.

Solid waste conditioning foresees incineration, pressing, decontamination, protective coating, packing into containers.

The fourth stage, namely interim storage of waste conditioned, may be caused by necessity to decrease waste radioactivity or lack of regional repositories for its disposal.

The fifth stage, namely transportation, subject to the regional repository has been created is the necessary stage of radwaste management scheme. Different containers of concrete, metal or their composition are used to transport waste package containers can be used not only for transportation but also for interim storage and disposal of waste.

The sixth stage, namely radwaste disposal, is intended to remove waste finally away of the area of human activities. Waste can be disposed into near-surface, buried and deep geological formations depending of their activity level and availability of long-living radionuclides.

Evaluation of the general situation in radwaste management in the Russian nuclear power displays that only first four stages of the above concept can be implemented today. Transportation and disposal of radwaste into regional repositories are examined today at the stage of technical and economical research applying to radwaste generated while operating and decommissioning Russian NPPS.

SOCIO-ECONOMIC ASPECTS OF NPP DECOMMISSIONING

A whole series of conditioning influence (Fig. 2) the decision to NPP decommissioning. The analysis of these conditions shows that economic consequences of the NPP decommissioning decision can be both direct and indirect.

EXISTENCE OF A LEGISLATIVE AND REGULATORY FOUNDATION FOR THE PROBLEM

DEGREE OF READINESS OF AN EMPLOYMENT PROGRAMME FOR RESIDENTS OF THE PLANT TOWN

DEGREE OF READINESS AND ELABORATION OF TECHNICAL DECISIONS FOR THE PROBLEM (PLANS FOR THE NPP)

RESOLVING POWER SUPPLY PROBLEMS FOR THE NPP AND NEARBY TOWN

DEGREE OF READINESS OF NATIONAL AND/OR REGIONAL SPENT NUCLEAR FUEL AND RADIOACTIVE WASTE REPOSITORIES AND STORAGE FACILITIES

RESOLVING FINANCIAL COMPENSATION PROBLEMS OF THE CITY AND REGIONAL BUDGETS

AVAILABILITY OF PERSONNEL PREPARED FOR WORKING THE PROBLEM (FOR VVER AND RBMK)

RESOLVING THE PROBLEM OF REPLACING NPP POWER CAPACITY FROM OTHER SOURCES

AVAILABILITY OF RELIABLE FINANCIAL RESOURCES FOR ALL STAGES OF THE PROBLEM (INCLUDING SCIENTIFIC, ENGINEERING AND PILOT RESEARCH PROGRAMMES FOR FEASIBILITY STUDIES OF DECOMMISSIONING NPPs)

ANALYSIS OF THE READINESS OF REGIONAL ELECTRIC GRIDS

DECISION ON DECOMMISSIONING THE NPP

Fig. 3 : Set of conditions to consider when making a decision on decommissioning a nuclear power plant

The economic loss from the NPP decommissioning has been broken down into elements to facilitate carrying out cost estimates. The loss from decommissioning any energy generated facility can be divided into two basic elements: technical/ economic cost and socio-economic cost. And of course these costs can be both direct and indirect.

The following costs are considered technical and economic:

- expenses for direct decommissioning activities;
- economic losses from electric power not generated (the maximum lost in the case of early decommissioning);
- expenses for indirect activities on decommissioning (including both activities at the NPP site as a preparatory activities);
- developing and carrying out scientifical engineering and research programmes);
- expenses for construction of a repository and storage facility for spent fuel and radioactive waste;
- travelling expenses;
- cost of energy (electric and heat) sources for NPP,NPP site, nearby plant town;
- cost of NPP power capacity substitution. Summary cost and labour intensity for one WWER type unit decommissioning are shown in Table 5.

Initial socio-economic factors to consider include;

- cost of a social rehabilitation programme for displaced NPP employees and population of nearby plant town;
- cost of financial compensation to the regional budget, nearby plant town budget and other budgets effected by NPP decommissioning;
- determine the fate of the near-by plant town and the local industrial infrastructure which services the NPP.

Indirect economic costs include:

- losses from sharp reduction or complete termination of use of a nuclear power engineering fuel base;
- losses from non-use scientifical technical and industrial base for development, construction and operation of nuclear power plants;
- losses, connected with loss of competent staff in the field of nuclear power engineering;

Table 5 : Summary cost and Labour Intensity for Decommissioning of One Unit with RBMK-1000 and WWER-440 Reactor Type

NN	Stage	Labour Intensity thous. person. hour		Cost thous. US$ / Specific cost $ kWt (el.)	
		RBMK-1000	WWER-400	RBMK-1000	WWER-440
1	2	3	4	5	6
1.	Pre-decommissioning Stage	21039.0	9326.0 128.0	127680.0 126.0	55331.0
2.	Stage 1–Unit preparation for safe storage under surveillance	12381.0	4645.0	169712.0 170.0	58948.0 137.0
3.	Stage 2 – Unit safe storage under surveillance	1632.5	910.0	114310.0 114.3	16815.0 38.2
4.	Stage 3 – Unit complete dismantling*			130000.0	44000.0
Total cost for pre-decommissioning and decommissioning activities from 1994 to 2039:				**541702.0** **541.7**	**175094.0** **397.0**

* — cost is shown only for dismantling of reactor and all residual radioactive parts of the unit

Table 6 : Summary cost result of economic losses from decommissioning of unit with WWER-440 (million $)

• Cost for socio-economic compensate measures for NPP personal and residents in case of NPP decommissioning	**–22.1**
• Cost of constructing and operating of regional radioactive waste repository	**–24.9**
• Cost of electric power needs for NPP decommissioning	**–9.0**
• Cost of heat	**–7.7**
• Cost of financial compensation to the decommissioning fund	**–3.0**
• Economic loss for industrial enterprises supporting the NPP activities	**–140.0**
	226.7
Loss to the gross national income from delivered electric power	**–5282.0**

- cost, of possible reconstruction of regional electric power grid to increase its capacity. Summary cost results of economic losses from decommissioning of NPP unit with WWER-440 is shown in Table 6.

Conducted valuation have shown that at a existing structure of a atomic engineering in Russia socio-economic consequences from nuclear power plants decommissioning without construction of replacing capacities on the existing sites can result in heavy socio-economic losses.

At planning NPP unit shut-down needs to be taken into account costs not only on fulfillment of decommissioning activities but also costs:

- On indemnification of consequences, connected with change of power and economic situation in region of NPP accommodation,
- On liquidation of social consequences from realisation of these activities.

As research show, maximally possible use of NPP infrastructure for construction of replacing capacities and manufacture of electrical energy permits not only essentially to lower the charges on NPP decommissioning but can be practically eliminated social problems.

SCIENTIFIC PROVISIONS

At present the NPP decommissioning activities are conducted pursuant to:

- The Decommissioning Concept, accepted in Russia;
- Scientific–Technological Programme "Decommissioning of NPP units, fulfilled the specified lifetime"
- Working programme "Management of NPP radioactive wastes"
- Annual plans of work on technological and industrial activity on Beloyarsk NPP and Novovorenege NPP

The programme structure provides realisation of activities on following directions:

- Development of the normative documents, determining and regulating decommissioning activities.
- Research and development of ways and means for deep decontamination high-contaminated and activated metal structures, materials and main NPP equipment.

- Research and development of ways and means for processing liquid radioactive waste, arising during operatioin and decommissioning
- Development of methods and means on maintenance of radiation safety on NPP and in a a environment at stages of preparation for decommissioning
- Development and manufacturing of the equipment for waste containerisation and packing, for transportation, long-term storage and dispoal of wastes arising during decommissioning.

ACTIVITIES CARRIED OUT ON THE SHUT DOWN NPP UNITS

1 and 2 units of Beloyarskaya NPP

The first and the second units of Beloyarskaya NPP have been shut down:

- The first unit with reactor AMB-100-in 1981;
- The second unit with reactor AIB-200 - in 1989.

Spent fuel are completely unloaded from the reactors and are located in water cooling ponds (BV - 1,2), total quantity of spent fuel assemblies is about 5000. The assemblies are placed in 17 and 35-place canisters. The ponds are in cooperation about 26 years.

The decision of spent fuel problems will be executed in three stages;

1 stage

- designing and construction on the NPP site of new storage for containers with spent fuel ensuring safety up to a moment of final removal of spent fuel;

2 stage

- designing and construction of a complex for SF conditioning, contiguous to storage;

3 stage

- removal SF on long-term storage into the storage.

At present the decommissioning activities on NPP are carried out pursuant to the authorised Decommissioning Programme.

Thus, at present on NPP a stage of preparation 1 and 2 units for decommissioning with temporary "endurance" of reactors is realised. This stage will be continued during 10-15 years before construction of spent fuel storage and complex for radioactive waste processing will be completed.

Simultaneously with this the following activities are executed:

- Development of complex for decontamination and processing of dismantled equipment:
- Design of a dry waste storage on NPP site;
- Development of the technical decisions on dismantling technology;
- Development of the options for management with graphite moderator;
- Development of technology for cooling pond waters cleaning from radionuclides;
- Development of the design for spent fuel storage;
- Development of the project of 1,2 units temporary endurance;
- Development of design for solidified liquid waste;
- Completion of complex engineering and radiating inspection of the equipment and structures of 1 and 2 units.

THE FIRST AND THE SECOND UNITS OF NOVOVORONEZH NPP

Units 1 and 2 have been removed from commercial operation:

- the first unit with a reactor VVER-210 in 1984.
- the second unit with a reactor VVER-365 in 1990.

Spent fuel assemblies are completely unloaded from reactors and located in cooling pond of the 1st unit, total quantity of assemblies is about 700. Other assemblies are sent on processing plant.

The decommissioning activities on 1 and 2 units are carried out pursuant to:

- NPP Decommissioning Programme;
- Decision on long safe storage of 1 and 2 units;
- Measures on transferring of 1 and 2 units in a safe condition at fulfillment of decommissioning activities;

On 1 and 2 units the following main activities have been carried out:

- the decision on using of separate reactor building rooms for storage of containers with conditioned waste is adopted;

- Technology for equipment decontamination is developed and installation is mounted.
- Experimental decontamination of outside surfaces of rooms and equipment is conducted;
- Dismantle of the clean equipment (pumps and fittings) for re-use is begun;
- the feasibility study of long term conservation of the units is finished;
- the project of a complex on metal waste processing is developed;
- Project of temporary site storage for drums with salt concentrate after deep evaporation of bottoms is developed;
- Concept of the waste management on the NPP site is developed;
- Technology of a steam generator decontamination is developed and an experimental decontamination of one SG is executed.
- Dismantle of the turbine number 1 on the first unit is executed.

Address:

ARMEN ABAGYAN
Vice-President
ROSATOMENERGO
119899 Moscow
Russian Federation

49

Conversion Problems at the Enterprises of Georgia

R. KHURODZE

The Republic of Georgia and other countries of the former Soviet Union participated in diverse work on a military theme.

Various scientific research and experimental constructional work for the production of special military techniques was conducted to the order of Military Institutions.

Work was mainly conducted in institutions founded by Georgian governmental ministries, such as the Scientific-Research Institute, Construction Bureau, and production enterprises, that were concerned with production in the defence field under the auspices of the Military Industry Committee.

Also, the different institutes of the Academy of Sciences of Georgia, which is the higher education establishment, were involved the war theme.

It is important to mention that scientists and other experts from the USA, Europe, Japan, and Russia were present at the May 1995 Symposium for Project Development and Conversion which was held in Tbilisi under the aegis of the International-Scientific Technological Centre, (founded by the USA, European Union, Japan and Russia).

The guests, after seeing results and future plans for defence conversion, fully supported the existing scientific-technological potential of the Republic.

Let me briefly dwell on the question of defence industry conversion in Georgia and the research conducted so far.

I am of the opinion that one of the brightest examples of conversion is the work performed by Georgian aviators. The aviation industry of Georgia which produced warplanes has developed the six-seat civilian aircraft designated JAK 58. Currently, the Georgian plane is being displayed at the renown

Le Bourget exhibition in France.

The Electric industry of Georgia has developed and serially-produced integral schemes for civil as well as military use. Scientific research and experimental-constructional work has been conducted for obtaining digital and line microschemes of minor and average degree integration.

More than 30 typonominals have been developed and implemented in the production.

A part of the scheme with standards technology only in Georgia were worked out original construction of the silicon type (on sapphire) distinguished with rapid activity and radioactive activity.

One of the important work conducted by the Electronic Enterprise on the decision was due to the increase of radioactive agility in IC. Complete microschemes were delivered to various enterprises of the Ministry of Defence and several branches of industry were applied to various equipment, military purposes in systems, (e.g. snowstorms), missiles, etc. Currently, such work is not being carried out. The Enterprise produces various equipment for the husbandry needs of the Republic.

During the formation of the Electronic Enterprise (particularly 1970-1990), optoelectronics as a prospective trend in solid electronics was being intensively developed.

Record parameters of radiated red light structures were obtained in the Republic.

Optoelectronic equipment (light diodes, scale, matrix, digital indicators) were created on the basis of the structures developed for the various applications, particularly for the information reflection system.

Basically, to the order of Military Departments, highly valuable and effective semiconductor light radiator indicators were created. Among them we can single out:

- Red light semi-conductor indicator scales with high brightness, for recording information on the photo film in the optical mechanical block of the board system in the night vision apparatus.
- Linear Scales of visual level to account for special techniques
- Semi-conductor mullet coloured table for the board system of military aircraft.

- Various light-emitting diodes for special techniques.
- Experience gained in this field is used for developing optoelectronic equipment of wide use for the creation of photo-electronic transformation of solar energy, etc.

Tbilisi instrument makers (particularly Scientific Production Unit (SPU) "Analyzpribor"), in accordance with their skills performed a considerable volume of work on a military theme, in particular, for the processes of production and delivery of weapons of mass destruction.

On the basis of scientific-technical potential gained while conducting work on a military theme, SOPU ""Analizpribor"" developed and produced a considerable number of apparatuses for civilian use, particularly photometrical analysers for defining the number and size of measured particles in liquids.

These photometers are used in civil aviation for control of fuel-greasing materials; children's nutrition; production of liquid medicines, metallic powder, natural water, and sewage; and for the control of technological contours of nuclear and thermal power stations.

The Institute of Metrology, Standardisation and Certification of the State Standard conducted work on high frequency range physometrical apparatus used in civilian applications of radio techniques communications, and radio-navigation.

Phase circulator of electric signals, the equipment for measuring the distance to the place of break in light, the equipment for measurement of time delays in fibre-optical components, and so on had a dominant place in the work of Scientific Research Institutes of Georgia. Work which was conducted for creation of the management systems and protection of transported nuclear power equipment, active media for lasers, systems for radiation protection of the staff, formation of nuclear materials, production of night-vision glass, monocrystals of cesium and arsenic mixture for directing laser rays, etc., has been halted.

The authorities have decided that the Physical Technical Institute of Sukhumi should learn war techniques production. At present the Institute is conducting research on the creation of equipment and material for ecological tasks, energy needs, communications and other technical fields.

The Institute of Cybernetics of the Academy of Sciences

is involved in conducting special bi-directional research in fields of theoretical cybernetics and physics. The first direction embodied questions such as identification of the kinds of (by the Soviet SOA programme) decision-making and mathematical modelling.

The main problem of the second direction was optical processing of information of semi-conductor structure, lasers, and conductors for the use of holography and fluid crystals.

Presently these two directions contribute to the Republic's formation and public service.

Light water pull-type nuclear reactor IRI 2000 had been in use in the Institute of Physics at the Academy of Sciences since 1959. Its initial capacity was 2000 kwt but after reconstruction in 1967 and 1974, capacity was raised to 5000 and later to 8000 kwt.

The reactor was equipped with low temperature channels which enabled a radiation pattern within the vicinity of active zone at low temperature (10° K) but inner channel measurements temperature at the temperature of liquid nitrogen.

Research work has been conducted in neutron physics, radioactive chemistry, biology, and in the analysis of neutron radiation activation methods. Solid body low temperature physics radioactive studies were of particular significance.

The Institute was the leading organisation in this field, and in the times of the former Soviet Union it conducted practical work at the order of the military industry complex, alongwith fundamental research.

The reactor was shut down in 1990 due to the exhaustion of resources. At present work is being conducted to bring the reactor in radio-safe state and is being researched on its basic technical possibilities and funds for the establishment of low capacity neutron source.

BASIC WAYS FOR CREATING POWER-SAVING MEANS AT THE GEORGIAN TECHNICAL UNIVERSITY

Recently intensive research has been conducted at the Georgian Technical University in cooperation with "Sakenergo" on improving the quality of the existing power systems as well as on search for new power sources and means for power saving. A number of significant projects have been developed, the implementation of which will considerably improve the crisis situation in the sphere of power supply in the Republic.

The project is based on the use of power-generation, drinkable, and service water free flow in the water pipe-lines (it means the pressure in water pipe-lines system). The problem is especially urgent in Georgia and Tbilisi, as, due to the slopes in some areas, water pressure is so high that it can put in motion a hydroelectric generating set with power up to one megawatt.

Analysis of the water supply system schemes in the Republic has shown that according to specific parameters it is possible to install up to 300 hydroelectric generating sets of different power and in a short amount of time to generate several hundred (up to 20 megawatts in Tbilisi alone) of total electric power.

It should also be noted that the electric power produced in this way is ecologically safe and considerably cheap as it becomes possible to minimise the construction and installation work costs which usually amount to 50 per cent of the general expenses.

A NUMBER OF LABORATORIES AND RESEARCH CENTRES OF GTU HAVE SUBMITTED THEIR PROPOSALS; LABORATORY OF CRYOGENIC AND COMPRESSOR MACHINE BUILDING HEAD OF THE LABORATORY—PROF. T. BEROSHVILI

Coal mining in Georgia and in many other countries is considered to be unprofitable as the greater part (more than 50%) of the coal is in fine granular fractions or in powder, the use of which is rather limited; It is, therefore, often impossible to sell as it, as in the case of Tkibuli coal.

Selling the coal powder (even at low prices) would make the coal industry profitable and if the powder could be used in household stoves this would create cheap resources for heating. Currently, 470,000 families in Georgia own private houses and use wood and electricity for heating, which both destroys forests and increases electric power shortages.

At GTU an original technology of the use of low cost coal fractions and coal dust in household stoves as well as older stoves with improved combustion processes and heat transfer from the evolved gas have been created.

Approximately 2 tons of coal are used by a stove with 3 kwt heat output during the whole season provided the stove functions 14 hours day.

An extremely simple and highly effective heat exchanger for heating water at the expense of internal combustion engine exhaust fumes and the cooling jacket heat has also been constructed. The heat exchanger synthesises barbotage and contact heat exchange; is highly compact, has a number of construction and technological advantages, and, above all, is characterised by high efficiency. For example, the use of such heat exchangers in Diesel generators having 200 kwt of electric power gives about 300 kwt additional heat which can heat an area of 4-3 thousand square meters.

A new generation of dry friction compressor, where the sliding friction of piston cylinders has been replaced by rolling friction, has been constructed. This together, with a number of other innovations, provides for the 20–25 per cent decrease in the electric power as compared to the analogs of the same class. A compressor modified specifically for compressing natural gas under the pressure from 1 to 250 atmospheres has been constructed as well.

RESEARCH-ENGINEERING CENTRE OF MICROELECTRONICS HEAD OF THE CENTRE—PROFESSOR R. CHIKOVANI

1. An original construction and technology for photoelectric solar energy converting cells has been worked out on the basis of silicon semi-conductors. Models of matrix type protected from mechanical influence are collected on the basis of photoelectric converting cells. The efficiency of the cells is approximately 10 per cent and the tentative cost of kilowatt electric power is $5.8. The above mentioned method is being elaborated for the full-scale production. Work is being conducted for creating highly effective (20–22% efficiency) photoelectric converter on the basis of Calcium-Arsenide semiconductor structure. Original construction of the photoelectric cells has been elaborated but owing to the high cost of the material needed the use of this type power source is rather limited and they are only used for special purposes (space research, etc.).
2. Microelectronics sensors for measuring liquid and gas flow have been constructed. The sensors can be used in the process of obtaining, processing, and utilising power carriers as well as in car engineering and domestic conditions. An electric signal is an output parameter of the above mentioned sensors; this makes it possible to match the sensors with a computer system.

This fact in turn makes it easier to automate the process of controlling the consumption, measurement, and use of power. To measure the consumption of gas flow, flow meters of thermo-pneumometric type have been constructed. For flows with speed of 0-30 m/sec measurement error does not exceed 3 per cent. For liquid flows a calorimetric flowmeter has been created; it is intended for precise measurement of minor flows with measurement error no higher than 1.5 per cent.

A laboratory model of natural gas flowmeter for domestic use with the range from 0 to 10 cubic metres per hour has been designed. A sensor for measuring the bearing of apparent of wind vector and speed has been constructed. This sensor will be of great importance for the optimum use of wind energy.

GENERAL AND THEORETICAL ELECTRONICS CHAIR #56; HEAD OF CHAIR PROF. L. MERABISHVILI

1. As the result of research, main parameter calculating methods of powerful impulse sources of current have been developed. Powerful magnetic devices for different purposes have been made.

The above mentioned devices can be used in such technological processes as: compression soldering, welding, and so on.

Using magnetic impulse devices for such technological processes greatly reduces the amount of energy necessary, and thus, it is an energy-saving technology.

2. Impulsive sources of aquatic systems for locating fault on powerful cables have been developed and made. These sources improve the effectiveness of locating aquatic systems greatly.

**GENERAL PHYSICS CHAIR, #95
HEAD OF CHAIR PROF. R. BABLIDZE**

The programmed automatic control system of optimum energy consumption.

The exterior lighting network and devices have bene designed. Instead of using traditional mechanical contractors, fast-acting symmetrical thyristors TC-171-320 were used for the first time. The load current in each phase made approximately 320 ampere. The system can provide the automatic switching on and off of exterior lighting at the selected level of natural programmed switching of certain sections of the network at a necessary successions. The devices has been examined in laboratories and it has proved highly reliable and economically effective.

THE VICE DEAN OF ENERGETIC DEPARTMENT PROF. O. VEZIRISHVILI

1. Since 1993 thermal pumping plants designed to answer technological purposes have been in use at Samtredia Tea Factory and Sagaredjo dairy integrated works. A number of devices are being installed now. The first results of exploitation have proved their effectiveness in saving energy and avoiding the environmental pollution. Thermal pumping plants are especially effective when used on the bases of thermal waste waters. The systems to provide the resorts of Tskaltubo, Borjomi and Simoneti with thermo cold have been designed also. It has been decided to supply Batumi with ecologically clean thermal pumping installations instead of building a new district fuel oil boiler house. This project is supported by Euro-Commission, and the programme considers use of the energy policy for Georgia. These measures will save approximately 2,3 ml. tonnes of equivalent fuel annually and improve the ecological situation too.
2. The perspective of using gas turbine thermoelectric blocks for combined production of heat and energy has been developed. This should save 25–30 per cent of fuel. On the basis of boiler houses functioning in Georgia, 30 gas turbine thermoelectric blocks of 1500 mgv total power can be constructed. They will save 3.3–4 million tones of equivalent fuel. It will reduce base shortage in the Republic and improve thermoelectric provisions to consumers.

THE REPUBLICAN CENTRE OF STRUCTURAL RESEARCH. THE HEAD OF THE CENTRE PROF. E. KUTELIA

1. The devices for protecting three phase-electric engines of diagnostic regime and emergency conditions.

The device can analyse the functioning of three phase engine at starting and operating conditions and also control a supply line (phase asymmetry, phase fall out). In case of emergency operation the device will switch the installation off and indicate the problem on the display panel.

2. The automatic stabilisers of wide range alternating current.

An electric network alternation within the limits of 150-270v stabilisers to maintain 220=50/v as the output power. The output power is additionally protected at more than 235v over-voltage. The output power is within the limits of 0.5–7.5 kw. Operating conditions—continuous.

ELECTRONICS ENGINEERING CHAIR 961
HEAD OF CHAIR, PROF. D. DVALI

1. Energy parameter (voltage, frequency, regulating, stabilising devices for reducing energy consumption of production equipment and household devices. Voltage 0.5–250 kW.
2. Compensators for reducing losses of devices which reduce reactive power consumption or reactive power compensators.

Address:

RAMAZ KHURODZE
Rector
Georgian Technical University
380075 Tbilisi
Georgia

50

A Proposal for a New Decommissioning Trend

N. CERULLO and **A. AMEGLIO**

ABSTRACT

The decommissioning of the nuclear plants is one of the main issues related to the use of nuclear power as energy source.

Radiological and technical problems are related to this topics and consist, not only in how to remove and to manage the reactors spent fuel elements and the process liquids, but also in how to disassemble all the reactor's activated and/or contaminated components and systems.

After a short review of the present law regulations and of the main technological activities related to the decommissioning, new decommissioning oriented reactor design criteria are proposed.

1. INTRODUCTION

The aim of this paper is to state and refresh the main problems and the operating steps related to the decommissioning of a nuclear power plant.

In fact, one of the most serious problems of our age and culture is the management of the wastes that every industrial and commercial activity, conventional and nuclear, produces. The nuclear ones pose the harder and the more serious problems.

In the nuclear reactors the radiological hazards are due to:

- the high radioactivity contents of the spent fuel elements;
- the contaminated materials and components;
- the presence of materials activated by interaction with the neutronic flux.

Therefore a radiological and technical problem arises, consisting, not only in how to remove and to manage the reactor's spent fuel and the process liquids, but also in how to disassemble all the reactor's activated and/or contaminated components and systems.

The decommissioning of a nuclear plant begins "when spent fuel elements and process liquids are removed from the shut down reactor"[1], and it is defined as "the whole of the theoretical, technical and legal dismantling's activities to achieve the unrestricted release of the site" in order "to free the territory from every radiological bond".

2: THE DECOMMISSIONING HAZARD

The hazard typology and magnitude is quite different between an active nuclear power plant and a nuclear power plant under decommissioning.

In fact a shut down reactor, when the spent fuel elements and process liquids are removed, is a "machine" unable to produce power. The hazard is connected with the residual radioactivity and moreover with the energy sources present to supply the dismantling operation.[5]

The energy sources are conventional and incomparably less powerful than nuclear ones; the radioactive materials are present in a physical shape that make a release very unlikely.

Three are the scenarios of a possible accident with environmental radiological impact[5]:

1) fire in a radioactive waste disposal;
2) loss of a radioactive liquid out of the waste tank;
3) flooding and reflooding of a radioactive waste storage.

A PSA analysis of these events, performed by Italian Authority, demonstrates a very low environmental radiological impact as shown in Table 1.

Table 1 : Results of PSA Analysis of the Italian Garigliano Nuclear Power Plant (BWR, 160 MWe)[5]

Event Number	*Radiological Impact*
1	10^{-9} Sv
2	20 mSv (to the critical group of people)
3	concentration, after the reflooding, in the near Garigliano river 10000 times lower than the Italian law limit

In normal or incidental conditions, the risks in a reactor in decommissioning are essentially connected with the operational staff exposure.

It is also important not to forget all the conventional hazards in the nuclear power plant decommissioning referred to the dismantling of complex structure (crush risks) and to the management of big amounts of conventionally toxic materials (for instance asbestos).

3. THE WESTERN COUNTRIES LAW REGULATION OF THE DECOMMISSIONING

In, perhaps, all the western countries the legal regulations of the nuclear power plants decommissioning have been established in the last 10–15 years and they are in the main lines similar[5].

These laws often consider the decommissioning starting when spent fuel and the process liquids are removed from the shut down reactor. Usually the decommissioning operations are a owner's task (except in Spain and Belgium).

When the nuclear power plant stops working, it is requested an operational plan that must be approved by the national nuclear regulation autocrity. Every stage and every single operation have to be previously approved and then operated under the supervision of the national safety autocrity. At the end of the activities, the owner has the duty to check and demonstrate the correspondence between the achieved and the requested goals.

It is important to note that, in many countries, exists a lack of stated "de minimis" law values for the unrestricted release of the activated or contaminated materials.

This is a serious problem because they represent very relevant parameters in the planning of the decommissioning process[5].

Table 2 : Expert EURATOM Group Recommended Values for Unrestricted Release[5]

Reference concentration values for the unrestricted release of steel wreckage and components:

β/γ emitters	Specific activity: 1 Bq/day (average over 1 lt, max. value: 10 bq/day). Surface contamination: 0.4 Bq/cm^2 (average over 300 cm^2).
α emitters	Surface contamination: 0.04 Bq/cm^2 (average over 300 cm^2).

4. THE TWO PRESENT OPERATING "PHILOSOPHIES"

To achieve the above written goal it is necessary to settle a decommissioning process that is a complex function of these different parameters[1, 5, 12, 13]:

- type and characteristics of reactor;
- site's geographical position;
- site's reuse urgency and importance;
- availability of final disposal site for uncleanable wastes;
- availability of adequate technologies;
- economical costs;
- legal problems;
- social and environmental interactions.

Although there are many different operational choices, according to the IAEA suggestions[1, 2], it is possible to identify three fundamental decommissioning stages:

I) Storage with surveillance: for a convenient time span the plant is conserved with a least but safe radiological, structural surveillance's plan.

II) Partial site's release: the plant is partially already disassembled and the disassembled zones are released without any radiological bond.

III) Total site's release.

The two decommissioning operating "philosophies"[10] are connected with the first stage's time span: one is to reach the final result in a relatively short item (order of few years) and the other is to reach the final result in many years (30–50 years).

An important factor to consider is that a great fraction of the radiation level in a just shut down reactor is due to nuclides that have a half life of the order of a few years: to delay the first stage for some decades means to reduce the radioactivity level by some orders of magnitude, with evident dismantling simplification's benefits.

The choice of the one or the other "philosophy" is influenced by many factors and, among these, it is possible to note:

- the kind and the power of the reactor;
- the availability of a final disposal site for the uncleanable wastes;
- the site's importance and the urgency of its reuse.

An example of the second philosophy is the decommissioning project[10] of the Italian "Garigliano" nuclear power plant (BWR of 160 MWe output): it involves safe storage of the reactor building for a period of some decades final dismantling. The plan of safe storage has the following main objectives:

- radioactivity containment for a long period with reduced surveillance and minimisation of active components
- safe storage operations achieved with extremely low occupational exposure
- reduced effect of all the necessary interventions on final dismantling.

As for long term contamination containment, use is made of three sequence barriers and to assure the integrity of the barriers the plan includes development of a system to keep inner humidity of the reactor building at a very low values.

5. THE MAIN TECHNOLOGICAL ACTIVITIES

The residual radioactivity in a just shut down reactor, after the removal of the spent fuel and process liquids, is essentially of two types:

I) The components activated;
II) The surfaces contamination.

All the reactor materials and components will be identified for[1, 2, 6, 7]:

- their specific activity that allows to evaluate the quantity of radioactivity;
- the radionuclides kind that affords to value the half time;
- the characteristics of nuclear radiation emitted (alfa, beta or gamma rays);
- the toxicity and the eventual future possibility of a new use.

After this characterisation, the materials and components are grouped in three classes[1, 6, 9]:

a) materials scheduled for unrestricted release: these materials have a specific activity level (Bq/gr) and a surface contamination level (Bq/cm^2) below the law limit. They can be freely released and reused.
b) materials scheduled for a restricted release: they have levels of specific activity and surface contamination that allow only a restricted and checked release in a non-nuclear environment or in a nuclear environment.
c) nuclear wastes: they need a particular management.

The radioactive waste's typology is: solid, semisolid and

liquid and semiliquid (e.g. sludge, ions-exchanges resins, filters, structural pieces, etc.).

Knowing the reactor's history, the first action to perform is to define which materials are in the just shut down reactor and then to calculate the supposed radiation level.

This characterisation and classification are done by direct measurement on materials or indirect measurement on picked samples: particularly indirect measurements are made for the components and materials that have unknown correlation factor between radionuclides easily calculated (beta-gamma emitters) or those that are hard to determine (alfa, X)[6].

The methods used in these measurements are, for instance[6]:

- gamma spectrometry
- X-gamma spectrometry
- liquid scintillation
- alfa spectrometry.

Measurements are done on the contaminated materials and component to investigate the possibility to remove the radioactivity: these measurements could be done, for instance, by the technique of the "smear test" and the total beta-gamma measure.

Usually residual activity values are defined in Bq/gr for activated materials and in Bq/cm^2 for contaminated materials.

The decontamination necessity is referred to two main goal[1, 5]:

- to reduce contamination to facilitate the dismantling operations;
- to permit the restricted or unrestricted release of structures, materials and components.

Obviously the decontamination for the unrestricted release is the most deep and complex.

Decontamination consists in removing the radioactive material from the surface where it is deposed[1, 6]. This objective is obtained in two ways:

I) solving the surface layer;
II) removing the contaminated layer.

The first technique is carried out chemically, by appropriate solutions, or electrochemically. In the second technique, material is removed physically, by water jet or wiping, etc.

The dismantling is segmentation of the interested parts to permit their removal[1, 12, 13].

The choice of the dismantling systems depends on many parameters of the considered component and, between these, it is possible to note: size, radioactivity level, constructive characteristics, costs, etc.[1, 6].

Dismantling implies the usage of these equipment's:

- cutting systems (mechanical, fire, laser, plasma, explosive, etc.);
- movimentation systems.

Usually to achieve the first decommissioning stage it is necessary to dismantle only those parts where the corrosion in long time-span could cause problems (release, static problems, etc.).

6. PROPOSAL OF AN INTERACTION BETWEEN THE DECOMMISSIONING EXPERIENCE AND THE NEW GENERATION REACTORS DESIGN CRITERIA

We think that it is possible to reduce the decommissioning difficulties using decommissioning oriented design criteria. This philosophy seems to allow a decommissioning costs reduction and particularly:

- to contain the radioactive wastes amounts;
- to minimise the workers absorbed dose;
- to achieve in a shorter time span the site unrestricted condition.

The proposed operating methodology is divided in three stages.

The first one consists in the preparation of a data-base.

We will research and assemble all the achievable informations about particularly the big size nuclear power plants. There are many cases of low power research reactors that achieved the site unrestricted release. The situation is different for the big size nuclear power plant: many gained the 1st stage (according to IAEA suggestions) but very few of big size nuclear power plants and usually with a particular exercise-life have arrived at the III.

We will organise the data considering essentially:

1) the type of the reactor (PWR; BWR, GCR, etc.);
2) the power output size;

3) the stage achieved.

The second stage of the proposing work will be an attentive and deep study and analysis of the decommissioning data bases. The aspects considered are essentially:

a) the reactor's radioactivity inventories;
b) the plant's lay-out impact;
c) the active technical disassembling operations.

Goal a) is to individuate the critical structures, components, geometry and design criteria under the aspect of radiological activation and/or contamination in the past reactors. *Goal b)* will be an analysis of the same type and similar size reactors: we will try to evidence how the components lay-out in the same type and size nuclear power plants influenced past decommissioning operations in terms of time, staff absorbed dose and costs. *Goal c)* is referred to the individuation of the critical active technical operations under the aspects of the complexity, the staff absorbed dose and the environmental impact.

The third stage will be an attempt to refine the design criteria of the new generation reactors for their future decommissioning. To achieve this, the work has to start from the results of the second stage. The objectives are essentially these:

A) to reduce the radioactive level of activation and contamination present at the end of the reactor life;
B) to simplify and speed up the dismantling operations optimising the components lay-out;
C) to facilitate and if it is possible to avoid the critical active technical dismantling operations.

The *A)* goal requires an attentive study of the neutronic activation and radiological contamination of the materials of the critical parts, structures and components (individuated in the first phase); it is due to respect all the design bonds of the considered components using a material with a low level of activation. It is to note that, in many cases, the component has to shield and then it is due to research a material presenting:

- high absorption cross sections values at the neutrons fission reactors energies;
- low activity of the generated activated nuclei, if any exists.

For instance one suggestion in this case could be to minimise the impurities presence (particularly of Cobalt) in the materials, to avoid their radiological activation.

The *B)* objective needs the past reactors components lay-out decommissioning impacts evaluating which parameters, bonds, and factors influence in this sense the disposition. For example probably we would find that it is very important to study the openings sizes to permit the future transportation of big components.

The *C)* goal requires, essentially, a study of the design choices that allows to modify all the past design solutions that caused the critical active technological dismantling operations. For instance, one of the most critical active technological dismantling operations is the cutting: the objective of this phase is to modify the characteristics, to avoid the cutting, of a component that in past reactors decommissioning has been cut (per example substituting, as it is possible, the soldering with screw-balls).

The studies of these two phases could give good results in terms of reducing decommissioning costs because they could permit to improve its velocity and to minimise the staff absorbed doses and overall the radioactive wastes amounts.

7. CONCLUSIONS

It is impossible to think to delay the problem of the nuclear power plants decommissioning and to leave it for the future generations.

It seems that the true problems of the decommissioning are not of technical nature: in reality they are of political and economical nature.

It is to emphasise the importance of the establishing "de minimis" law values for the restricted and the unrestricted release: they are vital for the choice of any decommissioning strategy.

The proposed research could give good results in terms of reducing decommissioning costs because they could permit to improve its velocity and to minimise the staff absorbed doses and the overall radioactive wastes amounts. A reduction of the reactors end-life diseases (particularly of the radiological wastes amounts) could give a benefit also to the social acceptability of the nuclear power plants.

BIBLIOGRAPHY

1. IAEA, "Methodology and Technology of Decommissioning Nuclear Facilities", Technical Reports Series no. 267-Vienna, 1986.
2. IAEA, "Safety in Decommissoning of Research Reactors", Safety series no. 74–Vienna, 1986.
3. IAEA, "Principles for establishing limits for the release of the radioactive materials into the environment", Safety Series no. 45–1979.
4. IAEA, "Factors relevant to the Decommissioning of Land-Based Nuclear Reactor Plants", Safety Series no. 52–1980.
5. Commissione tecnica per la sicurezza nucleare e la protezione sanitaria. "Rassegna e valutazione dell'attuale quadro nazionale ed internazionale delle normative, delle procedure e delle modalità per la disattivazione degli impianti nucleari: riferimenti per le attività istituzionali della commissione tecnica". ENEA, Roma maggio 1994.
6. UNI 9498/1, settembre 1989-Disattivazione di impianti nucleari. Criteri generali.
7. UNI9498/5, dicembre 1995- Disattivazione di impianti nucleari. Caratterizzazione radiologica.
8. *L. Sani.* "Problematiche connesse alle attività preliminari di disattivazione delle centrali nucleari". Energia Nucleare 11/2/ maggioagosto 1994.
9. *F. Marsilii.* "Disattivazione di impianti nucleari. Parte 6: caratterizzazione radiologica e classificazione dei materiali prodotti da operazioni di disattivazione ai fini della destinazione finale". Energia Nucleare 11/2/maggio-agosto 1994.
10. *M. Genova, F. Mantega, T. Vitiello.* "Disattivazione della centrale elettronucleare del Garigliano". Energia Nucleare 6/3/ settembredicembre 1994.
11. *M. Laraia.* "Decommissioning nuclear reactors in Italy: the unrestricted release issue". Energia Nucleare 8/1/gennaio-aprille 1994.
12. NUREG. "Technology, safety, and costs of decommissioning a reference Pressurised Water Reactor Power Station", NUREG/ CR 0130, NRC 1978.
13. NUREG. "Technology, safety, and costs of decommissioning a reference Boiling Water Reactor Power Station", NUREG/CR-0672, NRC 1980.

Address:

ADRIANO AMEGLIO and **NICOLO CERULLO**

University of Genva
16100 Genva
Italy

51

Assessment of Dismantling Procedures for Steam Generators of VVER NPP's Type with a View to Minimise the Generation of Waste

G. SCROFANI, R. LANZA and E. TAVELLA

1. INTRODUCTION

This Report investigates possible methods of safely decommissioning steam generators of VVER NPP's, including decontamination and re-melting procedures.

For the purpose of this report, it is assumed that the SGs are of a VVER 440 type but the study is valid for SGs of VVER-1000 NPP's too. The various units could differ in respect of the material used i.e. stainless steel cladding or unprotected carbon steel but there is little variation as far as oxide layers are concerned, however, because all VVER reactors are operating with the same water chemistry parameters.

As general strategy, an approach of SG decontamination followed by manual dismantling could be adopted. This is considered to be technically simpler and cheaper than remote dismantling. Once the sizes of the material coming from the dismantling of the SG have been reduced, it could be inserted into casks and melted to obtain the waste volume reduction and the decontamination below the re-using level of the melt products.

The following abbreviations will be used in the Report:

NPP	Nuclear Power Plant
SG	Steam Generator
ALARA	As Low As Reasonably Achievable
HP	High Pressure

2. OBJECT

The main object of the Report is the assessment of a decommissioning procedure for steam generators of VVER-440 NPPs. The SGs are horizontal cylindrical vessels about 11.5 m in length with a diameter of 3.5 m and total weight of about 160 tons (Fig. 1).

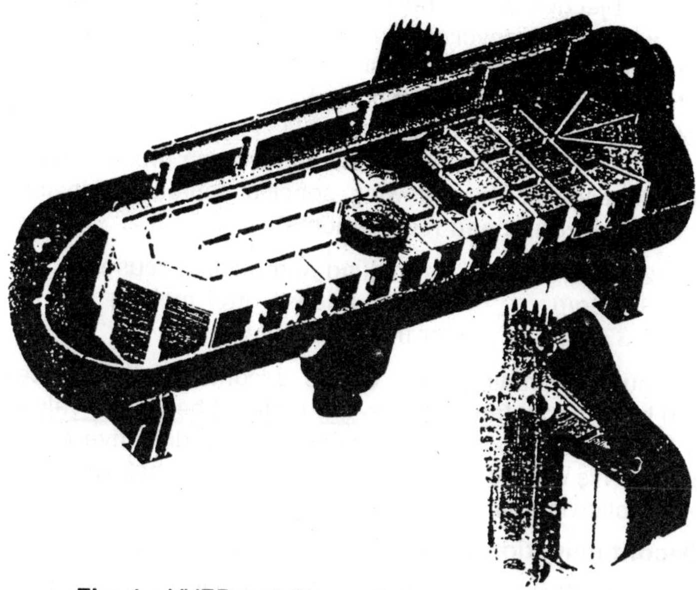

Fig. 1 : VVER-440 Steam Generator

The main phases of the dismantling strategy which will be assessed in detail in the following sections are listed below:

1. Decontamination of the whole SG
2. Dismantling, removal and sizes reduction of the SG
3. Melting of the material coming from the dismantling of the SG

The assessment will be based on the assumption that the SG has been already removed from service and it is now stored in the site of the NPP, outside of the Reactor Building.

3. DECONTAMINATION

The overall strategy for decontaminating the SG will be influenced by a number of factors such as the total activity associated and the volumes and types of secondary wastes generated.

Decontamination procedures of a SG prior to decommissioning for reaching at least re-melting conditions could be performed either on the whole component or on cut SG sections after partial dismantling or a combination of both operations. The general objective would be to establish a treatment scheme which will permit recycling of non-activated components while generating a minimum of waste. A further important requirements would be to reduce significantly the radioactive dose rate before the final dismantling of the SG.

The selection of a such decontamination process should take into account the following specific contamination limits (according to European requirements):

<200 Bq/g	for melting and nuclear reuse
<3.7 Bq/g	for free release after melting
<0.37 Bq/g	for free release

Thus, the main object of the decontamination process would be to decontaminate the SG material below the value of 200 Bq/g making possible its melting in a radioactive melting foundry. This value could be achieved if possible before dismantling or following partial dismantling of the SG.

i) Decontamination Methods

The material of the SG shell and tubes will not contain any induced radioactivity but heir surfaces will be covered with a contaminated thin corrosion layer. The removal of this corrosion is further complicated by the high packing density of the tubes in each tube bundle, the extremely large number of tubes and hence surface area which is corroded in each SG.

The decontamination can be carried out using either physical methods, chemical methods or a combination of both. Physical processes in general produce far less secondary wastes than chemical processes but they are much slower. It should also be pointed out that even with physical methods, some contamination goes into solution and so some chemical treatment of secondary waste is required in this case too.

The known methods of decontamination that could be used for decontaminating the SG are discussed below.

Physical methods

- Dry ice abrasion
- Dry grit blasting
- Explosion/shock/vibration
- High and low pressure water spraying
- Wet blasting
- Wet ultrasonic cleaning
- Hydrosonic cleaning

Of the above methods, *only high and low pressure water spraying* can be considered suitable for decontaminating the SG. A major advantage of using water is that it can be cleaned relatively easily to a level suitable for re-circulation. The total volume of secondary waste to store and dispose of at the end of the dismantling stage is thus quite small.

High pressure (HP) water at 250÷1000 bar is used to remove surface coatings and to clean back to bare metal and has a proved track record in nuclear applications. This method is very effective on flat surfaces where access is available (tanks or pipes of 50 mm diameter and above) but seems not applicable for decontaminating the SG tube bundles because of high packing density and small diameters of the tubes.

However, this system could be used for decontaminating the SG shell after partial decommissioning, if the specific contamination value suitable for melting is not achieved directly by decontaminating the single SG as well as to reduce doses on operators for next cutting and packaging operations. The decontamination of the sectioned SG shell would be carried out in a tented area with a small fixed spray system in a booth through which the SG section would be drawn by a monorail system. Because air is not involved in the process a relatively simple airborne contamination system is adequate.

Chemical Methods

- Inorganic acid
- Organic acid
- Electro-polishing
- Regenerative methods

Chemical methods have been used on a wide scale for decontaminating whole water reactor coolant circuits over a period of 30 years. Disadvantages are the large quantity of secondary wastes which would arise and the high costs associated with their treatment. However, some of the recent chemical processes which use ion exchange resins to regenerate the decontamination liquid are effective and thereby generate only small volumes of resin as waste.

Inorganic acids such as hydrochloric and nitric acids are very effective in dissolving and releasing scale from parent metal but they also attack the base metal vigorously resulting in unwanted secondary waste and excess use of reagents.

Organic acids such as oxalic, citric and Ethylene Diamine Tetra-acetic Acid (EDTA) are effective in dissolving iron oxide. The problem then is to put the contamination removed into some suitable solid form for long term storage. This usually involves a precipitation or ion exchange process.

The effluent would be collected into a tank and then pumped to the treatment plant. Finally the contents of the sludge tank and the ion exchange tank would be recovered and dewatered to at least 60 per cent by weight and solidified, probably in cement, in an approved package for storage.

ii) Selected Decontamination Procedures

Chemical processes seems to be the best solution to use for decontaminating the whole SG because it should require less technical efforts than the electrochemical and mechanical decontamination processes. As general rule, decontamination for low contaminated pieces could be carried out by only one cycle using an oxalic acid solution ($H_2C_2O_4$) with concentration of about 20 g/1. For higher contaminated pieces with a thick crude layer, it is necessary to perform several decontamination cycles to reach a low residual contamination.

To reduce the amount of secondary waste and to enhance the decontamination effectiveness, at the end of each decontamination cycle the aggressive chemicals could be pass through an organic decomposition module. By the action of focused ultrasonic waves the organic acids are completely decomposed into CO_2 and H_2O. With this treatment a large amount of resin capacity which is mainly required for the removal of the activity removed from the contaminated surfaces of the SG is saved.

Based on values gained by past experience, the activity levels (only Co-60) in the SG at the end of its service could be roughly estimated as follows:

Steam generator (total): 500 Bq/g
Steam generator (tubing): 2700 Bq/g

After the decontamination and 4 years of reduction due to decay, the following values could be achieved:

Contamination: 140 Bq/cm^2
Contact dose rate (outside): 20 mSv/h
Contact dose rate (inner surface): 37 mSv/h

With these dose rats values, the manual dismantling work and the re-melting will be allowed. For carrying out the decontamination, the SG should be connected at an external decontamination loop via the hot and cold-leg header (Fig. 2). The contamination set-up includes a re-circulation loop with a high pump rating which ensures as far as possible that decontamination chemicals are delivered to every heat exchanger tube in the SG. The decontamination loop, which incorporates a surge tank with degasification facility in addition to a heater and an organic decomposition module, is connected up in parallel to the recirculation loop. A chemical dosing system and a purification loop with ion exchanger columns which operate in parallel to the decontamination loop are included in the system.

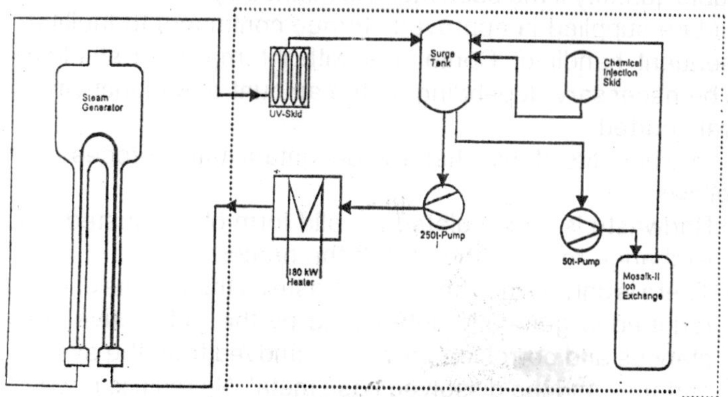

Fig. 2 : Chemical Decontamination of the Steam Generators

The modules necessaries for the decontamination process are resumed hereafter:

- Heater module
- Re-circulation pump
- Chemical dosing module

- Decomposition module for chemicals (UV-Skid)
- Surge tank with connection to the off-gas system
- Ion-exchange filter, possibly installed in approved waste containers.

Hoses are used to connect the separate modules to the SG and to each other. Each decontamination cycle involves the following stages:

- Chemical pre-oxidation step
- Reduction
- Decontamination
- Chemical decomposition

The number of decontamination cycles which are required depends on the level of contamination in the particular SG. At least three decontamination cycles are generally required.

The following data can be assumed:

- Volume 13 m^3
- Surface area 2180 m^2
- Thickness of crud 6 g/m^2
- Activity 5 kBq/cm^2

A three-cycle decontamination process on this scale produces approximately 400 liters of spent ion-exchange resins, or a double quantity if the base metal is thinned by 1mm. the resins should be supplied in approved storage containers to facilitate subsequent handling. Containers with internal lead shielding may be necessary depending on the amount of radioactivity to be transported.

The benefits of this chemical decontamination process are as follows:

1) Radwaste is accumulated in solid form of exhausted ion-exchange resin at the end of the process.
2) There is only a small amount of waste as the volume of resin required is generally determined by the radioactive substances and corrosion products removed from the system surfaces and the dissolved base metal. No resin is needed for removal of the decontamination chemicals which are converted into CO_2 and H_2O in an Ultrasonic Waves photoreactor.
3) The process is easy to implement and control from both chemical and technological point of view.
4) The process has already been applied successfully on

several occasions.

iii) Alternative Decontaminating Chemical Solution

Another decontamination process that could be used for decontaminating the single SG use a combination of nitric acid and Cerium nitrate at low pH and at ambient temperature. The Cerium is regenerated by injecting ozone during the operation. After neutralisation and precipitation of the decontaminating solution containing the removed activity, the residue will be dried in order to fix it in solid form for storage and disposal.

Main objectives that could be fulfilled with this procedure are:

1) Obtaining a high decontamination factor.
2) Minimising the volume of generated secondary wastes.
3) Performing the filtration of sludge and residues generated by the decontamination process using a standard process.

This procedure has been already applied but it is not experienced as the previous one.

4. DISMANTLING AND CUTTING

Once the SG has been decontaminated at least below a value that will allow manual operations, it will be segmented by cutting, removed and melted. Manually dismantling and cutting is considered to be technically simpler and cheaper than remote dismantling.

i) Cutting Techniques

The proper technique for cutting the SG shell depends on the selected dismantling procedure and sequence. Nevertheless, general criteria are:

- The cutting operation should be done in a closed and equipped space to better control the diffusion of the contamination.
- The cutting operations should be as simple as possible to allow the use of easy, common and rugged tools.
- The operators working time in areas with significant level of exposure intensity should be minimised as low as possible.
- The number of cutting operations should be minimised; the sizes of cut pieces should be as large as allowed by the transport cask or furnace capacity.

The following four methods could be used for cutting the SG shell:

1. Flame or Thermal Cutting

Main advantage is the high cutting speed (9 ÷ 12 m/h for 50 mm thickness), but it produces a large amount of aerosol with spreading of melted contaminated particulate which generates important containment problems.

2. Milling Cutting

This method produces a small amount of aerosol, but the cutting speed is low (2 ÷ 3 m/h for 50 mm thickness) as well as the tools require high forces.

3. Grinding cutting

The grinding cutting is faster (4 ÷ 6 m/h for 50 mm thickness) than the milling cutting but slower than the thermal methods, besides it has the disadvantages of both milling and thermal cutting.

4. High Pressure Water Jet Cutting

The application of HP Water Jet cutting technique requires the use of an abrasive to increase the cutting speed. Advantage is the low cutting forces on the tools. Disadvantages are the low cutting speed (0.5 ÷ 1.5 m/h for 50 mm thickness) and the large production of secondary waste (30 ÷ 80 kg of abrasive per metre, for a 50 mm thickness).

The quicker cutting system should be preferred to reduce the presence of operators in exposed area. As a result, the *flame cut* would be the best technique for cutting the SG shell, taking into account the need to controlling the airborne contamination produced during the cutting process. This method has also the lowest cost from both radiological and economical point of view. Also the milling technique could be used, with small problem about containment restrain.

As regard to the tube bundles of the SGs, various cutting methods could be used, once the shell has been cut: flame torches, grinding disks or hand lever shears.

ii) Dismantling Scheme

When the decontamination process has been concluded, the SG area should be prepared for the next dismantling and cutting operations. In particular, a containment tent should be installed around the SG to prevent airborne contamination. The main phases of the tentative dismantling scheme are as outlined below:

1. The two SG end sides are vertically cut by workers equipped with flame tools; when the cut is concluded, each SG end head is removed and transferred to another area for intermediate storage;
2. The free portions of tube bundles are cut by shearing the tubes (longitudinally) into smaller sizes in such a way that they are the correct length for packaging into standard 200 l transport drums and removed; the tube bundle ends and supports are cut in order to separate them from the shell.. This allows the remaining bank of tubes to be cut up into smaller bundles and removed. Then, all the cut tube bundles and supports are inserted into the transport drums;
3. Once all tubes have been cut and removed, the shell is vertically cut into sections which are then transferred to another area of the workshop and cut into a number of segments using flame cutting tool. The cutting area will be below a containment enclosure provided with a ventilation and filtration system to minimise the possibility of contamination of the workshop. Once cut, the segment is transferred to a HP water spray decontamination station;
4. The SG segments, which are effectively flat surfaces, are cleaned by passing through a small fixed HP water jets spraying system, to reduce contamination below the free-release level. A portable hand held HP water jet could be used in addition if some hot spot could be found. A further decontamination of the tube bundles is not foreseen since the free release level is not achievable anyway because of its complicated geometry and small diameters;
5. If the free release level is not achieved, the shell segments are further cut into suitable sizes for melting, using the same flame tools previously used for cutting the SG shell. The two SG end heads should be treated as the shell sections;
6. Once the sizes of the material has been reduced to approx. 80 cm of max. lengths, it could be packed into the drum. The drum is a cylindrical tank 595 mm diameter and 900 mm height with a load capacity of 260 kg. To reduce their

number, each drum should be filled up to get its load capacity in a short time and without requesting important manual activities. The segments of the shell and other bulk material could be melted directly without packaging;

7. The drums filled with the cut tubes and the segments of shell surfaces should be preliminary stored and then melted into a new melting Plant which will be purpose built in the NPP site.

The HP water treatment plant is shown in Fig. 3. The water used for the decontamination of the SG segments will be channeled into a sump, from which two pumps will transfer the water back to a sludge tank which will remove most of the coarse particulate by settling. Filters are provided to remove any that is recirculated before it reaches the HP spray nozzles. Most of the contamination will be in solid form of particles which should concentrate in the bottom of the tank. The contamination should be regularly monitored and, at suitable intervals, the sludge should be drained and stored into standard 200 litre drum which could contain about 40 kg of contamination (20% by weight) each. The lower part of the sludge tank will require shielding as well as each of the 200 litre drums and the facility used to store the drums.

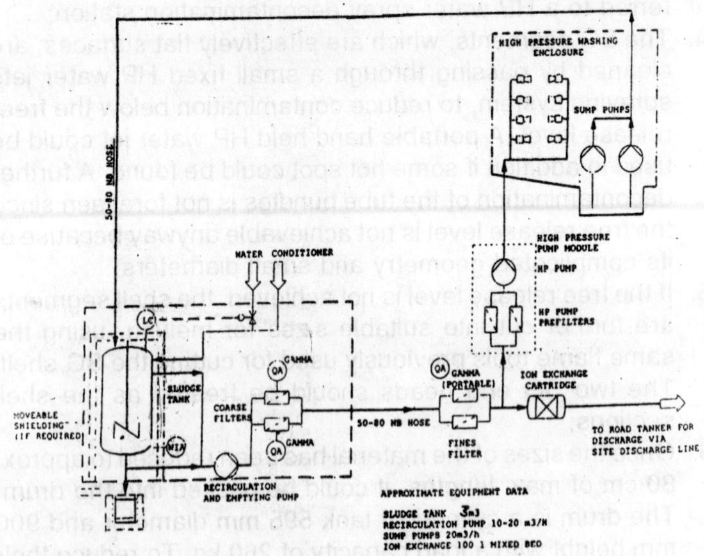

Fig. 3 : HP Water System

5. MELTING

The radioactively contaminated steel scrap coming from the dismantling of the SG (filled drums and bulk material) are melted after that it has been radiologically characterised using g-scanning method.

A new melting Plant should be purpose built in the site of the NPP taking into account hat significant quantities of contaminated steel scrap coming from the site itself or from other NPP could be melted. A good exploitation of a melting Plant, as experiences gained in Europe, foresees an annual throughput of scrap of about 2000 tons and a lifetime of 10 years.

In order to minimise the overall costs, the melted metal could be re-used for the manufacturing of casks and containers for storage and final disposal of radioactive waste, shielding equipment and other products used in the nuclear cycle. Recycling inside the conventional field can be realised, if the clearance level for unrestricted reuse, defined by the legal authorities and differing from country to country, can be fulfilled.

i) Melting Benefit

Melting technology is a method for the recycling of radioactively contaminated metallic material, based mainly on a separation effect of the nuclides during the melting at roughly 1500°C. This process leads to a significant transfer of these radio nuclides into the waste such as slag and filter dust, representing only a few per cent (less than 5%) of the overall mass.

Sr-90 and Pu-241 are nearly completely transferred into the slag while Cs-137 is equally distributed into the waste paths slag and filter dust. Co-60, Ni-63 and Fe-55, however, can be separated only in a small percentage because they have the tendency to remain in the metal structure of the melting product, the so-called ingot. Nevertheless, the residual activity in the ingots is homogenised inside the material and will be reduced during radioactive decay.

Main advantages of the melting technology can be summarised as follows:

- Substantial reduction of the waste volume compared to other standard handling;
- Decontamination of the basic metallic material;

- Homogeneous distribution of the residual contamination in the ingots;
- Easy classification by spectroscopy of the ingots qualifying them for the most favourable re-use.

The treated metal could be recycled and used as raw material for the manufacture of waste containers, shielding equipment and other products used in the nuclear cycle. Under favourable conditions of very high decontamination factors re-cycling inside the conventional field can be realised.

Even in the case of final disposal of low radioactively contaminated metallic waste the melting treatment shows a high cost advantage as shown in Fig. 4 Due to the max. obtainable volume reduction by melting the total costs of final disposal are about 50 per cent lower compared to treatment without volume reduction.

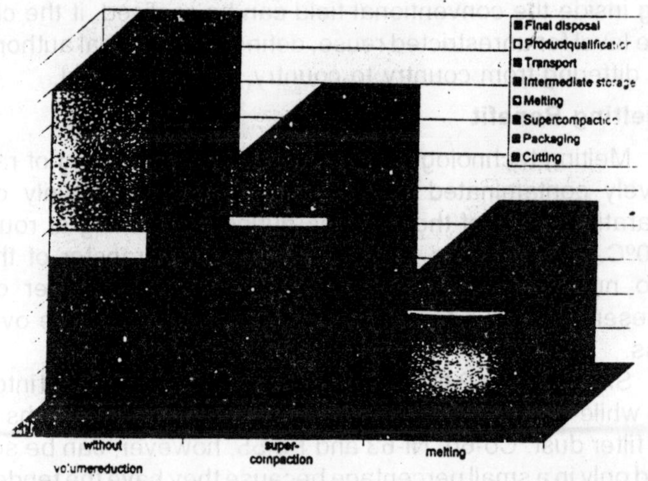

Fig. 4 : Cost Comparison for Various Waste Treatments

ii) Features of the Melting Plant

Basically, the special melting plant consists of a furnace with housing and various charging devices, exhaust and filter system including control and steering unit, control desk and storage areas for containers, drums, and moulds (see Fig. 5).

The main features of this melting Plant are given hereafter.

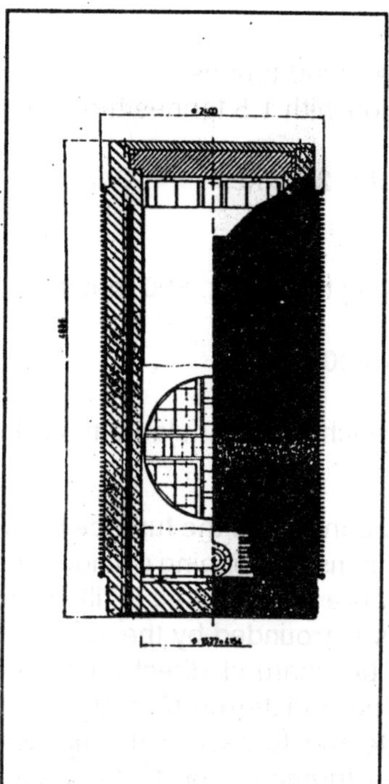

Transport-und Lagerbehälter zur Aufnahme von 21 DWR-Brennelementen (ca. 5 Jahre Abklingzeit)
Transport and storage cask for holding 21 PWR-fuel assemblies (approx. 5 years cooling time)

Belodekopazität Loading capacity	21 DWR-BE PWR-F/A
Außendurchmesser Outer diameter	2400 mm
Gesamthöhe Total height	4886 mm
Schachtdurchmesser Cavity diameter	1527 mm
Schachthöhe Cavity height	4154 mm
Behältergewicht, leer Cask weight, empty	92.3 t 92.3 m.t.
Behältergewicht, beladen Cask weight, loaded	106.0 t 106.0 m.t.

BE : Brennelement
DWR : Druckwasserreaktor

PWR: Pressurised water reactor
F/A: Fuel assembly
m.t. : Metric tonnes

Fig. 5 : Melting Plant for Radioactive Contaminated Waste

Main characteristics of the Plant:

- Overall Melting Performance: 2000 tonnes/y
- 1 Medium Frequency Induction with 1.5 tonnes/h nominal rate
- Furnace having a capacity of 3.2 tonnes
- Separate enclosure
- Joint control station
- Dimensions of input material up to approx. 800 mm length and 500 mm diameter
- Filter capacity: overall 2 × 15,000 m^3/h
- Power requirements: 2 MW
- Products: Ingots of 1 tonne each and granulated material

Melting Furnace

The induction furnace mainly consists of the furnace body, the furnace lid, the tipping mechanisms with tipping cylinder and the water cooling system. The charge to be melted is filled into a refractory lined crucible which is surrounded by the induction coil. Standard 200 l drums can be charged directly into the furnace while bulk materials can be transferred from the 200 l drums or from 20'-containers into the furnace with charging devices. The daily capacity of the furnace is up to 12 tonnes operating in one shift corresponding to 2,000 tonnes/year. A higher rate would require more shifts.

Casting

After melting, the material is cast into ingots or granulated. In the first case, the liquid iron from the furnaces is poured into the stacked and dried steel chills which are suspended from the furnace hoist.

In the second case, the liquid iron is filled into a transport ladle and brought to the granulating station. The main process of the granulating technique is given by the injection of the melt via a special nozzle into a water bath. The granules produced are filled into suitable drums and transported to the open-air site.

Samples of the liquid iron are taken for radiological and metallurgic analysis.

Filter Systems

The furnace is installed in a zone provided with a high efficiency ventilation system which consists mainly of an high efficiency cyclone, a bag filter, an HEPA-filter and an off-gas ventilator. The filter plant is designed for the required air throughput and exhausts the pure gas to the environment via stack.

Secondary Waste Treatment

Approximately 5 per cent of the melted scrap accrues in the form of secondary waste: slag from the melting process, filter dust, filter material, furnace lining, garbage. A planned annual throughput of 2,000 tonnes would result in approx. 100 tonnes of contaminated waste materials per year. As a rule, these are stored in tightly sealed 200 l drums. Bulky waste such as e.g. used filter cassettes require tightly sealed special containers with a min. content of approximately 1.5 m^3. After a period of intermediate storage, the secondary waste has to be brought to a final depository.

Auxiliary Systems

The following auxiliary systems are necessary: .

- Compressed air plant for supply of dry, oil-free compressed air (6 bar, 500 Nm^3/h)
- Emergency power station
- Cooling systems for the furnaces with a cooling rate of 5 m^3/h, cooling down via cooling towers.
- Space heating by radiators
- Repair and maintenance shop
- Special facilities such as laboratories and social facilities.

Radiation Protection Concept

The melting plant should be planned under consideration of the ALARA principle. Possible danger to the personnel of the melting plant exists in the form of external exposure from the contaminated scrap as well as the incorporation of radio nuclides during handling.

Special clothing and wearing of breathing masks are required for work areas with higher radiological impact. An impermissible distribution of activity is prevented by a staggered pressure system with a drop from higher to lower pressure level. Activity measurements of the inside air as well as smear tests to determine surface contamination will be taken regularly.

The activity which is released from the various work areas is piped to an air discharge off-gas stack via the above mentioned filter system. The release into the environment is done by continuous sampling as well as by periodical measuring and assessment. The residual activity in slag and filter dusts are measured and documented as well as the residual activity of the melting products.

Personnel Requirement

The melting plant can be operated by roughly 9 persons:

- Shift-leader: 1 engineer
- Operating of furnace: 2 melting operators and 2 unskilled workers
- Charging of furnace: 1 unskilled worker
- Transport: 1 crane operator
- Chills storage: 1 crane operator and 1 unskilled worker

Furthermore assistance by competent personnel regarding radiation protection tasks, radiological and chemical analysis, maintenance work and documentation are required. Typical maintenance work includes exchanging of bearings, fan blowers etc. The control and tests consist of, e.g. ultrasonic inspections of welding, visual controls and overload tests.

iii) Melted Material Reusing

In order to minimise the waste volume and therefore the overall costs the metallic waste material should be recycled. In principle, a broad range of applications exist for the re-use of the melted material. Depending on the residual specific activity in the ingots and the legal handling regulations, e.g. clearance levels, either recycling in the nuclear cycle or the conventional field, can be effected.

The following recycling paths within the nuclear cycle can be proposed:

- Manufacturing of casks and containers for storage and final disposal of radioactive waste.
- Generation of metallic granules to replace the natural aggregates in concrete structures (containers, shielding plates etc.).

As example, casks CASTOR V-21 and MOSAIK types could be manufactured. The CASTOR V/21 cask is designed for the transport and storage of 21 spent fuel assemblies from pressurised water reactors (Fig. 6). The MOSAIK cask is mainly

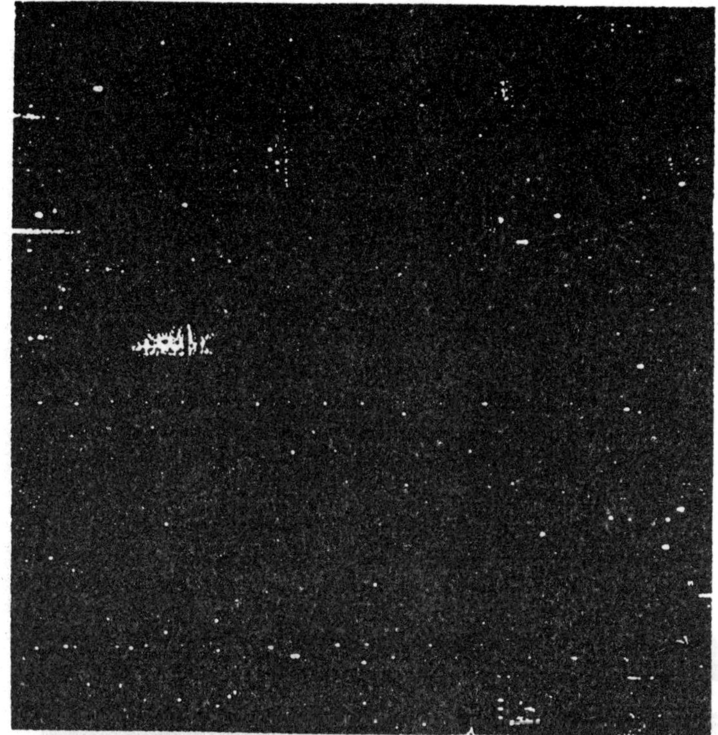

Fig. 6 : – CASTOR -V21 Cask

used for transport and storage of dried evaporator concentrators and ion-exchange resins. It is also used for holding core components (Fig. 7).

The total amount of containers of these types that could be manufactured using 1000 tonnes of radioactive scrap is given in the following Table.

Container Type	*Weight Empty*	*Recycling Portion*		*Total Amt.*
	(Mg)	*(Mg)*	*(%)*	
MOSAIK II–15	6	2.0	30	500
CASTOR V/21	92.3	27.7	30	36

The Table shows that a total amount of 36 containers CASTOR V/21 type could be manufactured using 1000 tonnes of radioactive scrap. Of course, there is a possibility to manufacture a certain amount of different containers by using parts of the radioactive scrap. Furthermore, the volume of final dis-

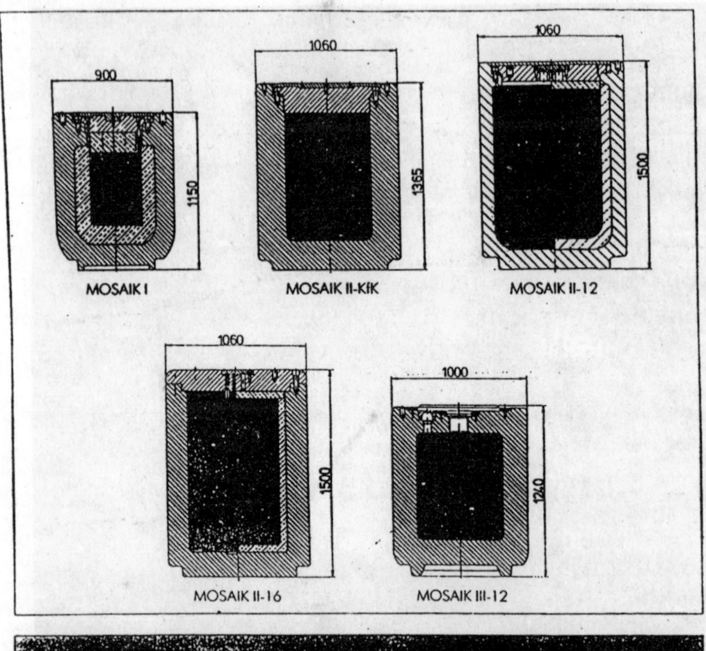

Außenhöhe Outer height	1150 mm	1365 mm	1500 mm	1500 mm	1240 mm
Innenhöhe max. Interior height max.	810 mm	930 mm	1145 mm	1170 mm	900 mm
Wanddicke Wall thickness	150 mm	215 mm	120 mm	160 mm	120 mm
Gewicht min. Weight min.	3350 kg	6400 kg	5000 kg	6530 kg	4290 kg

Fig. 7 : MOSAIK Cask

posal will be saved by using recycled material.

6. RADIOLOGICAL PROTECTION AND SAFETY REQUIREMENTS

Entry to and exit from all the areas in which dismantling work will be undertaken should be via a change areas, incorporating a physical barrier to prevent inadvertent access. Areas where containment is broken (e.g. cutting SG's into sections) or where contaminated surfaces are being disturbed (e.g. size reduction of tubes) will require containment and HEPA-filtered ventilation to prevent any potential spread of contamination.

In all areas where there is a potential for the generation of significant airborne contamination, personnel will wear appropriate respiratory protective equipment and hence maintain internal doses to operators ALARA.

Persons who intend to work in Radiation Controlled Areas must have attended all appropriate site induction courses, be a 'classified person', be issued with the appropriate written instructions, must wear a personal dosimeter. Persons who intend to work in Contamination Controlled Areas must also wear protective clothings.

No equipment may be removed from Contamination Controlled Areas without appropriate monitoring, packaging and certification. Radioactive waste packages must be checked for external contamination when they are removed from a Contamination Controlled Area.

The precise numbers of Health Physics personnel will be dependent on the number of work areas in use at any given time and the rate of generation of waste materials. However, for a project of this size, an appropriate Health Physics organisation would be:

- Health Physicist/Radiation Protection Adviser(1)
- Assistant Health Physicist/health Physics Supervisor(1)
- Health Physics Monitors/Changeroom Attendant(3)

The equipment required by this organisation would include:

- doserate meters;
- contamination meters;
- contamination air sampling equipment;
- gross alpha and beta/gamma counting equipment;
- personnel contamination monitoring equipment;
- consumable items, such as papers for airborne sampling, hardened ashless paper for smear sampling, sealable bags for holding samples, etc.;
- personal dosimetry (provided to all persons involved in the work);
- personal protective clothing, i.e. coveralls, overshoes, safety shoes, gloves, caps;
- respiratory protective equipment, i.e. dust masks and fullface canister respirators.

Address:

GIUSEPPE SCROFANI

ANSALDO

Genova

Italy

52

Environmental Challenges Resulting from Nuclear Waste Generation in the Former Soviet Union

D. BRADLEY

The Former Soviet Union (FSU) and the United States have developed the largest nuclear weapons complexes in the world. In doing so, they also created the world's largest inventories of radioactive waste. Although some of the waste has been stored in-safety managed systems such as tanks or converted into stable and storable forms such as glass, significant amounts of it have been discharged into the environment.

During the past 50 years, the FSU discharged substantially larger amounts of radioactivity into the environment than did the US, although the same types of discharges occured in each country: in the early years some wastes were discharged into surface water systems such as ponds and rivers, as well as being injected into the ground. Some of these practices are still being used today. In both the FSU and the US, the primary contaminated areas are located at or near facilities that reprocessed nuclear fuel from production reactors. In the FSU, essentially all components of the former Soviet weapons complex are in Russia. The reprocessing facilities are at Mayak (near Cheiyabinsk), Tomsk-7 and Krasnoyarsk-26, all of which are in Siberia. In the US, the major sites are at Hanford, Washington, at Savannah River in South Carolina, and at Oak Ridge, Tennessee.

The total amount of radioactivity released by both countries, roughly 1.7 billion curies in current radioactivity, is less than 0.4 per cent as much as exists naturally in the world's oceans which get their radioactivity primarily from potassium-40 (Carter 1993). However, the 1.7 billion Ci are concentrated in small areas, thus creating the potential for a substantial impact on the environment and human health. Of the 1.7 billion Ci released (not placed in some type of engineered container or structure) by both countries, about 3 million have been released in the US Fortunately, the technologies, data, models and scientific experience that have been developed and accumulated over the last 50 years are tremendous resources for finding solutions to the problems of environmental restoration and waste management. The US Department of Energy (DOE) and the Ministry of Atomic. Energy of the Russian Federation (Minatom) have been sharing these resources for the last seven years, creating a partnership that will extend far into the future.

RADIOACTIVE CONTAMINATION IN THE FSU

The event that increased worldwide awareness of radioactive contamination in the FSU was the Chernobyl accident in Ukraine in April 1986. International attention was also captured by the dumping of radioactive materials into the Kara and Barents Seas and the Seas of Japan and Okhotsk as a result- of naval operations. In both of these cases however, the amount discharged into the environment was small compared to that from the Russian production reactor and spent-fuel reprocessing sites.

By far the largest releases of radioactive waste into the Russian environment have occurred as a result of the inadequate management of radioactive waste resulting from the reprocessing of spent fuel. Since the late 1940s, some of this waste, primary fission products, has been released into surface-water systems since the 1960s, it also has been disposed of underground by means of deep-well injection. Together, both methods have led to Russia's most significant radioactive discharges to the environment at three sites in particular: Mayak (formerly Cheiyabinsk-65), Tomsk-7 and Krasnoyarsk-26.

Other activities that have released considerable amount of radioactivity into the environment of the FSU include nuclear weapons testing; peaceful nuclear explosions that were used for earth moving, storage cavity formation and oil and mineral recovery; and disposal of tailings and liquid waste from uranium mining and milling operations. Additionally, the storage/disposal of radioactive wastes and ionisation sources from research, medical, and industrial activities at the 'Radon' sites in the FSU may not afford adequate protection against releases to the local environment. Table 1 details the amounts of radioactive waste present in Russia as of 1995, not including those wastes dis-

Table 1 : Distribution of Radioactive Wastes in Russia for 1995 by Ministry

Source of Waste	*Type of Radioactive*	*Qty. (M^3)*	*Activity (Ci)*	*Storage Location*
A) Minatom				
Uranium mining and processing	Sludges and tails	1.0×10^8	1.8×10^5	Mines sites and other locations
Uranium enrichment & fuel element fabrication	Liquid and solid wastes	1.6×10^6	4.0×10^3	Storage at the plat
Nuclear power plants	Liquid concentrates	1.5×10^5	4.2×10^4	Waste storage
	Solid wastes	1.2×10^5	1.0×10^3	other storage
	Solidifies wastes	1.6×10^4	1.0×10^3	facilities onsite
Reprocessing plants for defence materials	Liquid wastes	2.5×10^4	5.7×10^8	Steel tanks at Mayak
	Vitrified wastes	9.5×10^3	2.0×10^8	
	Liquid wastes	4.0×10^8	7.0×10^8	Tanks, pools and reservoirs
	Solid wastes	1.0×10^8	1.2×10^7	
				Concrete storage at plants
B) Ministry of Defence				
Operation of nuclear submarines	Liquid wastes	1.4×10^4	1.8×10^2	On-land and floating storage facilities
	Solid wastes	1.3×10^4	8.0×10^2	Concrete storage facilities
C) Ministry of Transport				
Nuclear Icebreakes and container ship	Liquid wastes	3.9×10^2	0.6	On-land storage
	Solid wastes	1.4×10^3	2.1×10^2	On-land storage
	Solid wastes	1.0×10^2	2.0×10^4	On-land storage
D) State Committee of Defence Industry				
Construction and disposition of nuclear submarines	Liquid wastes	2.5×10^3	$5.\times10^2$	On-land and floating storage facilities
	Solid wastes	1.5×10^3	1.0×10^2	Storage at the facility
E) Ministry of Construction Industry				
Application of radionuclide sources	Laquid, solid and solidified wastes encapsulated spent ionization sources	2.0x105	2.0x106	Sixteen (16) "Radon" enterprises

posed of by deepwell injection, according to the responsible Russian Ministry (IAEA May 1995). Table 2 depicts the best estimate of the current activity of radioactive releases to the environment from production reactor and reprocessing sites, and other sources such as Chernobyl and naval operations. Figure 1 attempts to summarise a great deal of information into

Table 2 : Current Inventory of Radioactive Waste Releases to the Environment in the FSU

Site	*Current Radioactive Inventory Released to Environment, Ct*
Production Reactor and Reprocessing Sites	
Tomsk-7 waste injection	~1,000,000,000
Krasnoyarsk-26 waste injection	~450,000,000
Tomsk-7 reservoirs	~130,000,000
Mayak reservoirs, lakes, Techa River	122,200,000
Mayak production reactors coolant water	~132,000
Krasnoyarsd-26 production reactors coolant water	106,000
Tomsk-7 production reactor coolant water	37,000
Mayak-1957 HLW tank explosion	44,300
Krasnoyarsk-26 reservoirs	>19,000
Mayak-1967 release from Lake Karachai	~500
Sub total	**~1,702,538,800**
Other Waste Releases to the Environment	
Chernobyl	~1,575,000
Uranium mill tailings	600,000
Ocean near Sakhalin Island -90 Sr RTG, 1987	286,500
Submarine accident, Chazhama Bay, August 1985	~150,000
Kara Sea Reactor compartments w + w/o fuel	122,000
Barents, Kara and White Sea liquid wastes	20,750
Kara Sea solid wastes	15,500
Dimitrovgrad waste injection	90,000
To Irtysh Rives from weapons testing	88,000
Power reactor operations	36,000
Pacific Ocean liquid wastes	11,850
Pacific Ocean solid wastes	7,920
Discharges to marshes at Beloyarsk site	82
Sub total	**~3,003,600**
GRAND TOTAL	**~1,705,542,000**

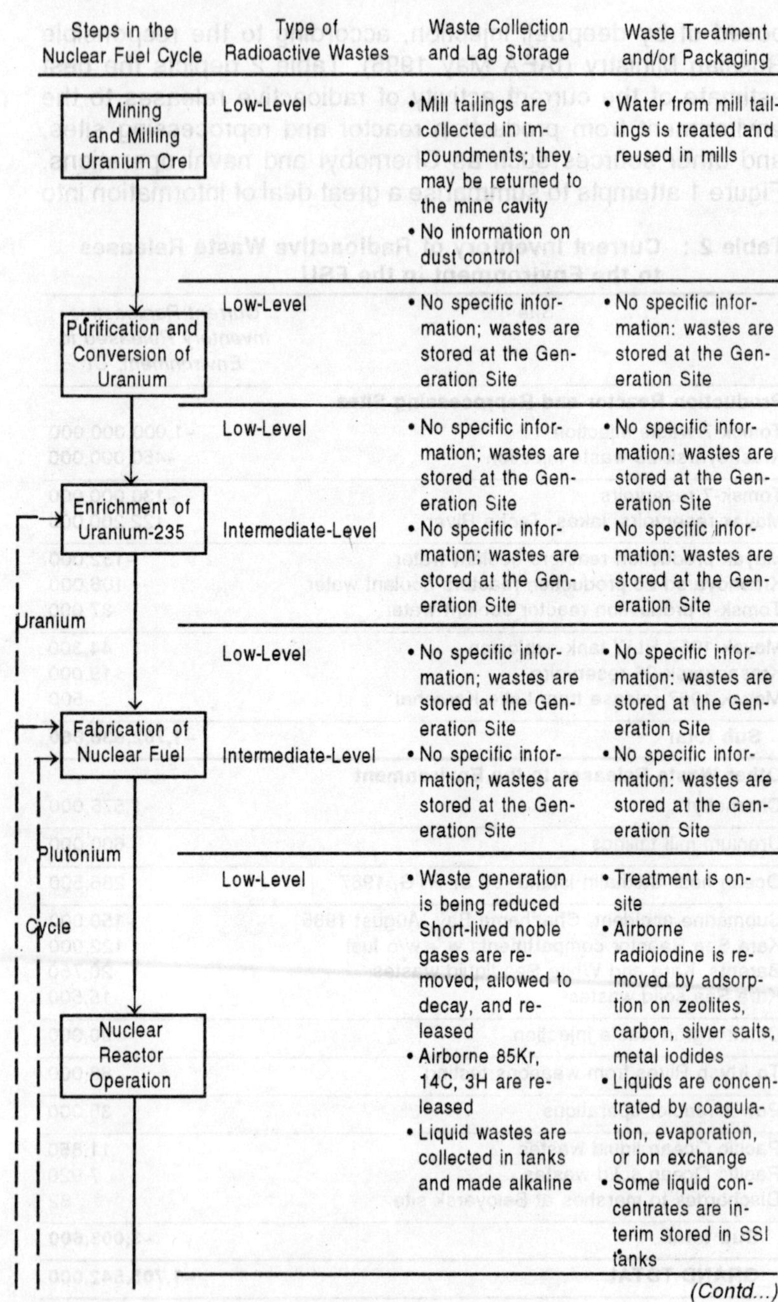

Steps in the Nuclear Fuel Cycle	Type of Radioactive Wastes	Waste Collection and Lag Storage	Waste Treatment and/or Packaging
Mining and Milling Uranium Ore	Low-Level	• Mill tailings are collected in impoundments; they may be returned to the mine cavity • No information on dust control	• Water from mill tailings is treated and reused in mills
Purification and Conversion of Uranium	Low-Level	• No specific information; wastes are stored at the Generation Site	• No specific information: wastes are stored at the Generation Site
Enrichment of Uranium-235	Low-Level	• No specific information; wastes are stored at the Generation Site	• No specific information: wastes are stored at the Generation Site
	Intermediate-Level	• No specific information; wastes are stored at the Generation Site	• No specific information: wastes are stored at the Generation Site
Fabrication of Nuclear Fuel	Low-Level	• No specific information; wastes are stored at the Generation Site	• No specific information: wastes are stored at the Generation Site
	Intermediate-Level	• No specific information; wastes are stored at the Generation Site	• No specific information: wastes are stored at the Generation Site
Nuclear Reactor Operation	Low-Level	• Waste generation is being reduced Short-lived noble gases are removed, allowed to decay, and released • Airborne 85Kr, 14C, 3H are released • Liquid wastes are collected in tanks and made alkaline	• Treatment is onsite • Airborne radioiodine is removed by adsorption on zeolites, carbon, silver salts, metal iodides • Liquids are concentrated by coagulation, evaporation, or ion exchange • Some liquid concentrates are interim stored in SSI tanks

(Contd...)

Steps in the Nuclear Fuel Cycle	Type of Radioactive Wastes	Waste Collection and Lag Storage	Waste Treatment and/or Packaging
			• Liquid concentrates are incorporated into bitumen or cement in steel drums, sometimes with solid waste incinerator ash • Some solid wastes are incinerated or compacted
	Intermediate-Level	• Waste generation is being minimised • Liquid wastes are collected in tanks	• Treatment is on-site • Some LLW concentrates are combined with ILW • Some wastes are incorporated into bitumen or cement in steel drums • ILW and alpha wastes may be vitrified in the future Some solid wastes are compacted
	Spent Nuclear Fuel	• Policy is to store VVER fuels in reactor pools for 3–5 years; however, delays in reprocessing are causing longer storage • RBMK fuels are discharged on-line and stored on-site	• No treatment
	Low-Level	• Some airborne radionuclides are collected upon generation • Liquid wastes are collected in tanks • Solid wastes are collected	• Radiokrypton may be removed; radionuclide is removed by mercuric nitrate scrubbing and adsorption using silver salts • Caustic scrubbing is also used for dissolver off-gases

(Contd...)

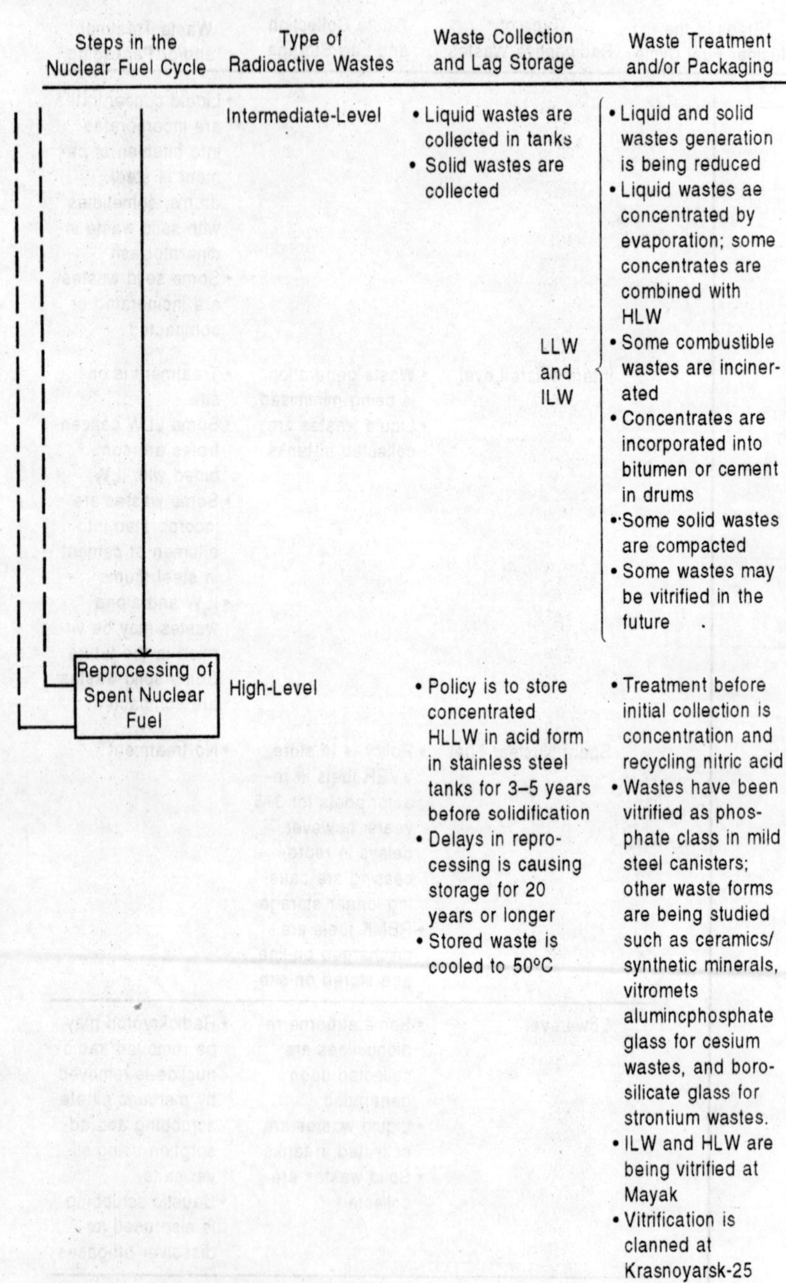

Steps in the Nuclear Fuel Cycle	Type of Radioactive Wastes	Waste Collection and Lag Storage	Waste Treatment and/or Packaging
	Intermediate-Level	• Liquid wastes are collected in tanks • Solid wastes are collected LLW and ILW	• Liquid and solid wastes generation is being reduced • Liquid wastes ae concentrated by evaporation; some concentrates are combined with HLW • Some combustible wastes are incinerated • Concentrates are incorporated into bitumen or cement in drums • Some solid wastes are compacted • Some wastes may be vitrified in the future
Reprocessing of Spent Nuclear Fuel	High-Level	• Policy is to store concentrated HLLW in acid form in stainless steel tanks for 3–5 years before solidification • Delays in reprocessing is causing storage for 20 years or longer • Stored waste is cooled to 50°C	• Treatment before initial collection is concentration and recycling nitric acid • Wastes have been vitrified as phosphate class in mild steel canisters; other waste forms are being studied such as ceramics/synthetic minerals, vitromets alumincphosphate glass for cesium wastes, and borosilicate glass for strontium wastes. • ILW and HLW are being vitrified at Mayak • Vitrification is clanned at Krasnoyarsk-25

(Contd...)

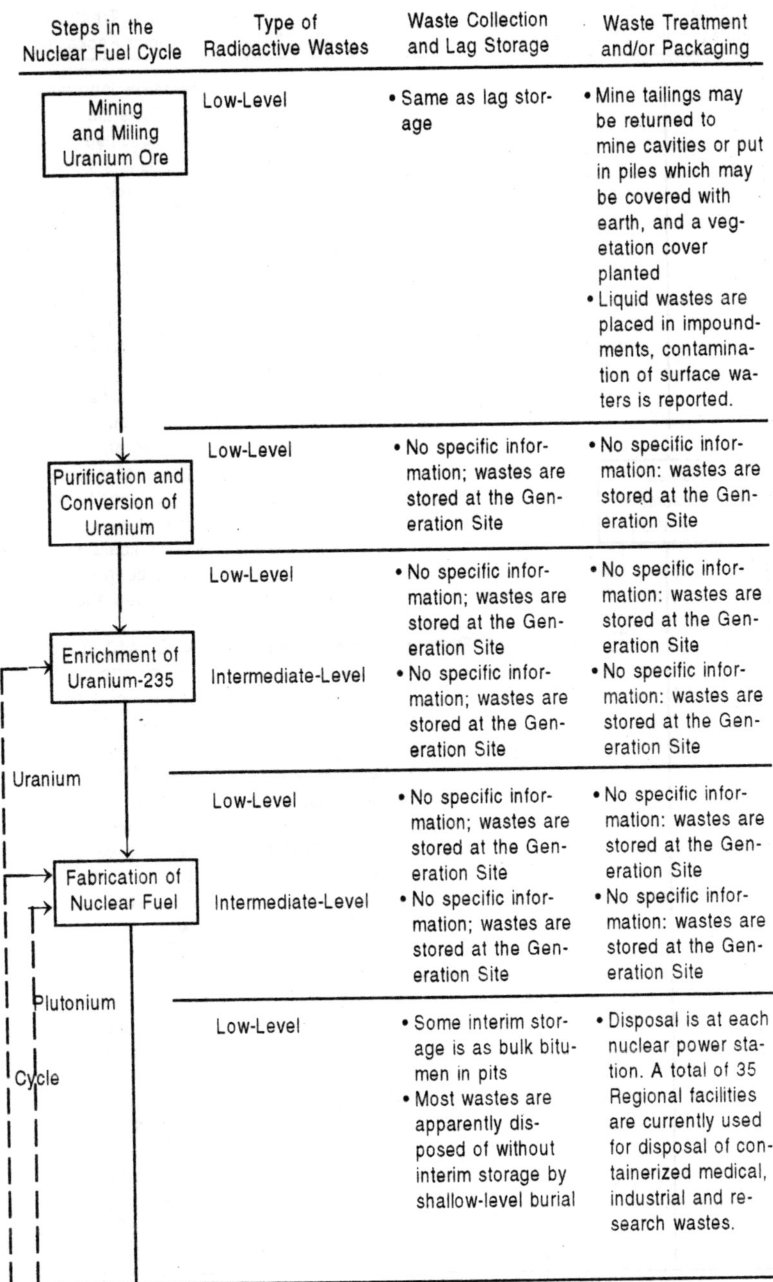

Steps in the Nuclear Fuel Cycle	Type of Radioactive Wastes	Waste Collection and Lag Storage	Waste Treatment and/or Packaging
Mining and Miling Uranium Ore	Low-Level	• Same as lag storage	• Mine tailings may be returned to mine cavities or put in piles which may be covered with earth, and a vegetation cover planted • Liquid wastes are placed in impoundments, contamination of surface waters is reported.
Purification and Conversion of Uranium	Low-Level	• No specific information; wastes are stored at the Generation Site	• No specific information: wastes are stored at the Generation Site
Enrichment of Uranium-235	Low-Level	• No specific information; wastes are stored at the Generation Site	• No specific information: wastes are stored at the Generation Site
	Intermediate-Level	• No specific information; wastes are stored at the Generation Site	• No specific information: wastes are stored at the Generation Site
Fabrication of Nuclear Fuel	Low-Level	• No specific information; wastes are stored at the Generation Site	• No specific information: wastes are stored at the Generation Site
	Intermediate-Level	• No specific information; wastes are stored at the Generation Site	• No specific information: wastes are stored at the Generation Site
	Low-Level	• Some interim storage is as bulk bitumen in pits • Most wastes are apparently disposed of without interim storage by shallow-level burial	• Disposal is at each nuclear power station. A total of 35 Regional facilities are currently used for disposal of containerized medical, industrial and research wastes.

(Contd...)

Steps in the Nuclear Fuel Cycle	Type of Radioactive Wastes	Waste Collection and Lag Storage	Waste Treatment and/or Packaging
			• Shallow land burial facilities are used for solidified waste Some liquid wastes have been injected into deep aquifers
Nuclear Reactor Operation	Intermediate-Level	• Some wastes are stored for up to tens of years at the reactor site • Many wastes are apparently disposed of without interim storage by shallow-level burial • Some interim storage is as bulk bitumen in pits	Some ILW may be disposed of with LLW at the respective power stations • Disposition of some ILW is to be determined, but may be vitrified and disposed of in geologic repositories in the future • Some liquid wastes have been injected into deep aquifers
	Spent Nuclear Fuel	• Some extended storage in the FSU in several at-reactor central storage pools • After storage in reactor pools, fuel is moved to central interim pool storage (20-30 yr) at reactor sites • Additional central storage is planned	• VVER-440, BN, research and naval spent fuel is reprocessed • VVER-1000 spent fuel will be reprocessed • RBMK fuel will not be reprocessed; disposal in a geologic repository is likely
	Low-Level	• No information on trapped airborne radionuclides • No extended interim storage of liquid and solid wastes	• Fuel disposition of trapped airborne radionuclides is unknown • Significant amounts of liquid wastes have been discharged to surface waters and/or injected underground

(Contd...)

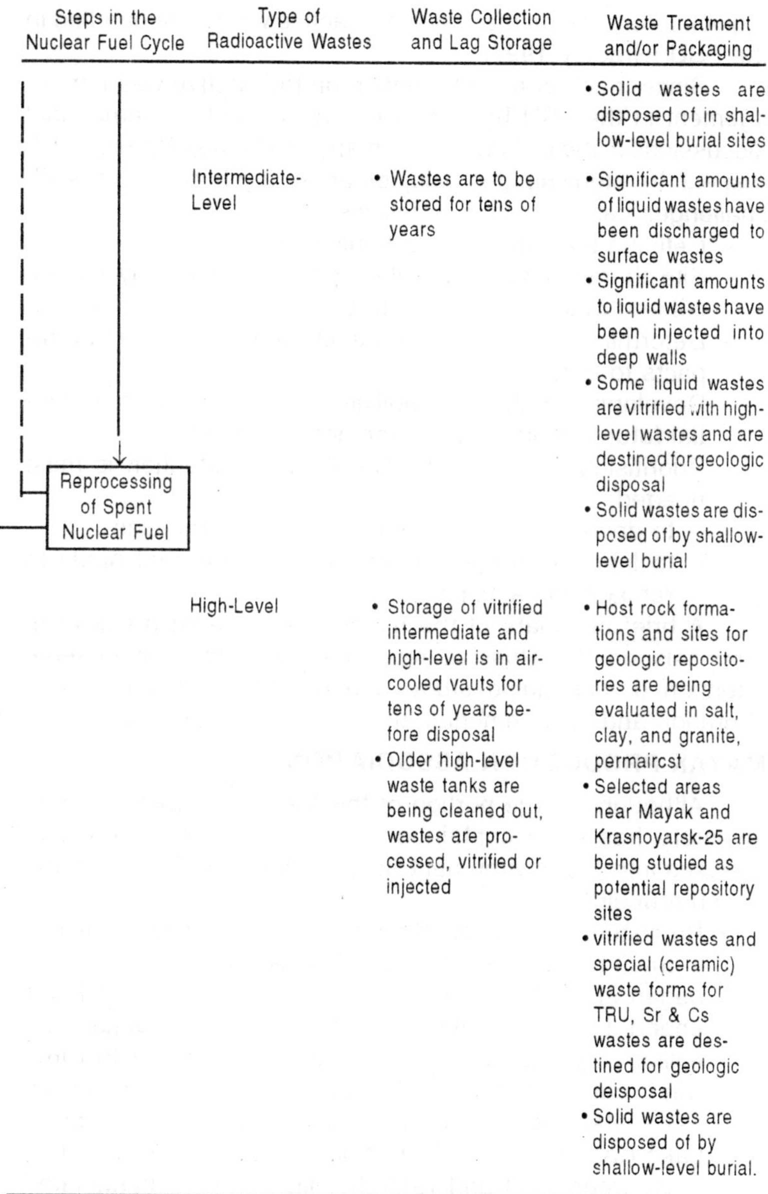

Steps in the Nuclear Fuel Cycle	Type of Radioactive Wastes	Waste Collection and Lag Storage	Waste Treatment and/or Packaging
			• Solid wastes are disposed of in shallow-level burial sites
Reprocessing of Spent Nuclear Fuel	Intermediate-Level	• Wastes are to be stored for tens of years	• Significant amounts of liquid wastes have been discharged to surface wastes • Significant amounts to liquid wastes have been injected into deep walls • Some liquid wastes are vitrified with high-level wastes and are destined for geologic disposal • Solid wastes are disposed of by shallow-level burial
	High-Level	• Storage of vitrified intermediate and high-level is in air-cooled vauts for tens of years before disposal • Older high-level waste tanks are being cleaned out, wastes are processed, vitrified or injected	• Host rock formations and sites for geologic repositories are being evaluated in salt, clay, and granite, permaircst • Selected areas near Mayak and Krasnoyarsk-25 are being studied as potential repository sites • vitrified wastes and special (ceramic) waste forms for TRU, Sr & Cs wastes are destined for geologic deisposal • Solid wastes are disposed of by shallow-level burial.

Fig. 1 : Overview of Uranium Nuclear Fuel Cycle and Radioactive Waste Management in the FSU

a quick reference on the ways radioactive waste are handled in the FSU (Bradley 1997).

Since much of the information on radioactive waste management in the FSU has only been available for international discussions in the last few years, many challenges still exist with respect to studying this subject area. A short list of such challenges can be stated as follows:

- Defining the extent of contamination
- Site characterisation activities (geology, hydrology, chemical and radioisotope inventory)
- Determining the extent of health and environmental impacts to date
- Developing analysis capabilities to gain insight into potential future health and environmental impacts
- Prioritising the extensive list of where attention is most needed:
  - to develop and implement mitigation strategies
  - to learn from the contamination events and apply to remediation actions.

A brief summary of the major contaminated areas (with emphasis on the Russian reprocessing sites) that deserve attention or continuing study with respect to contamination migration and environmental and human health impacts are:

MAYAK PRODUCTION ASSOCIATION

Although a large portion of the Mayak radioactive waste inventory has been contained in storage tanks, a series of releases into the environment have contaminated the site and the surrounding area:

- Radioactive waste discharges into Lake Karachi (which is a small natural lake near the Mayak reprocessing area), began in 1951 as large discharges into the Techa River ended. Lake Karachi is currently believed to contain 120 million Ci, primarily cesium-137 and strontium-90. In 1967, winds carried about 600 Ci or radioactive particles, mostly associated with dust from the dried exposed shoreline of Lake Karachi, up to 75 km from the site (Botov 1992).

  Between 1951 and 1989, 5 million M 30f contaminated solutions entered the water beating strata underneath the lake. The front of the southward-migrating plume of contaminated groundwater, which has a 90Sr concentration of

0,1 nCi per litre, has advanced 2.5 km over 40 years and is approaching the Mishelyak River. Because of its higher density, plume lies primarily in the lower part of the underlying rock fracture zone, at a depth of 70 to 100 m, and is under the discharge zone of the groundwater into the Mishelyak River (Drozhko and Romanov 1993). Since 1967 Russia has been filling in Lake Karachai to minimise the release of contaminants to the environment (Nuclear Waste News January 16, 1992).

The contamination of the Techa River, artificial reservoirs, and the Asanov Marsh area below the last reservoir, is associated with the direct discharge of high-level and other waste streams into the Techa River, primarily between 1949 and 1952. During this period, 76 million cubic metres of liquid waste were discharged, with a total beta activity of 2.75 million CI (Chukanov et al. 1991). The activity remaining today from these discharges is estimated to be about 200,000 Ci.

The five once-through cooled defence reactors at Mayak also discharged contaminated cooling water into Lake Kyzyitash and the Techa River/Reservoir system. Estimates of the decay corrected amount of activity remaining from these discharges (based on reactor data and operational experience at the Hanford reactors in the State of Washington, United States—which was also done for the single once-through cooled reactor at Tomsk-7 and the two at Krasnoyarsk-26), indicate that about 130,000 Ci remains from releases over the years from cooling water discharges at Mayak (Bradley and Jenquin1995).

In 1957, a waste tank containing about 20 million Ci of radioactive waste exploded, resulting in the discharge of about 2 million Ci into the atmosphere and surrounding environment, while the remainder fell in the immediate vicinity of the tank. Extensive efforts were begun shortly after the accident to remove surface contamination, relocate people and study the effects of contamination on agricultural products. Today, small areas still remain off-limits, but the entire inventory remaining from the accident has decayed to about 40,000 Ci (made up almost entirely of 90Sr), a far smaller amount than the millions of curies to be dealt with elsewhere. Therefore, the concerns at the Mayak site are centered on Lake Karachi and the Techa River-cascade (Bradley 1997).

TOMSK-7

The primary method of waste disposal at this site has been by discharges to man-made reservoirs or 'pits' and disposal via deep-well injection. While there have been a number of ongoing Russian-Western efforts to study contaminant migration and effect at Mayak, there have been relatively few at either this site or at Krasnoyarsk-26. The major waste management activities at Tomsk-7 that deserve further attention and study are as follows:

- Deep-well injection, which is the disposal of radioactive waste in underground formations that isolate the waste from the surrounding environment, has been used extensively at Tomsk-7 since 1963. The injection layers are reported to be at depths ci 240 - 340 m in sandy Cretaceous strata (Rybal'chenko et al 1994). Minatom officials have stated that the radioactive waste disposed of underground has a total radioactivity of billion curies (Moscow Interfax January 21, 1994).

  The reservoirs at Tomsk-7 contain about 130 million Ci of radioactive waste, (Nuclear Safety and Cleanup Report May 1994) and are thus similar in radioactivity levels to the discharges to surface water systems at the Mayak site. It appears however, that the pits used at this site for disposal of radioactive wastes generated from the reprocessing of production reactor fuel were lined. This action should help prevent the release of radionuclides to the environment—however it is not clear as to how effective they have been, and in fact there may have been breaches to the liner systems. To avoid windblown contamination releases, work was begun in 1991 to cover one of the reservoirs. During the winter months, sand is spread over the ice. Then, when the ice melts in the spring thaw, the radioactive particles are covered by a layer ci sand. (Bohmer and Nilsen August 22, 1995). About 75 per cent of the water surface area has been covered, most of the bottom sediments have been contained, and the liquids (decantate) are being disposed of by deep well injection (Tomsk Spravaka 1994).

The once-through cooled defence reactor at Tomsk-7 also discharged contaminated cooling water into the Tom River via canals. Estimates of the decay corrected amount of activity remaining from these discharges indicate that about 37,000 Ci remains from releases over the years from cooling water discharges (Bradley and Jenquin 1995).

KRASNOYARSK-26

The last of the three Russian production reactor and reprocessing sites, Krasnoyarsk-26 has apparently had far smaller discharges of radioactive wastes to surface water systems than either Mayak or Tomsk-7. Wastes have either been stored in tanks located underground [Krasnoyarsk-26 was built underground and reportedly has an internal volume 3.5 times that of the pyramid of Cheops (Pravda June 30, 1992)] or disposed of via deep-well injection. In summary, the following is known about radioactive wastes at this site:

Liquid radioactive waste from the reprocessing of production reactor fuel has also been pumped underground into the clay and aquifers sand strata beneath Krasnoyarsk -26 since 1963, and the practice continues today. Liquid radioactive waste is conveyed to the injection installation by a 15 km-long pipeline. The injection facility consists of a dual system of wells: those for the injection of waste and others for the withdrawal of water. As radioactive waste is pumped in, underground stratal water is drawn off, and a system of observation wells monitors the process. The injection site is located on a high terrace, 100 m above and 750 m from the Yenisey River. According to Russian scientists, the current activity of waste injected at this site is 450 million Ci out of a total activity injected of billion Ci. The volumes of radioactive waste that have been injected in two different geological strata are 2.5 million M^3 of low-level waste (at a depth of 190-225 m) and 2 million M^3 of intermediate and high-level waste (at a depth of 380-475 m). According to studies by Russian scientists, the wastes will not migrate significant distances from this injection site (or from those at Tomsk-7) prior to their decay to insignificant levels (Rybal'chenko et al 1994).

— The two once-through cooled defence reactor at Krasnoyarsk-26 also discharged contaminated cooling water into the Yenisey river. Estimates of the decay corrected amount of activity remaining from these discharges indicate that about

106,000 Ci remains from releases over the years from cooling water discharges (Bradley and Jenquin 1995).

— There are four reservoirs at Krasnoyarsk-26, with a total reported radioactivity content of >19,000 Ci. The Russians are currently 'liquidating' one of the reservoirs by adding earth and sorbents for cesium (NATO April 1995, DOE September 1993).

— A new installation for the underground injection of radioactive waste was planned for the western side of the Yenisey River to dispose of some of the radioactive waste from a new chemical separation plant called RT-2, which was designed reprocess reactor fuel from the Soviet-designed VVER-1000-type of pressurised water reactor. A tunnel was built under the river to transport liquid waste to the other side. Tritiated liquid reprocessing waste was to be injected into underground clay formations at this site, known as Site 27, which is 10 km from RT-2. Construction of RT-2 and Site 27 was halted in 1989, primarily because of public and environmentalist organisation protests (Nuclear Fuel January 4, 1993).

CHERNOBYL

The Chernobyl Unit 4 Nuclear Power Plant, an RBMK-1000 reactor located in Ukraine approximately 100 km north of Kiev, suffered a very severe accident with disruption of the core and the reactor building on April 26, 1986, at 01:23, Moscow time. The burning reactor became a source of major radioactivity escape to the environment. Initial estimates showed that about 3–4 per cent of the fuel was blown out from the unit either in the form of fragments or fine particles. To stop this process to mitigate the graphite fire, and to cool the core debris, a variety of heat absorbing and filtering materials were dropped onto the reactor area from helicopters (Rybal'chenko 1988). About 5,000 tons of materials, including 40 tons of neutron absorbers such as boron. 2,400 tons of lead, 1,800 tons of sand and clay, 600 tons of dolomite, tri-sodiumphosphate and polymerising liquids were thrown into the reactor area (Project Poiyn March 8, 1996). Table 3 shows the estimated radionuclide release at Chernobyl, calculated on May 6, 1986, allowing for decay of short-lived radionuclides (Nuclear News 10/86) and the amounts remaining after 10 years of radioactive decay. While this accident in certainly one of the most-studied from the aspect of radioactive waste management and contaminant migration, the current residual activity in the environment is actually quite small when compared to the releases noted previously.

Table 3 : Estimated Release of Radionuclides from the Chernobyl Accident

Nuclide	*Release Activity by May 6, 1986 Ci*	*Percetnage of Core Inventory Released by May 6, 1986 Ci*	*Decay Corrested to May 1996 Ci*
137 Cs	1,000,000	13.0	795,200
85 Kr	950,000	up to 100	498,800
90 Sr	220,000	4.0	172,900
241 Pu	140,000	3.0	86,370
134 Cs	500,000	10.0	17,300
106 Ru	1,600,000	2.9	1,800
240 Pu	1,000	3.0	999
238 Pu	800	3.0	739
239 Pu	700	3.0	700
144 Ce	2,400,000	2.8	330
242 Pu	2	3.0	2
133 Xe	45,000,000	up to 100	0
85m Kr	—	up to 100	0
131 I	7,300,000	20.0	0
140 Ba	4,300,000	5.6	0
95 Zr	3,800,000	3.2	0
103 Zu	3,200,000	2.9	0
99 Mo	3,000,000	2.3	0
141 Ce	2,800,000	2.3	0
89 Sr	2,200,000	4.0	0
132 Te	1,300,000	15.0	0
239 Np	1,200,000	3.2	0
242 Cm	21.00	3.0	0
Total	**80,930,000**		**1,575,140**

Note: Estimated error ±50%

URANIUM MINING AND MILLING

Another source of environmental contamination from FSU nuclear defence activities is from uranium mining and milling operations. In terms of volumes of waste materials generated and left essentially unprotected in the environment, uranium mining and milling operations have had a considerable environmental impact. As of 1990 the FSU had accumulated approximately 5 billion tons of uranium mill tailings, with an annual production of 6-7 million tons coming from uranium mining operations (Falci 1990). The total contaminated area from uranium mining operations in the FSU is estimated at 600 km^2 (IIASA 1993). The total activity of waste from FSU uranium milling and mining operations is 600,000 Ci (Nuclear Safety and Cleanup Report 1994) with the total for Russia given as 180,000 Ci (IAEA May 1995). Liquid radioactive waste solutions primarily consist of water used in mining operations, with total volume generated of up to 60 million m^3/yr (Kuznetsov September 1995). Plans have been made to cover tailings piles from uranium ore processing by the year 2000, returning 166,000 acres back to the national "economy". A radioactive release reduction of 300 fold may be achieved through the use of a 1.5-m thick clay and grass cover (Falci 1990).

DISCHARGES FROM NAVAL OPERATIONS

Prior to the early 1990s, the Russians disposed of liquid and solid radioactive wastes to oceans and seas from the operation of nuclear powered ships, primarily the Northern and Pacific submarine fleets. Solid Wastes, including reactor compartments, some with spent fuel left in the reactor cores, constituted by far the largest portion of the total inventory of radioactivity. Recent studies and decay correction calculations show that the current inventories of discharged solid wastes are about 140,000 Ci. The Northern Fleet stopped discharging LRW and SRW to oceans and seas in 1992, the Pacific Fleet in 1993. Results of the third Norwegian/Russian cruise to waste disposal sites in the Kara Sea in 1994 showed that radioactive waste-containing barrels were still intact. Although high concentrations of cesium were measured around the barrels, the concentrations were nearly normal 20 to 30 m away. Leakage from the solid wastes is expected to occur slowly, and there is tremendous dilution by the Kara Sea prior to contaminants reaching the Arctic seas and emerging between Svalbard and Greenland (Oslo Aftenposten October 18, 1994).

Tables 4 and 5 present data on SRW discharged to the Kara Sea, for reactor compartments and intermediate-and low-level wastes in large objects and containers (Sjoeblom and Linsley 1995). Table 6 gives some information in the Pacific Fleet's discharges of LRW and SRW in the Sea of Japan and Kamchatka regions (Danilyan and Vysotsky January 1995).

Table 4 : Summary Data on Low and Intermediate-Level SRW Dumped in the Kara and Barents Seas.

Area	Activity, Ci	No. of Dumpings	Years	Remarks
Novaya Zemlya Trench	3,320	22	1967–1991	3,174+C, 9LO, V
Sedov Fjord	3,410	8	1982–1984	1,108C, 104LO
Oga Fjord	2,027	8	1968–1983	472+C, 4LO, 1V
Tsivolka Fjord	2,684	8	1964–1978	1,600+C, 6LO, 1V
Stepovoy Fjord	1,280	5	1968–1975	5LO
Aprosimov Fjord	661	7	1966–1981	8+C, 6LO, 4V
Blagopoluchiya Fjord	235	1	1972	1LO
Techniye Fjord	1,845	3	1982–1988	146+C, 18LO, 1V
Off Kolguyev Island	40	1	1978	1V
	15,502	**63**		

Key to Remarks: C – containers; LO – Large objects; V – vessles

The largest radioactivity release from a Russian naval accident was due to an explosion in a nuclear-powered submarine on August 10, 1985, at a nuclear ship repair plant on Chazhma Bay in Primorskiy Kray. During a refueling operations of an Echo-II submarine, accidental removal of the control rods resulted in a reactor criticality and a steam explosion followed. Ten people were killed, fuel elements were damaged (one fuel element was actually blown out of the reactor) and substantial amounts of radioactivity were released to the surrounding environment. The combustion products along with radioactive materials fell within a radius of 50 m to 100 from the submarine.

A radioactive plume was produced that was several km wide, and fallout of aerosol particles occurred up to 30 km from the site of the accident. The release of radioactive substances to the atmosphere was calculated to be ~7.027 Mci, of which ~2.2 Mci were noble gases (NATO April 1995). A radioactive trail 6 km by 500 m was left on an adjacent hillside, and radioactivity is reported to have spread to nearby Konyushkovo, Abrek, and Razboynik Bays (Trud October 25, 1991).

NUCLEAR WEAPONS TESTS AND PEACEFUL NUCLEAR EXPLOSIONS

The former Soviet Union conducted nuclear weapons tests at the Novaya Zemlya islands in the Russian Arctic and at Semipalatinsk in Kazakhstan. In addition, it had an extensive

Table 5 : Information on the Naval Reactors Dumped in Novaya Zemlya Fjords and Kara Sea

			No. of Reactors				White Book[b]		Further Studies			
Site	*Year of Dump-ing*	*Factory No.*	*Without Spent*	*Con-tain-ing Spent*	*Dumping Depth*[a]	*Type of Solid Waste Form*	*At the Time of Dumping*		*At the Time of Dumping*		*1993–94*	
Abrosimov Fjord	1965	(No. 285)	1	1	20 (10–15)	Reactor compartment	30,000	(807)	11,600	(314)	634	(17.7)
		(No. 901)	--	2	20 (10–15)	Reactor compartment	15,000	(400)	2,950	(79.6)	727	(19.7)
		(No. 254)[c]	2	--	20 (10–15)	Reactor compartment	500	(14)	93.2	(2.52)	9.46	(0.256)
	1966	(No. 260)	2	--	20	Reactor compartment	500	(14)	43.7	(1.18)	5.05	(0.137)
Tsivolka Fjord	1967	(OK - 150)	3	0.6[d]	50	3 Reactor compartment with a box containing fuel on top	5,500	(150)	19,500	(530)	2,200	(59.5)
Novaya Zemlya Depression	1972	(No. 421)	--	1	300	Reactor in metal container with lead shell(e)	30,000	(800)	1,050	(28.3)	293	(7.91)
Stepovoy Fjord	1981	(No. 601)	--	2	50 (30)	Submarine	7,400	(200)	1,360	(36.8)	614	(16.6)
Tochenlye Fjord	1988	(No. 538)	2	--	35–40	Reactor in motal container	500	(14)	5.59	(0.152)	5.05	(0.137)
Total			**10**	**6.6**			**89,400**	**(2,400)**	**36,600**	**(993)**	**4,400**	**(122)**

a) Depths from the White Book (Yablokov 1993), those in brackets from Norwogian-Russian Cruiso data.

b) Fission products as in the White Book; activation products estimated on the basis of the White Book as follows: total content of activation products in reactors withoug spent nuclear fuel not more than 100 kCi (3700 TBq). 50kCi of which was in 3 reactors of OK-150. Thus the remaining 7 reactors contained not more than 7 kCi each.

c) Either 254 or 260.

d) The SNF was not contained in the naval reactor, but in a reinforced concrete and metal shell.

e) Probably no furfural.

Table 6 : Characteristics of Radioactive Waste Discharged in the Pacific Ocean Region by the Russian Navy

Region	*Sea*	*Distance from Coast*	*Depth(m)*	*Period of Discharge*	*Type of Radioactive*	
					Liquid (Ci)	*Solid (Ci)*
9	Japan	120	3,300–3,400	1974–1993	10,840	3,234
8	Kamchatka	25	2,000–2,600	1969–1992	—	2,993
6	Japan	50	1,900–3,300	1967–1990	—	969
10	Japan	120	2,900–3,300	1974–1984	536	721
7	Kamchatka	20	1,400–1,500	1962–1992	361	—
5	Japan	25	1,100–1,500	1962–1993	117	—
1	Japan	90	3,000–3,400		Not Used	
2	Japan	160	2,800–3,400		Not Used	
3	Okhotsk	160	1,200–1,300		Not Used	
4	Kamchatka	160	4,700–5,300		Not Used	

programme using nuclear explosions for civilian-related applications such as seismic studies of the earth's crust, creation of underground storage cavities, and enhanced recovery of mineral resources. These were called peaceful nuclear explosions, or PNEs. A total of 130 nuclear explosion were conducted at Novaya Zemlya, 456 at Semipatinsk, with 129 conducted elsewhere (primarily PNEs), for a sum of 715 nuclear explosions, including 124 PNEs. Summaries of the tests by location and type are given in Tables 7 and 8 respectively (Miniatom and MOD 1996).

Table 7 : Location of Soviet Nuclear Weapons Tests and Peaceful Nuclear Explosion

Nuclear Test and Peaceful Nuclear Explosions	
Nuclear Test Sites	
Semipalatinsk Test Sites	456
Northern Test Site (Novaya Zemlya)	130
Sub-total	**586**
Outside Nuclear Test Sites	
Russian Federation	91
European Part	59
Asian part	32
Kazakhstan	33
Ukrain	2
Uzbekistan	2
Turkmenia	1
Sub-Total	**129**

Total No. of Nuclear Tests and Peaceful Nuclear Explosions = 715

Table 8 : Summary of Soviet Nuclear Tests and Peaceful Nuclear Explosions by Type

Atmospheric, Underwater and Space Explosions	
Air Explosions	177
Surface explosions	32
Underwater and water surface explosions	5
Space explosions	4
High-altitude explosions	1
Sub-total	**219**
Underground Tests	
Tunnel Tests	245
Shaft tests	251
Including cratering explosions	5
Sub-Total	**496**
Total	**715**

Although there has been a significant amount of information published in the last few years on the FSU weapons testing programme, very little data have surfaced with respect to actual amounts of radioactive contamination entering surface water systems. Releases from underground tests to surface waters have likely been quite small; releases from surface tests were diluted and carried great distances by winds. Releases from PNEs were likely to be small to the Obnenisey river systems, because most PNEs were underground and had relatively small nuclear yields. A brief summary of what has been reported is as follows:

- **for Novaya Zemlya:**

The first nuclear test on Noyaya Zemlya took place on September 21, 1955, 100 feet below the surface of the water in Chernaya Bay on the southwestern coast of Novaya Zemlya. There are elevated levels of 137CS, 239Pu, and 241 PU in the lower sediments of this area today (Nilsen and Bohmer 1994). Two more tests were conducted in this area in 1957 and 1961. A decay corrected estimated upper limit in 1995 for the releases to the Barents Sea from these underwater tests equals ~4,050 Ci of 137Cs and 2,700 Ci of ^{90}Sr (NATO April 1995).

Data on surface contamination indicate that the radioactivity near Novaya Zemlya does not differ much from that in other Arctic Areas. In other words, the vicinity of the test site has not received significant more contamination than one would expect from the general global fallout (NATO April 1995). However, there are three areas having higher radiation levels, two of which were associated with atmospheric tests conducted prior to 1963. The third area was linked to an underground explosion in August 1987 that vented radioactive gases (Moscow Radio World Service January 20, 1992).

- **for Semipalatinsk:**

In the period of surface testing at Semipalatinsk, about 100 Cilk M^2 of 239 Pu, strontium and cesium, 14C, and other long-lived isotopes entered the surrounding environment. From 1961-1969, about 38,000 Ci of radioactive strontium and almost 50,000 Ci cesium were introduced into the Irtysh River (Kazakhstanskaya Pravada January 4, 1992). Five underground cavities, formed in salt layers by nuclear explosions, have filled with water and may represent a source for migration of radionuclides (US Embassy Alma-Ata March 15, 1993).

Based on the findings of a mission to Semipalatinsk in July 1994, the IAEA has preliminary concluded that the former nuclear weapons test site no longer constitutes a health risk to local populations. The IAEA did recommend that studies be done on plutonium dispersed by early surface explosions and potential contamination of groundwater by radionuclides that may have leached from contaminated, areas. Two areas were also noted as requiring administrative measures to limit access: the Ground Zero site about 50 km from Kurchatov city where 124 atmospheric explosions were conducted, and Lake Chagan. This lake was created from a nuclear explosion intended to create a dam in the Chagan River (Nucleoriics Week January 26, 1995).

The IAEA also reported that the remaining nuclear weapons tests at Semipalatinsk, 223 at a southern site and 123 at an eastern site, were underground within the Degelen Mountains, and were of no present radiation concern (Nucleonics Week January 26, 1995). Some 88 per cent of the steppe pastureland that used to be the closed site has been declared safe, another 8 per cent, requires special inspections, and 720 kM^2 are too contaminated for human use (Moscow Segodnya March 2, 1995).

OTHER RADIOACTIVE MATERIALS DISCHARGES OR RELEASES

- Deep-well injection also has been used to dispose of radioactive waste at the Scientific Research institute for Nuclear Reactors in Dimitrovgrad Rybal'chenko et al 1993). The site appears to have been used for the disposal of radioactive waste from research and reactor operations, with far smaller amounts of radioactivity injected into the ground than at Tomsk-7 and Krasnoyarsk-26.
- The largest single radionuclide source discharged to oceans and seas was a thermoelectric generator, powered by ^{90} Sr, accidentally dumped in the Sea of Okhotsk near Sakhalin Island in 1987. This hermetically sealed source contained 350,000 Ci (its activity as of 1996 would be 286,500 Ci). Annual studies in the area have failed to locate the source or any strontium signal (Danilyan and Vysotsky January 1995).

SUMMARY AND CONTAMINATED SITE PRIORITY

It can easily be seen, based only on the summary information presented above that the challenges from the legacy of the nuclear weapons complex of the FSU are many—and significant enough that they are not easily resolved. Some issues are severe enough—such as at Mayak that they will likely preclude "cleanup" in the Western sense. With such a broad scope of contaminated sites, however, it is necessary to prioritise potential characterisation and remediation activities. Based on what is known to date, an attempt at such a prioritisation is given in Table 9. By prioritisation of areas to be studies, we are likely to be able to provide the maximum benefit in terms of human and environmental health. The Mayak and Tomsk-7 sites are clearly at the top of the list for characterisation and remediation, followed by those areas with smaller contamination levels. Although listed as a "Third Tier" in Table 9, primarily because of our current lack of detailed knowledge of contaminant migradon at Tomsk 7 and Krasnoyarsk-26, efforts should be implemented and/or expanded to better understand contaminant migration from these deep-well injection sites as soon as possible due to the amount of radioactive materials disposed of there.

Table 9 : Prioritisation of Contaminated Areas in the FSU for Characterisation and Remediation

Site	*Contaminated System*	*Estimated Current Activity, Ci*
First Tier	**(Large source term, available for migration**	
• Mayak	Lake Karachai, reservoirs, Techa/Ob river system	122,000,000
• Tomsk	Reservoirs, marshes – Tom/Ob rivers	130,000,000
Second Tier	**(Small source term, available for migration)**	
• Chernobyl	Pripyat river, Kiev reservoir	~1,600,000
• Uranium mining	Reservoirs, streams	600,000
• Chazhma Bay	Chazhma Bay, Sea of Japan	~150,000
• Krasnoyarsk	Reservoirs, Yenisey river	>19,000
Third Tier	**(Large source term, unknown migration potential)**	
• Tomsk	Deep Well injection–underground strata/water	1,000,000,000
• Krasnoyarsk	Deep Well injection–underground strata/water	450,000,000
Fourth Tier	**(Small source term, unknown migration potential)**	
• Arctic seas	Solid wastes–Northern Fleet	137,000
• Dimitrovgrad	Deep Well injection–Underground strata	90,000
• Sea of Japan	Solid wastes–Pacific Fleet	< 7,000
• Weapons test sites	Contaminated runoff	?
• PNE's	Contaminated runoff	?
• Radon's	Contaminated runoff, materials availability	?

BIBLIOGRAPHY

1. *Bol'shakov V.N., R. M. Aieksakhin, L.A. Bol'Shov, V. N. Chukanov, L. A. Kochetkov, V. I. Petukhov and A. A. Tsyb.* 1991. "Conclusion of the commission for Evaluating the Ecological Situation in the Sphere of Influence of the Industrial Plant (IP) 'Mayak' of the USSR Academy of Sciences No. 1140501, on June 12, 1990, Moscow, Russia.
2. *Botov N.G.* 1992. 'ALWP-67: A Little-Known Big Nuclear Accident.' Proceeedings of the Third International Conference on High Level Radioactive Waste Management, April 12-16, 1992, Las Vegas, Nevada.
3. *Bradley D. J. and U. P. Jenquin.* May 1995. 'Radioactive Inventories and Sources for Contamination of the Kara Sea by Riverine Transport. "PNWD-2316, Richland, Washington.
4. *Bradley D. J., ed. D. Payson.* 'Behind the Nuclear Curtain:

Radioactive Waste Management in the Former Soviet Union." 1997. Battelle Press, Columbus, Ohio. 716 pages. ISBN 1-57477-022-5.

5. *Carter M. W.*, 'Radionuclides in the Food Chain," article by *R. J. Pentreath,* 'Radionuclides in the Aquatic Environment.' Springer-Vertag, Berlin-New York. 1988. pp. 99-119.
6. *Chukanov V. N., E. G. Drozhko, A. P. Kulignin, G. A. Mesyats, A. N. Panyagin, A. V. Trapeznikov, and P. V. Volbuev.* April 10-14, 1991. 'Ecological consequences of the Creation of Nuclear Weapons on the Example of the Atomic Industrial Complex Near the City of Kyshtym." Distributed at the Conference on Environmental Consequences of Nuclear Development, University of California, Irvine, and translated for Lawrence Livemore Laboratory by the Ralph McElroy Translation Co.
7. *Danilyan V. A. and V. A. Vysotsky.* January 1995. "Nuclear Waste Disposal Practices in-Russia-United States Study Group on Russia's Pacific Ocean Region. Presented at the Japan
8. Dumped Nuclear Waste in the Sea of Japan, Sea of Okhotsk, and North Pacific Ocean, January 12-13, 1995, Biioxi, Mississippi.
9. Department of Energy (DOE). September 1993. Third Meeting of the US/Russian Joint Coordinating Committee on Environmental Restoration and Waste Management (JCCEM), Washington, DC, September 20-21, 1993.
10. *Drozhko E. G. an G. N. Romanov.* 1993. 'Conceptions, Tasks, and Main Results of South Urals Region Radiation Rehabilitation,'Mayak' Production Association, Chelyabinsk-65, Russia.
11. Presented at the Fourth ex USSR Nuclear Society Meeting, June 28-July 2, 1993, Nizhniy Novgorod, Russia.
12. *Falci F. P.* 1990 Foreign Travel Report,'Travel to USSR for Fact Finding Discussions on Environmental Restoration and Waste Management, June 15-28, 1990,' DOE Office of Technology Development, July 27, 1990.
13. International Atomic Energy Agency (IAEA). May 1995. Prsented at the Seminar on International Cooperation on Nuclear Waste Management in the Russian Federation. Vienna International Center, May 15-17, 1995, Vienna, Austria.
14. IIASA. 1993. 'General Description of Nuclear Legacy of the Former USSR.' Background Document for Advisory Committee Meeting On Radiation Safety of the Biosphere, October 4-5, 1993. Luxembourg.
15. Kazakhstanskaya Pravada. January 4, 1992. "History of Radiation Safety at Semipalatinsk Detailed ,' p.l.

16. Minatom and MOD (Ministry of the Russian Federation for Atomic Energy and the Ministry of defence of the Russian Federation). 1996. USSR Nuclear Weapons Tests and Peaceful Nuclear Explosions: 1949 through 1990 Eastview Publications, Minneapolis, Minnesota.
17. Moscow Interfax. January 21, 1994. "Energy Ministry Denies Plutonium Dumped in Siberian Lake."
18. Moscow Radio World Service. January 20, 1992. "Novaya Zemlya Radiation Study Results Released.'
19. Moscow Segodnya. March 2, 1995. "Semipatatinsk Radiation Pollution Problems Highlighted ,' p.9.
20. NATO. April 1995. "Cross-Border Environmental Problems Emanating From Defence-Related Installations and Activities: Volume: Radioactive Contamination.' Final Report, Phase 1; 1993-1995. Report No. 204, North Atlantic Treaty Organisation.
21. *Nilsen T. and N. Bohmer.* 1994. Bellona Report Volume i. 'Sources of Radioactive Contamination in Murmansk and Arkhangel'sk counties.' Published by the Bellona foundation, Oslo, Norway.
22. *Nikipelov B. V., A. F. Liziov and N. A. Koshurnikova.* PRIRODA, February 1990, Experience with the First Soviet Nuclear Installation. Translated by Alexander Shiyakhter, Department of Physics, Harvard University.
23. Nuclear Fuel, January 4, 1993. "Russian Plutonium Programme at the Crossroads: A Special Report,' pp.4-7.
24. Nuclear News. October 1986. " Chernobyl: The Soviet Report. 59.
25. Nuclear Safety and Cleanup Report. May 1994. "East, Central European Radwaste Detailed in ANS Directory/Report." 3(5):43-45.
26. Nuclear Waste News. January 16,1992. "News Briefs,'p.25.
27. Nucleonics Week. January 26, 1995. "IAEA Clears Semipalatinsk Area Conditionally for Living," pp. 6-7.
28. Oslo Aftenposten. October 18,1994. "Rusty Nuclear Waste Barrels Leaking,' p. 12. Pravada. June 30, 1992. 'Implications of Early Krasnoyarsk-26 Closure Viewed,"p.i.
29. Project Poiyn. March 8, 1996. Hypertext Database: Chernobyl and its Consequences, Russian Research Centre "Kurchatov Institute.' Internet Address publication: http//polyn. net. kiae, su/ polyn/ manifest. html.
30. *Rybal'chenko I.* 1988. 'Main Principles of Safe Removal, Conditioning, Transportation andStorage/Disposal of Severely Damaged Nuclear Fuel and Other Accident Generated Waste.'Paper presented at an IAEA Advisory Group Meeting, December 12-15, 1988, Vienna, Austria, during an IAEA Project on Post Accident Man Agreement of Severely Damaged Nuclear Fuel

and Radioactive Waste Arising.

31. *Rybal'chenko A. I., M. K. Pimenov, V. M. Kurochkin,* Scientific Research Institute of Industrial Engineering, Moscow, Russia; *E.P. Kaimin, E.V. Zakharova,* Institute of Physical Chemistry, Russian Academy of Sciences, Moscow, Russia. 1993. 'The Environmental Effect of the Disposal of Liquid Waste Into Deep Geological Formations.' Presented at the *Fourth ex-USSR Nuclear Society Meeting*, June 28-July 2, 1993. Nizhniy Novgorod, Russia.
32. *Rybal'chenko A. I., M. K. Pimenov, P. P. Kostin, V.D. Balukova, A. V. Nosukhin, e. I. Mikerin, N. N. Egorov, E. P. Kaimin, I. M. Kosareva and V. M. Kurochkin.* 1994. "Underground Disposal of Liquid Radioactive Wastes.' Moscow Izdat. Translated from Russian by Ben Teague for Sandia National Laboratory, 1996.
33. *Sjoeblom. K. L. and G. Linsley.* January 1995, Presented at IAEA International Arctic Seas Assessment Project (IASAP) working group meeting, January 23–27, 1995, Vienna, Austria.
34. Trud. October 25, 1991. "Concern Over Post-Accident Conditions at Nuclear Submarine Repair Plant, ' No. 251, p. 2, cols. 1-6.
35. US Embassy, Alma Ata, Kazakhstan. March 15, 1993. 'Local Press Report on Nuclear Explosions in the Altay Oblast.'

Address:

DONALD BREDLEY
Pacific Northwest Laboratory
Richland WA 99352
USA

53

Conceptual Basic and Status of Nuclear Power Plant Decommissioning Effort in the Russian Federation

V. GLAZOUNOV and L. KHAMYANOV

NPP unit decommissioning, although being a usual phase in an industrial facility's life cycle, still has significant peculiarities. These are due to radioactive contamination of NPP unit major equipment and structural elements. The presence of contamination leads to significant changes in the technical policy of the process of facility decommissioning. The assessment of potential changes in this policy has necessitated the development of a special "Concept of NPP Unit Decommissioning" which has reflected the technical policy (philosophy) of the decommissioning process. This article addresses the conceptual base of this philosophy.

The concept stems from the principle that NPP unit decommissioning is undertaken after expiration of special lifetime or in case of technical impossibility or economic inexpediency of ensuring further safe operation. The concept provides for a possibility to extend nuclear unit lifetime, but presently there is no available procedure for such extension. A special programme on the procedure for nuclear unit lifetime extension should be developed and implemented.

Appropriate conditions for extending the lifetime may be possibly implemented starting from, serial (standardised) VVER-1000 reactor units. The concept is also based on the principle of full replacement of "spent" nuclear capacities through construction of a new unit at the same site. Such an approach is explained by the advantage of using the existing NPP structure, proficient personnel etc. Also of importance is the fact that the decommissioned unit will be mothballed for a long period of time and should be kept under, radiation monitoring. It is obvious that the optimal way here is to use the radiation monitoring system of operating nuclear units for monitoring the decommissioned unit as well. According to the basic principles the process of nuclear unit decommissioning consists of several phases which reflect technical, economic, ecological, social and even political aspects of the sophisticated decommissioning problem.

NPP UNIT SHUTDOWN

The essence of this phase is the termination of unit commercial operation and bringing it to nuclear-safe condition. At this phase complete core defueling and preparations or equipment mothballing are undertaken.

Radiation safety is assured by the normal radiation monitoring system. The duration of this phase is up to six months.

NPP UNIT MOTHBALLING

At this phase effort is complete for comprehensive engineering, and radiation review of the unit to justify particular concept of its decommissioning and finalise the work programme at this and subsequent phase. The zone of contaminated equipment for prolonged hold-up is identified and the isolating protective barriers around the zone are set up. Based on the normal system, radiation monitoring tools and techniques are established, and the monitoring procedure is developed. The duration of this mothballing phase is three-five years.

EXTENDED HOLD-UP OF DECOMMISSIONED UNITS

Extended hold-up is maintaining of contaminated equipment in the unit isolated zone in a radiologically safe condition. The aim of this phase is to reduce radioactivity through natural decay of radionuclides while monitoring contaminated zone isolation from the environment. The duration of extended holdup depends on many reasons and is determined in each specific case in the particular concept and in the project on nuclear unit decommissioning. Isotopic analysis of induced activity in the structural elements and components has shown that after twenty-thirty years of hold-up the main dose-contributing nuclides basically decay and activity level is determined by the long-lived nuclides. The total activity reduction during the twenty-thirty year period of hold-up is estimated to be sixty-hundred times. If there are other reasons, further unit holdup from the viewpoint of activity level decrease becomes effective.

UNIT DISMANTLING AND BURIAL

This phase completes the unit life cycle the major element here being the collection, reprocessing and burial of equipment and structural elements including the concrete. An important task at this and at the previous phase is the recovery of metal and concrete where the acceptable nuclide specific activity is ensured. The duration of this phase is estimated to be five-ten years, thus, for the project in question the total duration of all the phases of nuclear unit decommission amounts to thirty-forty years. For a multi-unit Plant and in case the replacement of "spent" nuclear capacities is assured, such a decommissioning project duration comparable with the unit operation life seems optimal. The concept of NPP unit decommissioning in Russia reflects all the major phases of the IAEA and REA-OECD concept.

STATUS OF NPP DECOMMISSIONING EFFORT IN RUSSIA

In 1988 Beloyarsk NPP Unit 1 completed its lifetime and was shut down, and in 1989 the same happened to Beloyarsk NPP Unit 2.

Both units are equipped with pressure tube reactors. In 1984 and 1990 Novovoronezh NPP units 1 and 2 VVER-type reactors were shut down respectively. Decommissioning experience with the above units has to a great extent enabled to identify the technical policy or the concept of decommissioning, which in 1991 was adopted by the Russia's Ministry of Atomic Energy in the form of a regulatory document. Further promotion of NPP decommissioning concept is also ensured by the R&D Programme which has been under way since 1987. The funding of research effort under the programme and of actual decommissioning work at Beloyarsb and Novovoronezh Plant sites is provided from a single NPP decommissioning fund, which has been established by the Rosenergoa on operating organisation cn the basis of deductions from the Russian nuclear electricity generation costs.

54

Nuclear Submarine Decommissioning and Related Environmental Problems

A. SARKISOV

1. GENERAL ASSESSMENT OF SITUATION ASSOCIATED WITH NUCLEAR SUBMARINE DECOMMISSIONING

Of the problems associated with nuclear wastes, the issues of nuclear powered submarine utilisation occupies a particular place.

Nuclear submarines (NS) that have been taken out of military service as well as those intended for utilisation present potential sources of both nuclear and radiation hazard.

The following data provide the evidence for the magnitude of the NS utilisation problem (Fig. 1).

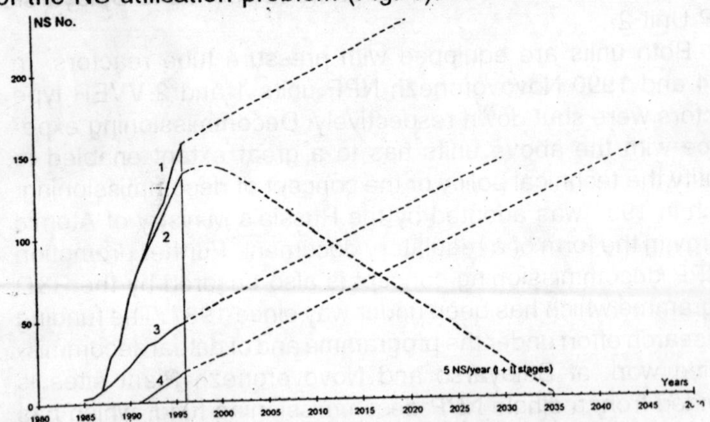

Fig. 1 : Dynamics of nuclear submarines decommissioning and utilisation

1. Nuclear submarine decommissioning dynamics
2. Dynamics of the nuclear submarine accumulation under account of dismantling
3. Dynamics of the submarine corres unloading
4. Accumulation dynamics of the three - compartment blocks.

By the beginning of 1966 more than one hundred and fifty nuclear-powered vessels had been decommissioned in Russia both on expiration of their service life and according to the treaties on the reduction of strategic offensive weapons. By the beginning of year 2000 this number is expected to increase to one hundred seventy-eighty units. With the onset of the next millennium, the second wave of the NS decommissioning is going to pick up its speed.

According to the data published, the number of NS that have been decommissioned in the USA to date exceeds twenty units.

Other NS possessing countries, Great Britain and France in particular, have also encountered the problem of their utilisation.

Taking into account both the great overall number of NS destined for utilisation and their high radiation potential, we consider the NS utilisation as one of the currently most important global environmental problems.

The following data illustrates the current state of NS utilisation works.

By the early 1990s, the United States managed to design and start a complex flow-line NS utilisation. The several transportation-technology scheme of this procedure has been described in office reports. However, to our regret, many a technical detail still remain not clear enough. Therefore, any estimations of the environmental acceptability of these utilisation procedure need to inevitably rely upon American reports that claim their sufficient agreement with national and international environmental safety standards.

The NS utilisation scheme accepted in the United States comprises fuel removal from the reactor vessel, submarine deactivation, cutting-out the reactor compartment followed by its combined transportation, i.e. by water and then by land, to the sites of its long-term storage, where they are buried in special trenches in a deserted region near Henford, Washington.

All the machinery and equipment installed in the submarine's remainder is then taken away, the devices feasible for further use undergo appropriate reconditioning, while the rest of the equipment as well as the submarine hull and superstructure are to be scrapped. Hazardous materials and wastes are localised and land-buried.

Following this procedure, some fifty nuclear propelled submarines have been utilised in the United States to date. Great Britain and France have embarked on similar programme, but with a limited NS number.

In solving this problem Russia appears tc be in the most difficult situation: the number of the nuclear submarines that it has to decommission and utilise by the year 2000 is three times and of the vessel nuclear power plants is five times that of the United States.

At the currently existing technology level, the NS dismantling remains both labour consuming and very expensive: the cost of a single NS dismantling makes up some fifty billion rubles (the prices refer to 1995). The Russian Government takes some more to rig the NS utilisation bases with necessary equipment but the lack of funds allows only some particular problems to be solved while the major task of the development of the complex NS utilisation technology remains out of proper consideration.

In the absence of the infrastructure providing for the complex NS utilisation, and in particular, with unprepared long-term storage sites the NS utilisation is being carried out now only according to the temporary scheme. To ensure the buoyancy of a detached NS block, the reactor compartment (RC) is cut out together with two adjacent ones, and this three-compartment unit is tugged to special shipyards allocated for the temporal NS reactor storage afloat.

By the beginning of 1996, eighteen nuclear submarines have been utilised with the three compartment block cutting-out, eight vessels have been prepared for the long-term storage afloat, and thirteen vessels were under the procedure of utilisation and preparation for the long-term storage.

2. MAJOR PROBLEMS ASSOCIATED WITH THE NS UTILISATION

The major difficulty of the NS utilisation problem lies not only in its great magnitude. Any attempt to solve the problem involves the necessity of taking into complex consideration a great number of diverse and sometimes contradictory factors which determine its the essence—political, economic environmental, social, and technological ones.

In our opinion the key points of the problem are the following ones:

- safety ensuring for the long-tern (15-20 years, at least) storage afloat of the out of service nuclear submarines;
- unloading and transportation of the spent nuclear fuel that is accumulated in naval bases and fuel servicing vessels in large quantities.
- cutting-out of the reactor compartments and development of the infrastructure for their long-term storage.

At this point, particular attention should be focused on treatment the reactor compartments containing emergency steam generating installations; processing and safe storage of large volumes of liquid radioactive wastes; scrapping the NS remainders according to efficient and environment saving technologies; establish the conditions that would rule out the possibility of the release and proliferation of fissing materials at all the stages of NS decommissioning and utilisation.

The time limits of this report allow me to briefly dwell only on the first three of these issues, the most acute ones to my opinion.

3. SECURITY MEASURES FOR DECOMMISSIONED NUCLEAR SUBMARINES

All the industrial and naval enterprises involved with the NS utilisation are unable to dismantle more than five-six submarines a year. At this utilisation performance the submarines that are currently destined to the dismantling will remain waterborne for at least ten-fifteen years more expecting for their turn to be utilized (Fig. 2). The decommissioned and anchored nuclear submarines present a serious potential nuclear and radiation environmental hazard. The critical security-related point here is that more than 60 per cent of these submarines still have stored spent nuclear fuel on board.

Estimates of the overall radiation potential of the fuel in their reactors give the number of 200-300 million Ci.

The vessels under decommissioning have, as a rule, substantially worked-out fuel in their cores. Taking into account the accumulated fission fragments activity in this fuel, the after effect of a hypothetical nuclear accident would correspond to 5-6th degree of IAEA nuclear accident's scale.

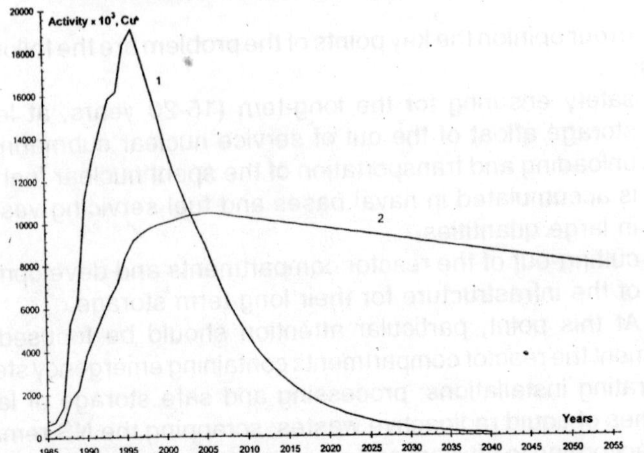

Fig. 2 : Integral Activity of Decommissioned Nuclear Submarines with Unloaded Fuel

1. γ-radioation activity (^{60}Co) accumulated in the constructive materials of reactor compartments of the decommissioned nuclear submarines.
2. Activity (63 Ni) accumulated int he constructive materials of reactor compartments of the decommissioned nuclear submarines.

Remembering that the nuclear accident in Chazhma Bay in 1985 occurred with "fresh" fuel (a new core had been just charged) we can conclude that the consequences of the accident with the completely burned-out core will be much vaster. The outburst of long-lived radionuclides in this case will exceed the one that took place at the nuclear accident with the "fresh" core by 3-4 orders of magnitude.

Such an accident may result in radiation contamination of water and shore areas that will far surpass the environmental consequences of the Chazhma event.

The possibilities of the submarines defueling are currently limited by a number of reasons, the main of which is that the storage sites for the spent nuclear fuel are full. Thus, the time period over which the fuel will reside in the cores of the decommissioned and anchored nuclear submarines is expected to be long enough.

The actual rate of the core unloading renders the on-board storage of the spent cores the keenest issue.

Therefore, while the complex technology line for the spent nuclear fuel of the NS is not developed yet, justification, urgent design and adoption of the measures ensuring nuclear and radiation security of a long-anchored waterborne NS are much needed.

At the same time, a serious effort should be indisputably made to accelerate the construction of facilities for defueling the decommissioned nuclear submarines.

Therefore, the construction of new sets of reloading equipment and of new sites for temporary expeditious storage of the spent nuclear fuel is the task of primary importance with no essential alternative.

The risk of the nuclear and radiation accident is aggravated by the poor technical condition of the anchored nuclear submarines. The lifetime of the major part of the decommissioned submarines are from twenty-five to thirty years, while 35 per cent of anchored ones were launched more than thirty years ago. Up to 40 per cent of the decommissioned nuclear vessels have not been under any maintenance or reparation for more than ten years.

The decommissioned submarines were anchored in the special shipyards being essentially unprepared to the long-term secure floating storage. Neither technical nor organisational measures were properly taken: partial dismantling of the material part was not accomplished, the water-tightness of the hull was not provided etc. As a matter of fact, once deco decommissioned, the nuclear submarine appears to be out of the routine NS operation and maintenance system: less than one forth of the staff crew remains on board.

The first generation nuclear submarines are now especially deteriorated. The most part of its machinery and equipment had worked out their resources long ago and as a result of a long idle standing became absolutely unfit for any use, including such vital systems as fire extinguishing and pumping-out ones. The corrosion of their light hulls reached a dangerous level, the main ballast tanks are damaged. Therefore, within a few years, tugging the vessels of the first generation from their anchoring places to the utilisation facilities may appear impossible without special transportation docks.

This jeopardises the security of the mere presence of decommissioned nuclear submarines on the float and demands the urgent development and adoption of the measures for their safe storage.

The measures should indisputably ensure water-tightness, unsinkability, explosion and fire safety, nuclear radiation and environmental safety. The number of the operating personnel should be sufficient for providing the secure long term storage of a waterborne NS.

4. WHAT TO DO WITH THE SPENT NUCLEAR FUEL IN THE DECOMMISSIONED NUCLEAR SUBMARINES?

The real magnitude of the potential nuclear and radiation hazard is primary conditioned by the backlog of the spent nuclear fuel accumulated by now in military fleet facilities. This is more than three hundred cores and more than seventy thousand fuel assemblies nearly half of which resides in suppressed reactors of the decommissioned nuclear submarines.

This resulted from a steady unbalance established in 1989 between the amount of SNF coming from the vessels and its transfer to Mayak utilisation base.

Three major reasons of this negative situation can be distinguished:

* Shortage of the transport means for the spent fuel transportation. Four spent fuel transportation casks that MINATOM has at its disposal can only transport 2.5 submarine cores per trip.
* The cost of shipping and reprocessing the spent fuel has sharply increased in the recent years. In 1995, the cost of transportation of one train from North fleet bases to Mayak processing plant made up eight billion rubles.
* The fleet bases are unprepared to accept and load the heavy special train: they have not got proper load-lifting mechanisms, and all the other equipment is in a poor tool condition.

The situation is further aggravated by the fact, that the large amounts of the spent fuel are accommodated in some inadequate compartments ashore hardly suitable for long-term storage of highly radioactive fissile materials.

Furthermore, by now. The capacity of these, storage sites as well as the capacity of waterborne expeditious storage facilities has virtually come to an end.

Even under the most beneficial conditions and normal financing, the MINATOM shipping resources allow the transportation to Mayak of only 18-20 cores per year, while the expected annual increase of the spent fuel from decommissioning and routine recharging of in-service nuclear submarines is thirty to thirty-two cores.

The above mentioned circumstances combine to virtually rule out any likelihood of transportation of the spent fuel backlog from naval bases and from decommissioned submarines before processing earlier than in the year 2025-2035, provided no extraordinary measures are taken, which seems even less likely.

The situation with the spent nuclear fuel hampers seriously the utilisation works, in fact it has paralyzed the just started up utilisation process.

5. TREATMENT OF THE REACTOR COMPARTMENTS OF NUCLEAR SUBMARINE UNDER UTILISATION

To provide the nuclear and radiation safety of the decommissioned NS, its reactors should be defueled, activity filters emptied of ionites and technological coutours drafted of water.

The defueling eliminates potential nuclear safety risks leaving high level of the radiation risk, sine the reactor compartment with no fuel in it remains highly radioactive on its own (app. 100 000 Ci) and requires special handling.

The problem of the reactor compartment treatment as one of the most difficult and costly one in the set of the NS utilisation related problems. It requires special techniques to provide radiation safety for the personnel and to eliminate relevant environmental risks.

This problem provided the focus for serious studies in Russia and abroad. In particular, a renewable conceptual approach is developed in a study carried out under NIKIET supervision, in which the reactor compartment is viewed as a container with compactly accommodated radioactive equipment. On cutting out and separating from the submarine hull, the reactor compartment should be kept (stored) over the time period sufficient for the natural decay of the dominant radionuclides that are contained in the equipment and the hull structure materials. (The most hazardous here is hard γ-radiation of ^{60}Co)

This approach to the reactor compartments (RC) treatment allows all the main equipment be left at its place and the most hazardous works associated with the exposure to radiation to be postponed.

There exists an alternative approach to the reactor compartment treatment that suggests the RC and its equipment scrapping simultaneously with that of the NS remainder. The aim is to lesson the amount of highly radioactive metal that otherwise should be stored for long or buried forever.

The latter approach is the connection of special ship-scraping complex facilities. The technology principles of the metal regeneration from the radioactive scrap have been well developed. However, the radioactivity dosage will not permit any personnel to attend to the line flow while, by rather conservative estimates, the cost of automatic processing line will make up several tens of trillions rubles.

In the opinion of our leading institutions and specialists that is in accord with that of the majority of foreign experts, indisputably preferable is the variant suggesting the RC long-term storage prior to any dismantlement.

Duration of the preliminary storage decisively determines the strategy of reactor compartment handling. The former is estimated based on the analysis of technologic-economical radiation and environmental factors.

The storage duration is naturally conditioned by the demand that it should last until the overall radioactivity and hard gamma-radiation in particular decays to the extent that the radiation exposure to personnel dismantling the compartment is lowered to an acceptable level offering the possible of using conventional technologies.

According to the data of NIKIET concerning the first generation submarines, the intensity of the gamma-radiation outside the hull, i.e. the one that detects the irradiation dosage, will approach an acceptable level due to the natural radionuclides decay (4.7 mR/h at the tolerance dose equal 2.8 mR/h) as early as over 70 years of storage. By this time the soft β-radiation associated With ^{63}Ni and ^{59}Ni decay will prevail in the overall compartment radioactivity.

The estimates of OKBM for the second generation nuclear submarines gave close results.

Apparently, the time period over which the gamma-radiation decreases to an acceptable level depends on the structure peculiarities of a particular reactor as well as by the regimen of its operation throughout the service life.

Time dependence of the overall intensity of the induced hard gamma-and soft beta-radiation for all the reactor compartments of the nuclear submarines intended to utilisation, is shown in Fig. 3.

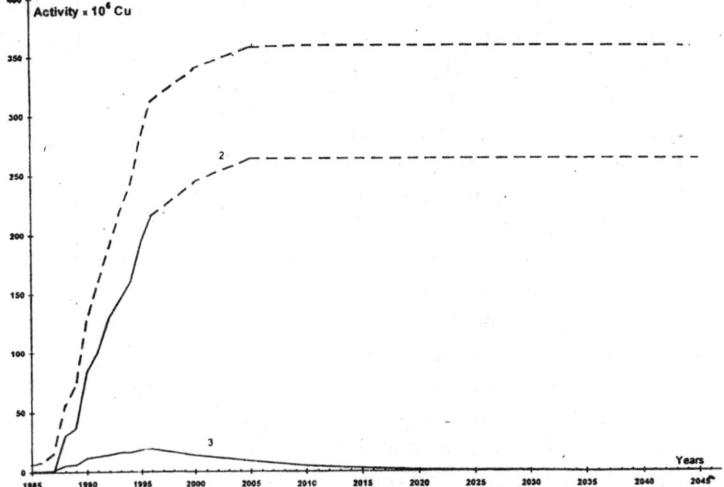

Fig. 3 : Integral Activity of Reactor Compartments of Decommissioned Nuclear Submarines

1. Accumulated activity of spent fuel of decommissioned nuclear submarine without load account.
2. Accumulated activity of the spent fuel of decommissioned nuclear submarine with account of loaded fuel.
3. γ-radiation activity (^{60}Co) accumulated in the constructive materials of the nuclear submarine reactor compartment

Speaking of the proper consideration for technical factors at determining the RC pre-storage duration we mean above all the state of the equipment, systems and hull superstructure. The most important here is the absence of through corrosion damages in the RC housing and maintenance of the overall RC state fit for the dismantlement.

The major economical factors that influence the choice of RC storage lapse are the cost of the upbuilding of special RC storage facilities, of their equipment, construction of the sites for RC temporal storage and development of the transportation infrastructure.

Taking into account all the aforementioned factors, the

duration of RC pre-storage for the first and second generation submarines, should make up, on the average, seventy to one hundred years.

Several variants of the long-term RC storage can be considered: in the sites ashore, in trenches, in underground workings, i.e. passageways of depleted mines, submerged in local waters, in the depth of the permafrost.

Whatever the variant, the PC housing should stay air or water-tight throughout the storage period. Special safety barriers should be constructed to block the pathways of the possible radionuclides migration and measures should be taken to make the hulls reliably corrosion-proof.

Moreover, all the storage variants listed above require the construction of an infrastructure to provide both continuous radio-ecological monitoring as well as attendance to the storage facilities throughout the whole RC storage lapse.

By now, there exists a decision of the Russian government to appropriately recommend and equip existing mine tunnels and to use them as PC long-term storage facilities. These works are virtually not started yet, and under the lack of adequate financing, it is most unlikely, that these facilities will be put into operation earlier than in 2000.

This decision is arrived at based on technology and economical investigations that do not seem neither reliable nor complete enough. In particular, dubious is the safety of such a storage throughout the whole planned period, if not for other, than for the following reasons:

- the possibility always exists of the tunnel vault or of the portal anticline destruction due to tectonic shoves, earthquakes or other geological adversary events;
- the tunnels that are situated 11 m below sea level can be waterlogged due to number of reasons.

The time period of the forced storage of the waterborne nuclear submarines offers the possibility for correcting the accepted concept and for search for some optimal variant of handling the separated reactor compartments.

It seem unable to design the study in more detail, based among other factors on recently refined data that we now have at our disposal, as well as on the instructive experience of other countries, first of all, on that of the United States.

CONCLUSION

In conclusion it is worthy to highlight the point, that the nuclear submarines decommissioning, dismantlement and utilisation, if carried out with reliable measures, providing for environmental safety, working personnel and population health security, presents difficult funds, technology and labour-intensive problems that in Russia becomes the issue of the State level due to the great number of submarines to be utilised.

Of the problems waiting for urgent solution, the security of the waterborne decommissioned submarines is, to our opinion the one of the primary importance, due to the inevitable long term (to fifteen-twenty years) storage afloat, the most part of the submarines leaving the spent nuclear fuel in them.

The serious obstacle that hampers the realisation of the NS utilisation programme, especially under the conditions of inadequate and irregular financing, is the absence of a head organisation responsible for its implementation. The responsibility is currently scattered among a great number of the State departments. At the same time, there is no institution some power that could coordinate all the works and investigations. It seems unable to recognise a Management board of the Russian nuclear submarines utilisation programme, that will be responsible for the funds distribution and for the implementation of all the utilisation works.

The first step in this direction, incidentally requiring no substantial organisational and financial efforts, would be the organisation of inter-departmental coordination council for nuclear submarine utilisation at the Russian Government, that would comprise responsible representatives of the Ministry of Defence, Ministry of Defence Industry, MINATOM, GOSKOMPRIRODA, and Russian Academy of Sciences.

The point of the particular importance is that due to apparent reasons, the NS utilisation is a large-scale global, rather then only local environmental problem,. The scientists and practitioners in this field in many countries acknowledge, that taking into account the current state and trends, the solution can be sought only in the urgent international cooperation of many countries, both in engineering-scientific and economic fields. An illustrative example is presented by a fruitful cooperation of Norway and

Russia in reducing environmental risks associated with utilisation of the Northern Fleet nuclear submarines.

Address:

ASHOL SARKISOV
Nuclear Safety Institute
Russian Academy of Sciences
113191 Moscow
Russian Federation

55

On the Possibility to Use Submarine Nuclear Power Units to Generate Peaceful Energy

A. BLAGOVECHTCHENSKI and F. D'AURIA

ABSTRACT

Preliminary information about principle possibilities of using nuclear power units of Russian submarines for energy supply is given in this paper. In the frame of the dismantling of the military power all over the world, this information can be usefull for specialists in atomic energy generation.

On the one hand the number of potentially available nuclear units, i.e. the overall power, makes attractive the proposal on the other hand, great technical and organisational problems make impractical to justify the use of nuclear submarines.

1. INTRODUCTION AND TECHNICAL BACKGROUND

Russian Submarine Nuclear Power Units (SNPU) include the following main components or systems: the nuclear reactor of PWR type: water is the primary coolant and the moderator of neutrons simultaneously;

- the once through Steam Generators (SG);
- the Main Circulating Pumps (MCP);
- various connections to the primary circuit.

In the first generation of Nuclear Submarines (NS), the main components of the primary circuit are connected by long pipelines as shown in Fig. 1(a). Such layout configuration has received the label "loop scheme".

In the subsequent generation of Nuclear Submarines, the so-called "lock construction" layout was adopted; the main

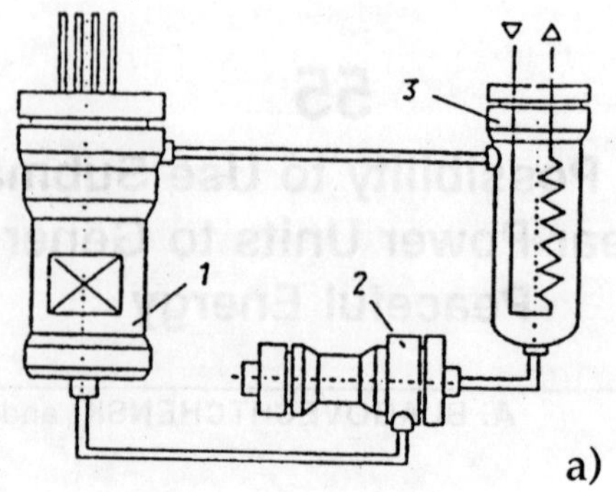

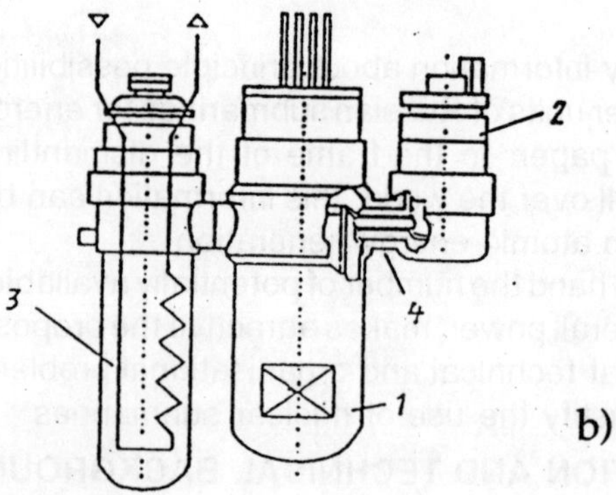

Fig. 1 : Characteristic Configurations of the Primary Circuit of Nuclear Power Units

a) Loop Scheme
b) Block construction

LEGEND
1. Reactor
2. Main Circulating Pump
3. Steam Generator
4. Branch of the 'tube-in-tube' solution

components of the primary circuit are connected by short branch and the 'tube in tube' concept is applied as shown in Fig. l(b). This configuration enhanced the reliability of the entire system. In fact long pipelines are eliminated and the in factory assembling of the system is possible and the consequent dispatch to the ship-building factory is ensured.

In both the configurations, the steam produced in the Steam Generators is directed to operate the steam turbine.

In the first generation of submarines the main turbine allows both the rotation of the screw and of the electro-generator (Fig. 2(a). In the subsequent generation, autonomous turbogenerators are installed: the turbogenerator is rotated by the own steam turbine as detailed in Fig. 2(b) The electrogenerators produce a three phase current with 380 V and 50 Hz frequency, ref. (1).

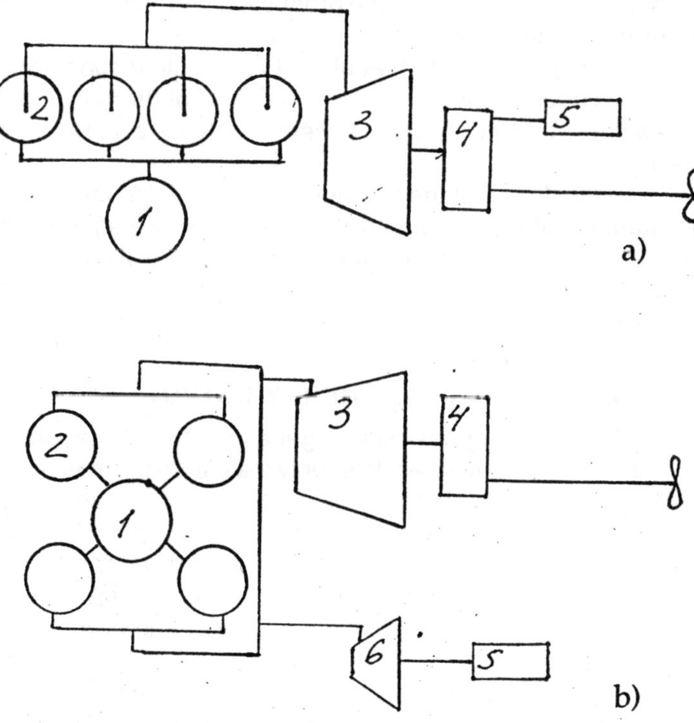

Fig. 2 : The skeleton scheme of submarines Nuclear Power Unit.

a) Single turbine scheme
b) Dual turbine scheme
LEGEND
1. Reactor
2. Steam Generator
3. Main turbine
4. Reduction Gear
5. Electrogenerator
6. Turbine of Electrogenerator

The current reliability and safety of the nuclear power units is ensured by:

- proven design technology and scientific background;
- use of PWR systems characterised by negative power/temperature coefficient;
- reliability of the main components and of the auxiliary components;
- availability of special systems for normal and emergency decay heat removal;
- high technical culture in the production and in the operation of the nuclear units.

It is important to note that almost 250 nuclear submarines are or have been in operation in Russia, most of these including two reactors, and in no one of them reactivity excursion occurred.

In connection with the standing problem of decommissioning a number of nuclear submarines, the consideration of the possibility and of the effectiveness of using the related reactors for generating peaceful energy, is of great interest. Some problems expected in achieving such goal, are emphasised in this technical note.

2. CURRENT ELECTRICAL ENERGY OUTPUT

The in-principle possibility to get a direct electrical output from the nuclear submarines, to supply the electrical grids on the coast or potential isolated coastal customers, does not evoke doubts. However, in the present system configuration, the total efficiency of using the nuclear reactor (i.e. ration between output electrical energy produced and total power or CURC =Coefficient of Use of Reactor Capacity) is very low and of the order of 10 per cent. This can easily be seen looking at the power of the Main Turbine and of the electrogenerator Turbine (Fig. 2 b), which are in the ranges 15-36 MW and 2-3 MW, respectively, ref. (1). In addition, it is necessary to consider that a significant part of the produced electrical power is already needed for the operation of the nuclear unit (pumps and other electrochemical devices) and other submarine equipment; thus, the actual CURC is even lower than the value that can be deduced from the above.

The total electrical power output can be obviously increased by the creation of a specific coastal energy platform that will consist of several submarine units. However, this solution looks expensive and does not result in the increase of the CURC.

In principle it would be possible to use the full power of the Nuclear Submarine, i.e. by installing additional electrogenerators rotated by the Main Turbine. Again, this solution may reveal impractical owing to the absence of free volumes in the submarine units suitable for accommodating the needed large sized and heavy components.

In this connection, it seem worthwhile to note that the Siemens electrogenerator installed in the Russian icebreaker "Taimir" (29 mwe power, n = 3000 rpm, Output Voltage = 6.3 kV), has the length equal to about 9.5m, the transversal size of about 4.5m and the mass of 71 tons, approximately. Other than such components, in order to transmit the electrical power to the coastal grid of to an isolated coastal customer, a large size electrotransformer is needed too.

A substantial reconstruction of the Nuclear Submarine with essential elongation of the vessel, installation of a special electric equipment, substitution of the Main Turbine reduction gear would be needed. All of this implies technical that are not difficulties justifiable from an economic point of view. The designer of the Submarine Units might give a more comprehensive view on this subject and draw proper conclusions from the analysis.

3. CURRENT THERMAL ENERGY OUTPUT

Superheated steam is produced as output of the steam generators of SNPU. Typical outlet conditions are:

- mass flow rate range: 110 - 250 Tons/hr,
- pressure range: 3.2-3.6 MPa,
- temperature range: 290-310'C.

In order to fully avoid the risk of radioactive pollution of the environment, in the case of a steamgenerator tube leakage event, the direct use of this steam is not to be allowed "on the coasts". A new system could be created including heat exchangers, pipelines and pumps allowing the heat transfer from the second circuit (II) to the third one (III), as shown in Fig. 3a. The need to have suitable pressure differences between the loops, (i.e. conditions pI > pII and pIII < pIV as its is foreseen in Russian design for Nuclear Stations of Heat Supply), for safety reasons, may prevent the possibility to produce steam in the third loop because of the dependence of the boiling temperature upon pressure.

Definitely, in the third loop, it is possible to have only hot liquid water at enough high pressure. Simple evaluations show that the heat transfer areas of the exchanger between the loops II and III may be larger than those typical of the steam generator (i.e. the heat exchanger between I and II in Fig. 3(a). The qualitative trends of the water temperature as a function of the amount of heat transfer from one loops of the other, are shown in Fig. 3(b).

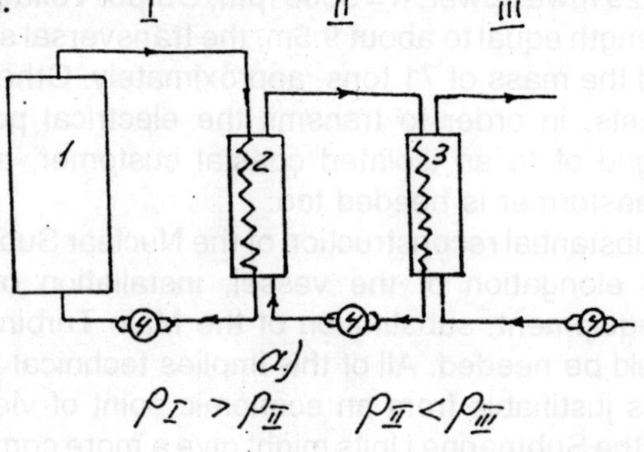

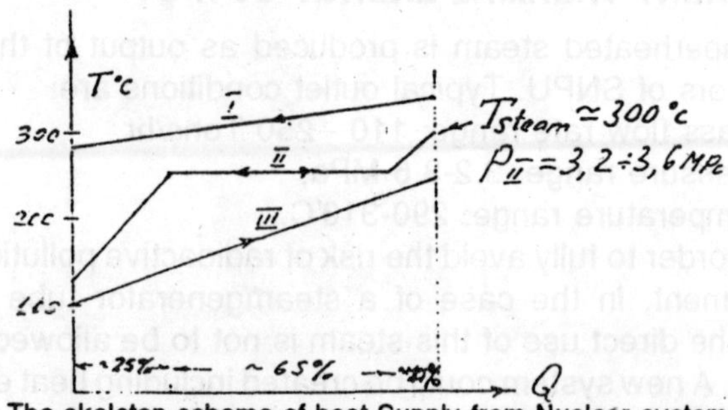

Fig. 3 : The skeleton scheme of heat Supply from Nuclear system.

a) Triple loop configuration

b) Spatial trends of liquid or steam temperatures

LEGEND

1. Reactor
2. Steam Generator
3. Heat exchanger
4. Circulating Pump

This discussion confirms the need of large reconstruction of the Nuclear Submarine. From a technical ground, it is impossible to consider as justifiable the fulfillment of this large and expensive task.

4. MAIN TECHNICAL AND INFRASTRUCTURE RELATED PROBLEMS

The achievement of a nuclear submarine working in the thermal or electrical energy production requires the solution of different technological and infrastructure problems in different areas, that are detailed in the following:

- ensuring radiation safety (dosimetric control, radiochemical analyses, etc.), ensuring the reliability of core transportation and replacement (including transshipment);
- acceptability of the storage of liquid and of high radioactive waste;
- availability of spare parts;
- maintenance;
- qualification of the personnel;
- etc.

Several highly qualified specialist in the different areas are needed.

5. CONCLUSIONS

The performed analysis shows that the features of the Nuclear Submarine make its use for producing either electrical either thermal power, unjustified on a technical ground. Specific cases including the solution of a sudden energy crisis in a remote region of the earth, e.g. following a natural disaster, may be more attractive.

It seems worthwhile to mention that in Russia, special new projects are devoted to the design of nuclear-heat cogeneration plants (NHES = Nuclear Heat Electrical Stations) which are based upon the existing scientific and technical know-how including the large experience in the ship nuclear power filed. This way is technically and economically more justified to solve the problem of energy supply, ref. (3). These Russian projects are already known and diffused abroad (e.g. the project KLT-40 of OKB mechanical Engineering, Nizhny Novgorod, or the NHES project according to the conception of the Krylov Ship Building Research Institute, St. Petersburg).

BIBLIOGRAPHY

1. *V. Burov,* "Russian ;Military Shipbuilding in the Third Century of its History", St. Petersburg 1995.
2. *L. T. Osipenko, L. Jiltzov, N. Mormul,* "Nuclear Submarine epoch."- Moscow 1994.
3. *A. Blagovechtchenski,* "On concept of use of technological potential in the field of marine nuclear power for decision of common power problems" - Symposiums Energetica, St. Petersburg 19-23 Nov. 1996.

Address:

ANATOL BLAGOVECHTCHENSKI
St. Petersburg State Technical University
195251 St. Petersburg
Russian Federation

FRANCESCO D'AURIA
University of Pisa
55126 Pisa
Italy

56

Stability of Transition to a World Without Nuclear Weapons: Technical Problems of Verification

V. ZHIGALOV

A serious psychological barrier to acceptance of the concept of achievability of a nuclear-free world by many politicians and public figures is fear of facing the prospect that one or more nations or extremist political groups might develop their own nuclear potential, which would result in a critical world situation. In actuality, this is a question of stability of a nuclear-free world. Every system in nature, in order to exist for a significant time, must maintain its state in the face of disturbances. The height of the barriers that enable the system to return to the initial state must not be less than the level of natural fluctuations.

Clearly, under conditions of a nuclear-free world, the advent of a group that has even limited nuclear potential will lead to a breakdown of this state.

From this standpoint, the most effective system of verification is an absolute necessity in this period for significant progress in formation of the concept of a nuclear-free world. This system must ensure detection of so-called undeclared nuclear activity on early stages.

Scientists of Russian nuclear centers are working today on this problem. We consider this paper a comprehensive attempt to look at the technical and organisational aspect of the problem of transition to a nuclear-free world, setting aside the difficulties of resolving purely political problems.

We can mention, two basic complex factors that determine the attainability and stability of a nuclear-free world. The first is the set of conditions that determine the level of efforts to have nuclear weapons. Actually, this is a factor that characterises the force of disturbances on the state of a nuclear-free world, it is a factor that might be called a kind of "temperature of nuclear trends" of society. It is determined in large measure by political aspects. Today, a lot is being done to get this "temperature" down. A consequence of this, in particular, is voluntary disarmament of the RSA, which had been secretly stockpiling nuclear weapons. But it would probably be a big delusion to believe that this "temperature" at any time will reach zero. There will always be attempts to have the nuclear bomb. Note that the growth of the nuclear power industry is, of course, a factor that will make monitoring difficult, and that to some extent will stimulate efforts to have nuclear weapons.

The other important factor is the presence of controls and deterrents. One such is the operation of such a major tool as the IAEA.

Having an effective verification system in place is a kind of barrier that prevents violation of conditions of a nuclear-free world, and ultimately determines the stability of these conditions.

The presence of nuclear forces in the world also acts as a deterrent.

Research has shown that there is an option for transition to a nuclear-free world that can be sketched as follows. With a drop in the "temperature of society" (i.e. the effort to have nuclear weapons) and a rise in the barrier of stability (development of an increasingly advanced verification system), there is a reduction in nuclear forces of deterrence, which will finally vanish altogether. In fact, the barrier of nuclear forces of deterrence is gradually replaced by an early warning system of undeclared nuclear activity. The central issue here is the question of what such an early warning system would be like, and how much it would cost.

An important point in designing an acceptable verification system is development of techniques and an analytical management model. It is also important to look at the problem of developing a verification system from the standpoint of maintaining the NPT. An essential component of the project is ascertainment of the main areas that will have to be developed primarily in the monitoring and control system, as well as determination of the volume of hardware that will have to be enabled to attain the goal of ensuring effective safeguards.

Analysis has shown that the option of a nuclear-free world has a realistic alternative that involves the existence of limited nuclear deterrence. In this case, the reduction of nuclear deterrence proceeds to a certain level. On the one hand, this level ensures sufficient ecological safety. This means that even if the entire arsenal were to be exploded, there would, with a large safety factor, be no global ecological changes. On the other hand, the combination of nuclear deterrence and a technical verification system would ensure international stability. The technical system of verification in this case would provide for safeguards sufficient to preclude secret development of a nuclear arsenal that could pose a serious threat to the nuclear deterrence potential of the society being-safeguarded. Of course, this is a simpler and less expensive system that would be the case if there were absolutely no nuclear deterrence.

The alternative choice is determined to a considerable extent by the answer to the question of how the technical system of verification would be set up, how much it would cost, and how much development time would be needed for different task levels.

It should be noted that a verification system can be developed for a finite number of scenarios in which efforts are made to possess nuclear weapons. Therefore, assessment of the extent to which safeguards can be trusted in systems under consideration requires special analysis and is a major factor that determines the feasibility of existence of nuclear deterrence. In working on this topic, we will set aside the question of how this nuclear deterrence will be set up and monitored. From the technical standpoint, the only thing that matters is the nuclear potential of this force. In our opinion this quantity is determined mainly by the effectiveness of the verification system. Let us say that the nuclear potential should be two orders of magnitude greater than what could be secretly developed. In this context, an event with controls in place should occur with sufficiently low probability (<1%).

The figure shows a schematic of the main units of the system for monitoring and control of undeclared activity.

The major problems that determine general verification systems may be formulated in the following points:

- Working out hardware requirements of the complex system

of monitoring and control to assure effective safeguards against attempts to develop nuclear weapons.

- Assessing sensitivity and effectiveness of technologies used in the monitoring system.
- Determining tasks and technical capabilities of international and national verification services.
- Analysing possible options of setting up information systems and expert support systems that incorporate processing of information on monitoring of fissile materials and undeclared activity.
- Providing equipment for verifications and inspections done in accordance with signals of the monitoring system.
- Developing methods of diagnosis and techniques for dealing with nuclear terrorism the influence of interaction of international and national security and control services.
- Studying possible methods of forceful response to cut off undeclared activity.

We hope that the advent of a version of a technical system of monitoring and control of nuclear activity will stimulate criticism improvement, proposal of alternative approaches, and sound economic assessments. In our opinion carrying out all-inclusive developments of this type in sufficient detail may answer the question of stability of a world without nuclear weapons (or with a limited number of such weapons), and what technical measures must be provided to make the transition to that world.

The participation of leading scientists of Russia's nuclear centres that have played an active role in development of nuclear weapons is surely advisable in this field of research. Having specific expertise in the area of determining possible paths of proliferation of information that is conducive to success in the development of nuclear weapons, and technical methods of detecting such activity, they could contribute greatly to solving one of the most important problems of our age.

In working on this problem use is being made of research by specialists of the IAEA and US national laboratories who are dealing with issues of non-proliferation of nuclear weapons.

At the RF Nuclear Centre All-Russian Research Institute of Experimental Physics, methods are being developed for analysing nuclear weapons proliferation that are based on a systems approach in which the entire process is broken down into corresponding stages in considerable detail. It is assumed in the model that the probability *p* of occurrence of a given event under specific conditions is a function of time, and that p –> 1 t –> ∞. The behaviour of the time function depends on several parameters that include economic indicators, the worldwide open source information status with regard to this issue, and hardware capabilities currently available in the nation.

The scheme of analysis includes such major units as fissile materials, technologies necessary for producing nuclear weapons, the design stage of development, stage of finalising and fabrication, the capability of series production, and the development and manufacturing if delivery systems. Processes unassociated with the cause and effect relation are parallel.

The most likely path of proliferation has the greatest probability and is chosen by shorting through different trajectories, probability being defined as the product of the probabilities of elements, since these events are assumed to occur independently:

$$P=P_1{}^*P_2....P_n = \Pi P_i$$

Time restrictions are important in analysis.

Time limitations are expressed as

$$T= \Sigma t; <T^o$$

Where To is the time for which the prediction is constructed.

The nature of economic limitations is the same.

Assumptions regarding decision making may considerably alter the structure and probability of the critical path of proliferation. In particular, when a decision is being made regarding development of one or more nuclear weapons for terrorist purposes, those units with development of delivery systems, questions of series production and experimental development, as well as special issues of nuclear safety and weapon reliability make a minor contribution to analysis of probabilities.

To evaluate the probability of each element of a trajectory of proliferation, approximating formulas are usually chosen in the form of a produce of functions of a number of specific parameters that are determined from information about a given nation with allowance for expert assessment by specialists (see diagram 1),

$$\text{Pi} = \varphi(t_i) \cdot \theta\ (Ci)$$

Where ti is research time, and *ci* is development cost. On the initial stage, formulas

$$\varphi(t_i) = e^{-q/t_i}$$

$$\theta(C_i) \begin{cases} C_i/C_o, & C_i < C_o \\ 1, & C_i \geq C_o \end{cases}$$

were used for very simple interpolations.

In some cases when doing an analysis, the probability of some elements may be assigned as *pi* = 1 because of specific conditions, as when there are no problems of series production when a single use is being planned.

Factors were analysed using the schemes given above to assess the role of the degree of provision with information regarding nuclear technologies. It was shown that development times, requirements for amount of fissionables, and economic expenditures on a project may be reduced by a considerable factor. In some cases, the time dependence shows "threshold" behaviour. Diagram 2 shows qualitative relations for the likelihood of development of a single nuclear weapon in third-world countries as a function of time with different assumptions about parameters.

Each element of the scheme is characterised by attributes of development associated with availability of personnel, special equipment, availability of materials, and in some cases the presence of specific ecological features.

The construction of an optimum critical path allows assessment of the issue of specific critical technologies by temporary inclusion of this element in the structure of the critical path.

Such a scheme of computation enables construction of a prediction in time at each specific instant for various States, and assessment of different scenarios of development of the situation.

From the viewpoint of applying this work to expand the operations of IAEA with the scope of Programme 93+2, the use of such an approach may be of interest for analysing undeclared activity and design of specific inspection control procedures.

Analysis of different proliferation scenarios enables two predominant directions to be singled out in considering risks of proliferation:

1. the State level—model of the totalitarian State. The main features that have to be noted in view of their influence on the option being considered:
   - rigid system of information security on the State level;
   - long-range planning of armament system design;
   - construction of State technological production facilities with certain requirements with respect to radiation safety;

   weapons engineering with allowance for requirements regarding nuclear and radiation safety.
2. Non-state level-model of terrorist group. The group may be nationalist in nature; its members may have fanatic tendencies. One option represents this group as an international underground organisation with the possibility of carrying out its activity in democratic States.

   Basic features:
   - rigid protection of information, using the institution of company trade secret security;
   - planning the production of several devices without developing delivery systems;
   - setting up special commercial structures for carrying out production;
   - ignoring radiation safety requirements;
   - using zealots for theft and transport;
   - requirements for nuclear and radiation safety considerably released in weapons engineering.

The analysis of proliferation scenarios is without question an extremely complex procedure, and is not limited to the aforementioned two scenarios. This is used in this report to illustrate the basic results that arise when evaluating risks of proliferation.

A key element needed for developing nuclear weapons is fissile materials. The schematic gives a simplified summary of some scenarios of proliferation depending on different approaches to the acquisition of plutonium. The complete probability of realising plans for development of nuclear weapons is understood to mean summation over kinds of scenarios with certain coefficients that determine the contribution of the scenario.

The probability of passing through a chain is defined as the product of the probabilities of the individual elements of the

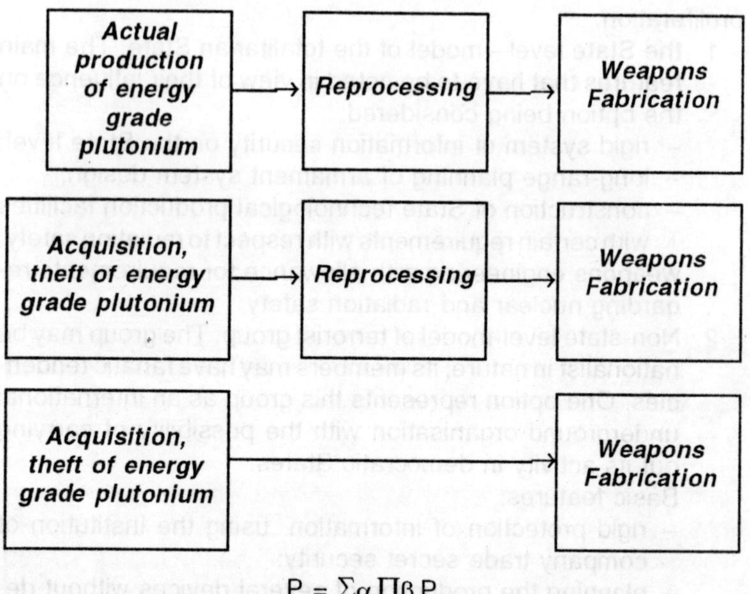

$$P = \Sigma\alpha_i \Pi\beta_i P_{ik}$$

where i is summation over chains involving weapons fabrication;
k is the product of probabilities within a chain;
αi are coefficients that weight the contribution of a scenario;
β_i is the set of coefficients for choice of the option of actions;
P_{ik} is the probability of realisation of an element of a chain.

chain, as it is assumed that they occur as independent events.

In our opinion the risk of proliferation is considerably increased by a process that uses energy grade plutonium. This is due to the following factors:

- the possibility of using reactor grade plutonium in nuclear weapons;
- the difficulty of radioactive waste handling is exaggerated in thefts, especially where there are zealots;
- there is a considerable contribution to formation of risk of proliferation associated with the actual procedure for storing fissile materials.

The monitoring service makes provisions for spotting signs of undeclared nuclear activity. These signs are concretely manifest factors that may accompany operations in the development of nuclear weapons (elevated content of certain elements in analyses, purchase of special equipment for vague purposes, and so on).

Three major units of the monitoring service are distinguished: remote monitoring stations (radiochemical analysis, ecological measurements, satellite monitoring, seismic monitoring, and so on);

- expeditionary monitoring (characteristic ecological samples, possibly within the framework of national services);
- information unit (chiefly data processing with respect to certain criteria of export-import operations and open-source publications).

In the case of occurrence of a signal from the monitoring system associated with violation of the established regime, the control verifications service is activated. Verifications (inspections) may be carried out both by the national services themselves, and by an independent international service.

A major part in management and coordination of actions with respect to nuclear activity belongs to a special center. The work of the center is planned within the scope of a controlling analytical programme. The main units in accordance with their functions are shown on the diagram:

- *Nuclear efforts analysis unit*

  basic task is to determine the major territories of 'hot spot" of the kinds of activity that have to be observed. Analysis requires a large body of information associated in particular with political objects;

- *monitoring facilities optimisation unit*

  basic task is to distribute monitoring facilities as a function of data acquired from corresponding nuclear efforts analysis units, and to assess the extent to which safeguards are to be trusted;

- *monitoring system signal analysis unit*

  basic task is analysis of anomalous signals with allowance for the entire body-of data on monitoring and control, and data output to the control verification generation unit;

- *control verification generation unit*

  basic task with respect to input data is to suggest a scheme of monitoring system signal verification, and if necessary to schedule routing and technical support, as well as procedures for cutting of undeclared activity;

- *safeguard confidence evaluation unit*

basic task is to assess the extent to which safeguards against undeclared nuclear activity are to be trusted on controlled territories with allowance for all acquired information through the channels of monitoring and control verifications systems. One of the tasks is to correct the operating conditions of monitoring facilities for the purpose of elevating the level of safeguards.

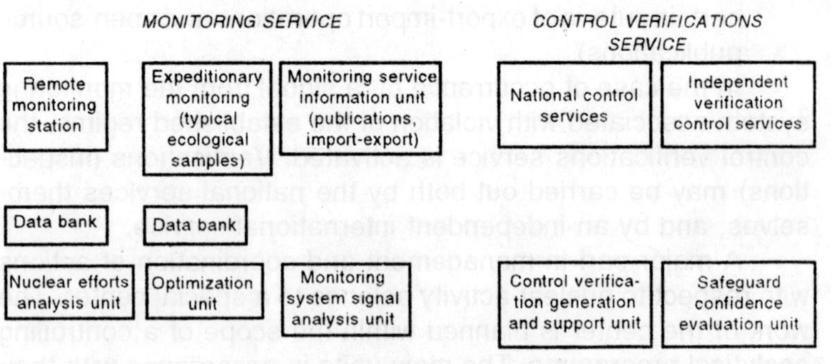

General block diagram of system for spotting undeclared activity

Address:

VLADIMIR ZHIGALOV
Russian Federal Nuclear Centre
Nizuni Novgorod Region 607190
Russian Federation

57

The Conversion of Military Science and Technology: Former Soviet Union Case

M. MARTELLINI

The end of the Cold War, which can be said to have coincided with the fall of the Berlin Wall in November 1989, has brought about deep changes in the very concept of defence. The cold War was characterised by the confrontation of two military blocks, which have granted peace by showing that their respective nuclear potentials were large enough to ensure the enemy's total destruction. Today, nuclear deterrence on a large scale has lost much of its relevance, and a progressive reduction in military spending is under way. Defence budgets at a world level have gone from about 1,300 billion dollars in 1987 to about 900 billion dollars in 1993 (both at 1993 rates).

While it requires deep changes in the defence strategies of all Nations, this new international situation is encouraging the conversion of the military sector for the benefit of the civilian economy. This process involves many companies that have so far been working mostly or exclusively on the basis of military orders. In particular, from the point of view of nuclear non-proliferation and environmental safety, some urgent problems arise, among which: the civilian management of military nuclear technologies; the management and storage of weapon grade materials, namely weapon-grade Plutonium and highly enriched Uranium resulting from the dismantlement of nuclear warheads; cleaning and reusing large areas which have so far been taken up for the production of weapon-grade Plutonium and Uranium enrichment (in the former Soviet Union some of the so-called "atomic cities" were entirely devoted to this task-see annex); retraining scientific personnel and technical experts who have worked in the nuclear military industry.

In the early 1990s, after the collapse of the Soviet Union, the idea had been mooted of a large Fund, to be set up with savings from military budget cuts. The Fund would finance large-scale projects for the creation of infrastructures in developing countries, as a way of accelerating their economic growth.

Today, after almost six years, this aspiration, known as the "Dividend for Peace", has turned out to be unrealistic: unfortunately, it has now become clear that in the first stages of the conversion process the financial resources to be allocated every year by all the countries involved will be at least as large as those spent on the arms-race during the Cold War. The "Dividend for Peace" objective can be effectively implemented only if it is combined with a global project aimed at the rational use of the Earth's primary resources (water, food and energy), in order to allow a sustainable development in the most severely depressed and degraded areas of the world and to protect the environmental heritage. Moreover, the conversion of the military industry into civilian industries and markets vary significantly from one country to another.

In this respect, the former Soviet Union is a very special case, and a critical one for many reasons. The role of the defence industry in the Soviet Union was enormous. The Cold War and the competition with the West lead to a huge concentration of resources in the military complex, which absorbed 80 per cent of human and material resources and employed over 16 million people. This military effort was reflected in some twenty "atomic cities". These were inaccessible "closed cities", where all the nuclear military R&D of the former Soviet Union was carried out.

The best human resources were all in the military sector not only because of better salaries, indirect benefits and a higher social status, but also because of the attractiveness of working with modern facilities and advanced technologies, which could not be found elsewhere. The military industry monopolised technological knowledge to such an extent that many civilian products and even commodities could be made only in military plants: civilian aircraft and radar, but also television sets, refrigerators and cookers were the exclusive prerogative of the military industry. What may have seemed a high level of integration between the military industry and the civilian economy was, in fact, the cause of deep technological backwardness in the civilian industry.

The Soviet military complex, just like military industries elsewhere, was not particularly keen on considering cost aspects, for the Army, its main client, was more interested in top quality than in cost effectiveness. However, unlike western defence industries, the Soviet military complex operated not in a market but in a planned economy, where it did not need to compete with other groups, which made its prices totally unrelated to and unaffected by production costs.

This is why the conversion of the military industry, the development of dual technologies, meeting at the same time military and civilian requirements, and the civilian-military integration have met with many more difficulties in Russia than in Western countries.

As early as 1988 Gorbaciov made a first attempt at conversion still based on centralized planning (Gosplan). The Gosplan established some general objectives, which were subsequently specified for each sector, elaborated by the relevant bureaucratic apparatus and transferred to single enterprises in the form of production goals. These production objectives had little to do with the capacities and competencies available in the factories. The results of this type of command conversion have been disastrous in most cases.

The next step has been an attempt at conversion based on globalisation, through the creation of joint ventures between Russian military plants and western civilian enterprises, with the former providing a selection of advanced technologies and low-cost labour and the latter offering a knowledge of the market and its demands and the organisation skills needed to run a cost-effective business. Unfortunately this second solution has not been very successful either, owing to Russia's economic instability and uncertainty and an almost total lack of the infrastructures needed to allow competitive and low-cost production (road and rail transport between production centres was so poor that people travelled by plane). One exception was a number of joint ventures set up between the US Department of Energy with some American enterprises and Russian "atomic cities" to reprocess and dismantle radioactive wastes. Joint activities also included the production of civilian radioisotopes, biorganic materials (biomass) and even particle accelerators for the medical industry and civil engineering.

The third attempt, suggested in late 1992, consisted in conversion through privatisation. The idea was that military companies that decided to operate on the civilian market would be listed for privatisation. However this solution has been hampered so far by the fact that military production in Russia can't yet be passed over to private hands. In addition, most Soviet military enterprises are grouped in large centers including—besides factories—residential buildings and services for the workers and their families. If these companies were to be privatised, these gigantic complexes would have to be dismantled, production units separated from other buildings, and overall sizes reduced to manageable proportions. This would not be an easy process and it may generate protests and social instability. It should also be considered that the managers and the employees of these huge industrial military complexes lack any market culture, and therefore any such project would have to include substantial retraining investments.

Some figures will help to understand the difficulties of military conversion in the former Soviet Union/Russian Federation. Due to cuts in contracts, the number of defence enterprises under the control of the Ministry for Defence Industries (Minoboronprom) went down spontaneously from 1,800 in 1991 to 500 in 1997 and their aggregate military and civilian output fell by 82 per cent (see P. Mann, "Economics Morass Foils Military Progress", AW&ST, May 26, 1997, pp. 7273). Very few of the 1,300 lost enterprises were converted efficiently for civilian production. Most were just disintegrated or prolonged their agony by selling out stocks and equipments, leasing space, surviving on small contracts and shifting their labour to a partial working week.

At the previous Como conference *"Science and Technology for the Construction of Peace"* (4-5 October, 1996) and the current meeting *"Nuclear and Biological Decommissioning: Management of Global Security Threats"* (27-28 June, 1997)—organised with the support of UNESCO Venice Office, Italian Ministry of Foreign Affairs, Lombardia Region and the Municipality of Como—various aspects of military conversion were analysed with the aim of laying the bases for a series of workshops and seminars on these issues, to be held in future in the frame of our UNESCO International School of "Science for Peace".

The preparatory workshops that preceded these conferences established a new way of tackling the conversion problem and achieving its main goal, namely to re-allocate the human and financial resources accumulated in 45 years of Cold War.

The key principle is to start from market demands and then to identify which macro-sectors in the military industry could meet those demands by using their know-how and technologies for the benefit of the civilian industry. Nevertheless, the selection of market demands can't be based only on strictly economic and market considerations, but it should also reflect some fundamental global aspects and paradigms, such as environmental protection and sustainable and durable development.

A particular topic is the decommissioning of "nuclear equipment". For instance weapon-grade materials from dismantled warheads could be used in civilian nuclear reactors to generate electricity, as envisaged by international disarmament treaties. However, while this is the best solution to assure non-proliferation of atomic weapons, it will require existing and potentially dangerous nuclear power plants to operate up to the year 2010 in the best of hypotheses! Consequently, if this solution is adopted, it will mean convincing the public opinion that such a solution is necessary to grant international security.

Another aspect that has emerged from the workshops in Como concerns some smaller-scale areas of the military industrial complex and their potentials to create market demands and to offer society the scientific and technological know-how accumulated during the Cold War. In this field specialised contracts could be signed with public and private companies. For example, together with the production of tanks and self-propelled guns, the former Soviet Union developed a strong military transport industry, which could be used for the development of the civilian transport sector.

And finally, one possibility to fulfill in the near future the goal of the "Dividend for Peace" mentioned above, would be to set up *ad hoc* committees, within UNESCO or other international organisations, designed to analyse various scientific and technical aspects of military conversion and to issue area studies for applications in civilian industries.

ANNEX: THE PLUTONIUM "CLOSED CITIES"

The main "Plutonium cities" of the former Soviet Union are at Chelyabinsk-65, Tomsk-7 and Krasnoyarsk-26. These centers, which were inaccessible in the age of the USSR, were used to produce fissile Plutonium for atomic bombs, recycle exhausted nuclear fuel and prepare some parts of the bomb. During the Cold War they developed about 100 tons of Plutonium and 500 tons of highly enriched Uranium. Before 1980, Chelyabinsk-65,Tomsk-7 and Krasnoyarsk-26 contained 5,5 and 3 water-cooled reactors for the production of Plutonium respectively. These reactors generate some 5-6 tons of Plutonium every year. After 1992, there remained only two operational reactors, one at Tomsk-7 and one a Krasnoyarsk-26, both producing 1,5 tons of Plutonium a year. They will continue operating until the year 2000. After the end of the cold War, the once steady government orders for weapon-grade Plutonium collapsed. Today the Russian government only partially covers operation costs for reactors and Plutonium recycling plants (20% less). Salaries are not paid regularly and operations are often suspended. Unemployment is growing among the workers and employees in the three Plutonium towns, where out of a total population of 280,000 as many as 37,000 work directly in the nuclear military complex. The estimated number of jobless at Tomsk-7 alone was 2500 in 1995 and is expected to rise to 9000-9500 by the year 2000. Plutonium production centres are going through a crisis and may face serious risks connected to the management and storage of weapon grade materials and the functioning of nuclear reactors. Yet, traditional Plutonium-cycle activities (defence production, storage and management, recycling of exhausted nuclear fuel and environmental cleaning of the sites) will no longer be the main source of income for these three areas. Alternative non-nuclear economic activities will necessarily have to be developed, somehow taking advantage of the existing know-how, technologies and highly specialised personnel, if the future of these towns is to be ensured and the risk of nuclear proliferation avoided.

Address:

MAURIZIO MARTALLINI

University of Milan
20133 Milan
Italy

PART—7

UTILISATION/DISPOSAL OF THE EXCESS FISSILE WEAPON MATERIALS: SCIENTIFIC, TECHNOLOGICAL AND SOCIO-ECONOMIC ASPECTS

58

Potential Contribution to the Recycling of Weapons Grade Plutonium in NPPs

D. HAAS, C. VANDENBERG and Y. VANDERBORK

Since 1986, more than 270 tons of MOX fuel rods have been fabricated in the BELGONUCLEAIRE Plant located at Dessel (Belgium) and, for the main part, loaded in commercial reactors in France, Switzerland, Germany and Belgium. Their excellent in-reactor performance up to 50 GWd/tHM has been demonstrated through surveillance programmes as well as specific experimental tests. That is the reason why BELGONUCLEAIRE advocates the recycling of weapons grade Pu as MOX.

This paper presents the main achievements of MOX fabrication in Europe and particularly in Belgium, and its commercial utilisation in Light Water Reactors. It describes also the important contribution that BELGONUCLEAIRE can provide to promote and implement the MOX option.

1. INTRODUCTION

A direct result of the Strategic Arms Reduction Talks (START) treaties is that the United States and Russia have a binding agreement to dismantle warheads into components. This will generate large amounts of high enriched uranium (HEU) and weapons grade plutonium (W-PU). Regarding the latter material, although it is difficult to predict precisely when and in which quantities it will be available, it is anticipated that 100 to 200 MT (50-50 to 100-100 for the USA and Russia respectively) will be stockpiled before its final disposal in the coming decades.

There is a consensus that these quantities of W-Pu must be transformed as soon as possible in order to fix plutonium in an environmentally inoffensive form such that it is inherently as difficult to recover the fissile material from the immobilised form as it is from standard spend fuel. Among various technical options studied at the moment, there are two main possibilities allowing to meet this goal:[1, 2, 3]

— either to use it as nuclear fuel for electricity production in civil power plants and to keep the remaining plutonium in association with the fission products in spent fuel in a once through strategy;

— or to blend the W-Pu with fission products to fabricate a highly radioactive but stable compound by vitrification, and to dispose of it in a deep geological repository.

The choice of using plutonium as nuclear fuel (MOX) is natural considering technical, economical and safeguards aspects. This paper describes the potential contribution of the Belgian industry to this solution which is also based on the broad experience of other European countries (mainly France and Germany) in the field of MOX fuel technology.

2. BELGONUCLEAIR'S EXPERIENCE WITH PLUTONIUM

Since its creation in 1957, BELGONUCLEAIRE has worked in the field of plutonium and has gained appreciable experience in the study, fabrication, fuel management and licensing of plutonium-bearing cores for FBRs, PWRs and BWRs. Already in 1963, it irradiated mixed oxide fuel (MOX) in small pressurised water reactor (PWR) in Belgium and since then, BELGONUCLEAIRE has fabricated fuel for fast breeder reactors (FBRs) and for light water reactors (LWRs). In putting into practice the use of this strategic material in nuclear reactors, we had to solve problems associated with fabrication, plant operation, safety, safeguards, security, transport and storage. This section briefly describes these topics[4,5,6,7].

Today, MOX technology is in industrial use in France, Germany, Switzerland and Belgium. About 400 tons of MOX for LWRs have been fabricated and irradiated successfully. Among those 400 tons, 270 were produced by BELGONUCLEAIRE between 1986 and 1995. This quantity contains 13 tons of W-Pu in LWRs.

i) An Early Beginning in MOX fabrication

After a period of R&D programmes performed jointly with

the National Research Centre (SCK.CEN) in its laboratories, BELGONUCLEAIRE now owns an industrial fabrication plant, called P-Zero (P0) which is operating since 1973. The plant is located in the Mol-Dessel nuclear site in Belgium.

The R&D programme covering all aspects of MOX fuel research was started in the 1960's initially within the framework of a co-operation between EURATOM and the US Atomic Energy Commission and later in various programmes carried out with the backing of the Belgian Government and the European Union.

The industrial plant P0 was first used to fabricated fuel for demo-programmes in LWRs and FBRs. From 1977 to 1985, BELGONUCLEAIRE fabricated more than 18,000 fuel rods for various FBRs and more than 225 MOX assemblies were irradiated in the LWRs of four European countries (cf. Table 1). As a result of these programmes, it was shown that MOX fuel can reach the same performance as standard uranium fuel or even behave better than the latter.

Table 1 : LWRs Loaded with MOX fuel from P0 plant—Dessel (Jan. 1996)

Country	*Demonstration Programmes*		*Commercial Deliveries*	
Belgium	PWR	BR3	PWRs	DOEL 1
France	PWRs	CNA CAP	PWRs	SLB 1 & 3 CRAVELINES 4 & 5
Switzerland	PWR	BEZNAU 1	PWR	BEZNAU 1
Germany			PWRs	GRAFENRHEINFELD BROKDORF PHILLIPPSBURG UNTERWESER
			BWRs	GUNDREMMINGEN 1 & 2
Sweden	BWR	OSKARSHAMN		
The Netherlands	BWR	DODEWAARD		
Italy	BWR	GARIGLIANO		

ii) The Industrial Phase

Since 1986, MOX fuel is produced on a commercial basis for LWRs of the countries shown on Table 1. Since then, the

MIMAS (MIcronised MASter blend) process has been used to obtain a fuel which is best adapted to LWR fuel specifications and large scale production, and which is soluble for the purpose of civil reprocessing. This process is described schematically in Fig. 1. The MIMAS MOX pellets are composed of a solid solution of UO_2 and PuO_2 homogeneously dispersed in a UO_2 matrix.

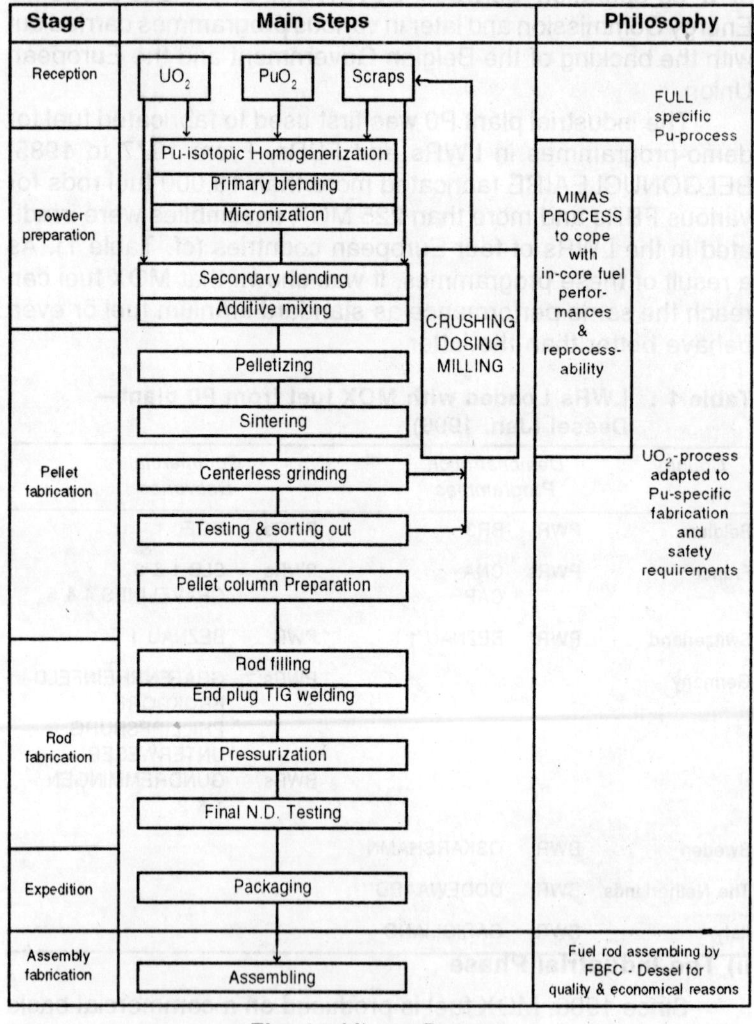

Fig. 1 : Mimas Process

This result is obtained through two blending steps: the primary (or master) blend obtained by ball-milling and the secondary (or final) blend.

The MIMAS process presents several advantages:

— the primary blend (30-40% Pu) can be stored immediately under usual safeguard control and used later, upon request, for fabrication of pellets after the final blend step;
— high flexibility for Pu isotopic composition homogenisation through cross-blending;
— demonstrated economical viability shown by full capacity production;
— mature technology.

By end of 1995, more than 150,000 fuel rods had been manufactured by BELGONUCLEAIRE which represents about 270 tons of MOX out of the 400 tons fabricated world-wide.

More than 90 per cent of the production is currently being irradiated in commercial LWRs of has already been unloaded from these reactors. Fabrication experience is presented in Fig. 2. Fig. 3. summarises the irradiation experience gained so far. The behaviour of the fuel has been demonstrated to be very similar of that of the corresponding UO_2 fuel.

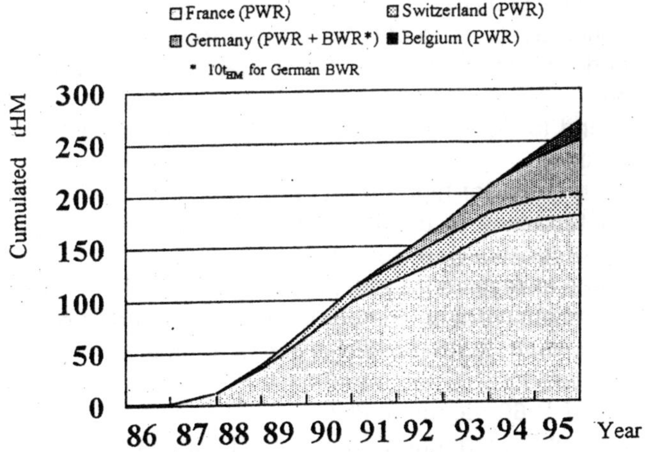

Fig. 2 : Cumulated Mimas Production at Belgonucleaire

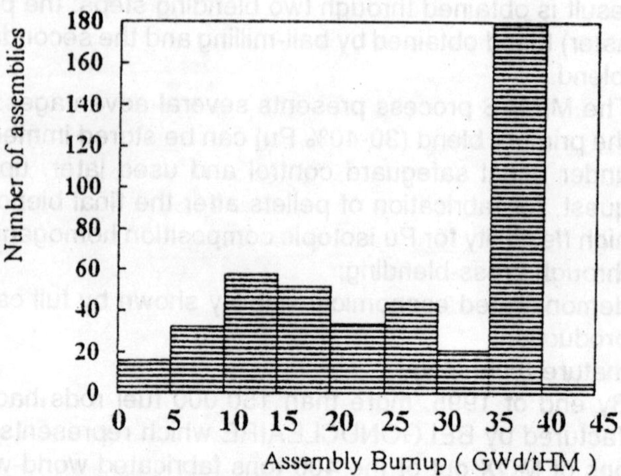

Fig. 3 : Mimas Fuel Commercial Irradiation Experience

On the basis of in-reactor behaviour and post-irradiation results (with fuel having reached up to 80,000 MWd/tM) it can be said that MOX is a proven mature industrial fuel for NPPs.

The future (1996-2005 decade) for BELGONUCLEAIRE's MOX activities shall be characterised by the following evolution:

— an increased share of BWR (8×8 or 9×9) fuels and thus a need for even more flexibility requested from our production;

— a trend to increase the Pu content in the product for two reasons:
  - a request for high burn-ups;
  - a tendency to decrease the fissile Pu content in the powder received from the reprocessors;

— a trend to receive Pu with higher activity, with an effect on the radiation doses and on the thermal heat dissipated by the powder, both having a direct effect on production efficiency;

— the ICRP-60 recommendations

To meet the main new constraints imposed by this evolution, adaptation of the MOX plant is necessary in the following directions:

— the decrease of the percentage of fissile Pu content and the higher Pu enrichment go along with a more important heat generation in the powders, requiring modification of the powder blending and transfer system;

— a significant increase is foreseen of the neutron emission by the even isotopes of plutonium (Pu 238, Pu 240 and 242): this is leading to the massive introduction of neutron shielding in the workshops;

— personal exposure is to be reduced also by remote processor control and by mechanisation of different process steps.

For the long term perspectives, additional improvements shall probably be requested depending too on the customer's particular requirements.

Finally, the MOX production experience shall also be improved by permanent analysis and feed-back of the fabrication data and by implementing the necessary qualification programmes e.g. to demonstrate the fabrication of new products (new powder types).

The same MIMAS process has been adopted by COGEMA for the Cadarache (since 1996) and MELOX plants. It will lead to a comulated production of more than 1,200 tHM of MIMAS fuel by the and of the century.

It should be noted that the production of the COGEMA and BELGONUCLEAIRE plants is commercialised world-wide through the COMMOX G.I.E. (Groupment d'Interet Economique).

iii) Plant Licensing and Construction Expertise

At the same time as developing the MOX product and MOX fuel fabrication, BELGONUCLEAIRE has progressively built up MOX fuel plant engineering expertise.

BELGONUCLEAIRE's capabilities in this field rely mainly upon three achievements, namely the existing Dessel Plant (P0), the project for the Dessel plant extension (P1) using the MIMAS process and participation in the MELOX project in France.

This expertise is being largely expanded and updated by the continuously growing operating data base of the P0 plant as it has been working at full capacity since 1989.

From the point of view of the environment, the sensitive areas of safety and proliferation have been thoroughly studied in Belgium which is a densely populated country located in the heart of Europe and which does not contain large uninhabited areas that can be isolated easily. These factors—which are of great importance in the context of the present discussion—have been treated at length in numerous publications and will be discussed briefly again below.

a) Environmental Effects

The MIMAS process is based on the dry processing of UO_2 and PUO_2 powders which are mixed, compacted and sintered. It includes direct recycling of most of the quantities of scraps produced which avoids storage of the latter and the need for expensive auxiliary facilities.

The waste is practically always in the solid state which reduces the amounts for disposal in the storage facilities.

b) Safety Aspects

The use of plutonium with relatively high contents of Pu 241, Am 241 and Pu 238 produced in LWRs has necessitated specific studies of radiation shielding problems since the plant is only partially automated to assure greater flexibility. and since the new ICRP recommendations shall decrease the admissible doses to the workers.

Over a total industrial operating period of more than 20 years, the men radiation dose has always been kept below the permitted value.

The P0 plant, together with the P1 plant, were subjected to other safety studies pertaining e.g. to criticality, fires, accidents due to external causes, in accordance with regulations imposed by the Belgian law and the EURATOM Treaty under the supervision of the Licensing Body.

c) Safeguards and Traceability

Belgium has been a party to the EURATOM Treaty since its initiation in 1958 and adheres to the Non-proliferation Treaty (NPT) with all its implications. It has been a strong supporter of an unconditional extension of the treaty with no limitation on duration.

The BELGONUCLEAIRE plant in Dessel is subjected to the strictest safeguards system. In addition to the national authorities, IAEA and EURATOM teams permanently inspect the plant. The inspection procedures are constantly rationalised by highly competent personnel in order to minimise the loss of capacity and to optimise, where necessary, the human and material resources.

The proper handling of fissile materials is continuously controlled and accounted by the two-fold international and independent monitoring systems. No relevant discrepancies have been identified to date.

d) Security

The physical protection of sensitive materials such as plutonium is defined by the Convention on the Protection of Nuclear Materials which is already over ten years old. This means that they are under constant surveillance by guards or electronic devices and are surrounded by a physical barrier with a limited number of points of entry under appropriate control. Access is restricted to person whose reliability has been ascertained.

In Belgium, the physical measures, rules and systems are determined by the Ministry of Justice. Most of these are not revealed to the public so as to increase their efficiency, while others are made visible as a dissuasive measure.

e) Transport of Fissile Materials

BELGONUCLEAIRE, through its daughter company TRANSNUBEL (set up in 1976), has acquired considerable experience with the transport of radioactive materials, especially spent fuel, plutonium dioxide and MOX assemblies. The companies work in close collaboration to solve various specific problems pertaining to safety, safeguards and the conservation and protection of the environment. So far, TRANSNUBEL has successfully carried out some 1,000 transports of radioactive substances every year without any major difficulties.

iv) MOX Fuel Engineering Expertise

Since it started developing MOX fuel fabrication technologies in the 1960s, BELGONUCLEAIRE has been directly involved with research and development programmes, including fuel fabrication, irradiation testing either in research and reactors or in LWR power plants and post-irradiation work in collaboration with competent laboratories in Belgium (CEN. SCKMol) and/or in other European countries.

The reactors used for such tests were:

- the BR3 PWR (15 MWe) located at Mol, shutdown in 1987
- the CNA PWR (located in Chooz), shoutdown in 1991
- the DODEWAARD BWR in the Netherlands
- the BESNAU 1 PWR in Switzerland
- the BR2 test reactor in Mol, using mainly the pressurised water loop designated CALLISTO (27 rods).

Since 1980, BELGONUCLEAIRE has organised international MOX fuel test an demonstration programmes, aiming at evaluating the mechanical and thermal behaviour of its fuel in LWR conditions. Mechanical testing includes ramp and transient tests. Thermal testing makes use of instrumented fuel rods. These programmes are financed by laboratories, utility companies of fuel manufacturers from several countries in Europe and Asia and from the US.

Particular examples of these programmes are:

- the PRIMO programme: irradiation and ramp testing of PWR MOX fuel rods in BR3 PWR up to 50 GWd/tM (1987 to 1993).
- the DOMO programme: irradiation and ramp testing of BWR MOX fuel segments in the DODEWAARD reactor up to 60 GWd/tM (1987 to 1996).
- the FIGARO programme: irradiation in the BEZNAU 1 PWR MOX fuel rods (up to 50 GWd/tm) followed by ramp testing including central temperature measurement and pressure detection (1995–1997).
- the NOK M109 & M308 programmes: irradiation in the BEZANU 1 PWR of MOX fuel rods and segments (up to 55 GWd/tM) followed by ramp testing for PCI (Pellet-Clad Interaction) behaviour determination.
- the ARIANE programme: this programme is aimed at determining the source terms from high burn up MOX and UO_2 fuels irradiated in DODEWAARD (NL), NEZANU 1 and GÖSGEN (Switzerland).

A status of recent experimental programmes as well as new proposals is presented in Table 2.

3. WHAT COULD BE BELGONUCLEAIRE'S CONTRIBUTION TO THE DISPOSITION OF W-PU?

BELGONUCLEAIRE can make available its experience of the MOX fuel fabrication process that uses PuO2 powder of

Table 2 : Mimas Fuel—Experimental Programmes

Fuel type	*Reactor*	*Burnup (GWd/ tHM)*	*Status*	*Programme*
17x17 rods	BR 3	41	Examination completed	PRIMO
17x17 rods	BR 3 & BR 2	60	Under examination	Bilateral
14x14 rods	BEZNAU-1	49	Under examination	FIGARO, NOK M 109
14x14 segments	BEZNAU-1	55	Extension of irradiation underway	NOK M 308 (not started yet)
8x8 segments	DODEWAARD	58	Examination completed	DOMO
8x8 segments	DODEWAARD	50	Under examination	ARIANE
14x14 Fuel rods	BEZNAU-1			
17x17 segments	MTR	30	New programme proposed	Helium release determination
8x8 segments	DODEWAARD and MTR	60	New programme proposed	High burnup behaviour and He release

different origins as well as its know-how regarding the construction or the operation of MOX fuel fabrication plants if it is decided to construct such a plant in the USA or in Russia[8, 9, 10, 11]. Moreover, its experience in MOX fuel engineering and irradiation programmes will be a valuable asset to support any licensing study[12].

i) Contribution to MOX Processing

As indicated above, after 1984 BELGONUCLEAIRE developed the MIMAS fuel fabrication process. This process which is essentially a dry process is a simple well-tried method. It has been used in the fabrication of more than 60 per cent of the rods produced so far for irradiation in LWRs over the world. This share will further increase in the future with the start-up of MELOX.

The MIMAS process produces excellent isotopic homogeneity of the Pu in the product, even with Pu of various origins (LWRs, GCRs, ex-W-Pu), of various forms (PuO_2 powder or a UO_2-PuO_2 powder mixture) and for various batch sizes. Therefore, it is now considered as world reference and could be easily adopted for MOX fabrication with W-Pu, for large scale demonstration programmes or later fuel reloads deliveries. Such deliveries would be treated commercially through COMMOX.

Of course, the solution, called EUROFAB by the USDOE, which consists in using existing facilities in Europe for the fabrication and US utilities for the burning, necessitates trans-continental transports of the materials and thus further licensing steps.

ii) Contribution to Licensing, Designing or Constructing a Fabrication Plant

BELGONUCLEAIRE has made the preliminary conceptual design of P1 as a possible extension to its existing P0 plant and has also included in tis studies the decommissioning aspects. The main features of P1 are given in Table 3. It has a nominal capacity of up to 60 tHM/yr. (for LWR fuel) and is of modular design. It is based on the MIMAS process and has been designed in such a way to permit its adaptation as a result of the experience acquired with P0.

Table 3 : P1—A New Medium Size Flexible MOX Fuel Plant

Capacity	*Design Capacity: 60 tHM/year*
Flexibility	**Line organisation adapted for high operation flexibility:** • P0 experience for small campaigns/multi-clients/multi-designs • two fabrication lines (NB: 4 lines with P0) • redundant/flexible handling **Plant lay-out adapted for flexibility along plant life:** • possibly progressive (de)-commissioning • prudent & progressive remote operation • in-life process & capacity upgrades.
Other characteristics	**Design for high Pu-contents (10–12%) and high burnup/age Pu (45 GWd/tM - 2 yrs) – Design complying with latest safety regulations:** • radio-protection: ICRP-60 • advanced fire protection • aircraft crash resistance • earthquake resistance

Fig. 4 shows schematically the lay-out of the plant. Fig. 5 is the general planning schedule for its construction and indicates a lead time before production limited to four years for construction once the authorisation has been obtained. This plant, relying upon an existing design integrating the experience gained in operating the existing plant as well as the know-how

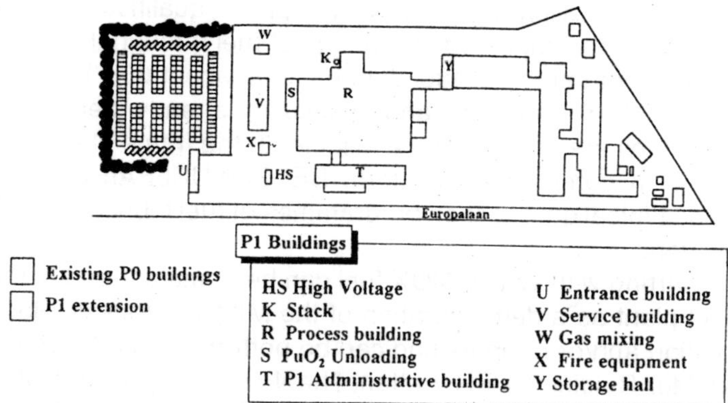

Fig. 4 : BN Dessel Plant—General View

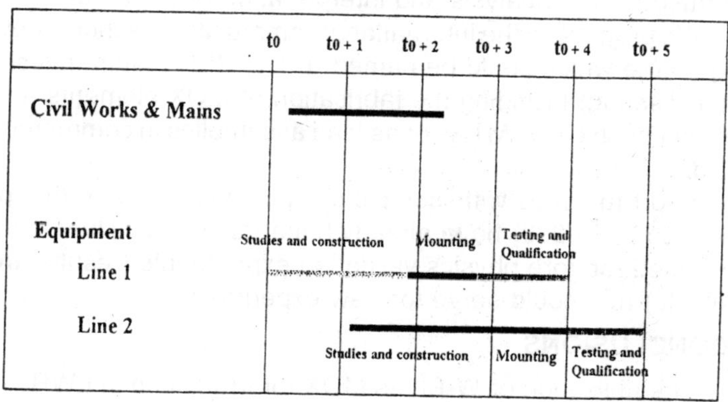

Fig. 5 : BN Dessel Plant P1 — General Planning

of the plant engineering, could be built in the countries (USA, Russia) where the W-Pu is available, after transformation into oxide powder. The plant could be accommodated in existing nuclear facilities, after a proper feasibility study. The advantage of this scenario versus EUROFAB is to avoid expensive and time consuming overseas shipments.

iii) Contribution to Fuel Engineering

BELGONUCLEAIRE has managed for more than 15 years R&D programmes on the behaviour of fuel (UO_2 and MOX) under irradiation. This gave the company the experience required to perform code calibration, licensing support, qualification of a new product and improved knowledge of fuel through the International Programmes. A fuel data base on about 60 PWR and 50 BWR MOX fuel rods and results of already performed programmes are immediately available. In the frame of the use of W-Pu in commercial reactors, BELGONUCLEAIRE could propose typical qualification programmes of fuel fabrication and irradiation.

Starting with W-Pu, MOX fuel can be manufactured in the Dessel plant as a demonstration of the MIMAS process. Base irradiation and/or ramp testing can be performed in a MTR like BR-2 (Mol, Belgium), HFR (Petten, The Netherlands) or HALDEN (Halden, Norway), or in a commercial NPP. Post-irradiation examinations can be organised in various laboratories with our expertise for the analysis and interpretation of the results.

We propose to the international community to launch a new programme which could be named "P.I.T." (Plutonium Investigation Testing) including the fabrication of MOX elements and their testing in either MTRs or as lead assemblies in commercial NPPd.

Short fuel rods with isotopic compositions close to that of W-Pu (95% Pu 239) have already been fabricated. They have been used for core physics validation experiments (results are available) and could serve for new experiments.

4. CONCLUSIONS

The utilisation of W-Pu as MOX for irradiation in LWRs is an attractive, mature and economical solution which answers the criteria issued in Dec. 1994 by the DOE for the "spent fuel standart".

The choice of this solution will allow a quick start of the programme if standard MOX fuel is used and if the selected fabrication technology is the one for which both fabrication and irradiation experience are maximum. MOX fuel can be manufactured either in Europe, in existing facilities (EUROFAB), or—as a further step—in new MOX plants to be built in the USA and Russia, taking advantage of existing facilities.

Both possibilities would benefit from the experience acquired in Europe with this type of fuel and in particular at BELGONUCLEAIRE:

— Mature fabrication technology (270 tHM fabricated with the same MIMAS process).
— Large irradiation experience in both PWRs and BWRs (240 fuel assemblies in the range 35–40 GWd/tHM burn up and 60 GWd/tHM as peak pellet in commercial reactors).
— Plant operation for more than 20 years and availability of an existing design for a plant of 60 tHM/yr. capacity.
— Post-irradiation results to support MOX licensing.

References

1. "Management and Disposition of Excess Weapons Plutonium", National Academy of Sciences Report, National Academy Press, Washington, D.C. (1994).
2. CANTER, H., "Status of U.S. Department of Energy Plans for the Disposition of Surplus Weapons-Grade Plutonium", U.S. DOE, 5th Int'l Conf. on Radioactive Waste Management and Environmental Remediation, ICEM'95, Berlin, Germany, Sept. 3–7 (1995).
3. EGOROV, N.N. & al., "Civil Use of Weapons-Grade Plutonium from Russian Nuclear Weapons—The AIDA/OX Programme, MINATOM, GLOBAL'95, Int. Conf. on Evaluation of Emerging Nuclear Fuel Systems, Versailles, France, Sept. 11–14 (1995).
4. GOLDSCHMIDT, P. & VERBEEK, P., "Plutonium Recycling: A Question of Timing", Current Issues on Nuclear Fuel Recycling, Kyoto, Japan, Feb. 6–7 (1995).
5. GOLDSCHMIDT, P. & VERBEEK, P., "The Disposition of Plutonium from Dismantled Warheads: A West European Electric Utility View", Nuclear Europe Worldscan no. 5-6 (1994), pp. 49–54.
6. VANDERBORCK, Y. & al., "MOX fuel Fabrication and Irradiation Experience in Belgium", 2nd Annual Int'l Policy Forum: Management and Disposition of Nuclear Weapons Materials, Lansdowe, VA., USA, March 21–24 (1995).
7. CORNET, G. & MICHEL, A., "MOX Industrial Reality and Political Acceptance", GLOBAL '95 Int. Conf. on Evaluation of Emerging Nuclear Fuel Cycle Systems, Versailles, France, Sept. 11–14 (1995).
8. DE DONDER, M. & VANDERGHEYNST, A., "BELGONUCLEAIRE and MOX Plant Engineering", MOX Fuel Fabrication Technology Symposium, Tokyo, Japan, Apr. 14 (1995).

9. HAAS, D. & al., "Experience with MOX fuel", The nuclear Energy Institute, FUEL CYCLE 95, San Diego, CA. USA, Apr. 2–5 (1995).
10. VANDERGHEYNST, A. & PAY, A., "MOX Fuel Fabrication Plants in Diesel: Operation Experience with P0–P1, Second Generation Plant", RECORD, London, UK, Apr. (1994).
11. VANDERGHEYNST, A. & DEBAUCHE, M. "Key Issues of MOX Fuel Plants Engineering", ICONE 3-Nuclear Power and the Energy Future, Kyoto, Japan, Apr. 23–27 (1995).
12. HAAS, D. & van VLIET, J., "Status of MOX Fabrication for LWRs at BELGONUCLEAIRE", KTG Fuel Conference, Karlsruhe, Germany, Dec. 5–6 (1995).

Address:

DIDIER HAAS
Commercial Director
Belgo Nucleaire
1200 Brussels
Belgium

59

Experience and Activities in the Field of Plutonium Recycling in Civilian Nuclear Power Plants in the European Union

ALBERT DECRESSIN, DIDIER J. GAMBIER, JEAN-PAUL LEHMANN and **DIETER ERNST NIETZOLD**

The European Union industry has established a world-wide leadership position in manufacturing and exploiting plutonium bearing fuel (Mox). About 15 to 20 tons of plutonium have been manufactured in the Mox fuel fabrication plants of E.U. companies.

The current capacity of about 60 tons of Mox fuel per year is being upgraded to reach 400 tons/year by the year 2000. As a result, the excess amounts of separated plutonium, presently stored in the European Union, should no longer raise but should steadily decrease to converge to zero.

Studies by the European Commission have indicated that the best use at present of weapons-grade and reactor-grade plutonium is to burn it in operating and future planned nuclear reactor. Disposing of plutonium by blending it with fission products or immobilising it into synthetic matrices appears to be far from being an industrially viable option. Following this path would mean to continue storing the excess plutonium of both military and civilian origin for an unknown, but very long period of time.

For these and other reasons, the European Commission is striving to foster international cooperation between the European Union companies, having a long industrial experience accumulated in the field of recycling plutonium, and so, far, the Russian Federation and the Newly Independent States. This cooperation is aiming at supporting projects that could be mutually beneficial to all parties involved.

To meet this objective, several programmes have been established either bilaterally or multilaterally, in particular within the framework of the International Science and Technology Center (I.S.T.C.) in Moscow. Some examples of such collaborations will be described.

1. INTRODUCTION

Under the terms of the START Treaty, warheads should be dismantled at an assumed rate of 2,000 per year. This activity should free a large quantity of highly enriched uranium and weapons-grade plutonium that add to the burden of the countries concerned.

From a technical viewpoint and focusing on plutonium management, disposal of weapons-grade material at a meaningful industrial scale already represents a major problem in the short term. The problem of storage become even worse in the long term and could be at odd with the very spirit of the disarmament initiative.

The option of plutonium disposal through blending with fission products or immobilisation into synthetic matrices is far from being a viable industrial option. Following this path would mean to continue storing the excess plutonium for an unknown, albeit very long period of time.

Studies of the European Commission have concluded that the use of weapon-grade and reactor-grade plutonium in operated and planned nuclear reactors is viable option, in economical and industrial terms but also with regards to physical control and safeguarding.

The pioneering and long standing experience acquired by European Union's organisations in the field of civilian use of plutonium allows these organisations to play a pro-active role in the implementation of several international cooperation and assistance programmes of the Union, either on a bilateral or multilateral basis.

This article will therefore concentrate on describing the present capabilities and programmes of European Member States, mainly in Belgium, France, Germany and the United Kingdom; the experience and activities of the general directorates of the European Commission will then be reviewed in particular in the area of research and development, nuclear safety, radiation protection, safeguarding, transport, combating illicit trafficking, and in the area of international cooperation and assistance; finally, ISTC projects in the field of plutonium recycling that are supported by the European Union will be presented as a pioneering example of future collaboration with the Russian Federation and the Newly Independent States.

2. EXPERIENCE AND ACTIVITIES IN THE EUROPEAN UNION

The **European Union** industry has acquired a thorough knowledge and experience of plutonium and Mox fuel handling, transport, physical protection and safeguarding, based on actual industrial achievement.

Mixed oxide (Mox) fuel, made by blending reactor-grade plutonium with natural or depleted uranium, is currently in industrial use in light water reactors of nuclear power plants in Belgium, France and Germany. More than 400 tons of Mox fuel have been fabricated and have permitted the recycling of more than 20 tons of reactor-grade plutonium. Thirty-four LWRs are fully licenced in Europe for using Mox fuel and fifteen reactors are actually loaded with plutonium fuel.

On average 21 tons per year of plutonium will be separated from LWRs and AGRs fuel in the reprocessing plants of La Hague (F) and Sellafield (UK) for the next 10 years; of this total, 15 tons per year will be European plutonium. The reprocessing of fuels from the Magnox GCRs is currently supplying a further 2.5 tons/year of separated plutonium.

With 15 to 20 tons of plutonium so far manufactured in E.U. Mox fuel fabrication plants and a production capacity of 60 tons of Mox fuel per year, soon to be upgraded to 400 tons/years, the European Union industry has placed itself in a world leadership position in the field of plutonium bearing fuel (Mox) manufacture and utilisation. As a consequence, the excess amounts of separate plutonium presently in store in the European Union will not raise any more but will converge to zero in the near future.

The European experience accumulated in the field of reactor-grade plutonium (RG Pu) with various isotopic compositions is a valuable basis for the application of these recycling technologies to the use of weapons-grade plutonium (WG Pu) in power reactors.

In **Belgium**, a high level of expertise has been reached. Of the 400 tons of Mox fuel or so fabricated thus far in the world, about 270 tons (equivalent to the recycling of 13 tons of Pu) were produced in Belgium. Mox fuel has been used in the first European PWR (the 11.5 MW Br-3) at the Belgian Nuclear Research Centre at Mol already in 1963.

The P0 Mox fuel fabrication plant of Belgonucleaire, at Dessel, has been operated since 1973 with a capacity of 35 t HM per year; a PI extension of 60 t HM/year being foreseen. For the fast reactor programmes, Belgonucleaire has fabricated more than 18,000 fuel rods using more than 6t HM for different customers during the period 1977-1985, in particular for 40 per cent of the core of the German SNR-300 at Kalkar. for the light water reactor programmes, Belgonucleaire has fabricated 150,000 rods (i.e. about 800 fuel assemblies and 270 t HM) between 1986 and 1995 for many various clients.

In 1995, 16 Mox fuel assemblies were loaded in Belgian PWRs, 8 in the 900 MW Tihange -2 unit and 8 in the 970 MW Doel-3 station; this is the first time that Mox fuel elements are being used at a commercial scale in power reactors in Belgium.

In **France**, plutonium was first separated in December 1949 from an irradiated rod of CEA's Zoé reactor at Fontenay-aux-Roses and its production started in 1958 at Marcoule and 1966 at La Hague. Then the fabrication of plutonium bearing fuel was developed to feed the fast neutron reactors Rapsodie at Cadarache, Phénix at Marcoule and Superphénix at Creys-Malville.

For economical reasons, the mid-term interest focused on the Pu recycling in water reactors. From 1974, Mox fuel use experience has been developed in the Chooz 310 MW PWR in the Ardennes; the plutonium recycling in civil power plants continued at an industrial scale in 1987 with the fifth reload of Saint-Laurent-des-Eaux-B-I PWR. Now, nearly 500 Mox fuel assemblies have bene loaded in the seven EDF's PWRs operating with this fuel at Saint-Laurent, Gravelines, Dampierre and Blayais. EDF has 16 PWRs licenced to use Mox fuel and intends to burn Mox fuel in 28 PWRs of 900 MW as from year 2000. For the future, a plutonium version of the European Pressurised Reactor EPR is currently studied by the consortium EDF-EVUNPI (Framatome+Siemens).

Today the plutonium content in the 1100 tons of spent fuel unloaded yearly from EDF's PWRs is well over 10 tons, but only 850 tons/year of spent fuel are reprocessed leading to 8.5 tonnes of separated Pu.

By the end of 1994, 460 Mox fuel assemblies had been supplied by Fragema to seven French 900 MW PWRs and abroad. Since 1989 the fabrication of Mox fuel is done in the Cogema's CFCa plant at Cadarache which was previously dedicated to fast reactor fuels and has a capacity of 30 t HM/year; in the future this plant will be dedicated to fabrication of Mox fuel for the German LWRs. In 1995 the new 120 tHM/year Melox plant at Marcoule started industrial fabrication of Mox for supplying EDF's PWRs requirements; this plant could easily be upgraded to a 160 tHM/year capacity. Construction of a new Mox plant is foreseen at La Hague; this facility will be able to accommodate various types of Mox fuel and its capacity will be in the range of 80 tHM/year.

As far as cooperation with Russia is concerned, the French Government signed in November 1992 a bilateral agreement on collaboration in the field of nuclear weapon destruction in Russia and peaceful use of the released fissile materials. The Aida programme is aimed at facilitating the reduction of the considerable amount of Russian weapon-grade plutonium and at defining an efficient method for its utilisation. French and Russian scientists are studying the feasibility of building the so-called Tomox facility in Russia to transform some 1.3t WG Pu/year into approximately 1.5 tons of Mox for the 600 MW Bloyarsk FBR and 20 tons of Mox for the 4 Balakovo 1000 MW VVERs.

In **Germany**, plutonium fuel recycling in thermal reactors dates back to 1966 with the insertion of Mox assemblies into the VAK reactor at Kahl, followed in 1970 by the loading of Mox in the 350 MW PWR at Obrigheim. Several commercial power PWRs and BWRs were then loaded with Mox. The Alkem 25 tHM/year Mox fabrication plant started operation at Hanau in 1965; it was handed over later to Siemens, but was shut down in 1991. A new 120 tHM/year Mox fabrication plant was built at Hanau's site next to the Alkem plant, but was in turn abandoned in 1995.

Presently the 21 LWRs, in operation with an installed capacity of 23.6 GW produce 450 tons/year of spent fuel containing approximately 4.5 tons of plutonium. So far about 18 tons of plutonium were separated, of which 7 tons have been fabricated to Mox fuel elements and recycled in thermal reactors mainly PWRs, the balance being stored in different forms; about 1.5 t Pu have been used for the manufacturing of the core loading for two fast reactors, the Kalkar SNR and the small KNK-II prototype. Up to year 2003 it is expected that 42 tons of separated plutonium are available for re-use in German rectors. At present 11 PWRs, and 2 BWRs are licensed to use Mox elements and 6 other NPPs have applied for the use of Mox.

A total of 250 Mox assemblies have been manufactured and loaded in Germany LWRs; the overall German experience with these Mox fuel elements is excellent.

Regarding cooperation between Germany and the Russian Federation, an initiative was started in 1993 to transfer equipment and technology from Siemens AG to Russia for processing weapons-grade plutonium into Mox fuel. The German Foreign Ministry has funded a feasibility study on Mox use with WG Pu, the result of which being a proposal to build a pilot fabrication plant at Chelyabinsk on the basis of Siemens know-how.

In the **United Kingdom**, the processing and storage of plutonium at BNFL's Sellafield site started in the early 1950's and included civil plutonium from 1964 onwards. Thermal Mox fuel has been made for a number of reactor types including GCRs, BWRs and PWRs. Five Mox assemblies were loaded in the first AGR a Windscale with excellent results. Also about 20 tons of fast reactor Mox fuel has been made in the period since 1970; this was used in the prototype PFR operated by UK AEA at Dounreay, Scotland. Altogether the UK stockpile of civil plutonium is substantial with some 80 tons RG Pu, half as irradiated fuel in storage ponds, and half as separated PuO_2 stored under rigorous safety, security and safeguards measures. UK has an enormous potential for recycling Pu from reprocessing of natural uranium Magnox fuels, the obvious reactors to burn Mox fuel being the AGRs; but this has not yet been developed on an industrial scale. To demonstrate Mox fabrication, in 1993 BNFL brought into operation its 8 tHM/year MDF facility at Sellafield with an annual capacity of 20 PWR assemblies and its first customer, the Swiss NOK, received delivery of Mox fuel in 1994. The 120 tHM/year SMP Mox fabrication plant is currently under construction adjacent to the Thorp reprocessing plant which came into operation at Sellafield in 1994; SMP, designed to fabricate Mox for various types of reactors, is due for active commissioning in late 1997.

Concerning cooperation with Russia, BNFL's 1991 collaboration agreement with the former Soviet ministry was extended until 1997 with the Ministry of Atomic Energy of the Russian Federation (Minatom) and covers the whole of the nuclear fuel cycle from R & D, through design, construction, operation and decommissioning. Specific Mox technologies are part of the programme and a BNFL-Minatom working group is investigating a potential process to transform Pu into PuO_2.

3. EXPERIENCE AND ACTIVITIES OF THE EUROPEAN COMMISSION

Disposing of excess plutonium of any origin is a major concern for the public, but even more so when considering surplus military plutonium. Given the currently implemented policy of disarmament and non-proliferation, the European Union has a vital interest that nuclear weapons states, and Russia in first instance, realise quickly the disposing of its plutonium stock. This approach would help contributing towards reduction of excess stockpiles of military fissile material; it would strengthen the enforcement of the international non-proliferation treaties and also foster the implementation of the Russian-American disarmament agreement. The fastest and more economical solution for disposing of the surplus WG Pu would be to recycle it in existing nuclear power stations, in line with the Russian Federation statement on plutonium being a precious source of energy for its population.

Russian plans to date have focused on recycling plutonium in fast neutron reactors, such as the BN-600 operating at Beloyarsk (Siberia) and BN-350 at Schvevchenko (Kazakhstan), or the few BN-800s presently at the planning stage. Russian research centres have begun studying the use of Mox in VVER pressurised light water reactors using technologies similar to those used in the PWRs in operation in eight European Union Member States. Russia currently has seven 1000 MW VVER in service at Balakovo, Kalinin, and Novovoronezh, and seven others under construction; more 1000 MW VVERs are operating or under construction in neighbouring countries.

The huge quantities of fissile materials transferred from the military sector to the civilian sector pose a new challenge. Development of new techniques and adapting existing ones, requires an important **R & D** effort as well as investing into industrial plants. While European Union industry and Member States are already contributing a major effort to that end in Russia, the European Commission's Directorate General for Sciences, Research and Development—**D.G. XII**—and **Joint Research Centre—JRC**—have also taken significant steps in this direction. The Euratom Framework Programme includes a research and technological development and demonstration scheme for nuclear fission safety covering the period 1994–1998. Much of the JRC's effort in the field of nuclear materials is developed in relation with the European Safeguard Research and Development Association ESARDA, grouping all the European Member States, the JRC being in charge of the association's permanent secretariat. A strong collaboration is also pursued between this association and the American Institute of Nuclear Materials Management INMM. As soon as September 1994, the Commission issued its COM(94) 383 communication to the European Council and European Parliament for an assistance programme to Russia to monitor the fissile materials from disarmament; this programme is based on the JRC's works.

Several projects were approved during the Summit of the heads of the E.U. Member States at Essen and are in progress and financed by the Commission's Directorate General for External Relations: Europe and the Newly Independent States, Common Foreign and Security Policy, External Service—**D.G. IA**—Technical Assistance to C.I.S.—TACIS—budget. Priority in the nuclear fission safety was given to training with the main project of creating the Russian Methodological and Training Centre RMTC at Obninsk, near Moscow, where more than 1000 experts are expected to be trained.

In addition, the JRC is currently developing research aimed to design new plutonium fuels; European scientists devoted to this task work in the JRC's Institute of Transuranium Elements—ITU—in Karlsruhe, in close cooperation with experts of the most concerned Member states (D, F, UK). The ITU currently investigates a novel scheme for burning weapons–grade plutonium (WG Pu) in conventional light water reactors with a view to optimising the destruction rate of plutonium and increasing the proliferation resistance of the resulting spent fuel. The main feature of this scheme is the use of a fuel based on WG Pu and weapons-grade high enriched uranium (WG HEU) in an inert matrix. Ex-military nuclear non-fissile materials such as tritium and beryllium need also to be disposed of in the civilian sector and research is underway to using them in fusion programmes.

• In the field of **nuclear safety**, the European Commission's Directorate General for Environment, Nuclear Safety and Civil Protection—**D.G. XI**—has developed considerable experience in the definition and application of regulations and conventions for plutonium handling and transfers. Required safety standards are based upon a philosophy of defence in depth which minimises the probability of incidents or accidents during fabrication, transport and reactor use of nuclear fuels. Any design of Mox fuel assemblies and Mox loaded reactor core has to obey the same safety requirement as UO_2 elements and core; PuO_2 fuel rods and assemblies have to meet the same thermal and mechanical limits as specified for UO_2 fuel. Transient and accident simulations (loca, lofa) for PWRs and BWRs cores containing Mox fuels show only small differences compared with those for pure UO_2 cores. Altogether, no significant changes occur by the introduction of plutonium bearing fuel as long as adjustments are fulfilled to fit the Mox neutronic properties : higher neutron absorption requiring greater neutron flux, higher temperature coefficient, lower proportion of delayed neutrons, behaviour under load-follow conditions and at high burn-up levels.

Responsibility of D.G. XI includes the **radiation protection**. Fresh fuel assemblies containing plutonium have a higher dose rate than the standard UO_2 elements: this entails special handling radiological protection measures, including use of a dedicated examination ring for in-plant handling. For radioprotection reasons as well, it is better not to store Mox fuel for a long period of time, as plutonium progressively turns into radioactive americium.

Materials **transport** comes under the authority of the Commission's Directorate General for Transports—**D.G. VII-**. Shipment of plutonium must meet the same standards used for other radioactive materials; they are based on the integrity of the package under any accident conditions, thus ensuring the same level of protection for the workers, population and environment independent of the selected transport mode. Supplementary tests, for beyond the present IAEA criteria, can be imposed by national authorities on the containers selected for special shipments, such as was done for the sea transport of plutonium between France and Japan. Other conditions above and beyond the safe container are also imposed (e.g. shock absorbers for road and rail shipment) so as to minimise and to avoid complete loss of the cargo or involvement of the cargo in case of extreme fire. This is the case for the transport of plutonium between La Hague and Rokhasho Mura where the following measures are in force: ship with double hulls, compartmental holds, redundant fire prevention and suppression systems, monitoring of cargo's temperature, routing and time-scheduling navigation.

Transport of radioactive materials is managed for preventing accidents with the normal traffic and other precautions are inspired by security reasons like physical protection (escorts, radar systems). It is necessary to consider possible causes of package failure: for such cases emergency response planning and preparation is foreseen. The strengthening of regulations and practices are currently under review and discussion for a wide range of activities and responsibilities related to transport of sensitive nuclear materials; the opportunity is considered to issue specific regulation notably for the development of a new Type C safer package for the air transport of plutonium. The European Commission's Standing Working Group on Safe Transport of Radioactive Material, which includes representatives of the Members States' safety authorities, is actively involved in these activities for the implementation of which the European Parliament has earmarked a significant 1996 budget. This programme stresses as well the need for assistance and training of the staff of the competent authorities of the Russian Federation and the Newly Independent States.

The production, movement and processing of nuclear materials, including plutonium bearing spent fuel and separated plutonium, are subject to strict **safeguards** procedures aimed to verify that their use continues to be as declared, that is to ensure that they are not illicitly made into nuclear weapons or explosive devices. The Commission's regional safeguards system operated by the

Directorate General for Energy **D.G. XVII-E**'s Euratom Safeguards Directorate deals with the individual nuclear operators within the territory of the European Community Member States and the Euratom system constitutes a regional safeguards system which complements the different national systems of accountancy and control, and the UN's international safeguards authority IAEA. The Euratom safeguards system is since 1957 founded in European law (Euratom Treaty's Chapter VII) and it applies to all civil nuclear materials; its mission is to detect diversion from peaceful use, to check that specific uses are correct and that obligations imposed by suppliers are respected.

Since 1970, Euratom has carried out continuous inspections at plutonium bulk facilities. Given that the existing industrial reprocessing plants and Mox fuel fabrication facilities, as well as number of Mox burning power stations, are all located with in the territory of the European Community, Euratom has obviously gained substantial experience in safeguarding these parts of the fuel cycle, a fact which is reflected by the amount of inspection effort dedicated to that area. This trend will lead in the future to increasing activities related to sensitive nuclear materials of which plutonium takes a prominent role. The European Commission is open to share the experience gained in this field with those countries embarked on the disposition of weapons-grade fissile materials.

Illicit trafficking in nuclear materials continues to pose a global risk and a potential danger to public health and safety. The European Commission recommended in 1994, the COM(94)383 communication to the European Council and the European Parliament on the illicit trafficking of nuclear materials and radioactive substances, with recommendation of a series of measures for combatting this practice. Several Commission's General Directorates are involved, including DG 1, DG 1A, DG XI, DG XII, JRC, DG XVII and DG XXI. The nature of the materials involved in this trafficking is quite various and in some occurrences only minor quantities of plutonium were at stake, as was the case of the well publicised seizure at Munich international airport. A close cooperation has been implemented between the Commission and the concerned organisations of the Russian Federation: Gosatomnadzor, Minatom and Kurchatov Institute. Four projects were initiated to improve the nuclear material accounting and control system in Russia and seminars were organised for the training of the personnel. A long term action in this field is developed at the RMTC centre of Obninsk with Tacis financing. The achievement in Russia of an efficient system of accounting and control of fissile material in the industrial nuclear facilities requires important resources and an international cooperations indeed needed; the European Commission is joining efforts with the USA and Japan to reach this objective.

4. PARTICIPATION OF THE EUROPEAN UNION IN THE INTERNATIONAL SCIENCE AND TECHNOLOGY CENTRE (I.S.T.C.)

The **International Science and Technology Centre** (ISTC) is an inter-governmental organisation established in March 1994 by an international agreement with four initial parties, the European Union, Japan, the Russian Federation, and the United States, now joined by Armenia, Belarus, Georgia, Kazakstan and Kyrgystan. ISTC recently broadened participation with admission of a first funding partner, the European Organisation for Nuclear Research CERN in Geneva. The task of ISTC is to develop, approve, finance and monitory science and technology projects addressing the following objectives:

- to provide weapons scientists and engineers in Russia and other interested states of the CIS opportunities to redirect their talents to peaceful activities;
- to contribute to the transition to market-based economies responsible to civil needs;
- to support basic and applied research and technology development for peaceful purposes;
- to promote the integration of scientists and engineers from Russia and other CIS states into the international scientific and engineering communities.

The ISTC operates with two administrative bodies: the Governing Board and the Secretariat, located in Moscow, with an executive Director and three Deputees all from the initial Parties to the Agreement. To manage its affairs, the Governing Board is advised by a Scientific Advisory Committee and is helped by a standing Coordination Committee.

The ISTC currently supports 202 projects for 81.6 M$, providing opportunities for more than 11,000 scientists and engineers, previously involved in the development of weapons of mass destruction, to engage in peaceful activities. The European Union is funding or co-funding 133 of these projects for a total amount of 29.6 M$, provided through the Tacis programme. The European Commission is the E.U. management body for the ISTC, with DG IA in charge of administrative and political aspects, and DG XII in charge of the evaluation and monitoring of projects.

ISTC projects span a wide spectrum of technology and research domains, including a little more than 10 per cent of projects pertaining to nuclear fuels and storage. The European Commission supports the involvement of European organisations at all levels from the definition of projects to the development and exploitation of results, although the latter must be dealt outside the ISTC framework.

As an example, when the ISTC was set up in 1994, Commission's Directorate General for Energy—**D.G. XVII**—took the initiative of proposing a project, now known as ISTC Project **N.369,** on the **"study of the technical and economic feasibility of the use of ex-weapons plutonium and civil plutonium as fuel for both fast and thermal reactors".** It received an encouraging response from an number of European Members States already active in that field, Belgium, France, Germany and the United Kingdom, and submitted a draft study for examination by the Ministry of Atomic Energy of the Russian Federation (Minatom). This initiative resulted in Minatom officially submitting in 1995 the Project N.369 to Governing Board of the ISTC, which approved its funding by the E.U. for 0.58 M$.

In this example of practical collaboration, E.U. industrialists involved in plutonium recycling technology expressed their interest in collaborating with Russian research through the project development. The companies involved are **Belgonucléaire** (Belgium), **BNFL** (United Kingdom), **Cogéma** (France), and Siemens (Germany).

On the Russian side, Minatom appointed seven Russian research and engineering institutions to conduct the feasibility study, making one of them responsible for directing and coordinating the project: the **Institute of Physics and Power Engineering IPPE** in Obninsk. The other participants are the All-Russia Scientific Research and Design Institute of Power Technology A-RSRDIPT in St Petersburg, the "Bochvar" All-Russia Scientific Research Institute of Inorganic Materials A-RSRIIM in Moscow, the "GidroPress" Special Design Bureau SDB GP in Podolsk, the "Chlopin" Radium Institute RI in St Petersburg, the Experimental Design Bureau of Machine Building EDBMB in Nizhnij Novgorod, and the Governmental Specialised Project Institute GSPI in Moscow.

The N.369 **feasibility study**'s aim is to put emphasis on the first stage of Pu recycling in Russia with large amount of separated civil plutonium being accumulated at P0 Mayak and quantities of released weapons plutonium, and with few reactors actually using plutonium. The study is based on a comparative system analysis of the technico-economic parameters of various scenarios for using WG Pu and RG Pu.

A major concern within the public opinion is in first instance to make sure that the excess separated plutonium from military origin can be disposed of as soon as reasonably possible. The material derived from the dismantlement of the warhead pit consists of metal plutonium and the first step towards denaturation is to convert this metal into plutonium oxide PuO_2; this is to be achieved inside the military facilities workshop and the study of this step is not part of the Project N.369. But before it can reach an industrial stage, this first phase will require process development, time and investments.

The second step of denaturation of the ex-military plutonium is to use it in fabricating Mox fuels and this is to be done inside civilian, internationally safeguarded facilities where PuO_2 can be handled. The third and final step towards denaturation is the irradiation of the Mox fuel assemblies inside the core of the power reactor where plutonium is burnt and mixed with highly radioactive fission products and trans-uranium elements, making it resistant to diversion or use in explosive devices.

Given the above mentioned time incurred in developing industrial scale facilities necessary to deal with ex-military plutonium and considering also the fact that civil separated plutonium in stockpile is as well a public opinion concern, it seems reasonable to take opportunity of this delay to accelerate, as soon as possible, familiarisation of the operators with the valuable experience of recycling excess in inventories of RG Pu as Mox fuel in operating power reactors.

The Project N.369 feasibility study considers three different options:

- fast neutron reactors to be built on a single integrated site with the appropriate fuel cycle facilities;
- pressurised light water reactors VVER already in service, under construction and planned, partly loaded (one third of core) with Mox fuel assemblies, as well as in fast reactors;
- pressurised light water reactors of VVER type in operation modified to handle full Mox cores.

The comparison of all options is to be performed on the basis of a uniform methodology using data obtained in a coherent manner. The work is done taking into account the conditions existing in the nuclear power development and the economic context of Russia, as well as available experience in Mox utilisation abroad, inter alia in Belgium, France, Germany and the United Kingdom.

The economic assessment of these options will be made within the framework of a system analysis integrating the following parameters:

- efficiency and time scale of inventory drawdowns;
- non-proliferation concerns;
- people's health and environmental protection;
- resistance of materials to diversion;
- nuclear safety.

The programme comprises two phases, each lasting a year:

- Phase 1: Characterisation of the principal fuel parameters for fast and thermal reactors (NB-600, BN-800, VVER-1000, VVER-500) having full or part Mox cores, taking into account the most stringent safety requirements. Calculation of investment and operating costs for successive stages of plutonium use in the reactors and in fuel cycle and storage facilities.
- Phase 2: The above mentioned results will serve as input data to a system analysis of the various scenarios, using a standard methodology, so as to determine their economic efficiency, with a view to the impact assessment from the point of view of radiological protection, waste management and evaluation of measures to prevent misuse of nuclear materials.

The human resources mobilised, over two years, for the Project N.369 at the seven Russian institutions amount to 1300 men × months, corresponding to an average 50 scientists and engineers. The basic data will be calculated using neutronic codes widely used in the similar studies carried out by IAEA and OECD, taking into account the balance of energy produced and changes in isotopic composition; mathematical models will be used to determine the technico-economic feasibility of the options and scenarios. The Russian research institutes will make their data banks, software and data-processing capability available to the project. Progress reports will be produced and the findings of the study will be published in the specialised national and international press and at international scientific meetings.

The Project N.369 is guided by a **Joint Steering Committee** (JSC) composed of experts from the Russian Federation and the European Union. Members of this JSC are representatives of the seven Minatom's specialised institutes, the European Commission and the four nuclear fuel industrial companies of Belgium, France, Germany and the U.K.

The JSC or its Secretariat, which is assured by representatives of the IPPE and Commission's DG XVII, meets twice a year either in Russia or in the European Union. The last JSC meeting was held in Saint Petersburg on October 30–31, 1995, and 1996 meetings will take place in Brussels and Sellafield. The trilateral organisation of the JSC is a very typical aspect of the Project, its fundamental object being to create a dialogue and to assure an efficient collaboration between operators from the East and the West of Europe.

With the definite intention of avoiding repetition works and encourage complementarities and cross-fertilisation between different programmes and different institutes, it has been decided that links be established between Project N.369 and **other projects** related to **plutonium use** and also funded by the **ISTC**. Several projects, either already approved or still under review, have been identified as bearing a potential complementarity with N.369; their results or conclusions could represent a significant and beneficial input to take into consideration.

The approved ISTC projects deserving to be mentioned are:

– **N.116**, funded by the E.U., pertains to the development of computational and methodological techniques for verification of nuclear data bases used in calculations of neutron-physical characteristics and used in analysis of nuclear reactor safety. This reactor physics project will last 3 years, and is focused on VVER reactors. It involves also simulation and experimental works at the ROMB stand in Chelyabinsk-70. The leading institute is the NIKIET-RDIPE in Moscow and E.U. collaborating industry includes GRS (Germany), Framatome (France) and Belgonucléaire (Belgium).

- **N.273,** funded by the E.U. for 2 years, focus on radiation aspects in nuclear fuel cycle based on reprocessed uranium Repu and Mox fuel, including: calculation of hazardous nuclides; study of radiation characteristics; analysis radiation conditions during pellets and assemblies fabrication; determination of spent fuel characteristics and wastes during production and reprocessing. The Russian institutes involved are the Bochvar and the Kurchatov, and the E.U. collaborators involve the Institute of Transuranium Elements of the JRC in Karlsruhe (Germany), Cogéma (France) and BNFL (U.K.).
- **N.290** concentrates on civil and military plutonium utilisation in fast and thermal reactors, and management of wastes generated in the process. This 3 year project is being funded equally by the E.U. and Japan for a total of 0.45 M$. The project specificity is to transform plutonium alloys from dismantled warheads to dioxide and to develop technologies of low dust-producing granules of Mox fuel. The leading institute is the Bochvar in Moscow, but the project involves other Minatom's divisions in Dimitrovgrad, Obninsk, Nizhniy Novgorod, Sverdlosk and Chelyabinsk. International collaborators to this project are Siemens and the JRC (Germany), BNFL (United Kingdom) and PNC (Japan).
- **N. 332** is concerned with developing safe methods for storing large quantities of plutonium and uranium from dismantled warheads. ISTC funding was granted for a preliminary feasibility study; if continued, the 2 year project will consist of computation methods and test experiments (conversion into oxide, powder mixture, sintering). The leading institution is the VNIIEF in Arzamas and the sup-

porting organisation is the European Commission's Institute for Transuranium Elements ITU in Karlsruhe.

Further related projects are currently under review by the ISTC.

5. CONCLUSION

The European Union nuclear industry is a world leader in the industrial development and commercial exploitation of mixed plutonium-uranium oxide fuels technology in pressurised light water reactors and is well placed to guide the use of such Mox fuel in the VVER-1000. The experience and activities in this field involves the European Community as a whole, several of its Member States with their national nuclear research organisations, their regulatory authorities and their specialised industrialists, several divisions of the European Commission including its Joint Research Centre and its Euratom Safeguards Directorate. The European Union is one of the funding parties of the International Science and Technology Centre ISTC in Moscow which provides mass destruction weapons scientists and engineers in Russia opportunities to redirect their skill to peaceful works; several projects funded by the ISTC are dealing with the use of plutonium of military or civil origin as a fuel in nuclear power plants.

At international level, the experts were given the task to study options for the long-term disposition of fissile materials, particularly of plutonium, taking into account the issues of non-proliferation, environmental protection, safety, and technical and economic factors. the most rewarding issue is the use of plutonium for electricity generation and a significant contribution can be brought into the picture by the cooperation between scientists and engineers of the European Union and the Russian Federation. The various actions taken i the European Union will contribute to provide an indispensable reference for decision making concerning the disarmament issue. They will pave the way for the orientation of new R & D projects and identify areas of industrial cooperation between countries having experience in this field.

60

Uranium-Plutonium Fuel for Fast Reactors

S.A. ANTIPOV, V.A. ASTAFIEV, A.E. CLOUCHENKOV
K.I. GUSTCHIN and T.S. MENSHIKOVA

INTRODUCTION

First efforts to utilise plutonium as nuclear fuel date back to 50s.

Those investigations allowed to create in the mid 60s pilot section at PU "Majak", where the cores for the reactors BR-5, IBR-2, IBR-30 and experimental fuel assemblies for reactor BOR-60 were produced.

The subsequent R&Ds were focused on the mixtures of uranium and plutonium oxides _ the main fuel for fast reactors of the first generation.

The initial investigations of uranium-plutonium fuel were based on mechanical mixing of individual dioxides of uranium and plutonium. In 1980 a "Paket" facility was developed at PU "Majak" with the sections of pelleting and fuel element outfitting. The capacity of the facility is 350 kg/year by the sum of uranium and plutonium oxides. Already in 1980-1981 the first 10 full-length fuel assemblies were manufactured (assembling of fuel elements into fuel assemblies was conducted at the Elektrostal plant) and charged into the BN-350 reactor.

In 1969 the decision was taken to establish a pilot-industrial U-Pu fuel production—a complex 300 at PU "Majak".

Mechanical mixing of metal dioxides was laid as a basis for the technology of fuel manufacture.

The ammoniac technologie of fuel fabrication having been developed at Russian Scientific and Research Institute of Inorganic Material since 1986 alongside with the objective to develop ecologically acceptable (producing little dust) processes, proceeded from the condition to obtain mixed oxides via chemical co-precipitation of the corresponding compounds of uranium and plutonium.

At the beginning of 1987 a pilot_ industrial facility "Granat" was designed and constructed at PU" Majak", which starting from April 1988 has begs to produce U-Pu oxides.

Currently 2 technological processes are at the stage of R and D:

- mechanical mixing of individual oxides (MMO—process);
- method of ammoniac granulation ("Granat").

The above technologies allow to fabricate powders that can be conventionally presented by 2 classes: dispersed, granulated powder class.

"Granat" method provide for the fabrication of granular with the size of 50-1000 mkm (mainly 200-800 mkm). Individual oxides being used in the technology mechanical mixing have the particles of the size of up to 20 mkm.

When choosing one or another pattern of obtaining the initial powders, not only dusting but a member of other factors as well should be taken into account. These factors are the following: the possibilities of producing powders with stable properties; simplicity of the technology of core fabrication; absence of additional operations to improve the powder technological properties (for example, granulation, additional calcination, grinding, etc.).

The development of the technology to produce MeO_2 core was commenced in 1963-65.

The initial products were uranium dioxide obtained via ammonia precipitation, and plutonium dioxide obtained via preliminary precipitation of plutonium - ammonium pentaoxalate, providing for fuel ceramic properties of plutonium dioxide.

To provide for the yield the MMO powder require the granulation that includes preliminary compaction, grinding, sieving and pelletising. The operations enumerated are dust-producing and complicated in view of equipment.

Individual oxides of uranium and plutonium, being used for mechanical mixing, are the most dust-producing at the given level of the technology development.

Currently the characteristics and types of initial powders, heat treatment modes, mixing and dispersion modes are deter-

Table 1 : Pellet Fabrication Schemes Prepared from Different Oxide Powders

Operation	*Zol-Gel*	*Granat*	*AU(PU)T*	*PCC*	*MMO*
Reduction of the founding solution	+	+	+		
Dosage up to require composition of Pu(U+Pu)			+		+
Blending with binder for average composition	+	+	+	+	+
Dispersing of the mixture					+
Preliminary pressurising				+	+
Ball milling and sieving				+	+
Grinding and mixing with the binders				+	+
Pressing	+	+	+	+	+
Sintering	+	+	+	+	+

Table 1 presents the technological patterns of core fabrication of various types of initial materials. Only pressing and sintering—the two technological operations—are common for all the powder types.

mined. In a member of work this technology is shown to have an extensive possibility to affect in the required direction the powders and cores characteristics.

The strength of the powder granular produced by MMO technologie, is caused by the pressure preliminary compaction and is easily controlled during the process of granulation and, therefore, is no need to control this parameter for the given initial powder.

Besides that, the technology of fuel fabrication from MMO powders anticipates the dispersion operation that is carried out with the goal of thin PuO_2 particle's distribution is the UO_2 matrix to form solid solution at the sintering stage.

The degree of grinding and dispersion (uniform mixing of fine dispersed components) influences considerably the physicochemical and technological characteristics of powders, the content of sintered solid solution after sintering, fuel structure and solubility after chemical reprocession and solution of prod-

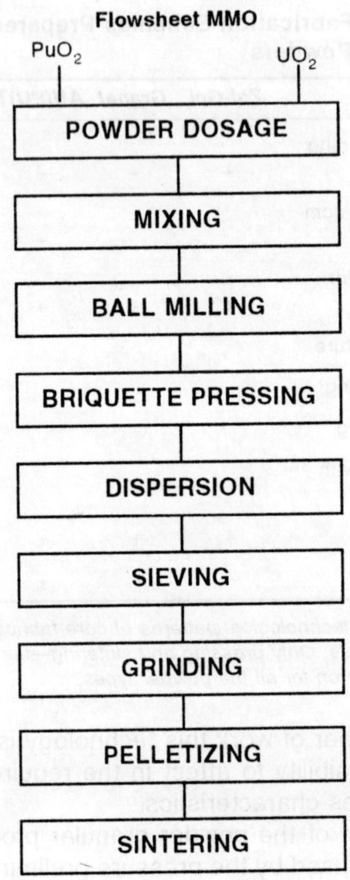

Fig. 1 : The Methods of UO_2 and PuO_2 powders mechanical mixing

ucts rejected. Grinding leads to decrease in size of particles (agglomerates) and shift in their morphology, which causes changes in specific surface, fluidity, bulk density, mouldability and sinterability. The influence of dispersion and grinding on technological characteristics is ambiguous. Dispersion and grinding modes are critical to the process. Inactive grinding and mixing with the use of grinding bodies can somewhat increase fluidity and bulk density, and decrease specific surface. On the contrary, active grinding and dispersion with fine and extra fine grinding processes increases specific surface and decreases fluidity and bulk density. In mixed fuel production, the degree of dispersion determines the size of solid solution zones with different content and allocation of Pu inside fuel pellets.

Presence of Pu-enriched zones in sintered pellets influence the characteristics of Doppler coefficient and solubility in reprocession of rejected products and irradiated fuel.

Because of this, specifications for reactor cores contain tolerances for sone size and Pu content.

Thus, the mixing, dispersion and grinding processes are the base of mixed fuel production, as they provide for uniform allocation of components and the physicochemical and thermal characteristics in the fuel column. Powders of different origin and character are used in core production; because of this, one of the key aspects is to optimise the mixing, dispersion and grinding processes.

Introduction of a AVS-150 type eddy layer apparatus into the technological process has allowed to improve to a large extent the homogeneity.

METHOD OF AMMONIAC CO-PRECIPITATION IN THE PRESENCE OF SURFACE SUBSTANCES "GRANAT"

As a basis for this process the method of uranium and plutonium hydroxide's precipitation in the presence of surface-active substances (SAS) was used.

"Grant" powders provides for a low level of dusting at all the stages of technological process of fuel core fabrication.

As a result of the investigation conducted by the Russian Scientific and Research Institute of Inorganic Material and then jointly with PU 'Majak' during 1985-1988, positive information was obtained both in view of the product quality and possibilities of the process waste utilisation, and the perspectives of the equipment development.

For a short period (about one year) the design of the 'Granat' facility was developed, the basic technology equipment was manufactured, the mounting and start-up-and-adjustment work were performed. The mixed oxide's facility production rate is 0.6-0.8kg/day.

With the 'Granat' technology of product fabrication from powders, the properties of initial granular are not changed. They are predetermined at the chemical technological stage of oxide fabrication. At the stage of product fabrication it is impossible to

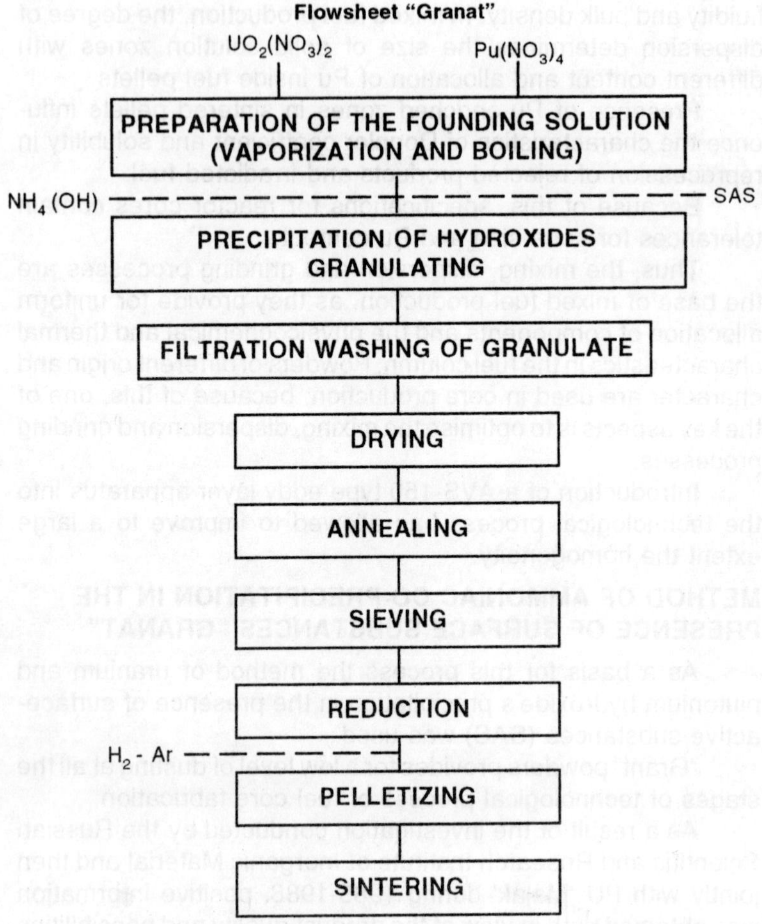

Fig. 2 : Method of Ammoniac co-precipitation in the presence of surface substances "Granat"

control the granular strength, therefore, this parameter of the initial powder must be controlled to its correspondence with the required value (0.3-0.5kg/m).

The carried out investigation have shown that the granulated of the 'Granat' facility allows to obtain sleeves with the required structure and composition.

During the period of December 1988 - February 1989 about 50 kg Of mixed oxides with plutonium content of 23.5±0.5%

mass was produced at 'Granat' facility, and about 700 kg of mixed oxides were produced by the beginning of the second quarter of 1993.

Recommendations given on the base of laboratory investigations, were fully confirmed at all the process stage when experimental fuel assemblies were produced in semi-industrial scale.

SINTERING

Sintering is the basic stage of production of pelletised fuel. In the process of sintering all physicochemical and technological peculiarities of the initial fuel are revealed, and the most important fuel characteristics are formed, namely density geometry, structure, degree of homogenisation, oxygen co-efficient.

The nature of the initial oxide influences considerably the characteristics of sintered pellets, as it determines the morphology of particles, the specific surface of the powder, bulk density, presence of impurities, etc.

Below are shown the peculiarities of sintering process for U and Pu dioxides.

There exist fairly strict tolerances on MOX pellets for fast reactors. Many of those are analogical to tolerances on U pellets in the list of criteria. Utilisation of Pu as the splitting material brings forward new requirements which influence considerably the fuel's performance under irradiation, the working capacity of fuel elements and neutron-physical characteristics of the reactor in general. The basic difference between MOX fuel and U fuel is that for MOX fuel either a homogeneous structure (solid $(UPu)O_2$ solution) is obtained, or fine dispersed Pu inclusions are uniformly allotted within a U counter die, and, which is more, determined O/Me ratio and U/Pu ration are obtained.

As a rule, we use batch-operating furnaces with molybdenum/tungsten heaters to investigate and fabricate experimental cores, which makes it possible to attain a sintering temperature up to 2000∞C, charge mass up to 2000 g and to use H_2-Ar mixture as the sintering atmosphere. It is known that to attain a pre-stoichiometric constitution of MOX fuel, a gas atmosphere with a corresponding oxygen potential (DGo_2) is needed. But with the use of batch-operated furnaces a regulated oxygen potential is practically unattainable, because moisture in the pellets and in the atmosphere (sorbing on the water-cooled body, shields, heaters) interferes into the furnace.

In batch-operated furnaces humidity does not practically vary, and moisture forming in the process of reduction of pellets influences considerably the o potential. In this case, the o potential of the sintering atmosphere depends on the O/Me ratio of initial oxides, the speed of gas flow, the speed of boats moving through the sintering area, and temperature gradient in the axial and radial directions. All these parameters are controllable and capable of regulating which permits to obtain stable and determined O/Me ratios after sintering, as it was corroborated by the experience of foreign companies in MOX fuel production.

At the Mayak plant cores are sintered in batch-operated shaft high-temperature furnaces.

To provide for control over the sintering atmosphere, a molybdenum ampoule was inserted into the furnace to isolate the sintering area from the heaters and the body of the furnace. In this case, the moisture of the outcoming gas also depends on temperature, but to a considerably smaller extent.

Completion of this sintering mode made it possible to obtain a regulated O/Me ratio.

POTENTIAL WAYS OF EFFECTIVE UTILISATION OF PLUTONIUM, WEAPON'S GRADE ONE INCLUDED

In Russia reactor scientists and designers have a unified opinion that plutonium can be most effectively utilized in fast reactors. As distinct from the concept that was shared by us 10 or 15 years ago which envisaged the maximum plutonium breeding we are now engaged in resolving the directly opposite task, namely producing fuel that would drastically reduce or fully eliminate plutonium breeding.

Several directions are under study. One of them is to produce fast reactor MOX fuel containing much more plutonium, up to 45 per cent against 20-25 per cent. The second one is to prepare plutonium oxide fuel containing the minimum allowable uranium and an inert diluent additive, i.e., fuel that would significantly reduce plutonium breeding.

The two directions of the most effective plutonium utilisation through the use of oxide fuel seem promising since, first, much experience has been gained in oxides, and, second, the stored civil plutonium is in the oxide form.

The plutonium content of the fuel now in use in fast reactors is 20-25 per cent; the rest being depleted uranium. However, this fuel inevitably results in significant plutonium breeding. The plutonium breeding ratio can be much reduced if the plutonium content of the fuel is increased while that of uranium is decreased. We decided in favour of the fuel containing 45-50 per cent plutonium. Use was made of both ammonia co-precipitation and mechanical blending. The specifications for the pellets were the same as those placed on the standard product.

The improved process is capable of producing pellets having the stable density in the narrow range of 10.5 -10, 7 g/cc. X-ray spectral and x-ray diffraction analyses corroborate the formation of an oxide solid solution (Fig. 3).

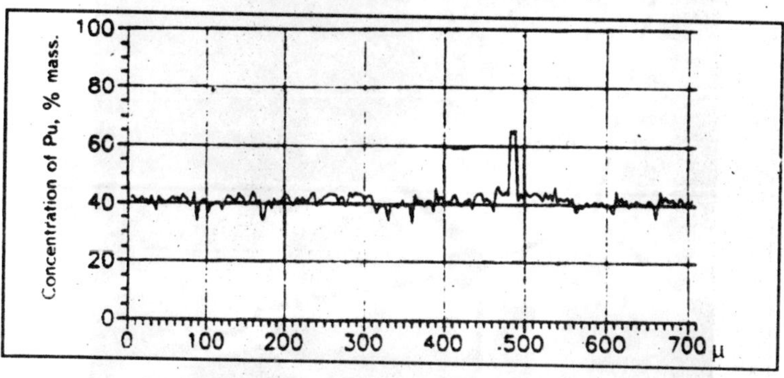

Fig. 3 :Plutonium distribution in ($U_{0.55}$ $Pu_{0.45}$) O_2 specimens prepared by chemical co-precipitation

The nitric acid solubility of higher plutonium content pellets depends on the method of the mixed oxide preparation. The pellets of co-precipitated oxides as two-fold processed are adequately dissolvable, i.e., 99.4 per cent. The solubility of mechanical blended oxide pellets is but 93.4 per cent.

The second direction of the potential substantial reduction of the plutonium breeding ratio in fast reactors is related to a development of fuel having the minimum permissible content of uranium through partial substituting the latter by and inert diluent.

The work in this field has been initiated. Cerium and aluminium oxides are under study as potential inert dilutents;

also zirconium oxide is suggested for studies. The choice of those materials has been made based on general implications, since no phase diagrammes of the respective oxide systems or other data are available. To generate some benchmark data the first experiments were conducted using uranium oxide and diluents at the 10 per cent content of the latter. Pellets manufactured from mixed dioxides prepared by the ammonia precipitation method have the density of 10.3-10.5g/cm^3 and fine uniformly distributed pores. The diluent content is as specified; is distribution is adequately uniform (Fig. 4).

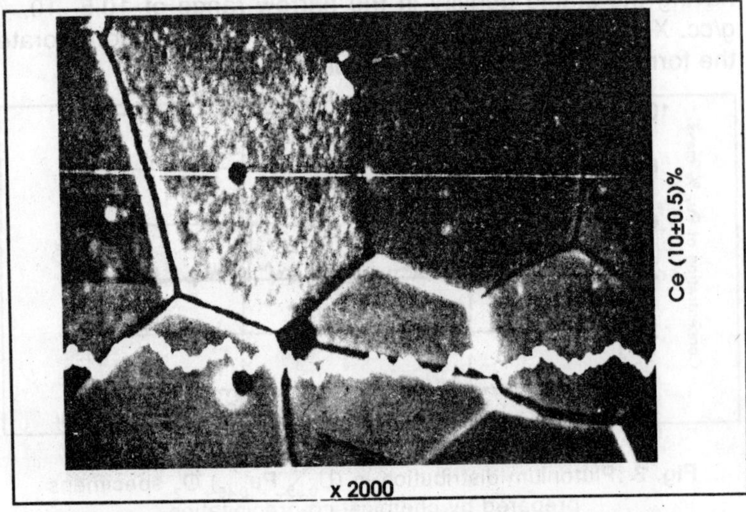

Fig. 4 : Microstructure and distribution Ce

SUMMARY

Thus, the technology was worked out for fabrication of MOX fuel pellets from co-precipitated and mechanically blended mixed oxides. Both the processes ensure the homogeneous structure of pellets readily dissolvable in nitric acid upon reprocessing.

To increase the plutonium charge in a reactor-burner the process was tested for producing MOX-fuel having higher plutonium content and an inert diluent.

It is shown that it is feasible to produce fuel having a homogeneous structure at the content of plutonium upto 45 per cent mass.

Address:

SERGEY ANTIPOV
Chief of Laboratory
All Russia State Scientific Centre of Inorganic Materials
Moscow
Russian Federation

61

Performance and Characteristics of a Small-Sized Oxide-Fuelled Fast Reactor (PRISM) for the Burning of Excess Plutonium

CARLO ARTIOLI, GEORGIOS GLINATSIS and FRANCA PADOANI

The increasing availability of fissile material and the world-wide perception of the need to reduce or control such material make a new approach towards nuclear reactors necessary. The Nuclear Fission Division at ENEA is presently interested in evaluating various options for the burning of excess plutonium, among which the LWR reactors using rock-like inert matrix fuel in a once-through cycle and fast reactors with recycling. This paper deals with the latter option. A fast system, rich in neutrons, can produce plutonium by breeding, or it can burn it (together with Minor Actinides) with a high degree of effectiveness: this flexibility makes the fast system worth considering.

In the framework of a co-operation project between GE-USA and ENEA, the PRISM MOD D oxide-fuelled reactor (small size, 840 MW) was studied as a burner. A complete set of studies was carried out covering several fields: neutronics, thermo-hydraulics, dynamics and safety.

A negative void coefficient was achieved thanks to a particular arrangement of the fuel subassemblies (leaky core) and to the enrichment distribution. The central zone, loaded with shielding subassemblies, accounts for a negative void coefficient even in the inner zone. Moreover, this zone could become a suitable place for burning long-lived fission products. As a burner, using a standard fuel, the expected performance is a consumption of about 60–70 kgPu/TWhe. ULOF and TOP analysis show a begin transient evolution. Only half of the GEMs are sufficient to close the ULOF transient; the reactor power reaches its maximum at 1.35 of the nominal figure during a TOP accident.

A method of detecting a failure in a subassembly and identifying the subassembly concerned was developed. It was demonstrated that it is possible to create a sufficient number of tags for them to be recognised as different by the detector system, by suing "cocktails" of only four tag gases. The total gas required to overcome the uncertainties, but only the mutual rations. By processing the results of the detector system with a simple code, the effectiveness of the method can be significantly increased. Moreover, experimental measures can be used to reduce the required quantity of gas.

1. INTRODUCTION

Significant quantities of weapon-usable fissile material, namely plutonium and highly enriched uranium, are now considered as surplus to requirements and this excess is doomed to further increase as the arms reduction agreements between United States and Russia are implemented. Moreover the growing and even larger accumulation of plutonium from the civil nuclcar programme is another formidable element and means that a new approach toward fissile material is required.

The use and disposal of civil plutonium is a difficult matter to handle, as its value is not the same for all countries: greater availability and more recycling may be seen either as a manna from heaven or as a plague. On the other hand, for military plutonium it is perhaps easier to find a common ground and the use of the keywords "surplus" or "excess" is already evidence of an intention to get rid of it.

But at which cost, and within time frame? As soon as all the variables of the proble are taken into account, including security, safety and environmental impact, it is clear that an approach similar to that for civil plutonium is required. Despite the differences, an analysis of the options for the disposal of the weapon material can cover a large part of that for civil plutonium and vice-versa.

Several options have been proposed, varying from vitrification, the launching into space or its use as a fuel, hence exploiting its economic potential. They could be outlined as follows:

– options already technological available;
– options not yet available, but that would be made available with present technology in the short term;
– long-term options, that in some cases have still do demonstrate their efficacy.

The time schedule should not however hamper further studies of the most promising options, but at the same time the urgency of the plutonium disposal problem compels us to turn towards systems that may be available in the short term.

The burning of plutonium in nuclear reactors in one of them. Looking at the various systems proposed, it should be noted that the 'optimum' system probability does not exist, but more probably that a combination of potentially 'optimum' systems will be the practical and best solution (for instance LWR+LWR-MOX+Fast-Burner). Independently from the chosen system, two issues have to be duly considered: first, the need for intrinsically safe reactors in order to be publicly acceptable; second, the problem has to be fully addressed, hence considering the degradation of plutonium in the reactor and the presence of minor actinides (Np, Am, Cm) and long-life fission products (I129, Tc99).

If giving too high a value to plutonium has been one mistake of the past decades, the opposite mistake should if possible avoided. System dedicated to mere destruction of plutonium, such as accelerators, could be seen as a panacea now, but perhaps as a terrible waste in the future. The flexibility of the system should be a strategic issue. Regardless of economic considerations a fast system, that can act as a breeder or as a burder (to some extent) according to the loaded core, could assure this flexibility.

However the problems of plutonium disposal should not lead us to neglect the problem of waste. A correct ethical approach toward posterity requires systems that neutralise, now, potentially dangerous products.

2. AN OXIDE-FUELLED FAST REACTOR (PRISM)

The Advanced Liquid Metal Reactor (ALMR) is a fast reactor design based on the Power Reactor, Innovative Small-Module (PRISM) concept originated by General Electric, with the objective of obtaining a competitive fast reactor system with improved safety, enhanced plant licensability and simplified plant operations. The ALMR Programme was sponsored by US-DOE until fiscal year 1994-95 and is still under development with GE resources.

The solution analysed by GE assumes a metallic alloy of Uranium, Plutonium and Zirconium (U, Pu, Zr) as reference fuel, in both breeder and burner solutions. Innovative elements have been introduced into this system in order to enhance safety:

- the Gas Expansion Modules (GEMs) operate in the event of a loss of primary coolant flow. They are hollow assembly ducts, closed at the top and filled with over gas trapped in the upper part by the compression of the sodium: at normal operating pressures the sodium level is above the top of the active core. Reduction in pump head made the gas expand and expel most of the sodium; being located towards the border, this fact increases the neutron leakage and hence inserts a negative reactivity;
- the Ultimate Shut-down Device (USD) is a manually activated device with the means of bringing the reactor to cold subcritical conditions in the event of a complete failure of the normal scram system;
- the Reactor Vessel Auxiliary Coolant System (RVACS) is a passive system for the removal by natural convention for the shutdown heat for loss-of-cooling events.

Moreover, the PRISM concept presents a high flexibility that made possible important changes in the core with no effects on the rest of the plant and its performance.

ENEA OXIDE-FUEL CORE BASED ON PRISM MOD.B (840 MW_{th})

In the framework of a co-operation project between GE-USA and ENEA (PECOS Programme), an alternative core design was developed, to set up an optimised small-size oxide-fuelled core. In particular the tasks performed were:

- optimisation of the breeder solution and minor actinides (MA) transmutation;
- development of a burner core with a high plutonium consumption rate;
- development of a Passive Monitoring Device (PMD) to detect clad failure valid for both options;

- study of a methodology to identify the failed subassemblies, valid for both the options.
- seismic analysis.

As the present topic is the disposal of plutonium, we now refer only to the optimised burner option studied for the PRISM Mod. B (840 MW_{th}).

This optimisation is based on the break-even core geometric configuration (Fig. 1). In a break-even core the breeding ratio must be nearly equal to one, and to achieve a net consumption rate of plutonium and eventually of minor actinides (MA), fertile material must be removed from this configuration. Thanks to the flexibility of the PRISM design, only a few other modifications were necessary to transform the break-even core into a burner core, assuming the following requirements as guidelines:

- *Reactivity burn-up swing as small as possible:* it assures better loading factors and safety performances in the core. A high reactivity burn-up swing reduces the available CR-Scram reactivity north at the Beginning Of the Equilibrium Cycle (BOEC).
- *Negative, or low positive Na-void effect:* as it strongly related to safety, it provides for a benign closure in most accident situations.
- *Doppler effect:* notwithstanding the high fissile content it should be sufficiently negative;
- *Breeding ratio significantly less than one:* the Pu quantity-burned per cycle must be consistent with the requirement for Pu and MA economical burning.
- *Classical design solution:* the additional cost of further developments—for instance new material and/or fuels with a high Pu content (>35% for oxide fuel)—must be limited in order to compensate the additional cost of the fuel cycle-length reduction, intrinsic to the burner concept.

Past experience on oxide fuelled core designs as well as the above considerations account for the assumed design limit:

reactivity burn-up swing	$\leq$ 10$
peak fuel burn-up[(1)]	180 MWd/kg
peak fast fluence[(2)]	3.8 (+23) n/cm^2
peak linear power[(3)]	$\leq$ 300w/cm
Transuranic content	$\leq$ 35%

(1) to assure limited cladding strain and fuel pin integrity;
(2) to assure acceptable swelling behaviour in the clad and the core structural material (HT9);
(3) to assure limited fuel pin peak temperature;

A Pu content limited to about 35 per cent, as well as assuring the use of the present fuel fabrication and reprocessing

Figure 1
840 MWth PRISM Mod. B Metal Breakeven Core

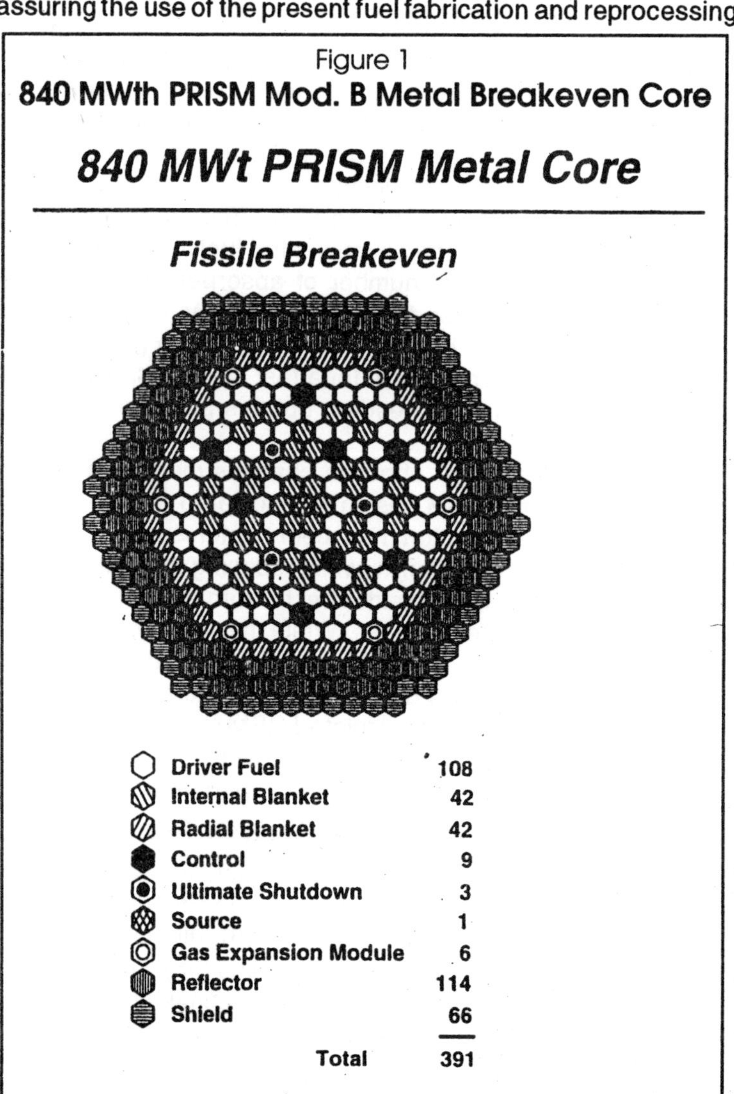

Driver Fuel	108
Internal Blanket	42
Radial Blanket	42
Control	9
Ultimate Shutdown	3
Source	1
Gas Expansion Module	6
Reflector	114
Shield	66
Total	391

technologies, provides well known irradiation performances. The chosen reference isotopic composition of the feed fuel was:

Pu-239/pu(*)	58.003%
U-235/U	2.0%

(*) 10 year old LWR discharged fuel: Pu/TRU = 89.26%

The core divided into two enrichment zones (i.e. different Pu content) and the inner one has the lowest enrichment. A higher enrichment towards the border was found to play a positive role between the place where sodium boiling can occur and favourable Na-void coefficients.

Moreover, to improve the Na-void effect in the inner fuel rings at low enrichment, the seven central fuel assemblies were substituted with an equal number of absorber assemblies. A possible interesting option for this central zone could be issue as a suitable place for burning long-lived fission products. As their transmutation is negligible for a fast neutron spectrum, an appropriate moderation is needed, but this problem has to be investigated.

Another of the most effective changes made possible by the flexibility of the PRISM design is associated with the GEMs. Their number and position has been optimised in order to maximise their reactivity worth, while reducing the mutual interference effects and the 'shield' effects from control rods, that greatly affect their effectiveness. While the metallic solution by GE assumes the GEMs are distributed all along the same ring, for the oxide solution and distribution over more than one ring was adopted. The required calculated GEM reactivity worth was estimated to be about 3$; this figure provides for a reasonable margin, even taking into account any experimental correction.

Neutronic and Thermohydraulic Analysis Results

Neutronic breakeven core was not modified with regard to the total number of assemblies (391) (i.e. the core diameter) and to the number of the control assemblies (9), ultimate shutdown assemblies (3) and radial shield assemblies (66). Different core configurations were considered for the optimisation analysis, varying one or more than one of the following parameters:

- active core height;
- fuel assembly number;
- fuel pin number per assembly;

- fuel assembly distribution in the two enrichment zones;
- batch number;
- cycle length.

In particular, three families were analysed (Fig. 2.1 and Fig. 2.2):

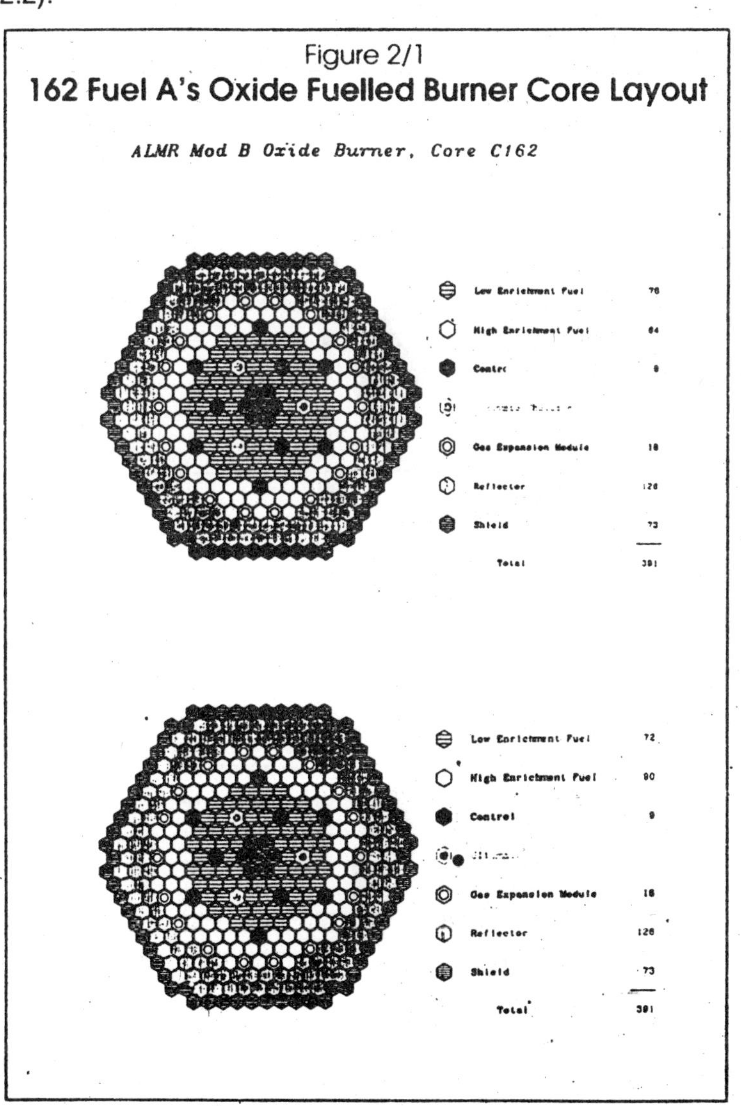

Figure 2/1

162 Fuel A's Oxide Fuelled Burner Core Layout

162 fuel assemblies } at 271 pins
180 fuel assemblies } at 271 pins

180 fuel assemblies at 331 pins

For each family different sub-families were studied, modifying the other parameters. Each configuration was character-

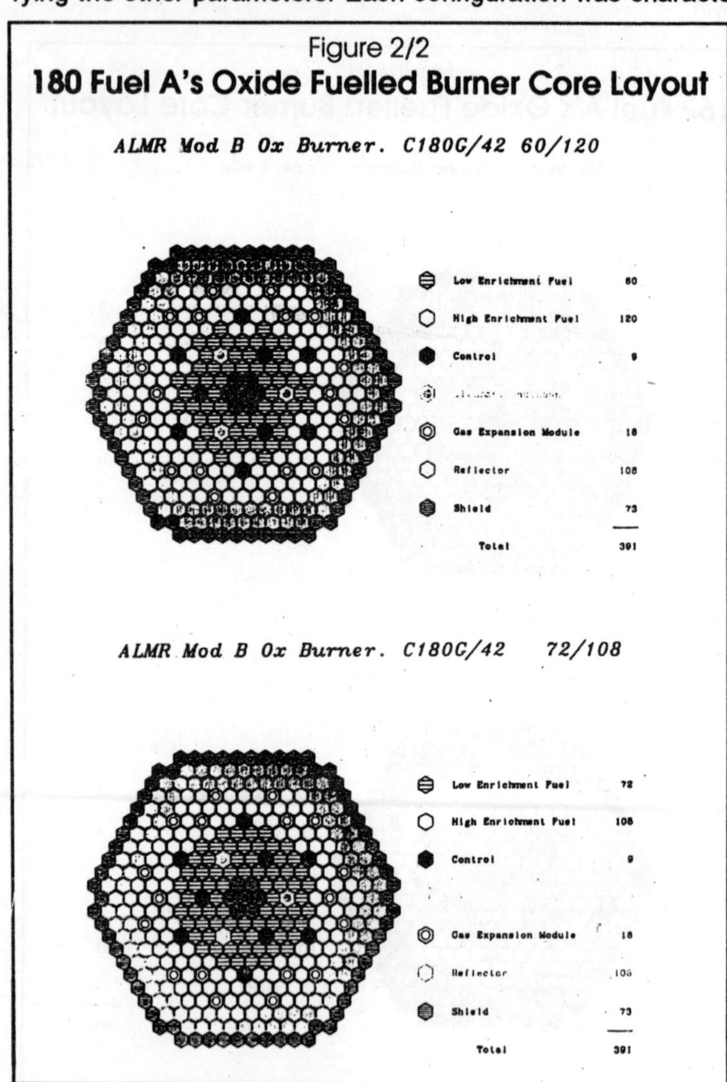

Figure 2/2
180 Fuel A's Oxide Fuelled Burner Core Layout

ised by its own neutronic performances and safety parameters, related to the design assumptions, and some results are given in Tab. 1, where only the most interesting options are shown.

Table 1 : Neutronic Performances for Some Options for the ALMR Prism Mod. B Burner Cores

Configuration	*C162–271*	*C180–331*	*C180–331*
Height (inc)	42	42	42
Pins number	271	331	331
Pellet Type	Full	Full	Hollow
Inner/Outer	72/90	72/108	72/108
Batch Number	5	6	4
Cycle Length (months)	12	8.5	8.5
GEM A's Number	18	18	18
TRU-Enrichment	28.50	35.38	39.33
Pu-Enrichment(1)	26.27	32.86	36.69
TRU Inventory (kg/year)	3314.3	3406.1	3262.8
Consumption Rate(2)			
kg/Core/year	102.0/97.3	138.8/137.4	158.8/158.1
kg/TWhe	45.2/43.0	61.4/59.9	70.3/70.0
% Inventory/year	3.078	4.075	.967
Breeding Ratio	0.629	0.504	0.429
BU-Swing ($)	–10.81	–10.12	–11.80
Peak BU (MWd/kg)	17.90	181.01	165.51
Aver. Bu(MWd/kg)	110.63	114.01	104.69
Peak Fast Fluence	3.05E23	2.24E23	1.76E23
Peak Linear Power	322.2	240.2	229.3
Doppler Coeff. ($)	–1.416	–1.408	/
GEM's worth ($)(3)	–2.848	–3.143	–3.462
CR's worth ($)	–35.05	–37.29	–40.70
Na-Void Worth ($)			
Fuel	–0.780	–2.542	–3.050
Total	–6.135	–8.815	–10.138
β-effective	0.00305	0.00290	/

(1) Pu/(Pu+U)

(2) TRU/Fissile Pu

(3) 18 GEMs fur all the configurations

For instance the 162 fuel A's/42 inch – 271 pin configuration appears as the one with highest peak linear power, peak fuel Burn-up and peak fast fluence. The thermohydraulic analysis on the peak fuel assemblies was performed for this configuration and the results are shown in Tab. 2 and Fig. 3. The design temperature constraint is satisfied in both Beginning and End of Life (BoL and EoL); Fig. 3 shows that the total flat-to-flat dilation for the peak fuel assembly is lower than the design limit.

Table 2 : T/H-Results for the C162G18/42-72/90 Peak Fuel Assembly

	BOL	*EOL*
Peak Pin Linear Power (W/cm):	323.4	247.7
Nominal Thermal Analysis (°C)		
- Outlet Temperature	523.8	487.1
- Peak Subchannel Temperature	552.8	509.5
- Peak Cladding Midwall Temp.	561.6	516.2
- Peak Fuel Surface Temperature	760.7	668.7
- Peak Fuel Centerline Temperature	2225.6	1789.9
+2 Sigma Thermal Analysis (°C)		
- Peak Subchannel Temperature	596.7	544.1
- Peak Cladding Midwall Temp.	613.7	554.9
- Peak Fuel Surface Temperature	837.9	726.1
- Peak Fuel Centerline Temperature	2453.6	1980.8
- Peak Fuel Surface Temp. (at SCRAM)	899.7	773.3
- Peak Fuel Centerline Temp. (at SCRAM)	2748.0	2191.1
Cladding Target Temperature: 616 °C		
Oxide Fuel Melt. Temperature: 2845 °C (approx.)		

Its attention is paid to the Pu-Consumption rate it will be noted that: on the one hand, in order to increase the plutonium consumption, the solutions with higher enrichment in Pu content would have been preferable; on the other hand higher Pu content poses more technology and safety problems and it is more difficult to satisfy design and economic constraints. Among the families studied, the 180 fuel assemblies-331 pins, full pellet, made possible high enrichments satisfying design constraints, though reducing the cycle length. This configuration was then defined as the 'reference configuration' and was used for further safety considerations.

Safety parameters for the reference configuration are shown in Tab. 3. Generally speaking, favourable safety parameters were found and an intrinsic reactivity feedback was assured.

Figure 3

Flat-to-Flat Dilation (inches) for the Peak Fue A's of the C162/42 - 72/90 Solution

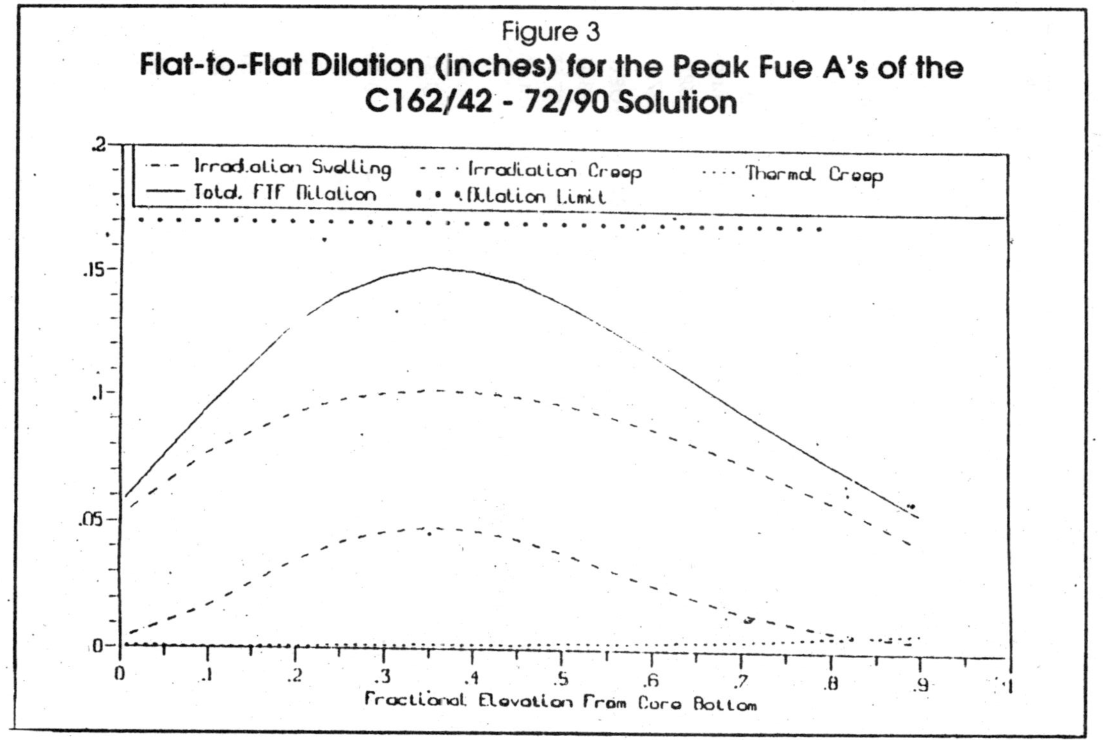

Table 3 : Neutronic Results and Performances for the Reference Solution

Core Configuration	*C180G18/42-B-331*	
Core Height (inc/cm)	42 / 106.68	
# of Fuel A's / # of Pins per A's	180 / 331	
Inner/Outer	72 108	
Batches Number / Cycle Length (months)	6 / 8.5	
TRU-Enrichment (%)	33.17 / 36.86	
Consumption Rate (kg/Core.year)[1]	138.8 / 135.3	
% Inventory/year	*4.075*	
Breeding Ratio	*0.504*	
BU-Swing ($)	*–10.12*	
Peak / Aver. Burnup (MWd/kg)	*181.00 / 114.01*	
Peak / Aver. Fast Fluence (10^{23} n/cm^2)	*2.52 / 1.52*	
Peak Lin. Power (W/cm; BOEC/EOEC)	*240.2 / 227.7*	
	BOEC	**EOEC**
GEM's Worth ($)	–3.114	–3.171
CR's Worth ($)	–36.63	–37.95
USD's Worth ($)	–10.70	–11.10
Uniform Axial Expansion		
Net Effect	–0.22479	–0.22552
Geometry effect	0.21461	0.21487
Uniform Radial Expansion		
Net effect	–0.66454	–0.66566
Geometry Effect	0.21426	0.21512
Doppler Coefficients		
Inner Fuel	–0.00178	–0.00183
Outer Fuel	–0.00224	–0.00231
Fuel Density Coefficients	0.41010	0.42746
Structural Density Coefficients	–0.00394	–0.00609
Sodium Density Coefficients	0.01116	0.01054
Sodium Void Reactivity ($)		
Inner Fuel	–0.73394	–0.62633
Outer Fuel	–1.89110	–1.83268
Others	–6.18756	–6.35789
Total	–8.81260	–8.81690
Total β-effective	2.9084E-3	2.8887E-3
Prompt Gen. Time ? (sec)	4.8537E-7	5.0901E-7
Delayed Neut. Lifetime ? (sec)	3.4303E-2	3.4006E-2

(1) TRU / Fissile Pu

- Net axial expansion negative coefficients provide a significant negative feedback during a transient event associated with axial fuel expansion as the temperature rises. In the same way, net radial negative coefficients provide a negative feedback during thermal expansion of the grid plate depending on sodium inlet temperature.
- The Doppler coefficients (TdK/dT) were reasonably good considering the high fissile content and the absence from the burner core of fertile assemblies and axial blanket zones.
- The density coefficients are expressed in term of fractional variation for core reactivity, in the presence of a change of density. Because of the calculation methodology, a positive value implies a negative feedback as material density decreases with increasing temperature; conversely a negative value provides a positive feedback. The high value of the fuel density coefficient and its high temperature gradient assure a large negative feedback.
- In particular negative sodium void effects were achieved, as shown in Fig. 4 and Fig. 5. Fig. 4 shows the Na-void worth per fuel assembly, averaged over the ring, at the beginning and at the end of the fuel cycle. The negative effect in the inner rings is due to the presence of the absorber zone at the centre of the core. Fig. 5 shows the Na-void worth distribution per assembly for the reference configuration and highlights the achievement of negative values for all assemblies but a few with low-positive values. Negative void effect increases (as absolute value) as the sodium volume fraction increases—for instance reducing the pin diameter—or augmenting the fissile content, while keeping other volume fractions constant. Negative void effect decreases as fission products increase (i.e. at EOEC); moreover it also decreases introducing MA into the core.
- Regard to the β-effective value, it may be noted that it decreases as the fissile content increases. Though a small value implies a reduction of the prompt-criticality martin, safety analysis of accidental situations always shows a benign closure, thanks to all the other parameters.

It is worth nothing that locating GEM assemblies in high neutronic importance zones it make possible to have high GEM reactivity worths with a relatively small number of assemblies (18), of which 6 distributed in the seventh ring and 12 in the ninth.

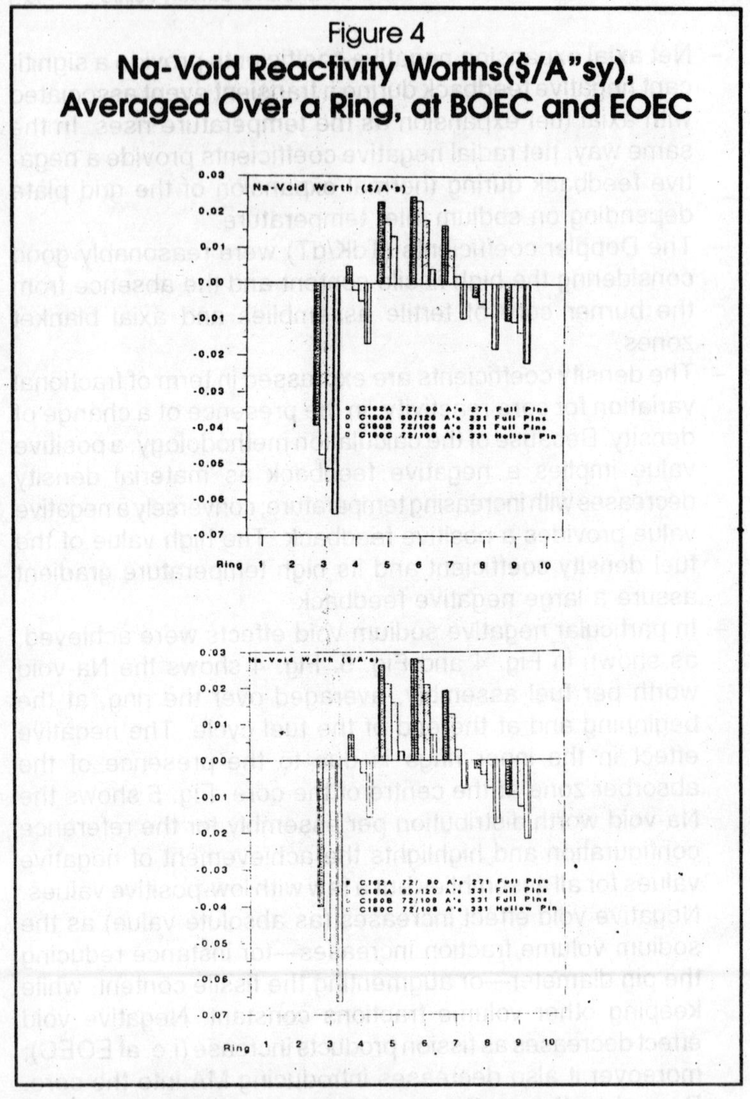

Figure 4
Na-Void Reactivity Worths($/A"sy), Averaged Over a Ring, at BOEC and EOEC

Safety Analysis

ALMR safety design goal is to accommodate Anticiate Transient Without Scram (STWS) events, in particular:

- Unprotected Loss of Flow (ULOF)
- Unprotected Transient Over Power (UTOP); and

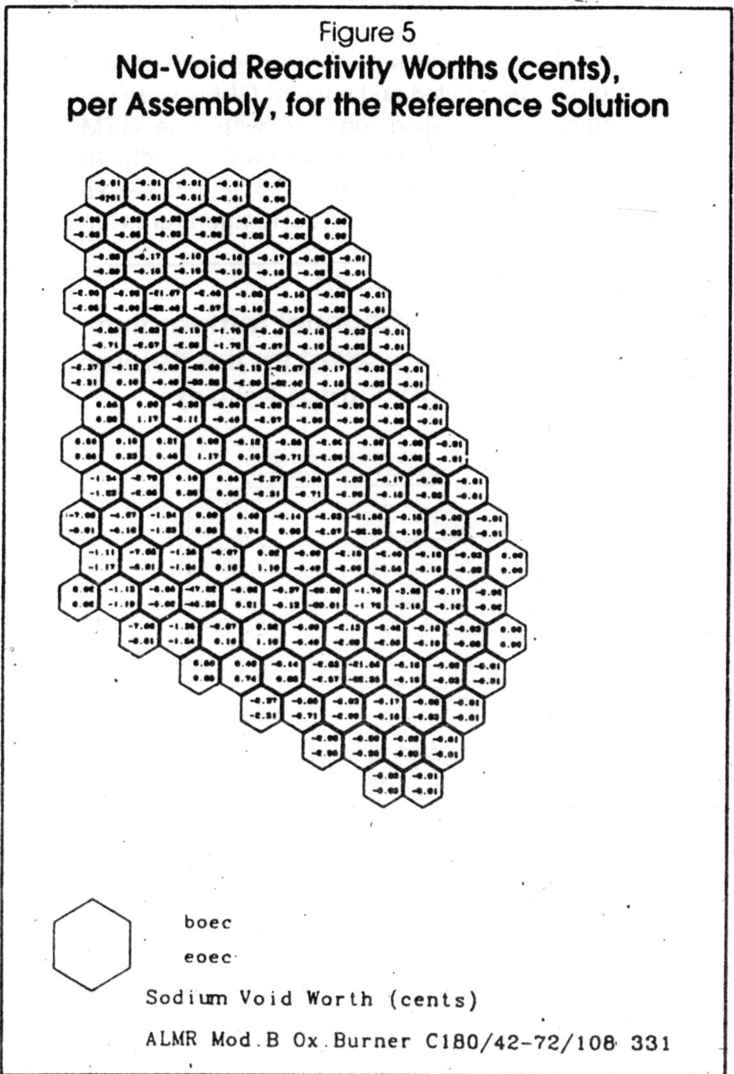

Figure 5
Na-Void Reactivity Worths (cents), per Assembly, for the Reference Solution

– Unprotected Loss Of Heat Sink (ULOHS).

Analysis of ATWS events was performed only for the reference solution at BOEC, the events being initiated with the reactor at its nominal power and the conservative assumption at the non-availability of the passive RVACS.

- The ULOF analysis was performed for two cases: all the four pumps were failed and none of GEMs were failed; the four pumps were failed together with nine GEMs. The GEMs reactivity worth considered was the effective one, corrected by experimental factor ??? Fig. 6.1–2 show the calculated reactivities and the fuel temperatures for the hot

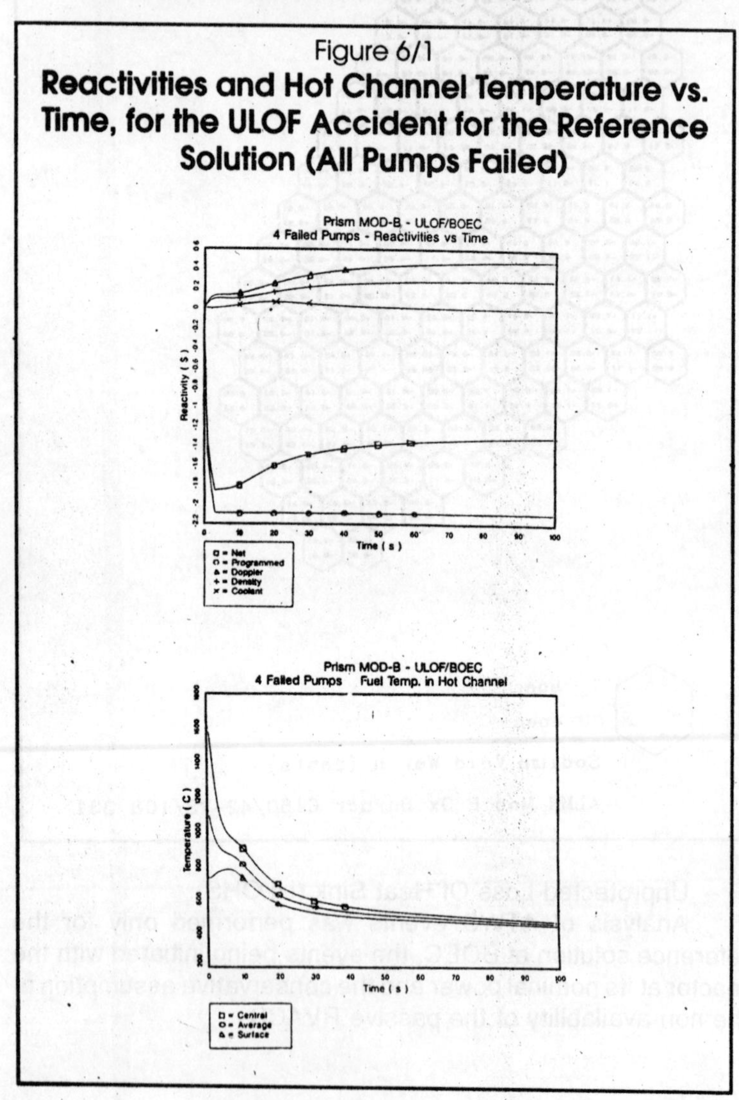

Figure 6/1

Reactivities and Hot Channel Temperature vs. Time, for the ULOF Accident for the Reference Solution (All Pumps Failed)

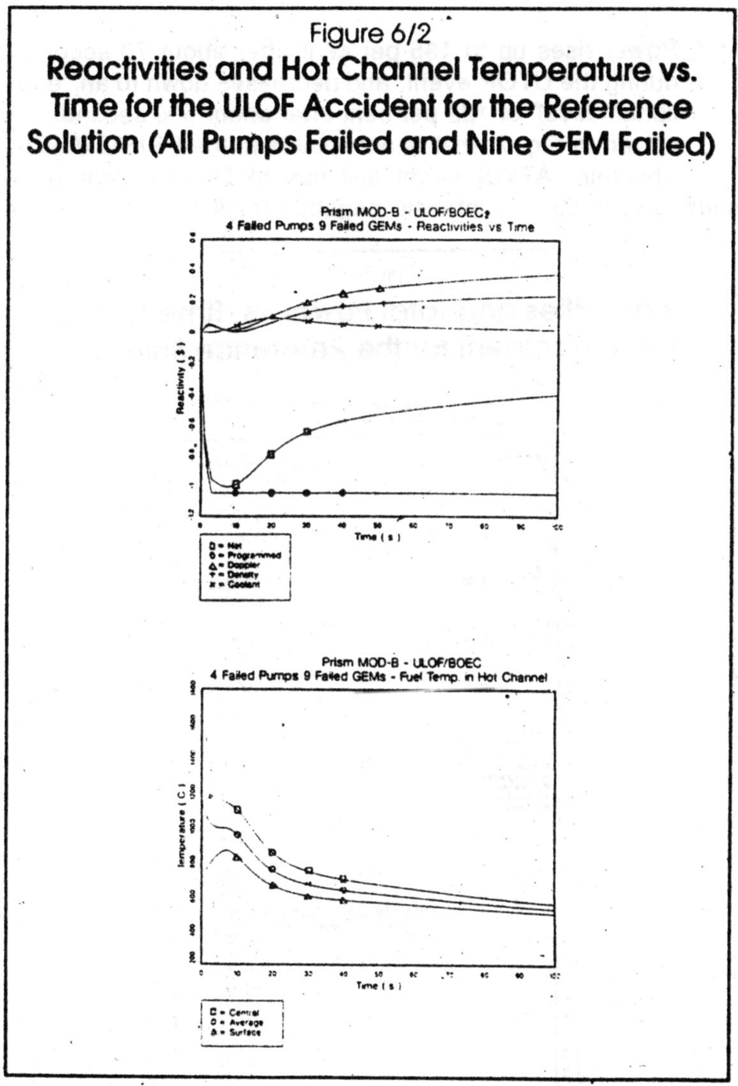

Figure 6/2
Reactivities and Hot Channel Temperature vs. Time for the ULOF Accident for the Reference Solution (All Pumps Failed and Nine GEM Failed)

channel: after about 100 seconds the net reactivity and fuel temperature reach stability conditions for both cases.

- The UTOP analysis was performed with the insertion of a positive reactivity up to 5.0$: this value was determined by the mechanical device that controls the CR's withdrawal.

Power rises up to 135 per cent after about 70 seconds, during the UTOP event, and decreases down to an "equilibrium level" of 122 per cent after about 500 seconds, as shown in Fig. 7, that is acceptable in the PRISM design.

The third ATWS event still has to be investigated. In summary, thanks to intrinsic negative feedback and passive

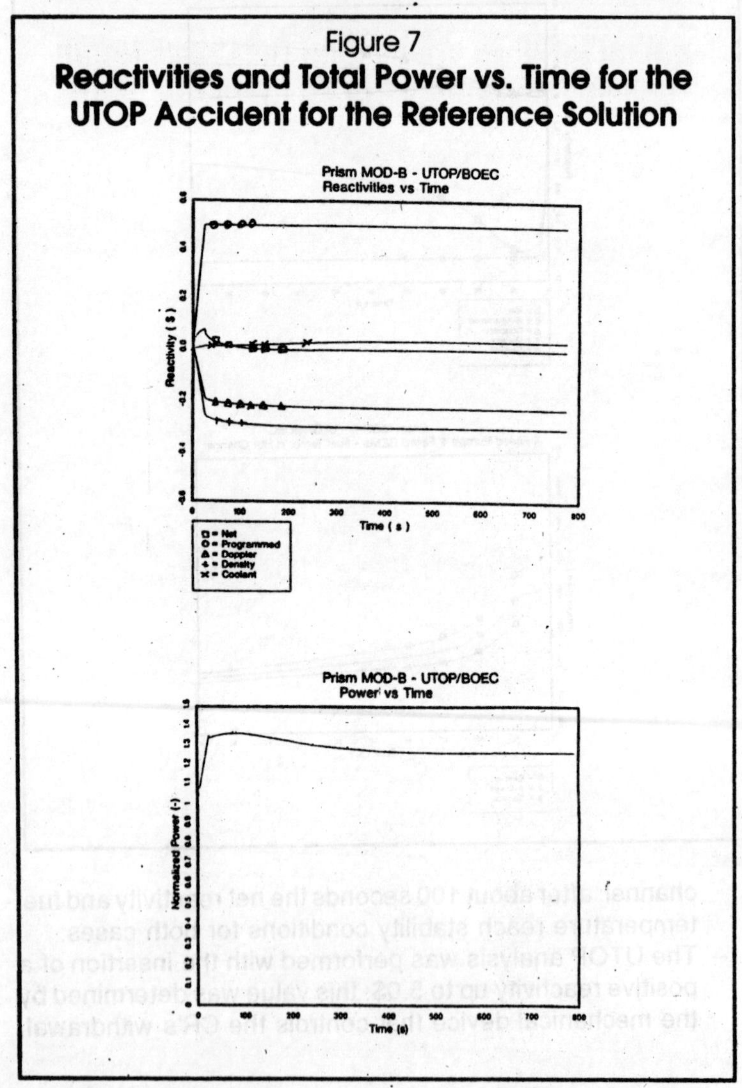

Figure 7
Reactivities and Total Power vs. Time for the UTOP Accident for the Reference Solution

safety devices, the analysed ATWS events were proved to be "controlled" and satisfying safety constraints.

Plutonium Consumption and Conclusions

Plutonium consumption performances for the reference core are commarised in Tab. 4. A Pu-consumption rate of about 60 kgPu/TWhe was found, as expected with Pu enrichments of about 35 per cent, equivalent to about 140 kg/core/year.

Higher Pu-consumption are could be possible introducing other modifications, for example increasing the Pu-content using

Table 4 : Fuel Inventory and Mass Balance (in kg) for the Reference Solution

Core Configuration	**C180G18/42-B-331**	
Batches Number	6	
Cycle Length (onths)	8.5	
TRU-Enrichment (%)	33.17 / 36.86	
Average Enrichmnt (%)		
TRU	35.38	
Pu/(Pu+U)	32.86	
TRU-Inventory (kg/Core)	3406.1	
Consumption Rate[(1)]		
kg/Core/year	138.8 / 137.4	
kg / TWhe	61.4 / 59.9	
% TRU-Inventory/year	4.07	
	BOEC	**EOEC**
Fuel Inventory (kg)		
Pu238	72.7	81.0
Pu239	1682.0	1603.0
Pu241	236.7	219.7
Total Pu	3037.4	2949.9
NP237	141.8	131.9
Am241	171.0	166.3
Am243	34.9	35.9
Total TRU	3406.1	3307.8
U 235	10.5	9.5
U 238	6466.5	6378.6
Total U	6477.6	6388.9
Total F.P.	489.0	675.5

(1) TRU / Fissile Pu

hollow pellets as shown in Tab. 1. Unfortunately this, or any other solution, introduces additional costs, apart from worsening safety performances (Doppler).

Moreover a burner core introduces economic cost penalties anyway compared to breakeven or breeder cores. While previous neutronic performances remain in changed, key parameters for a comparison could be:

– amount of TRU that must be processed each year: this is greater for a burner core than for a comparable breakeven core or even more for a breeder, because of the breeding ratio factors and the peak-to-average fuel burn-up ratios:
– cycle length: it is reduced for a burner core;
– reshuffling/reloading strategy and number of fuel assemblies involved; it penalises burner cores because of the reduced plant availability, due to the additional down time each year.

Nevertheless the acceptability of an over-cost in presence of a high Pu-consumption rate has to be valued weighting political issues rather than economic and/or technical ones.

The most interesting outcome of all the analyses reported in this paper is the conclusion that an optimised small-size oxide-fuelled burner core can provide noteworthy safety performances, including negative Na-void effects and reasonable Doppler, associated with high Pu-consumption rates, about 60 kgPu/TWhe. Moreover, while satisfying all technological, safety and economic design constraints, this system could be set up even within a "Classical Design Solution", thus representing a practical option with immediate applications.

3. STUDY OF A METHODOLOGY TO IDENTIFY A FAILED ASSEMBLY

The study of a methodology to identify a failed assembly is presented in detail as it is an important safety and economic issue; moreover this methodology has a general value that makes it applicable to a number of fields.

A Passive Monitoring Device (PMD) to detect clad failure was theoretically studied and its prototype was tested at ENEA experimental laboratories. This system can be briefly described as cylinder acting as a delay-line, whose task is to reduce the strong noise produced by Na activation and affecting the detection of fission products.

The PDM is purely addressed to safety and makes possible the mere detection of failure, while an economic approach would require more detailed information: i.e. the identification of the failed assembly. The basic idea of this identification system was to attach a kind of blueprint to each assembly. This could be possible by inserting a "cocktail" of tag gases into all the pins of each assembly and detecting their loss by means of their daughter activity. The tag gas combination and their mutual rations will be different and unique for each assembly, so that the spectrometric analysis of the energy and number of gamma emitted by the escaped daughter can identify the assembly concerned.

The choice of tag gases has been restricted to a few materials, the characteristics of which meet the following requirements:

- parents and daughters have to be distinct from cover gas isotopes or fission products;
- parents and daughters have not to be chemically reactive with any reactor components, i.e. they have to be noble gases;
- parent activation cross-section on the one hand has to be low enough to consider the parent stable or almost stable compared to the subassembly life, but on the other hand it has to be high enough to produce a significant quantity of daughters;
- daughter decay constant has to be low enough to allow a significant equilibrium concentration, but high enough to present a detectable activity.

Four gases have been selected—Krypton-78, Xenon-124, 126 and 128:

PARENTS			DAUGHTERS		
	fast σ (barns)	half-life(*) (years)		λ (s^{-1})	half-life
Kr-78	.007	3100	Kr-79	5.5E-6	1.455 d
Xe-124	4	5.5	Xe-125	1.13E-5	17 h
Xe-126	.05	440	Xe-127	2.2E-7	36.41 d
Xe-128	.01	2200	Xe-129m	9.0E-7	8.89 d

(*) Assuming the maximum flux reactor

An original methodology was developed was developed at ENEA to address this problem in a general way. The problem can be divided in four stages:

1) definition of the number and the type of parents;
2) definition of the critical scenarios that rule the minimal usable quantity of gas;
3) definition of the "cocktails" of tag gases so that:
   - there is a sufficient number of tags, one per assembly;
   - they are identifiable:
   - the required gas volume has to be consistent with that available in the pins;
4) taking the radial flux distribution properly into account.

1) The first stage has already been defined, but the method does not depend on these particular gases, and if a choice of non-activated gases could be made, the methods would be even simpler.

2) Restant and End of Life (EOL) are critical times; in the latter, the parent concentration is the smallest; at the restart the daughter's concentration is only beginning to build-up. Taking into account the parameters influencing both scenarios:

EOL {
- S detector sensibility (10Bq)
- $\varnothing$ max flux (for depletion) (10^{15} n/cm^{2}s)
- T lifetime (for depletion) (10y)
- τ breaching - detection delay (10h)
- D gas fraction detected (10^{-5})

} RESTART (all of the above and:)
- $\varnothing_R$ restart flux (10^{15}n/cm^{2}s)
- T_{BO} black-out time (1d)

An acceptable counting time sets the detector sensibility; the other values, though reasonable, could be modified (for example after testing the delay between the moment the gas slips out of the failed assembly and the moment it is detected or the fraction that actually reaches the detector) without affecting the methods. Their use is to define the minimum quantities of gas to be inserted in one pin in order to have a measurable signal at the detector of the same intensity. These minimum quantities have been named 'quanta' and for the previous parameters are as follows:

Parent gas	quanta volume (normal cm^3)
Kr-78	3.3
Xe-124	0.03
Xe-126	0.4
Xe-128	1.9

3) The identification of the failed assembly is made possible by the analysis of the cocktail of gases reaching the detector, that is unique for each assembly. Nevertheless it is not possible to measure the absolute quantities because of the uncertainties of the migration model and then on the effective fraction of gas reaching the detector. The only valid measurement could be made for the mutual ratio of gas quantities.

Uncertainties

As a background noise is inevitable, the greater the difference among cocktails, the more certain the regression from the detected signal to the correct cocktail and hence to the failed assembly. In order to avoid huge quantities of tag gases inserted into the pins, it is of primary importance to identify the uncertainties affecting the attribution to a given assembly and then the minimum mutual quantities of gas required. Assuming conventional uncertainties on the independent variables within a confidence interval of 2σ, the uncertainty affecting the calculated gas ratios are:

Scenario		Factor	Description		Scenario
RESTART	22%	R_o	quantities of gas charged into the pins	5%	EOL
		R_e	parent gas activation cross-section	15%	
		R_d	parent gas depletion	26%	
		R_a	daughter decay during flight toward the detector	4%	
	7%	R_t	R_s instant of the gas loss after Restart while daughter concentration is not at equilibrium		

The global uncertainties for the two scenarios were:

RESTART	28%
EOL	31%

The latter was the reference value.

The next step is how to create different cocktails that could be identified unambiguously, notwithstanding the uncertainties.

Let us define:

$$d = \frac{1+u}{1-u}$$, u = uncertainty (for u = 31% → d = 1.9)

Hence two close binary cocktails are recognised as different if their ratio changes by a factor d.

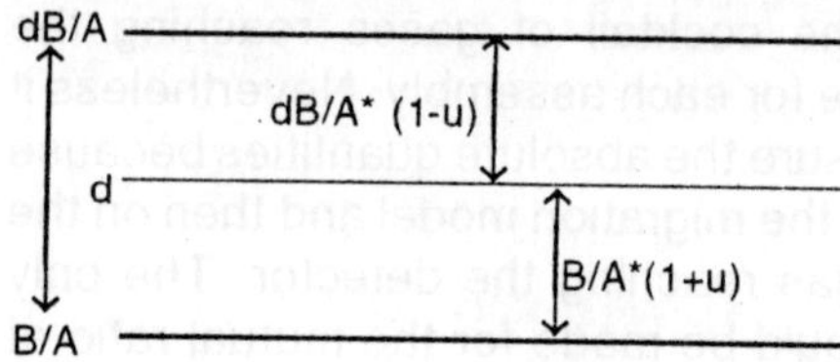

In the example there are two cocktails of gases A and B. In the second one the gas quantity B was increased by d. To identify the cocktail, there must be no overlapping between B/A*(1+u) and dB/A*(1–u)

$$\rightarrow d = \frac{1+u}{1-u}$$

The parameter *d* can be called the 'uncertainty diameter' and represents the minimum difference between two gas ratios necessary to differentiate them. As seen before, the minimum quantity of parent gases to give rise to the same signal at the detector was defined as quantum. In order to recognise a cocktail as different, the gas quantity increase has to take place by steps equal to *d*, so that the n_{th} variation for a given gas, the total gas quantity is the (quantum*d^n).

With a little imagination a simplified way to find and represent these quantities was found.

PARENT GAS QUANTITY IN A COCKTAIL

A cocktail can be defined with its own blueprint (BP) as:

$$BP\ (x_1, x_2, x_3, x_4)$$

where

$$x_i = \log_d \frac{Q_{ai}}{Q_{qi}}$$

Q_{ai} = parent gas quantity loaded into the pin

Q_{qi} = quantum, previously defined as the minimum quantity of a given parent gas

$d = \frac{1+u}{1-u}$ "uncertainty diameter" u = uncertainty

When the loaded quantity is equal to the quantum for that gas it is x=0 and because the increase is by step equal to *d*, the next minimum quantity has to be *d* times Q_{qi} (x=1) and for $Q_{ai} = d^n Q_{qi}$ it is x=n. BP (x_1, x_2, x_3, x_4) is then a sequence of full, positive numbers. For economic reasons at least one gas has to be present as a quantum, so that at least one number of the sequence is a zero.

For instance BP (2, 0, 1, 3) means that the concentration are exactly (or proportional to);

1st gas (Kr-78): quantum × d^2 = 3.3 ncm^3 × 3.61 = 11.91 ncm^3
2nd gas (Xe-124): quantum × d^0= 0.03 ncm^3 × 1 = 0.03 ncm^3
3rd gas (Xe-126): quantum × d^1 = 0.4 ncm^3 × 1.9 = 0.76 ncm^3
4th gas (Xe-128): quantum × d^3 = 1.9 ncm^3 × 6.86 = 13.03 ncm^3

and it is completely equivalent and indistinguishable, for instance, from BP (3,1,2,4) because only mutual ratios are considered.

GRAPHICAL REPRESENTATION

The xi can be viewed as spatial coordinates in a 4-dimension hyperspace, whose axes are the four tag gases and where any cocktail BP (x_1, x_2, x_3, x_4) is a point. Since, as already stated, it is more convenient to have at least one x_i=0, the permitted gas combinations will lie in the four positive 3D volumes bounding 1/16 of the hyperspace.

Because the allowed quantities have to increase by step = d, the correspondent increase in x will be Δx=1. Therefore uncertainties can be represented as pseudo-spheres around the BP(x_1, x_2, x_3, x_4) point of diameter equal to one, regardless of the position of the representative point. This fact makes possible the creation of a regular lattice that defines the permitted positions and hence the permitted gas combinations. In an easier 3D representation (only 3 tag gases) the point will lie on the three planes delimiting the positive 1/8 volume, as shown in

Fig. 8. The uncertainties will be a pseudo-circle of diameter=1, defining the permitted gas combinations in a square lattice equal to 1.

If the cocktails are very different from each other it could be possible to identify them overcoming even higher uncertainties. For example the spatial distance between BP(1, 1, 0) and

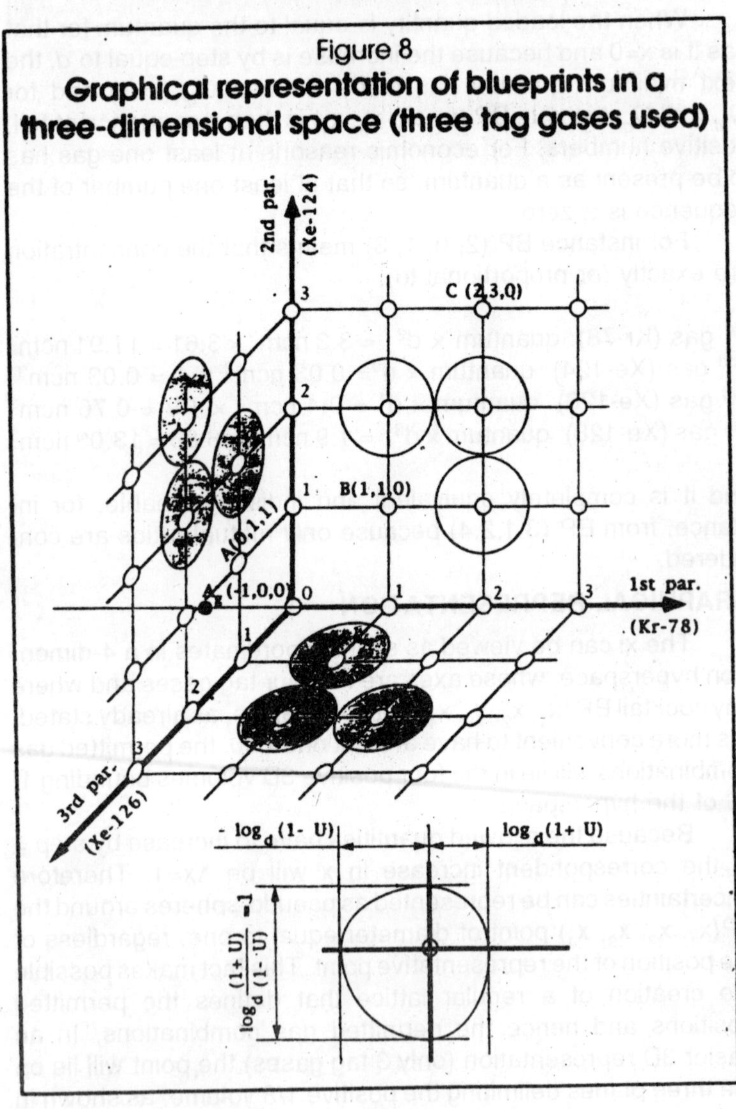

BP(2, 1, 0) will be 1 (d=1.9, uncertainty equal to 31%), while the distance from the first point and BP (2,3,0) will simply be the conventional geometric distance (in that case 2.24, equivalent to an uncertainty of

$$1.9^{2.24} = \frac{1+u}{1-u} = 62\%).$$

The choice of possible cocktails is flexible and can be made on the basis of volume or cost considerations. Fig. 9 shows an optimisation based on parent gas volumes.

4) Flux and spectrum strongly influence many parameters (parent depletion, daughter production...) and the choice of the assumed uncertainty of 10 per cent cannot take into account by any means the strong radial distribution variations. On the other hand, the choice of a higher uncertainty on the flux will increase the "uncertainty diameter" therefore dramatically increasing the required gas volumes.

A better solution could be the division of the core into different zones, where the flux has a limited spread. For each zone a peculiar "family" of blueprints could be created with the methodology already described. The number of families, and hence the number of zones in which the core could be divided, depends on the number of combinations allowed. Having 4 tag gases we can create:

- 6 different couples AB, AC, AD, BC, BD, CD
- 4 different triplets ABC, ABD, ACD, BCD
- one quartet ABCD

The core was then divided up to 11 zones, where the flux uncertainty was combined with its spread and a total flux uncertainty of 20 per cent was assumed. Considering only the number of cocktails that can be created in a volume of 15 ncm^3, the number of the permitted blueprints is:

	uncertainty diameter d	*blueprint number*
AB	1.4	10
AC	1.4	15
AD	1.6	16
BC	1.4	17
BD	1.6	17
CD	1.6	21
ABC	1.4	24
ABD	1.6	36
ACD	1.6	2
BCD	1.6	97
Total		*325*

Figure 9

Optimisation of parent gas volumes in a three-dimensional space (three tag gases used)

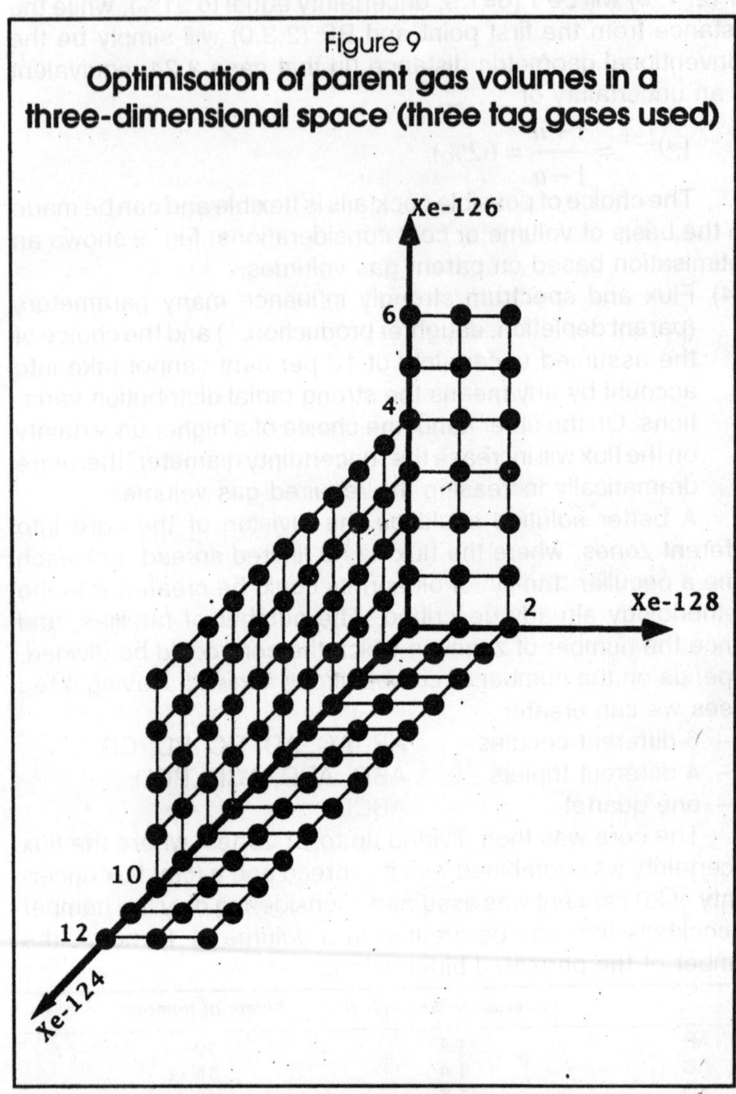

The total number, only using couples and triples, is 325. There are enough blueprints to allow a selection: the only constraint is the number of families or, in other words, the maximum number of flux zones.

Processing the result of the detector system by a simple dedicated Monte Carlo code, the effectiveness of the method can be significantly increased. The code must analyse the detector signal and indicate all the possible cocktails involved and associate to them the probability of having a correct attribution. For instance the code answer could be:

95% probability that the failure occurred in the assembly characterised by BP(0, 1, 3, 2);

1% in the assembly characterised by BP (0, 2, 3, 2);

1% in the assembly characterised by BP (0, 1, 2, 3);

0.5%

0.001% probability of having a simultaneous failure in the assemblies characterised by BP (0, 1, 3, 1) and BP (0, 1, 3, 3); etc.

Ad hoc experimental measures could make possible an even greater efficiency and smaller gas quantities.

BIBLIOGRAPHY

1. C.L. Cockey *'Actinide Transmutation in the Advanced Liquid Metal React (ALMR)'* IAEA/IWGFR Specialist Meeting on Use of Fast reactors for Actinide Transmutation, Onbinsk, Russian Federation, September 22–24 1992
2. G. Glinatsis *'Oxide-Fuelled Burner Optimised ALMR Cores'* "ENEA-FT-FBC-00001/0 Technical Report Bologna, Italy November 16, 1994.
3. A.J. Lipps, G. Glinatsis *'Actinide Transmutation and Pu Burning in the PRISM Modular Reactor'* Second International CAPRA-Seminar Karlsruhe, September 21-22, 1994.
4. G. Glinatsis *'Optimised Oxide-Fuelled ALMR Burner Cores"* Second International CAPRA-Seminar Karlsruhe, September 21-22, 1994.
5. C. Artioli *'A methodology to identify a failed assembly'* ENEA, forthcoming.

Address:

GEORGIOS GLINATSIS

Franca Padoani

ENEA

40129 Bologna

Italy

62

Getting the Plutonium Disposition Job Done: the Concept of a Joint-Venture Disposition Enterprise Financed by Additional Sales of Highly Enriched Uranium

MATTHEW BUNN

There is a growing international consensus that the hundreds of tons of excess weapons material left over from four decades of Cold War arms competition pose, in the words of the U.S. National Academy of Sciences, "a clear and present danger" to international security. Ensuring secure control over weapon-usable nuclear materials, an action plan to stop nuclear materials, an action plan to stop nuclear smuggling, and international cooperation in disposition of excess weapons plutonium will all be central items on the security agenda for the P-8 nuclear summit in Moscow next month.

In this paper, I will attempt to outline a concept which has the potential to provide both the substantial financing needed for plutonium disposition and the stable long-term management structure required to implement the effort. But first, it is important to address the earlier steps in the process.

URGENT FIRST STEPS: SECURE, MONITORED STORAGE

Despite the growing international sense of the urgency of fissile material disposition, the fact is that no matter what option we choose, the job won't be finished for decades to come. Therefore, ensuring safe and secure storage in the meantime along with some from of international monitoring to ensure that the arms reduction now underway are truly irreversible—is by far the most urgent item on the fissile material agenda.

The United States, the states of the former Soviet Union,

and other countries are engaged in a far-reaching cooperative programme to modernise systems for security and accounting of weapons-usable materials in the former Soviet Union. The United States and Russia are also cooperating to build a modern, safe, and secure storage facility for fissile materials from dismantled warheads at Mayak. Nothing could be a more critical investment in international security than ensuring that the essential ingredients of nuclear weapons do not fall into the hands of terrorists or rogue states. I urge everyone in this room to consider whether their own government is doing enough to solve this critical problem—and if not, to urgently recommend that additional steps be taken.

The problem of placing these materials under some form of international monitoring is also urgent, but there, less progress has been made. The United States has announced that over 200 tons of its weapons-usable material is excess to its defense requirements—roughly 175 tons of highly-enriched uranium (HEU), 38 tons of weapons-grade plutonium, and 13 tons of fuel-grade plutonium—and has begun placing this material under International Atomic Energy Agency (IAEA) safeguards. We have encouraged Russia to do the same; while there have been a few encouraging statements on the Russian side, no concrete action has been taken so far. Moreover, much of the U.S. excess material (and probably the Russian material as well) is in the form of weapons components, to which traditional safeguards techniques cannot be applied without giving away information that would contribute to proliferation—and much of the rest of the excess material is difficult—to-measure scrap. Thus, before the majority of this material can come under IAEA safeguards, either greatly modified safeguards approaches will have to be developed or materials will have to be processed to inspectable, unclassified forms, which will take a considerable time.

To fill that gap, the United States and Russia have agreed in principle to undertake a bilateral programme of mutual, reciprocal inspections of the plutonium and HEU from dismantled weapons—including inspection measures that would reveal a limited amount of classified weapons information whose exchange between advanced nuclear weapon states does not pose a security risk to either side. Our Presidents have also agreed to an unprecedented exchange of data other for the first time how may nuclear warheads and how much plutonium and HEU we really have, and to consider other measures to confirm the dismantlement of nuclear warheads. Unfortunately, however, last fall, when an agreement on exchange and protection of classified nuclear information that would have provided the legal basis for these measures was almost complete, the Russian side suspended the negotiations, and the talks have not yet resumed.

Let me emphasise again: these measures to ensure secure, monitored storage—preventing proliferation and ensuring the irreversibility of arms reduction—are more urgent in the near term than is plutonium disposition, the principal subject of this conference.

FISSILE MATERIAL DISPOSITION: ALSO URGENT

Nevertheless, it is also critical to move as quickly as practical to actual disposition of excess fissile materials, transforming them into forms that no longer pose urgent security risks. As Russian Minister of Atomic Energy Victor Mikhailov has said, disarmament will only be truly real when we have gotten rid of the materials from which the weapons could be rebuilt. Precisely because it will take a long time, it is urgent to begin: I am reminded of the French marshal, who, when told by his gardener that it would take thirty years for trees to grow along the boulevard leading into his estate, told the gardener he had better start planting tonight, rather than waiting until tomorrow. I am confident that there will be consensus on the urgency of moving forward on fissile material disposition at the nuclear summit next month.

As you all know, HEU disposition is relatively straightforward in a technical sense, an in fact is already being accomplished: the United States and Russia have agreed that the United States will purchase 500 tons of HEU from dismantled Russian weapons over the next 20 years, blended to proliferation resistant low-enriched reactor fuel. In one stroke, this agreement reduces the stockpile of weapons-usable material at risk of theft, provides a direct financial incentive for continued weapons dismantlement, helps ensure irreversibility of the ongoing arms reductions, pumps needed hard currency into the Russian economy, and provides a valuable commercial product to the U.S.—all without requiring direct Federal subsidies from the U.S. budget.

I can report that this agreement is now working-six tons of HEU was blended, delivered, and paid for last year, that figure is increasing to 12 tons this year, and we are now finishing the specific transparency arrangements needed to assure that the U.S. is buying what it thinks it is, and using it only for peaceful purposes. Three weeks ago in Moscow, the business manager of the Ministry of Atomic Energy described this arrangement to me as 'a shining example of U.S.-Russian cooperation". And the United States is planning to do much the same with its own excess HEU.

The plutonium disposition job is much more difficult-in part because in today's market, fuel made from plutonium cannot compete with cheap uranium fuels, even if the plutonium is "free".

While plutonium disposition is a harder job, technologies do exist that can get that job done over the next few decades, while meeting the requirements of non proliferation, arms reduction, environment, safety, and health. Existing light-water of CANDU reactors can burn plutonium as MOX fuel; existing technologies for immobilisation of liquid high-level wastes could also incorporate plutonium, making it roughly equally proliferation resistant. Virtually every week there is another proposal for some new technology to apply to this mission—but all these ideas share the severe drawback of requiring a substantial R&D programme to prove them out definitively, delaying the day when disposition of this dangerous weapons material can begin.

THE KEY QUESTIONS: FINANCING AND MANAGEMENT

Since the technologies needed to do the job already exist, and new technologies would only delay matters, technology is not the key issue. The key issues revolve around how plutonium disposition can be managed and financed on the immense scale required.

This is a big job—tens of tons of material, hundreds of millions of billions of dollars in required investment, decades to complete. It also involves the unpopular nuclear industry, and a toxic weapons—usable material that the public in many countries has come to fear. In all likelihood, virtually any of the opinions will face protests by hundreds of thousands of people, not to mention lawsuits and other legal interventions. No one country will be able to get the job done alone: to pull it off will require far-reaching international cooperation, considerable creativity, and an immense investment of political will. Managing a sustainable programme lasting decades in such a controversial area will require institutional arrangements of unprecedented durability.

Financing the operation is a particularly difficult problem. For an plausible option, initial investments of hundreds of millions or even billions of dollars will be required. In the United States, it is likely—though by no means certain—that Congress can be convinced to fund these substantial investments, if—and only if—a parallel disposition programme is underway in Russia. If Russia simply leaves its excess plutonium in storage in readily weapons-usable form, congress simply will not fund a U.S. plutonium disposition programme. Even if Russia is moving forward, ensuring that the financing from Congress is sufficiently stable over the long term to accomplish the mission will he a difficult challenge. A U.S.-Russian agreement calling for reduction in stockpiles of fissile materials—as will be discussed later by other speakers—could be an important factor in ensuring stable implementation over time.

But with Russia's current economic circumstances, the Russian government is not likely to be able to finance the initial capital investments needed for plutonium disposition anytime in the foreseeable future. Thus, if the plutonium job is to get done, either the international community will have to subsidise it, or some creative financing approach—such as some from of counter trade, where the operation would be financed by exports of some other valuable commodity—will be needed.

What kind of money are we talking about? That depends in part, of course, on the specific disposition technology to be implemented. Russia has made it clear that it will only consider reactor options, and the reactor options involving the lowest initial capital investment are those involving reactors that already exist-primarily light-water reactors—so that only storage, MOX fabrication, and pit-processing facilities would have to be provided, not additional reactors. (Realistically, Russia cannot expect the international community to help finance expensive new reactors in the name of plutonium disposition when existing reactors could do the job.) The fissile material storage facility now being built at Mayak, with U.S. assistance, is estimated to cost roughly $300 million. The up-front capital cost of a MOX plant big enough to do the job would be of the order of $600 million—if it were built in the West, and did not face the sort of delays that have often forced up the cost of nuclear projects in recent years. Construction in Russia might cost less particularly if Germany were to provide the components from the MOX plant at Hanau (the so-called "Hanau East" option). However, anything that might be saved on the 4600 million figure for the MOX plant would probably be added black in the cost of licensing and any necessary modifications to existing reactors. Adding an industrial-scale capability to convert plutonium "pits" to oxide might add another $100 million, if the facility could be installed in an existing plutonium-handling facility (such as the MOX plant), or more if a new facility had to be build from scratch. So in round numbers, the up-front capital investment required is $1 billion.

This is not an inconceivable amount to be provided through international assistance. If, for example, the G-7 countries and Russia agreed to split the cost over a five-year period, this would mean an on-budget subsidy from each of these states of only $25 million per year. I strongly believe this would be an excellent investment in international security, worth far more than its cost.

But after promising to provide $2 billion to help Ukraine shut the Chernobyl reactor, and $4 billion to help building light-water reactors in North Korea, the international community is not particularly eager to immediately undertake another billion-dollar project of this kind. A certain "nuclear compassion fatigue" has set in.

Is there a way such an operation could be financed an managed without requiring such sustained on-budged subsidies from the major powers? I believe the answer is yes, and that the answer lies in the highly-enriched uranium that also comes form dismantled weapons, which is a product with substantial commercial value, whose full market potential is not yet entirely exploited.

A JOINT-VENTURE "ENTERPRISE FOR NUCLEAR SECURITY"B

Let me offer the following as an approach to addressing, in part, both the financing and management problems:

Imagine establishing a semi-commercial joint venture between Russia's Ministry of Atomic Energy (MINATOM) and Western fuel cycle and construction firms—the "Enterprise for

Nuclear Security". This enterprise would be charged with the task of completing and operating the plutonium storage facility now under construction, and building an operating a pit-to-oxide conversion facility and a MOX plant in Russia.

The entire operation would be managed by the international partners in the joint venture, including internationally-managed guard teams and accounting systems at the storage and the processing facilities. Participation by both MINATOM and Western firms would ensure that the MOX produced was appropriate for Russian reactors while meeting the quality standards Western utilities demand—certified by the Western fuel cycle firms the utilities are already accustomed to working with. All the facilities would be under IAEA safeguards, modified as necessary to avoid compromising proliferation—sensitive information related to plutonium pit design. (Non-nuclear-weapon states such as Germany and Japan would certainly demand such safeguards as a condition for their participation). The United States would accept parallel international monitoring measures at the storage and processing facilities involved in its parallel plutonium disposition programme.

The billion-dollar initial capital investment required would be financed, in this concept, by additional sales of HEU above and beyond the 500 tons the United States has agreed to purchase. Russia would transfer to the joint venture's ownership approximately 100 tons of additional HEU, which, at the prices currently prevailing in the U.S.-Russian purchase agreement, would be worth roughly $2 billion. Rather than releasing that material onto the market all at once—which would crash the uranium price—the joint venture would use this material as collateral against which it could borrow, and meter the uranium into the market slowly, consistent with demand, to pay back the loans. Given discounting for risk and the cost of money, along with the cost of blending the HEU for sale, the total up-front capital available to the joint venture from HEU worth $2 billion would probably be of order $1 billion—roughly what is required to finance the initial investments. (Needless to say, perhaps, all the number in this paper are intended to be illustrative, not definitive.)

After the initial capital investment, paid for with HEU sales, operating revenues form sales of MOX fuel (at price equal to or lower than those of LEU fuel of equal energy value, to give utilities an incentive to by) would be sufficient to pay the continuing operating costs of the facilities, allowing the joint venture to survive as a stable, ongoing commercial enterprise over the long period required—a critical point. That is to say, the fact that MOX costs more than uranium fuel arises largely from the need to pay off the capital cost of the MOX plant: MOX sold at LEU price can more than pay for the operating costs, if the capital costs have been covered by other means. Any additional profits that might by generated could be split between the joint venture partners, or used to finance other critically needed nuclear security improvements, such as additional modernisation of security and accounting systems for nuclear material.

The joint venture could contract the job of blending the HEU for eventual sale to MINATOM's facilities that are blending HEU for the U.S.-Russian deal, if they had excess capacity, or other firms in the West might get the business. Similarly, the joint venture might determine that until a new MOX plant could be provided, it made sense do contract with European firms for initial MOX production.

Russia does have additional stocks of HEU it no longer needs, and has made some informal inquiries concerning possible additional sales. But no definite arrangements have been reached, in part because of the tight trade restraints Western countries have imposed on Russian Uranium exports. The United States has a stringent anti-dumping agreement in place limiting Russia's access to U.S. markets; the current unwritten Japanese policy is not to buy any Russian uranium at all, primarily because of the islands dispute; and EURATOM has also adopted an informal policy of limiting Russia to a small portion of the European uranium market. Thus, in this approach, the action the G-7 countries would have to take instead of providing on-budget subsidies would be to modify their trade restraints enough to allow uranium blended from the HEU belonging to this MINATOM-Western joint venture to enter their markets. One approach would be to simply count the material belonging to the joint venture as coming not from Russia but from a separate entity—the joint venture—which would not count against Russia's restraints. With uranium prices now recovering to $12 a pound a still headed upward, and world uranium demand greatly outpacing current production, such a step would be less painful than it might have been a few years ago.

Thus, the "Enterprise for Nuclear Security" concept could potentially make it possible to finance plutonium disposition; create a management structure for implementing plutonium disposition that can sustain itself over the long term; ensure international, rather than merely national, guarding and accounting at the plutonium storage and processing facilities, as well as IAEA safeguards; eliminate an additional 100 tons of HEU; and provide substantial business to both MINATOM's desperate nuclear cities and to Western firms—all at little or no direct on-budget cost to the countries involved.

Russia would have to provide 100 tons of HEU, and would get in return a plutonium storage facility and a MOX plant, along with a long-term commercial relationship with Western fuel cycle firms. The Western countries would have to modify their trade restraints to allow uranium from that additional HEU into their markets, and in return the Western countries and Russia would both get substantial security dividends from disposition of plutonium under stringent non-proliferation controls.

If the international community is going to help Russia finance a MOX plant—whether through direct subsidies or through offering increased access to their uranium markets—some non-proliferation conditions are likely to be required. To insure that the disposition operation itself does not increase risks of proliferation, extremely stringent standards of security and accounting for the material throughout the process will be essential. International safeguards will have to be applied as early in the process as practical. There would have to be commitments not to export the MOX fuel or other plutonium products to countries of proliferation concern; perhaps there would be an agreement that the MOX would be used only by Russia, or only by the nations participating in the joint venture. As the international community's interest is in disarmament and non-proliferation—not in assisting Russia's development of a broad plutonium fuel cycle, or creating fierce commercial competitors for the West's fuel cycle business—some commitment would be needed that any MOX plant built with Western participation would be used only for weapons plutonium, or only for plutonium that is already separated. That might not be a very onerous commitment, as most of the potential operating lifetime of such a plant would be over in any case by the time the large existing stocks of excess weapons plutonium and separated civilian plutonium were processed. And there would likely have to be some arrangement—however difficult it might be to arrive at—to ensure that new excess stocks of separated plutonium were not built up as fast as the old stocks were built down, or there would be little point in the enterprise.

Of course, if Russia did not want to agree to such conditions, it is always free to do whatever it likes with its plutonium, with whatever financing of its own it can make available—but the prospects for such financing in the near term do not appear very good.

AN AMBITIOUS AGENDA

To sum up:

First, we must move quickly to modernise security and accounting systems for all weapons-usable nuclear materials, particularly in the former Soviet Union.

Second, we must bring excess plutonium and HEU under international monitoring, to ensure the irreversibility of nuclear arms reductions.

Third, we must move as quickly as practical toward actual disposition of excess plutonium and HEU. Technology already exists to get the disposition job done. The key issues will be how to finance and manage the operation, particularly given its immense scope and controversial nature. An international joint venture "Enterprise for Nuclear Security" that would build and operate plutonium disposition facilities under stringent non-proliferation controls, financed through additional sales of HEU, is a potentially promising approach to addressing the most difficult issues facing the disposition problem.

This is an ambitious agenda—but to be any less ambitious, with the immense challenges we face, would be to shirk our responsibilities. Working together, the international community can put together the means to get these jobs done. The future of efforts to reduce nuclear arms and stem their spread depends on our success.

Address:

MATTHEW BUNN
Study Director
US National Academy of Sciences
Washington DC 20418
USA

63

Management of Russian Military Plutonium

C. PIERRE ZALESKI

The objective of this paper is to propose and to discuss a solution which makes it possible to store, as quickly as possible, all weapons-grade plutonium no longer used in the Russian military programme, in a way that makes diversion extremely difficult, and the re-use in weapons form by the Russian government difficult and visible.[1]

Two main conditions apply to this solution. First, it should be achieved in a way acceptable to the Russian government, notably by preserving plutonium for possible future energy production uses; second, the economics of the total system shall be as good as possible and there should be no charge, or a limited one for the storage of plutonium.

I will propose a solution already outlined in Ref. 1: to store plutonium in a specially designed fast reactor, or at least a specially designed fast reactor core. I will attempt to demonstrate that this solution compares favourably to other possible solutions, applying the criteria set out in the goal and the two conditions mentioned above.

In addition, this solution should have the following side advantages:

- utilising available personnel and installations of the Russian nuclear military complex;

---

1. It seems however clear that whatever durable solution may be developed, an initial temporary centralised storage of weapons and plutonium (and even HEU) from weapons will be necessary as the delays of implementation of more permanent solutions shall be at least of the order of a decade (except perhaps for HEU). This centralise safe storage is probably the most urgent and the most important task for those who deals with the management of excess Russian Military Plutonium.

- providing possible basis for decommissioning of older and less safe Russian reactors;
- giving a quantitative experience of construction and operation of a series of sodium-cooled fast reactors;

It also, however, presents a major problem: the need for rather large capital investment, with the risk of not getting the appropriate return on investment due to the generally difficult political and economic situation of Russia.

REVIEW OF OTHER POSSIBLE SOLUTIONS

The solution of mixing plutonium with fission products and disposing of this mixture in irretrievable deep geologic formations would, in our view, not meet the first condition, that is, the approval of the Russian government. In addition, this solution also presents a not negligible net expense, and may be questioned from the non-proliferation point of view, as in the long term one cannot exclude diversion of plutonium even from supposedly irretrievable storage, and the separation from fission products and re-use of plutonium may not pose major difficulties.

Placing the plutonium in the retrievable storage under Russian national or international safeguards may also not be very economic because of the cost of safeguards and the degradation of plutonium quality during the storage period, but more importantly, it present some diversion risks and it would not prevent any future Russian government from very easily acceding to the plutonium for military use. Therefore this solution, even if it is probably the only available for the initial temporary storage, shall not be extended beyond the minimum necessary period.

A more promising solution may be to "burn" this plutonium in the form of mixed-oxide fuel in existing light water reactors, modifying its isotopic composition in addition to mixing it with fission products. Here, however, two options are to be considered. One is to export the plutonium to other countries, for example Germany or France, which already have programmes for MOX use, the second is to burn it in Russia. It seems to me that the first option has the following drawbacks:

- the difficulty for the Russian government to accept this export, losing its sovereignty over it (Ref. 2);
- a possible issue of public acceptance in host country, whose public may be reluctant to deal with a problem it considers outside its responsibility.

- the issue of the value of this plutonium; presently, countries using MOX are not limited by availability of plutonium but rather by a shortfall in fuel manufacturing capacities and the availability of reactors licensed to burn MOX fuel. For example, Electricite de France has recently decided that plutonium will have a zero value on its books. In fact, the Russian plutonium will displace French civilian plutonium, which will have to be placed in storage, with accompanying costs and degradation in quality. Therefore, the Russian plutonium will have to bear this cost, and will have a negative value. This may not be acceptable to the Russian government.
- the substitution of the Russian plutonium for French civilian Pu would in fact also be contrary to present French policy of having a minimum quantity of separated Pu in storage.
- the proliferation benefit of replacing weapons-grade plutonium in storage with civilian Pu would have to be clarified, including the question of weapons-worthiness of plutonium depending on isotopic composition and the technical capabilities of potential proliferators, an issue on which there is no agreement (Ref.3).

In summary, it is true that burning Russian plutonium in foreign reactors, for example French reactors, may appear an attractive solution; indeed, EDF's 28 reactors potentially able to burn the plutonium (with 1/3 MOX core), 16 of which are already licenced to do so, could burn the 100 tonnes of available Russian military plutonium in less than eight years. However, the abovementioned six drawbacks mean that it is not a practical solution.

Therefore, one should rather consider burning or storing Russian plutonium in Russia itself. Burning it in existing Russian LWRs is theoretically possible. However, only some of these reactors of present design may accept plutonium, and no one are licensed for this. In addition, it would take many decades to burn the 100 tonnes in existing Russian LWRs (Ref.4).

Another possible solution which seems relatively attractive is to burn the plutonium in newly built reactors. Fast breeders, which are clearly better for this purpose than LWRs (Ref. 4 and 4bis), have been proposed as a possible solution. Indeed, by completing two BN-800 type reactors in Russia, construction of which was begun some years ago but frozen for lack of funding, in a once-through cycle one could transform into spent fuel all

100 tonnes of ex-military plutonium in 30 years of operation (Ref. 4 and 4bis).

CONCEPT PROPOSED: FNPSR OR CAPTURE

Finally, the concept that I am proposing, which I have called Fat Neutron Plutonium Storage Reactors (FNPSR) and has also been dubbed CAPTURE (Ref. 5), has as an objective to store a maximum quantity of plutonium in the reactor core itself for a relatively long fuel cycle duration.

To achieve this, I suggest to seek a reactor core for which the typical design objectives will be:

- large fuel rod diameter - about twice that used in present breeder designs;
- low specific power, in range of 1 MWe per 10-15 kg of Pu;
- high internal breeding ratio, in the range of 0.9-1.0; and
- long fuel residence time, in the range of 10-20 years.

These design goals are consistent with the idea that plutonium has practically no value for the coming decades (there is an excess of Pu) and that in case of ex-military Pu, one may even envisage a fee being charged for its storage, giving it in practical terms a negative value.

If one achieves this 10-15 kg Pu per MWe, between 10 and 7 GWe of CAPTURE type reactors would be needed to store all 100 tonnes of Russian plutonium as soon as the reactors are built. Another solution may also be considered, i.e., to built only half this capacity and irradiate the 100 tonnes of plutonium in two batches. However, this brings us closer to the previous solution with BN-800 reactors, and the differences become less clear-cut.

Assuming construction of CAPTURE reactor series totalling some 8 GWe, after maximum burn up is achieved, in some 30-35 years (15 for development and construction, and 15-20 for incore residence time), the following options would be available;

- exchange Russian denatured plutonium for U.S. weapons-grade Pu, and use the latter for the second fuel load of the CAPTURE power stations;
- reprocess the spent fuel and re-use the Pu in the same reactors;
- store the denatured Pu in safeguarded, retrievable storage for future use, and use civilian Pu for the second fuel loading.

After two or three fuel cycles in CAPTURE plants, when the reactors have reached the end of their useful life, the following options would be available:

– if nuclear energy is to be discontinued and no development of breeders considered at this time, one could dispose of Pu definitively in underground repositories, or if environmental and proliferation concerns so dictate, burn the Pu in fast burner reactors (cf. French CAPRA programme, Ref. 5).
– if nuclear energy is developed as a long-term solution, as is expected by some today (notably the French, Russian and Japanese governments), one could use the Pu in fast neutron breeder reactors.
– if the conditions are still not clear 55-60 years from now, one could store the plutonium in a new series of CAPTURE power stations. This would preserve all options, and the eventual decision could wait for a clearer context.

One of the important aspects of this kind of project is timing. The following schedule may be imagined for implementing the CAPTURE idea:

– 10 years for the development and validation of the concept, notably fuel irradiation experiments; however, start of construction on the first reactors could be envisaged before all experimental results are in hand, for example, seven years after the start of the project.
– construction time 5-6 years, with construction starts every year on three reactor units.

With these rather optimistic assumptions, the total project would take some 15 years, by which time all military plutonium will be stored safely in CAPTURE reactors. Indeed, there is an implicit assumption in this optimistic schedule that the series of CAPTURE reactors, which will be rather simplified version of BN600/800 (with lower specific power, simpler fuel handling equipment) will not require construction of a prototype.

Given that dismantlement of plutonium warheads will likely follow dismantlement of 235 U warheads, and that plutonium would be needed for fabrication of first fuel loads for CAPTURE about eight years after the beginning of the project, the above schedule is rather consistent with safe disposal of military plutonium.

TECHNICAL ASPECTS

With a plutonium value that is zero or negative-indeed, the owner of plutonium may be obliged to pay a storage fee, covering, for example, storage and safeguards costs—fast neutron breeder cores must be re-optimised.

Therefore, the natural idea is to use as much plutonium as possible in each CAPTURE core, so as to store more Pu per MWe, and also to increase the diameter of fuel rods, thus decreasing the relative share of fuel fabrication in the total cost of a kilowatt-hour. It is also natural to try to increase fuel residence time in order to simplify fuel handling equipment without penalising availability. It is therefore desirable to seek higher internal breeding ratios in order to allow longer residence time without large reactivity swings.

Meeting these objectives will lead to a higher Doppler coefficient, which plays a positive role in controlling power excursions, but also to a more positive sodium void coefficient, which has an adverse effect on safety.

The core will, therefore, have to be optimised to ensure overall safety. This can, for example, be achieved via a relatively flat core design, with appropriate upper plenum design ensuring that any sodium voiding of a core section will inevitably lead to sodium voiding of the corresponding section of the upper plenum, where the reactivity effect can be designed to be negative.

For reasons associated with core design safety, it seems reasonable to limit the size of the power plant. A logical size would be between 600 MWe and 800 MWe, as Russian technology for this size of breeder reactor plant is well-developed.

The BN-600 FBR has been operating very successfully for over 10 years, with availability that is not only among the best in the Russian nuclear programme, but also among the best in the world, all type of nuclear plant's considered. The plants average availability over 10 years was 97.5 per cent with a capacity factor of 71 per cent.

In addition, Russia has developed detailed projects for the BN-800 fast breeder reactor, directly inspired from BN-600 technology (Ref.4). Therefore, with some cooperation from western Europe (Phenix, Superphenix, EFR projects) and Japan (Monju), Russian scientists and technicians should be able to design and build a safe 800-MWe fast neutron power station without too much development.

The above-mentioned core optimisation and the development and fabrication of Pu-bearing fuel with large-diameter rods will require close collaboration with western countries, notably with France, which has more experience in Pu-bearing fuel than Russia.

Core and fuel aspect represent the most innovative aspect of the CAPTURE design, and will probably require the most R&D. In fact, the idea that the author expressed in November 1994 was taken up for study by A.A. Kamaev of the Institute of Physics and Power Engineering in Obninsk, Russia and presented in Cadarache, France, in June 1995 (Ref. 6). His conclusions are that if one limits the core dimension by the diameter of the BN-800 vessel, the FNPSR core could not achieve the design objectives proposed in this paper.

The limiting values that Kamaev found were: specific power of 7,4 kg of Pu/ MWe, inner diameter of fuel element cladding, about 8,8 mm; fuel residence time of 4,1 effective years; and breeding ratio of about 0,84. However, in his conclusion Kamaev stated: "Achievement of FNSPR type reactor characteristic values is possible in the new reactor plant design developed on the basis of design solution used for the BN-1600 heat removal system. In this case R&D work should be carried out to substantiate design solutions used for the first time in the practice of home nuclear reactor building industry".

This clearly shows that the design goals mentioned above are not irrealistic, but will probably need more R&D. In fact, in Kamaev's preliminary review, he did not consider the possibility of increasing the diameter of the reactor vessel beyond that of existing projects (BN-1600) and changing the design of the fuel handling system. His preliminary results show that it would be necessary to study a rather flat core, with even larger diameter, larger fuel volume fraction, larger fuel element diameter and probably larger section of subassembly. This pancake type core, with low specific power (kilowatt per liter of core), evidently calls for much simplified design of reactor vessel, vessel closure, and fuel handling system, the last perhaps inspired by those used for light water reactors (Ref.7).

The relatively large vessel diameter required for this core

design should not present major construction difficulties or increased costs, as we are talking about a relatively short stainless steel vessel with thin walls (no need to contain high-pressure liquid). The closure, the internals and the fuel handling mechanism should be much simplified compared to the classic breeder design. There may, of course, be some difficulties with the design, as for example the problem of intervention on failed fuel assemblies. One would have to determine by study if it is possible to accept a long waiting time before such intervention, inherent in the type of fuel handling I am suggesting, considering the low probability of this kind of event.

Generally speaking, this simplified vessel-wide and short-and simplified closure with no rotating plug but rather a type of leak tight cell (Ref. 7); associated with BN-800 type components—such as pumps, intermediate heat exchangers, steam generators—may hopefully be considered, with proper design and test efforts, as reasonably proven.

ECONOMIC ASPECTS

A rough economic evaluation of the potential of this project may be done in two ways.

One way of evaluation is to use the 1993 EFR (European Fast Reactor) study carried out by utilities and manufacturers from France, Germany and the U.K., which seems the most recent, serious and pertinent study of the subject, and to transpose it to Russian conditions.

This study (Ref. 8) shows that in a western European context, and assuming zero plutonium value, a series of 1,500-MWe FBRs can be in the range of economic competitivity with LWRs at least with some uncertainty margin.

The EFR design was not optimised for plutonium storage; therefore, some gains can perhaps be anticipated thanks to:

- simplified fuel handling equipment design (fuel handling every 15 to 20 years);
- much lower cost fuel cycle: the large fuel rod diameter makes it possible to produce more energy per rod (for example, four times as much), and the cost of fuel rod fabrication should not be very sensitive to diameter; and
- potential fees to be paid by plutonium owners (for storage function).

These gains should thus lead in the EFR context to a situation very competitive with LWRs. This result should be transposable to the Russian context, assuming well-managed construction of a series of identical plants (10 × 800-MWe units).

There may, however, be one penalising point, namely, the 800-MWe size, which is suggested for core design and local pragmatic reasons (the existence of BN-600 and of a detailed project for BN-800). However, this size only slightly smaller than the largest modern Russian LWR of 1,000 MWe (VVER-1000). Therefore, the economic size-related penalty should probably be significantly smaller than the advantage due to the re-optimisation of the EFR core (see above). In fact, there is a certain tendency in Russia towards smaller reactors (for example, VVERs of 500 MWe and 630-MWe are in current Russian plans); thus, any economic penalty stemming from relative reactor size could disappear or even be reversed.

The second way to evaluate the economic potential of the project is to use the Russian internal comparison. According to V. Kagramanian, head of laboratory of systems analysis of nuclear power at the IPPE (Obninsk), a highly favourable experience has been gained in Russia from fast reactors: the BN-600 has the highest load factor among the country's reactors, and the BN-800 design modified (BN-800M) meets the latest, more stringent safely requirements. The economics of this reactor is equivalent to that of the medium-size thermal reactors (VVERs) or fossil-fired power plant (Ref. 9)

This, combined with gains from core optimisation, gives a very positive indication about the economics of the concept.

One can therefore expect the CAPTURE project to have a good potential for competitivity with LWR projects in Russia, that is, that it could produce electricity at a lower price.

FINANCIAL ASPECTS

The need for new electric power plants in Russia is quite evident. Even if the domestic demand is not growing, because of the developments of energy saving and more efficient energy use, the possibility of exports to neighbouring countries and the need to replace older power plants, nuclear or not, justifies some new constructions.

The very difficult economic situation of Russia, however, makes the issue of financing difficult. Therefore, if this project is to go forward, international financing seems necessary, at least for most of the investment required. It seems to me that the international community, and especially OECD countries, should be interested in facilitating safe storage of weapons-grade plutonium. Some of the OECD countries also should be interested in maintaining world expertise and increasing operational experience in fast neutron reactor power plants, and the entire international community in helping the Russian economy reconstruct.

A potential additional motivation for the potential lenders may appear if the Russian authorities accept to link the building of these new plants with decommissioning of older and less safe nuclear power plants.

This being said, the important question is how Russia can reimburse the money borrowed for these projects.

What will certainly reassure potential lenders is a contract expressing the reimbursement in a commodity exported normally by Russia which has well-established international value and is in demand in the lending countries, for example, natural gas.

The other advantage to link the reimbursement to gas exports is that Russia can consider that it saves gas when producing electricity with new plant, and reimburses only part of the saved gas.

As the Russians may have a problem to market all their available gas, it would probably be necessary to conclude a separate contract to purchase gas for reimbursement of the CAPTURE project, in addition to normal commercial gas supply contracts.

It is quite clear that financing is the largest obstacle to the entire CAPTURE project. To finance the study and development phase of the CAPTURE project, one can think of all governments interested in the main objective of CAPTURE - managing excess Russian military plutonium-as well as governments interested by the side benefits outlined at the beginning of this paper.

Using as far as possible the expertise and available time of Russian scientists and engineers, as well as their installations, development expenditures for this project should not be excessively high.

Indeed, as some eminent Russian scientists have indicated

(Ref. 10), that in the present situation and for a given amount of hard currency, the Russian nuclear industry may perform much more work than western industry. For specific examples related to the upgrading of old Russian reactors, they suggested that Russian industry would be 16 times more efficient than western industry.

To be conservative, it would not be extraordinary to consider a factor or two to characterise the relative efficiencies of the two industries (Russian and western). This may be due to the relatively low cost of labour in Russia expressed in hard currency, as well as the high contribution of labour costs in the total cost of nuclear plant construction.

Assuming further that at least 80 per cent of the construction work on CAPTURE stations could and would be done by Russian in-dustry, the cost, in hard currency and as a percentage of typical western costs for the same work, can be expressed as 80 per cent/2+20 per cent=60 per cent.

If in addition the development phase is successful, construction of a series totalling some 8 GWe of CAPTURE reactors should not exceed about $10 billion. In these conditions, the project should be self-supporting, with no, or very modest, fee for military plutonium storage. The risks—political, economic and technical—may discourage normal financing One can, however, imagine that the OECD governments interested in the project may subscribe some sort of guarantee for the risks involved.

CONCLUSION

The CAPTURE concept seems potentially sufficiently attractive to at least deserve further investigation. The next, relatively inexpensive step, could be the study of cores without the constraint imposed by vessel diameter. If this study leads to satisfactory results, a slightly more expensive, but still not too costly, step might be contemplated, to study and design a simple, large-diameter, low-height, vessel with a very simplified closure, internals, and fuel handling equipment.

Thereafter, a decision should be taken if more costly steps involving mock-ups, tests, fuel irradiation, and other design and experimental efforts should be launched. It is so be hoped that at least the first step will be launched in the near future.

Of course the decision to go along with the CAPTURE development should be made by Russian Government, which has to decide if it is in Russia's interest to build a series of Fast Neutron Reactors for a total capacity of 8 GWe, and if it wish to commit it Gas resources to finance this investment.

If the decision is positive then Russian Government shall negotiate with interested OECD Governments the term of a loan.

If on the contrary the decision is negative, or if the development of CAPTURE is not successful, the next best solution for Military Plutonium Management seems to be the construction of 1,... 2... or more BN 800 Fast Neutron Reactors. BN 800 is already designed but may be reoptimised for low or negative plutonium value.

For example of two BN 800 are build, they will transform the 100 tons of Military Plutonium in spent fuel during some 30 years of operation.

REFERENCES

1. Zeleski, C.P., November 1994, *Fast Neutron Plutonium Storage Reactors: An example of the utilization of the military nuclear complex for peaceful purposes.* UNESCO ROSTE, Venice.
2. Remarks of N. Yegorov, first deputy minister, Ministry of Atomic Energy, Russian Federation Proceedings of seminar on Back End of nuclear Fuel Cycle, CGEMP, Université de Paris Dauphine, May 1993.
3. Remarks by C.P. Zaleski, annual conference of Japan Atomic Industrial Forum, Tokyo, Japan, April 1995.
4. Kagramanian V. et al., "Aida-MOX" Proceedings of Global 95, Versailles, France September 1995.

4bis. Mikhailov, V. N. et al., Ministry of Atomic Energy-Russian Federation, Moscow, V. Mugorov et al., Institute of Physics and Power Engineering, Obninsk, Russia; I. N. Avrorin et al., NNIITF, Moscow, "Plutonium in Nuclear Power of Russia". OECD-NEA Expert Group on the Management of Plutonium, Paris, March 1995.

5. Zaleski, C.P., 1995, "Fast neutron reactors: development in future decades" *International Journal of Global Energy Issues,* Oxford, U.K., Vol. 8, Nos 1-3, pp. 133-142.
6. Kamaev, A.A. "Technological Aspects of FNPSR Development on the Basis of Russian LMFRs. "French-Russian seminar, Cadarache, June 1995.

7. *Inter-departmental Fast Neutron Reactor Studies.* Electricité de France, Direction des Etudes et Recherches, Clamart, France 1971-72.
8. Lefèvre, J. Hubert, G., and Aubert, M. "Le project de réacteur rapide européen EFR: état actuel et perspectives." *Revue Générale Nucléaire*, No. 6, p. 504-516, Paris, 1994.
9. Ermakov, N.I., Ministry of Atomic Energy-Russian Federation; V. N. Murogove and V. M. Poplavski, Institute of Physics and Power Engineering, Obninsk, Russia. "Role of Fast Reactors in the Future of Power Engineering, fuel supply and the Environment. " International Topical Meeting on Sodium-Cooled Fast Reactor Safety, Obninsk, September 1994.
10. Ponomarev-Stepnoy, N. N. and Adamov, E. in Proceedings of the MIECCGEMP seminary, *La Sécurité de l'Approvisionnement en Energie d el'Europe: Rôle de la Russie.* CGEMP, Université de Paris-Dauphine, Paris, 1994.

Address:

PIERRE C. ZALESKI
Delegue General
CEGMP University of Paris-Dauphine
75116 Paris
France

64

Non-Fertile Fuels for Burning Weapons Plutonium in Thermal Fission Reactors

C. LOMBARDI, A. MAZZOLA and F. VETTRAINO

In the last few years, the excess plutonium disposition has become ever more a topical and critical issue. As a matter of fact, more than 200 MT of Plutonium coming from spent fuel reprocessing have been already stockpiled and over the next decade, under the already ratified agreements, another about 200 MT of weapon-grade plutonium are expected do be available from nuclear weapons dismantlement. On this basis, an ever growing plutonium production is no longer the goal and the already stored quantities should be burnt in power reactors by taking care that no new plutonium is generated under irradiation. This new outlook in considering plutonium has led many designers to reassess the Fast Breeder Reactors (FBR) role and shifting from breeder to burner machines perspective. Several solutions for burning plutonium have been so far proposed and discussed from the safeguards, proliferation resistance, environmental safety, technological background, economy and time schedule standpoint. A proposal for plutonium burning in commercial Pressurised Water reactors (PWR) by using a non-fertile oxide-type fuel consisting of PuO_2 diluted in an inert matrix is reported hereafter. This solution appears to receive an ever growing interest in the nuclear community.

In order to not produce new plutonium during irradiation an innovative U-free fuel is being researched, based on an inert matrix which will consist in a mixed compound of inert oxides, such as ZrO_2, Al_2O_3, MgO, CeO_2 where the plutonium oxide is dispersed in. The matrix will fulfil the following requirements: good chemical compatibility, acceptable thermal conductivity, good nuclear properties, good stability under irradiation, good dissolution resistance. The plutonium relative content will be comparable to that used in MOX fuel. The fuel is expected to be characterised by a high chemical stability (rock-like fuel), so that after discharge from reactor and adequate cooling time, it can be considered a High Level Waste (HLW) suitable for final disposal in the deep geological formations without requiring any further reprocessing treatment (once-through solution). The fuel pellets, similar to those currently employed in the commercial LWR's will come from the ceramic mixed powders technology or from Gel Supported Precipitation (GSP) microsphere process.

The neutronic calculations show that commercial PWRs, partially fed with Pu-inert matrix fuel and operated in a once-through cycle, have good plutonium elimination capabilities. The plutonium still remaining in the spent fuel is quality-poor, difficult to be recovered and then highly proliferation resistant.

Radiotoxicity levels versus time in inert matrix spent fuel, do not show any increase with respect to standard unreprocessed spent fuel. A most relevant open issue is, however, the technological development and qualification of this new fuel.

1. INTRODUCTION

The end of the cold war and the recent agreements between USA and the states former USSR will make available relevant quantities of weapon grade plutonium (WG Pu) and of highly enriched uranium (HEU); the nuclear weapons reduction established by the START II treaty, signed in 1993 by Bush and Eltsin, and ratified by the American Senate just few weeks ago, should result in about 200 MT of WG Pu and 1000 MT of HEU. These figures are only estimated amounts, since is not been known precisely how large is the military stock in both parties and there is also an high uncertainly about the size of the residual storage of the fissile materials, i.e. those materials not yet introduced in the nuclear weapons; therefore, most people believe that the quantities which we are dealing with are far higher than the reported values. The international scientific community is faced by the problem of finding a reliable methodology for the disposition and/or management of relevant quantities of strategic materials, in a economic manner, by: eliminating the diversion risk, reducing the environmental impact and the radiotoxicity hazards. It is wishful that the identified solutions will allow a limited recovery of the economic resources employed in the production of nuclear weapons.

2. HIGHLY ENRICHED URANIUM (HEU)

The problem of disposition of HEU is affordable in a rather simple way: it could be completely reconverted to civil utilisation by diluting it with natural and/or depleted uranium, down to enrichment values of 3-5 per cent, suitable for being used as fuel in the commercial LWRs. To this purpose, the USA have signed an agreement in summer 1992 with the Russian Atomic Energy Ministry (MINATOM) for purchasing 500 MT of HEU; this agreement has given rise to heavy disappointment among American industrial and commercial operators of nuclear energy, worried about the perturbation effect that this action might bring to the uranium market. However, strictly speaking from the technical viewpoint, the HEU question is well defined. Some relevant doubts do however exist about the former Soviet HEU stock dimensions; Secretary Mickhailov has revealed in September 1993 that the 500 MT HEU transferred to USA, represent only 30-40 per cent of total HEU existing in Russia[1]. Should these news were right, only for Russia would be there a total amount of 1200-1500 MT, much higher than the total 1000 MT estimated for both Super-Powers, confirming in this way the doubts about the real entity of the existing stocks.

3. WEAPON GRADE PLUTONIUM

The management of WG plutonium is more complex since it is possible to reduce its potential hazard by simple dilution operations or other chemical processes. At moment, the WG Pu is stored by Super-Powers waiting for a definitive solution, nevertheless this situation seems to be not sustainable any longer. A quite large number of solutions have been proposed for the WG Pu disposition, some of them are viable, others more or less exotic or unfeasible.

A first solution outlooks the introduction of Pu in glass bodies, possibly mixed to high level radwastes; this option though decisively safer than the simple storage and giving the possibility of permanent storage of the Pu, appears rather costly and still doubtful about its complete irreversibility, i.e. about its definite effectiveness against the proliferation risk.

Another solution consists in the *spiking* process: the Pu is irradiated in reactor for a short time, but sufficient to render it highly radioactive, in order to preclude any easy access to the material. It is not eliminated however, the problem of the continuous safeguard control being not definitely prevented the possibility of Pu recovering through chemical process; in addition, the associated radwaste management would represent a rather delicate feature for this solution.

Some options consider the WG Pu utilisation as nuclear fuel; only by fissioning Pu for producing energy there is in fact the possibility of recuperating a part of the economic investments made for the production of this material. The fission of 200 MT of Pu would produce something like ~ 200GWe-yr of electricity, that is to say about 80 per cent of the world year production of electricity made by nuclear power plants. The entire quantity of estimated HEU and WG Pu would produce through the fission process ~1200 GWe-yr of electricity, somewhat slightly less than the entire world electrical energy consumption in a year. In terms of equivalent-oil, this huge energetic potential represents 2.3 GTep which would require to have going around over the seas 4600 loads of a 500.000 tonne oil super-tanker; this equivalent-oil quantity corresponds to about 7 GT of carbon dioxide (CO_2) that would not be discharged into the atmosphere.

During the fissioning process Pu is altered and denatured; its isotopical composition is modified moving up towards higher atomic weight isotopes. This worsens the quality at a such an extent that makes the recuperation strongly unattractive either for utilisation in thermal reactors or for diversion purposes. Moreover the irradiation generates highly radioactive fission products, which render highly difficult the accessibility to residual Pu. Although there still exist the theoretical possibility of recuperating Pu remaining in spent fuel, the associated complexity and the high technological level that would be necessary, would facilitate the safeguard role of the proposed international organisations (IAEA). Finally, through fission the most of Pu can be transmuted and therefore physically eliminated; one should consider that this is the really safe way for preventing diversion attempts ad for complying with the necessary safety requirements.

4. THE DEBATE ON THE CIVIL PU ROLE

The discussion about methodologies for WG Pu disposition is connected to the more general issue pertaining to civil or reactor-grade Pu (RG Pu) exploitation. Pu has been ever since considered as a precious material, substitute of natural fissile and therefore relevant efforts have been addressed to increment its production in the perspective of using it especially in fast reactors. Compared to thermal reactors, fast reactors allow, in fact, for a better utilisation of U by enlarging of about a factor 100 the energy that can be extracted from a given quantity of mineral material, simply by converting U-238 (99.3% in natural uranium) to Pu-239. During the last few years, however, some facts have modified this kind of approach. One first point is that the U price has undergone a rapidly decreasing trend as an effect of the reduced demand and increased availability of natural U on the world market. Moreover, many experts do think that the reserve exhaustion do not represent a serious problem since new survey activities and technological improvements should render accessible even lager amounts of material. The experience derived from exploitation of other rare materials, such as zlnc, tin and copper confirms this consideration, being their price in current money, undergone a continuous decrement during the last decades. Other considerations also play an important role:

- the U stock pile is an increasing function of the accepted extraction cost;
- the natural U price affects only marginally the energy production cost (~10%) and this turns out to be a very limited penalty in case of extraction cost increase;
- the ever more urging actions linked to nuclear weapons dismantling will make available large quantities of HEU and WG Pu.

Furthermore, FRs at commercial scale have been delayed in the far future, on the basis of technical, political and economical issues. Moreover, the repeated malfunctions, even though of not relevant entity, happened to Super-Phénix in France and the very recent incident to Monju reactor in Japan, which had been connected to the grid only in August 1995, beyond putting in evidence the management difficulties of those reactors, do bring the public opinion to a negative attitude towards this solution. Pu is presently utilised, even if at a limited extent, in thermal reactors in form of U-Pu mixed-oxide (MOX), whose economic advantage as we will see later on, is more and more questioned. To this respect it appears from the very recent news, that Electricité de France (EdF) holds that Pu recycling in PWRs finds justification only if a burn-up level of at least 60 GWD/T is reached, which stands as a rather hard challenge anyway. On this basis, during last few years a new way of considering Pu has come progressively to taking shape: Pu is not any longer looked as an useful material, but is rather considered as a waste product in UOX cycle, just to get rid of. Given its relevant energetic content, it should be exploited under this respect at the best it can be done, while trying to avoid any further production of it. In this perspective WG and RG Pu can be defined as "excess Pu".

5. THE PU ELIMINATION BY FISSION

To the end of Pu disposition, either WG or RG, by fission process, different solutions have been proposed and discussed from the environmental protection standpoint, proliferation resistance, technological background and economic aspects and planning[2–6]. Among these, the following hypothesis appear as the most interesting:

- utilisation in FRs;
- utilisation in LWRs;
- utilisation in HTGRs.

FRs would be the best candidates for Pu burning if the blanket would be replaced by a reflector. This hypothesis has driven the attention of the French designers in order to find a new strategic position for super-Phénix reactor; in fact the R&D programme approved in 1994 assumed to utilise the reactor as an incinerator for Pu and Minor Actinides (A). The third core, already approved and expected to be loaded during 1998-99, will reach a Pu burning capability of ~20kg/TWhe; in the meanwhile a test fuel assembly should be loaded with the purpose of verifying the possibility of reaching burning capabilities of 70-100 kg of Pu/TWhe (CAPRA Programme) and burning at the same time some minor actinides such as Am and Np[7]. An analogous function has been outlooked for the Japanese prototype 280 MWe fast reactor Monju, which should be utilised for a minor actinides burning research programme[8]. It is evident, however, that due to the very limited number of available FRs, the Pu burning capability through this line would result rather limited. Anyway, the utilisation of a FR in a "burner mode" would result in an exactly opposite direction with respect to the original objective. That would go in the sense of a really reduced role of this kind of reactors, perhaps arriving to put under question even their presence in the near-mid future of nuclear energy panorama. Alternatively Pu can be utilised in LWRs as MOX fuel; among the afore mentioned solutions this is the only one which has been thoroughly tested and that starts to have real application at industrial scale. The MOX fraction loaded in the present PWR cores lies in the range 30-40 per cent of the total core. This fraction ensures that power distribution and safety parameters fall within the design margins of UOX cores, in order to not bring any change to the reactor control system. France, Germany and Japan are getting engaged in research programmes aimed at increasing the MOX percentage in the cores up to 50 per cent or even 100 per cent[10], but that will require careful evaluations since significant modifications to the control system could be necessary. The so far available data demonstrate, however, that the economic benefit in utilising MOX fuel is not unanimously acknowledged.

Recently, P. Leventhal, Director of the Nuclear Control Institute, has claimed that the Pu recycle as MOX is undoubtedly more expensive than simple storage of spent fuel; nevertheless, many non-technical reasons induce electric utilities to follow this strategy[11]. A significant confirmation of that comes from the fact that Electricité de France (EdF), leader in reprocessing and Pu recycling, and resolute supporter of this strategy, for the first time is going to attach a zero value to Pu in its 1995 budget thus reflecting a comparative cost evaluation of MOX and UOX fuels[12]. While other electric utilities, especially in Germany, are operating since several years with a Pu zero value, EdF has maintained a positive value until an economic justification has been possible. Practically, EdF is now admitting that MOX fuel is so highly expensive with respect to UOX fuel that no margins for a positive component of Pu value are available any more. Nevertheless, according to the conclusions of the Seaborg Report, which concluded that Pu should be considered an energy source to be exploited in power reactors as a MOX fuel,

EdF has announced at the Global '95 conference that authorisation for 12 ore PWR units has been asked to French Nuclear Safety Authority, beyond the already authorised 16 units, to utilise MOX fuel. There is a set of technical reasons which make the MOX management more expensive than UOX. One first issue is that the fabrication of Pu fuel is much more difficult than the only U based fuel, due essentially to safety reasons which require remote installations operation to be used. A second point is that the Am-241 coming from the Pu-241 decay, is a strong neutron absorber and, together to the Pu-236 decay products, is an emitter. This causes the Pu fuel to be manufactured immediately after reprocessing, in order to avoid ray shieldings and minimise undesirable poisoning elements which would alternate the core neutronic economy. A third important point deals with the allowable number of recyclings in a sequence. During irradiation the Pu composition is modified in such a way to pass from the initial high Pu-239 content to an isotopic mixture where the Pu-240 and Pu-242 contents are increased. The presence of even isotopes degrades the Pu quality at such a low level that Pu coming from a second recycle would be already of so poor quality to render it unusable in a LWR core; to cope with that the Pu could be mixed also with a higher quality Pu, but that not be a decisive solution either. Conversely, the Pu composition is not degraded at such low level as to surely eliminate proliferation risks.

The MHTGR reactor, which has been proposed for Pu disposition is a modular concept of the HTGR and could profit also of the experience achieved in the military facilities dedicated to tritium production. The fuel for this reactor consists in UOX or MOX microspheres coated by inert ceramic material (SiC) and put in form of small cylinders by an hot pressing process, and finally located in hollow graphite blocks. Some irradiation test have been performed in the past by using Pu fuel coated particles, but either the Pu fuel fabrication or the characteristics of a reactor operated with MOX, would need further and deep verifications. The remarkable difficulties in reprocessing this kind of fuel would constitute a strong barrier in case of an attempt aimed at residual Pu recovery. There exist a main handicap however for this line, that is no reactor of this kind is available at moment and the very few old HTGRs have been shut down.

The solutions discussed above, do not completely meet, however, the requirements of a disposition which must be safe, economically viable and above all definitely proliferation resistant. It is, therefore, fully justifiable the research for alternative solutions based on the utilisation of nuclear power reactors presently available and operating. In this perspective the PWRs, due to their worldwide diffusion and consolidated technology, appear as the best candidates for Pu disposal by introducing some innovations in their fuel cycle.

6. INERT MATRIX FUEL FOR PU BURNING

In order to not produce new plutonium during irradiation, an innovative *U-free* inert matrix fuel concept is being researched from very recently[13–20]. The inert matrix consists in a mixture of inert oxides such as zirconia, aluminia, spinel and ceria, where the plutonium oxide is dispersed in. The objective of such a fuel is plutonium burning in commercial PWRs by avoiding the breeding effect, i.e. new Pu-239 generation which takes place when U-238 is present. The plutonium relative content will be about 5 per cent mol, comparable to that one used in the MOX fuel. The matrix will hold characteristics such as: good chemical compatibility, acceptable thermal conductivity, good nuclear properties (low prasitic capture rate), good stability under irradiation.

Moreover, the inert matrix fuel should comply with the following basic requirements:

- proliferation resistance: very low solubility in nitric acid so as to discourage any possible recuperation attempt either from fresh or irradiated fuel by current dissolution techniques;
- economics: the fuel pellets must be manufactureable through the already well established MOX fuel technology (mixed powders technology) or Gel Supported Precipitation (GSP) microsphere process; the inert oxides have to be abundant and easily available;
- burn-up performances: it is expected that this type of fuel shall be able to withstand a batch average burn-up same as for standard fuel, according to commercial PWR requirements for next decade;
- low environment impact: on the basis of the required high chemical stability, after discharge from reactor and ad-

equate cooling time, the spent fuel should be considered as a HLW suitable for final disposal in the deep geological formations without requiring any further reprocessing treatment (*once through solution*).

As already found by other authors[6–18], the phase relations of some sintered mixed compounds suggest that the fuel should be characterised by a particularly high chemical stability (*roch-like fuel*) which persists in the irradiated fuel. It is remarkable moreover, that such a type of fuel, after being irradiated will hold only a marginal Pu-239 content and such an high concentration of even isotopes of Pu, which coupled to its high chemical stability should render this fuel "hinerently safe" from the proliferation standpoint. The wet GSP microsphere process is proposed for fuel manufacturing, in order to avoid the not negligible Pu fines contamination problem during fabrication, being that a most relevant drawback in mixed powders technology. The second positive point for GSP application, is the potential utilisation of coated microsphere, same as in HTGR fuel, which would enhance the fission gas retention capability provided that coating integrity is conserved during pellet fabrication.

Candidate inert matrix fuel systems are:

- PuO_2 oxide dispersed in: ZrO_2 (stabilised)–Al_2O_3–MgO or ThO_2-Al_2O_3-MgO. These compounds are ternary systems with ThO_2 or stabilised ZrO_2 in solid solution with PuO_2; spinel (Mg_2AlO_4) and Al_2O_3. They are particularly investigated at the JAERI, Japan.
- PuO_2 oxide dispersed in ZrO_2 stabilized with CaO or Y_2O_3, in order to have the highly stable cubic phase zirconia. The use of zirconia, as only matrix material, could require annular pellets instead of solid pellets, due to its low thermal conductivity.
- PuO_2 oxide dispersed in spinel (Mg_2AlO_4). Spinel is a highly promising ceramic product foreseen for MAs burning in FRs and already studied for fusion technology applications. Annular pellets should be required in this case also, because the rather low melting point of spinel. The inert matrix fuel is also suitable for burning HEU in PWRs: in this case the major advantage with respect to UOX (obtained by denaturing HEU with depleted U), would be that no new Pu from neutronic captures in U-238 is generated.

A R&D activity on the above inert matrix fuel system is presently being carried out as a cooperative effort between ENEA and Polytechnic of Milan.

7. CORE ARRANGEMENT AND PU BURNING IN PWRs

Three core configurations are possible for burning Pu via intert matrix fuel in PWRs:

a) core 100 per cent loaded with inert matrix fuel;

b) core partially loaded with inert matrix fuel assemblies (fraction ranging from 1/5 to 1/3), being the remaining part loaded with standard fuel;

c) core uniformly loaded with all fuel assemblies containing a suitable number of inert matrix fuel rods.

The first configuration would be highly challenging due to the quite large amount of plutonium that could burnt in this case. Because of different neutronic features of Pu with respect of U, such type of reactor would have, however, a dynamic behaviour rather dissimilar from the current PWR one: this fact might affect the reactor controllability in such a grade that new control systems, addition of large amounts of burnable poisons and other significant countermeasures should be considered. On the other hand, the urgency in facing the plutonium issue and the need to find simple and more readily viable solutions would suggest that configurations *b.* and *c.* more easily applicable to commercial PWRs, being the characteristics of these layouts much closer to the currently MOX-fuelled cores.

Anyway, possible countermeasures could be the addition of some small fractions of depleted U or even using ThO_2 as a main component for all matrices. In the last case some depleted U could be still necessary in order to reduce the proliferation potential of U-233 produced during irradiation from the Th breeding. In a more general case, Th utilisation could constitute an additional option of the solution here proposed which permits the U-233 recoverability; this possibility seems to bring a renewed interest even though it will introduce more complexity in the fuel cycle management[21].

Despite reduction on Pu loaded in each fuel batch, solutions *b.* and *c.* which result equivalent in principle, do not show relevant problems for Doppler and void coefficients, for being rather similar to current PWR MOX-fuelled cores. In fact, the reactivity coefficients of solution c. are just slightly lower than those in a conventional PWR[16–22].

A wide preliminary investigation on configuration c. has been carried out at Nuclear Engineering Dep. of Politecnico di Milano[16–17]. For the analysis an AP-600 type core using 3.2 per cent U-235 enriched fuel, was selected as Reference Reactor (RR). It is worth to remind that this reactor is characterised by a lower power density with respect to standard PWRs, so that higher power peaks can be allowed without attempting to fuel rod integrity. The Burner Reactor (BR) was similar to the RR with the difference that in the fuel bundle, 56 rods over a total of 264 are substituted by inert matrix fuel rods (IMF).

The IMF rods layout was chosen in such a way to obtain a uniform dispersion of these kind of rods among the standard UO_2 rods, but no optimisation was done up to now to this respect.

The fissile Pu mass content has been assumed equal to the U-235 mass content in the 3.2 per cent enriched RR fuel. This amount allows the UO_2 rods within the BR assembly to achieve the same burn-up level as that in the RR[24].

The results related to configuration c. show that, no matter the Pu source (RG or WG), more than 98 per cent of Pu-239 is fissioned during irradiation. Also the fissile fraction of Pu is quite high:

~90 per cent for the RG Pu and ~93 per cent for the WG Pu. Furthermore, about 73 per cent of all Pu loaded is burnt in the case of RG Pu and about 81 per cent in the case of WG Pu. An indicator of the proliferation potential of a Pu isotopic mixture is given by fissile to fertile atoms ratio, which goes from 2.45 in fresh fuel to 0.37 in spent fuel if dealing with RG Pu, while goes from 15.1 to 0.5 if dealing with WG Pu. These low fractions of fissile Pu in the spent fuel, make highly unattractive an ever possible recovering attempt. Finally, the all Pu mass is reduced by a factor 3.7 for RG and 5.4 for WG Pu.

It is worthwhile to compare the previous figures of inert matrix fuel to those typical of a MOX fuel[23]. Because the presence of U-238 which generates Pu under radiation, the all Pu mass still present in spent MOX after a first cycle, ranges between 60 and 70 per cent of the loaded mass, depending on burn-up level; the burnt Pu fraction goes then from 30 to 40 per cent of the Pu mass which has been loaded in the reactor. During irradiation the Pu isotopic composition moves towards higher mass number isotopes, nevertheless the fissile isotope fraction is still prevailing; in fact, the fissile Pu percentage in MOX goes typically from 71 per cent in fresh fuel to 52–56 per cent in the spent fuel. The above figures show an interest for the IMF solution either in terms of overall Pu reduction or in terms of proliferation resistance. It must be recognised, however, that the inert matrix fuel solution still requires an important research activity for its complete technological qualification.

The significance of the proposal here presented is confirmed by the many parallel research programmes on Pu burning by inert matrix fuel, which are presently under way at JAERI-Japan, CEA-France, PSI-Switzerland and ENEA/Politecnico di Milano-Italy. Also US research institutions are showing interest of this kind of fuel. An information exchange among these organisations has been already undertaken. The work being carried out at ENEA and Politecnico di Milano on this topic, contributes to a 3-year activity programme on Pu and minor actinides burning in PWRs, performed in cooperation with other European partners and supported by the EU within the EAEC-4th Framework Programme 1994–98.

8. RADIOTOXICITY ISSUES

Although the Pu mass reduction and the increased proliferation resistance are assumed as key requirements, an analysis has been performed in order to assess whether the option of burning Pu in an inert matrix would produce any increase of actinide radiotoxicity. Radiotoxicity is defined in terms of number of ALI (Annual Limit of Intake):

$$\text{Nb. of ALI} = \Sigma\{Ai(t)/ALIi\}$$

where Ai(t) is the radioactivity of each radionuclide at time t after irradiation and ALIi is the Annual Limit of Intake through ingestion of the public. As reference values, 1 MT of 3.2 per cent enriched uranium spent fuel irradiated up to 34.4 MWD/kg was considered in the RR and the actinides contained in this unit mass have been labelled as mix M1. In the Pu burning option, the Pu in mix M1 was assumed to be recovered with a 2 per cent loss rate, manufactured as inert matrix fuel and irradiated in the BR. The actinides contained in the inert matrix fuel after irradiation, together with those left in the reprocessed spent fuel, after

Pu separation, were marked as mix M². The performed calculations have shown (see curves) that the radiotoxicity of mix M2 was larger than that of mix M1 for a period of about 25 years, because of the larger initial contributions of Am and particularly Cu. The Pu direct contribution was clearly reduced by the fissioning process. After this period the radiotoxicity of mix M2 fell below that of mix M1, under the effect of the lowered direct contribution of Pu and the reduced Am build-up. What is worthy to be stressed is that the significant reduction of Pu overall mass and increase of its proliferation resistance, did not involve any increment in actinide radiotoxicity compared to UOX spent fuel. Finally, if we do refer to radiotoxicity per energy unit, the actinide radiotoxicity in mix M2 becomes about 20 per cent lower than that in mix M1.

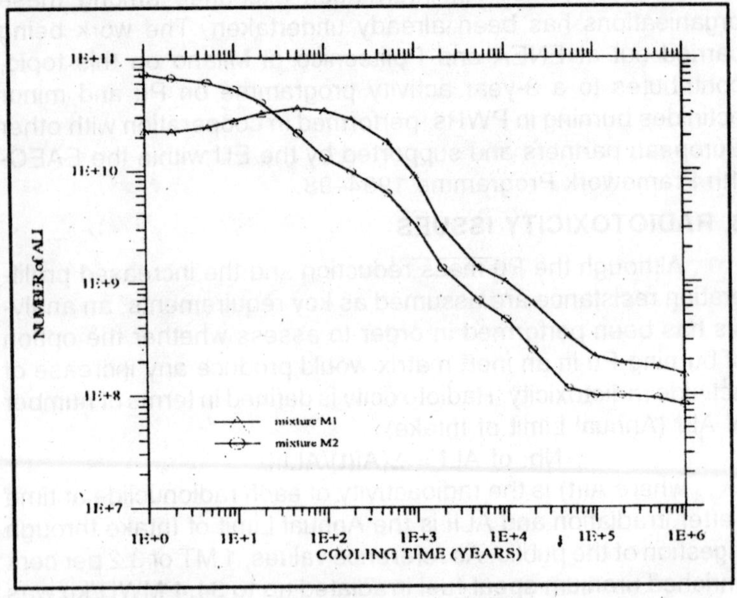

9. CONCLUSIONS

The Pu coming from nuclear stockpile dismantling represents a true challenge to which the present nuclear technology can find a rational solution.

The weapons Pu holds a not negligible energetic potential, possible to be converted to useful electricity in the already existing commercial PWRs.

In this way the huge economic investments made during the last decades on nuclear weapons, could be recovered at least at limited extent and anyway converted to a peaceful goal.

This objective can be achieved under the optimised conditions of system safety, proliferation resistance, environmental impact and rational use of the technical and scientific competences already existing, by especially implementing innovative solutions in the nuclear fuel cycle, such as for ex. the once-through "U-free" inert matrix fuel devoted to current commercial PWRs.

References

1. M. Hibbs, "*Mikhailov Told US It Was Buying Less than Half of Soviet HEU=20 Stock*", Nucleonics Wick, September 30th, 1993.
2. M.R. Buckner, "*Comparison of Options for Plutonium Disposal Reactors*", 20 WSRC-TR-92-554, U.S. Department of Energy, 1992.
3. M.R. Buckner and P.B. Parks, "*Strategies for Denaturing the Weapons-Grade Plutonium Stockpile*", WSRC-RP-92-1004, U.S. Department of Energy, 20, 1992.
4. D.F. Newman, "*Burning Weapons-Grade Plutonium in Reactors*", 4th 20 Annual Scientific and Technical Conf., June 28-July 2, 1993, Nizhni Novgord, 20 Russia.
5. C.E. Walter and R.P. Omberg, "*Disposition of Weapon Plutonium by Fission*", Proce. Int, Conf. and Technology Exibition on Future Nuclear Systems: Emerging Fuel Cycles and Waste Disposal Options (Global '93), Seattle, Washington, September 12-17, 1993, 2, 846, American Nuclear Society
6. D. Diswas et al., "*Weapons-Grade Plutonium Disposition in Pressurised Water Reactors*", Nuc. Sci.. and Eng, 121,1
7. R. Capitini, "*New Objectives for Superphenix*", Plutonium, 7,20,1994
8. N. Usui, "*Monju Finally Generates Power To Grid But Cost Issue Remains*", Nucleonics Week, September 7, 1995
9. K. Uematsu, "*Technological and Economic Aspects of Pu Utilisation in Fission Reactors*", Proc. International Symposium on Conversion of Nuclear Warheads for Peaceful Purposes, Rome, Italy, 1992
10. P.M. Chantion and J. Finucane, "*Plutonium as an Energy Source: Quantifying the Commercial Picture*", IAEA Bulletin, 3, 35, 1993
11. P. Leventhal, "Pause for Thorp", The Guardian, November 26, 1993
12. A. MacLachlan, "*EDF To Erase Positive Pu Value In 1995 Accounts*", Nucleonics Week, November 2, 1995.

13. E. Cerrai and C. Lombardi, "*Burning Weapon-Grade Pu in Ad Hoc Designed Reactors*?", Proc. Int. Symp. on Conversion of Nuclear Warheads for Peaceful Purposes, Rome, Italy, June 15–17, 1992, II.
14. C. Lombardi and A. Mazzola, "*The Uranium Exploitation and the Plutonium Dilemma*", Energ. Nucl., anno 11, 1, 27.
15. C. Lombardi and A. Mazzola, "*Weapon-Grade Plutonium: Annihilation Via Thermal Fission in Unconventional Non-Fertile Matrices*", Int. Congress on Nuclear Weapons and Underdevelopment: Effective Projects from Italy, Rome, Italy, July 4–5, 1994.
16. C. Lombardi and A. Mazzola, "*Plutonium Burning in Pressurized Water Reactors Via Non-fertile Matrices*", to be published on Nucl. Sci. and Eng.
17. C. Lombardi and A. Mazzola, "*Exploiting the Plutonium Stockpiles in PWRs By Using Inert Matrix Fuel*", to be published on Ann. Nucl. Energy.
18. H. Akei et al., "*A New Fuel Material for Once-Through Weapons Plutonium Burning*", Nucl. Tech., 107, 182, 1994.
19. Proc. *IAEA TCM on Unconventional Options for Plutonium Disposition (With Main Emphasis on Their Technical Description and Status of Development)*, Obninsk, Russia, November 7–11, 1994.
20. J.M. Paratte and R. Chawla, "*On the Physics Feasibility of LWR Plutonium Fuels Without Uranium*", Ann. Nucl. Energy, 22, 7, 471, 1995.
21. C. Rubbia et al., "*A Realistic Plutonium Elimination Scheme With Fast Energy Amplifiers and Thorium-Plutonium Fuel*", CERN/AT/95-53 (ET), European Organisation for Nuclear Research, 1995.
22. A. Puill and J. Bergeron, "*Improved Plutonium Consumption in a Pressurized Water Reactor*", Proc. Global '95-International Conference in Evaluation of Emerging Nuclear Fuel Cycle Systems, September 11–14, Versailles, France, 1, 897, 1995.
23. H.W. Wiese, "*Investigation of the Nuclear Inventories of High-Exposure PWR Mixed-Oxide Fuels With Multiple Recycling of Self-Generated Plutonium*", Nucl. Technol., 120, 68, 1993.
24. C. Lombardi, A. Mazzola, F. Vettraino, "*Plutonium Annihilation in PWRs Via Non-Fertile Inert Matrices*", GLOBAL '95, Versailles, France, Sep. 1995.

Address:

CARLO LOMBARDI

Dept. of Nuclear Engg.
Polytechnic of Milano
Milan
Italy

65
A Clever Use of Ex-Weapons Material

JEAN PIERRE ROUGEAU

INTRODUCTION

According to START I & II Agreements, the United States and Russia are committed to remove thousands of nuclear warheads from their arsenals and, once dismantled, those warheads make available significant quantities of fissile materials: Tritium, Highly Enriched Uranium (HEU) and weapons grade plutonium (wgPu). Part of this material, not needed for future defence programmes, will be permanently withdrawn from the military stockpiles and will have to be treated in order to make sure it could never become weapon-usable again, neither for the owner country nor for potential terrorists.

Many studies have been and are still conducted to evaluate options for the disposition of warheads origin fissile materials. But the outcome does vary a lot according to each different material. Tritium is currently being recycled and is still needed in the military industry. The disposition of HEU is well ahead of wgPu treatment since implementation of HEU recycling has already started. As far as wgPu is concerned, although civilian grade plutonium is currently recycled in commercial reactors in Europe, there is no consensus yet on the benefits of adopting the same strategy for this material.

A concrete approach of what could be a realistic and clever disposition programme is presented to underline the real benefits of the recycling option.

AMOUNTS OF MATERIALS INVOLVED

The quantities of fissile materials involved are considerable, especially when compared to civilian industry current needs or even forecasts.

In Russia, stockpiles are estimated to be above 1000 Mt HEU and 150 Mt wgPu. 500 Mt HEU are going to be withdrawn under the so-called "HEU agreement" signed with the U.S., and most of the wgPu is supposed to be withdrawn from the military industry'; nevertheless since many different figures have been released either by MINATOM officials or by journalists for both materials the picture is not fully clarified: In the United States, a recent DOE report on military plutonium production and inventories has shown that there were about 85 Mt wgPu in the US military stockpiles. These amounts are actually in various forms and spread over several DOE sites. HEU stocks are unofficially said to be above 500 Mt. But the whole stock is not actual surplus. Those materials will not all be "available". President Clinton officially declared in March 1995 that 200 Mt of military fissile materials were surplus and going to be put under IAEA safeguards. Among the 200 Mt, 165 are HEU and 38 are wgPu according to Hazel O'Leary in a statement last September. A DOE HEU EIS draft released in October 95 assumed a maximum surplus value of 200 Mt HEU, with an estimated 50 per cent enrichment level, which proves that 200 Mt are probably not a definitive figure. It is not the point here to pretend revealing the actual figures in place of official bodies. For practical purpose it is assumed hereafter that the total "available" quantity of wgPu amounts to 100 Mt in Russia, which is the most often quoted figure, and to 50 Mt wgPu in the US, quantity which is commonly used in the different DOE scenarios studies.

As far as HEU is concerned, this paper takes into account the currently most reliable figures, that is the 500 Mt Russian HEU from the HEU deal, and 200 Mt US HEU as stated by the DOE.

RECYCLING OF HIGHLY ENRICHED URANIUM

The process used in order to recycle HEU consists in diluting it and thereby transforming it into a low enriched fuel that can be loaded in commercial reactors.

Such blending has two advantages: it is a way of taking HEU away from military purpose and at the same time provides a commercial product.

Energy Resource

Russian and American HEU are not precisely the same

materials, and both countries do not plan to blend it exactly the same way.

Whereas Russian material surplus consists essentially in HEU enriched at level above 90 per cent, American HEU is on average 50 per cent U235. Moreover, the isotopic assay of the material used for dilution and of the resulting dilution is different from one country to the other.

Taking into account the isotopy and blending process differences between US and Russian stocks, 200 Mt US HEU would produce 1200 billion kWh and result in saving 32 000 Mt NATU and 19 million SWUs.

In Russia, resulted LEU would generate 4800 billion kWh in place of 150 000 Mt NATU and 90 million SWUs.

Global surplus has a saving potential of 3 years of NATU requirements and 3.5 years of enrichment need in the world.

These comparisons help to figure out why the recycling option of surplus HEU is not a mere option, but rather the solution. On the other hand, it is a way of underlining the negative impact that those materials could have on the enriched uranium industry, which is obviously not in a position to absorb suddenly this new source of supply.

Recycling Implementation

US and Russia signed in January 1994 an agreement whereby Russia would dilute 500 Mt HEU and sell it to US for commercial use. More precisely, diluted HEU was to be transferred during 20 years within annual quantities between 10 and 30 Mt per year. It was afterwards decided that more than 20 Mt was not conceivable: it would be beyond absorption capacity of the market and also above current Russian dilution capacities.

The first delivery of blended LEU occurred in spring 1995, after completion of a purification process which was said to be more difficult and expensive than expected.

The United States has also released in October 95 and EIS draft on HEU disposition that recommends the dilution of American HEU for commercial use in nuclear power plants. But like in Russia, part of the US surplus is at a level of enrichment and impurity that makes commercial recycling unfeasible or unprofitable, especially because of high concentration in isotopes U234 and U236. The recyclable part of US HEU surplus is estimated to be 65 per cent to 85 per cent of the "available" quantity.

The rationales for HEU recycling did not take long to convince. Recycling is actually less a technical issue than a commercial one. The real question is: how to manage the entry of a huge amount of substitution product into the market without long term damage to the enrichment industry? Both parties, US and Russian governments, have agreed that the HEU deliveries should not disturb the Uranium and Enrichment markets and have therefore planned a 20-year completion period.

RECYCLING OF WEAPONS GRADE PLUTONIUM IN LWRs

As far as military plutonium is concerned, the challenge is more complex and the industrial solution is not yet selected.

Whereas the military industry is used to handle such a material, civilian industry is not familiar with separated wg-plutonium. Civilian nuclear industry is in fact used to deal with reactor-grade plutonium which is a natural by-product of nuclear energy generation; this is the case in those countries that have already chosen the so-called "closed fuel cycle", and have developed plutonium recycling. On the contrary, in the U.S. many still prefer to keep plutonium included in spent fuel rather than separated and "available".

But in the military plutonium issue, discussions on whether plutonium is to be separated or not are irrelevant simply because wgPu is already separated, and the recycling experience accumulated by the industry will serve to process it safely.

Non-Proliferation

The major goal of wgPu disposition is to prevent anyone of being able to use it as a bomb component. Three actions can reduce the chance that someone could some day easily recover a sufficient amount of such material:

– to reduce total amount of fissile material
– to decrease its concentration in isotope 239
– and to restrict its possible retrieval.

It has been demonstrated, through several years of experience in Europe, that the recycling of civilian Pu in mixed oxide (MOX) fuel for LWRs fulfil these 3 functions.

Illustrative figures quoted hereafter are related to a PWR 900 MW with fuel irradiated at 43 GWd/Mt.

MOX and Pu Balance

The point here is to demonstrate from actual figures the

ability of MOX fuel to consume plutonium through energy generation.

One UO_2 fuel assembly does not contain any plutonium when loaded in a commercial reactor but contains about 1 per cent of this material after irradiation. In concrete words, a 500 kg fuel assembly contains 5kg plutonium when unloaded. Energy generation in an UO_2 assembly, which is due to Plutonium for up to 40 per cent, results in a net positive balance of plutonium.

In comparison, civilian MOX fuel is about 7 per cent Pu when fresh for an expected burn-up of 43 GWd/t. Of the 35 Kg plutonium initially loaded, only 25 kg plutonium remains after irradiation. In a MOX assembly, energy generation consumes plutonium.

A LWR power unit, loaded with 1/3 core of MOX fuel, consumes an amount of plutonium in MOX assemblies that equals what is created in UO_2 elements. Higher fractions of MOX assemblies will transform this reactor in a plutonium burner. Reactors of the next generation are currently designed to receive 50 per cent or 100 per cent MOX fuels.

MOX and Isotopic Degradation

Recycling plutonium through MOX fuel is the only way of achieving isotopic degradation, because irradiation in LWR actually down grades the material.

Plutonium recovered by reprocessing of spent 43 GWd/Mt UO_2 fuel is about 66 per cent fissile. Its content in fissile isotopes is 55 per cent Pu 239 and 11 per cent Pu241.

Incorporation in fresh MOX fuel and irradiation result in an increase in Pu241 content but a greater decrease in Pu239 content.

When unloaded, the plutonium is only 54 per cent fissile, which makes it clearly less suitable for any weapon utilisation.

Plutonium Retrieval

Last of the three actions to be undertaken is to make the material as less accessible as possible.

The irradiation of MOX fuel in LWR produces a matrix of spent MOX fuel from which plutonium cannot be easily recovered.

To retrieve plutonium in a spent MOX fuel matrix is feasible, like to extract plutonium from a spent UO_2 matrix is feasible. But the retrieval is possible only through a costly and technically demanding industrial process, because of fission products releases. This could not be easily done.

Recycling wgPu through MOX fuel in LWRs, fulfils the goals of the disposition programme in terms of non proliferation.

Industrial Background

Process

To recycle plutonium through utilisation of MOX fuel requires to implement a very specific fuel fabrication process, that European industry has developed. Even if the raw material is not absolutely similar, civilian industry has enough experience to expect that the process transposition to wgPu will not raise any major obstacle.

The first step is to convert plutonium metal into plutonium dioxide which is the usual input of MOX fabrication line. Such a process does not industrially exists yet and would have to be developed for the disposition programme. But detailed studies are currently conducted in cooperation with the surplus owners to design a suited conversion process. Even if all results of these studies can not be released, being still restricted to defence activities area, it is known that several different experiments on wgPu samples were conducted successfully. Obtaining sinterable PuO_2 from wgPu seems reasonably feasible.

Beyond the conversion, the fabrication line itself will be very similar to a traditional civilian MOX fabrication line, with some differences related with the properties of wgPu compared to civilian Pu.

But these process modifications will not induce huge difficulties. Low content in isotopes 238 and 241 would even make the process easier in terms of thermal power, which is 7 to 10 times lower than with civilian plutonium and in terms of health physics and safety, because, and neutrons emissions are also notably lower.

The only constraint would come from criticality, leading to the need for some specific adaptation of the civilian MOX technology, for example smaller size equipment in the first part of the facility. Designing an adapted process in terms of criticality is already done for civilian MOX. More criticality will simply result in taking into account new parameters for the plant, but is not harder to achieve than current design.

European MOX Industry Experience

In Europe, several countries have decided to implement the reprocessing/recycling route for years and have therefore created the related industry. Reprocessing pilot plants have been built in Germany and in Belgium and industrial scale reprocessing plants are operating in France and in the UK. Several European countries like Germany, Belgium, Switzerland, United Kingdom or France do have been involved in the MOX industry, either by operating MOX fabrication facilities or by loading commercial LWR with MOX fuel.

About 400 Mt MOX have already, been loaded in commercial reactors since first test in Belgium in 1963, in order to produce more than 100 billion kWh.

18 European reactors, PWR and BWR, have already been loaded with this fuel, without any problem,.

Three industrial MOX fabrication plants are operated in France and Belgium totalling more than 200 Mt MOX/y nominal capacity. The UK is also involved in a pilot scale facility and is embarked on a large industrial fabrication plant with completion expected at the end of the century.

After more than thirty years of experience in MOX technology, the European industry is ready to offer a mature fabrication process and to adapt its design for military origin material, which will not induce difficulties.

Economics

Beyond its non-proliferation aspects and its technical feasibility, the wgPu recycling option is also a way of making the disposition programme profitable. A great amount of a very energetic material will in such way produce electricity.

Energy Resource

The amount of material, first, is considerable; 150 Mt of wg-Pu is about three times the world annual production of fissile plutonium in the LWRs. And when compared to the reprocessing industry output, the figure is seven times the quantity currently recovered by reprocessing LWR fuels.

Considering the conditions under which plutonium is loaded in commercial reactors, it is conceivable to assume 4 to 5 per cent wg Pu in Mox fuel. If burned this way in commercial LWR, 150 Mt wgPu would produce more than 1000 billion kWh, which is about 5 times the annual electricity requirement of a country like Italy for example.

Resulted fresh material savings would be around 25000 Mt of natural uranium and 16 million SWUs, which helps to understand the interest demonstrated by several utilities towards this programme.

Costs

Building wgplutonium stockpiles was part of the defence programmes and related costs were included in military budgets. It is hard to anticipate the conditions under which civilian industry will receive this new fuel supply. Anyone knows that this point is essential for reactors owners motivation.

As far as the material form is concerned it is assumed that industry will receive sinterable PuO_2 powder from governmental organisations. In such a case, from a strict MOX industrial point of view, and provided that the conversion costs are supported by governments budgets, the fabrication costs of wgPu MOX will be in the range of civilian MOX fabrication costs.

Provided it is transferred to industry in a suitable form, that is sinterable PuO_2 powder, the wgPlutonium can be processed in order to produce energy at a competitive price.

WgPu in LWRs Recycling Implementation

The rationales for wgPu recycling option are clear. Provided that needed facilities are available and that entities involved in the programme are motivated, the implementation of wgPu disposition will be achieved together with electricity production.

Host Reactors

Three kinds of facilities are needed to implement such recycling: the first one includes the conversion plants that will produce PuO_2 powder. The MOX fabrication plants are the second. Those facilities do not exist yet in the US nor in Russia, but would be easily constructed with European cooperation, like it is mentioned above. Thirdly the power plants are needed.

Since wgPu is a very sensitive material in terms of proliferation, it has to be constantly protected while stored, transported or processed. Therefore the best way of anticipating theft risks is probably to keep it inside its original territory. This argument is true not only for power plants selection, but also for all facilities involved in the disposition process.

In addition to proliferation issues, it is quite logical to consider that former USSR and US, having spent huge amounts of money producing wgPu, now get some return on investment by using this valuable material in their own reactors. And those host reactors exist in both countries.

A recent report states that at least 40 US LWRs are technically suited to burn MOX fuel. And according to current studies and calculations, in particular those conducted by Russian Institutes, the VVER 1000 reactor type is also well suited to be loaded with MOX fuel. About 20 of those currently exist or are under construction in Russia and Ukraine.

Those reactors will have to go through licensing procedures before being able to receive the new fuel. But these procedures should not take any longer than time needed to build MOX plants, that is about 5 years.

Motivation

Last condition in a concrete approach of such a disposition programme is the motivation of entities involved in the process.

The actors are on the one hand the DOE and MINATOM, public entities which major goals are to implement safely the governments commitments and to spend the least for the disposition programme.

On the other hand, utilities and operators of the reactors must also find their interest in the programme, which is to benefit from a new and competitive fuel supply without major core modifications.

The situation is a little different in the US compared to Russia. In Russia where plutonium has always been considered as a "national treasure" that should never be treated as waste and recycling seems to be the prevailing option. In the US, there is a current evaluation of two different options for the disposition programme, which are wgPu vitrification and wgPu recycling. In fact, 10 utilities have responded to a DOE request of interest for wgPu recycling in February 96, which means a real change in the U.S. way of considering plutonium.

What Could be a Realistic and clever Pu Disposition Programme

Is the recycling of 50 Mt wgPu in the US and 100 Mt in Russia realistic over a reasonable period?

Considering a 1000 MW LWR 30 per cent MOX loaded every year with 8 Mt MOX containing 320 to 400 kg wgPu, 50 Mt US wgPu could be consumed by 12-15 reactors within 10 years. The quantity of MOX needed would amount to 100-120 Mt/y.

As far as the former USSR stock is concerned, quantity is more important, but 9 to 11 VVER 1000 loaded with 30 per cent MOX could burn 100 Mt wgPu within 34 years. Such a programme would require 70 to 85 Mt MOX/y depending on the Pu content in MOX fuel.

The number of reactors involved in each country gives obviously the rhythm of the recycling programme. And an increase in the MOX part of each reactor core could also influence greatly the disposition timing. But considering that those stocks have been produced for 50 years, a 10 to 30 disposition years approach seems not to be out of the picture.

Quantities of wgPu MOX to be fabricated, from 70 to 120 Mt/y, are in the range of existing or under construction industrial MOX fuel plants, which confirms that MOX industry competence is well suited for cooperation in the disposition programme.

French involvement in Russian Disposition Programme

An example of the cooperation programmes between surplus materials owners and recycling industry is given by the current cooperation between France and Russia in this field.

French and Russian Governments have signed in 1992 an agreement on collaboration towards peaceful utilisation of exweapons material, called AIDA MOX.

More precisely, the joint programme aims at studying the feasibility of all steps of the Russian wgPu recycling process, from conversion to MOX reprocessing. A comprehensive summary of the programme was presented at the Global 95 Conference last September.

Several scenarios are currently evaluated, depending on the type and number of reactors involved. As far as MOX fabrication is concerned, the feasibility of constructing a first facility to incorporate Russian wgPu into MOX fuel for VVER and fast neutrons reactors is currently under study. The TOMOX plant would aim to transform 1300 kg wgPu per year in order to produce around 20 Mt MOX for 4 VVER 1000 at Balakovo and 1.5 Mt fuel for a BN600 fast breeder.

It would be the first step towards a large facility able to produce MOX for all reactors involved in the disposition programme.

It would also be the first step towards the effective start of Russian wgPu disposition.

At the end of 1996, it will result in the definition of an industrial programme for wgPu MOX fabrication and in-reactor recycling, preferably in cooperation with other European countries.

CONCLUSION

WgPu recycling is politically safe, technically feasible, economically profitable, and realistic.

Parties involved do have the means of completing such a programme, and they can benefit from the civilian MOX industry experience and know-how. Some cooperation programmes have already started.

The only one ingredient missing is a strong political initiative, which is needed to turn, in the decades to come, military plutonium into kWhs. As far as the industry is concerned, the implementation of such a programme could start soon.

A CLEVER USE OF EX-WEAPONS MATERIAL

Jean-Pierre ROUGEAU-COGEMA

Centro Di Cultura Scientifica A. VOLTA - COMO - ITALY
March, 18/20, 1996

1. AMOUNTS OF MATERIAL INVOLVED

	HEU	wgPLUTONIUM
RUSSIA	1,000 Mt ⇒ 500 Mt	1,50 Mt ⇒ 00 Mt
US	(500 Mt) ⇒ 200 Mt	85 Mt ⇒ 50 Mt

2. RECYCLING OF HIGHLY ENRICHED URANIUM

	RUSSIA	US	TOTAL
• Surplus	500 Mt	200 Mt	700 Mt HEU
• % U235	90%	50%	
• Blending Material	1.5%	1.4%	
• Resulted LEU	4.4%	4%	
• Billion kWh	4,800	1,200	6,000 billion kWh
• Savings Mt Natu mILLION SWUs	1,50,000 90	32,000 19	3 years WORLD NATU NEED 3.5 years WORLD SWU NEED

3. PROLIFERATION : MOX IN LWRs

			Fresh Fuel	Spent Fuel	Balance
• Plutonium Balance (43 GWd/t - 4 Years storage)	2 UO2 FA	1000 kg	0 kg Pu	10 kg Pu	+ 10 kg
	1 MOX FA	500 kg	35 kg Pu	25 kg Pu	– 10 kg

⇒ With 30% MOX Core, Plutonium Balance = 0

			Pu239	Pu241	Fis Pu
• Isotopic Degradation (43 GWd/t - 4 Years storage)	MOX	Fresh Fuel	55%	11%	66%
	⇒	Spent Fuel	37%	17%	54%

• Access & Retrieval — Heavy and costly Industrial Process

TRANSPOSITION TO WG PLUTONIUM

4. WGPU IN LWRs RECYCLING : INDUSTRIAL BACKGROUND

PU METAL TO PUO2 — Industrial Process still to Develop

MOX PLANT ADAPTATION

- THERMAL POWER — 7 to 10 times lower
- HEALTH PHYSICS AND SAFETY — α, γ, neutron Activity Reduction
- CRITICALITY — More Constraints

⇒ BENEFIT FROM EUROPEAN MOX EXPERIENCE

5. WG PU IN LWRs RECYCLING : ECONOMICS		
SURPLUS	**150 Mt wgPu**	
⇔	3 years Fissile Pu Production in LWRs 7 years Fissile Pu Recovered by Reprocessing LWR Fuels	
GENERATION	1000 billion kWh	
SAVINGS	25,000 Mt NATU 15 million SWUs	
COST	wg PuO_2 Cost wg Pu MOX Cost	Included in Governmental Budgets In the Range of civil Pu MOX Cost

6. WGPU RECYCLING : IMPLEMENTATION	
HOST REACTORS	wgPu Producting Country = wgPu Recycling Country
MOTIVATION	• RUSSIA : Pu = "National Treasure" • US : Vitrification OR Recycling ⇒ 10 US Utilities Interested in Recycling Option

PROGRA-MME	**Surplus wg-Pu**	**Reactors Loaded**	**MOX FABRICATION NEEDS**	**YEARS**
RUSSIA	100 Mt	9–11 VVER 1000 30% MOX	70–85 MtHM/y	34
US	50 Mt	12–15 LWR 30% MOX	100–120 MtHM/y	10

Address :

JEAN PIERRE ROUGEAU
Vice-President
Corporate Street
COGEMA
78141 Velizy
France

66

Utilisation of Already Separated Plutonium in Russia: Consideration of Short and Long-term Options

ANATOLI S. DIAKOV

Today there are a growing Russia's stocks of separated plutonium recovered from dismantled nuclear warheads and from military and civil reprocessing plants that present an international security problem. In the long term, the basic Russian approach for disposition of this plutonium is to burn both weapons plutonium and civil plutonium in the fast-neutron reactors, but due to current political and economical situation in Russia this plan can not be realised any time soon. Under such conditions the first priority should be placed on the set up of a plutonium storage regime under bilateral or international control.

INTRODUCTION

The overriding objective of U.S. and Russian nuclear disarmament initiatives is to dispose of excess of nuclear fissile materials released from weapons in such a way that they can not be reused in the country of origin or stolen by terrorists group, for military purpose.

The natural disposal method for the weapons-grade uranium is to blend it with natural uranium to produce a nuclear reactor fuel. In February of 1993 the governmental Russian -US agreement have been signed in accordance with 500 tons of Russian weapons uranium which will be recovered from dismantled warheads and mixed with depleted uranium will be sale in the US as raw material to fabricate a reactor fuel. This agreement is being implemented. In 1995 Russia delivered to the United States LEU derived from 6,1 tons of HEU, and in 1996 will deliver the LEU derived from 12 tons of HEU obtained from nuclear weapons.

In the case of weapons-plutonium the situation is much more complicated. The utilisation of already separated plutonium, particular of excess plutonium from retired nuclear weapons, is rising a complex set of technical, economical, environmental and political problems. All these problems are closely related and mutually reinforcing. The determination of optimal option for disposition of excess weapon plutonium based on such criteria as technical viability, resistance to theft or diversion, economic, timeliness, environmental protection and others is being studied by Russian and foreign experts.

FSU/RUSSIAN STOCK OF WEAPONS PLUTONIUM

During the Cold War the FSU/Russian nuclear military production complex produced about of 125 tons of weapons-grade plutonium.

As the results of nuclear arms reduction most of this plutonium will become "surplus". Currently Russia is dismantling nuclear weapons and plutonium components at rate of some 7 tons per year which are being shipped to storage at the disassembly plants near Seversk (Tomsk-7), Ozersk (Chelyabinsk-65) and Arzamas-16. It is expected that some 100 tonnes of plutonium will be released from weapons in Russia[1].

Furthermore, Russia will continue to produce significant amounts of weapons-grade plutonium. Only 10 of the 13 Russian plutonium-production reactors have been shut down. Although, the three remaining reactors are now operating principally to supply heat to cities Tomsk and Krasnoyarsk, they continue to produce weapons-grade plutonium at a rate about 1.5 metric tonnes each year. The Russian government has obligated that as of October 1, 1994 all newly produced plutonium be not used in weapons and wiil be stored in the oxide form.

RUSSIAN STOCK OF CIVIL PLUTONIUM

Table 2 presents the amounts of fuels discharged from Russian power reactors[2] and estimates of the amounts of reactors-grade plutonium.

Russia is reprocessing spent fuel from domestic and Soviet-built reactors VVER-440, BN600 power reactors, naval reactors, and research reactors. At present, about 30 tons of separated reactor-grade plutonium[3] in the form of the plutonium dioxide is being stored at Chelyabinsk-65. It is supposed to

Table 1 : Estimated Weapons Plutonium Production in FSU/ Russia by 1996

Type of reactor	*Power, MWt (designed/upgraded)*	*Period of operation production MT Wpu*	*Estimated*
A	100/90	06.19.48/06.16.87	6.5
IR-AI	50/500	12.22.51/05.25.87	3.4
AV-1	300/1200	04.01.50/12.08.89	8.9
AV-2	300/1200	04.06.51/06.14.90	9.0
AV-3	300/1200	09.15.52/11.01.90	6.3
1–1	600/1200	11.20.55/09.21.90	8.5
1–2	600/1200	09.58/12.31.90	8.2
ADE-3	1600/1900	07.61/08.14.90	11.9
ADE-4	1600/1900	02.26.64/in operat	12.7
ADE-5	1600/1900	06.27.65/in operat	12.1
AD	1600/1800	08.25.58/06.30.92	13.5
ADE-1	1600/1800	61/08.29.92	12.3
ADE-2	1600/1800	63/in operat	13.2
Total	**126.2**		

Table 2 : Russian Civil Plutonium Production Data

Reactor's type	*Mass of Spent fuel MT*	*Estimated mass of Pu MT*
RBMK	6100	38
VVER-1000	1000	11
VVER-440	1250	17
BN-600	65	6
Total	**72**	

fabricate this material into reactor fuel before beginning to use weapons plutonium because the growth of the radioactivity in the civilian plutonium due to decay of Pu241 makes it difficult to handle.

RUSSIAN APPROACH TO PLUTONIUM DISPOSITION

The Russian Ministry of Atomic Energy (MinAtom) views plutonium as a valuable energy sources[7]. Its current concept, of how to utilise plutonium, is based on that approach which was developed two decades ago when there was a great energy demand, and entails the following stages:

- near-term stage—secure storage of both surplus weapons and civil plutonium;
- medium-and long-term stages—utilization of civil plutonium and excess of weapons-plutonium in the fast neutron

reactors and in the thermal reactors.

Another possibility for the disposition of Russian plutonium is to use it for MoX fuel fabrication and sell this fuel on the world market. The Canadian government as well as Canadian nuclear industry has expressed support to the idea of transforming Russian excess weapon plutonium into MOX fuel and burning it in CANDU reactors. The feasibility study of CANDU option is currently in progress.

Although Russian experts are studying of non-reactor options within the frame of the joint U.S./Russian Plutonium Disposition Study, there is currently little enthusiasm in MinAtom for this approach to plutonium disposition. MinAtom officials repeatedly stated that the priority is given to the use of weapon grade plutonium in nuclear fuel for power production industry and its immobilisation or geologic disposal. Use of the existing vitrification technology has always been perceived as unsafe[4].

A research and development programme was adopted by MinAtom to coordinate efforts on implementation of technology and construction of equipment to use of weapons plutonium in the MOX-fuel fabrication for fast and thermal reactors. This programme includes:

- processing of metal plutonium to plutonium oxide in the procedures of its dissolving, filtration, purification, precipitation and heating;
- fabricating of the fuel elements and fuel assembly;
- processing of the radioactive wastes resulted from the conversion procedure;
- production of the containers for secure and safety storage of plutonium dioxide and for its transportation;
- construction of storage facility.

MinAtom requested $8.5 million as a part of the FY 1995 funds to support this programme, but to the end 1995 it has received about 50 per cent of these planned funds.[5]

STORAGE

Evidently, the time required for nuclear weapons dismantlement is much lesser than even the time to decide on optimal option for disposition of plutonium surplus. Therefore, it is necessary to store this surplus in safe and secure manner. In this context, the construction of storage facility which would provide storage for fissile material from dismantled nuclear weapons, and support the schedule on weapons dismantlement becomes a first priority.

The construction of a fissile material storage facility were begun by MinAtom, with the U.S. assistance, at the Mayak site last Spring, and it's completion is scheduled on 1997. The storage capacity is 50 thousand containers. This facility will correspond to all modern international standards for safe, secure and accountable storage. The cost of the construction is $150 million. The U.S. is providing the essential part of financing for material and labour cost, $75 million was allocated for this project.

Also, (January 13, 1995) the Russian Government adopted the programme of immediate measures on implementation of the national nuclear material protection, control and accounting system (MPC&A). The strategy for realisation this task includes the development of:

- the draft of new legislation to regulate the state activity in this field;
- the concept of national MPC&A system;
- the base documentation to regulate an agency activity;
- the federal programme to introduce MPC&A system;
- the technical project of the state information system for MPC&A
- the structure of the inspection service.

Several governmental agencies including MinAtom and GosAtomNadzor are involving in ongoing work. The close collaboration on these issues with Western countries is established. The GosAtomNadzor as a general manager of this programme requested $22.2 million to support programme activities but up to now only about of $1 million have been received for the development of new legislation[6].

BURNING PLUTONIUM IN REACTORS: STATE AND PROSPECTS

In order to examine what utilisation option is the most promising if Russia insists on its approach, a number of criteria were selected in this paper:

1) the technical viability;
2) timeliness;
3) resistance to theft or diversion;
4) cost.

Because the lack of information, the cost presented here represents only an estimate of capital investments. Definitely, these set of criteria is not complete and sufficient but it allows to evaluate the merits and disadvantages of the different options.

FAST REACTORS

Technical Viability

Russia has began experiments with plutonium for fuel fabrication in the middle of 50-th however the systematic studies of plutonium fuel started with BOR-60 reactor in 1970[1]. Although, a number of different kind MOX fuel elements were tested in Russia, the prototype of industrial power production fast reactors BN-350/600 have been fueled mostly with enriched uranium fuel.

The construction of first two fast reactors BN-800 was started at the South Urals site but then has been suspended in the initial stage due to financial problems. The new BN-800 fast reactor is designed and has passed all required examinations. No problems are expected with plutonium of various isotopic compositions. Also, the fast-neutron reactors could process larger amounts of Plutonium than LWRs of equal power output, and the radiotoxicity of its spent fuel would be significantly less.

There are three pilot installations in Russia to produce MOX fuel for fast reactors: two at the Mayak association in Osersk and one at the RIAR in Dimitrovgrad. The capacity of "Paket" installation at the Mayak allows to produce 10-12 fuel assemblies annually (300 kg of MOX fuel with about of 20 per cent of reactors-grade plutonium) for BN-600/350 reactors. The modified installation "Packet" has a MOX production capacity up to 40 fuel assemblies (1 tonne of MOX fuel). The capacity of "Granat" installation at the RIAR is about 1 tonne of MOX fuel. The design and technology of these pilot installations are not corresponding a modern requirements[11], and their use for plutonium utilisation is doubtful.

The construction of industrial scale MOX fuel fabrication plant Complex-300 to fabricate fuel for BN-800 reactors was started at the Mayak but has been suspended lately in the 50 per cent complete due to financial difficulties. The proposed capacity of this plant is 900 fuel assemblies annually (60 tonnes of MOX fuel).

Timeliness

To implement this stage of plutonium utilisation, MinAtom proposes to build four 800-Megawatt fast-neutron reactors: three near Chelyabinsk-65 and one at Beloyarskaya site and to complete the construction of Complex-300 MOX plant at Ozersk site[7]. The Bn800 reactor design allows to irradiate 1,6 tons plutonium per year. If implemented, four reactors would consume 100 tons of plutonium during 15 years. But taking into account that a substantial period of time would be required to built reactor (about ten years) and that each next unit would be installed following a delay 5 years therefore the process of plutonium disposition would start by 2010 and finish by 2030/35.

Resistance to theft or diversion

The chemical processing of metal plutonium to plutonium oxide and mixing operations take place in compact facility within closed area with tight security and monitoring. When MOX fuel is cladded and assembled into subassemblies it becomes rather difficult to steal it because the reactor subassembly is very heavy. The irradiation of MOX fuel results in the creation of the intense radioactivity and increases difficulties for theft. Additionally, an implementation of this plan will diminish risk of diversion and thefts because not only fuel fabrication, but also fuel and plutonium transportation will be under comprehensive control within a closed site.

Cost

Because the lack of information, the cost presented here represents only the estimation of capital investments. The estimated cost for this project is about of $3.8 billion: $800 million to complete construction of one BN-800 ($765 million[9]) and "Shop -300" plant ($35 million[2]), and $3 billion for construction three additional BN-800s.

LIGHT WATER REACTORS

Technical Viability

In the past in Russia, use of plutonium in thermal reactors was viewed to be ineffective. For this reason on one existing Russian LWR reactors (VVER-440, VVER-1000) was designed for use of MOX fuel and in Russia there is no experience as well as facility for MOX fuel fabrication for water reactors. Although some Russian experts believe that four modern VVER-1000 units at the Bolakovskaya NPP could be modified and loaded with MOX (1/3 core), but experts from the GosAtomNadzor (Russian Nuclear Regulation Agency) and from Institute of Physics and Power Engineering in Obninsk expressed their doubts[9] that even modern VVER-1000 reactors can be easily modified at moderate cost and licensed to accept plutonium fuel.

It is well known the Western Europe has the experience of using MOX fuel in thermal reactors, though it does not apply directly to the use weapon plutonium. Currently MinAtom in collaboration with France, Germany and the US is conducting technical and economical evaluation of plutonium utilisation in the existing and future Russian commercial LWRs. These studies should be completed this year.

Based on the preliminary results each of four current existing VVER-1000 reactors at Balakovo would recycle 250 kg. plutonium per year, assuming 1/3 loading of the core, and 850kg per year with 100 per cent core loaded with MOX.

Timeliness

To estimate the period of disposition with use of the water reactors the next scenario is considered. Four VVER-1000 units at the Balakovo are reconstructed and loaded 1/3 of the core by MOX after the year 2000. Two partially completed VVER-1000 units are redesigned and introduced with 100 per cent core load by Mox after the year 2005. Also three additional of advanced design VVER-640 will be constructed as replacement power sources for three operating production reactors and put in operation after the year 2010. Their annual plutonium consumption is approximately 370 kg with 100 per cent MOX core[10]. When this scenario will be implemented completely the annual plutonium consumption would be about 3,8 tonnes.

Obviously, some time is required to carry out research, experimental, design, licensing and construction work on use of MOX fuel in the Russian water reactors, and construction of the MOX-fuel Fabrication plant. In accordance with current estimation the pilot installation could be introduced on line by 2000, with the capacity 1 tonne plutonium per year or 20 tonnes of MOX fuel. Assuming the introduction in operation of the full scale MOX fuel fabrication plant with a capacity 120 tonnes MOX per year by 2005 the disposing of plutonium excess would be finished by 2032-2035,

Resistance to Theft or Diversion

If implemented this option will essentially broaden the geography of plutonium and MOX fuel transportation. Presumably, the chemical processing of metal plutonium to plutonium oxide powder and fabrication MOX fuel elements will take place at the Mayak site, where is the construction of the storage facility is going on. The Chemical and Mining Combine at Zheleznogorsk (former Krasnoyarsk-26) will produce fuel assemblies while the nuclear power plants dispersed within the great region. Therefore, due to the broadening of the area of plutonium and MOX fuel transportation the risk of theft or diversion is growing.

Cost

To realise this option it is necessary to built three new reactors with a cost up to $0,9 billion per reactor, to complete the construction of two reactors with estimated cost at the level of $ 1,5 billion. The cost of construction of pilot MOX-fuel fabrication plant is estimated about $60 million[12] and cost of full scale plant is about $250 million. In addition some funds require to reconstruct the old four VVER-1000 units. So, the total cost will be at the level of about $4,5 billion.

CANDU

Atomic Energy of Canada and Ontario Hydro propose to dispose of up to 100 tonnes of weapons plutonium resulting from disarmament programmes in Russia and the U.S. by utilising it as MOX fuel in the Bruce A Reactors, four 825 Mwe CANDU reactors operating in the Canadian Province of Ontario. The outline of the proposal looks as follows: chemical conversion of weapons plutonium components to plutonium oxide at a Russian facility; fabrication of MOX fuel and production of fuel elements for CANDU at a Russian facility; transportation of fuel elements from Russia to Canada; irradiation of elements in two CANDU reactors.

The spent fuel resulting from the process the Canada plans to store on its territory.

Technical Viability

Because of the unique flexibility of the CANDU design to adapt itself to many different fuel cycles, preliminary conclusions indicate that MOX fuel can be incorporated in the design with no changes to the reactor hardware, and within the existing licensed performance envelope. The plutonium concentration in the fuel is about 1.2 per cent and the existing pilot scale fuel fabrication facilities or Complex-300 could be converted for CANDU MOX fuel fabrication purpose.

Timeliness

In accordance with preliminary estimation the full-programme of MOX fuel production would begin by 2002. Each 825 Mwe CANDU reactor at the Ontario Hydro's Bruce A Station is capable to utilise about 1 tonne of plutonium. Therefore, it is possible to utilise 50 tons of Russian plutonium and 50 tons of the U.S. plutonium at a single station within 25 years.

Resistance to Theft and Diversion

The converting plutonium to plutonium oxide and the mixing operations take place in compact facility located within closed area amenable to tight security and close monitoring. After mixing the plutonium with the depleted uranium oxide, the volume is increases substantially and it makes difficult to steal it.

Certainly, the transportation of plutonium fuel for such long distance creates a significant risk to theft and diversion. However, any diversion should be detected quickly. Also, this risk would be diminished by using a specially designed vessel and supporting transport vehicles, as well as by implementation a special safeguards and security system.

Cost

A key elements of economic evaluation in this case is not the cost of capital investment but is the cost of MOX fuel fabrication, the cost of MOX fuel production facility conversion, and the cost of transportation.

The fabrication of unenriched uranium fuel for CANDU reactors is considerably cheaper with regard to LWR and its about \$65 per kg U[11]. On other hand, in current plants MOX fuel fabrication cost is higher than fabrication of uranium fuel. Because there is no data on the MOX fuel fabrication for CANDU, the cost presented here represents only approximations. It is supposed that cost of CANDU MOX fuel including fabrication, cost of depleted uranium and conversion of metal plutonium to plutonium oxide is a three times higher than an ordinary cost, so it is about $400 per kg. The plutonium is assumed to be free. The fuel transportation cost is assumed at the level of spent fuel transportation cost $50 per kg. The cost of conversion of complex-300 for CANDU fuel fabrication assume on the level funds needs to complete its construction. Under these assumptions the total cost of disposing 50 tons of Russian weapon plutonium would be at the level $2,25 billion.

CONCLUSION

Although Russia has some experience with fast-neutron reactors, and the fabrication of plutonium fuel, but due to the lack of funds it is questionable that this option will be realised any time soon. The utilisation of plutonium in LWR would be initiated in a shorter period of time and from this point of view this option looks more promising. But taking into account that Russia has an over capacity for production of low-cost LWR uranium fuel, it will be difficult for MinAtom to justify and get a large-scale subsidy to implement LWR MOX-fuel disposition concept. A decisive role on realisation of CANDU variant will play a cost of MOX fuel elements production at Russian facility, but also including cost of weapon grade plutonium. Currently, the economic merit of each options is not easy to predict because the lack of information, but evidently a full scale implementation of any option and its realisation will take a substantial period of time.

These observations indicate that the real question that needs to be answered is what priority needs to be placed in the strategy of dealing with the problem of weapon plutonium. This question is easily answered when one considers the current turbulent political and economic situation in Russia. The priority that makes the most sense is to concentrate efforts on short-term options. The main concern, and highest priority for now, must be to create a regime that will prevent the reuse of retired weapons grade material in new weapons and prevent its diversion to the black market. This will create a base for irreversibility of nuclear-weapons reductions and confidence in the international community that no proliferation of nuclear weapons is taking place.

It seems there is only one way to realise this goal. That is to make a determined effort to set up a reciprocal regime of storage of both Russian and the U.S. excess plutonium under bilateral or international control.

REFERENCES

1. V.N. Mikhailov, V.V. Bogdan (MinAtom), V.M. Murogov, V.B. Lytkin, V.S. Kagrarnanyan (IPPE). E.N. Avrorin, V.I. Chitaikin (VNIITPh) Plutonium in Russian Nuclear Power Industry, presentation at the Workshop on the Accumulation of Plutonium in Russia: Technical, Socio-Economical, Ecological, and Political Problems, Moscow, 27–28 April, 1995.
   V.N. Mikhailov, V.M. Murogov, at al. Plutonium utilisation in nuclear power of Russia, paper presented at the International Political Forum on Weapons Grade Fissile Material Management, March, 1994, Pittsburgh, USA.
2. Programme of the Radioactive Waste Management in the Russian Federation.
3. Interview with E.G. Dzekun (PO "Mayak").
4. G.G. Borisov, Scientific and Technical Aspects of Plutonium Transition into Glass-Matrix, presentation at the Workshop on the Accumulation of Plutonium in Russia, Moscow, 27-28 April, 1995.
5. V.I. Kuzmenko (PO "Mayak"), Research and Development Programme on Plutonium Utilisation at PO "Mayak", talk given at the Workshop on the Accumulation of Plutonium in Russia, Moscow, 27-28 April, 1995.
6. Interview with S.D. Lutsev (GAN), August, 1995.
7. The Concept of Development of Nuclear Power in Russian Federation. 14 July, 1992, the Council of the MinAtom RF
8. V.F. Menchikov, Ecological cost for Different Type of Nuclear Fuel Cycle, talk given at the Workshop on the Accumulation of Plutonium in Russia, Moscow, 27–28 April, 1995.
9. Victor M. Murogov, Vladimir S. Kagramanian, Alexander N. Chebeskoov, Scenarios of Separated Plutonium Utilisation in Russian Thermal and Fast Reactors, paper presented at the ICEM'95 Conference, Berlin, Germany, September 4–8, 1995.
10. N.N. Egorov, E.G. Koudriavtsev, X. Ouin, and B. Sicard, Civil Use of Weapons-Grade Plutonium from Russian Nuclear Weapons: The AIDA/MOX Programme, paper presented at the GLOBAL '95 Conference, Versailles, France, September 11–14, 1995.

11. The Economics of the Nuclear Fuel Cycle, OECD report, Paris, 1994.
12. The technical study of MOX fuel production with weapons plutonium, Joint Report of MinAtom and Siemens, 1995.

Address:

ANATOLI S. DIAKOV
Center for Arms Control, Energy and Environmental Studies
141700 Moscow
Russian Federation

67

The Minatom Concept of Surplus Weapons Plutonium Utilisation in Russia

N.N. YEGOROV, V.V. BOGDAN and
V.S. KAGRAMANIAN

INTRODUCTION

After the end of cold war period when East-West mutual faith is increasing, the process of real disarmament has started and rapidly gained in scope; The number of warheads to be dismantled on both sides is equal to tens thousand. It is expected that hundreds tons of high enrichment uranium and tens tons of weapons grade plutonium will be released from the military field as a result of this process. Accumulation of weapons grade plutonium released from the nuclear weapons as well as civil plutonium produced by the modern power reactors gives rise to the well-grounded concern of the world public. The problems of plutonium management are not only technological, environmental and economical, these are also political problems since they are closely connected to the nuclear weapon non-proliferation issues and public acceptability of nuclear power.

There are different opinions on what should be the fate of accumulated plutonium stocks in the future. Some are considering plutonium mainly as the wastes of military and civil nuclear industry and studying various ways of deliverance from these wastes, reduced in the end to their disposal in geological strata in one or other form. These options are still only theoretical and they require large scope of studies to be made on their feasibility, environmental safety and cost-effectiveness.

Others consider plutonium first of all as a raw material for power engineering, which can be utilised in power reactors thus enhancing significant increase of value is inadmissibly low, i.e.

less than 1 per cent for modern reactors. This viewpoint is based on the experience gained in many countries on the use of civil plutonium as mixed uranium-plutonium fuel in two reactor types: sodium cooled fast reactors and water cooled water moderated thermal reactors.

It is well known, that plutonium can be utilised in a most effective way in fast reactors. These reactors allow using plutonium if any isotope composition, and they can operate both as plutonium breeders (BR>1) and surplus plutonium burners. When operating in breeder mode on condition that repeated plutonium recycle is provided, unlimited amount of depleted uranium is involved in it: power production process. This can be used for increasing by hundreds times the efficiency of natural uranium utilisation and assuring long term development of nuclear power in the future without any limitations on the power resources.

Plutonium utilisation in thermal reactors is significantly less effective. It is only possible to burn plutonium in these reactors. Plutonium repeated recycling is hampered by considerable amount of non-fissionable Pu-242 isotope and long-lived radiotoxic isotopes of neptunium, americium and curium (so called minor actinides) accumulated in the spent fuel. At large, efficiency of natural uranium utilisation can be increased only by 50 per cent as compared to the existing value.

However even such increase can be of certain interest for instance under conditions of fast reactor development delay.

From the very beginning of nuclear era in our country, the opinion existed that the long terra wide-scale nuclear power development would not be reasonable without significant increase of utilisation efficiency of natural uranium resources and hence of plutonium utilisation for energy production. Urgency of this problem for our country has even increased by now owing to the USSR disintegration since most natural uranium resources are located beyond the Russia borders.

Proceeding from this, Minatom Concept has been elaborated on management of released surplus of weapons grade plutonium and accumulated civil plutonium stocks. The essence of this Concept consists in the following postulates:

- in the long-term aspect: cost effective and environmentally safe efficient realisation of plutonium power generating capability by means of its use as a fuel for power reactors;

- in the near-term aspect: during the period of reactor technology preparation for plutonium utilisation reliable storage of this plutonium should be provided;
- the most important condition for this Concept realisation both on the stage of plutonium storage and on the stage of its utilisation, is assurance of non-proliferation of fissile materials.

Below brief description of state-of-the-art on this Concept realisation is given.

STORAGE

The storage of civil plutonium released as a result of RT.-I chemical plant operation is now provided at the PA MAYAK Works. The total amount of this plutonium is about 30 t. In the recent years, plutonium production as a result of chemical reprocessing of spent fuel received mainly from VVER-440, BN-600 and BN-350 reactors has been about 1 t per year. Some part of accumulated plutonium is of weapons grade, i.e. that obtained as a result of reprocessing of BN-350 and BN-600 fast reactor blanket subassemblies.

At the PA MAYAK site, there are special storage premises adapted for all materials coming as a result of dismantling of nuclear weapons. The amount of materials delivered for storage is determined by the dismantling process rate achieved, i.e. about 2000 warheads a year.

The design of specialised storage has been developed for fissile materials released in the process of nuclear weapons dismantling. This storage construction has been started on PA MAYAK site. Now containers for the fissile materials storage are being received at the site.

WEAPONS GRADE PLUTONIUM UTILISATION FOR POWER GENERATION

Demonstration Stage

New Russia is not ready yet for the wide scale utilisation of power generating potential of both civil and weapons grade plutonium. Two main users of plutonium are considered: sodium cooled fast reactors of BN type and water cooled water moderated thermal reactors of VVER type.

The validation of plutonium utilisation in both fast and thermal reactors has been initiated bearing in mind further use of civil plutonium in these reactors. Of course, there are certain features of weapons grade plutonium utilisation however they are not crucial. Besides, when developing MOX fuel manufacturing technology and irradiating this fuel in the reactors just weapons grade material is used.

BN Type Reactors

Considerable experience has been gained in our country on fast reactors and in particular on the reactor technology. BN-350 (start-up in 1973) and BN-600 (start-up in 1980) fast reactors have been built and successfully operated. Also, certain experience has been gained on manufacturing technology and using of mixed uranium-plutonium fuel in fast reactors. There are pilot lines for manufacturing trial MOX fuel subassemblies at PA MAYAK Works and in Melekess, and over 2000 MOX fuel elements have been irradiated in BN-350 and BN-600 reactors. Post irradiation analysis results have demonstrated fuel characteristics being preserved for required burn-ups.

Now fast reactor technology in our country is almost ready for implementation of demonstration stage on weapons grade plutonium utilization. This demonstration can start in this century. It requires development of so-called hybrid core design with partial MOX fuel loading for the BN-600 reactor. This reactor now operates in converter mode using highly enriched uranium. When changing over to the new core design it would be reasonable to eliminate radial blanket, now used for production of weapons grade plutonium. Development of hybrid core design and its realisation would require certain additional analytical and experimental studies for safety validation.

Possible rate of weapons grade plutonium utilisation in BN-600 reactor hybrid core is 500 kg/year. Taking into account the fact that the reactor life time would expire in 2010, the total amount of utilised plutonium would be about 5 t.

In order to provide BN-600 hybrid core with the fuel, pilot plant should be constructed for weapons grade metallic plutonium conversion into oxide and manufacturing MOX fuel. This plant can be constructed on PA MAYAK site in two or three years using mainly technologies already developed in our country and abroad.

Increasing of plutonium utilisation rate in the BN-600 reactor by means of changing over the BN-600 core to MOX fuel is possible in principle, however it would require considerably more studies because of the necessity to observe the strict requirements in force in Russia on sodium void reactivity effect, which should be negative.

VVER Type Reactors

As far as the second direction, i.e. VVER reactors concerned now only analytical and experimental studies are underway to validate the possibility of MOX fuel utilisation in operating VVER-1000 reactors. This work has been started in our country because of some slowing down of fast reactor development programme and intention to limit accumulation rate of extracted civil plutonium when RT-2 plant will be put in operation in the future for chemical reprocessing of VVER-1000 reactor spent fuel.

First MOX fuel elements for VVER reactors have been manufactured and irradiated by now in test thermal reactor MIR. Critical facility SUPR construction has been started in Obninsk for analytical and experimental validation of neutronic and physical parameters of the VVER type reactor core with MOX fuel. Analytical studies on operational safety, reactor control, accidental parameters and VVER core design improvements issues related to the plutonium utilisation are under way. Some 30 or 40 experimental MOX fuel subassemblies for VVER-1000 reactor are going to be manufactured and irradiated.

It seems possible to combine tasks on MOX fuel preparation both for fast reactor technology demonstration stage and for analytical and experimental validation of possibility of plutonium utilisation in the VVER-1000 reactors. One pilot plant could be used for both metallic plutonium conversion to the oxide and MOX fuel manufacturing for fast and thermal reactors. Now the design of this plant is being developed on the basis of about 1 t of weapons grade plutonium utilisation per, year in the BN-600 and VVER-1000 reactors.

Large-Scale Plutonium Utilisation

BN Type Reactors

On the basis of the positive design and operation experience of the BN-350 and BN-600 fast reactors the BN-800 fast reactor has been designed and construction was initiated of two units of this reactor. Presently this construction process is actually frozen for the lack of money. According to plans it was assumed earlier to built three such units at the South-Urals site, PA MAYAK and one unit at the Beloyarsk site instead of BN-600 to the moment of its decommissioning. Initially fast reactors were aimed at the utilisation of civil plutonium being stored at PA MAYAK as a result of spent fuel reprocessing at RT-1 plant. The developed reactor BN-800 design has plutonium breeding ratio value of about 1 i.e. these reactors were supposed to use plutonium produced by thermal reactors as their initial loading, passing then to their own plutonium.

For this purpose it was supposed to arrange reprocessing of BN-800 spent MOX fuel and further repeated plutonium recirculation. To provide these reactors with MOX fuel "Complex-300" shop has been designed and its construction began at the PA MAYAK as well.

At present "Complex-300" construction is suspended because of financial difficulties and the delay in South-Urals NPP with BN-800 reactors construction. Evaluated level of completion is about 50 per cent. The stoppage in "Complex-300" construction results in the technical aging of the design in terms of technology, process equipment, control instruments and automatic devices provided. Today the technical decisions need to be revised.

Studies performed demonstrate that the large-scale weapons plutonium utilisation in the quantity of 50 t for 30 years can be realised on the basis of one BN-800 reactor functioning in "once through" cycle, i.e. without spent MOX fuel subassemblies reprocessing. Fuel fabrication for this reactor at a rate of 1.6 t plutonium per year, can be realised at the first line of the "Complex-300". Being financed the first BN-800 reactor and the first line of the "Complex-300" could be started before the year 2005. Utilisation of BN-800 reactor at the South-Urals site, where weapons plutonium storage and MOX fuel fabrication plant are being constructed, will allow us, when weapons plutonium is utilised, to exclude any transport of materials containing weapons plutonium beyond the PA MAYAK site. Such an arrangement of the weapons plutonium utilisation allows to meet the most strict requirements in terms of fissile materials non-proliferation insurance.

The second BN-800 reactor, which can be built at Beloyarsk site, can be used for civil plutonium utilisation, which is being separated and stored at the RT-1 plant. Fuel for this reactor can be fabricated at the first line of the "Complex-300" as well.

VVER Type Reactors

Before the completion of the programme on calculational and experimental substantiation of possibility MOX fuel use in VVER-1000 reactors development of any programmes on utilisation of weapons plutonium on the VVER reactors basis would be premature. Nevertheless preliminary estimates show, that the existent VVER-1000 reactors and in a larger degree new design reactors of increased safety VVER-640 can, in principle, be considered as a spare version for weapons plutonium utilisation in the case if BN-800 construction will not be completed for either cause. In the case with VVER it will be difficult to arrange utilisation of weapons plutonium within a single site. Here one have to arrange a safe transportation of the fuel with weapons materials from the MOX fuel fabrication plant near Krasnoyarsk to a number of sites in the European part of Russia with operable VVER-1000 or with new VVER-640.

INTERNATIONAL COLLABORATION

The work on substantiation and realisation the concept on utilisation of excessive weapons plutonium is being conducted in Russia in close collaboration and with certain financial support by Western countries.

With European countries joint calculational neutron physics studies are being conducted on Russian fast and thermal reactors on MoX fuel utilisation. Some works are carried out on the substantiation of technology on conversion of metallic plutonium into oxide. Design studies are conducted on pilot installation aimed at the MOX fuel fabrication for the hybrid zone of BN-600 reactor and VVER-1000 reactors.

In collaboration with the USA specialists evaluations are conducted of various options of excessive weapons plutonium management in Russia and in USA. On the basis of these studies perspective directions of collaboration are supposed to be determined as far as optimal options of weapons materials management for each country are concerned.

Russia is very interested in continuation and enlargement of the collaboration. It is connected with the large scale of the task we have encountered in Russia, complex solution of which needs hundreds millions of dollars and today's heavy economical situation in the country, very strong complicating fulfillment of the task.

CONCLUSION

The fuel cycle industry of Russia has necessary base and experience to begin solution of problems on ensuring safe utilisation of weapons plutonium.

Russian concept of plutonium management (both civil and weapons) is based on the fuel cycle closing in the nuclear power industry to increase the efficiency of the fuel use and decrease long lived waste activity.

On the basis of short term programme of plutonium management in Russia lies safe and reliable storage of weapons and separated civil plutonium till they will be used in reactors.

An effective option of separated civil plutonium and weapons plutonium being released with conversion the latter in the spent fuel form can be realised in the frames of a Nuclear Power Center being created at PA MAYAK (RT-1, "Complex-300" and BN-800 reactors).

Further studies are needed on calculational and experimental substantiation of optimal use of MOX fuel in fast reactors BN and in thermal reactors VVER type having in mind nonproliferation aspects, nuclear and radiation safety, economics and ecology.

It is worth while widening and coordination of the international collaboration with the aim of practical realisation in Russia the most prepared for conditions of Russia option excessive weapons plutonium, being released, utilisation on PA MAYAK basis. This is completion of the "Complex-300" and of one BN-800 reactor, as well as construction on an installation for metal to oxide conversion.

Address:

VLADIMIR S. KAGRAMANIAN
Institute of Physics and Power Engineering
Moscow
Russian Federation

68

Get SMART: The Case for a Strategic Materials Reduction Treaty, and its Implementation

NEIL J. NUMARK

Inventories of weapons plutonium removed from nuclear warheads should be reduced as quickly as possible to prevent large scale rearmament by the United States or Russia and to minimise the risk of theft or sabotage by a sub-national group. The U.S. and Russia should agree to a Strategic Materials Reduction Treaty (SMART) establishing an aggressive timetable for manual reduction of national security needs and schedule the final disposition of this material.

An aggressive disarmament timetable will require an aggressive implementation programme. This should take advantage of available resources within the U.S. and Russia as well as in third countries, including potentially both reactor and immobilisation options, as long as stringent safeguards and security can be guaranteed at all participating facilities. Many existing light water reactors in the U.S. are well suited to the purpose, and several private operators of these plants have formally expressed interest to the U.S. government in providing such service. Russian fast and light water reactors appear to be less readily available to burn weapons plutonium. Russia, the United States and other G-7 countries should develop international programmes to facilitate the most rapid possible reduction in weapons plutonium inventories, consistent with SMART. Such international cooperation would add credibility and transparency to the nuclear disarmament process in the spirit of the Non-Proliferation Treaty, and could add momentum toward the conclusion of both a Comprehensive Test Ban Treaty and a Fissile Material Production cutoff Treaty.

This international programme should take advantage of existing global infrastructure for the use of plutonium fuel as well as other capability (e.g., in the U.S. and Canada) that could supplement existing civilian Pu use plans, as well as vitrification or other immobilisation facilities. In combination, it is reasonable to forecast a global capability to disposition at least 10 tons of weapons plutonium per year. In addition to providing plutonium disposition capacity, third countries should participate in financing disposition programmes and offering safeguards technologies to the global disarmament effort.

Weapons plutonium which the U.S. and Russia declare to be in excess of national security needs and place on the SMART timetable should be held under their ownership and under international safeguards until sent to various vinal disposition sites. However, it would be advantageous to establish an international entity with representatives of Russia, the U.S., other G-7 countries and possibly other states to play a role concerning the flow of weapons plutonium to third country disposition sites, based on the resources available in each country. Such entity could also play a role in preventing the further accumulation of stocks of separated civilian plutonium.

1. INTRODUCTION

The dismantling of weapons delivery systems by the United States and Russia has already yielded large quantities of material now unneeded for defence purposes, and extensive research is under way concerning the best way or ways to get rid of these materials irreversibly. But while these evaluations progress and the selection of disposition methods remains a subject of great interest, a question that often seems forgotten is: How much of the dismantled materials will the U.S. and Russia actually *decide* to disposition? Will we declare the vast majority of the dismantled materials to be in excess of national security needs, *and* will we in fact successfully disposition this quantity? Clearly, as successful as disposition programmes may be, their benefits to society are limited by whatever amounts of fissile material the U.S. and Russia include in these programmes. The United States has declared about 175t of HEU, but only about 40t of Pu, to be in excess of national security needs; to my knowledge no such declaration has yet been made on the

Russian side. Furthermore, even the amounts that *have* been declared to be excess are of course not yet irreversibly dispositioned, and there is the risk that they could be returned to weapons use.

I would like to state as the underlying premise of my talk something which is very simple and upon which I think most or all of you would agree: that it is healthier for the planet if both sides downsize our nuclear arsenals as quickly as possible, keep them small, and establish a system of international control to oversee this process. By arsenals, I am including all weapons-grade fissile material that is stili in the possession of each country, even it has already been declared to be in excess of national security needs and relinquished by the military, i.e., anything that has not yet been irreversibly converted to a non-weapons usable form. The point is, it is strange, and inadequate, that we have almost come to accept the dismantling of *delivery* systems as "disarmament", rather that the dismantling of the bombs themselves. If the bombs still exist, we have not disarmed them.

Let us assume steady progress towards START mile-stones and, looking just at plutonium, a fairly optimistic scenario in which 50 MT of Pu per side - i.e., roughly half - is declared surplus to national security needs, released for disposition, subjected to IAEA safeguards, and successfully dispositioned over the next two or three decades. Even with such a rosy forecast, a rough calculation shows that there would still be enough remaining weapons-grade Pu thirty years from today to arm 20,000 warheads (not just the 3,000 or so allowed under START-II).

Clearly a 50 MT reduction in W-Pu per side is totally inadequate. Even if we were confident that this remaining inventory was extremely well guarded, it would still be in the national arsenal, leaving open the possibility of an all-too-easy resumption of the arms race. Plus, if we accepted indefinite long-term storage of such a large remaining inventor of "Loose nukes" (no longer in weapons but not yet dispositioned), we also run the risk of major changes in government and periods of loss of control over the material, potentially leading to diversion.

In any case, it is not even clear that we now on our way to shedding even this inadequately small portion of our arsenals. We forget in good times that there is some risk that the process will lose momentum. There are still those voices in each country who would bring us back into a spiraling arms race.

2. SMART

What can we do to ensure that the U.S. and Russia will indeed eliminate the vast majority of the "loose" inventories? What we need is a *programme* and an *agreed reduction timetable* that both sides must honour if they expect the order to do it. The U.S. and Russia need a treaty-level agreement—a Strategic Materials Reduction Treaty (SMART)—establishing a joint framework and timetable for the disposition of fissile materials, and should jointly declare the majority of existing inventories to be in excess of national security needs and schedule its final disposition. SMART should have the following features:

- Goal of achieving a minimal residual inventory within 20 years, from the current inventory of about 100 tons W-Pu perside to perhaps 10 tons each;
- Compliance by whatever method or combination of methods each country prefers—including domestic and third country options, employing either reactor burning or immobilisation or other approved methods - as long as the aggressive treaty goal is met; and
- As a first step, an agreed timetable for deforming plutonium pits, which should be relatively quick and inexpensive. Pits would then need to be refabricated in order to use them in weapons.

According to the U.S. National Academy of Science[1], this last point—pit deformation—would only introduce a delay on the order of months to a rearmament programme. However, it would have great symbolic importance, and if we could also reach agreement to decommission all pit fabrication facilities, we would introduce an even greater barrier to a rearmament programme.

Implementing SMART would ensure the irreversibility of disarmament. Without such an agreed schedule, it is unlikely we can make rapid reductions because neither side can possibly implement significant cuts without a reciprocal move on the other side. We might instead be engaged, at best, in a game of stop-and-go disposition, with each side constantly concerned that they maintain near-parity with the other, and at worst, in a stalemate where both sides sit on large inventories indefinitely.

Of course there are voices in both countries arguing that we should hold on to these inventories as they might be needed again for military purposes. This is a narrow and outdated Cold War mentality (and indeed, one of the major reasons that we should get rid of the inventories quickly). As long as the U.S. and Russia reduce inventories *in parallel*, nuclear parity will endure; large reductions are possible before we reach the fissile material inventory level of the other declared nuclear weapons states, China, France and the U.K. Furthermore, once the U.S. and Russia begin to make substantial progress, the other nuclear powers should reduce their inventories as well.

How Smart Fits Into Disarmament Framework

In addition to START II, which the U.S. Congress recently ratified, President Clinton and Yeltsin made a joint statement on the Transparency and Irreversibility of the Process of Reducing Nuclear Weapons at their summit meeting in Moscow in May 1995. Their statement declared that:

"Fissile materials removed from nuclear weapons being eliminated and excess to national security requirements will not be used to manufacture nuclear weapons;

No newly produced fissile materials will be used in nuclear weapons; and

Fissile materials from on within civil nuclear programmes will not be used to manufacture nuclear weapons".

These three elements-no reuse, no new production for weapons, and no transfers from civilian to weapons-are all the right elements, but they are not irreversible. SMART would make the first item irreversible. As an agreement to get rid of existing inventories, SMART would go hand-in-hand with START and with the commitment not to produce more fissile material for use in nuclear weapons (which has yet to be formalised internationally in a Fissile Material Production Cutoff Treaty).

Negotiations have been ongoing to implement other aspects of the May 1995 agreement, concerning increased transparency and irreversibility and there need for an agreement for cooperation allowing data exchanges. Hopefully these items will be resolved in connection with April 1996 summit. But regardless of progress on this front, SMART can proceed independently.

3. AN INTERNATIONAL PROGRAMME FOR ACCELERATED DISPOSITION OF W-PU

How can we achieve the aggressive disarmament timetable spelled out under SMART? I will quickly summarise the current prospects for W-Pu disposition within the United States and Russia, and then turn to the potential contribution of third countries towards meeting aggressive disposition milestones.

United States

On the U.S. side, things are moving too slowly and not one kilogram has yet been dispositioned despite various existing technologies that could already have been employed to demonstrate their suitability for the purpose. The reason is that U.S. policy has not yet been decided and unfortunately even pilot-scale efforts will not begin until it has been. The development of disposition options is on hold until at least the end of 1996 when the U.S. Department of Energy expects to announce a decision.

On the other hand, DOE deserves credit for its recent steps to release inventory data, as part of Secretary of Energy Hazel O'Leary's openness drive. The report released by O'Leary on February 6 reveals that the U.S. inventory of Pu totals 99.5 metric tons, including 66.1 MT at the Pantex site in Texas[2]. It is DOE's intention that the release of these details will prompt Russia, China and others to make similar disclosures, and I hope that will occur.

Returning to DOE Pu disposition studies, Fig. 1, from their February 1996 draft environmental impact statement, indicates that DOE is evaluating options in three categories: Deep Borehole Disposal; Immobilisation Followed by Repository Disposal; and Reactor Options. (Also shown in the "No Action" alternative, which would of course be the worst possible outcome and one which DOE will certainly not select, but is a standard element in the U.S. government's review of the environmental impact of actions it proposes to take).

The use of existing reactors has substantially more support than building new reactors or completing partially-constructed ones, mainly for the obvious reason of economics. (This includes the option of employing Canadian CANDU reactors, which I will discuss in a few minutes). In December 1995, DOE asked the nation's electric utilities to indicate whether they had an interest in offering one or more reactors for the purpose, and last month

Fig. 1 : Surplus Plutonium Disposition Alternatives

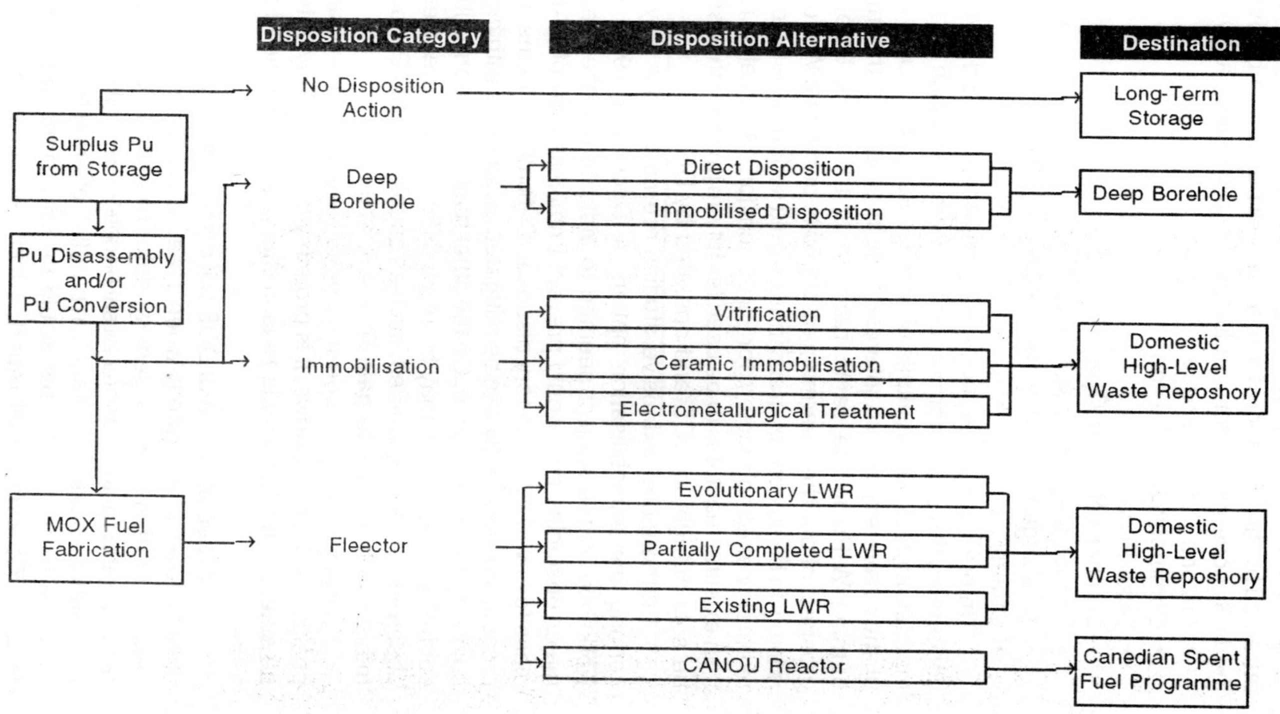

received positive responses from several utilities (who are attracted by the prospect of free fuel and a possible subsidy from the U.S. government for conducting Pu disposition activities):

- Arizona Public Service
- Commonwealth Edison
- Duke Power
- Southern Nuclear Operating Co.
- Tennessee Walley Authority
- Washington Public Power Supply System
- Others

Unfortunately DOE tied this request to a related invitation concerning the use of commercial reactors to produce tritium for maintenance of our stockpile of nuclear weapons. Some of the utilities have expressed interest both in producing tritium and burning W-Pu in the same reactors at the same time. The prospect of our the disarmament benefits of burning W-Pu, and is likely to be so controversial that it would seriously delay efforts to employ existing reactors for Pu disposition. May feeling is that if an existing reactor is to be used for tritium production, it should be sold outright to DOE and operated by DOE.

Furthermore, extensive studies are now under way concerning the immobilisation option. A December 1995, DOE workshop on the subject seemed to indicate that there are no major obstacles that would prevent immobilisation of W-Pu in glass followed by repository disposal. Critically concerns during processing apparently can be mitigated and a glass with high Pu solubility can be designed. On the other hand, long-term critically control in a repository may be more difficult to assure. Neutron absorbers are being studied, such as gadolinium, but Pu loading may have to be low to gain licensing approval.

Finally, while the borehole option seems to be receiving positive technical reviews, it is probably the least likely option to be selected for the simple reason that a site would need to be found.

My expectation is that DOE will not narrow down these nine choices to a single strategy when it reaches its decision later this year, but rather make a general determination to develop and demonstrate reactor and immobilisation options in parallel. Potentially both options could be pursued in parallel, offering the assurance of some redundancy. This would be consistent with the NAS Reactor Options Panel report, which recommended "binding both processes on-line by the end of the century of as shortly thereafter as possible". I would note that we already know that some Pu residues are unsuitable for MOX fuel and a waste management solution will be necessary, so the main question really is whether DOE will also pursue the MOX method of disposition. My hope is that they do and that they develop both options as expeditiously as possible, consistent with SMART objectives.

Russia

Russian experts at this conference have already described possible concepts for utilisation of excess weapons Pu in Russia. My understanding is that using MOX in LWRs. At the September 1995 ASME conference in Berlin, Drs. Murogov, Kagramanian and Chebeskov described three possible scenarios: 1) building 3–4 BN-800 fast reactors, which could transform both civil and ex-weapons Pu into spent fuel by 2030 or 2040; 2) using only the existing BN-600 and VVERs, which can burn only 25 tonnes of ex-weapons Pu in their remaining life; and 3) constructing new VVER-1000s (11 units burning 1/3 MOX could consume 3.3 tons/year)[3].

At the conference, Dr. Rybatchenkov of the Ministry of Foreign Affairs of the Russian Federation, noting the financial difficulties in executing these concepts, spoke favourably of the option of disposing Russian W-Pu in existing foreign reactors. Focusing in particular on a proposal from Canada, Rybatchenkov stated that preliminary study "shows that it may be of interest to Russia from political, economic and social points of view." He noted the benefits of the concept for assuring transparency and irreversibility of W-Pu disposition; offering hard currency to Russia; and creating new work places in Russia's nuclear industry. Speaking also of other possible partners such as the U.K. (Sizewell), France, Germany, Belgium and even the U.S., Rybatchenkov concluded that "the very idea of using foreign reactors for disposition of Russian weapon grade plutonium seems to be sufficiently productive and deserves a thorough investigation side by side with traditional projects, involving Russian reactors"[4].

Third Country Scenarios

Taking advantage of available resources not only within the U.S. and Russia but also in third countries, including potentially both reactor and immobilisation options, might make possible a more aggressive disarmament time-table under SMART. Under the condition that stringent safeguards and security must be guaranteed at all principating facilities, there could be significant disarmament benefits to broadening our disposition programmes in this manner. In addition to assuring accelerated disposition, international cooperation would add credibility and transparency to the nuclear disarmament process.

As I noted earlier, the Canadian nuclear industry has already offered proposals to both the United States and Russia to accept MOX fuel made from W-Pu at Ontario Hydro's Bruce station. Atomic Energy of Canada Limited projects that four reactors at the 8-unit Bruce station could consume 50 tons of W-Pu in just 12 years (or, for 8 units, 100 tons in 12 years). Ontario Hydro is committed to this proposal, and the Canadian Government is currently addressing it at the highest levels and is expected to announce a position very soon[5].

Clearly, the Canadian proposal is an important third country scenario, and indeed could be one of the most significant on *any* of the options for U.S. or Russian W-Pu disposition. But additional third-country scenarios should be considered as well for *their* potential benefits to the disarmament process. While it may be desirable to limit the total number of disposition *sites* in order to guarantee security arrangements, third country sites should be among those considered. Fig. 2 identifies a range of conceivable domestic and third country MOX scenarios.

Fig. 2 : Conceivable mox scenarios, domestic and third country

Activity	*Group I scenarios*	*Group II scenarios*	*Group III scenarios*	*Group IV scenarios*
Origin	Russia	Russia	USA	USA
Fabrication	Russia	Belgium, France, Japan, U.K.	USA	Belgium, France Japan, U.K.
Burning	Russia; All Thrid Co's[1]	All Third Co's	USA Canada	USA, All Third Co's
SF Disposal[2]	Russia; All Thrid Co's	Russia; All Third Co's	USA, Canada	USA; All Third Co's

Unlike disposition scenarios involving U.S., Russian or Canadian reactors, which would generally be above and beyond existing civilian MOX use plans, these other third-country options, in Europe or Japan, which would take advantage of existing worldwide infrastructure for civilian Pu utilisation, could either be *above and beyond* existing MOX use plans, or could potentially *substitute* W-Pu for reactor-grade Pu that is otherwise scheduled to be utilised in these countries. (France, Germany, Switzerland, Belgium and Japan are either already using MOX fuel in LWRs or plan to do to within the next few years).

These options are complicated by the fact that there is already a large surplus of separated civilian Pu awaiting MOX fabrication and disposition, and it will be several years before this backlog can be worked off. (A large portion of that material happens to be nearly weapons-grade itself, resulting from the reprocessing of low-burn-up fuel from gas-cooled reactors).

Nonetheless, the feasibility of certain scenarios should be carefully examined as possible further contributions *beyond* what happens in the U.S., Russia and Canada. I think these can be divided into two classes:

1. *Above and beyond existing Pu use plans:* It is conceivable that third-country reactors could consume some W-Pu beyond existing civilian Pu utilisation plans.

The main question is whether existing MOX fabrication capacity would be sufficient to accommodate such additional plutonium, or when it would be available, without building additional capacity just for this purpose.

Reprocessing capacity and MOX fabrication capacity are expected to reach equilibrium over the next decade or so. In the event that reprocessing contracts are cancelled, however, it is conceivable that, after the backlog of separated Pu is worked off, there could be surplus capacity at existing MOX fabrication plants.

Another way would be to some weapons—Pu ahead of already—separated civilian Pu to burn it sooner. But due to the problem of americium buildup in higher-burn-up fuel, this could only practically be applied to low-burn-up fuels that are nearly weapons-grade. Thus, the benefits would be minimal-mainly just symbolic.

One further scenario for increasing Pu consumption rates

in these countries has been proposed by Professor Atsuyuki Suzuki, of the University of Tokyo, under which Japan's government-owned reactors Fugen (Advanced Thermal Reactor) and Monju (Fast Neutron Reactor) would alter their operation to achieve faster Pu depletion, permitting the consumption of up to 2 tons/year of W-Pu[6]

A problem common to all of these "addition" option in that utilities right now appear to be increasingly disinterested in Pu fuel.

2. *Substitution for existing Pu use plans:* If W-Pu was sent to existing (or already-planned) MOX fabrication facilities *instead* of an equivalent amount of civilian Pu, reprocessing activities necessary to separate that amount of civilian Pu could therefore be deferred until it was no longer necessary to burn W-Pu. This would require complicated arrangements depending on:

- Political decisions by the U.K., France and others that accord enough importance to the disarmament benefits of weapons disposition in existing MOX programmes to justify a disturbance to existing commercial fuel cycle service arrangements;
- Future evolution of utility contracts for reprocessing and MOX fabrication services;
- The schedule for reactor loading of already-separated civilian Pu; and
- The willingness of utilities to accept weapons-grade rather than reactor-grade MOX fuel.

Blending Option—There are limits to how long reactor-grade Pu can be stored following reprocessing, due to the building of Americium-241, if it is to be acceptable for MOX fabrication and recycle in reactors. This is especially true for Pu arising from more recent irradiations because of the higher burn-up (higher burnup causes a lower grade of Pu including more Pu-241). W-Pu does not have this problem. Thus, a fifth possibility would be to mix W-Pu with separated civilian Pu to dilute the Am-241 problem.

Such blending would not allow as high a throughput rate as possible and is thus less than ideal from a disarmament point of view. But, on the other hand, as a *supplement* to the major disposition pathways, which will probably be in the U.S., Russia and possibly Canada, this could absorb some W-Pu. The question concerning this option is whether reprocessing could be slowed down to match the amount of W-Pu added to the stream. This would allow total inventory reduction; furthermore, the alternative—i.e., no change in the rate of reprocessing—would require more utility burning of Pu, which may be difficult to arrange.

Another blending idea I have heard is to use W-Pu to deal with second generation spent fuel (i.e., spent MOX), which has a still-lower grade (which is for the most part being stored for the time being). In this case it is not an issue of Am-241, because the fuel is not yet reprocessed, but rather a question of 239 content—second generation MOX fuel would perform better if blended with high-grade W-Pu. However, to me this is not attractive because you might actually cause *more* separation of Pu (spent MOX fuel might not otherwise necessarily be reprocessed, or at least not yet, because there is a backlog of separated first-generation Pu to work off).

The Table below summarises the above-mentioned disposition scenarios in countries with existing MOX infrastructure.

Opposition to third country reactor disposition could be expected from two principal quarters:

1. *Reprocessors:* To the extent that third country disposition in countries already using MOX fuel would substitute for existing plans, it would defer planned reprocessing activities. Possibly these same companies would gain additional MOX fabrication contracts as a result of the disarmament effort, but this would only partially offset the loss of important export sales and the associated employment benefits in France and the U.K. (An analogy can be made to the disposition of highly-enriched uranium (HEU), which will potentially cause economic dislocations to the uranium industry). If substitution is to be at all feasible, it seems that some form of compensation might have to be offered to make up for each ton-year of deferred business, and this must be considered in estimating the costs of this disposition method.

However, we should also keep in mind, in the event these arrangements are too difficult, that W-Pu burning that is above and beyond existing Pu plans would also have disarmament benefits, equivalent to W-Pu disposition in the U.S., Russia or Canada.

Table 1 : Specific Third Country Scenarios in Countries with Existing MOX Infrastructure

Additions
1. MOX fabrication capacity becomes available due to expiration and/or cancellation of reprocessing contracts
2. Put W-Pu ahead of burning already-separated low-burnup civilian MOX
3. Alter reactor operations to allow higher Pu depletion rates (esp. Fugen and Monju)
Substitution
4. Slow down reprocessing by same rate that W-Pu is introduced into existing MOX programmes
Blending
5. Blend with already-separated high-burn-up civilian Pu, diluting Am-241

2. *Opponents of Civilian Pu Utilisation:* Some organisations that have opposed civilian Pu use also oppose the MOX option for disposition of W-Pu, whether it is in the U.S. and Russia or in third countries, out of concern that this would ultimately lead to greater separation and use of plutonium. However, it should be possible to structure international efforts towards Pu disposition such that we can have the best of both worlds: achieving reductions in the total global inventory of separated Pu, and sending the signal that this inventory *should remain* low and should be under international control. As discussed below, this requires an international framework that has the goal not only to improve controls over all separated Pu but also to gradually reduce inventories. This would necessarily imply establishing some controls over reprocessing rates–base on demand. W-Pu disposition policies should be announced with a clear statement concerning the need to reduce *total* separated inventories, and to put remaining civilian and military inventories under greater international control.

Disarmament has increased the Pu glut, and to deal with the glut we must first of all increase the use and disposal of Pu, which is what U.S., Russian and Canadian disposition options would do. In addition, I think this is not just the responsibility of the U.S. and Russia, and other nations with existing MOX infrastructure who can contribute should do so. This also responds to the glut, either by increasing the use of Pu or bu slowing the separation of additional Pu.

The National Academy of Sciences concluded in its 1994 report that "substituting excess weapons plutonium for reactor-grade plutonium in existing civilian plutonium fuel programmes. Would be the quickest practical means of disposition for excess weapons plutonium if the complex international agreements required could be achieved." The international arrangements that would be needed should be further examined to ascertain which specific scenarios, if any, could offer significant benefits in terms of accelerated disposition, and which therefore should be pursued.

Table 2 : Summary of Third Country Disposition

- Contributes to accelerated disposition
- Adds credibility and transparency to the disarmament process
- Could be above and beyond existing MOX use plans, or a substitution for existing MOX use plans
- Opposition from reprocessors and opponents of civilian Pu utilisation.

4. INTERNATIONAL CONTROL

Consistent with these thoughts on managing the total supply and demand of Pu, we should establish a process ensuring that we do not accumulate excessive stocks of separated civilian Pu, and that the inventories that do exist are kept under international control.

Extensive groundwork has already been laid concerning international Pu storage. I would like to quickly summarise these ideas and then add my further thoughts on international controls over Pu stocks.

An international control system for all fissile material was clearly foreseen at the dawn of nuclear age, but has never been implemented. Based on a U.S. proposal, Article XII of the IAEA's statute, entitled "Agency Safeguards", specifies, among other things, the agency's right:

"to require deposit with the Agency of any excess special fissicnable materials recovered or produced as a by-product over what is needed [for peaceful purposes under continuing Agency safeguards] "in order to prevent stock-piling of these materials, provided that thereafter at the request of the member of members concerned special fissionable materials so deposited with the Agency shall be returned promptly to the member or members concerned [for peaceful uses under Agency safeguards]."

The IAEA initiated a study in 1976 of ways to implement International Pu Storage to fulfil this provision, and experts from members states met from 1978 to 1982 to prepare further proposals for IPS schemes, but the concept has lain dormant.

New proposals in recent years concerning weapons material have included: 1) a call for physical protection by the host country and verification by the IAEA, and subsequent extension of this arrangement to civilian material (Scheinman and Fischer)[7]; 2) creation of an international Pu depsoitory with an international guard force, and a "banking" system in which payments would go to depositors and withdrawals for safeguarded peaceful used would require payment of a fee (Carter and Cote)[8]; and 3) establishing a storage regime first for U.S. and Russian civilian and military inventories, which international regime (NAS)[9]. The NAS also urged agreement on cooperative international approaches to managing the reprocessing and use of Pu to avoid building up excess stocks. This latter idea would more or less internationalise the commercial use of Pu.

A specific way to manage Pu on an international basis so as to avoid the build-up of excess stocks would be for a group of countries, probably the G7 plus Russia, to charter an organisation with the specific but vital role of making *agreed determinations* and *arrangements* for Pu separation and disposition. Such organisation could be very valuable with respect to arranging third country disposition options for W-Pu as outlined above. Furthermore, it would seek to maintain a demand-driven system for Pu use. Although the reprocessing industry in particular might not welcome such international controls, such controls would at the same time provide a framework for internationally-agreed further industrial use of the closed cycle, which could have benefits to the industry in the future.

5. CONCLUSION

The conversion of surplus fissile materials over which the governments of the United States and Russia have relinquished military control will have important benefits for all nations of the world. But only if we declare a large portion of our total defence inventories to be surplus to national security needs, and permanently and irreversibly disposition them, will we achieve disarmament benefits that are more than simply symbolic. We need SMART, to take advantage to current warm relations between the U.S. and Russia and get on the track of disarmament. We also need an aggressive implementation programme including third country participation, which should make possible the disposition of at least 10 tons of W-Pu per year, and an international control, system over all Pu supply and demand. These efforts will add great credibility and transparency to the nuclear disarmament process in the spirit of the Non-Proliferation Treaty, and possibly momentum toward the conclusion of both a Comprehensive Test Ban Treaty and a Fissile Material Production Cutoff Treaty.

Note:

1. National Academy of Sciences, Committee on International Security and Arms Control, *Management and Disposition of Excess Weapons Plutonium*, National Academy Press, Washington, D.C., 1994; also NAS-CISAC, *Reactor-Related Options*, 1995.
2. U.S. Department of Energy, *Plutonium: The First 50 Years*, February 1996.
3. V.M. Murogov, V.S. Kagramanian and A.N. Chebeskov, Institute of Physics and Power Engineering, "Scenarios of Separated Plutonium Utilisation in Russian Thermal and Fast Reactors," *Fifth International Conference on Radioactive Waste Management and Environmental Remediation*, American Society of Mechanical Engineers, Berlin, Germany, September 3–8, 1995.
4. V. Rybatchenkov, Ministry of Foreign Affairs of the Russian Federation, "Possible Involvement of Foreign Partners in Resolving the Problem of Russian Weapon Grade Plutonium Disposition", *Fifth International Conference on Radioactive Waste Management and Environmental Remediation*, American Society of Mechanical Engineers, Berlin, Germany, September 3–8, 1995.
5. Personal Correspondence with Philip Compbell, President, AECL Technologies, Inc., March 11, 1996.
6. A. Suzuki, University of Tokyo, "Japan's Civil Use of Foreign Military Plutonium, "Proc. International Conference on Evalua-

tion of Emerging Nuclear Fuel Cycle Systems (Global '95), Versailles, September 1–14, 1995, pp. 661–668.
7. L. Scheinman and D.A.V. Fischer, "Managing the Coming Glut of Nuclear Weapon Materials", *Arms Control Today*, March, 1992.
8. A.B. Carter and O. Cote, "Disposition of Fissile Materials", Chapter 3 in *Cooperative Denuclearization,* G. Allison et al. (ed.), Center for Science and International Affairs, Harvard University, January, 1993.
9. National Academy of Sciences, Committee on International Security and Arms Control, *Management and Disposition of Excess Weapons Plutonium*, National Academy Press, Washington, D.C., 1994.

Address:

NEIL J. NUMARK
President
Numark Associates, Inc.
Washington DC 20036
USA

69

A Proposed Methodology for the Analysis and Selection of Alternatives for the Disposition of Surplus Plutonium

JAMES S. DYER, THOMAS EDMUNDS,
JOHN C. BUTLER and JIANMIN JIA

The nuclear states are currently involved in the development of comprehensive approaches to the long-term storage and management of fissile materials. A major objective of this effort is to provide a framework for prevention of the proliferation of nuclear weapons. The evaluation should include non-proliferation, economic, technical, institutional, schedule environmental, and health and safety issues.

The ANRCP has proposed that an evaluation of alternatives be guided by the principles of decision analysis, a logical and formal approach to the solution of complicated problems that are too complex to solve informally. This approach would consist of four steps:

1) identification of alternatives and objectives,
2) estimation of the performance of the alternatives with respect to the objectives,
3) development of values and weights for the objectives, and
4) evaluation of the alternatives and sensitivity analysis.

In order to facilitate the evaluation process, the ANRCP proposes the use of nine objectives grouped into the following categories:

1) Non-proliferation objectives (which includes resistance to theft and diversion by unauthorised parties, resistance to retrieval and reuse by the host nation, schedule and fostering progress and cooperation with other nations and Russia)

2) Operational effectiveness (which includes cost, technical viability, and other benefits)
3) Environmental, safety, and health considerations

In order to evaluate alternatives on the basis of these objectives, they have been clarified through the definition of secondary objectives in some cases. Once the objectives were defined, the next step is to develop measures of performance associated with these objectives. Some of these measures of performance use natural scales, such as cost (dollars), time (months), and environmental impacts (cubic metres of secondary waste). Other measures require specially constructed verbal scales and the performance of each alternative is assessed based on expert judgment.

1. BACKGROUND

The end of the Cold War and subsequent arms limitation and reduction agreements have led to a surplus of weapons-usable plutonium in the United States and Russia. In order to prevent the proliferation of nuclear weapons, steps must be taken to manage this plutonium in a manner which takes into account non-proliferation, economic, technical, institutional, schedule, environmental, and health and safety issues.

The purpose of this paper is to define a model and the methodology that could be used to support the selection of alternatives for the disposition of surplus plutonium. There are a number of methods that have been proposed to model preferences and support decisions, and each of them may be used constructively in some contexts. However, we believe that the significance of decisions regarding the disposition of plutonium requires the use of a methodology that can evaluate alternatives involving risk and multiple performance measures, and that is practical, theoretically sound, and transparent to external reviewers and interest groups. The one methodology that meets these requirements is multi-attribute utility theory (MAU), which has been supported for use in similar situations by the National Research Council, an agency of the United States National Academy of Sciences[1].

MAU (Keeney and Raiffa, 1976) is one of the major analytical tools associated with the field of decision analysis (Clemen,

1. National Research Council, letter to Ben Rusche, DOE/OCRWM, dated October 10, 1985.

1991; Holloway, 1979; McNamee and Celona, 1990; Raiffa, 1968; von Winterfeldt and Edwards, 1986). Simply, decision analysis is a logical and formal approach to the solution of problems that are too complex to solve informally. In the past, decision analysis has been applied to problems such as siting electricity generation facilities (Keeney, 1980), choosing among vendors for the evaluation of alternatives for the commercial generation of electricity by nuclear fusion (Dyer and Lorber, 1982), and selecting a nuclear waste clean up strategy (Keeney and von Winterfeldt, 1994).

The MAU methodology for the evaluation of alternatives for the disposition of plutonium consists of the following steps:

1. Identification of alternatives and objectives
2. Estimation of the performance of the alternatives with respect to the objectives
3. Development of values and weights for the objectives
4. Evaluation of the alternatives and sensitivity analysis

As a first step, reasonable alternatives for the disposition of plutonium must be identified along with the objectives that are used in the analysis. The alternatives and the objectives from a matrix in which each row corresponds to an alternative and each column represents an objective. The cells of the matrix contain estimates of the performance of each alternative on each of the objectives. When these estimates are uncertain, it is often appropriate to quantify them with ranges or with probability distributions determined using risk analysis methods (e.g., Clemen, 1991; Keeney and von Winterfeldt, 1991).

Typically, it is possible to gain a number of insights regarding the alternatives simply through a careful inspection of this matrix. For example, one or more alternatives may be "dominated" by another alternative, meaning that the dominating alternative performs as well or better on every objective than the dominated alternative. Alternatives that are dominated can often be eliminated from further consideration in the decision process, which may significantly simplify the remaining steps in the analysis.

Step three creates a value model based on the objectives by defining value functions, if necessary, on the measures of the performance of the alternatives, and by assigning weights to the objectives. This process is carried out with decision makers or their designated representatives, and allows the measures of performance on each objective to be aggregated into a single figure of merit. Finally, this value model can be used to determine ranking of each of the alternatives, and a sensitivity analysis is typically conducted to determine if this ranking is robust relative to reasonable changes in the weights or the other parameters that determine the value mode. This sensitivity analysis may include changes in the value model that are suggested by interactions with representatives of other interest groups or stakeholders.

This process should summarise the critical information needed for an evaluation of alternatives, and provide the insights that both support and explain the basis for this evaluation. However, it is important to emphasise that the decision analysis process does not lead to a computerised model that actually determines the decision for a complex problem. Rather, this process highlights the strengths and weaknesses of alternatives, the implications of trade-offs among these strengths and weaknesses, and the sensitivity of the evaluation to the underlying assumptions so that better informed choices can be made.

Any model of a physical process or of subjective preferences will omit some details in the abstraction from the real-world in order to crystallise the essence of the problem. Some of these Omitted details, may be relevant in the final selection of alternatives by a decision maker or decision makers, particularly when the alternatives are determined to be "very close" in the formal analysis. Further, the appropriate value model for use as a guide to public policy is, in general, not sharply defined. As a result, the decision analysis process will emphasise the support of the decision makers charged with the responsibility for the selection of alternatives, and will attempt to clarify the consequences of each choice. We subscribe to the philosophy that the result of using models should be insights, not numbers.

Sections 2-5 of this report will describe these four steps of the MAU methodology in more detail. Section 6 summarise the discussion.

2. IDENTIFICATION OF ALTERNATIVES AND OBJECTIVES

i) Alternatives

The evaluation process begins with the identification of the set of reasonable alternatives that are appropriate for serious consideration. This screening process may be aided by reference to a set of criteria that identify the most important considerations guiding this preliminary selection process. Examples of the use of screening processes to determine reasonable alternatives for the disposition of surplus plutonium are provided by the studies conducted by the National Academy of Sciences (1994) and by Office of Fissile Materials Disposition of the United States Department of Energy (OFMD, 1995).

The reasonable alternatives for plutonium disposition determined in these studies fall into three categories: reactor alternatives, immobilisation alternatives, and borehole alternatives. The reactor alternatives would use surplus plutonium to fabricate mixed oxide (MOX) fuel for nuclear reactors that generate electrical power. The spent fuel from these reactors would ultimately be transferred to a national waste management system for ultimate disposition. The immobilisation alternatives combine the surplus plutonium materials in borosilicate glass or ceramics; additional radionuclides may be added to provide a radiation barrier to inhibit recovery and reuse. This material could also be transferred to a national waste management system for ultimate disposition. The borehole alternatives involve the placement of the plutonium in a deep borehole, possibly after the material is immobilised in an inert matrix.

Other alternatives may eventually be considered by the United States and Russia. However, the general methodology for the evaluation of these alternatives should be flexible enough to evaluate and compare any reasonable approach to the disposition of the surplus plutonium.

ii) Objectives and Measures

The first step in the application of MAU is the development of a "hierarchy" of objectives, criteria, and measures. Objectives are often broad statements of goals. Typically two or more criteria are associated with objectives at the next level of the hierarchy to provide more specific statements of desirable characteristics of alternatives, and to help define the objectives in more detail. In complex decision problems, these criteria may be decomposed further into sub-criteria, and so on, until a sufficient level of detail is reached to allow measures to be identified.

In some cases, these measures may be quantified as estimates on a natural scale, for example, net present value of cost, time, travel miles, etc. In other cases, it may be necessary to construct scales that are more descriptive in nature, and that may require estimates for the alternatives based on expert judgment. In many cases, these measures are surrogates for higher-level issues.

Useful reference points for the identification of measures for evaluating plutonium disposition alternatives include measures proposed for previous studies involving technology choices (e.g., Keeney, Lathrop, and Sicherman, 1986; Keeney and von Winterfeldt, 1994; Merkhofter and Keeney, 1987), for previous studies concerned with the management and disposition of surplus plutonium (National Academy of Sciences, 1994), and for evaluations of technologies and sites for tritium supply and recycling.

Objectives. The objectives for any decision provide the basis for evaluating the relative desirability of available alternatives. For the purpose of illustrating the methodology, we present the objectives recommended by the National Academy of Sciences (1994) and used by the Office of Fissile Materials Disposition (OFMD, 1995) for the purpose of a preliminary screening. The objectives used by the OFMD for screening the alternatives for the disposition of plutonium were the following:

1. Resistance to theft and diversion by unauthorised parties
2. Resistance to retrieval, extraction, and reuse by the host nation
3. Technical viability
4. Environmental, safety, and health
5. Cost effectiveness
6. Timeliness
7. Fostering progress and cooperation with Russia and other nations
8. Public and institutional acceptance
9. Additional benefits

For this illustration of the methodology, these nine objectives have been reorganised to emphasise the commonality among some of them, and to provide additional detail regarding others. This reorganisation is shown in the form of a hierarchy of objectives in Figure 1.

At the highest level of this hierarchy, we have identified three major categories of objectives:

1. Non-proliferation which includes resistance to theft, resistance to reuse, international cooperation, and timeliness (objectives 1, 2, 6 and 7 from the original list of nine)
2. Operational effectiveness which includes and cost effectiveness (objectives 1 and 5 from the original list of nine)
3. Environmental, safety, and health (objective 4 from the original list of nine) which has been decomposed into human health and safety, environmental protection, and socio-economic effects at the next level in the hierarchy

Such a reorganisation of the nine objectives would simplify the task of creating a value model, and particularly the assessments of weights on the objectives, as we discuss in Section 4. In addition, this simplified structure would provide a natural means for transferring the insights from the model to the decision maker.

It should also be noted that objectives 3, 8 and 9 from the original list, technical viability, public and institutional acceptance, and additional benefits, have been dropped from the proposed hierarchy. Technical viability refers to the level of technological development associated with the alternative, and is essentially a surrogate for the risk of possible delays and cost overruns. These concerns can be captured in an evaluation through the use of probability distributions on measures of time and cost, if necessary.

Public and institutional acceptance is a major concern in any screening process, and the basis for the elimination of many of the alternatives that may originally be considered. However, the other objectives that have been selected for this illustration are based on meeting public concerns. Therefore, we believe that an alternative selected based on the other eight objectives will be one that would also be ranked highly on the objective of public acceptability. In addition, the economic impacts of the alternatives on local communities have been included in the proposed measures of the Environmental, Safety, and Health objective, as we shall discuss.

Also, we have deleted the objective of "additional benefits" from this hierarchy. Some of the alternatives may offer the possibility of producing useful by-products, such as the production of electric power by nuclear reactors or the possibility of sharing costs with other programmes. However, the most significant examples of these "other benefits" can be captured as offsetting costs, and will be effectively measured by the cost effectiveness objective.

As previously mentioned, Figure 1 represents the highest level of the objectives for selecting a plutonium disposition alternative. Figure 1 a, 1b and 1c provide the details for the three main objectives of the analysis: Non-proliferation, Operational Effectiveness and Environment, Health and Safety respectively.

Two comments are in order regarding Figures 1, 1a, 1b and 1c. First, the major purpose of these diagrams, particularly Figure 1, is to assist the decision makers in "making sense" of an evaluation of alternatives based on thirty eight detailed performance measures. The reorganisation of the objectives as shown in this hierarchy is neither unique nor fixed. It could be altered based on feedback from those involved in determining policy toward plutonium disposition. Second, the fact that one objective or sub-objective appears at a "lower level" in the hierarchy than another does not imply that it is less important, or that it should receive a smaller "weight" in the analysis than another objective.

The objectives categorised as providing assurance against non-proliferation (Figures 1 and 1a) indicate five distinctly different areas of concern. The first objective is to minimise the opportunities for theft of the materials by unauthorised parties. Generally, an alternative will be more resistant to theft during the processing steps required to transform the material from weapons-usable plutonium into its final form for permanent disposition if these steps are relatively simple and transparent, if the form of the material is not "attractive" to potential thieves because of size, radioactivity, or other concerns, and if effective safeguards and security can be applied.

The second objective is to maximise the resistance of the disposition alternative to the diversion of the plutonium by the host nation during processing, and to provide an internationally verifiable and acceptable process. Providing adequate accessibility safeguards, and measurement capability will allow an alternative to satisfy international inspection standards and provide assurance that diversion by the host nation is not taking

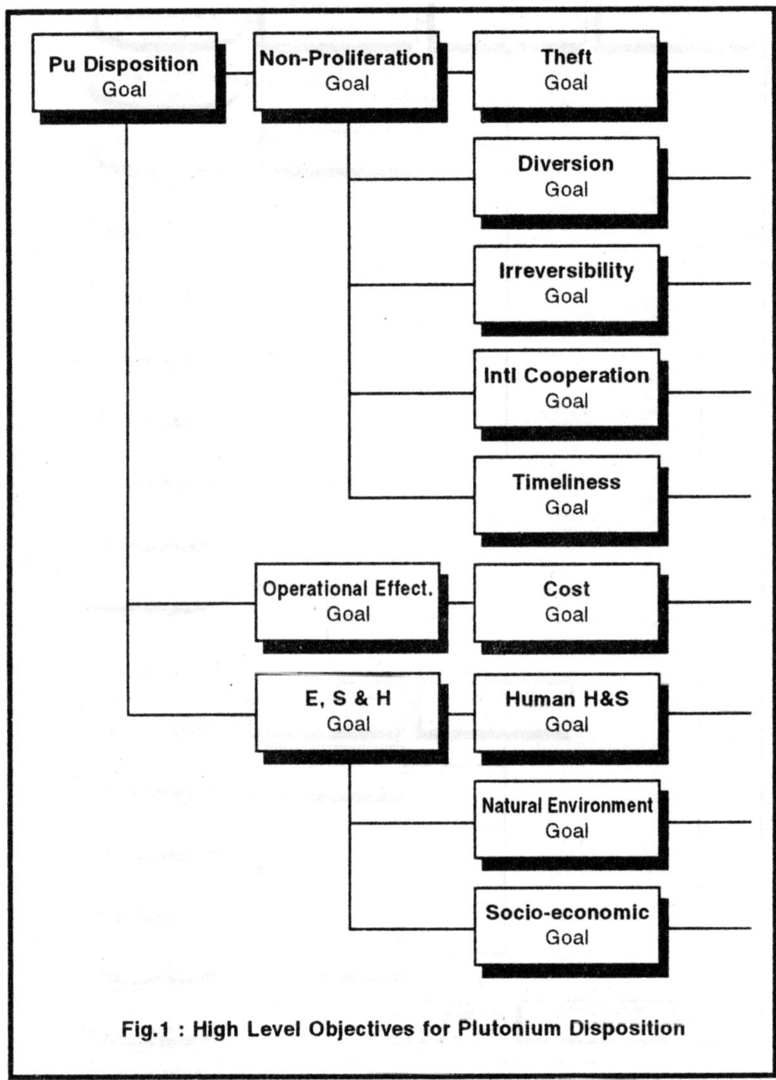

Fig.1 : High Level Objectives for Plutonium Disposition

place. Many of the factors considered in the theft sub-objective can also apply here.

The third objective is to maximise the difficulty of recovering the disposed material after processing has been completed. The disposed material will be less attractive for reuse by the host nation if it meets the "spent fuel standard", or would be as costly,

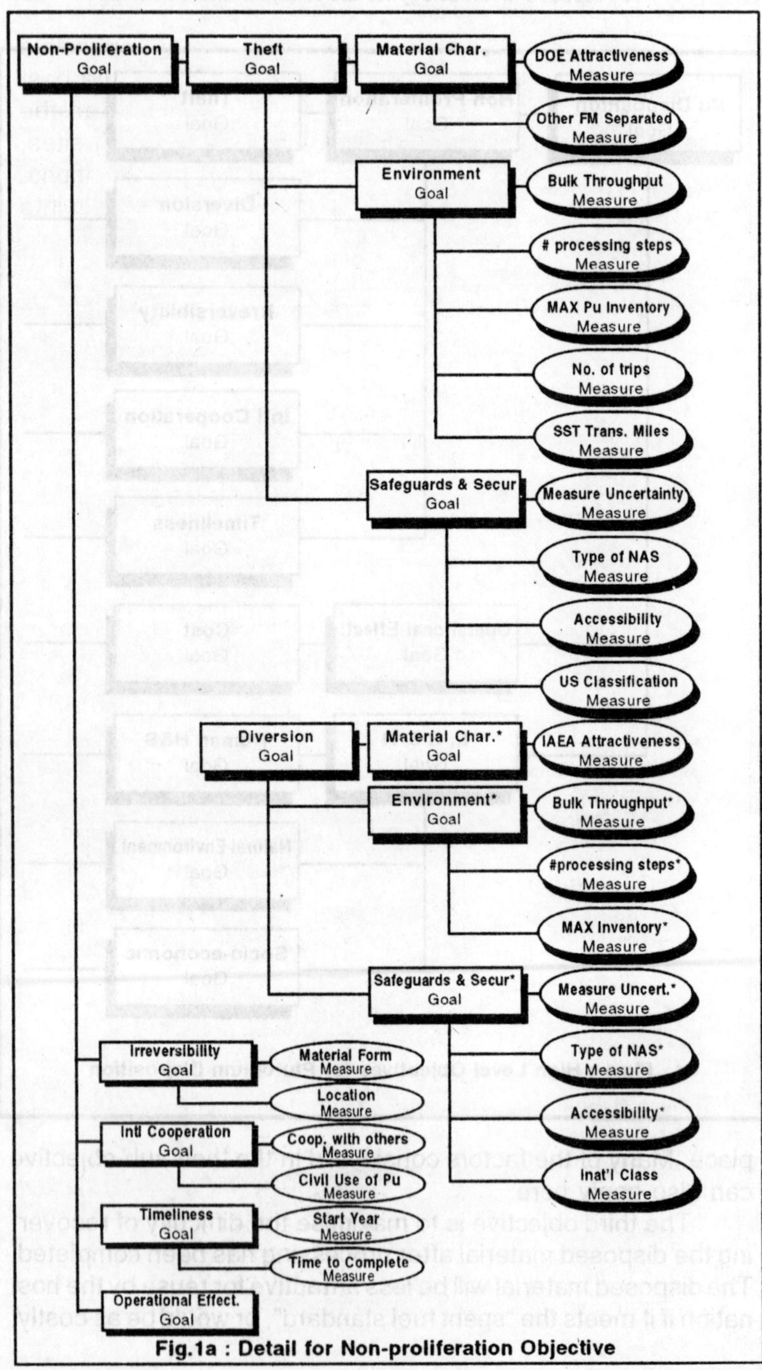

Fig.1a : Detail for Non-proliferation Objective

Operational Effect.
Goal
Cost
Goal
Investment Cost
Measure
Life-cycle Costs
Measure

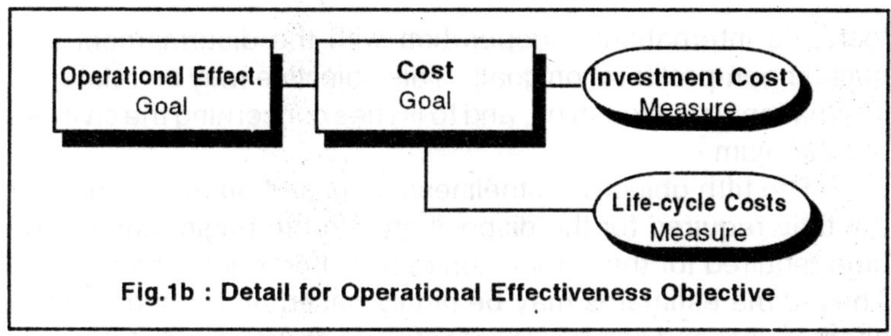

Fig.1b : Detail for Operational Effectiveness Objective

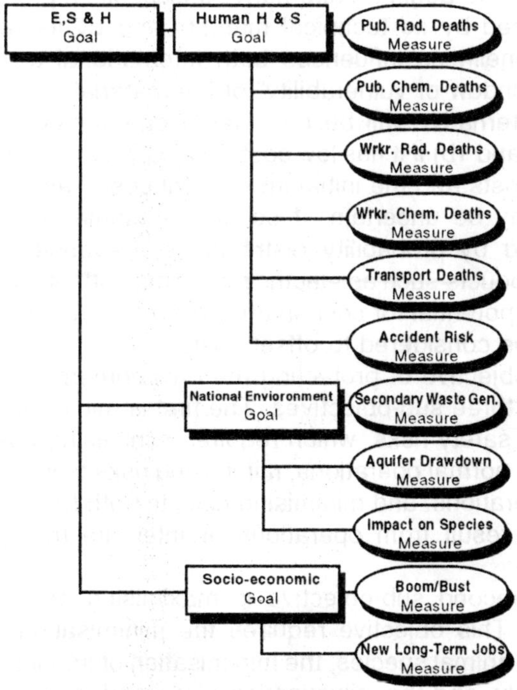

E,S & H
Goal
Human H & S
Goal
Pub. Rad. Deaths
Measure
Pub. Chem. Deaths
Measure
Wrkr. Rad. Deaths
Measure
Wrkr. Chem. Deaths
Measure
Transport Deaths
Measure
Accident Risk
Measure
National Envioronment
Goal
Secondary Waste Gen.
Measure
Aquifer Drawdown
Measure
Impact on Species
Measure
Socio-economic
Goal
Boom/Bust
Measure
New Long-Term Jobs
Measure

Fig.1c : Detail for Environmental Safety and Health Objective

detectable, and time consuming to retrieve and fabricate into weapons as the recovery of plutonium from spent commercial reactor fuel. The final form and location of the disposed material will determine its long-term resistance to reuse.

The fourth non-proliferation objective is concerned with

fostering international cooperation with the disarmament and nuclear non-proliferation goals. This objective may be related to international relationships, and to issues concerning the civil use of plutonium.

The fifth objective, timeliness, is based on an estimate of the time required for the disposition effort to begin, and on the time required for the completion of the effort once it has begun. These time estimates may be highly uncertain for some of the alternatives, and can be represented as probability distributions when necessary. The assessment of the uncertainty associated with the time to begin an alternative's disposition process may be influenced by its technical maturity and by its regulatory history. Timeliness influences both international cooperation and the "window of vulnerability" of the material.

An alternative will be considered operationally effective (Figures 1 and 1b) if it has low cost. The cost may consider both life-cycle costs and the initial investment costs, and estimates of both may be uncertain. If so, these estimates should be represented by probability distributions. Revenues resulting from by-products such as electric power may offset some of the costs. The potential for cost sharing with other related projects may also be considered to offset costs.

The objective of protecting the environment, safety, and health has three sub-objectives. The first is minimising human health and safety risks, which requires minimising risks to the public from normal operations, minimising risks to workers from normal operations, and minimising risks to both from accidents that could result from operations or inter-site transportation activities.

The second sub-objective is maximising environmental protection. This objective requires the minimisation of direct impacts on animal species, the minimisation of impacts on local water supply, and the minimisation of secondary wastes.

The third sub-objective is related to the socio-economic impacts of the alternatives. The short-term socio-economic disruptions by the alternative should be minimised, while any long-term economic and social benefits should be maximised. These socio-economic impacts also relate to the screening objective of encouraging public acceptance of the alternative, particularly in the local communities that would be affected by construction and operation.

Measures. In order to evaluate the alternatives, a measure or a set of measures is needed for each of the objectives, as shown in Figures, 1a,. 1b and 1c. These measures should be selected so that each alternative can be evaluated on each of them, and so that each measure is then logically linked to one or more of the objectives.

The measure or set of measures associated with an objective should cover all aspects of the objective. In some cases the selection of an appropriate measure may be clear. For example, it is customary to measure the life-cycle cost of an alternative in terms of discounted net present value dollars. Similarly, concerns regarding the timeliness of the disposition activities associated with an alternative may be captured by measures of the "time to start the disposition activities" and the "time to complete the disposition activities". However, when no relevant and/or natural scales are closely linked to an objective, such as maximising the likelihood of international cooperation, it may be necessary to work with experts to construct a measure to indicate different levels of achievement.

3. ESTIMATION OF THE PERFORMANCE OF THE ALTERNATIVES ON THE OBJECTIVES

Given the identification of the alternatives and the definitions of the measures, the next step is to obtain estimates of the performance of each alternative on each measure. This step defines the alternative-by-objective (and measure) matrix that summarises the overall performance of each alternative on the relevant measures. An example of such a matrix is provided in Figure 3, where performances of three hypothetical alternatives are evaluated on five measures used for illustration purposes only. The entries in the cells in this matrix may be in the form of point estimates, ranges, or in the form of probability distributions. For example, a probability distribution might be represented by a simple three point distribution of the form (0.05 fractile, media, 0.95 fractile), that reflects the uncertainty associated with the estimates of performance. Probability distributions are included for the life cycle and investment costs of an alternative in Figure 2.

	MEASURES				
	Life Cycle Costs ($B)	*Investment Cost ($B) Fatalities*	*Expected Worker (# species)*	*Impacts on Species (yr.)*	*Completion Time*
Alternative A	(1, 2, 5)	(2, 2.5, 3)	.001	1	2010
Alternative B	(2, 4, 10)	(2, 3, 5)	.002	0	2025
Alternative C	(-1, 0, 5)	(2, 3, 5)	.001	3	2025

Fig. 2 : Example of Alternatives by Objectives Matrix

A careful inspection of this simple matrix may provide some rich insights regarding the alternatives. For example, one or more alternatives may be identified as clearly inferior because of their poor performance on most if not all of the relevant objectives. Others may obviously "rise to the top' because of superior performance on many of the objectives.

In order to obtain the performance estimates with respect to these measures, a series of assessment meetings may be necessary to focus on the major objectives. For example, experts in the area of safeguards and security (S&S) may be asked to evaluate the performance of the alternatives on the non-proliferation objectives. Other teams may be involved to develop estimates of cost and time, while still other groups mayfocus on the analysis necessary to develop the measures of environment and health impacts.

4. DEVELOPMENT OF VALUE FUNCTIONS AND WEIGHTS

Once the performance of each alternative on each measure in the alternatives-by-objectives matrix has been obtained, the next step in the analysis involves assembling the measures into a "super-measure" of the desirability of each alternative. The aggregation procedure is complicated by the diversity of the types and scales of the individual measures. As evident in Fig. 2, some measures may be represented by probability distributions while some are expressed as point estimates. Some measures units are dollars and some are cubic metres of secondary waste, while others are defined over constructed scales, further complicated the aggregation procedure.

Utility theory provides the basis for the appropriate approach to aggregate the seemingly disparate measures. It is a logically consistent and tractable means of representing the degree to which each alternative fulfils the objectives shown in Figure 1. The use of utility theory ensures that any recommendation reflects:

- the relative attractiveness of a specific level on a measure
- the relative attractiveness of performance on different measures and objectives
- the interactions, if any, between objectives.

These three issues will be addressed in the following sections. For a more detailed presentation of these topics see Keeney and Raiffa (1976) and von Winterfeldt and Edwards (1986).

i) Single Attribute Value Functions

The relative attractiveness of performance outcomes on a measure is captured by a single attribute value function. A value function is constructed or assessed so that it incorporates a decision maker's preferences for performance on a measure in a utility value or score; a superior objective measure will score higher on the value scale. Value functions can be linear or non-linear as dictated by both normative concerns and the nature of the decision maker's preferences. Once constructed, value functions can be combined with probability distributions to ensure that the risk associated with an alternative is properly evaluated.

Figure 3 illustrates two *hypothetical* value functions. It is particularly important to emphasise that these value functions are used here only for the sake of exposition. The first value function represents the value associated with different levels of the Life Cycle Cost of an alternative. The function is decreasing because lower cost is preferred to higher cost; hence, lower costs receive higher value scores. The function is linear because the range of dollar amounts being considered may be small in comparison to the national budget, so the marginal value of each incremental dollar over this range is assumed to be equal.

The second value function representing "Type of Nuclear Accounting System" (defined per facility as the percentage of time in the facility that "item" accounting is used) is a bit more complicated. Intuitively, the idea facility would utilise 100 per cebt item accounting and receive the highest value. Due to the

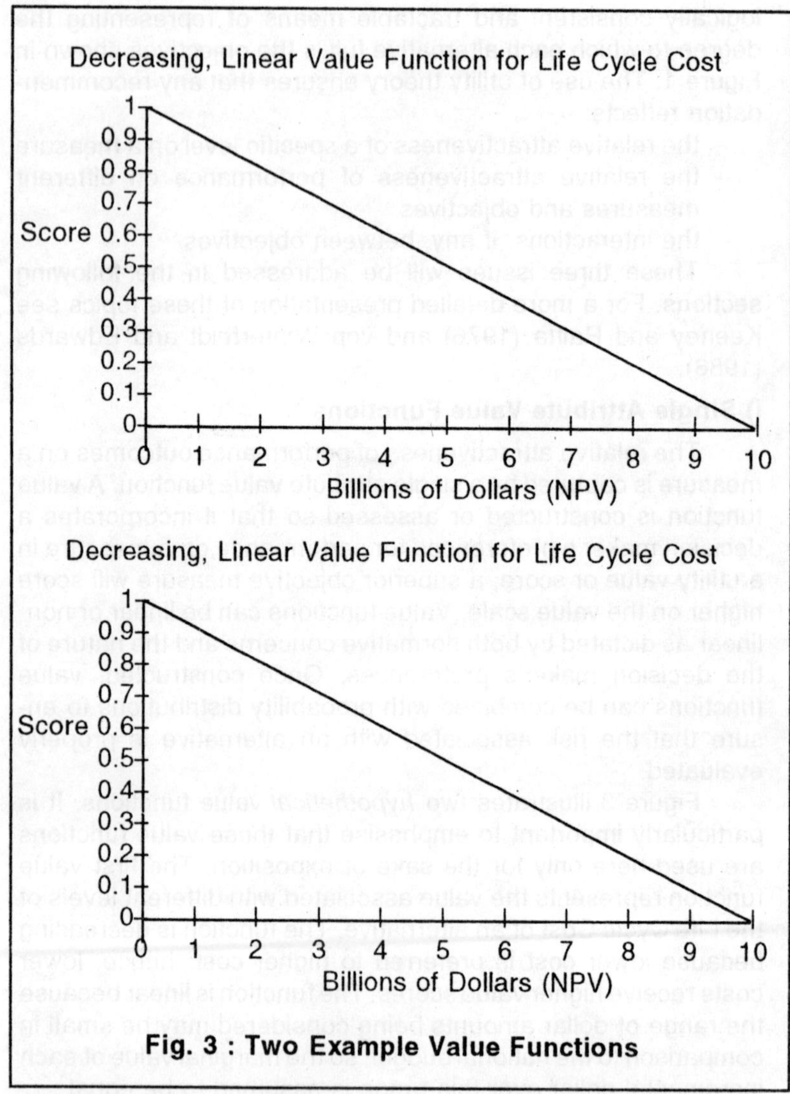

Fig. 3 : Two Example Value Functions

comparative ease of measuring material that is classified as item, even a small decrease from 100 per cent item accounting receives a stiff penalty. Looking at the scores for facilities that use very little item accounting (these facilities rely heavily on "bulk" accounting), it is also clear that moving from 0 per cent item to 10 per cent does not receive a substantial increase in score relative to moving from 90 per cent to 100 per cent. The scale for Type of Nuclear Accounting System is "exponentially biased toward item accounting".

ii) Weights

Each objective, sub-objective, and measure in the attribute hierarchy is given a weight. These weights reflect the value trade-offs among objectives (or sub-objectives and measures within objectives), and are dependent on the ranges of the outcomes considered in the analysis.

As a simple example, consider the problem of choosing among disposition alternatives based on the objectives of cost, ES&H, and non-proliferation. Suppose that three alternatives are under consideration with costs of \$2.2, \$2.4, and \$2.5 billion, respectively, and with representative values on the other two objectives. Now, suppose that a fourth alternative is added to the list with a cost of \$3.0 billion, and with values on the other two measures that lie within the ranges of values determined by the original three alternatives. Utility theory prescribes that the weight on cost in choosing among the original three alternatives (where costs range from \$2.2 to \$2.5 billion) should be *smaller* than the weight on cost in choosing among the four alternatives (where costs range from \$2.2 to \$3.0 billion). Intuitively, this is because a wider range of costs is considered in choosing among the four alternatives; i.e., cost is more of a discriminating factor in choosing among the four alternatives than in choosing among the original three.

As a result of this insight, it should be clear that weights on objectives are not simply measures of the "relative importance" of each objective. Loosely speaking, they are measures of the importance of the *increase* from the worst to the best level of performance on one objective compared to the increase from the worst to the best level of performance on another objective. Therefore, weights must be assessed carefully to ensure that the results of the evaluation are consistent with the preferences of the decision maker or decision makers.

This assessment procedure can be based on a dialogue with a decision maker (or a group of stakeholders) that can take the following form. First, we assume that we have specified the ranges over which the performances of the alternatives can vary on each objective; that is, we have identified the "worst" and "best" feasible levels of performance on each objective. Next, we assume that an alternative achieves only the worst levels of performance on each of two objectives, say objective A and objective B. Holding its levels of performance constant on all of the other objectives, we ask the decision maker if it would be appropriate to pay more to increase the performance of this alternative from the worst to the best level on objective A, or to increase its performance from the worst to the best level on objective B?

Suppose that the decision maker responds, "I would pay more to increase objective A from its worst to its best level of performance. "Next, we would ask her to identify a level of performance on objective A so that she believes it would be appropriate to pay the same amount to increase objective A from its worst level to this level of performance as to increase objective B from its worst to its best level of performance. The response to this question determines the ratio of the weights on objectives A and B, and additional questions comparing the other objectives provide sufficient information to specify the numerical values of these weights.

In some, cases, it may appear that responses to questions of this type would be extremely difficult to make. However, the assessment process can be aided by the skills of a trained analyst, and a variety of " consistency checks" can be used to ensure that the responses are meaningful. These assessment protocols are also scripted to minimise biases in the responses, systematic errors that are known to occur as a result of the limitations of human information processing capabilities. For additional details and examples of assessment dialogues, see Keeney and Raiffa (1976) or von Winterfeldt and Edwards (1986).

Weights can be used to combine objectives and measures at different levels of the hierarchy, and the individuals who provide the judgments required to develop these weights may be different, depending on the level. For example, the judgments required to combine the measures for the sub-objectives "minimise number of processing steps" and "minimise attractiveness of material" for the objective Maximise Resistance to Theft may be more appropriately obtained from S&S experts who help evaluate alternatives as described in Section 3.

At the highest level of the hierarchy of objectives, the weights are less related to expert judgment, and much more to questions of policy. These higher level weights should be obtained in interviews with persons representing national policy makers.

iii) Aggregation Methods

In order to obtain an overall evaluation for each disposition alternative on a higher level objective, we may use an aggregation model that can combine different measures into a single value. The model also must show the results of "sub-aggregation" at lower levels of the objectives hierarchy so that decision makers can better compare the attractiveness of alternatives. Since the decision for plutonium disposition involves both multiple criteria and risk, it is appropriate to use multi-attribute utility models for this study (Keeney and Raiffa, 1976).

If stakeholder preferences are consistent with some special independence conditions, then a multi-attribute utility model $u(x_1, x_2, \ldots x_n)$, where x_1 represents the level of performance on measure *i*, can be decomposed into an additive, multiplicative, or other well-structured form that simplifies assessment. An additive multi-attribute utility model can be represented as follows:

$$(1) \qquad u(x_1, x_2, ..., x_n) = \sum_{i=1}^{n} w_i u_i(x_1)$$

where u_1 (•) is a single-attribute value function over measure *i* that is scaled from 0 to 1, w_i is the weight for measure *i* and

$$\sum_{i=1}^{n} w_i = 1$$

If the decision maker's preference structure is not consistent with the additive model (1), then the following multiplicative model may be used, which is based on a weaker independence condition:

$$(2) \qquad 1 + ku(x_1, x_2, ..., x_n) = \prod_{i=1}^{n} [1 + kk_i u_i(x_i)]$$

where $u_1(\bullet)$ is also a single-attribute value function scaled from 0 to 1, the k_1's are positive scaling constants satisfying $0 \leq k_1 \geq 1$, k is an additional scaling constant that characterises the interaction effect of different measures on preference. The value of k can be determined from one additional question similar to the questions used to determine the objective weights. As a special case when

$$\sum_{i=1}^{n} k_i = 1$$

the multiplicative model (2) reduces to the additive model (1).

The choice of the appropriate model for aggregation will be based on information collected from interviews with policy makers. For approaches to the assessment of an additive utility model and a multiplicative utility model, see Keeney and Raiffa (1976).

5. EVALUATION OF THE ALTERNATIVES AND SENSITIVITY ANALYSIS

i) Evaluation and Ranking

Once the single measure value functions have been completely defined, the data from the alternatives-by-objectives matrix (see Figure 2) are converted to component utilities. For measures that are known with a high degree of certainty, this process amounts to supplying the measure as an argument to the value function to obtain a score for each alternative on each measure. If a measure has been defined with a probability distribution, the appropriate value function is applied to the distribution to provide an expected utility value for the measure.

The component value function scores are aggregated, using the correct multi-attribute utility function, within each of the major objectives, and within each of the categories of objectives identified by the decision maker as illustrated in Figure 1. During this aggregation, the weights are used to reflect the trade-offs between measures, and are multiplied by the corresponding scores. This stage of the evaluation process is important and useful for decision makers as it provides scores for each alternative for the major objectives of the plutonium disposition problem, and on the three categories of objectives identified in

	MAJOR CATEGORIES		
	Non Proliferation	*Operational Effectiveness*	*Environmental Safety and Health*
Alternative A	0.7	0.3	0.7
Alternative B	0.5	0.8	0.1
Alternative C	0.4	0.4	0.4

Note : Scores are from 0 (least preferred) to 1 (most preferred). Scores are purely hypothetical.

Fig. 4 : Example of Phase I Aggregation

Figure 1. At this stage it is possible to examine the relative strengths and weaknesses of the alternatives. A hypothetical example of the results of this phase of the analysis is provided in Figure 4.

It is often possible to obtain important insights from an inspection of this table of scores. In addition to highlighting the relative strengths of the alternatives on the major objectives or the objective categories, alternatives that are dominated may also be identified. For example, these hypothetical scores indicate that Alternative A dominates the No Action alternative since its scores are as good or better on every major category. Note that this table could be created at a "lower level" in the hierarchy as well, highlighting the nine objectives used in the screening process. Comparisons among objectives and sub-objectives at different levels in the hierarchy may also be used to provide additional insights.

Weights may be assessed to represent trade-offs between the major objectives. This will allow another level of aggregation to provide a measure of the overall utility of each alternative. This step will allow for quick comparisons regarding the relative desirability of the alternatives, and should provide an excellent means of ranking the field of contending disposition alternatives.

ii) Sensitivity Analysis

Before final disposition recommendations are made, the analysis must be tested to see if the evaluation of alternatives is robust. This sensitivity analysis basically amounts to making changes in the performance on the measures and/or weights and observing changes in the resulting evaluations and rankings.

It is impossible to predict exactly what sensitivity analyses would be performed as the evaluation process is problem specific and iterative. The following types of analyses have been useful when examining similar problems.

- Change the weight of an important objective while leaving the ratios between weights on other objectives unchanged. This will highlight the effect of changing the emphasis placed on an objective.
- Investigate specific entries in the alternatives-by-objectives matrix (over a reasonable range). In cases where ranges of values or probability distributions are provided as estimates, values other than the mean (average) may be selected for the sensitivity analyses (e.g. the 10th and /or 90th percentile levels). This process helps test the robustness of the ranking to the assumptions and facts that form the basis of the analysis.

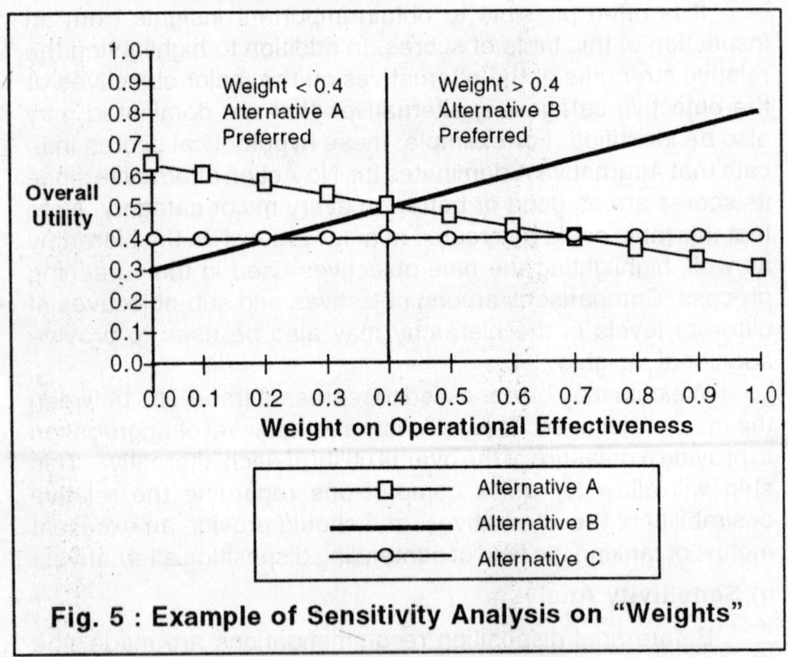

Fig. 5 : Example of Sensitivity Analysis on "Weights"

- Manipulate the measures and/or probability distributions to reflect specific viewpoints. For example, it may be appropriate to investigate the implications of an optimistic perspective about the life cycle costs of an alternative. This analysis will demonstrate the effect of different perspectives about the alternatives.

Any sensitivity analyses performed will be summarised in several formats. Numerical results are presented and, where appropriate, graphical representations are also provided. These results should also be explained intuitively to ensure that all participants understand the implications and are able to contribute to discussions.

Fig. 5 provides an example of the first type of sensitivity analysis based on the hypothetical scores in Fig. 4. The weight placed on Operational Effectiveness is varied from 0 to 1 holding the ratio of all other weights unchanged. This analysis indicates that if the weight on Operational Effectiveness is less than 0.4 (holding the ratios of other weights constant), then Alternative A will be preferred; if it is greater than 0.4, then Alternative B is preferred. Similar analyses could be performed on all other objectives and sub-objectives.

6. SUMMARY AND CONCLUSIONS

This presentation presents a proposed methodology for the analysis and selection of alternatives for the disposition of surplus plutonium. The approach is intended to be general, and could easily be modified to address specific issues and concerns unique to a country or a stakeholder group.

This approach to the evaluation of alternatives has several advantages over the presentation of a great deal of technical information in the form of discussion and tables, and then a verbal argument regarding the selection of the preferred alternative. First, the approach brings some order and structure to the evaluation process. It helps to focus the different teams of personnel who are responsible for generating information regarding one or more aspects of the complex alternatives required for plutonium disposition. Second, it provides a "score card" that can be used by policy makers and stakeholders to understand, relatively easily, the strengths and weaknesses of the various alternatives. Third, the evaluation and sensitivity analysis can easily reduce the set of "reasonable alternatives" to a smaller subset that may be viable candidates for the final choice, depending on the implied "weights" that are assigned to the objectives.

The selection of alternatives for the disposition of plutonium is a critical issue that requires simultaneous consideration of many conflicting objectives. We believe that this approach can help countries to make these decisions in a logical and informed manner, and to communicate with each other more effectively regarding the rationale behind these choices.

References

Clemen, R.T. (1991). *Making Hard Decisions.* Boston: PWS Kent Publishing.

Dyer, J.S. and Lorber, H.W. (1982). the Multi-attribute Evaluation of Programme-Planning Contractors. OMEGA, 6, 673–678.

Hertz, D.B., and Thomas, H. (1983). *Risk Analysis and its Applications.* New York: John Wiley.

Holloway, C.A. (1979). *Decision Making Under Uncertainty: Models and Choices.*

Englewood Cliffs, N.J.: Prentice-Hall.

Keeney, R.L. (1980). *Siting Energy Facilities.* New York: Wiley.

Keeney, R.L., Lathrop, J., and Sicherman, A. (1986). An Analysis of Baltimore Gas and Electric Company's Technology Choice. *Operations Research*, 34, 18–49.

Keeney, R.L. and Raiffa, H. (1976). *Decisions with Multiple Objectives.* New York: Wiley.

Keeney, R.L. and von Winterfeldt, D. (1991). Eliciting Probabilities from Experts in Complex Technical Problems. *IEEE Transactions on Engineering Management*, 38, 191–201.

Keeney, R.L. and von Winterfeldt, D. (1994). Managing Nuclear Waste from Power Plants. *Risk Analysis*, 14, 107-130.

Krantz, D.H., Luce, R.D., Suppes, P., and Tversky, A. (1971). *Foundations of Measurement.* New York: Academic Press.

McNamee, P. and Celona, J. (1990). *Decision Analysis with Supertree.* 2nd edition. San Francisco: Scientific Press.

Merkhofer, M.L. and Keeney, R.L. (1987). A Multi-attribute Utility Analysis of Alternative Sites for the Disposal of Nuclear Waste. *Risk Analysis*, 7, 173-194.

OFMD (1995). Summary Report of the Screening Process to Determine Reasonable Alternatives for Storage and Disposition of Weapons-usable Fissile Materials, March 17, 1995.

National Academy of Sciences (1994). *Management and Disposition of Excess Weapons,* Washington D.C.: National Academy Press.

Raiffa, H. (1968). *Decision Analysis.* Reading, MA: Addison Wesley.
Von Winterfeldt, D. and Edwards, W. (1986). *Decisional Analysis and Behavioural Research.* New York: Cambridge University Press.

Address:

JAMES S. DYER
Amarillo National Resource Centre for Plutonium
University of Texas
Austic Texas
USA

70

Economic Assumptions for Evaluating Reactor-Related Options for Managing Plutonium*

GEOFFREY ROTHWELL

This paper discusses the economic assumptions in the U.S. National Academy of Sciences' report, *Management and Disposition of Excess Weapons Plutonium: Reactor-Related Options* (1995). It reviews the Net Present Value approach for discounting and comparing the costs and benefits of reactor-related options. It argues that because risks associated with the returns to plutonium management are unlikely to be constant over time, it is preferable to use a real risk-free rate to discount cash flows and explicitly describe the probability distributions for costs and benefits, allowing decision makers to determine the risk premium of each option. As a base-line for comparison, it assumes that one economic benefit of changing the current plutonium management system is a reduction in on-going Surveillance and Maintenance (S&M) costs. This reduction in the present value of S&M costs can be compared with the discounted costs of each option. These costs include direct construction costs, indirect costs, operating costs minus revenues, and decontamination and decommissioning expenses. The paper also discusses how to conduct an uncertainty analysis. It finishes by summarising conclusions and recommendations and discusses how these recommendations might apply to the evaluation of Russian plutonium management options.

* I have been aided by discussions with D. Hale, R. Knecht, D. Korn, M. May, M. Rothwell, A.D. Rossin, R. Smith, J. Taylor, and W. Weida, the Centre for International Security and Arms Control Technical Seminar. However, I accept full responsibility for the errors that remain. Also, I acknowledge the financial aid of R. Noll through the Programme on Regulatory Policy at the Centre for Economic Policy Research, Stanford University.

1. THE NET PRESENT VALUE OF REACTOR-RELATED OPTIONS FOR PLUTONIUM MANAGEMENT

Since the end of the Cold War and the start of decommissioning nuclear weapons by the U.S. and Russia, a problem has presented itself: What should be done with the weapons-grade nuclear materials?

There are two materials to consider: highly enriched uranium (HEU) and weapons-grade plutonium (WPu). Because HEU can be blended down to concentrations appropriate for nuclear power plant fuel, attention has focused on WPu. In the U.S. Several organisations have addressed the issue of excess WPu, including the National Academy of Sciences' Committee on International Security and Arms Control.[1] This Committee produced *Management and Disposition of Excess Weapons Plutonium* (see NAS, 1994). Also, the Committee's Panel on Reactor-Related Options for the Disposition of Excess Weapons Plutonium wrote *Management and Disposition of Excess Weapons Plutonium: Reactor-Related* Options (NAS, 1995).

My paper discusses many of the economic assumptions in NAS (1995) for evaluating reactor-related plutonium management options. On the value of economic analysis for this problem, see NAS (1995, p. 280): economic considerations are less important than security in reaching a conclusion about the relative attractiveness of alternative options for WPu disposition. But the study of costs is nonetheless worthwhile, both to assist in ranking options that are not distinguishable on security grounds and to facilitate planning for the investments that will be required for one option or another.

To facilitate an economic analysis, my purpose is to provide a unified method, based on the current economics and finance literature, to guide future evaluations of reactor-related options. Because I will not discuss all economic assumptions in NAS (1995), this paper should be read as an addendum to that report.

To aid nuclear-weapons states in determining how to efficiently management plutonium, economists would suggest that decision makers use the Net Present Value (NPV) method of economic decision making[2]. Under this method all costs and benefits (real and monetary) are discounted to the present[3]. I discuss the discounting procedure in Section 2.

The uncertainties associated with reactor-related economics complicate the NPV calculations. At least three options are available. First, contingencies (usually in percentage terms) can be assigned to cash flows. Second, the discount rate can be adjusted to reflect project risk. However, determining the appropriate adjustment requires strong assumptions. Third, an uncertainty analysis can be preformed to approximate the probability distribution associated with the NPV of each option. The decision maker can then calculate the risk premium associated with each probability distribution. I argue in Section 3 that the third option is most consistent with an economics approach to decision making under uncertainty.

Although the primary *social* benefit to careful plutonium management is a reduction in the risk of proliferation, this is difficult to quantify[4]. As a starting point, one quantifiable *economic* benefit of a proliferation-resistant plutonium management system is the reduction in on-going Surveillance and Maintenance costs. I discuss this base line benefit in Section 4. I discuss discounted costs in Sections 5 through 8. Section 9 lists my conclusions and recommendations and discusses their applicability to Russian plutonium management. I conclude by reviewing the limitations of NPV analysis. I begin by discussing the discount rate.

2. THE APPROPRIATE DISCOUNT RATE IS THE REAL RISK-FREE RATE

A difficult parameter to determine in any economic analysis is the appropriate discount rate. Under the NPV method all cash flows for all investment alternatives are discounted to the year when investment choices are made. (Non-monetary flows also should be discounted, but this requires careful consideration; see Section 10.) To simplify this discussion I assume that these choices are made in the present (however, cash flows can be discounted to any base year). An appropriate discounting rate must be selected for each cash flow. Regarding project costs, the discount rate should be such that funds set aside in the present would cover all future costs.

Two types of discount rates must be distinguished: nominal and real. The nominal rate is equal to the real rate plus the inflation rate[5]. Further, the real rate is equal to the risk-free rate plus a risk premium. Higher rates of return through a risk premium are required by investors for higher risk projects. The risk-free rate is used to discount investments with near certain returns; for example, investments in short-term (U.S.) Government securities. In the first quarter of 1996 the nominal risk-free rate in the U.S. was about 5 per cent per year. If the anticipated inflation rate in early 1996 over the period of the short-term risk-free investment was 3 per cent per year, the anticipated real rate was about 2 per cent per year during 1996.[6] (NAS, 1995, P. 75, uses a 3 per cent inflation rate throughout its analysis.) I discuss risk premiums here and appropriate inflation rates in Section 5.2.

If project returns are well defined and if risk is constant over the life of the project, then a risk-adjusted discount rate is appropriate for discounting cash flows. If both assumptions hold, risky *costs* should be discounted at *less* than the risk-free discount rate and *benefits* should be discounted at *more* than the risk-free discount rate. This is because for risky costs (negative cash flows), we must set aside funds in the present to cover expected future costs plus a risk premium. A rate lower than the risk-free rate provides for this risk premium. For risky benefits (positive cash flows), funds in the present are equivalent to risk-free rate provides for this risk premium. (See Copeland and Weston, 1980, Chapter 10). Because risks associated with the returns to plutonium management are unlikely to be constant over time, it is preferable to use a real risk-free rate to discount cash flows and explicitly describe the probability distributions for costs and benefits. The decision maker can then determine the risk premium that equates a risk-free (certainty equivalent) NPV to an uncertain NPV calculated from risky cash flows.

3. CHOOSING AMONG PLUTONIUM MANAGEMENT OPTIONS REQUIRES UNCERTAINTY ANALYSIS

Many cost estimates include a factor (typically a percentage increase) that represents the uncertainty associated with the estimate. Unfortunately, as noted in NAS (1995, p. 82), there is no uniform method of applying contingency factors to cost or benefit estimates. The Electric Power Research Institute (see EPRI, 1986, or later editions) relates the contingency factor to the level of the estimate's detail:

Simplified estimate	30 %	to	50 %
Preliminary estimate	15 %	to	30 %
Detailed estimate	10 %	to	20 %
Finalised estimate	5 %	to	10 %

Therefore, if all reactor-related options are at the same level of detail, the same contingency factor would apply. Also, if the economic benefits have been made at the same level of detail, the determination of those options with positive NPVs is unchanged when all costs and benefits are increased by the same contingency factor. If this is true, no contingency factor is required in determining whether an investment option has a positive NPV.

If different uncertainties apply to different cash flows, there should be an analysis of the risks associated with estimating costs and benefits, particularly if risk-free discount rates are used. An appropriate approach to decision making under uncertainty is outlined in Nordhaus (1994, Chapters 6-7). The goal is to determine an option's probability of having a positive NPV. The decision maker can then compare the NPV distributions of the different options. How can this probability distribution be approximated?[7]

The first stage of uncertainty analysis is to determine which economic parameters have the greatest impact on the NPV. This is done through sensitivity analysis by varying each parameter, for example, by plus and minus 50 per cent, and calculating the sensitivity of the NPV to this change. Those parameters that have the greatest influence on the NPV are retained for further analysis. As Nordhaus sates (1994, p. 105), "It should be emphasised that at this initial stage the purpose is to begin with order-of-magnitude estimates of the uncertainties... to determine the important ones and screen out those that contribute little to the uncertaint(y)..." For example, in evaluating plutonium management options, the influential variables are likely to include (see NAS, 1995, p. 376): the discount rate, the reactor construction cost, the market price of electricity, and the cost of plutonium Surveillance and Maintenance, see Section 4.

The second stage is to approximate the distributions of the most influential parameters. Because these distributions can have any shape (normal, lognormal, triangular, uniform, etc.), the probabilities should be described non-parametrically, for example with quintile values, i.e., 20 per cent of the distribution is less than the first quintile value and 20 per cent of the distribution is more than the fifth quintile value. These probabilities can be determined either (1) by examining historical data (for example, data on the historical risk-free discount rate), (2) by expert opinion through panel surveys, or (3) by polling the users of the NPV analysis, e.g., decision makers.

The third stage involves either Monte Carlo or Latin Hypercube analysis. Under Monte Carlo analysis, random values of each parameter are generated from the underlying probability distributions and the NPV is calculated. This procedure is done hundreds or thousands of times by computer, generating a probability distribution of the likely outcomes for the NPV. Latin Hypercube analysis involves two steps. First, the probabilities of sets of parameters are calculated from the underlying probabilities. Second, samples based on the probabilities of these sets are selected randomly and the NPV is calculated for each randomly selected set, generating a probability distribution.[8]

These methods for incorporating uncertainty into the NPV analysis give decision makers more information regarding the distribution of an option's NPV, but can reduce transparency in the analysis. If transparency is highly valued and "the value of more complicated approaches is questionable" (NAS, 1995, p. 82), contingency factors should be avoided unless they are directly associated with the detail of the cost estimate or unless options have different levels of detail. On the other hand, if probability distributions for the most important parameters can be transparently modeled, uncertainty analysis allows a more complete comparison of reactor-related options.

4. ECONOMIC BENEFITS: SECURITY AND REDUCING THE COST OF PLUTONIUM STORAGE

Secure plutonium management gives many benefits. One quantifiable economic benefit of changing the present plutonium management regime is the reduction of on-going Surveillance and Maintenance (S&M) costs. These costs include (1) maintaining plutonium handling facilities, (2) transportation services, (3) inventory and accounting charges, and (4) security and national and international regulatory costs. Implicitly, these activities are a "Monitored Retrievable Storage" programme. (See Rothwell, 1993b.) Is these costs must be paid in every future year and not increasing over time, their present value can be calculated as a perpetuity, equal to the product of annual costs and the inverse of the discount rate.[9]

For example, if in the U.S. plutonium could be stored for $2-$4 per gram per year (NAS, 1994, p. 122) and there are about 38.2 metric tons (MTPu) of excess weapons plutonium (DOE, 1996), S&M costs would be between $ 76.4M (million) and $152.8M per year in perpetuity.[10] At a 2 per cent real discount rate (and assuming no increase in S&M costs) the present value of storage is 50 (equal to the inverse of 2/100) times $ 76.4M to $152.8M per year, or $3.8B (billion) to $7.6B. Therefore, to be economically beneficial (under this benefit only) and reactor-related option should cost no more than $3.8B to $7.6B.[11]

These values should be adjusted for the speed at which plutonium S&M costs would decline as the plutonium was disposed. For example, if the stock of plutonium declines by 3.3MTPu per year for 30 years starting in 10 years, storage costs during the disposal process would be about $1.5B to $3.1B, discounted to the present. Because reactor-related options require long disposal periods, their economic benefits are reduced by $1.5B to $3.1B.

Of course, the quantification of *social* benefits should be added to this one *economic* benefit. If a reactor-related option provided greater security than could be provided by the S&M option, but at higher price, decision makers must determine whether the negative NPV would compensate an increase in security. For example, if an advanced reactor was built to consume 38.2 MTPu during 30 years at a (net of electricity sales) PV cost of $4.5B (i.e., within the range described in NAS, 1995, p. 325), then at $2 per gram per year the NPV would-be negative $2.2B (i.e., $3.8B - $4.5B - $1.5B for storage during disposal) or at $4 per gram per year the NPV would be $0 (i.e., $7.6B - $4.5B-$3.1B for storage during disposal). At the lower storage cost, decision makers must decide whether the extra security would be worth $2.2B. Therefore, to understand projected costs for reactor-related options, I discuss economic assumptions affecting these costs in the next three sections. These costs include direct and indirect costs of construction and net operating costs.

5. CAPITAL COSTS: CONSTRUCTION AND FINANCE CHARGES

The direct costs of constructing facilities for plutonium

management (including plutonium handling facilities, fuel fabrication facilities, nuclear reactors, waste handling facilities, decommissioning, and disposal facilities) include engineering, licensing, site selection costs, site acquisition costs, site preparation costs, structures construction, and equipment acquisition. Besides land, these direct costs can be categorised as labour (both engineering and construction), materials (including energy), and equipment (both construction equipment and purchased components). To discount these costs into the present involves determining (1) total resource requirements for each stage of construction, (2) unit costs, (3) the time profile of expenditures, and (4) inflation rates over the construction period.

In this section I discuss total resource requirements. I ignore issues in determining unit costs (particularly international extrapolations). I focus on nuclear power plants (NPPs) in the plutonium fuel cycle, however my comments also could apply to plutonium handling and fuel fabrication (or vitrification) facilities. I will not review costs of particular reactor-related options. (See NAS, 1995, pp. 306-329, for an extensive review.) Instead, I discuss two economic assumptions, then I review two issues associated with financing costs during construction.

i) Real Construction Costs

First, consider the assumption that bigger plants have lower unit costs.[12] Although this is often true, this assumption masks the increase in total costs and financing costs that accompany large plants. For example, while there is a decrease in total resource requirements with the doubling of a nuclear power plant from 600 megawatt-electric (MW) (e.g., 1,650/kilowatt, kW, 1992$, see Braun, 1992) to 1,200 MW (e.g., 1,450/kW 1992$), total costs increase from $1B to $1.75B. As finance capital becomes harder to acquire, smaller units with lower total costs become more attractive. (See my discussion of the influence of increasing competition in electricity generation on the size of new NPPs in the U.S. in Rothwell, 1995, or the review of this discussion in ANS, 1996.) At the limit, modular nuclear units, even with higher costs per kilowatt, become more financially viable than larger nuclear units.

Second, consider the assumption that (because of learning and the ability to spread fixed development costs over more than one unit) total resource requirements decrease with more plants (beyond the obvious savings in building more units at a single site). (See NAS, 1995, p. 318, discussion of first-of-a-kind costs and learning factors.) In the nuclear power industry, this has been empirically demonstrated by the French programme. However, most estimates of NPP direct construction costs do not explicitly account for this decrease. These estimates are implicitly either average costs for all projected units or average costs for the later units in a series. Construction cost estimates should explicitly forecast the increase in first (or single) unit costs over expected long-run average costs, (One can interpret the high NPP construction costs in the U.S. as representing the costs on the first or second units for a series of plants that were never built.).

ii) Discounting Construction Costs

Two other issues associated with total resource requirements involve the time to complete construction. They are (1) financing costs (known as "interest during construction") and (2) the increase in nominal and real costs over the construction period (known as "escalation during construction"). First, "Interest During Construction" (IDC) in levelised annual cost analysis accounts for the "time value of money." IDC accounting is equivalent to discounting construction costs to the time of plant completion (which is then levelised over the plant's life). Under NPV analysis, it is not necessary to include IDC because discounting to the present accounts for the "time value of money."

Second, there is the problem of adjusting for different inflation rates for different cash flows during construction. If the nominal values of all project costs are expected to increase at the same rate and this rate is equal to the inflation rate anticipated by financial markets, then discounting can be done at the real rate. But all project costs (and benefits) do not increase at a uniform rate. It is easy to assume that if there are competitive labour markets (with the free movement of labour and no shortages of skilled workers), wages will rise at the anticipated inflation rate. Therefore, labour costs can be discounted at the anticipated real rate. On the other hand, the cost of instrumentation and control equipment (now largely based on microprocessor and other advanced technologies) is likely to decline. If the anticipated inflation rate for these costs is negative, the real discount rate would be greater than the nominal rate. For example, if these costs are nominally declining at 5 per cent per year, the appropriate real discount rate in the U.S. in early 1996 would be 10 per cent per year. In contrast, waste management costs are increasing over time. One could imagine that these costs are nominally increasing at 5 per cent per year. Therefore, the real discount rate for these costs would be zero, i.e., income from funds set aside today would just cover future increases in these costs.[13]

If some costs are increasing more rapidly or more slowly than the anticipated rate, these costs must be segregated and discounted separately. If the real discount rate is used, real inflation rates must be used for the segregated cash flows. The real inflation rate is the difference between the anticipated inflation rate and the nominal inflation rate for the segregated cash flow. Following the examples above, the real inflation rate for labour is zero per cent (i.e., labour costs are moving with the anticipated inflation rate). For instrumentation it is-8 per cent (i.e., these costs are declining 8 per cent per year when compared to labour costs). For waste management the real inflation rate is 2 per cent (i.e., these costs are increasing 2 per cent faster than labour costs). These inflation rates allow one to use the real discount rate and account for differences in inflation for specific unit costs. "Escalation during construction" (EDC) inflates construction cash flows to the time of plant completion. Under NPV analysis, it is not necessary to account explicitly for EDC if cash flows have been properly discounted.

6. INDIRECT COSTS

Indirect cost markups are applied to direct costs for both construction and operation.[14] There are several types of indirect cost, including those identified in NAS (1995, p. 80): "construction facilities, equipment, and support services, safety and environmental engineering, inspection and other quality-assurance activities, project administration, and the like." These cost categories are often added as a percentage of direct costs. These indirect costs (particularly project administration that can include programme integration, construction management, and project management fees) can be as large as direct costs.[15]

Although the Department of Energy (DOE) has been trying to reform its contracting procedures (see DOE, 1994), there are few guidelines dictating appropriate indirect cost markups. Therefore, it is not surprising that the DOE found non-uniform treatments of indirect costs among the contractor studies that it had commissioned for the evaluation of reactor options for plutonium management (see NAS, 1995, p. 81).

This non-uniform application of indirect cost percentages affects more than advanced reactor contracting, in response to GAO (1993) and Independent Project Analysis (1993), and other studies, DOE is initiating "Department-wide bench marking of various indirect-cost categories against the best in class of public and private business" and initiating plans "for specific goals for reducing indirect costs." (DOE, 1994, p. 52).

These initiatives were started before the current Congress and its attempts to balance the U.S. federal budget. Because projects within the DOE compete with one another for funding (for example, funds for advanced reactor projects compete for funding with environmental restoration at DOE laboratories), there will be fewer funds for plutonium management options. Therefore, in the first stage of evaluating reactor-related options care should be taken (1) to identify redundant indirect cost categories and percentages and (2) to equate non-redundant indirect costs across options. Later stages of the evaluation process should consider whether the indirect cost totals and percentages are reasonable under more competitive contracting practices. Where these are not reasonable, guidelines should be determined before specific proposals are requested.

7. OPERATING COSTS AND REVENUES

This section reviews economic assumptions in evaluating fuel and non-fuel operating costs net of electricity revenues. I argue that the NPV approach should be applied to these costs as its is applied to the benefit of decreasing plutonium S&M charges, direct and indirect construction costs, decommissioning costs, and environmental externalities.

Externalities exist when consumption or production by one party affects another party's well-being or ability to produce. Reactor-related plutonium management causes positive and negative externalities. Some are positive because careful management increases the security of citizens in all nations, not only Americans and Russians. Some are negative such as environmental externalities. If externalities are not properly internalised, inefficiency can result.[16] Because producers of positive externalities do not enjoy all the benefits of production, too little is produced. Because producers of negative externalities do not face all the costs of production, too much is produced. If externalities cannot be internalised, for example, through price mechanisms, activities with positive externalities should be subsidised to bring production to efficient levels. Activities with negative externalities should be charged an implicit price, which is counted as a cost of operation. Of course, implicit prices are unknown and only forecast with uncertainty. As with all operating costs, implicit costs should be included in an uncertainty analysis.

i) Fuel Costs

While NAS (1995, pp. 280-306) review of the cost of mixed plutonium-uranium oxide (MOX) and low-enriched uranium (LEU) fuels is extensive, the same economic assumptions applied to reactor-related facilities also should be applied to fuel facilities. In particular, there is little discussion of the environmental externalities, including decommissioning, of fuel-cycle facilities. For example, use of LEU increases uranium mine and mill tailings. The decommissioning of uranium mines and mills is not fully reflected in uranium oxide prices. Also, using depleted uranium in MOX fabrication, decreases the S&M costs of uranium tailings at the enrichment facilities.[17] Therefore, future evaluations of fuel costs should include the costs of environmental externalities, particularly decommissioning costs.

ii) Non-Fuel Operating Costs

Following DOE/EIA (1995, p. vii), non-fuel operating costs at U.S. NPPs include routine operating and maintenance (O&M) costs and capital additions during operation: "Approximately 67 per cent of the reported O&M costs are labour-related, and the remaining 33 per cent are for expenditures for maintenance material and supplies." There are three types of capital additions: (1) plant retrofits required by the Nuclear Regulatory Commission (NRC), an example being there design of the control room;

(2) repairs needed to keep a plant operational, such as the replacement of the steam generator; and (3) changes to the plant that will improve performance and productivity. Some recently completed case studies found that for the plants studied, roughly 50 per cent of the capital additions were the result of regulatory compliance actions. The other 50 per cent were largely due to repair or replacement of plant components. Only a small fraction of the capital additions costs were undertaken to improve plant performance.

In the early 1990s, for all U.S. NPPs above 400 MW in operation by 1993 average annual total O&M costs were about $95 per kW (1993$) and capital additions were about $25 per kW. The average total O&M costs are comparable with O&M costs for new reactors in Table 6-15 of NAS (1995, p. 318), but NAS did not estimate capital additions. Assuming no changes in regulation during operation, capital additions are likely to add 10 per cent to annual costs.

Administrative and general expenses (including insurance) and taxes (other than income taxes) have proved troublesome to incorporate into the operating cost analysis. In particular, NAS (1995, pp. 78-79): "There is controversy about the existence, magnitude and appropriateness of the 'subsidies' for government activity associated with government's freedom from property taxes and insurance costs, and about whether evaluations of the costs of government projects should be adjusted to cover these factors." NAS (1995) resolves this problem by performing its cost calculations with and without an adjustment of 2 per cent to the fixed charge rate. For new reactors this increases the fixed charge rate about 20 per cent. Although this simplification might not influence the conclusions in NAS (1995), private electric utilities do not capitalise property taxes and insurance, so adjustments to the fixed charge rate are inappropriate. These are administrative expenses and should be treated as annual costs.

Beyond the question of how to model these costs is the problem of whether government facilities should pay property taxes and insurance. Traditionally, DOE facilities have contributed "payments in lieu of property taxes" to local governments. These payments, or property taxes, represent a contribution to the provision of local public services. These services include maintenance of roads and other public properties used by the facility and social services, including schools, used by the facility's employees. In this sense, these payments internalise the costs of externalities generated by the facility. On the other hand, local governments collect taxes on incomes generated at the facility (for example, through sales taxes) and usually collect larger taxes on properties that increase in value because of the facility. Therefore, it is difficult to determine *a priori* whether local governments subsidise DOE facilities. However, to avoid the unfair DOE competition with the private sector in electricity sales, it would be safe to assume that a DOE-financed facility should pay the same property taxes as a private electric utility.

The appropriate insurance charge can be similarly determined. There are several types of insurance to consider. It is convenient to assume that health and safety insurance payments for facility employees are adequately incorporated into O&M costs. (It is possible that health and safety externalities have not been completely internalised because of lack of information; it is also possible that these costs are higher than their efficient levels.) It is safe to assume that the facility would be covered under the Price Anderson Act, so off-site damages would be covered after a catastrophic accident.[18] Finally, insurance to cover losses at the facility equal to payments required by the financial community and regulators of private electric utilities should be added to annual administrative expenses for a DOE-financed facility.

iii) Operating Revenues

One of the benefits of managing plutonium with NPP is that electricity can be sold to offset costs. These benefits can be represented as negative operating costs. Net operating costs (costs minus revenues) can be discounted in the NPV analysis. To evaluate reactor-related options, three forecasts are required: (1) the future price of electricity, (2) the NPP's capacity factor, and (3) the NPP's operating life. All three forecasts require uncertainty analysis.

Forecasting the price of electricity is particularly difficult now in the U.S. given the possibility of deregulating the electric utility industry during the next decade. (See ANS, 1996.) Assuming a competitive environment, electricity produced at a plutonium-consuming NPP will compete with all other producers of electricity in regional markets for long-term contracts and spot sales. There will be no protection of monopoly rights in captive service territories under rate-of-return regulation. This also implies that the low cost of capital enjoyed currently by the electric utility industry is not likely to continue. Therefore, financing costs assumed in earlier analyses are likely to be too iow for future generators and future prices are more likely to be on the high side of estimated ranges.

Because of the stability of production at NPPs, reactor operators will compete for long-term contracts. This reduces the uncertainty associated of forecasting prices.[19] Long-term contract prices will be determined by the low-cost long-term producer. During the life of a future NPP, this is likely to be the gas-fired combined-cycle base load power generator. Therefore, assumptions made in NAS (1995, p. 108) are reasonable: "For our analysis we choose a reference value of $0.05/kWh with a judgmental 70-per cent confidence interval of ±$0.015/kWh."[20]

The second issue in forecasting operating revenues is the annual level of electricity generation. This forecast is typically modeled with the projected capacity factor, i.e., the ratio of average projected generation to the potential maximum generation. But capacity factors vary over the life of the NPP; see Rothwell (1990). Further, this variation with age also varies over time and across reactor manufacturers. Because annual electricity generation is subject to random variation, the capacity factor is uncertain.[21] Therefore, while it is convenient for comparisons to assume the same base-case capacity factor for all NPPs (for example, 75 per cent), the sensitivity of the NPV to the capacity factor should be investigated.

Similarly, the economic operating life of a power plant is uncertain; on modeling this uncertainty, see Rothwell and Rust (1996). While the licensed lifetime could be 40 years, a power plant might not be economically competitive throughout its licensed lifetime. If the primary purpose of a new or modified NPP is to dispose of plutonium, once the excess weapons plutonium has been covered to spent nuclear fuel, decision makers must reevaluate the economics of continued operation at that time treating earlier expenditures as sunk. Because there is the possibility that discounted operating costs of continued operation would exceed discounted revenues, analysis should treat the operating life as uncertain.

8. DECONTAMINATION AND DECOMMISSIONING

Decontamination and Decommissioning (D&D) a nuclear fuel facility or power plant involves several steps: (1) safe shutdown, including the removal of process materials, (2) waste disposal, (3) Surveillance and Maintenance (S&M) until D&D is complete, (4) D&D planning, (5) site contamination characterisation, (6) decontamination, (7) demolition or refurbishment of remaining facilities for future use, and (8) environmental restoration. See DOE/EM (1994). But DOE/EM (1994) guidelines post-date cost estimates for plutonium management options. Also, these guidelines have not followed Nuclear Regulatory Commission requirements for decommissioning NPPs. Until there is more D&D experience with NPPs and a detailed correspondence made between this experience and DOE/EM (1994), DOE's Nuclear Energy Cost Data Base formula for NPP D&D is unlikely to reflect future D&D costs for DOE facilities.

For example, there is much confusion surrounding Nuclear Regulatory Commission regulation of NPP decommissioning. From the late 1970s through 1995 the Nuclear Regulatory Commission sponsored studies to estimate decontamination costs for generic Pressurised and Boiling Water Reactors (PWRs and BWRs). See references in Wood (1991) and Konzek, Smith, Bierschbach, and McDuffie (1995). These are approximately \$130M (1992\$) for PWRs and \$170M (1992\$) for BWRs.[22] Further, these estimates assume low-level waste management at Hanford with price regulation. If prices are deregulated and allowed to rise to those at Barnwell, decontamination costs could rise by \$94M (1993\$). See Konzek et al (1995, v2). These estimates are for radioactive decontamination because the Nuclear Regulatory Commission is only concerned with releasing the NPP site from its licensing requirements. Electric utilities operating NPPs are now required by the Nuclear Regulatory Commission to maintain Nuclear Decommissioning Trust funds to provide for the eventual *decontamination* of these facilities.

However, costs of other decommissioning operations are not reflected in the decontamination cost estimates. Other D&D costs depend on the requirements of other regulatory agencies.

Some of these costs are reflected in estimates done for state public utility regulatory commissions. See Strauss and Kelsey (1991). For example, the estimated cost of a more complete D&D of Trojan (the unit that served as the model for the Nuclear Regulatory Commission's generic PWR decontamination estimate) in 1986 was $173M (1986$), or $216M (1992$). Also the original construction cost of Trojan (see DOE/EIA, 1986, p. 109) was $860M (1982$) or $ 1,237M (1992$), so Trojan's estimated D&D cost will be at least 17.5 per cent and could be as high as 25 per cent (with higher radioactive waste costs) of its construction cost.

While it is instructive to compare projected decommissioning costs with construction costs, most NPP designs were not optimised for decommissioning. With 10-year design and construction periods, 40-year operating lifetimes, and up to 60 years of safe storage before decommissioning, the discounted cost of D&D had little relevance in decision making when NPPs were ordered. Recent experience with D&D has lead to nuclear facility design that considers decommissioning. See DOE/EM (1994, p. 2–2): A design trade-off is considered when a design can be changed to facilitate maintenance or decommissioning activities. For example, a storage tank located in a poured reinforced concrete room would be difficult to access for decommissioning. An alternative design that would provide easier access would be preferable from a decommissioning standpoint, for instance one non-load-bearing wall could be built out of easy-to-remove block. Further discussions of this technique, called facilitation, can be found in NUREG/CR-0569 (1979).

Therefore, estimated construction costs could increase to lower future decommissioning costs. Increases in construction plus decommissioning costs likely will be closer to 20 per cent of the earlier estimated construction costs than to the rule-of-thumb "that the at-shutdown cost of decommissioning is 10 per cent of the construction investment as of start-up." (NAS, 1995, p. 84).

9. CONCLUSIONS AND RECOMMENDATIONS

This section lists the conclusions (•) and recommendations (→) made in this paper and discusses their applicability to Russia.

i) The Net Present Value Method and Discounting

- NPV analysis has been accepted by economists as the guide for making investment decisions.

→ Plutonium management and disposition options should be ranked by their non-proliferation benefits, and within each security class, options should be ranked by NPV.

- Because uncertainties associated with the costs and benefits of plutonium management will change over time, risk-adjusted discount rates are not appropriate for discounting.

→ The most appropriate discount rate to evaluate plutonium management is the risk-free rate with explicit modeling of uncertain costs and benefits.

Although NPV analysis would also be appropriate to evaluate Russian plutonium management options, the selection of the appropriate discount rate for both costs and benefits is more problematic than in the U.S. There has been little historical observation of real interest rates in post-Soviet Russia. One could argue that the real risk-free rate is much higher in Russia than in the U.S. because the social rate of discount is much higher, i.e. in the present situation the Russian people place a higher premium on consumption today than on consumption tomorrow than in the U.S. Further, financial research is required to calculate (1) the appropriate inflation rates for the cash flows in this context, (2) the appropriate risk premiums, including the exchange-rate risk for the nuclear power industry, and (3) the appropriate risk-free discount of future consumption by the Russian people.

ii) Uncertainty and Risk

- The costs and benefits of plutonium-management alternatives are uncertain.

→ All economic variables should be included in an uncertainty analysis. The first step is to determine the most influential variables with sensitivity analysis. Second, specify probability distributions for each of the influential variables. Third, simulate the NPV distribution based on the probability distributions of the influential variables. Decision makers can then evaluate the potential risks of each option.

Uncertainty analysis should be applied to the costs and benefits of Russian plutonium management options. Due to the lack of historical data, probability distributions should be based on expert opinion. Due to the lack of experience of Russian decision makers in evaluating the resulting probability distribution for NPVs, the analysis must be more transparent than in the U.S. The first step is use sensitivity analysis whenever possible.[23]

iii) Economic Benefits

- One economic benefit to changing the current plutonium management regime is a reduction in on-going plutonium Surveillance and Maintenance (S&M) costs.

→ The present value of costs for alternative management options should be compared with the present value of S&M costs and any other economic benefits.

Civilian and excess weapons plutonium is stored in dozens of sites in Russia. Before the economic benefit of reducing S&M costs in Russia could be calculated, the storage and transportation system should be rationalised and upgraded. Some of this work is underway with U.S. and European assistance. Because S&M costs are presently so uncertain, they are unlikely to provide an appropriate baseline with which to compare plutonium management costs. Therefore, a more appropriate benefits baseline must be defined for the Russian situation.

iv) Construction and Operating Costs

- Although larger facilities have lower average unit costs, total costs are higher.

→ Budget constraints require consideration of facilities that might be smaller than minimum efficient scale.

- First-of-a-kind costs in the nuclear industry have been larger than expected costs.

→ Cost estimates should explicitly model costs increases for first-of-a-kind construction projects.

- The risk-free discount rate is equal to the real risk-free discount rate plus the anticipated inflation rate. In 1996 the real risk-free discount rate was about 2 per cent and the anticipation inflation rate was about 3 per cent.

→ Costs that are expected to increase at the anticipated inflation rate should be discounted to the present at the risk free rate. Cash flows that are expected to increase (or decrease) at rates above (or below) the anticipated inflation rate should be discounted at rates below (or above) the risk-free rate.

- In NPV analysis appropriate real discount rates account for escalation during construction and interest during construction.

→ If appropriate discount rates have been used, it is not necessary to include escalation or interest during construction in the NPV analysis.

- Indirect cost estimates have not been uniform across studies of plutonium management.

→ Guidelines should be established to describe appropriate tasks to be included in indirect costs, appropriate percentages for the ratio of indirect to direct costs, and appropriate management fees.

- Estimates of front and back-end nuclear fuel cycle costs have not been made at the same level of detail as reactor-related costs.

→ The same assumptions for the NPV analysis of reactor-related costs applied to fuel costs.

- One half the capital additions nuclear power plants in the U.S. have been for regulatory compliance; historically capital additions were about 20 per cent of total non-fuel operating costs.

→ Assuming no changes in regulatory requirements, annual non-fuel operating costs should be at least 10 per cent higher to reflect repair and replacement of plant components.

- Accounting for administrative expense (including insurance) and taxes (other than income taxes) have proved troublesome to incorporate into operating cost estimates.

→ Estimates of insurance and tax payments for federal facilities should be equal to these payments for electric utilities operating similar facilities. Insurance premiums implicitly paid by local communities should be included as costs.

- Projecting electricity revenues involves forecasting three uncertain variables: (1) the future price of electricity, (2) capacity factors in each year of operation, and (3) the economic life of the NPP.

→ Base line values should be selected for these variables and the sensitivity of the NPV to probable realisation of these variables should be investigated.

- Decontamination and Decommissioning costs based on decontamination estimates are likely to be too low because of the requirements by non-nuclear regulatory agencies.
- → Decommissioning and waste management costs, including changes in construction design to facilitate decommissioning, are likely to be close to 20 per cent of construction costs.

Recommendations regarding construction and operating costs largely apply to Russia. However, I did not discuss international extrapolations of U.S. cost estimates. There is a near universal assumption in Western economic analysis that Russian labour rates are low and will continue to be low compared to labour rates in the U.S. Although Russia could be deindustrialised and living standards could fall to those in non-industrialised nations, this is unlikely to happen. More likely is a gradual rise of labour rates: first to rates comparable to those in Eastern Europe (e.g., Poland) soon and later (with Eastern European wages) to those in Western European over the time horizon under consideration. Further, because of the complex structure of the nuclear power industry in Russia, indirect costs are also complex. Guidelines for maximum indirect to direct ratios should be established to clarify the cost estimation process.

10. LIMITATIONS OF NET PRESENT VALUE ANALYSIS

Although NPV analysis under uncertainty provides the best available technique for economic decision making (because it forces users to make their assumptions explicit), there are two problems with it. First, it is biased towards costs and benefits that can be quantified and monetarised (cash flows). In plutonium management there are many considerations that elude monetarisation. The most important is the anticipated loss from a proliferation-related incident, e.g., burning plutonium to cause panic. Other costs that are difficult to quantify are environmental and health and safety effects. Second, NPV analysis is biased toward the present generation. Discounting across generations requires the assumptions that future generations will have the same values. This is a strong assumption. Costs and benefits to generations beyond the present must be carefully evaluated. Therefore, NPV analysis under uncertainty should be considered as only the best starting point for further research.

End Notes

1. Other committees at the National Research Council considering related problems include the Committee on Separations Technology and Transmutation Systems (STATS) of the Board on Radioactive Waste Management. See, for example, National Research Council (1995). Also see the American Nuclear Society's 1995 special panel report, ANS (1995).
2. NAS (1995) also considered the levelised annual cost (LAC) approach. While this might be appropriate for comparing kilowatt hour costs for electricity generation, the NPV method is more general and is more flexible for comparing non-electricity generating and non-reactor related options. On the NPV method and its superiority over other approaches see, for example, Brealey and Myers (1991, Part 2: Value).
3. The NPV approach must be modified when there are expenditure constraints, e.g., budget or foreign currency constraints. These modifications can be handled by linear programming techniques.
4. NAS (1995, p. 374) considers three classes of management options: current reactor options, vitrification, and advanced reactor options: "The main factors besides timing that affect the comparative security of disposition options are (1) the extent of exposure to theft or diversion in the processing and transportation steps that an option entails and (2) the theft and diversion risks posed by the plutonium in its final form and location." Except for timing, all three classes provide similar non-proliferation benefits. On advanced reactors, see May and Avedon (1994).
5. These are approximations. More correctly, (1 + nominal rate) is equal to (1 + real rate) times (1 + inflation rate).
6. Here, I avoid the question of what the real rate should be by suggesting that the empirical real risk-free rate be used for discounting. Ibbotson and Brinson (1987) and Nordhaus (1994, p. 126) found that real returns between the mid-1920s and the mid-1980s clustered between 0 and 2 per cent per year for fixed-interest instruments for premium borrowers. For a summary of the debate over real rates, see Nordhaus (1994, pp. 122-135). For an extend discussion, see Lind (1982).
7. The process described here can be done with the Lotus or Excel spreadsheet add-in, @RISK, available from Palisade Corporation, telephone 607-277-8000 or e-mail palisade@palisade.com.
8. Nordhaus (1994, Chapter 8) describes a final stage of uncertainty analysis: determining the consequences of rejecting a project that has a high probability of being negative, but could have a positive realised NPV. For a more complete analysis of this stage, see Dixit and Pindyck (1994).

9. S&M costs need not be paid forever to be approximated as a perpetuity. For example, at a 2 per cent discount rate the present worth factor is 25 (one-half of the asymptotic value) after 35 years. At 7 per cent the present worth factor is 10 (approaching an asymptotic value of 14.3) after 18 years.
10. These are costs of storage at new or international facilities built to International Atomic Energy Agency standards. Marginal costs at existing facilities with no other use would be much lower. Because marginal costs are so low and fixed costs are so high, consider the $2–$4 per gram to be an average cost, covering both fixed and variable costs.
11. If the discount rate is 7 per cent per year, the present value of S&M costs reduces to 14.3 times $200M to $400M, or $2.9B to $5.7B.
12. This is expressed in NAS (1995, p. 291) as "we apply the widely used rule of thumb that the construction costs and operating costs for such facilities increase as the 0.6 power of output (meaning unit costs will decrease as the 0.4 power of output)." NAS applies this rule to conversion of plutonium metal to oxide.
13. I have argued that a zero discount rate is appropriate for evaluating the cost of decommissioning U.S. nuclear power plants. The costs of decommissioning appear to be rising as fast as returns on Nuclear Decommissioning Trust funds because the scope and standards for decommissioning are increasing. See Rothwell (1993a). Once regulatory requirements and low-level waste costs stabilise, the discount rate would rise, lowering the present value.
14. Other indirect cost categories include government sponsored research and development expenses and government project management costs. These are identified as "other preoperational costs" in NAS (1995, p. 81).
15. As a member of the NAS committee that produced NRC (1996), I am most familiar with cost estimates for the decontamination and decommissioning (D&D) of the U.S. uranium enrichment facilities, see NRC (1996), particularly Appendix J, "Review of Existing Cost Estimates." Ebasco Environmental prepared a series of cost estimates for this D&D project between May and September 1992. In their final estimate, the direct cost of D&D operations, decontamination facilities, and waste management was $6.5B (1992$). Indirect costs, Programme Integration, Construction Management, and Management and Operations (M&O) fees were $6.3B and contingency was $3.2B. The total cost was $16B. The indirect cost markup was 43 per cent, equal to 26 per cent for field indirects, 3.3 per cent for home office overhead, and 10 per cent for contractor's profit (1.43=1.26 ×

1.033 x 1.10). Programme Integration ($2.6B) included costs associated with project oversight by the M&O Contractor, the Construction Manager, the Remedial Design Engineer, and the costs of preparing environmental impact statements. The Construction Management markup was 5 per cent ($468M) on all direct and indirect costs, not including Programme Integration. The M&O Contractor markup was 5 per cent ($491M) on all direct and indirect costs and construction management, not including Programme Integration. This included surveillance and maintenance, additional security, contractor design and review, contractor construction engineering, health physics, overhead adders, and markups on construction management. See NRC (1996, p. 268).

16. Another solution to the problem of externalities is to redefine property rights. See Coase (1960). For an application of the Coasian solution to nuclear power plant decommissioning externalities, see Pasqualetti and Rothwell (1993).
17. NRC (1996, Chapter 7) recommends that depleted uranium hexafluoride be converted to uranium oxide for long-term disposition. Use of depleted uranium oxide in MOX fabrication reduces the cost of this disposition.
18. Dubin and Rothwell (1990) calculate that the implicit subsidy to an average NPP of the Price Anderson Act after the 1988 amendments is about $22M ($1985). This calculation was based on Nuclear Regulatory Commission estimates that the probability of a loss above $10B was 1 in 1,250,000. These estimates should be revised given the safety record of the nuclear power industry since 1985. To compare reactor-related options with non-reactor related options, similar implicit subsidies should be calculated. In particular, implicit subsidies to S&M storage options should be determined.
19. The variance of spot prices is considerable. The Wall Street Journal now quotes the weighted average of a megawatt hour sold at the California-Oregon and Nevada-Oregon borders for non-firm power on (6am to 10pm) and off (10pm to 6am) peak. The temporal standard deviation of these prices is higher than the reported weighted average, suggesting a high variance with a skewed distribution.
20. The reasoning in NAS (1995, Appendix C) is different from that here because the authors assumed rate-of-return regulation. Their reasoning involves determining (1) the appropriate avoided cost (e.g., as measured by the cost of replacement power as in Che and Rothwell, 1995) and (2) the appropriate cost of capital for government-financed electricity generators. These complications are avoided here by assuming a competitive electricity

market for long-term contracts.

21. Rothwell (1996) presents a stochastic model of electricity generation: "Under the appropriate assumptions, one can show that maximising expected present value for (a nuclear power plant) is equivalent to maximising the probability of producing output."
22. These values are updated from Nuclear Regulatory Commission estimates of $105M (1986$) for PWRs and $135 (1986$) for BWRs in Wood (1991, p. 48) with inflation factors in NAS (1995, p. 78). Although Konzek et al (1995) found that the inflation rate for NPP *decontamination* has been lower that the general inflation rate, these estimates are lower than those quoted in NAS (1995, p. 84), i.e., a base cost of $145M (1992$) for PWRs and $185M (1992$) for BWRs plus a factor related to the thermal size of the reactor in megawatts (MWt). For the generic reactor of 3,400 MWt this factor would increase cost by $44M (1992$).
23. Although high and low values are given for cost estimates, there is no explicit analysis of the sensitivity of costs and benefits to specific causes in US AID (1995).

References

American Nuclear Society (ANS). 1996. "Deregulating the U.S. Electric Industry," *Nuclear News* (La Grange Park, IL: American Nuclear Society, January).

American Nuclear Society (ANS) Special Panel Report, 1995. *Protection and Management of Plutonium* (La Grange Park, IL: American Nuclear Society).

Braun, C. 1992. "Economic Rationale and Prospects for Advanced Reactors," presented to the Energy Information Agency Technical Workshop on Nuclear Issues (Washington, DC: November 12).

Brealey, R. and S. Myers. 1991. *Principles of Corporate Finance*, 4th Edition (New York, NY: McGraw-Hill).

Che, Y.K. and G.S. Rothwell. 1995. "Performance-Based Pricing for Nuclear Power Plants," *The Energy Journal* 16, 4 (December) pp. 57-77.

Coase, R.H. 1960. "The Problem of Social Cost," *Journal of Law and Economics* 3 (October) pp. 1-44.

Copeland, T.E. and J.F. Weston. 1980. *Financial Theory and Corporate Policy* (Reading, MA: Addison-Wesley).

Department of Energy (DOE). 1996 "Department of Energy Declassifies Location and Forms of Weapons-Grade Plutonium and Highly Enriched Uranium Inventory Excess to National Security Needs," *DOE Facts* (Washington, DC: U.S. DOE, February).

Department of Energy (DOE). 1994. *Making Contracting Work Better and Cost Less: Report of the Contract Reform Team* (Washington, DC: U.S. DOE, February) DOE/S-0107.

Department of Energy, Energy Information Administration (DOE/EIA). 1995. *An Analysis of Nuclear Plant Operating Costs* (Washington, DC: DOE) SR/OIAF/95–01.

Department of Energy, Energy Information Administration (DOE/EIA). 1986. *An Analysis of Nuclear Power Plant Construction Costs* (Washington, DC: DOE) DOE/EIA-0485.

Department of Energy, Office of Environmental Restoration (DOE/EM). 1994. *Decommissioning Handbook* (Washington, DC: U.S. DOE, March) DOE/EM-0142P.

Dixit, A.K. and R.S. Pindyck. 1994. *Investment under Uncertainty* (Princeton, NJ: Princeton University Press).

Dubin, J.A. and G.S. Rothwell. 1990. "Subsidy to Nuclear Power through the Price-Anderson Liability Limit," *Contemporary Policy Issues* 8 (July) pp. 73–79.

Electric Power Research Institute (EPRI). 1986. *Technology Assessment Guide* (Palo Alto, CA: EPRI) P-4463s-SR.

General Accounting Office (GAO). 1993. *Department of Energy Mangement Problems Require a Long-Term Commitment to Change.* (Washington, DC: GAO, August) GAO/RCED 93-72.

Ibbotson, R.G. and G.P. Brinson. 1987. *Investment Markets* (New York: McGraw-Hill).

Independent Project Analysis, Inc. 1993. *Project Performance Study* (Washington, DC: U.S. Department of Energy, November) DOE/EM-12396-H1.

Konzek, G.J., R.I. Smith, M.C. Bierschbach, P.N. McDuffie. (1995). *Revised Analysis of Decommissioning for the Reference Pressurised Water Reactor Power Station.* (Washington, DC: U.S. Nuclear Regulatory Commission, November) NUREG/CR-5884.

Lind, R.C., editor. 1982. *Discounting for Time and Risk in Energy Policy* (Washington, DC: Resources for the Future).

May, M.M. and R.E. Avedon. 1994. "The Future Role of Civilian Plutonium: Summary Report of a Workshop Held at Stanford University, March 29–30, 1994." (Stanford, CA: Center for International Security and Arms Control, Stanford University).

Moses, L.N. and D. Lindstrom, editors. 1993. *Transportation of Hazardous Materials: Issues in Law, Social Science and Engineering* (Boston, MA: Kluwer Academic Publishers).

National Academy of Sciences (NAS). 1995. *Management and Disposition of Excess Weapons Plutonium: Reactor-Related Options* (Washington, DC: National Academy Press).

National Academy of Sciences (NAS). 1994. *Management and Disposition of Excess Weapons Plutonium* (Washington, DC: National Academy Press).

National Research Council (NRC). 1996. *Affordable Cleanup? Opportunities for Cost Reduction in the Decontamination and Decommissioning of the Nation's Uranium Enrichment Facilities* (Washington, DC: National Academy of Sciences, March).

National Research Council (NRC). 1995. *Nuclear Wastes: Technologies for Separations and Transmutation* (Washington, DC: National Academy of Sciences, pre-publication copy).

Nordhaus, W.D. 1994. *Managing the Global Commons: The Economics of Climate Change* (Cambridge, MA: The MIT Press).

Nuclear Regulatory Commission. 1979. *Facilitation of Decommissioning of Light Water Reactors.* (Washington, DC: U.S. Nuclear Regulatory Commission, December) NUREG/CR-0569.

Pasqualetti, M.J. and G.S. Rothwell. 1993. "Who Pays to Close a Nuclear Power Plant?," *Public Utilities Fortnightly* (January 15).

Pasqualetti, M.J. and G.S. Rothwell, editors. 1991. *Nuclear Decommissioning Economics: Estimates, Regulation, Experience and Uncertainties*, Special Issue of The Energy Journal.

Rothwell, G.S. 1996. "Organisational Structure and Expected Output at Nuclear Power Plants," *Review of Economics and Statistics* (forthcoming).

Rothwell, G.S. 1995. "New Nuclear Power Plants Require Changes in Utility Structure," presented at the American Nuclear Society 1995 Winter Meeting (San Francisco, CA: November 2).

Rothwell, G.S. 1993a. "Costs for Dismantling Nuclear Power Plants," *The Wall Street Journal* (February 23) p. A19.

Rothwell, G.S. 1993b. "An Economic Review of Monitored Retrievable Storage for Spent Nuclear Fuel," in Moses and Lindstrom (1993).

Rothwell, G.S. 1990. "Utilisation and Service: Decomposing Nuclear Reactor Capacity Factors," *Resources and Energy* 12, p. 215–229.

Rothwell, G.S. and J. Rust. 1996. "On the Optimal Lifetime of Nuclear Power Plants," *Journal of Business and Economic Statistics* (forthcoming).

Strauss, P.M. and J. Kelsey. 1991. "State Regulation of Decommissioning Economics," in Pasqualetti and Rothwell (1991) pp. 55-72.

U.S. Agency for International Development (US AID). 1995. *Joint Electric Power Alternatives Study: Appendix G—Joint Parallel Nuclear Alternatives Study for Russia.* (Washington, DC: National Technical Information Service, June).

Wood, R. 1991. "Federal Regulation of Decommissioning Economics," in Pasqualetti and Rothwell (1991) pp. 45-54.

Address:

GEOFFREY ROTHWELL

Centre for Economic Policy Research

Stanford University

Sanford CA 94305-6072

USA

Table 1 : Economic Assumptions for Evaluating Reactor-Related Options for Managing Plutonium*

Geoffrey Rothwell, Ph.D.
Senior Resarch Associate
Department of Economics
Center for Economic Policy Research
Stanford University
Standord, CA 94305-6072 USA
(415) 725-3456, fax = (415) 723-8611
rothwell@leland.stanford.edu

Utilisation of the Excess Weapons Plutonium: Scientific, Technological and Socio-Economic Aspects
Como, Italy
March 20, 1996

Table 2 : The Economic of Plutonium Management

- NAS Committee on International Security and Arms Control (CISAC)
- NAS. "Management and Disposition of Excess Weapons Plutonium" (1994)
- NAS. "Management and Disposition of Excess Weapons Plutonium: Reactor-Related Options" (1995)

Table 3

NAS (1995, p. 280): "economic considerations are less important than security in reaching a conclusion about the relative attractiveness of alternative options for WPu disposition. But the study of costs is nonetheless worthwhile, both to assist in ranking options that are not distinguishable on security grounds and to facilitate planning for the investments that will be required for one option or another".

Table 4

NAS (1995, p. 374) considers three classes of disposition options: current reactor options, vitrification, and advanced reactor options: "The main factors besides timing that affect the comparative security of disposition options are (1) the extent of exposure to theft or diversion in the processing and transportation step that an option entails and (2) the theft and diversion risks posed by the plutonium in its final form and location". Except for timing, all three classes provide similar non-proliferation benefits.

Table 5 : Economic Evaluation of Plutonium Management

- Net Present Value (NPV)
- Stages of NPV Analysis:
- 1st: Preliminary evaluation of the costs and benefits: which options are likely to yield positive NPVs?
- 2nd: Rank options by NPV
- 3rd: Evaluate specific proposals

Table 6 : What is NPV?

- Net Present Value is discounted benefits (B) minus discounted costs (C):

$$NPV = \sum_{t=1}^{T} \frac{B_t}{(1+r)^t} - \frac{C_t}{(1+r)^t}$$

Table 7 : Appropriate Discount Rate

- Nominal Discount Rate + Real Rate + Inflation Rate
- Real Discount Rate = Risk-free rate = Risk Premium
- For example, in U.S. in 1996
- Nominal Rate on short-term government securities = 5%
- Anticipated Inflation Rate = 3%
- Real Risk–Free Rate = 2%

Table 8 : Risk Adjusted Discount Rates ?

- If risk is constant over the life of the project, then a risk-adjusted discount rate is appropriate
- But risky costs discounted at less than the Risk-Free Discount Rate
- And risky benefits discounted at more than the Risk-Free Discount Rate

Table 9 : NPV with Risk Adjusted Discount Rates

- A premium is added to the discount rate

$$\mathrm{NPV} = \sum_{t=1}^{T} \frac{B_t}{(1+r')^t} - \frac{C_t}{(1+r'')^t}$$

Table 10 : Are Plutonium Management Cost and Benefit Risks Constant Over Time?

- This is unlikely
- Solution:
- (1) Use a real risk-free rate to discount cash flows and
- (2) Explicitly describe the probability distributions for costs and benefits.

Table 11 : The Present Value of Economic Benefits

- One quantifiable economic benefit
- Reduction in ongoing Surveillance and Maintenance (S&M) costs: the quicker the reduction in S&M costs, the greater the economic benefit
- Therefore, time to completion matters.

Table 12 : Contingency

- EPRI's Technology Assessment Guide:
- Contingency factor related to level of estimate's detail
    - –30% – 50% : Simplified estimate
    - –15% – 30% : Preliminary estimate
    - –10% – 20% : Detailed estimate
    - –5% – 10% : Finalized estimate

Table 13 : Uncertainty Analysis

- (1) Sesitivity Analysis of economic assumptions
- (2) Approximate probability distributions of economic variables
- (3) Monte Carlo approximation of NPV distribution
- (4) Evaluation of stochastic distribution to dtermine risk premium

Table 14 : Comparing Two NPV Distribution

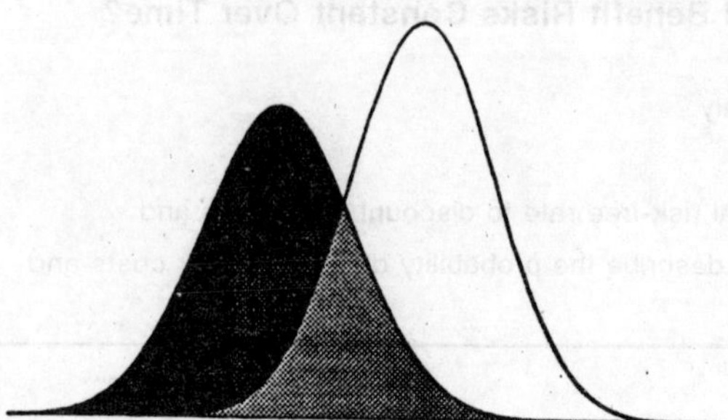

- One distribution has lower mean but a higher variance
- The other has a higher mean but a lower variance

Table 15 : Risk Aversion and Risk Premiums

- (Risk Aversion)
- When the value of the expected NPV is greater than the expected value of NPV
- Risk Premium:
- When investors are indifferent between the risk-free outcome and the risky outcome plus a risk premium

Table 16 : NPV and Certainty Equivalence

- Risk premiums are added to Benefits and Costs before discounting

$$\text{NPV} = \sum_{t=1}^{T} \frac{B_t + RP'}{(1+r)^t} - \frac{C_t + RP''}{(1+r)^t}$$

Table 17 : A Risk Averse Investor

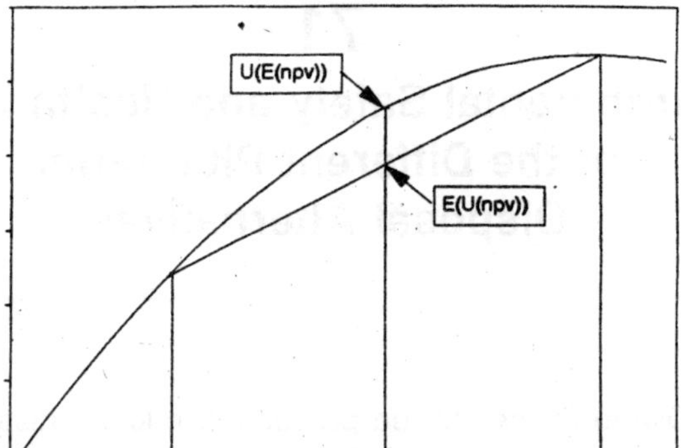

- The value of the mean NPV is greater than the average value of two risky NPVs.

Table 18 : Comparing Different Adjustments to NPV

- NPV without adjustment

$$\text{NPV} = \sum_{t=1}^{T} \frac{B_t}{(1+r')^t} - \frac{C_t}{(1+r'')^t}$$

- NPV with risk-adjusted discount rates

$$\text{NPV} = \sum_{t=1}^{T} \frac{B_t}{(1+r)^t} - \frac{C_t}{(1+r)^t}$$

- NPV with risk premiums

$$\text{NPV} = \sum_{t=1}^{T} \frac{B_t + RP'}{(1+r)^t} - \frac{C_t + RP''}{(1+r)^t}$$

Table 19 : Economic Assumptions for Evaluating Russian Options

- Use Net Present Value analysis with Russiân-appropriate economic assumptions
- Determine Russian real risk-free discount rate, risk-premiums, and inflation rates for labour, materials, equipment, and waste management.
- Calculate fixed and variable costs for upgrading plutonium managemnet system, compare with costs of disposal
- Conduct sensitivity analysis to determine critical variables outcome plus a risk premium.

71

Environmental Safety and Health Risks of the Different Plutonium Disposal Alternatives

L.A. BOLSHOV

1. Owing to the started process of nuclear disarmament, a problem of the future of the excessive weapon plutonium arose. The solution of the problem should satisfy the requirements of non-proliferation of nuclear weapons, providing of people's health and safety, and protection of the environment. Thus, the first task is to convert weapon plutonium into form that would exclude any possibility of military usage of this plutonium. One of such forms is spent nuclear fuel.

There are two technical methods to reach the standard of the spent fuel. The first way, known as reactor choice, is based on the conversion of weapon plutonium into mixed dioxide fuel (MOX fuel) to be processed in heat reactors. In this case, plutonium is a valuable power resource. An alternative way, known as glazing, assumes mixing of plutonium with high-active as glazing, assumes mixing of plutonium with high-active waste followed by glazing and burying. In this case, plutonium is merely a kind of waste, which seems to be unjustified extravagance from the economical point of view.

The reactor variant of the weapon plutonium conversion call be implemented faster than glazing. In principle, it provides more effective protection from the possible repeated military usage of plutonium. The reason is that the isotope content of plutonium is changed and reduced in the process of fuel burning, while in the process of glazing it remains constant.

According to the data of the Uranium Institute, the total amount of the weapon plutonium in the former USSR reaches 105-130 tons. The estimates of the plutonium power resource made by specialists of the, ENTEK show that 100 tons of plutonium can produce from 1000 to 8000 TWt-h of the electric power, depending on the, applied power technology (heat reactors with opened or closed fuel cycle, fast reactors with BF=1.00-1.05). Tables 1 and 2 display the estimated data oil the possibility to use plutonium in Russian reactors. It is clear that the most effective "furnaces" are first reactors, the next is VVER-1000 consuming the mixed fuel.

It follows from the above, that the main direction of the weapon plutonium conversion is most likely to be its using as nuclear fuel. In this connection, IBRAE RAS performed assessment of the possible radiological consequences of a severe accident with core melting on the NPP with fuel enriched by Pu-239. These investigations involved:

- model calculations of the possible activity of Pu-239 release into atmosphere resulting from the accident on VVER-1000 reactor operating with fuel enriched by Pu-239;

Table 1 : Annual Balance of General Plutonium Isotopes. Open Fuel Cycle, kg/[GW(el)–year]

	(A) VVER-1000 *UO_2*	*(B) VVER-1000* *PUO_2+UO_2+Zr*	*(C) BN-800* *PuO_2+UO_2*	*(D) BN-800* *PuO_2+Th*
Input:				
239 Pu	--	764.2	1910	2421
240 Pu	--	49.4	124	94
241 Pu	--	4.8	12	7.5
242 Pu	--	0.3	0.7	0.3
Pu	--	818	2047	2522
Output:				
239 Pu	124	130	1798	1690
240 Pu	55.2	144.2	262	186
241 Pu	30.9	73.2	22.8	12.5
242 Pu	12	37.2	2	0.8
Pu	222	384	2085	1889+770 kg 223U
Output *Input:*				
239 Pu	124	–634	–112	—731
240 Pu	55.2	95	139	92
241 Pu	30.9	68.4	10.8	5
242 Pu	12	36.8	1.3	0.5
Pu	222	–434	39	–633+770 kg 223U

Table 2 : The Possibility of Utilization of Plutonium Li Russian Reactors (Data Presented by Physics and Energetics Institute from the Report "Engineering Analysis of Production of Uranium-Plutonium Fuel from Weapon Plutonium, and its Possible Utilisation in Nuclear Energetics" Minatom of Russia–Siemens – GRS. 1995)

Reactor kg/yl.	*Loading Pu, kg/yr*	*Yield Pu,*	*Balance, kg/yr*
1 VVER-1000	0	223	+223
2 VVER-1000	254	308	+54
3 VVER-1000	364	395	+31
4 BN-600	1141	1053	–88
5 BN-800	1637	1508	–129

Notes : Reactor VVER-1000 is loaded with: (1) standard poor-enriched $U0_2$ fuel (4.3% of 235U, taken as a reference), (2) 1/3 of the reactor is loaded with MOX-FA (concentration of plutonium is 3.5%; poor uranium dioxide with 0.25% of 235U serves as a carrier, and 2/3 of reactor is loaded with uranium FA(0.25% of 235U), and (3) 1/3 of the reactor is loaded with MOX-FA but with higher concentration of plutonium in the mixture of oxides (5% Pu) and with lower level of enrichment of uranium (3.7% of 235U) in uranium FA. Reactors BN-600 and BN-800 are completely loaded with uranium-plutonium fuel.

– calculation of the Pu-239 contamination density of the landscape, resulting from such an accident;
– calculations of the population and personnel exposure doses resulting from Pu-239 release in an accident;

The used base of information comprised:

– actual data on contamination levels of the environment objects by plutonium isotopes after the Chernobyl accident;
– experimental data on plutonium isotope content in human body in the regions surrounding the Chernobyl NPP and fuel-processing plant "Mayak" in Chelyabinsk region;
– modern dosimetry data on the radiation impact of plutonium isotopes on human body.

2. The model calculations of the consequences of major accidents at NPP's with PWR-type reactors are performed in IBRAE using the modified computer code "MELCOR", which allows for specific features of the design of Russian reactors [1]. Figure 1 represents the nodalisation of VVER-100 used in the calculations with "MELCOR". The basic scenario of the severe, accident on VVER-100 with Pu-239 enriched fuel was based on the event chain that gives the most severe consequences. The

rupture of the main circulation pipe in one of four loops at the hot segment with equivalent diameter Du850 and bilateral outflow of coolant was chosen as an initial event. In addition, the presence of hidden defects in the plant equipment was assumed, namely: complete failure of power feeding from internal and external sources with respective failure of the emergency core cooling system (ECCS), and failure of the area spray system of steam condensation inside containment.

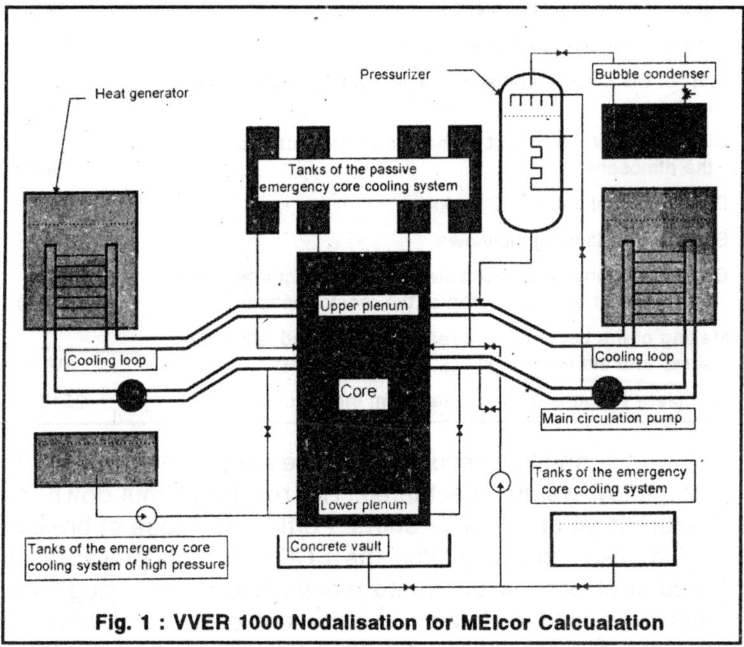

Fig. 1 : VVER 1000 Nodalisation for MElcor Calcualation

The calculations assumed that the fuel load at the moment of the accident was equal to 75,000 kg of UO_2 and 1,500 kg of PuO_2. the fission products are released inside containment and then leak to the external space through the project leaking in the containment. The resulting chain of the accident events is represented in Table 3.

The rupture of a loop of the main circulation pipe is followed by the intensive coolant outflow into the reactor pit. At the beginning, the Outflow coolant is in the liquid state (water); later on, when the water level is reduced lower the edge of the emergency pipe, it turns into steam. The process is accompanied by the abrupt decrease of pressure and coolant level in the

Table 3 : The Chain of Main Events of the Accident Process

Event	*Time from the start of the calculation, s*
The initial event - the rupture of the main circulation pipe	10.0
Actuation of the emergency protection system; start of the MCP run down	12.0
Connecting of tanks of the emergency core cooling system; start of bore solution injection into reactor	15.0
Start of core drying and hearing	17.0
Start of the steam-zirconium reaction with hydrogen release in the core	20.0
Cladding failure and start of the fission product release into the atmosphere	26.0
Devastation of the ECCS tanks	174,.0
Start of the cladding meltdown	347,.0
Complete vapourisation of water from the reactor and start of heating of the core scrap at the reactor bottom	1300.0
Melting of the bottom of the reactor vessel and fall of the scrap into the reactor pit	1660.0
Complete vapourisation of water from the reactor pit	25100.0

first circuit, and two seconds after the start of the accident, the emergency system is actuated. The reactor is shut down, and the run down of the MCP starts. Further decrease of pressure should result in connecting the ECCS of high pressure and injecting of bore solution into reactor. However, owing to the mentioned hidden defects, it does not occur, and when the pressure decreases lower 5.9 Mpa, the passive tanks of ECCS are connected. The action of the tanks only slightly reduces the velocity of the level decrease in the reactor, and by the seventh second the drying of the core starts. Water is vapourised in the lower (poured) part of the core and outflows in the form of stream into the reactor pit, and then in the adjacent volume. Approximately 125 tons of water, from the total amount of 220 tons in the first circuit appear in the reactor pit, the rest amount spreads in the from of stream over other compartments of the reactor building.

Power removal from the dried part of the core is insufficient for the effective cooling of fuel rods, and by the 10th second from the accident start, the temperature of the upper part of fuel rods reaches the lower boundary of the stream-zirconium reaction (approximately, 1000 K). Hydrogen release and spreading over the plant compartments start. The reaction is accompanied by large heat production, which results in still more intensive heating of the core. Sixteen seconds after the start of the accident, when the cladding temperature reaches approximately 1300 K, cladding failure occurs, and gaseous fission products from the inter-rod space are released into the atmosphere. Further on, the release of the volatile fission products from the solid matrix is intensively developed, thus reducing the residual heat production in fuel. By the 337 the second from the accident start, the rod temperature reaches the zirconium melting temperature, and the destruction of the core starts. The cells of the second, the most power-stressed zone are destroyed first, and several seconds later, the central zone is destroyed. The unmelted fuel tablets also fall down at the vessel bottom. The remaining water rapidly boils and cools the falling scrap. By this moment, the amount of water in the first circuit is sufficient for keeping the scrap at the bottom of the vessel poured. However, the water vapourisation is enhanced, and approximately 1300 seconds later the start of the accident water is completely absent from the reactor vessel. The debris of two central zones heats the bottom, and it is melted 1600 seconds after the start of the accident. The melt appears in the reactor pit, and intensive vapourisation of water from the pit and concrete destruction by the action of high temperature start. Further destruction of core occurs inside the reactor vessel, and only 5% of its total mass remain solid by the moment of the calculation end. The complete vapourisation of water from the reactor pit takes 25,000 seconds (or, approximately, seven hours).

Figure 2 illustrates the thermo-hydraulic processes that occur in course of the specified scenario. It shows the time dependence of pressure inside the VVER-1000 containment and the value of mass of UO2 released into environment.

3. Using the obtained parameters of the emergency release, we calculated the trajectory of propagation of the radioactive cloud and made a prediction of the probable radiation situation in the close proximity of the plant.

Such calculations are performed in IBRAE RAS on the

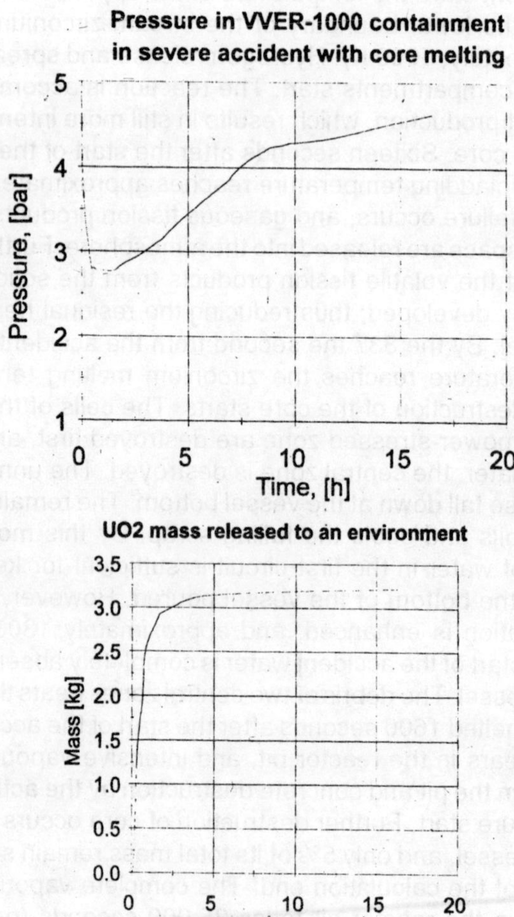

Fig. 2 : Thermohydraulic Processes for the Specified Accident Scenario

basis of customary Gaussian models of the release propagation, which have shown good applicability within the distance of several kilometers from the release source (programme code "TRACE" [1]), and on the basis of more complicated Lagrange models (programme code "NOSTRADAMUS" [1]), which allows one to perform calculations for the distances of several hundred kilometers from the plant and accounts for changes in meteorological situation. The mentioned computer codes can predict the concentrations of radionuclides in the surface layer of the atmosphere; density of the territory contamination by radioactivity after the cloud pass; exposure doses of gamma radiation in the open air, and possible doses of external and internal radiation of people. The results of such calculations are intensively used in Russia when preparing and conducting practical games, Which emulate major radiation accidents, for assessment of the expediency of implementing various population-protecting measures (people housing, population evacuation, iodine prophylactics, etc.) in the region of the NPP, and for the analysis of the required and sufficient means and forces for such measures [2-4].

The performed calculation show that under usual weather conditions and within the specified accident scenario, the territory contamination density by Pu-239 at the distance 5 km from the NPP can reach 1.3 mBq/m^2, or 60 times larger than global average soil contamination density resulted from the nuclear weapons tests, according to estimates reported by the UN scientific committee for Nuclear Radiation Effect.

4. In order to transform the territory contamination density to human exposure doses, we performed the analysis of actual and calculated data on people exposure doses by plutonium isotopes after the Chernobyl NPP accident. It was shown that 10 years later the accident, the ratio of the territory contamination density by Pu-239, 240 in kBq/m^2 and the annual effective exposure dose in μSv/year can reach 2.7, and 70 years after the accident this value could betion density by Pu-239, 240 in this case is 0.66 mSv.

Table 4 displays the data on the comparison of radiation burden upon the population due to plutonium isotopes and other radionuclides for the occurred accidents (Chernobyl NPP and fuel plant "Mayak" in the Chelyabinsk region) and the accident described by the considered scenario.

In conclusion, we can emphasise that, if the post-accident contamination density of the territory at the distance 5 km from the NPP that uses Pu-239-enriched fuel would not exceed 1.3 kBq/m2, then average cumulative people exposure doses resulting from the Pu-239 release would not exceed 0.86 mSv for 70 years of residence in the zone. This value is 1.2 per cent of the exposure due to natural background for the same time interval. The dose for the personnel in the zones of intensive dust production can be 5-10 times large, but even in this case they are not dangerous for the human health. Thus, using the Pu-239-enriched fuel for the NPP with VVER-1000 reactors, even in the case of the severe accident with core meltdown would not aggravate the consequences of the accident for the population residing in tile zone of the NPP.

References

1. Command and Headquarters Exercise "Polyarnye Zori-95". Apatity, Murmanskaya oblast', May 29- June 2, 1995. Final Report. Appendix A. Documents Prepared for the Participants of the Practical Game. Moscow, NSI RAS, 1996.
2. Arutunjan R.V., Linge I.I., Pavlovski O.A. et al. French-Russian Role-Play on Decision Making in the Event of Radiological Contamination of the Large Areas of Land. Portsmouth 94 Proc. of the IRPA Regional Congress, Nuclear Technology Publishing, 1994, p. 329-332.
3. Command and Headquarters Training "Kalinin NPS-94". Moscow. November 22-24, 1994. Final Report, Moscow, NSIRAS, 1995.
4. Command and Headquarters Exercise "Polyarnye Zori-95". Apatity, Murmanskaya oblast', May 29 - June 2, 1995. Final Report. Moscow, NSI RAS, 1995.

Address:

LEONID A. BOLSHOV
Director
Nuclear Safety Institute of the
Russian Academy of Sciences
113191 Moscow
Russian Federation

72

The Disposition of Weapon Grade Plutonium: Costs and Tradeoffs

WILLIAM J. WEIDA

This paper explores some of the economic issues surrounding a major area of expenditures now facing the nuclear powers: the disposition of weapon-grade plutonium either through 'burning' in nuclear reactors for power generation or by other means.[1] Under the current budgeting philosophy in the United States, programmes managed by the Department of Energy (DOE) tend to compete with one another for the total funds assigned to that agency. For example, in the FY1995 DOE budget a trade-off was made between increased funding for nuclear weapons and reduced funding for site cleanup. No matter which disposition alternative is chosen, if disposition funds are controlled by the DOE in the US or by a government agency in any other country, disposition is likely to complete directly or indirectly with other alternatives for energy funding. And if they are subsidised by any government, research into plutonium as reactor fuel or the operations associated with such use are likely to consume funds that might otherwise be available to support sustainable energy alternatives.

When all costs are considered, final waste disposal costs will be incurred whatever disposal option is taken. These costs could potentially be offset by doing something profitable with the plutonium prior to final storage, but this paper has shown that finding a profitable use for plutonium is unlikely. Thus, the more probable case is one where the costs of basic waste storage are increased by whatever costs are associated with the disposition option chosen. The factors most likely to significantly increase costs appear to arise from four areas:

(1) The level of subsidisation in the "profitable" parts of the

disposition programme.

(2) Those items (such as reprocessing) that increase the volume of waste and thus, the cost of waste disposal.

(3) The cost of security and its direct relationship to the number of times plutonium is handled or moved.

(4) The cost of research and development of new and unproven methods of disposition.

INTRODUCTION

Over the last three years, the uneconomical aspects of burning plutonium have been made abundantly clear by a number of studies, In spite of this, of all the materials, systems, facilities, and laboratories involved in the design and operation of nuclear weapons, the most readily available assets for reuse are usually identified as being the plutonium from warheads. Over the last two years, quasi-private consortia have put considerable effort into convincing the US government to embark on such a programme. These efforts have either

(1) assumed that there was an economical way to burn plutonium for power,

(2) proposed the construction and operation of new reactors specifically built to burn plutonium as part of a regional conversion plan for old nuclear weapon sites, or

(3) claimed that even if power generation itself was uneconomical, it would still provide a way to dispose of the large stocks of plutonium that was economically sound in the long run and was worthy of government support.

At the same time, other technical solutions for the plutonium problem have also been proposed. Many of these are transmutation techniques that would require large amounts of federal research and development money to construct facilities to turn plutonium into shorter-lived elements.[2] Others, such as shooting plutonium into the sun, are equally expensive. With the exception of the Integral Fast Reactor (IFR), which has also been marketed under category (2) above, transmutation has generally been proposed as a pure government research project.

In this paper, comparisons between plutonium and other forms of nuclear power generation will be made using the general "industry model". In these comparisons, the costs associated with the wastes generated during the creation of nuclear power will not be explored because these costs are approximately identical no matter what kind of nuclear operations are undertaken. However, a full accounting of these costs would be necessary before any form of nuclear power generation is compared to coal, gas, hydroelectric, or solar generation schemes.

As a further issue, it should also be remembered that most nations are currently struggling with nuclear proliferation issues. Recent problems with North Korea have demonstrated that because plutonium is normally produced as a by-product of reactor operations, civilian nuclear power generation may be in fundamental opposition to proliferation goals in spite of international safeguards installed at most plants. Reactor grade plutonium is about 25 per cent Pu240 as opposed to 6 per cent in weapons-grade plutonium.[3] However, the use of reactor-grade plutonium in a nuclear weapon was successfully demonstrated at the Nevada Test Site in 1962.[4] Further, actually burning plutonium for power legitimises the reprocessing of spent fuel and the possession of plutonium, both of which vastly complicate the proliferation issue. When evaluating any disposition option, one should keep in mind that the major obstacle to building a bomb is getting plutonium. When that obstacle is overcome, the rest is much simpler.

THE ECONOMIC VALUE OF PLUTONIUM

Value is normally established through a market mechanism in which a buyer and seller negotiate a price viewed as fair by each. At the present time, no market for plutonium has formed, due partially to a lack of demand, partially to plutonium's role as a heavily controlled substance, and partially to adverse public reaction over the shipment and use of plutonium. The only sizeable market for civilian plutonium in recent years was one created by Japan's purchase of plutonium from France for future use in its breeder reactor programme. Pricing in that market was not public, but Japan's unique lack of alternative energy sources make its determination of the value of plutonium inapplicable to most other countries. Further, adverse publicity generated by the plutonium shipments on the Akatsuki-maru in 1992-1993 are likely to prohibit similar purchases by Japan in the future—thus terminating the market. It is probable that there is another, illicit market for plutonium, but prices in this market are surely much higher than the actual value of plutonium because of the risk involved. Hence, neither previous experience nor the illicit market provide much guidance as to the actual value of plutonium.

If all costs of plutonium are considered, it would be some of the most expensive material ever created by man. Further, the true costs of reusing plutonium generated through dismantlement of nuclear weapons would also have to include:

The research costs accumulated in developing the materials.

The initial costs to extract uranium, to purify the materials and to make plutonium in reactors.

The cost to fabricate the materials into weapons.

The cost to maintain plutonium in weapons.

The cost to dismantle weapons and free plutonium for other uses.

And finally, the future costs of waste disposal would have to be included.

Accounting for the past costs of plutonium would make it too expensive for any alternative use and, whether legitimately or not, these costs are usually counted as the costs of doing business during the Cold War. As a result, alternative uses of plutonium are usually considered under the assumption that all past costs are sunk costs and future decisions are based only on the future costs of disposition.

Russian Perceptions of the Value of Plutonium

Russia's approach to valuing plutonium appears to be at least partially based on the costs expended to create it. Viktor Mikhailov, Russian Minister of Atomic Energy, has said that plutonium cost the Former Soviet Union six times as much to make as HEU so it is unacceptable to destroy it.[5] He has also stated that "we have spent too much money making this material to just mix it with radioactive wastes and bury it."[6]

However, many Russians recognise that if plutonium has any real economic value, is only as a future energy source. The Director of the Obninsk Institute has acknowledged that plutonium does not have any economic value in the near term, but he noted that doesn't mean that the economy won't eventually change to favour the use of "products for future technology, which we cannot use today."[7] In spite of this, Russian engineers are planning to use existing supplies of reactor-grade plutonium long before such use becomes economically viable—possibly to avoid the handling difficulties encountered as components of reactor-grade plutonium break down and become more radioactive over the next few years. This will allow Russia to keep its weapons plutonium in storage for the next few decades.[8]

It is probable that the near-term attempts by Russia to use plutonium as reactor fuel are partially based on two perceptions—both of which affect assessments of value: first, such use allows more oil and gas to be exported, and second, shipping reactor fuel is easier than shipping coal, oil, or gas in a country as large as Russia. Viktor Mikhailov has proposed using part of the $10 billion in hard currency Russia will generate by selling 500 metric tons of highly enriched uranium to the United States to help build new breeder reactors to produce more plutonium. Completing the first two reactors and a MOX factory to turn the plutonium into reactor fuel would cost $2.5 billion, according to experts at the Institute of Physics and Power Engineering.[9]

Assuming these potential uses of plutonium do actually represent an assessment of its value instead of bureaucratic inertia or some other rationale, these assessments are clearly not universally accepted in Russia. Aleksei Yablokov, an adviser to President Boris Yeltsin and a former environment minister, claims it is not clear that existing nuclear reactors, let alone new ones, make economic sense when Russia could replace all its reactors with natural gas and coal-fired power plants for an estimated $6 billion to $7 billion. In contrast, he claims the International Atomic Energy Agency has stated that upgrading existing Russian reactors to Western safety standards would take between $26 billion and $120 billion.[10]

US Perceptions of the Value of Plutonium

As was the case in Russia, the US Department of Energy has usually calculated the value of US plutonium based on the costs to manufacture it. And as it was in the Russian case, the fallacy in the proposition that something is valueable if it is expensive to produce is obvious. Further, in a country such as the US which is not currently considering the use of reprocessing or breeder technology, the prior costs used to value plutonium are neither market-based nor necessarily rational. As a result, all calculations of the value of US plutonium by independent sources have come to the same conclusions: the value is negative—at least for the foreseeable future.

As an added factor, when one evaluates the economics of plutonium burning in commercial reactors as a method of disposition, basic physical rules apply: first, reactors using plutonium generate approximately the same amount of power as they would if only uranium was used. Second, the quantity of material put into a reactor becomes the quantity of spent fuel generated by the reactor. Thus, only one cost comparison between plutonium and uranium is necessary to show if plutonium can be burned with an economic benefit and therefore, has a positive value:

(1) The costs of processing and fabricating reactor fuel—and whether these costs would be higher or lower when plutonium is used.

At prices of ($20/kg) for natural uranium and ($70/SWU) for separative work, low enriched uranium (LEU) costs about $750/kg. This is about half the estimated cost of $1300–$1600/kg for new MOX fuel. At these prices, disposing of 200 tons of plutonium would cost and additional $1.5 billion over burning LEU (or $2–$10 billion using a reasonable range of future LEU and MOX prices).[11]

Also of interest is whether the costs of disposing of plutonium might be lowered by burning it in a reactor, or whether the overall costs of disposition can be reduced by simply disposing of plutonium without burning it. These disposition costs will be explored later in this paper, but with no operating market for plutonium, and with existing prices for plutonium that have been set by governments, it is fair to say that no one has established an economic value for plutonium. The lack of market information causes pricing problems that cascade through any commercial operations when plutonium is introduced into a power-generating regime where market analysis and cost control govern which power sources are exploited.

THE NATURE OF THE COMMERCIAL NUCLEAR INDUSTRY IN THE UNITED STATES

Since its inception, subsidies have been a way of life in the US nuclear power industry. A 1992 report found that over the period 1950 to 1990, 20 per cent or $96 billion of the $492 billion (in 1990 dollars) spent to develop and obtain nuclear power was provided by the US federal government.[12] According to the DOE, of total subsidies to the energy sector provided by the federal government in 1992, nuclear energy received $899 million of $4.88 billion expended—or about 18 per cent. However, while most other sources of energy (oil, coal, etc.) received either tax subsidies to lower prices or direct subsidies to encourage consumer use—both of which acted to stimulate demand for the product—nuclear energy received almost all of its subsidies ($890 out of $899 million) in Research and Development. In fact, in the US, nuclear energy received 44 per cent of all energy R&D subsidies in 1992.[13]

Over the last forty years, funding of nuclear energy research has continued with little actual implementation of research results. As construction of new reactors in the US ceased—the last US commercial nuclear reactor was started in 1978—a few large companies stayed in the reactor research and development business without having to sell economically viable reactors. In such a situation, there has been no need for commercial products—instead, the emphasis has been on selling and maintaining large research and development programmes. Each new research and development (R&D) proposal has been further removed from the last project private industry and the public were willing to accept and fund. One result of this situation has been to create an industry interested in the development of sources of power, not the economics of producing power.

This helps explain the nuclear industry's continuing research into the use of plutonium burning reactors in the face of overwhelming evidence that such reactors would be economically unfeasible. As time has passed, the economic viability of even standard nuclear reactors has deteriorated. This is unlikely to improve in the future when plans to generate power from plutonium are proposed to take place. Evaluating the current status of nuclear power in the US, Shearson Lehman reports that: "Evidence suggests the average operating costs of nuclear power plants are now higher than those of conventional plants and other power supply alternatives."[14] And Moody's investor's service has stated that:

"Given increasing competition from other types of generating facilities and renewed efforts via conservation and demand side management programmes to reduce the need for new capacity additions, nuclear power's economics must be comparable with alternative fuel sources and energy efficiency and conservation options. In a deregulating environment, the pressure to maintain competitively low rates will compel utilities to select the most economic option. And given the challenges outlined above, we do not think that nuclear plants are likely to provide such economic benefits."[15]

The questionable ability of commercial nuclear reactors to compete with other power generators in the US has been further complicated by recent increases in the price of uranium. Many uranium producers ceased operations in the 1980s and the supply of down-blended Russian uranium available to western markets has been below expectations. Since January 1, 1996, the price of uranium has risen 18 per cent to $14.75 a pound, and prices have risen by 55 per cent since the start of 1995. Uranium prices peaked at $40 in 1979 and were at $5 per pound in 1991 when the industry's competitiveness problems began.[16]

Chow and Solomon estimated that thermal cycle plutonium use[17] in commercial reactors will not be feasible until the price of uranium-bearing yellowcake reaches $100/LB and they estimated that this will not occur for 50 years.[18] They further projected that fast reactors will not be profitable until yellowcake prices reach $220/LB in about 100 years.[19] It should be noted that the costs of burning plutonium are always compared with the costs of burning plutonium are always compared with the costs of burning highly enriched uranium (HEU) or low enriched uranium (LEU) in reactors. Since the inherent costs of regulation, operating efficiency, waste disposal, and contamination associated with nuclear operations are approximately the same for both uranium and plutonium operations, these costs are never discussed in this kind of comparison. However, these costs, as well as the increasing price of uranium, are all factors when a country, city, or power supplier considers the price of alternative nuclear and non-nuclear power sources. And in this type of comparison, the very factor that makes the cost of using plutonium compare favourably with the cost of using uranium—a sharp increase in the price of urnanium—also makes it more likely that neither fuel will be competitive with non-nuclear methods of power generation.

BURNING PLUTONIUM

The economic realities of commercially operating breeder reactors have been faced by the US, Japan, Germany, and France over the last twenty years, and the result has been basically the same in all four countries.

The US:

The United States spent $1.7 billion on a breeder reactor in the 1970s and early 1980s when uranium was expensive and the number of nuclear power plants in the US and abroad was expected to continue to increase. As the price of uranium dropped from $40 a pound in 1982 to a little over $5 a pound in the early 1990s the US dropped its breeder reactor plans. The US cancelled the Integral Fast Reactor research programme several years ago and US experts continue to believe that there will a long-term glut of uranium.[20]

France:

In France, Superphenix, the first commercial-scale fast breeder reactor, has had continuous problems since it began operations in 1986. Leaks in the liquid sodium coolant system have persisted, the reactor was shut down in 1990, and it operated only 174 days in eight years. France's Atomic Energy Commission (CEA) claimed the reactor could burn 200 kg of plutonium a year and in March, 1994, it said it will make modifications to achieve the goal. It has not said what these modifications will cost. The French government says it has abandoned the idea that the Superphenix will ever make money.[21]

The French civilian power industry owes bondholders billions of francs and the French government no longer guarantees nuclear bonds. An independent study by the International Project for Sustainable Energy Paths claims that the French nuclear power industry is "the world's most indebted corporation and cites continual over-estimation of demand and under-estimation of costs. Official estimates of costs were based on an increase of 1.5 per cent per year. In actually, costs rose at 5–6 per cent per year. Overall, government nuclear power costs have probably been underestimated by at least 60%.[22]

Germany:

Construction of Germany's Kalkar breeder reactor was

halted in 1986 before the reactor became operational. The entire project was abandoned in 1991 after total expenditures of over $4 billion. In 1995 the 40 acre reactor site was sold to an amusement park developer from the Netherlands who plans to market it as the "Nuclear Water Wonderland."[23]

Japan:

Japan has delayed from 2010 to 2030 several breeder reactors and a large reprocessing plant. The delay was partly due to citizen pressure and partly due to the sharply increasing financial risks. The Monju breeder reactor was activated in March, 1994, at a cost of $5 billion and has been so expensive to run that a second reactor has been delayed until the next century.[24] On December 8, 1995. Monju experienced a sodium leak in the piping of the secondary cooling system that did not involve the release of any radiation. The accident was the largest ever to occur in the piping of an operating reactor and also the worst in terms of the leakage rate.[25]

The Japanese programme is now about 30 years old. Early forecasts about how cheap this reprocessing/breeder programme would be, have been wildly wrong—Japanese breeder reactors are now estimated by some sources to be 5 to 15 times more expensive to run than conventional nuclear power plants.[26]

Specific Cost Additions Arising from the Use of Plutonium in Commercial Reactors

The use of mixed oxide fuel (MOX) containing plutonium in Light Water Reactors (LWRs) is technically proven. Reactors that use LEU can have 1/3 of their core in MOX. Three US reactors of the System 80 type at the Palo Verde Nuclear Generating Station are pressurised light water reactors (PWRs) that could handle a full core load of MOX. Using these reactors, it would take 30 reactor years—or 10 years for all three reactors—to convert 50 tons of plutonium into spent fuel.[27]

A National Academy of Sciences (NAS) study estimated that a new MOX fabrication facility would cost between $400 million and $1.2 billion and would take about a decade to complete.[28] The cost of MOX fuel fabrication is estimated to cost over $2000 per kilogram of heavy metal, about six times the fabrication cost of LEU fuel.[29] At MOX fabrication costs of $1300–$2000 per kilogram, the cost of uranium would have to rise to $123–$245 per kilogram just to equal MOX fabrication costs even if the plutonium used was free.[30] In addition, it is not clear that the NAS study included the additional costs of security and handling that would be certain to accompany any plutonium processing or storage.

Cost estimates for geologic repository disposal of spent fuel from commercial power reactors are about $300,000 per ton of heavy metal (in 1988 dollars). However, the cost of disposal of a ton of plutonium would be higher because it must be diluted to make re-extraction difficult. Assuming a cost on the order of several million dollars per metric ton of plutonium, total disposal costs would range from $100 million to $300 million for 50 metric tons of plutonium.[31]

As was previously noted, the economics of plutonium burning have been investigated and rejected. Chow and Solomon looked at five options for using plutonium in reactors:[32]

1. Use plutonium as fuel in existing fast reactors without reprocessing. Using weapon-grade plutonium in this manner would cost $18,000/kg.
2. Use LWR's with 1/3 or partial MOX fuel without reprocessing. The cost for this is $7,600/kg with weapon-grade plutonium.
3. Use LWR's with full MOX fuel loads without reprocessing. The cost for this is $5,600/kg with weapon-grade plutonium.
4. Store plutonium for 20 or more years. Cost: $3,800/kg.
5. Mix plutonium with waste and dispose of it as waste. Cost: $1,000/kg in marginal costs over storing the waste alone—which would lead to costs of about $4,800/kg.

None of these options has any commercial value. In the first three, the extra costs of handling plutonium because of its radioactivity, toxicity, and potential weapon use outweigh any benefits. Further, storage sites will not be ready until 2010 at the earliest, and when storage costs are taken into account, they raise the cost of burning plutonium in LWRs by $4000 to $10,000/kg. Because of these significant cost additions, the use of plutonium in civilian reactors creates no economic benefits.

CONVERSION AS A RATIONALE FOR PLUTONIUM DISPOSITION IN THE US

Once rationale for burning plutonium has been that even

if a specific burning operation may not be economically viable, the use of the burning programme as a tool of economic conversion[33] in a region or state justifies its implementation. A proposal of this nature raises the issue of whether a conversion programme of similar impact be provided more cheaply.

The Triple Play Reactor, proposed for the Savannah River Site (SRS), and Project Isaiah, proposed for the old Washington Public Power System (WPPS) reactors around the Hanford site, have both been suggested as conversion programmes where new or refurbished reactors would burn plutonium. Both programmes have claimed they would be privately financed and, by implication, profit-making.

As a general principle, economic conversion is both site and sector based. On a site basis it preserves the local economic community by changing the base of economic support for the site. In an economic sector, it frees resources to be used in other ways for the benefit of the nation at large. Thus, the purpose of conversion is not to substitute one government-funded programme for another, it is to change the economic base (the source of funds) for the region or sector. This cannot be achieved unless conversion generates economic benefits, and the Isaiah and Triple Play options demonstrate how the conversion approach to disposition has tried to adapt to the economic realities of plutonium burning.

The Isaiah Project

Proposed in 1993, this project involved burning plutonium in MOX and producing electricity by completing the WPPS #1 reactor at Hanford, WA and the #3 reactor at Satsop, WA. It has been claimed this would create 9,000 direct construction jobs, 2,500 permanent operations jobs and 13,500 secondary jobs in the region. Each plant would produce 1,300 MWe.[34]

In 1993 dollars, completion costs for WNP-1 were $1.7 billion and for WNP-2 they were $1.6 billion. Operating costs were estimated at about $21 million/year, and O&M costs at about $123 million/year including the spent fuel disposal fee. When financing costs were included, the $1.7 billion completion cost for WNP-1 rose to $2.8 billion. However, private financing was supposed to cover all project completion costs and return $4 billion to the Federal government.[35]

While these financial arrangements sound promising, the poor economics surrounding this project were summed up by a clause in the Project Isaiah contract that stated that DOE would "enter into a long term contract......[with] *a federal obligation to make debt service payments if revenues from the sale of steam [power is] not adequate."*[36] (author's italics)

Triple Play Reactor

The System 80+ "triple play" reactor was proposed by a quasi-private consortium to burn plutonium, produce tritium and generate electricity at the Savannah River Site. The triple play reactor is an advanced, pressurised light water reactor that would use a 100% MOX core and produce 1350 MWe. Two units would consume 100 metric tons of plutonium in 30 years and would cost $6.25 billion to build and deploy.[37] The reactor's Programme Plan displayed considerable "uncertainty in costs" in MOX fabrication[38] and it proposed that the US federal government provide $50 million in up-front financing.[39] The private consortium offered to pay back the $50 million if DOE ultimately decided to proceed with the proposal at the end of the three-year study phase.[40]

In addition, the Triple Play reactor required an extensive list of other subsidies:

The federal government had to provide a site and infrastructure at no cost to the consortium.[41]

The consortium pays disposal fees for waste, but then passes them through to the government, not to the consumer of the power.[42]

The government supplies plutonium oxide, depleted uranium oxide, and the site lease, all at no charge, and it further agrees to sole-source irradiation services from the plant.

The "annual fees" required from the government were estimated at $78 million for plutonium burning alone—about a 10% subsidy.

An annual fee would also be assessed for tritium production based on revenue losses and other factors.[43]

The government shared liability for any increased costs due to regulatory changes or any other factors over which the consortium had no control.[44]

Similar subsidies are likely to be required by project Isaiah because a majority of the proposed revenues from both projects are from electrical generation. An electricity-producing, plutonium-burning light water reactor is not economically feasible because of the additional facilities and security procedures required for plutonium handling. MOX fabrication will also add hundreds of millions of dollars to normal operating costs. Each of these factors increases the financial risk associated with the project.

In an October 10, 1995, DOE briefing on its 'Dual Track' strategy for producing tritium, Secretary O'Leary essentially abandoned the Triple Play option while proposing to "examine the policy and regulatory issues associated with purchase of a commercial reactor or irradiation service."[45] The cost of this option was estimated by the DOE to be between $.2 billion and $4.5 billion[46]—a cost to the US government that can be assumed to be less than the cost of subsidising a Triple Play reactor. In 1988, the GAO estimated a total construction cost of $2.2 billion for converting the WNP-1 light water reactor at Hanford to produce tritium.[47] This would equate to a cost of about $2.9 Billion in 1995 dollars.

DISPOSITION REQUIREMENTS

Total Quantities of Plutonium

In 1991, the US had about 19,000 nuclear warheads and the Former Soviet Union (FSU) had about 32,000. Under START I and START II, the US and FSU agreed to reduce to 3,500 US and 3,000 FSU strategic warheads by 2003. Numbers of remaining tactical warheads may vary, but a good estimate would be about 1,500 US and 2,000 FSU tactical warheads. Thus, each side will have about 5,000 nuclear warheads in 2003. About 2,500 warheads could be dismantled each year in the US, but only about 1,170 will be dismantled if parity is maintained with the FRS's rate of 2,250 per year.[48] The supply of plutonium available for reuse in the United States as of September, 1994, is shown in Table 1.

The outlook for Japanese plutonium inventories has changed considerably with the recent accident at Monju. Table 2 shows the Japanese plutonium inventory at the end of 1994. According to the long run supply and demand balance first presented by the Japanese Atomic Energy Commission in June, 1994, and revised in August, 1995, the annual supply of plutonium from the

Table 1 : US Plutonium Inventories and Locations

Site	*Pu Inventory (in Metric Tons)*	*Pu in Wastes (in Kilograms)*
DoD & Pantex	66.1	N/A
Rocky Flats	12.7	47
Hanford	11.0	1,522
Argonne Lab-West	4.0	2
Los Alamos	2.7	610
Savannah River	2.0	575
INEL	.5	1,106
Lawrence Livermore	.3	N/A
Others	.2	N/A
Oak Ridge	N/A	41
Nevada Test Site	N/A	16
Total	99.5	3,919

Source: Secretary of Energy Hazel O'Leary, "Plutonium: The First 50 Years–United States Plutonium Production, Acquisition, and Utilization from 1944 to 1994," Department of Energy, Washington, DC, February 6, 1996.

Table 2 : Japan's Separated Plutonium Inventory (as of end of 1994)

Facility	*Kilograms of Pu*	*Stockpiled(s) or in use/ready for use(u)**
Reprocessing plant 836		
As nitrate	710	s
Stored as oxide	126	s
Fuel Fabrication plant	3,018	
Stored as oxide	2,032	s
Under test or processing	948	u
Completed fuel	38	u
Reactor sites	498	
Joyo	6	u
Monju	15	u
Fugen	53	u
Critical assemblies	425	u
Overseas reprocessors	8,720	
UK(BNFL)	1,412	s
France (COGEMA)	7,308	s
Total	13,072	11.588(s)+1.484(u)

Takagi, Dr. Jinzaburo, Citizens' Nuclear Information Center, 1–59–14–302, Higashi-nakano, Nakano-ku, Tokyo 164, Japan, January 10, 1996.
*Attribution to u and s by Dr. Takagi.

Tokai reprocessing plant is about 0.4 ton. This was short of the expected consumption of 0.6 ton by Monju, Fugen and Joyo, and this shortage was expected to be offset by plutonium imported from France. The Japanese government now says that plutonium stored for use in Monju will be used instead in Joyo, but the expected annual demand by Joyo and Fugen together is less than 0.2 tons. Thus, the total Japanese plutonium surplus in Japan and Europe could amount to around 25 tons by the turn of the century.[49]

50 or more metric tons of excess weapon grade plutonium are now available in Russia.[50] While specific inventory breakdowns are not available for Russia, France, Great Britain, and Germany, based on the assumptions that there are less than 4 kg of plutonium in each warhead and that there are 20 metric tons of plutonium in the military inventories of other nuclear weapons powers, the global inventory of plutonium is approximately as shown in Table 3.

Table 3 : Global Inventies of Plutonium

Source of Plutonium	*Metric Tons*
Military plutonium	248
Separated civilian plutonium	122
Unseparated plutonium in civilian spent fuel	532

Source: Makhijani, Arjun, and Annie Makhijani, *Fissile Materials In A Glass, Darkly*, IEER Press, Takoma Park, Maryland, 1995, p. 11.

Aside from these inventories, the total amount of plutonium available for use in non-weapon applications is directly dependent on dismantlement of old warheads. The pit of a nuclear weapon is composed of plutonium and weighs, on average, about 3–4 kilograms. The secondary is HEU that is over 90 per cent U235 and weighs about 15 kg.[51] Surplus US warheads contain about 50 tons of plutonium and up to 400 tons of HEU. Soviet Warheads contain about 100 tons of plutonium and more than 500 tons of HEU.[52]

Beginning in October, 1992, the DOE set a dismantlement goal of 2000 warheads per year for the US. As Figure 1 shows, this goal has never been met, and, at the current pace, dismantlement work at Pantex is likely to continue for three or four more years. On January 19, 1994, the DOE announced a decision to increase pit storage at Pantex from 6,000 to 12,000 pits. As of May, 1995, there were 7239 pits stored at Pantex, meaning that there will be about 8500 pits stored in May of 1996. Given the remaining storage space—3500—it is likely that there is insufficient space for the remaining warheads scheduled for dismantlement and this can be expected to increase pressure to find some other means of disposing of these materials.[53]

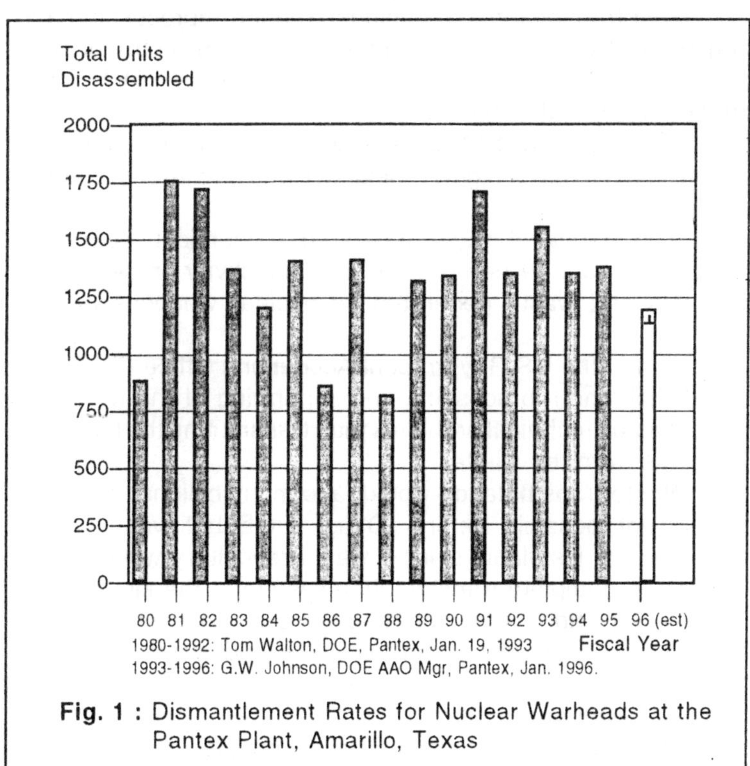

Fig. 1 : Dismantlement Rates for Nuclear Warheads at the Pantex Plant, Amarillo, Texas

THE COSTS OF TRANSMUTATION AND OTHER NON-BURNING OR TECHNICAL FIXES

Complete elimination of plutonium is only possible through two means: first, wait until the natural radioactive decay destroys it—this would take thousands of years. Second, transmute plutonium by using some technique to bombard its nuclei and split them into fission products. Option two can only occur through a nuclear reaction in a reactor or in a particle accelerator.[54] Most elements created by transmutation would have much shorter half-lives than plutonium. Thus, the potential benefits of transmutation could be:

1. A reduced volume of material.
2. Reduced radioactive life of materials.
3. Less risk of human intrusion into storage areas.[55]

Most transmutation techniques require reprocessing and, hence, are likely to be unacceptable in the US on the basis of both proliferation and waste generation concerns.[56] DOE's own studies show that in the US.

"Over a period of 40 years, [reprocessing] plants generated 105 million gallons of highly radioactive and hazardous chemical waste."[57]

"The high-level waste generated in chemical separations contains almost 99 per cent of all radioactivity present in by-products and waste generated by nuclear weapons production."[58]

In fact, the US Government Accounting Office (GAO) has noted that "the reprocessing and separating of the waste are more difficult technical problems than transmuting the long-lived elements from the waste."[59]

Waste transmutation would take many billions to develop and is not possible before 2015.[60] DOE believes it is not economically justifiable since a waste repository would still be needed. A complete transmutation system of the kinds shown in Table 3 includes a reactor or accelerator to transmute reprocessed fuel, a spent fuel reprocessing and waste separation facility, a fuel fabrication facility, and storage facilities for spent fuel and residual wastes.[61]

In addition to the transmutation techniques shown in Table 3, the following proposals have been subjected to enough investigation to either allow the estimation of costs or to identify similar process for which costs are generally known:

1. Monitored Surface storage

In 1990, a monitored storage facility for 50 tons of plutonium had an estimated capital cost of $170 million (1990 dollars) with an operating cost of $28 million pér year.[62] Preliminary

Table 4 : Potential Transmutation Technologies

Potential Programmes	*Sponsor*	*Units & Time to Destroy 90% OF LWR Actinide Waste Expected in 2010*	*Schedule/ Cost ($1993)*	*Destroys Actinides*	*Destroys Fission Products*
Advanced Liquid Metal/ Integral Fast Reactor (ALMR/IFR)	DOE, GE Argonne	19 Units and 200 years Ops Cost: $32 B	$5B (1 reactor) +4 B/Unit for remainder Start: 2015 Operate: 200 yr.	Yes	No
Accelerator Transmutation Project (ATW)	LANL	19 Units 40 years	Develop: $5B Start: 2016 Total: $120B	Yes	Some Including Pu and U
Phoenix Accelerator Pu nor U	Brookhaven National Lab	1 or 2 units 25 years	Develop: $29B Development Time: 15-20 yr.	Yes	Some Neither
Particle-Bed Reactor (PBR)	Brookhaven National Lab	20–70 Units 40 years (150 yr. for Pu)	Develop: $1.3B Development Time: 16 yr. No cost estimate	Yes	Yes
Clean Use of Reactor Programme (CURE)	Hanford/Westinghouse		Rsch: $74–160 M No cost estimate	No	Yes

Source : Developing Technology to Reduce Radioactive Waste May Take Decades and Be Costly, GAO/RCED-94-16, United States General Accounting Office, Washington, DC, December, 1993.

estimates by Fetter were that storing plutonium would cost about $1 per gram per year. Thus, storing 200 metric tons would cost roughly $200 million per year for a net present value cost of $2 billion.[63]

In 1994, civilian plutonium reprocessors charged between $2 and $4 per gram for storing separated plutonium. At these prices, storing 50 metric tons of plutonium for a decade would cost from $1 to $2 billion.[64]

2. Deep Geologic Disposal /Seabed Disposal

The cost is essentially that for vitrification and for burial in Yucca Mountain—i.e., the cost of both operations. See the vitrification option below.

3. Launching Plutonium into the Sun

A 1982 NASA study estimated the cost of this option at $200,000 per kilogram of plutonium. Several hundred kg could be handled at a time. This is probably not feasible due to public fears about the potential for a crash and resulting dispersion of plutonium from one of the rockets.[65]

4. Underground Nuclear Detonation

In one Russian proposal, 5000 warheads would be destroyed in a single explosion of a 100-kiloton warhead. A US option proposed using small shafts to destroys 5 warheads at a time (about 3000 detonations would be required). Even if one destroyed 50 warheads at a time, 300 detonations would be required—almost half of the 730 US underground tests conducted to date.[66] The costs of this option would be similar, although not identical, to the costs to conduct underground tests of nuclear weapons which ranged from $20 million to $60 million per test. Sixty-three past US underground tests involved more than one explosive device, and one test used six devices.[67] Thus, the technical aspects of this option are fairly well known. Costs would be lowered to the extent that the extensive telemetry and experiment design required for successful underground tests would not be required. However, costs would be increased to the extent that underground pollution of aquifers and other resources occurred.

5. Vitrification

By 1994, the DOE had spent over $1 billion trying to vitrify liquid wastes and had not yet succeeded. However, plutonium may not share these problems and it could be formed into blocks weighing thousands of pounds to make theft more difficult.[68] However, while vitrification of plutonium alone is an option, it is unlikely to present a sufficient barrier to reuse.[69] For this reason, prior to vitrification, plutonium will most likely be mixed with other materials that would make repurification more difficult.[70]

There are three general vitrification options with potential for plutonium disposition:

1. Vitrification of plutonium mixed with gamma-emitting fission products so the resulting glass logs meet the spent fuel standard.[71] These fission products have much shorter half-lives than plutonium. For example, the half-life of Cesium 137 is only 30 years as opposed to 24,000 years for plutonium. Thus, the mix would become less resistant to proliferation over time. This is likely to take longer since vitrification plants are not prepared for this task.[72]
2. Vitrification of plutonium with depleted uranium or some other alpha-producing element.
3. Vitrification of plutonium with a non-radioactive element, such as europium, that would render the mixture unsuitable for weapons without reprocessing.[73]

According to one proposal, the US could incorporate high level waste (HLW) like plutonium into 25,000 tons of glass at a rate of about 1000 tons of glass per year. This would allow the disposal of 100 tons of plutonium in five years if the glass contained only 2 per cent plutonium. A recent analysis by Pacific North-west Laboratories estimates the total additional cost at $100 million to convert 100 tons of plutonium metal to oxide and mix it with other HLW—ten times cheaper than storage, and ten to fifty times cheaper than MOX.[74] One could also place a barrier to misuse by subnational groups by making the canisters in which vitrified plutonium is stored highly radioactive.[75]

Critics note that vitrification leaves weapon-grade plutonium in a recoverable form that is not isotopically contaminated and has low probability of going critical in storage. Plutonium is relatively insoluble in water, but boron, which is the medium that absorbs neutrons, is not. Thus, they claim vitrified waste stored in the presence of water could go critical.[76] However, the MIT engineering group that has performed the research on the "clean glass" borosilicate vitrification option found that extracting the plutonium from this medium would present a formidable barrier to subnational groups. Further, tests conducted by this group show the borosilicate glass to be quite durable in the presence of water.[77]

6. Mix and Melt

Proponents of this method, such as Argonne National Laboratory, claim that reactor grade plutonium cannot easily be used to make weapons because it has 5 times the isotopic contamination of weapons grade plutonium, its radioactivity makes it more difficult to work with, and weapons made from this material would have an unpredictable yield.[78]

The mix and melt alternative proposes to melt plutonium together with spent fuel resulting in a mix that is too isotopically contaminated to easily make weapons without a time-consuming and expensive process. After this procedure is performed, the separation of isotopes would be so difficult that anyone who could do it would likely prefer to use "fresh" weapons grade plutonium instead.[79] The cost of this option would probably be similar to the $4 billion cost of the pyroprocessing reprocessor recently proposed as part of the US Integral Fast Reactor project.

Wolfgang Panofsky, chair of the NAS Plutonium Study, notes that the mix and melt process, as a variant of pyroprocessing, is largely unproved and would require a plant capable of handling over 2000 tons of heavy metal and 50 tons of plutonium—a plant that would be a giant reprocessing facility that has undergone neither engineering development nor any aspect of the regulatory process. The reactor grade plutonium in question is of approximately the same isotopic concentration of Pu239 as weapon grade plutonium. Further, no proposal for decreasing the isotopic concentration of Pu239 makes it less desirable to terrorists since "it has been amply and convincingly documented that, while reactor-grade plutonium has not been the material of choice for past weapons builders, an explosive device with an assured yield of one to two kilotons could be built from reactor-grade plutonium by relatively elementary methods.....In the hands of a terrorist or proliferator such devices would be formidable indeed."[80]

7. Burning plutonium in unconventional matrices

This solution has been proposed by members of the Nuclear Engineering Department of the Centro Studi Nucleari Enrico Fermi in Italy. Japan, Russia, and Switzerland are considering the unconventional matrix approach for their PWRs, as is Canada in its D_2O reactor.

One of the authors of the proposal has noted that the economic implications of this proposal are not markedly different from those for burning MOX. Discharged fuel would not undergo reprocessing or chemical treatments, leading to a one-time pass through only. The inert matrices have not undergone systematic testing and the development of these fuels still remains to be accomplished.[81]

CONCLUSION

Several studies on alternatives for disposition of plutonium have noted that due to potential proliferation problems and the danger these pose for all people, disposition issues should be decided based on expediency and safety, and economic considerations should not play a major role in this process.[82] However, a student of the military budgeting process or the budget considerations surrounding a major infectious disease such as AIDS will realise that there is no precedent for real-world decisions—even those that concern threats to large numbers of people—being made in an environment free of economic considerations. In fact, in making such decisions it is not unusual for economic costs and benefits to be considered first, not last. For this reason, it is necessary to identify those factors involved in the disposition area that will create common costs across all options, and to specify those areas where specific factors are likely to be major cost drivers that could discriminate between the various disposition options.

This paper has shown that burning plutonium in power generating reactors is not economical at the present time and is unlikely to become economical at any time in the near future. As the recent National Academy of Sciences study stated,

"Exploiting the energy value of plutonium should not be a central criterion for decision making, both because the cost of fabricating and safeguarding plutonium fuels makes them currently uncompetitive with cheap and widely availably low-enriched uranium fuels, and because whatever economic value this plutonium might represent now or in the future is small by comparison to the security stakes."[83]

However, even if burning plutonium is not economical, is it still cheaper than other methods of dealing with or disposing of plutonium? This question incorporates both proliferation risk and economics, and the following framework is suggested as a way in which it might be considered:

First, it is obvious that increased handling of plutonium leads to increased costs and increased proliferation risks.

Second, any proposal to burn plutonium in reactors to reach a spent fuel standard might also be accomplished more simply and cheaply by mixing plutonium with waste to a spent fuel standard to start with.[84] As an isotopically different element, plutonium can always be chemically separated form spent fuel whether it was generated inside a reactor or simply mixed with existing spent fuel, although the difficulty associated with this operation can be increased by adding other elements to the mix.

Third, waste storage costs, irrespective of the method of storage chosen, are based on volume and radioactivity and will be the same for all burning and non-burning options. In any process that requires putting material in a reactor, whether for power generation or simply to dispose of the material, the volume of material will remain constant throughout the process and the radioactivity of the spent fuel will be approximately the same for storage considerations. The only exception to this rule occurs when reprocessing is involved. Then both waste volume and costs rise dramatically. And fourth, for transmutation, costs are altered because one is handling hotter material for relatively shorter periods of time—but these time periods are still so extensive that discounted cost comparisons between alternatives cannot show significant differences. In addition, transmutation technologies still require reprocessing and they still must absorb the cost of research and development. Other options do not have either of these negatives.

Viewed in this light, final waste disposal costs will be incurred whatever disposal option is taken. These costs could potentially be offset by doing something profitable with the plutonium prior to final storage, but this paper has shown that finding a profitable use for plutonium is unlikely. Thus, the more probable case is one where the costs of basic waste storage are increased by whatever costs are associated with the disposition option chosen. The factors most likely to significantly increase costs are the major cost drivers that create differences among the various options for plutonium disposition. At this point, major costs of plutonium disposition appear to arise from four areas:

i) The level of subsidisation in the "profitable" parts of the disposition programme.
ii) Those items (such as reprocessing) that increase the volume of waste and thus, the cost of waste disposal.
iii) The cost of security and its direct relationship to the number of times a material is handled or moved.
iv) The cost of research and development of new and unproven methods of disposition.

These four costs outweigh all other costs of disposition by many orders of magnitude and, as a result, minimising their impact should be the major consideration in choosing among disposition options.

Notes:

1. For example, see
   Chow, Brian G. and Kenneth A. Solomon, *Limiting the Spread of Weapon-Usable Fissile Materials*, National Defence Research Institute, RAND, Santa Monica, CA, 1993, and *Management and Disposition of Excess Weapons Plutonium (Pre Publication Copy)*, Committee on International Security and Arms Control, National Academy of Sciences, National Academy Press, Washington, DC, 1994.
   "Burning" is the techno-slang word for using plutonium in nuclear reactors by mixing it with uranium to form a mixed oxide fuel (MOX) that can be burned in light water reactors (LWRs).
2. Elements with half-lives of 50 to 100 years instead of the 24,000 years possessed by plutonium.
3. von Hipple, Frank, "Fissile Materials Security in The Post-Cold-War World," *Physics Today*, June, 1995, p. 32.
4. Makhijani, Arjun, and Annie Makhijani, *Fissile Materials In A Glass, Darkly*, IEER Press, Takoma Park, Maryland, 1995, p. x.
5. Garwin, Richard L., *Critical Question: The Value of Plutonium*, presented at Cornell University, October 1, 1993.
6. Wald, Matthew L. And Gordon, Michael, "Russia And US Have Different Ideas About Dealing with Surplus Plutonium," *New York Times News Service*, New York, August 19, 1994.

7. Ibid.
8. Ibid.
9. Ibid.
10. Ibid.
11. Fetter, Steve, *Control and Disposition of Nuclear Weapons Materials*, Working Papers of the International Symposium on conversion of Nuclear Warheads for Peaceful Purposes, Rome, Italy, June 15, 16, 17, 1992, pp. 144–148.
12. These figures significantly understate the current estimates of the costs to bury nucelar wastes and decommission reactors. Source: Komanoff Energy Associates, *Fiscal Fission: The Economic Failure of Nuclear Power,* 270 Lafayette, Suite 400, New York, NY, December, 1992.
13. *Federal Energy Subsidies: Direct and Indirect Interventions in Energy Markets*, SR/EMEU/92–02, Energy Information Administration, US Department of Energy, Washington, DC, November, 1992, p. 7.
14. *Electric Utilities Commentary,* "Are Older Nuclear Plants Still Economic?, Insights from a Lehman Brothers Research Conference," vol. 2, no. 21, May 27, 1992, p. i.
15. *Nuclear Power*, Moody's Special Comment, Moody's Investors Service, New York, NY, April, 1993, p. 7.
16. Heinzl, Mark, "Uranium Prices Rise on Scarcity, Steady Demand," *The Wall Street Journal*, February 26, 1996.
17. Reprocessing plutonium and U from spent fuel and using plutonium-bearing mixed-oxide (MOX) fuel in thermal nuclear power plants.
18. Chow and Solomon, Op. Cit., pp. xvi, xvii.
19. Ibid., p. xvii.
20. Wald, Op. Cit.
21. Rothstein, Linda, "French Nuclear Power Loses its Punch," *The Bulletin of the Atomic Scientists*, July-August, 1994, pp. 8,9.
22. Ibid.
23. "A Great Fixer-Upper," *The Bulletin of the Atomic Scientists*, Vol. 52, No. 2, March/April, 1996, p. 8.
24. Sanger, David E., Japan, "Bowing to Pressure Defers Plutonium Projects," *The New York Times*, February 22, 1994.
25. Takagi, Dr. Jinzaburo, Citizens' Nuclear Information Center, 1–59–14–302 Higashi-nakano, Nakano-ku, Tokyo 164, Japan, January 10, 1996.
26. Sanger, Op. Cit.
27. Makhijani and Makhijani, Op. Cit., p. 26-27.
28. *Management and Disposition of Excess Weapons Plutonium*, Op. Cit., p. 159-160.

29. *Nuclear Fuel*, January 26, 1992.
30. Feiveson, H.A., *Plutonium Fuel: An Assessment*, OECD, Paris, 1989, p. 69.
31. Makhijani and Makhijani, Op. Cit. p. 66.
32. Chow and Solomon, Op. Cit., pp. xxi, xxii.
33. Conversion programmes attempt to replace an obsolete or formerly federally funded economic base in a community or region with a new industry or base that does not rely on the old funding sources. In the case of nuclear plant locations, conversion progrmmes have proposed to replace older DOE plants with new, privately-funded nuclear facilities.
34. Letter from Robert Wages, President, OCAW, to Elmer Chatak, President, Industrial Union Department, November 3, 1993.
35. Letter from John R. Honenkamp, SAIC, to Dr. Matthew Bunn, National Academy of Science, November 9, 1993.
36. Communication from Lauren Dodd, Battelle Institute, "The Isaiah Project" Pacific Northwest Laboratories, October 1, 1993.
37. *Programme Plan for deployment of a System 80+ Multi-purpose Nuclear Facility at Savannah River Site*, System 80+ Team, Savannah River Site, Aiken, SC, March 31, 1994, pp. 1–12.
38. Ibid., p.8.
39. Ibid., p.9.
40. Personal communication between Brian Costner and George Davis of ABB Combustion Engineering in May, 1995.
41. *Programme Plan for deployment of a System 80+ Multi-purpose Nuclear Facility at Savannah River Site*, Op. Cit., p. 68.
42. Ibid., p. 70.
43. Ibid., p. 75 and personal communication between Brian Costner and George Davis of ABB Combustion Engineering in May, 1995.
44. Ibid.
45. O'Leary, Hazel, Briefing on Dual Track Strategy, Department of Energy, October 10, 1995.
46. Ibid.
47. GAO, 1988 and Barker, Rocky, "Experts Weigh Four Designs," *The Idaho Falls Post Register*, April 24, 1988.
48. Chow and Solomon, Op. Cit., pp. 9,10.
49. Takagi, Op. Cit.
50. *Management and Disposition of Excess Weapons Plutonium*, Op. Cit., p. 1.
51. von Hippel, Frank, and Marvin Miller, Harold Feiveson, Anatoli Diakov, Frans Berkhout, "Eliminating Nuclear Warheads," *Scientific American*, August, 1993. p. 46.
52. Ibid., pp. 46-47.
53. Norris, Robert S. and Arkin, William M., "US Nuclear Weapons Stockpile, July 1995," *The Bulletin of the Atomic Scientists*, Vol.

51, No. 4, July-August, 1995, pp. 78-79.
54. *Developing Technology to Reduce Radioactive Waste May Take Decades and Be Costly*, GAO/RCED-94-16, United States General Accounting Office, Washington, DC, December, 1993, p. 11.
55. Ibid., p. 10.
56. Makhijani and Makhijani, Op. Cit., p. 98-100.
57. *Estimating the Cold War Mortgage—The 1995 Baseline Environmental Management Report*, US Department of Energy, Office of Environmental Management, DOE/EM-0232, March, 1995, pp. B.4 to B.5.
58. Ibid., p. B.5.
59. *Developing Technology to Reduce Radioactive Waste May Take Decades and Be Costly*, Op. Cit., p. 13.
60. Ibid., p. 3.
61. Ibid., p. 4,5.
62. Bloomster, C.H., P.L. Hendrickson, M.H. Killinger, and B.J. Jonas, *Options and regulatory issues related to disposition of fissile materials from arms reduction,* PNL-SA-18728, Pacific Northwest Laboratories, US Department of Energy, Washington, DC, 1990, pp. 12-13.
63. Fetter, Op. Cit.
64. *Management and Disposition of Excess Weapons Plutonium*, Op. Cit., pp. 126-127.
65. International Physicians for the Prevention of Nuclear War and The Institute for Energy and Environmental Research, *Plutonium—Deadly Gold of the Nuclear Age*, International Physicians Press, Cambridge, MA, 1992, pp. 130–138.
66. Ibid., pp. 130–138.
67. "Known Nuclear Tests Worldwide, 1945–1994," *The Bulletin of the Atomic Scientists,* Vol. 51. No. 3, May/June, 1995, p. 70.
68. For a discussion of potential problems and benefits associated with vitrification, see comments by Wolfgang Panofsky, Kevin Wenzel et al, and Alex De-Volpi in "Letters," *The Bulletin of the Atomic Scientists,* vol. 52, no. 1, January/February, 1996.
69. Makhijani and Makhijani, Op. Cit., p. 4.
70. Wald, Matthew, "Encase Excess Plutonium in Glass, US Urged," *The New York Times*, November 17, 1994.
71. The spent fuel standard proposes to make plutonium as difficult to retrieve as it would be if it was in the form in which it exists in nuclear reactor fuel that has been irradiated (used) to the extent that it can no longer effectivley sustain a chain reaction and thus, has been removed from the reactor for disposal. This irradiated fuel contains fission products, uranium, and transuranic isotopes.

72. Makhijani and Makhijani, Op. Cit., p. 88.
73. Ibid., p. 4.
74. Fetter, Steve, *Control and Disposition of Nuclear Weaspons Materials,* Op. Cit., pp. 144–148.
75. Makhijani and Makhijani, Op. Cit., p. 89.
76. DeVolpi, Alex, "Fast finish to plutonium peril," *The Bulletin of the Atomic Scientists,* Vol. 51, No. 5, September/October, 1995, p. 20-21.
77. Wenzel, Kevin W.; Sylvester, Kory W.; Cerefice, Gary, "Reply to 'Fast finish to plutonium peril'," *The Bulletin of the Atomic Scientists,* Vol. 52, No. 1, January-February, 1996, pp. 59-60.
78. DeVolpi, Op. Cit.
79. Ibid.
80. Panofsky, Wolfgang K.H., "No Quick Fix for Plutonium Threat," *The Bulletin of the Atomic Scientists*, Vol. 52, No. 1, January-February, 1996, pp. 3, 59.
81. Letter from Carlo Lombardi, Dipartimento Di Ingegneria Nucleare, Centro Studi Nucleari Enrico Fermi, to William Weida, November 30, 1994.
82. For example, see Makhijani and Makhijani, Op. Cit.
83. *Management and Disposition of Excess Weapons Plutonium,* Op. Cit., p. 3, 4.
84. For a discussion of potential problems and benefits associated with "mix and melt" approaches to plutonium disposition, see comments by Wolfgang Panofsky, Kevin Wenzel et al, and Alex DeVolpi in "Letters," *The Bulletin of the Atomic Scientists,* vol. 52, no. 1, January/February, 1996.

Address:

WILLIAM J. WEIDA
The Colorado College
Colorado Springs CO 80903
USA

73

Utilisation of Excess Weapon Plutonium: Scientific and Technological Aspects of the Conversion of Military Capacities for Civilian Use and Sustainable Development

HANS-PETER WINKELMANN

1. THE IMPERATIVES OF SCIENTIFIC AND TECHNOLOGICAL ASPECTS OF CONVERSION OF MILITARY CAPACITIES FOR CIVILIAN USE AND SUSTAINABLE DEVELOPMENT

Throughout history there has always been some reduction in military production in times of détente following wars. However, the end of the Cold War, in the aftermath of one of the biggest military build-ups and arms races ever, has left in its wake problems of unexpected magnitude: Threats to nations or to the world community are no longer limited to military aggression. While the decline of a military threat in the recent years resulting from the East-West détente is evident, the increasing environmental degradation caused by the high status of existing military potentials continues to be a source of deep global concern—despite all disarmament efforts.

Military activities have caused such formidable damage to the environment and to human health that their consequences will be felt for decades and, in instances of forms of radioactive contamination, for much longer. Unfortunately, hardly any of world's nations are devoid of contaminations caused by military activities, devastated landscapes, polluted groundwater and injured biota. Furthermore, damage has been spread throughout the oceans and atmosphere. The catastrophic nature of environmental contamination becomes more apparent as additional sites of contamination are discovered in the course of the ongoing disarmament process and as the effects of the damage at known areas manifests itself fully. The true extent of the environmental damage resulting from military activities during the Cold War is both massive and only now becoming apparent. The seepage of chemical or radioactive contamination into water systems and the corrosion of containers dumped at sea, containing chemical weapons or radioactive material are but two examples of problems that grow more serious with the passage of time. The scope of the problem and the technological needs can only be hinted at. A particularly important concern is the large number of nuclear weapon test areas that witnessed many of above-ground nuclear explosions and subjected surrounding populations to decades-long exposure to high levels of radioactivity. Thus, pollution from radioactive waste stored on land and at sea is a subject of great concern. There are many areas where nuclear waste has been stored in disregard of internationally accepted standards established to prevent environmental damage. The location and assessment of such sites is particularly urgent. In the first place, efforts must focus on pollution from radioactive materials, primarily those affecting oceans and waterways. In this context, problems of radioactive contamination arising from the accidental loss and decommissioning of nuclear submarines have received considerable public attention in the recent past. It is necessary to assess the effects on the marine and human environment of present levels of radioactivity, including the risks of increased levels of contamination owing to corrosion.

In dealing with the environmental legacy of the Cold War the dimensions of the effort needed to correct these problems is truly massive. A comprehensive and fast solution is beyond the economic means of today's society. Even spread over several generations, satisfactory clean-up can only be managed by prioritizing the problems, tackling or containing those that are most urgent and developing much more cost-effective technologies as soon as possible. International cooperation to share experience and resources, and to plan for the necessary efforts is imperative not only because of their similar nature and transboundary character. In the long range, alternative technologies, capable of drastically reducing clean-up costs, will have to be developed in order to bring the whole problem, within affordable financial resources. International cooperation is needed to speed up the development of these technologies and to use the best efforts of the cooperation partner scientists.

The end of the Cold War and the evident decline of military threats between major powers of the world on one hand, and the wide spectrum of increasing environmental concerns including that are defence-related underscore the need for conversion of military scientific and technological capacities as an effective strategy to accompany national and international disarmament efforts and as a contribution to a development path in harmony with the natural carrying capacity of our planet. For the first time, the need to transform large-scale scientific, technological and industrial capacities formerly devoted to military purposes has become a worldwide issue affecting all nations, major powers as well as many smaller countries.

At the same time, the challenges of global environmental change have led the world community to adopt—at the United Nations Conference on Environment and Development - the new paradigm of sustainable development. Sustainable Development has broadened the traditional understanding of security: Global environmental change as the main destabilizing effect is also perceived as a vital risk to nations. In addition, environmental concerns are often blocking the dismantlement of nuclear weapons or their transport or dismantlement sites resisted by local populations and political bodies.

The question of how to respond to the need to transform large-scale scientific, technological and industrial capacities designed to serve the military-industrial complex by redirecting them towards environmental protection and accelerating the development process has increasingly become part of the discussion of a new understanding of "security".

The new understanding of security, the industrial restructuring for sustainable development, the human resources issues, the cleaning up of the Cold War legacy and developing timely alternative use plans for military facilities constitute the frame work for the issues of concern. This framework should be looked at more detailed in following.

2. THE FRAMEWORK OF SCIENTIFIC AND TECHNOLOGICAL ASPECTS OF CONVERSION

The scientific and technological aspects of the conversion of military capacities for civilian use and sustainable development concerning the utilisation of excess weapon plutonium consist of the following main issues:

1. **The new understanding of "security":** Threats to nations are not simply military threats. Today, the future of nations is equally threatened by global environmental change and economic or social instability.
2. **Industrial restructuring for sustainable development:** Conversion of the military-industrial complex into clean, market - and consumer-oriented production facilities is part of overall efforts of industrial restructuring in the pursuit of sustainable development.
3. **Human Resources Issues:** In countries where the military sector (industry, national R&D institutions, and the armed forces themselves) is particularly important as compared to overall economic activity, unemployment—including that of scientific and technological staff-resulting from disarmament, could lead to an outflow of skills to nations still expanding military R&D and production.
4. **Cleaning up of the Cold War legacy:** Abandoned military sites and facilities are often among the worst areas struck by toxic waste pollution. Military activities have caused such formidable damage to the environment and to human health that their consequences will be felt for decades and, in instances of forms of radioactive contamination, for much longer. This is task which requires scientific and technological inputs. Closely related to this aspect of conversion is the environmentally sound disposal of existing military hardware and surplus weapon systems including nuclear weapons.
5. **Developing timely alternative use plans for military facilities:** This is part of the development strategies in economies heavily dependent on the presence of such facilities. It is essential to avoid large-scale unemployment initiating brain-drain and to secure public support for conversion measures. It requires also scientific and technological assessment and economic projection.

The notion to diversify resources traditionally devoted to the military and released in the process of disarmament to address such risks is intriguing. Scientific and technological aspects, including the conversion of military research and development (R&D) with the objective to develop environmentally sound technologies are of particular interest in the context.

Such a "trade-off" has been termed in a broad discussion to "peace-dividend". Different terms are being used to describe the process of conversion, for example "reinvestment", and strategies to utilise the expected peace dividend for sustainable development efforts. Such a pay-off as it is implied in the objective of the utilisation of excess weapon plutonium cannot simply be described in financial terms. While the release of actual financial resources from military to civilian endeavours is questionable, the peace-dividend essentially exists in terms of human resources as disarmament is freeing scientific and technical resources to pursue profitable R&D, and in terms of a long-term economic impact resulting from the restructuring of production from a narrowly-based military one to a more efficient production serving a broader civilian market. The issue is after all, how it can achieved to count the civilian market. The issue is after all, how it can be achieved to count the civilian utilisation of excess weapon grade plutonium under such understanding of conversion for sustainable development.

Closely related to this aspect of conversion is the environmentally sound disposal of existing military hardware. Such conversion for sustainable development is an exercise where tremendous short-term costs have to be balanced against long-term benefits. Cleaning up the Cold War legacy is in particular costly and a complex venture when it is connected with the dismantlement of nuclear weapons. There are various options for how to solve the disposal problems of these materials, each involving different cost levels for realization and eventual economic and political benefits. In the search for an optimal method of conversion of weapon-grade plutonium it is necessary to bear in mind the grave potential hazards of storing plutonium even during a limited period of time, both in respect to the environment and human health, as with respect to non-proliferation and the prevention of terrorism. In terms of the discussion on conversion it should be clearly distinguished between the civilian utilisation of excess weapon plutonium and nuclear disarmament.

3. NUCLEAR DISARMAMENT: THE ROLE OF THE MILITARY FOR CLEAN UP, MANAGEMENT AND PREVENTION AS THE MAJOR OPTION

Therefore it should taken into consideration that modern military establishments have developed highly sophisticated techniques and technologies that could be applied by the military in an active conversion role to protect, restore and improve the environment. Military establishments can assist in achieving these objectives by contributing their technical expertise, advanced equipment and communications and surveillance systems. The range of scientific and technological areas which could be used for such purposes include the handling and disposal of highly radioactive substanaces as well as the destruction of weapons.

The high pace of nuclear disarmament, on the one hand, and the unpreparedness of the current nuclear fuel cycle to adopt and process such large quantities of new nuclear materials, as well as the actual lack of demand for these materials lead to the need for a political decision on the management of the nuclear disarmament process. It is in this context that a conversion policy to manage effectively the nuclear disarmament process in full compliance with all international norms and safety standards is of utmost importance. Therefore the need for a close international cooperation on this issue is evident. Such a decision would provide additional time for a reexamination of all options and should include a new role for the military sector in the overall nuclear weapons safety, security and dismantlement process. This prospective to concentrate on nuclear disarmament instead of real conversion for civilian application could lead, in a first step, to a safer world and in it, to a sustainable future.

I would like to underscore that the military sector, including military R&D and parts of the armed forces, are ideally suited for dealing with these and other defence-related environmental problems. What is needed now in terms of dealing with the nuclear disarmament management is the development of an overall prioritised plan of action that will result in a systematic assessment of the problem, an analysis of associated risks, a selection of clean-up technologies and the development of a rational basis for deciding on the order in which problems should be addressed. It would be not needed to establish a new international agency or the allocation of huge resources in any one centralized body. But is it indispensable to create coordinated, inter-governmental efforts, relying heavily on existing agencies and appropriate organisations, to share experiences and resources and to develop a rational plan to deal with this pervasive threat to the security of us all. This would be the most pressing recommendation which should be forwarded to the forthcoming G-7 Summit on Nuclear Safety in Moscow from the environmental point of view associated with the scientific and technological aspects of the conversion military capacities related to nuclear disarmament for civilian use and sustainable development.

4. CONCLUSIONS FOR SUSTAINABLE DEVELOPMENT

All the main issues and problems mentioned above have severe consequences for the entire nuclear disarmament management process. Such problems of the nuclear disarmament management which are linked to sustainable development are part of chapter 22 of the Agenda 21 which is entitled "Safe and Environmentally Sound Management of Radioactive Wastes". The programme area *Promoting the safe and environmentally sound management of radioactive wastes* has the objective to ensure that radioactive wastes are safely managed, transported, stored and disposed of, with a view to protecting human health and the environment, within a wider framework of an interactive and integrated approach to radioactive waste management and safety. Special emphasis is laid on the international and regional cooperation and coordination of states and, in cooperation, with international organisations as the main basis for action. The particular objectives of such measures are to promote policies and practical measures to minimise and limit, where appropriate, the generation of radioactive wastes and provide for their safe processing, conditioning, transportation and disposal. The further aim is also to promote the safe storage, transportation and disposal of radioactive wastes, as well as spent radiation sources and spent fuel from nuclear reactors destined for final disposal, in all countries by facilitating the transfer of relevant technologies and/or the return to the supplier of radiation sources after their use, in accordance with relevant international regulations or guidelines. With regard to the scientific and technological means the chapter 22 of the Agenda 21 suggests to promote research and development of methods for the safe and environmentally sound treatment, processing and disposal, including deep geological disposal, of high-level radioactive waste as well as to conduct research and assessment programmes concerned with evaluating the health and environment impact of radioactive waste disposal.

Address:

HANS-PETER VINKELMANN
Managing Director
INFU-Institute for Environmental Research
University of Dortmund
Italy

74

The Prolification Risks of Plutonium Mines

HAROLD A. FEIVESON

Several recent observers have called attention to the risk that mined geological repositories, the planned destination of at least some of the plutonium recovered from nuclear warheads and additionally of a large fraction of the world's spent commercial fuel, could eventually become low-cost sources of fissile material for nuclear weapons. This paper argues that the risk of these so-called "plutonium mines" as a source of weapons-usable material depends on the accessibility of plutonium in repositories relative to other sources of fissile material. A preliminary analysis suggest that the range of circumstances under which plutonium mining will be significantly more attractive than alternative routes to fissile material are fairly narrow. Furthermore, the proliferation risks of alternatives to spent fuel disposal in repositories, in particular schemes to partition the plutonium in the spent fuel and to burn it in reactors or accelerators, appear greater and far more immediate than the risks of repository mining. Nevertheless, the risks of such mining are significant enough that safeguards should be maintained on spent fuel repositories indefinitely, even after the repositories are closed and the contained spent fuel made "irretrievable".

INTRODUCTION

The U.S. appears committed to convert surplus plutonium recovered from dismantled warheads to a form meeting the "spent fuel standard"—that is, to a form similar to spent fuel from commercial reactors. This would be done either through immobilisation of the plutonium in some waste form (such as vitrified high-level waste) or by burning the plutonium in the form of mixed-oxide (MOX) fuel in reactors to produce spent MOX fuel. Presumably then the resulting waste would be disposed of as would commercial spent fuel. For the U.S. (along with some other countries, including Sweden and Canada), the intention has been to dispose of this spent fuel , after some period of retrievable storage, by emplacing it into mined geological repositories. After this period, all access tunnels and ventilation shafts will be backfilled and sealed, and all supporting facilities will be decommissioned and dismantled. The repository will then be considered "irretrievable", in the sense that recovery of the emplaced material would require much greater cost and effort.

But is a mined geological repository a safe destination for the recovered plutonium (and for the spend fuel from commercial reactors)? Some analysts think not. They fear that the repositories will become overtime "plutonium mines" because of three factors. First, the penetrating radiation barrier that renders spent fuel extremely hazardous to handle will decay to a very long level after few centuries of cooling, so that the material can be acquired and reprocessed at much lower cost than spent fuel of more recent vintage. Second, the time and effort necessary to recover the fuel from the repository will decrease as mining technologies improve. Third, the isotopic quality of the plutonium in commercial spent fuel will approach (although never reach) that of "weapons-grade" with time.[1]

It is the purpose of this paper to examine such claims and whether the prospect of future" plutonium mines" really poses an unacceptable long-term risk.

STANDARD OF COMPARISON

The critical question in defining the risks of plutonium mining is how much easier (if at all easier) will such mining be compared to other ways in which countries can acquire fissile material. This comparison depends foremost on the future of commercial nuclear power. If a country has an operating nuclear fuel cycle, it will always have a ready supply of spent fuel available, either in retrievable storage or in the reactor cores. (If the nation, as well, operates commercial reprocessing plants of contracts for reprocessing, it will also process stockpiles of separated plutonium and, if it has its own reprocessing plant, the ability of reprocess quickly any spent fuel). Under these circumstances, it is seems likely that spent fuel in a sealed geological repository would be relatively unattractive with respect to both sate-sponsored and sub-national diversion, assuming that repositories were safeguarded at a level consistent with other stages of the fuel cycle.

The relative attractiveness of spent fuel in a repository would be greatest in the context of a "nuclear-free" future, in which nuclear power had been phased out and neither operable reactors nor retrievable spent fuel storage facilities existed. In this case, the only means of acquiring spent fuel other than mining the repository would be the construction and operation of production reactors and associated front-end facilities (e.g. uranium mining and fuel fabrication) from scratch. For both repository mining and new production, it will still be necessary to reprocess the spent fuel to obtain the plutonium.

Alternatively, a proliferant could seek to acquire weapons-usable material by enriching natural uranium bypassing the spent fuel route altogether. How difficult this alternative would be in the hundreds or thousands of years in which the security of a repository might be an issue is a crucial, but undecidable, question. For clarity of presentation, we disregard the enriched uranium route to weapons-usable fissile material, and focus exclusively on the plutonium route in a non nuclear world. Can plutonium be acquired more readily through new production or through mining a repository?

RESOURCES NEEDS FOR NEW PRODUCTION

The development cost of a dedicated plutonium production capability from scratch depends on the size of programme desired. Consider three categories of plutonium production (all costs are given in 1992 dollars):

i) A "minimum acquisition" plutonium programme, based on a gas-graphite production reactor rated at 30 MW-thermal (MWt) and capable of producing around 10 kg of plutonium per year. This programme is estimated to require a capital cost of $120-$300 million, of which $35-$100 million is the construction cost of the reactor[1]. The average cost of the reprocessing component is $15-$40 million, or about 12 per cent of the total. The time for construction of this project is estimated to be 3-4 years, with a crew of 100.

ii) An "intermediate acquisition" programme capable of producing around 100 kg of plutonium per year. One programme, based on a 400 MWt reactor, was estimated to require a capital investment in the range of $1-2 billion, including $400 million to $1 billion for the reactor alone, with a cost overrun of up to 100% possible in the event of delay. The construction time for this reactor is estimated to be 5-7 years, requiring a staff of 200-300[2], although the time could be shortened through a crash effort. The cost of the reprocessing plant in the case was not given; scaling from the previous example (40% of the reactor cost) yields a value of $160-$400 million.

iii) A "maximum acquisition" plutonium programme, in which the desired production rate is limited only be the resources available. For example, during the Cold War, multiple 2150 MWt reactors were constructed at the Savannah River Site, each capable of producing about 600 kg of plutonium per year. Today, total capital costs of such a reactor would be between $1.5-3.0 billion; and a large reprocessing plant might cost an additional $1 billion.

CHARACTERIZATION OF A MINED REPOSITORY

No geological repository has been fully designed, let alone built. However, authorities in the U.S., Sweden and Canada have developed conceptual designs which can be used to describe a "nominal" repository. In particular, repository designs, which have been developed by the U.S. Department of Energy for Yucca Mountain, and by Atomic Energy of Canada Limited (AECL), can be used for this purpose[1].

We may imagine the repository to consist of two principal parts;

- above-ground facilities, for unloading and packaging fuel for final disposal, for crushing rock to be used as backfill, for mixing concrete for vault seals, for handling and storing low-level radioactive wastes, and for various administrative and service functions;
- an underground vault, consisting of a series of tunnels about 300-1000 metres below ground, capacious enough to allow various heavy equipment to move.

A small number of shafts or ramps would connect the two parts of the repository.

In general, we may imagine two periods of concern after the repository is built:

- a pre-closure transition of 50-100 years before the repository is closed, during which period spent fuel would be placed into the repository;
- the period after the repository is closed.

During the pre-closure period, the spent fuel packages would be placed in containers and then sent underground for emplacement in the storage areas by a transporter truck. In the Canadian design, the vault would be filled room by room. After each room is filled (in the AECL design requiring about one month), it is backfilled with mixtures of clay, and sand.

Once the vault is filled, the surface and subsurface facilities will be dismantled; the tunnels, shafts, and exploratory boreholes will be sealed; and the secondary wastes disposed. Seals will be backfilled with clay and concrete and rock. In the final closure step, measuring instruments would be removed from exploratory boreholes and the boreholes would be sealed. Conceivably, the vault location could be identified with on-site markers.

RESOURCES AND TIME REQUIRED FOR REPOSITORY MINING

Retrieval of spent fuel from a closed repository would entail assembly of the necessary equipment on-site, construction of new surface support and material handling facilities, the drilling of new shafts into the geologic formation, and the use of grappling and other equipment to extract the spent fuel. We may imagine that the repository is filled either with canisters containing something like 5 kilograms of plutonium per canister, as in the Canadian design, or with casks, containing perhaps 100 kilograms of plutonium per cask, as in the Yucca Mountain concept. A “minimum acquisition” programme would require the recovery of 2 canisters per year on one cask per decade. An “intermediate acquisition” would require the recovery of 20 canisters or one cask per year.

For the extraction of from a few to several canisters to one or two casks, corresponding to the first two levels of diversion introduced above, the most direct way to recover material might be to drill a tunnel to level of repository, contact a canister or cask, and then somehow bring the spent fuel to the surface. For the extraction of some substantial number of canister or casks, it might be more efficient to reexcavate a part of the repository through the drilling of large shafts or trenches in which large equipment could be introduced into the repository much as the case when the repository was first filled.

Let us consider first the effort to obtain small amounts spent from a Canadian-type repository utilizing canisters containing about 5 kilograms plutonium per canister. Assuming that the driller had a general but not precise map of the burial spots, he would, in all likelihood, have to dig several tunnels to hit a canister. In a study done for the International Nuclear Fuel Cycle Evaluation (INFCE) in 1979, the authors estimated that it might take on the order of twenty one-metre diameter holes to recover a single canister. After contact, the canister wound have to be attached and hoisted through the shaft or tunnel. For a nominal tunneling cost of $3000-5000 per metre and tunnel length of 2-4 kilometres, a single tunnel drilled might cost $6-20 million. Twenty such tunnels would then cost $6-20 million. Twenty such tunnels would then cost $120-400 million, comparable to the cost of a small production reactor scenario. At a rate of billing of (say) 20-50 metres per day, and assuming that the tunnels are drilled in parallel, it might take a few months to contact and extract a canister. This is consistent with the INFCE study which concluded that the "recovery of one of a few canisters might be accomplished using ten drilling rigs in two months time; more probable recovery times are six months or longer"[1].

For recovery of spent fuel from a repository similar to Yucca Mountain, the direct tunnel approach would be far more difficult. The nominal waste cask envisioned for Yucca Mountain weighs 40 tons, with diameter about 1.7 metres and length about 5.2 metres. In this case, the tunnel drilled would have to be much bigger than one metre diameter, of course; and the grappling equipment needed to bring up the cask, or alternatively open it and remove the fuel assemblies, would be formidable. Again, it might be necessary for the driller to drill many tunnels before hitting a cask. Given time for initial preparation, bringing drilling equipment to the site, and then the actual drilling and extraction of the spent fuel, an effort of many months to a year or more and a cost of several hundred million dollars would probably be required.

For a massive reentry into a repository, reexcavation of the repository would probably be the preferred mode, with the divertor excavating out very shafts and trenches and emplacing specialised equipment and transporters down into the repository. This would involve a very substantial mining operation. Large underground mining operations today typically require capital investments on the other of a few hundred million to well over one billion dollars (with the higher figure reflecting isolated sites and/or difficult climates), and development times of 2-5 years before production can begin[2]. Excavation of a spent fuel repository would require a similar level of investment. For example, the development costs of the original (1988) design of the Yucca Mountain repository include $200 million for preparation of the site, $320 million for constructing the shafts and ramps, initial excavations at the repository level, and underground service systems, and $510 million for construction of surface facilities, for a total of over $1 billion (all costs in undiscounted 1992 dollars).

Although the advance rates of modern drilling methods can be quite high, on the order of 50 metres per day, the minimum time necessary to gain access to a repository will probably be determined by the significant effort involved in the on-site assembly and preparation of equipment. For example, the tunnel boring machine (TBM) being used to drill Exploratory Studies Facility (ESF) at the Yucca Mountain site was brought there in 52 separate truckloads. One engineer has estimated that one to two years would be needed to assemble a large TMB and auxiliary equipment, and to prepare a platform to launch the drilling.

Although the effort and time required to excavate a repository would thus not be trivial, once completed, the rate of removal of spent fuel from a repository could be on the order of the rate of emplacement. This could be quite high. For example, the rate of loading of the planned Yucca Mountain repository is currently anticipated to be 3000 tHM of spent fuel a year. Assuming equal rates of retrieval and emplacement, a large mining operation could produce 25 tonnes of plutonium per year. More than forty large Savannah River-type reactors would have to be built to attain a level of production equivalent to the mining route, at a much greater capital cost. Therefore, current repository designs would allow accumulations of plutonium equivalent to a rate of thousands of nuclear weapons per year. However, it remains to be determined whether this scenario is sufficiently credible to warrant concern.

COMPARISON OF MINING AND PRODUCTION

No reliable and precise comparison between the production and minings routes to the acquisition of weapons-usable material is possible over the hundreds or thousands (1) of years in which repository spent fuel must be a concern and during which period one must expect both mining and fissile-material production technologies to improve markedly. Moreover, aside from the vast uncertainties in future technology, other factors frustrate any exact comparisons, for example: How many tunnels will be necessary to contact a cask? Once a cask is contacted, will it be relatively easy to contact another cask nearby? Once the major capital investments in reactors or drilling rigs are made, what be the relative marginal effort required to obtain a continuing stream of fissile material?

However, despite all these and other uncertainties, the above analysis suggests that for the "minimum" and 'intermediate' acquisition scenarios, both the spent fuel mining and the new production route would require capital investments up to several hundred million dollars, and a development time of six months or more. It may be that, in some circumstances, the mining path will be quicker and cheaper, and since commercial spent fuel in a repository will be approximately ten times more concentrated in plutonium than production-reactor spent fuel, the separation of plutonium from the fuel could be done more rapidly. Spent MOX fuel fabricated from weapon-grade plutonium would be still more concentrated in plutonium. Nevertheless, mining will require substantial resources and activity, allowing the possibility of effective safeguards, a subject which is addressed after the next section. For a very large acquisition of weapon-usable material, reexcavation of a depository would appear to provide a more efficient route than the construction of new production reactors. But again such an effort would be highly visible and would allow effective safeguards or internationalisation of the repository.

THE ATTRACTIVENESS OF AGED SPENT FUEL

Relatively fresh spent fuel is strongly self-protecting from casual appropriation and processing. For example, assuming a nominal cooling period of 10 years and fuel burnup of 40 Mwd/kg burnup, the dose rate perpendicular to a PWR fuel assembly at the centre is about 4.000 rem/h at 1 metre, about 300 rem/hr at 5 metres. Since the LD-50 dose is 450 rem[2], the spent fuel must be handled and processed remotely behind heavy shielding. Specialized equipment, such as shielded machines with grappling hooks, casks to place the fuel assemblies, and transporters strong enough convey the casks are required to remove the fuel from reactor storage pools. In addition, once the spent fuel assemblies are obtained, many of the key steps to separate the plutonium will also have done remotely behind shielding. The equipment requirements are, therefore, extraordinarily daunting for terrorist groups and represent a hurdle even for states.

Unfortunately, however, the "self-protecting" field of penetrating radiation emitted by spent fuel decreases steadily with time, and effectively disappears after several hundred years of cooling. Thus, in about 300 years, both the total radioactivity and gamma and x-ray activity of the spent fuel will have decreased by a factor of one thousand.[1] After this period, the handling of spent fuel, once obtained, will be possible without heavy shielding; and the processing of the spent fuel will be possible in a glove-box facility, rather than in a shielded, remotely-operated reprocessing plant.

These potential attractions of aged spent fuel, however, do not appear overwhelming. First of all, it is evident that many states already have the capacity to handle and process fresh spent fuel. They have routine access to the specialised equipment required; and several of these countries already remove the fuel assemblies and transport them to reprocessing or away-from-reactor stores. Similarly, countries with reprocessing plants already in place would be able to separate the plutonium from the spent fuel readily.

If a proliferant did not have access to an already constructed separations plant, it would have to construct one. Facility requirements would depend strongly on the scope of plutonium production the proliferant was undertaking and its overall industrial sophistication. But, in general, the scale of the separations endeavour would not have to be great. Each half-ton fuel assembly from a PWR contains about one bomb's worth of material; so a plant capable of reprocessing one fuel assembly per week could in a short period assemble material for tens of bombs. The lead-time necessary to construct a separations plant of this capability has been estimated to be from six months to four years.[2]

The barrier to spent fuel processing is even less if a country is dealing with fuel from a new production reactor, rather than a commercial reactor. The burn-up of spent fuel from dedicated gas-graphite weapons-grade plutonium production reactors is typically quite low, below around 800 Mwd/t. This is around fifty times smaller than the typical burn-up of commercial fuel.

For low and intermediate production rates, low-burnup fuel can be processed in small, rudimentary reprocessing cells known as "caves", provided the operators are willing to accept high but not debilitating radiation doses. These cells, which date from the 1950s, utilise primitive, mechanically operated remote-handling devices known as ball-joint manipulators. The difference in cost between a small cave and a glove-box facility would probably not be great enough to influence the decisions made by a proliferant group.[1]

At higher production rates, small, locally-shielded caves would no longer be feasible for reprocessing production-reactor fuel, and larger plants, utilizing remote operation and maintenance, would have to be employed. Thus the comparative attractiveness of repository-grade spent fuel is maximised under these circumstances. However, at high production rates, the cost of a facility for reprocessing agent spend fuel would increase as well, because of the need to provide greater containment of alpha particles and to mitigate the increased risk of a criticality accident.

Over time, because the half-lives of Pu-240, Pu-241, and Pu-238 are 6540 years, 14.4 years, and 87.7 years respectively, compared to 24,100 years for Pu-239, the fraction of Pu-239 in the plutonium in a repository will gradually increase, making the plutonium slightly more weapon-grade. But the change is indeed very slight, the Pu-239 fraction increasing from about 58% for spent fuel one-year out of reactor to just under 70% in 1000 years.[2] Not only is the change in isotopic quality slight, but several sources have now confirmed that reactor-grade plutonium could be used to make both crude nuclear devices such as may be sought by terrorist groups or sophisticated nuclear weapons.[3]

DETECTABILITY AND THE ROLE OF LONG-TERM SAFEGUARDS

The safeguards aspects of spent fuel disposal in repositories have been examined by the IAEA over the last several years, with particular reference to the question of whether safeguards on spent fuel could ever be terminated. Under INFCIRC/66 and INFCIRC/153 guidelines, safeguards on fissile material cannot be terminate unless by determination by the Agency that the material is no longer usable for any nuclear activities or has become "practically irrecoverable" due to lack of access to the material. Before closure of a repository, the plutonium in the spent fuel would clearly not be irrecoverable and safeguards would have to be applied. Various studies have examined how this might be done through application of containment and surveillance and materials-accountancy measures at a repository site. Prior to closure, these measure should be as effective at a repository as elsewhere in the fuel cycle.

An advisory Group Meeting on safeguards related to final disposal of spent fuel and nuclear wastes, held at the IAEA in 1988, found that spent fuel does not qualify as being "practically irrecoverable" at any point, even after closure of a repository, and recommended that the IAEA should not terminate safeguards on spent fuel.[1] The IAEA has been re-examining this question, and is expected to formally determine that safeguards must be maintained on repositories containing spent fuel.[2]

Although the maintenance of long-term safeguards on spent fuel repositories is inconsistent with one of the fundamental goals of geologic disposal—namely, that a repository, after it is scaled, should not require active monitoring for any purpose—the IAEA determination to maintain safeguards does appear warranted. Mining a repository would not be a quick of quiet operation, but if could be done; and, therefore, if is reasonable that it be kept under safeguards. Safeguards could be strengthened if the repositories were also put under international ownership, so that no nation had carte blanche to access them.

The safeguards challenge must be considered in some perspective. As already noted, the threat posted by mining of a geologic repository is maximised in the context of a "nuclear-free" world in which there are no operating nuclear reactors or retrievable spent fuel storage facilities. However, if once assumes means for detecting clandestine production is in place, then it is sensible to assume as well that such techniques could be extended quite simply to safeguard geologic repositories. Indeed, the task of monitoring a series of known sites would be far more straightforward than the task of verifying the absence of clandestine activities, which could occur anywhere (and in particular, in industrial areas where the visual and thermal signals could be camouflaged). This suggests that in the presence of a monitoring regime, clandestine production would be more likely to escape detection than repository mining.

ALTERNATIVES

Comparison to alternative ways for a country to acquire spent fuel is one standard of comparison to evaluate risks of plutonium mining. But there is another standard of comparison which should be kept in mind. If the spent fuel is not put into a repository, it will have to go somewhere else, and the ensuing risks of that somewhere else will have to be considered. Disregarding exotic or underdeveloped alternatives, such as shooting plutonium and high-level waste separated from spent fuel to the sun, deep bore-hole emplacement, and sub sea-bed disposal, there appear but two practical long-term alternatives to geological repositories for the disposal of spent fuel:

(1) Indefinite, monitored, retrievable storage of spent fuel.

(2) The partition of plutonium from other constituents in spent fuel and its transmutation in a reactor or accelerator, with the high-level waste treated and then sent to a geological repository. The spent fuel created by burning the plutonium would be reprocessed repeatedly with the plutonium (and other actinides) gradually burnt-down. The repeated reprocessing and recycling is termed "partition and transmutation" (P-T). (A truncated process, in which separated plutonium is fashioned into mixed-oxide (MOX) or other reactor fuel and recycled only once or twice rather than repeatedly, would reduce the value of plutonium in spent fuel. But this would merely delay the spent fuel disposition choices, since the spent MOX or metallic reactor fuel would have to be disposed).

Pretty clearly, the first alternative would allow more ready access to spent fuel than an underground geological repository. What about the second?

PARTITION AND TRANSMUTATION

Partition and transmutation systems have been described and analysed recently by the Lawrence Livemore National Laboratory[1] and by the Panel on Separations Technology and Transmutation Systems (STATS panel) of the National Research Council.[2] Both reports drew on earlier work by Thomas Pigford and collaborators.[3] Both the Livemore study and the STATS panel consider a range of concepts, including the Integral Fast Reactor programme developed at the Argonne National Laboratory, the PRISM fast reactor concept developed by General Electric, and accelerator concepts developed by Los Alamos National Laboratory and Brookhaven National Laboratory.

In all the cases, partition and transmutation operations are designed to burn a large part of the plutonium (and other actinides) in spent fuel, with the produced high-level wastes, containing most of the fission products, after they were vitrified or otherwise treated, sent to a geological repository. The proliferation risks of P-T compared to a spent fuel repository then involve a weighting of the reduced risks of plutonium mining in a repository and the increased diversion risks associated with the above-ground processing and recycling.

This processing and recycling would have to be on a grand scale indeed. For example, consider what would be involved in treating U.S. spent fuel—approximately 2000 tonnes of heavy metal per year. Reprocessing of the fuel would require the equivalent of three plants the size of the THORP reprocessing plant in Great Britain (whose capacity is 700 tonnes of heavy metal per year). Since there are about 10 kg per tonne of transuranics in the LWR spent fuel, the reactor or accelerator deployment will have to be able to handle 20 tonnes of tansuranics per year. Assuming for specificity a fast breeder reactor of capacity 1.4 GWe, with a conversion ratio of 0.76 and lifetime of 40 years (the PRISM reactor with the lowest practical conversion factor), such a reactor could fission in its lifetime about 37 tonnes of transuranics.[1] Since, over forty years, the LWRs will generate about 800 tonnes of transuranics, the transmutation effort will involve some 20 reactors. After 40 years, the job won't be finished because the transuranic inventories in the reactors will still be substantial and will have to burned down still further. Depending on how completely one wishes to destroy the transuranics, the transmutation effort could take from hundreds of thousands of years.[2]

This scale of activity would be extremely difficult to safeguard and physically-secure. Even where separations technologies can be implemented such that the plutonium nominally stays mixed with most of the other actinides and thus would not be weapons-usable (such as is promised by the pyroprocessing technique developed at Argonne National Laboratory), it would almost certainly always be possible for a country to reconfigure the separations process to allow the separation of plutonium.[2] Safeguards would have to be applied in perpetuity as with safeguards on spent fuel repositories. So, even if the long-term risks of plutonium mining in a repository are significantly reduced through P-T, they may be overwhelmed by the short and medium-term risks of processing and recycling.

Transmutation of already separated weapon-grade plutonium, of course, would not require the reprocessing step. But the "spent fuel standard" implies that extraordinary efforts to dispose of weapons plutonium are not justified unless simultaneous efforts are undertaken for the disposal of commercial spent fuel.

Even if the risks of repository mining do not look excessive compared to P-T or on the basis of and MPS, they may be large enough to justify a serious reinvestigation of other permanent storage solutions, such as deep borehole or sub sea-bed.[1]

CONCLUSIONS

Although the diversion scenarios examined do not cover all possibilities, they do suggest that the range of assumption under which a repository will look attractive compared to other routes to plutonium acquisition is extremely narrow. It is true that after a couple of hundred years cooling, repository spent fuel will become less self-protecting than relatively fresh production-grade spent fuel. However, this fact does not mean that repository spent fuel will be valuable to mine.

In circumstances where a country has a working nuclear fuel cycle, mining a repository is not likely to look attractive compared to more direct ways to obtain weapons usable material. In a nuclear-free world, for small and intermediate acquisition programmes, repository mining might be quicker and cheaper than the production route, although give uncertainties in technology advances in the next several hundred years, it is hard to make any definitive comparison. At a minimum, it seems likely that mining will take at least several months and will be readily detectable if there are reasonable safeguards applied at the repository sites.

Mining a repository might look attractive to a country contemplating a large scale breakout if it did not already have access to many power reactors. In this case, the quantity of plutonium in a repository would be the main attraction, not so much the cooler fuel; for in the time necessary to establish a mine at a repository or to obtain large quantities of spent fuel from an ongoing civilian power programme, a country would probably want to construct a large, heavily-shielded reprocessing plant even for fuel cooler for several hundred years or more.

If anything, comparing the risks on mining a spent fuel repository to those of new production probably overvalues the repository as a plutonium mine. This is first of all because the construction of new production reactors and reprocessing plants is an appropriate standard of comparison, an appropriate standard only for instances where there is no ready access to an operating nuclear fuel cycle. For proliferants with access to operating reactors, other fuel-cycle facilities, and/or inventories of spent fuel in monitored, retrievable storage, the acquisition of plutonium would be much more direct and inexpensive.

Secondly, if the spent fuel accumulated in a nation's civilian power programme is not put into a repository, it will likely instead be put either into monitored, retrievable facilities or processed in a P-T programme. Each of these alternatives will allow plutonium to be mined more readily than would a repository.

For these reasons, recent suggestions to scale back efforts to develop a spent fuel repository in the United States are to be regretted. A scale-back would send a strong signal to other countries now contemplating what to do with their spent fuel. If could, in consequence, delay for decades the movement of spent fuel out of retrievable storage in this country and abroad, and lead also to partition and transmutation activities, with all their attendant hazards.

It seems likely that disposal of spent fuel in a mined repository in tuff of granite can be made an acceptable long-term option, at least with respect to plutonium mining. Under this option, by necessity, the fuel will be retrievable for some period; and although most repository schemes envision the repository to be closed, and the fuel made practically irretrievable, in 50–100 years after the repository is first opened, that decision could itself be changed over time, with the closure of the repository delayed still longer if doubts persisted concerning the long-term safety of geologic disposal. In the pre-closure period, a mined geological repository will effectively constitute a monitored, retrievable store—but one underground.

No matter how the competition among repository disposal, long-term storage, and partition and transmutation is finally decided, it is critical to recognize that, no matter what is done with spent fuel, all countries with nuclear power programmes will have access to spent fuel containing substantial quantities of plutonium. A one gigawatt-electric reactor will contain an in-reactor inventory of plutonium of about 600 kilograms and will discharge about 200 kilograms per year. Even if spent fuel was removed from reactor site after a very short period after discharge (say two years), a country with a single 1 gigawatt reactor would have continually on hand something like 1 tonne of plutonium in its spent fuel, enough for perhaps 200 warheads. By the time the country had built a reprocessing plant, it would thus have a substantial backlog of spent fuel from which to produce plutonium, and in the future it would be able to produce about 200 kilograms of plutonium per year. Resolution of the quandary of what to do with the long-term disposal of spent fuel can reduce pressures for reprocessing and it could speed the emplacement of spent fuel underground; but it cannot choke off the route to nuclear weapons through civilian nuclear programmes based on current reactor types.

BOWMAN AND VENNERI HYPOTHESIS

Finally, a brief comment may be made on another objection to a geologic repository which has recently been advanced. This is the hypothesis by two Los Alamos scientists, Charles Bowman and Francesco Venneri, that plutonium in glass or other medium underground would eventually become susceptible to critically and explosive events long after emplacement in a repository.[1]

... subcritical fissile material underground might reach criticality that is autocatalytic or self-enhancing. This criticality could come about upon dispersion into the surrounding medium by either natural or unnatural process, or by the fissile material being carried to other sites where it can collect into different autocatalytic critical configurations. Underground, where the material is confined and there is an abundance of moderating medium around it, the consequences of such super critical excursions could range from modes energy releases to the generation of explosive nuclear yields up to a few hundred gigajoules [about 100 tonnes] of high explosive equivalent from a single event... In varying degrees, all categories of waste containing fissile material appear to be susceptible to these criticality excursions, including vitrified weapons plutonium, research reactor and DOE spent fuel, commercial and MOX spent fuel.

The B/V hypothesis is now under challenge from several quarters and yet may be shown to be inconsequential. For example, it might be that waste forms could be developed which preclude criticality. But I would like to focus on the extremely long time frames involved. The B/V excursion couldn't occur until the canister shell loses integrity (hundreds to thousands of years), the nuclear poisons such as boron leach away (thousands to tens of thousands of years), and the underground fissile material gets concentrated through some mechanism. Such a mechanism is more likely for the decay product of plutonium—uranium-235—so that it may be that the explosive excursion could only happen after much of the plutonium has decayed to U-235. This would put the time of concern a couple to a few plutonium half-lives away or (say) 50–100 thousand years. The B/V analysis is not invalidated by the long-time frames involved, and may, in fact, suggest some useful fixes in repository geometry or in materials packaging. But, it would be foolish to let a speculative risk many thousands of years in the future or longer determine in any fundamental way how we dispose of weapons plutonium over the next few decades.

This is exactly the conclusion to draw also, I believe, on the speculative and distant risk of plutonium mining.

Note

1. See, for example, Johan Swahn, "The Long-term Predicament", Technical Peace Research Group, Institute of Physical Re-

source Theory, Chalmers University of Technology, Goteborg, Sweden 1992; and Pers Peterson, "Long-term Safeguards for Plutonium in Geologie Repositories, *Science & Global Security*. Spring 1996 (in press).

1. U.S. Congress, Office of Technology Assessment (OTA), *Technologies Underlying Weapons of Mass Destruction*, OTA-BP-ISC-115, U.S. Government Printing Office, Washington, D.C., December 1993, 156.
2. *Ibid*, 158

1. AECL, *Environmental Impact Statement on the Concept for Disposal of Canada's Nuclear Fuel Waste*, AECL-10711, COG-93-1, September 1994; TRW Environmental Safety Systems, "Strategy for Waste Containment and Isolation for the Yucca Mountain Site", October 9, 1995.

1. "Safeguards for Geologic Repositories", Prepared for Working Group 7 of the International Nuclear Fuel Cycle Evaluation, by the U.S. Delegation, INFCE/DEP/WG.7/18, September 1979, pp 18–19.
2. W. Peters, *Exploration and Mining Geology*, 1987, 162.

1. W.R. Lloyd, M.K. Sheaffer, and W.G. Sutcliffe, "Dose Rate Estimates from Irradiated Light Water Reactor Fuel Assemblies in Air", UCRL-ID-115199, January 31, 1994.
2. The LD-50 is the dose rate expected to be fatal to 50 percent of the exposed population.

1. Swahn, 69, 147. The factor of one thousand in 300 years is due principally to the passage of ten half-lives of Cs-137.
2. Memorandum from D.E. Ferguson to F.L. Culler, *Simple, Quick Processing Plant*, Oak Ridge National Laboratory, August 30, 1977.

1. After two years of cooling, each tonne of uranium gas-graphite reactor fuel, irradiated to a burnup of 800 MWD/t, will contain about 720 grams of weapons-grade plutonium and about 4000 curies (Ci) of hard (greater than 0.4 MeV) gamma activity. In order to produce 100 kg of plutonium per year, about 0.5 tonnes of fuel would have to be reprocessed per day (assuming 80% plant availability). This quantity of fuel could be subdivided into five batches per day, each containing about 400 Ci of hard gamma activity. The unshielded exposure from a point source of this strength at a distance of 50 cm is over 5 Gy/hr; to reduce this to (say) 1.5 mGy/hr, a target value that a proliferant could conceivably tolerate, would require attenuation by a factor of 3300. This can be achieved with 8 cm of lead or 15 cm of dense lead glass. These dimensions are well within the constraints of process cells of the type described.

2. Swhan, 131.
3. See, for example, National Academy of Sciences, Management and Disposition of Excess Weapons Plutonium, National Academy Press, Washington, D.C. 1994, Box: "Reactor-grade and Weapons-grade Plutonium in Nuclear Explosives".
1. International Atomic Energy Agency, Advisory Group Meeting on Safeguards Related to Final Disposal of Nuclear Material in Waste and Spent Fuel, STR-243 (Rev. 1), Vienna, 1988. Also see A. Fattah and N. Khlebnikow, "International Safeguards Aspects of Spent Fuel in Permanent Geological Repositories". Journal of Nuclear Materials Management, October 1990/January 1991.
2. G. Seneviratne, "IAEA Developing Safeguards Conditions for Vitrified Waste Repositories", Nuclear Fuel, September 12, 1994, 9.
1. Lawrence Rampspott, et al. *Impacts of New Developments in Partitioning and Transmutation on the Disposal of High-level Nuclear Waste in a Mined Geologic Repository*, UCRL ID-109203, Lawrence Livermore National Laboratory, Livemore, CA, March, 1992.
2. National Research Council, *Report of the Panel on Separations Technology and Transmutation Systems*, 1995.
3. See, for example, Thomas H. Pigford, "Actinide Burning and Waste Disposal", Invited Review for the MIT International Conference on the Next Generation of Nuclear Power Technology", UCB-NE-4176, University of California Berkeley, October 5, 1990.
1. Ramspott, et al, p. 4–3.
2. Ramspott, et al, p. 4–21.
3. Edwin Layman, "Interim Storage Matrices for Excess Plutonium: Approaching the Spent Fuel Standard Without the Use of Reactors", Research Report 286, Center for Energy and Environmental Studies, Princeton University, August 1994.
1. Swahn, op. cit, argued in his pioneering study that serious attention be given to deep borehole emplacements.
1. See Charles Bowman and Francesco Venneri, "Underground Supercriticality from Plutonium and Other Fissile Material", and several critiques, in *Science & Global Security*, Vol. 5, No. 3, 1996.

Address:

HAROLD A. FEIVESON
Princeton University
USA

PART—8

INDIAN AND PAKISTAN NUCLEAR TESTS

75

Technology, Strategy and Politics: An Inevitable Past Cold War Cocktail?

C. UDAY BHASKAR

"So far as I can see, the atomic bomb has deadened the finest feeling that has sustained mankind for ages."
Mahatma Gandhi

The unveiling of the atomic weapon at Hiroshima and Nagasaki in August 1945 and the impact of this new technological capability since then on international relations on the one hand and on the human sensibility on the other, is a reality that cannot be wished away. The weapon symbolizes the reality of ultimate deterrence, while disarmament remains the rhetorical flourish—the elusive Holy Grail a normative goal that has inexorably receded from the grasp of contemporaniety.

The nuclear issue has been driven by a combination of technological determinism and strategic innovation locked together in a symbiotic manner—the resultant techno-strategic determinant being embedded in the prevailing political ambiance of the period. Thus while the Cold War decades were anchored in bipolarity and an intense ideological battle that nurtured the edifice of mutually assured destruction/ deterrence (MAD) through the nuclear weapon, the post Cold War period is marked by an attempt at WMD (weapons of mass destruction) regulation led by the dominant powers. Considerations of realpolitik and self-interest currently govern international relations and over the last few years—or rather in the first few years after the end of the Cold War— a MWD pattern has evolved that validates the concept of nuclear deterrence in a selective manner and seeks to protect and nurture the power existing matrix.

Thus we have a system whereby the dominant powers of the post Cold War world are the five declared nuclear weapon powers, who are also the five permanent members of the UN Security Council—the PN-5. If one accepts the contention that WMD capabilities impinge on the macro-security of nations, then a dominant pattern emerges whereby the macro-security of the PN-5 and their allies who are an overlap of the G-7/8 and a host of politico-military partners have ensured their security by selective nuclear deterrence. This structure has been buttressed by various regimes such as the NPT, the MTCR, the CTBT, the NSG et al. which when pared to the essentials are a case of according a normative sheen and the purported pursuit of the collective good, to what is basically an attempt at protecting the exclusivity of the dominant powers/groups.

In this arrangement we note the manner in which technology and research and development (R&D) have been managed to further these dominant interests and it is the contention of this paper that given the relentless march of technological advances in the WMD sphere shaped by prevailing political-economic-military forces, equitable and total nuclear disarmament will remain an elusive goal—and that the current trend of activities in the national laboratories of the major powers is inherently against the spirit of disarmament. The paper will further argue that there is a degree of inevitability about such a development given the compulsions of late 20th century technological determinism, the determined attempt of the major powers to preserve the existing power matrix, and the manner in which the dominant 'ethical' narrative is being shaped and interpreted.

The concepts that merit scrutiny in the current debate are (a) the primacy accorded to the principle of "deterrence" - by either nuclear /WMD or conventional/technological means; (b) the incredible entropy of technology/ R&D in nuclear/missile related issues; (c) the impact of such synergism on the politico-military stratagems adopted by the major powers; and finally (d) the need to incorporate a nuclear management philosophy for the post Cold War world that would allow the disarmament square to contain the deterrence circle.

The post Cold War world is deemed to be one of considerable uncertainty as regards the attainment of security for the nation-state and there are certain realities that cannot be ignored. These include the primacy of the US led western alliance in the global military security arrangement and the synonymy of *core security* with the concept of "deterrence" as achieved by this grouping of the "we" on the one hand, and the engendering of *insecurity* in this cross-section on the other, when WMD capabilities are perceived to have leaked to the "them"—those outside the dominant groupings. Thus core macro security for the "we" is sought to be assured by the assuaging of any form of WMD insecurity through credible nuclear deterrence-and the denial of the same to "them".

Today there is an interesting conceptual debate where some sections of the strategic community opine that deterrence is imperative but can be achieved through means other than the nuclear, thereby making a plea for nuclear disarmament. This well-meaning spectrum includes retired generals and admirals from the world over—particularly from the USA—along with better known names such as Mr. Robert MacNamara, the former US Secretary of Defence. The counter-view to this radicalism is evidenced in the old Cold War theology that argues that total nuclear disarmament is neither strategically desirable nor technologically feasible and that nuclear deterrence is here to stay for eternity-the genie-cannot-be-put-back-into-the-bottle line. What is germane is the convergence that deterrence cannot be invalidated and the debate is more of means than the end itself.

It is here that technology and related R&D exude an incredible quality. Illustrative of this quality is the recent report that modern science and technology have allowed human thought to be photographed as it flashes across the brain—and the kind of quantification that has been deduced should also introduce a degree of humility when pitting the human sensibility against the nuclear genie. The technique of potentiometric probing associated with the University of Pennsylvania has also concluded that the human brain contains 100 billion neurons—comparable to the number of galaxies in the visible universe—which generates enough electricity to illuminate a light-bulb.

At the related technological plane in the WMD realm one notes an interesting inversion when examined through the MLT dimensional analysis framework. While in earlier decades, the major powers and their technological accomplishments focused on the enhancement of Mass - or kilotonage and megatonage of the nuclear explosive device with enhanced parameters of Length/range and CEP, currently the nuclear weapon is becoming more precise and more miniaturized- that is we have a phenomenal calibration and compression of Mass and the concomitant refinement of Length/precision. Thus we now have a family of mini and micro nuclear weapons and this combination will alter the frame of reference for disarmament and deterrence.

It may be shown empirically that the current trend among the major powers is to fund R&D in what is euphemistically referred to as FGN—or fourth generation nuclear weapons—nuclear weapons that will have yields of 1-10 tons of TNT equivalent. These weapons can be designed and tested in what are called ICF facilities—Inertial Confinement Fusion and super-laser laboratories that will allow the desired compression of fissile material. It is pertinent to note that these designs and tests will not be a contravention of existing regulatory mechanisms such as the NPT, the CTBT et al. and hence the nations that have the technological know-how can proceed in the matter to further their own perceived security interests. In short the dreaded nuclear weapon of monstrous lethality is becoming more precise, sophisticated and miniaturised—what in diabolical jest has been termed the designer variant of a weapon that is no longer of mass destruction quality—and hence politically more correct? In this manner the semantic packaging is also being transmuted to make the "new" nuclear weapon more benign and acceptable in the power game—but only in the hands of a select few.

As per current reckoning and according to some media reports, the superlaser labs in the world are now restricted to the following nations: USA, UK, France, Japan and China with some suggestions that similar facilities/know-how (?) either already existing or are soon to be located in Russia. The political grouping of these states is self-explanatory and can be covered under the rubric of the UNSC/PN-5/G-7 and represents the powerful "we" of the late 20th century/early 21st century, as opposed to the subaltern "them" - the technological have-nots of the world.

Reference may also be made to the current stockpile stewardship programme of the USA that by Presidential directive mandates the US Department of Energy (which funds the nuclear weapon labs and their research activities in this area) to maintain the credibility of the US nuclear arsenal. This is a complex and classified area and while there are conflicting views on the subject, this paper argues that there is an inevitability about "where the labs are going"—they will move towards higher and more effective levels of nuclear weapon related technological capability. Whatever be the public line on these R&D activities—it may be presumed that the fine-print of the stewardship programme and the politics that govern it—not merely in the USA but in all states that perceive a similar WMD insecurity—will move towards new weapon designs—and this activity is axiomatically detrimental to the movement towards disarmament. This is the inevitability that will have to be acknowledged in the current deliberations on the subject and not to do so would be tantamount to ignoring the reality that characterises a politically contentious issue.

The current global debate on WMD is also marked by an uneasy awareness that ballistic missiles are here to stay and that the only known defence against them - that is credible Ballistic Missile Defenses will go against the principles of the ABM Treaty that served to anchor strategic deterrence in the Cold War. This debate is being fuelled by domestic US political compulsions and a certain public perception about WMD threats and this technology—when realised—will shape the strategy of the major powers and the end result may not always be conducive to global stability. Thus again we come full circle to the inexorable linkages between technology, strategy and politics both domestic and realpolitik - and there may be an inescapable need to introduce an element of ethical principle and the accommodation of the interests of that part of the world whose strategic interests are not necessarily convergent with the "we" of the post Cold War world.

The practice of pursuing discriminatory policies that favour one section of the globe and then attempting to accord a normative sheen to what is essentially an explanatory framework that recognizes a selective reality may not augur well for nurturing global peace and stability in the next century. In an era of increasing global interdependence, where security for the nation-state, civil society and the individual have undergone a complex transformation, is it valid or feasible to prescribe the assuaging of WMD 'insecurity' in a discriminatory manner? While governments and the official *party-line* on WMD issues have an inherent constraint in not straying from the predictable groove, the challenge is for those who constitute the Track-2 community-academics, scientists, analysts and professionals now out of harness - to harmonize innovative thinking with a basic humanist integrity that will allow the squaring of the circle.

Address:

C. UDAY BHASKAR
Deputy Director
Institute for Defence Studies and Analysis
New Delhi—110067
India

76

Nuclearisation of South Asia

AHMAD NISHAT

The Indian nuclear blasts on 11 and 13 May at Pokhran close to Pakistan's border finally, exploded the myth of India's peaceful nuclear programme. India had insisted that its nuclear explosion of 1974 was for peaceful purposes. It had tried to mislead the world against all available evidence for the past twenty-four years, till Mr. Raja Rammana who was the Chairman of the Indian Atomic Energy Commission in 1974 publicly stated some weeks ago that the 1974 explosion was a bomb test.

Indian bureaucracy had used the code phrase "Buddha is smiling" to inform its Prime Minister in 1974 of the success of its nuclear test. It is ironic that the name of Lord Buddha the greatest apostle of peace in the sub-continent's entire history was chosen for such a destructive event. And his birthday the 11th May was selected for India's first explosion. India decided to repeat its performance on a much larger scale again on Buddha's birthday on 11 May 1998. What a great tribute to his legacy by the land of his birth.

The nuclear explosion of 11 and 13 May shook the very foundations of International community's non-proliferation policy. It was caught completely off-guard. The CIA admitted that despite its vast human and extensive satellite assets it was duped. Many in the sub-continent were rightly skeptical about this colossal intelligence failure. India's neighbours whose security was seriously threatened by its actions could not take such a laid back view. Pakistan has to keep a very close watch on what goes on in its neighbuorhood, which might have negative consequences for its security. Pakistan was getting some very strong and disturbing signals about India's nuclear plans. When it felt convinced, its Prime Minister wrote a latter to the President of United States on 8 April 1998 informing him of Pakistan's apprehensions that India was about to conduct nuclear tests. The fact that this timely warning was completely ignored is now of course a matter of public knowledge. I personally attended a conference in New Delhi on 6 March 1998 in which Mr. Jaswant Singh the leading strategist of the BJP addressed a large gathering of very senior retired military officers and diplomats on India's nuclear policy. I came away from the conference with a distinct feeling that India was launched on the road of overt nuclear weaponisation.

Why did India carry out the tests in such secrecy and in such great haste? Was there an imminent external threat that forced India's hand. We can look at the causes that the Indian Prime Minister has referred to in his letter written to various heads of state. First the threat from Pakistan. India attacked Pakistan three times in the first twenty-four years of its history, but we haven't had a conflict for the past twenty-seven years. We did not have any high level dialogue to discuss our differences, between 1994 and 1997, but even that was put right when Mr. Nawaz Sharif came to power in February 1997. He met his Indian counterpart and both agreed to immediately start a dialogue at the foreign secretary's level. We had three rounds of talks, in which some progress was made and then the Indian government fell. We were now hoping the new government to pick up the thread from where the old government had left off. Pakistan was seeing some improvement in its bilateral relations, though not as much as it would have liked. Both Pakistan and the rest of the world thought that a beginning had been made, however tentative, towards resolving the contentious issues that divide us, including the issue of Kashmir. Obviously the new Indian Government thought differently.

The second reason given by the Indian Prime Minister for the tests was the threat from China. To all neutral observers India-China relations appeared to be on the mend for the last ten years, after Mr. Gandhi's maiden visit to China in 1988. A number of CBMs had been instituted to improve the climate along the Sino-Indian border. A large number of troops had been transferred from India's Himalayan border to Kashmir to crush the internal uprising there. Not even a minor skirmish had taken place along the Sino-Indian border for many years. China was, therefore, not wrong in assuming that it had finally established a stable working relationship with India.

When the Chinese President came to Pakistan in December 1996 after his visit to New Delhi, he addressed the joint session of our Parliament and advised us to concentrate on our economic development and put the contentious issues aside for resolution later. He said, that was what China had also done. The Indians never tire of quoting that speech to us. So, the Chinese were quite shocked to be cast as the main villain in India's nuclear drama. I saw their disappointment personally when I visited China in the first week of June 98.

There were some internal compulsions as well that the Indian Prime Minister did not mention. The last time the Bhartia Janata Party got invited to form the government in 1996 its tenure lasted only thirteen days. This time again it did not have a majority and was forced to accept allies who had no ideological affinity with it. Some of them started to push their own agenda from the very beginning and continuously created difficulties for the government. Three ministers had to be dismissed within the first two weeks. The BJP government was facing a crisis almost on a daily basis and was desperately looking for something that will give it stability. Nuclear issue appeared to be an obvious choice. BJP had been an adherent supporter of India's nuclear weapons capability. Its election manifesto categorically stated that it would conduct nuclear tests and induct nuclear weapons into the Indian armed forces. It calculated that nuclear tests will create large popular support for it and also limit the smaller allies ability to blackmail it. And if it was forced to call early elections, it could win a majority and form the government on its own.

It is now public knowledge that a very small group of only six people were involved in taking this momentous decision. Even the cabinet was not consulted. How does one justify such violent change of course on a major strategic issue by a small coterie. Was there such a dramatic change in India's strategic environments, that it forced the Prime Minister to take immediate action. I would like to quote the statement of Mr. I. K. Gujral who was the Prime Minister only six weeks before India's tests. He says, "The sub-continent has been pushed into a nuclear arms race for purely political reasons. There was no security threat when I laid down the office. Mr. Deve Gowda, the Prime Minister before him, has attacked, "the political and partisan nature of the tests". Mr. Sangma who was unanimously elected speaker of the Indian Parliament under the United Front Government and belongs to the Congress party has said and I quote, "No convincing answers have come from the Government with regard to the reasons and timings of the tests".

It appears that the Hawks have carried the day to the exclusions of all voices of moderation and reason. The strategic elite in India considers the acquisition of nuclear weapons vital to its big power status. Mr. Jasjit Singh the Director IDSA New Delhi which is so ably represented by my colleague in this conference, stated in a seminar in Islamabad in 1996, that nuclear weapons were the modern currency of power which India must possess in order to play its rightful role commensurate with its size and stature. Its ambition of becoming a world power and permanent member of the UN Security Council, he feels could not be achieved without this modern currency of power.

Its leadership after the tests openly boasted that India was now the sixth nuclear weapons power and the world community had no option but to recognise and accept this fact. It believes it can crash through to the Security Council's permanent seat on the strength of its new found capability. The general assessment is that India on the other hand has seriously hurt its chances of achieving its ambition by flagrant disregard of a strong world trend towards non-proliferation.

The conduct of nuclear tests created a very serious security dilemma for Pakistan. The wide spread celebrations, the fire-works and the dancing on the streets in India was perhaps understandable. But what was not understandable was the very threatening and bellicose statements that started to emanate at the highest level of Indian political leadership. The Prime Minister set the tone by stating that India had the big bomb and it was now a nuclear weapon state. He was followed by the Indian Home Minister Mr. L. K. Advani who is considered to be more powerful that the Prime Minister. He warned Pakistan that it should recognise the changed strategic realities in the sub-continent and modify its policies with regard to both India as well as Kashmir, otherwise it should be prepared to face the consequences. Its Parliamentary Affairs Minister Mr. Madan Lal Khurana while addressing a public gathering at Jammu challenged Pakistan to nominate the time and place where it wanted to fight the fourth war with India. To top it all, the Chief Minister of Indian Kashmir publicly advised the Indian Government to launch a very strong and decisive battle against Pakistan. He also taunted Pakistan to detonate a bomb if it possessed one. It is not difficult to imagine the psychological climate that these utterances created in Pakistan.

I would now like to briefly deal with the factors that Pakistan took into account for dealing with this radically altered strategic environment. It first and foremost looked at the world community's response. A united, strong and unequivocal response by the international community to deal with this flagrant disregard of the strong consensus against nuclear testing would have helped Pakistan to take decision against testing. But what Pakistan observed was a G8 in complete dis-array. US talked of sanctions because its law made the sanctions mandatory. U.K., France and Russia disagreed and said that in their view sanctions would not be helpful. This wishy washy response by some important members of the G8 caused great disappointment in Pakistan.

The real problem for Pakistan was the new threat that Pakistan now perceived to its security and the manner in which to deal with it. Pakistan had believed that a stable strategic environment existed in the subcontinent before the blasts of 11 May 98. Pakistan's perceived nuclear capability which was variously called ambiguous, recessed or non-weaponised deterrence had provided it an effective safeguard against aggression. The fact that we had no conflict in the sub-continent for the last twenty seven years was in our view a consequence of our nuclear deterrence. We were quite willing and ready to keep the existing nuclear capability at this low and stable level. India took that option out of our hand by its unilateral action.

Many friends of Pakistan advised us not to follow the Indian example and take the high moral road of not testing. Pakistan took this advice seriously and initiated an intense open national debate on its options. Pakistani press, radio and television discussed the issues extensively for the next three weeks. A very broad spectrum of views were covered. A special debate was initiated in our Parliament. The Government consulted all political parties in the country both its allies as well as the opposition. It obtained the views of all the four provincial governments. It wanted to reach as broad a national consensus as possible on this vital national issue. Throughout this national debate, our consultations with important members of the international community continued. Pakistan did not want to take a pre-mature or hasty decision. The needs of our national security finally determined the course of action that we followed. We took the decision to test with great reluctance and in great sadness. It did not come as a surprise to the world. Many countries accepted the logic of our decision and knew that Pakistan would have liked to avoid taking this step if it had a choice. We believe strategic balance in the sub-continent which had been seriously disturbed has been restored. Our decision was taken in the most democratic tradition and consequently there is complete national consensus and resolve to deal with the economic and strategic consequences of our decision.

The important question now is how to deal with this new dangerous situation. The dooms day clock has been moved another five minutes closer to mid-night after the events of last May. Are these fears misplaced? Many defence analysts in the sub-continent compare the situation with the one that prevailed between USA and the Soviet Union during the cold war. I believe there are important differences. Soviet Union and the United States were separated by vast distances. They had no unresolved bilateral disputes. And of course they hadn't gone to war three times in just twenty-four years. On the technical side they had very reliable command, control and safety arrangements in place.

In my view there is a great deal to worry about from the growing nuclearisation of the sub-continent. The first and foremost is to look at the issues that have bedeviled the relations between the two neighbours. The major unresolved issue is that of Kashmir, which has been the reason for two of the three wars in the sub-continent. It cannot be wished away. The subjugated people of Kashmir will not allow it to be wished away. It is an issue that was taken before the Security Council by India and the Security Council passed a resolution on it which India accepted and has refused to honour for the last fifty years. It is not a territorial dispute. It is basic human right of a people for self-determination who refuse to live as an occupied colony. It is the moral obligation of the international community to help them gain their freedom as promised to them in the Security Council's resolution. Without a just solution of Kashmir, the prospect of peace in the sub-continent will always remain elusive.

The nuclear issue itself needs to be addressed simultaneously. First, step is that of further nuclear testing. Pakistan has already declared a qualified moratorium on nuclear testing. The qualified moratorium implies that Pakistan will not carry out any further tests unless there is grave provocation and threat to our national security. We are willing to sit down with India and the international community to consider a full moratorium on further testing.

And finally the CTBT. As you are aware Pakistan did not oppose the treaty and was willing to sign it, if India agreed to do the same. India which had co-sponsored the original resolution with the United States backed out when the moment of truth arrived. It came up with an entirely new argument demanding from the P-5 a time bound programme for total and complete disarmament and when that was not accepted, it went ahead and opposed the treaty. Pakistan did not sign the treaty because we were responsibly sure of what was likely to happen in our neighbourhood. The events of 11 and 13 May proved that our apprehensions were well-founded. We are giving the CTBT a fresh look and are willing to discuss it with a positive attitude.

Address:

AHMAD NISHANT
President
Institute of Regional Studies
Islamabad
Pakistan

77

Nuclear Proliferation After the Indian and Pakistan Tests

V. GILINSKY

This Forum is on reducing U.S. and Russian nuclear weapons, and how weapons scientists and laboratories can help to make that happen. Since the Forum was organised, nuclear explosions in South Asia have undermined assumptions about the future of nuclear weapons. Suddenly, weapon scientists were on the front page—not the rueful ones in Russia and the United States, but exuberant ones in India and Pakistan. They became national heroes for their explosions, as did their political leaders. India and Pakistan are not reducing their nuclear weapons; they are accumulating them.

The explosions, and the muted international reaction to them, mark an important turning point. But in what direction? It is difficult to know what to do about the South Asian nuclear tests, so it is tempting to play down their significance. To put the best face on it, the tests merely confirm what has been true for many years: that India and Pakistan are building nuclear weapons, as Israel did before them. India may have decided to get some thermonuclear testing done before a worldwide ban will make that much more difficult, just as France and China did earlier. If that is so, the tests were a backhanded recognition of the strength of the international consensus against further testing. India, Pakistan, and Israel are the only significant Non-proliferation Treaty holdouts. No other countries are comparably close to getting nuclear weapons. Did the May tests then signal the end of nuclear proliferation, at least for a long-time?

I think the opposite is more likely. The May tests—and even more so the mild international reaction to them—confirm the weakness of the bars to nuclear proliferation. To my mind they ease the way for more nuclear states in the Middle East and East

Asia. One thing is sure: this new circumstance, and the world reaction to it, will be watched closely by other countries interested in these weapons.

The situation calls for a re-evaluation of prevailing assumptions in arms control circles about the spread of nuclear weapons and what can be done about it. A good place to begin is by clearing away some of the make-believe that fills discussions on this subject.

ARE NUCLEAR WEAPONS REALLY USELESS?

It has, for example, become an arms control cliché that nuclear weapons are useless and, once countries realise this, they won't want them. Delegations of arms controllers from Western countries trooped to India and Pakistan for years to explain this. The Indian nuclear tests in May made clear the Indians did not find the arguments convincing. They think nuclear weapons are useful and important to have, not least for enhancing their place in the world. Can one disagree that these weapons do enhance a country's importance?

The five nuclear powers legitimised by the Non-proliferation Treaty obviously regard their own nuclear weapons as valuable and important. The United States and Russia have taken large steps to reduce their arsenals, but they are in no rush to give them up entirely. Neither Britain, France, nor China have agreed to reduce their nuclear arsenal. Nor has Israel, obviously.

These countries and others treat any move by a state to get nuclear weapons as a matter of the utmost importance, as it is. How else can one consider the possibility of Iraq or North Korean nuclear weapons? Significantly, North Korea got much more lenient treatment, as compared with Iraq. The United States, South Korea, and Japan agreed to pay North Korea billions in blackmail in return for freezing its small nuclear operations and construction, at least for the time being. The difference was that North Korea really did have the explosives for a bomb and so got more respect. North Korea proved that a few nuclear warheads can, if nothing else, be a powerful bargaining chip.

In a certain sense, it is true that their owners cannot do much militarily with their nuclear weapons. At the same time, as is characteristic of so many propositions about nuclear weapons, the opposite is also true: countries with nuclear weapons have more weight in the world. Years ago, Neils Bohr used the term "great truths" for such propositions whose opposite was also true. The subject of nuclear weapons is full of them. Bringing nuclear weapons into a power equation is the political equivalent of dividing by zero.

WILL A NUCLEAR "OPTION" SATISFY WOULD-BE NUCLEAR STATES?

Other prevalent assumptions in arms control circles were that countries like India and Pakistan would not manufacture and deploy nuclear weapons; that they would be satisfied to have components that they could assemble if needed; or that they would be satisfied to maintain the "option" to do so; and that they would not do full-scale testing because of the possible severe response that would bring.

As for the notion of an "option", anyone who has thought about it seriously knows that such a technological "option" is meaningless unless one had mastered every single technical step, down to the last one. One simply cannot foresee what technical problems one may run into. In other words, you only have an "option" if have actually done it.

It has been evident for years that both India and Pakistan had developed and manufactured nuclear weapons, as Israel had before them. Yet for years it was thought indelicate to say so, as if pretending that these countries did not have nuclear weapons would maintain the psychological barriers to test explosions and weaponisation.

India showed those psychological barriers were no more than plate glass, and Pakistan followed quickly through the opening. The tests served an important purpose: analysis and simulation may convince technical people but, in the world of affairs, a successful full-scale test is much more convincing. The recent misjudgment by observers regarding the likelihood of Indian tests shows that the outlook of arms controllers is a poor guide to the actions of players in the international power game. There are now reports out of India that the country is thinking about submarine-based warheads.

This experience has to make us wonder about other prevalent assumptions—such as the unlikelihood of actual military use of nuclear weapons. These assumptions rest on similar psychological propositions that in the heat of action and misinformation may not amount to more than plate glass, either. It doesn't help that in South Asia targets and launch pads are only a few minutes apart.

WHAT HAPPENS TO THE FICTION OF "FIVE NUCLEAR STATES"?

The Non-proliferation Treaty was designed to avoid precisely such dangerous situations. The idea was that it was in the interest of all countries to limit the spread of nuclear weapons. One had to draw the line somewhere, even if arbitrarily, so the idea was that all countries but the first five nuclear states would agree not to make nuclear bombs. In return, the five approved nuclear states agreed to eliminate their nuclear arms in time.

It didn't completely work out that way. The nuclear states are not contemplating eliminating their nuclear arsenals, but the United States and Russia have made important reductions in recent years. At the same time some non-nuclear countries obtained nuclear weapons (Israel, India, Pakistan, and South Africa). So long as these states did not test their weapons, it was possible to maintain the fiction that there were still only five nuclear powers. The elimination of the nuclear weapons stockpile in South Africa, and the giving up of nuclear warheads by several former Soviet republics, even suggested a reversal in the proliferation process.

The May nuclear tests in India and Pakistan have thrown this progress into doubt. The five grandfathered nuclear states have said in response they will not give India and Pakistan the status of nuclear weapon states under the NPT. The Five understandably want to avoid the impression that a country can shoot its way into club. But India and Pakistan are nuclear weapon states all the same, probably with dozens of bombs each (as is Israel, with many more). In time, the Five will want India's and Pakistan's cooperation in agreeing to an end of testing and in controlling the flow of dangerous technology. To achieve this, the Five will have to acknowledge India's and Pakistan's new nuclear status outside the NPT "regime."

WEAKNESS OF NPT: THE CASE OF NORTH KOREA

As it stands, that NPT "regime" is a pretty fragile thing. It rests on a Non-proliferation Treaty that allows for withdrawal upon a 90-day notice. The Treaty allows countries to accumulate nuclear explosives so long as countries promise to use the materials as fuel. (I am speaking, of course, about separated plutonium and highly enriched uranium.) Compliance with the Treaty is verified by a weak inspection system, and there is no explicit Treaty enforcement mechanism. The possibility of enforcement in any specific case relies on ad hoc actions of major countries, in practice led by the United States.

At least two member countries—Iraq and North Korea—have violated the Treaty. The North Korean case, the more serious of the two, is especially instructive. North Korea refused in 1993 to allow IAEA inspectors to check on whether it secretly separated bomb quantities of plutonium for nuclear weapons. When pressed, the North Korea threatened to exercise the 90-day provision to withdraw from the NPT. This panicked the organisers of the upcoming 1995 NPT review conference who feared the effect of a withdrawal on the prospects for Treaty extension. The Treaty custodians were more concerned about extending the Treaty than enforcing it. They also feared triggering North Korean violence against South Korea and Japan. As a consequence, the United States and other countries did not impose sanctions, nor did they insist on elimination of Treaty violations by North Korea. Rather, these countries offered North Korea two large nuclear reactors, worth about $5 billion, in return for freezing its small indigenous nuclear programme. North Korea agreed, but still refuses to allow IAEA inspectors to check on past plutonium diversion. No one is pressing them to do so, or to clarify North Korea's ambiguous status under the NPT. The most flagrant violator of IAEA Treaty safeguards received the most generous treatment.

WEAKNESS OF IAEA INSPECTION

For years, arms controllers regarded any criticism of IAEA inspection as heresy. To give them credit, in the North Korean case the IAEA inspectors at least tried to do their work. But in the earlier Iraqi case, the Agency missed the country's nuclear weapons programme completely. The Vienna Agency wasn't interested in finding anything outside of its narrow blinders. When, after the Gulf War, defectors revealed Iraq's secret weapon programme, the Security Council still had to prod the Agency pretty hard to get it to get tough with Iraq.

The Agency claims to have changed after that experience and says it is now taking a more serious approach to its responsibilities. There are some signs that this is happening. The Agency has, however, always suffered from an intrinsic conflict of interest in that its instinct is to protect anything nuclear its member states are doing. This includes fuel cycle activities involving separated plutonium and highly enriched uranium. The Vienna Agency is not going to get ahead of its political support.

WEAK INTERNATIONAL REACTION TO NUCLEAR EXPLOSIONS

Which brings us back to enforcement. Of course, India and Pakistan (and Israel) didn't sign the NPT and so their tests do not violate the Treaty. Still, it was widely thought that violating this international norm entailed serious penalties. It appears that despite all the talk deploring the Indian and Pakistani explosions, there isn't really much that the United States and other countries are going to do in response. This weakness is bound to make an impression on other would-be nuclear bomb-makers.

In many ways, the current reaction to the May test series is a replay of the events of twenty-four-years ago. The United States continued supplying nuclear fuel to India after the 1974 explosion and has been accommodating in a variety of ways since then. (Similarly, Russia has announced the May explosions will not affect the sale of two Russian nuclear power reactors to India).

The current U.S. administration made clear that it would take an even softer line on India's nuclear programme than that of its predecessors. The U.S. administration does not want nuclear issues to get in the way of what it regards as more important economic ones. Having turned a blind eye to nuclear developments in South Asia for so long, the United States, for one, is not left with any good options for dealing with them. A U.S. Secretary of Defence once said the government's anti-proliferation policy could be handled by two officials—one to count the new nuclear states and the other to wring his hands.

It is awkward, to say the least, for the Untied States to impose economic sanctions on a relatively democratic India for testing several warheads while embracing an undemocratic China with heavy nuclear armament. The reason the United

States has imposed some sanctions on India and Pakistan after the May tests is that existing law forced the President to do this. He has, however, interpreted the law's requirements—bars to certain financial transactions and credits—in the most lenient fashion. It is easy to rationalize the futility of economic and political sanctions in the current circumstances. But what does it say to other would-be nuclear states if there is no penalty for breaking accepted international nuclear norms? If we can so easily rationalize an Indian bomb, and then a Pakistani bomb, we will also rationalize other ones.

In some ways, even though Pakistan was responding to India, the Pakistan's explosions are the more worrisome ones. Pakistan is more representative of a wider class of countries that could plausibly attempt to get nuclear bombs, whether by their own means or through others, by purchase or theft. The psychological barriers against this have lowered, too.

The effect on North Korea will be especially interesting. There are reports out of Russia that a few years ago, North Korea asked to be allowed to test in China, but was turned down. That raises the question whether North Korea had a bomb to test. It seems likely that North Korea had a nuclear weapons programme which did not end with the 1994 U.S.–DPRK agreement. The recent tests by India and Pakistan are likely to reinforce the DPRK's determination to carry on with its secret nuclear programme. If a large state like India sees the need for nuclear weapons to guarantee its own national security, then North Korea may well see it as even more important. A weak world reaction to the South Asian tests may suggest to North Korea that it could get away with even more than it had imagined. Recall that, under the 1994 U.S.-DPRK agreement, North Korea retains its indigenous plutonium production facilities in a mothballed state. It is allowed to do so until the United States supplies the second of the two large power reactors promised to North Korea, which means North Korea will retain the means for continued nuclear blackmail for a long time. (The NPT, by the way, did not bar the accumulation of plutonium, only doing so out of sight of the IAEA inspectors—which says a lot about the NPT).

CONTROL OVER "CIVILIAN" TECHNOLOGY TRANSFERS STILL VITAL IN KEEPING THE LID ON

It has also become fashionable to say, just as about

sanctions, that controls on technology exports don't really work. In fact, such controls have played an important role in stopping or slowing nuclear weapons developments and they continue to be important. There is no need to make it too easy for countries around the world to make nuclear warheads. The key technologies are those for plutonium separation and uranium enrichment.

Would-be nuclear states emphasize the indigenous nature of their nuclear development. In reality, there is in almost every case a strong dependence on external technology. That is true for both India and Pakistan. Pakistan stole the uranium enrichment technology from the Netherlands and Germany. An enrichment plant at Kahuta is the source of highly enriched uranium for bombs. Pakistan was able to complete its work on the plant only with the active cooperation of European firms. They were happy to have the business and looked the other way, as did some of their governments. The United States did, too, during the Afghan war when helping Afghans to defeat the Soviets took precedence over stopping the Pakistani nuclear bomb programme.

It has of course been obvious for years that Pakistan's uranium enrichment programme was dedicated to military use, despite the protestations by Pakistani leaders to the contrary. Pakistan had no reactors that used enriched uranium.

The Pakistanis are widely believed to have gotten actual nuclear weapons assistance from China. If that is so, it would put China in violation of the NPT. The only ones who are calling the Chinese on this are the Indian, who aren't even member of the Treaty. The subject abounds in ironies.

MUCH OF INDIA'S WEAPONS-GRADE PLUTONIUM COMES FROM A CANADIAN REACTOR INDIA PROMISED TO USE ONLY FOR "PEACEFUL PURPOSES"

The Indians claim their bombs are based entirely on indigenously produced materials, free of entangling international obligations. That is not so.

India's unsafeguarded supply of weapons grade plutonium is thought to come from two small reactors at the Bhabha Centre near Bombay. These are the 40-megawatt CIRUS reactor, built in the 1960s, and the more-recently constructed 100-megawatt Dhruva reactor. Current estimates credit India with about 400 kilograms of weapons-grade plutonium—enough for over 80 warheads. About half is from CIRUS.

India got the CIRUS reactor from Canada in the early 1960s and the United States supplied the essential heavy water. The contracts included an Indian promise of "peaceful" uses. Despite this, India used the plutonium from this reactor for its 1974 explosion and then said this did not violate agreements because it was a "peaceful" explosion.

The Indian scientist in charge of the 1974 test has now said that was intended to be a bomb all along. There are recent reports in the nuclear trade press that after the 1974 Indian explosion, India continued to use CIRUS to produce weapons grade plutonium and manufactured warheads. Such actions would have been in direct violation of India's promises to Canada and the United States, as of course would be any use of such material in the current military test series.

The Indian ambassador to the United States said in Washington on June 17 that India would address this point promptly in an official communication. If it has done so, it has not made the statement public. The U.S. State Department would rather forget about the issue. Its Asian specialists think that the nuclear issues get in the way of more important opportunities to "engage" India. The trouble is, there is not much point in trying to "engage" India in new agreements if one does not insist on its compliance with existing ones.

CONCLUSION: ONE RULE FOR ALL

In launching a series of nuclear tests, India has created a new situation. What the effect will be on Indian-Pakistani relations, and on other countries that would also like to have the bomb, is an open question. The hopeful view is represented by the *Bulletin of the Atomic Scientists*, which has not moved the arms of the doomsday clock on the cover of its last issue. The same cover, however, carries a picture of Indian women cheering the bomb, and that seems more ominous.

To keep the nuclear lid on over the long run, the Five NPT nuclear states will need the support of the rest of the non-nuclear countries. To keep that support for measures that make it more difficult for more states to get nuclear weapons, the Five are going to have to take steps to reduce nuclear arms themselves beyond those already planned. These steps cannot just involve the United States and Russia.

However, much the official five nuclear powers don't like it, they will have to make larger reductions if the overall proliferation process is to be brought to a halt. That is so, again, not because the reductions would sway would-be nuclear states. Rather, it is because such reductions are essential for retaining the support of the rest of the non-nuclear countries for tough anti-proliferation measures. However much the majority of non-nuclear want to control proliferation of nuclear weapons, they also resent the dominance of the nuclear countries and will not suffer it forever.

The recent test explosions, and the dilemma of what to do about them in a world of sovereign states, tell us that in the end there is not tenable position except one that applies the same rule to all countries. Either we will have more nuclear reductions or more nuclear states. That means that the work of this conference is, if anything, more important than before.

Address:

VICTOR GILINSKY
Nuclear Consultant
48 Vellesley Circle
Glen Echo MD 20812
USA

78

India's Nuclear Planning, Force Structure, Doctrine and Arms Control Posture

BRAHMA CHELLANEY

India's emergence as a declared nuclear-weapons state has only sharpened its challenges to build a minimal but credible nuclear deterrent. India can build only a very small deterrent force because its stockpile of weapons-grade plutonium and its financial resources are limited. New Delhi will need to have a high level of confidence in the reliability and survivability of its arsenal so that the smallness of the deterrent is not a handicap. Such confidence can come about only by rigorously meeting the technical and policy requirements of a deterrent. Potential adversaries will have to be left in no doubt the retaliatory prowess of India if it were to come under attack.

The key challenge for India is to adequately deter two hand-in-glove nuclear adversaries, China and Pakistan. India's threat situation is unique: It is the only state in the world to share disputed borders with two closely aligned nuclear foes. India has yet to acquire the deterrent capacity to end China's nuclear threat. It will have to plug this vulnerability at the earliest opportunity.

New Delhi's decade-long policy of rapprochement with Beijing since 1988 has not won it Chinese friendship, but enabled China both to engage and to contain India, with engagement serving as a front for accelerated containment. Covert Chinese transfers have neutralised India's technological advantages over Pakistan. China also has sought to bring India under strategic pressure on another flank by building listening posts on Burmese islands and deepening collaboration with the military junta in Rangoon. Without being able to stand up to China, India will never be able to persuade Beijing to halt its containment of India or its clandestine nuclear and missile transfers to

Pakistan.

It took almost 24 years for India to lift the veil of ambiguity from its strategic posture and go overtly nuclear. All it needed was political courage and conviction. By detonating five nuclear devices based on different warhead configurations in a space of less than 45 hours, India sent an unequivocal message to the world: It is determined to secure itself with its own capabilities. Only a secure India can be at peace with itself and its neighbours. The nuclear tests thus will go a long way in shattering India's image as a meek, vacillating nation and in building a safer environment for it to focus its attention on economic modernisation.

These tests by a country, which has never committed aggression, were designed to ensure peace through deterrence. Over the centuries, India was repeatedly invaded but never sought to raid or overrun another civilization. From the time of Alexander the Great, it was attacked, plundered and at times even subjugated by foreign invaders, be it the Arabs or Central Asians or Europeans. Even tiny Portugal managed to set up a colony in India. After independence in 1947, India faced four wars in the first 24 years, one with China and three with Pakistan.

The fact that Prime Minister Atal Bihari Vajpayee's Government detonated five sophisticated devices within two months of assuming office showed that India already had the basic materials, equipment and skills for nuclear deterrence. While independent India's first Prime Minister, Jawaharlal Nehru, laid the foundation of India's nuclear programme, his daughter, Prime Minister Indira Gandhi, opened up the nuclear option in 1974 and his grandson, Prime Minister Rajiv Gandhi, helped establish a nuclear-delivery capability through the 1989 Agni intermediate-range ballistic-missile test. Mr Vajpayee was the leader India was waiting for to take it over the nuclear threshold.

In one go, India showed its capability to manufacture and test the most modern nuclear weapons-thermonuclear, boosted fission and low-yield types. No country has ever demonstrated such a range of weapon capabilities in one shot. In fact, no nation has conducted multiple tests of the kind India did. This was deliberate. It was intended not only to herald India's arrival as a nuclear-weapons state with deterrent capabilities but also to deal with external pressures. If India had conducted one test at time to certify its warhead models it would have come under swirling coercive pressures, possibly hindering its movement forward. By doing five bangs over two days, India gate-crashed the nuclear club, presenting a fait accompli to an astounded world. Nothing can undo this development. Unlike 1974 when a crude fission device was detonated without being configured as a warhead, the 1998 tests all involved warhead prototypes.

The tests were a natural corollary to India's firm opposition to the Comprehensive Test Ban Treaty (CTBT), which was taken through the backdoor to the UN General Assembly for endorsement after India vetoed its adoption at the Geneva negotiations. For the second time since the CTBT veto in 1996, India showed ti could stand up for its rights even if it means swimming against the international tide. Sceptics never believed that the Vajpayee Government would take India down the nuclear road, just as they never expected New Delhi not only to oppose the CTBT but to actually veto its adoption in Geneva. By opposing the loopholes-marred CTBT tooth and nail, India had signalled to the world that the testing option was essential for its security. If the Vajpayee Government had not gone ahead, India would have got stuck as a threshold state, bearing the burden of an open option but not reaping the benefits.

India's long-held nuclear-weapons option had come to symbolize its holier-than-thou attitude. The rest of the world had watched India sitting on the nuclear fence for decades, mocking the West self-righteously, but unable to make up its own mind. For more three decades since starting to produce plutonium, India had been debating whether to "go nuclear". Its indecisiveness had given it the distinction of being the only country to openly conduct a nuclear test and yet shy away from a military posture based on deterrence. India had to settle its biggest unresolved issue because it could not go on paying escalating international costs for merely retaining its nuclear option. It had remained a key target of all the export-control cartels. Even if the cumulative costs it had paid were ignored, the international costs over the next decade of its old approach would have exceeded the total costs of building a small, credible deterrent force.

Few seems to be aware that India has one of the world's oldest nuclear programmes. Its first prime minister, Jawaharlal

Nehru, laid the foundation of that programme by setting up the Atomic Energy Commission in 1948 to produce "all the basic materials" because of nuclear power's "strategic nature". The Commission was set up with Homi Bhabha as chairman and Nobel laureate C.V. Raman as an adviser.[1] Nehru had said even before assuming office that as long as the world was constituted on nuclear might, "every country will have to develop and use the latest scientific devices for its protection". By the mid-1950s, India had built Asia's first atomic research reactor and set in motion a broad-based nuclear programme.

After the Cirus reactor built with Canadian assistance started up in 1960, Nehru declared, "We are approaching a stage when it is possible for us ... to make atomic weapons". That stage was reached unquestionably in 1964 when India completed a facility to reprocess the Cirus spent fuel, making it the fifth country to be able to produce plutonium.

When the Chinese conducted their first nuclear test in 1964—almost five months after Nehru's death—India's top nuclear scientist, Homi Bhabha, declared that India could build a nuclear bomb within 18 months if it so decided. China's first nuclear test, barely two years after its invading forces inflicted a crushing defeat on India, sharply heightened New Delhi's insecurity. The following year, Pakistan, taking advantage of India's security travails, infiltrated its army troops into Kashmir, triggering a full-scale war.

It was Prime Minister Lal Bahadur Shastri who initiated the Indian nuclear-explosives programme in 1965. But a series of events put a brake on that programme. These included the passing away of Shastri, Bhabha's own death in Europe in a mysterious plane crash, and the political instability triggered by an initially weak government under Prime Minister Indira Gandhi. When India eventually conducted a nuclear detonation in 1974, it astounded the world. U.S. intelligence was caught unaware, even though Mrs. Gandhi had told Parliament in 1972 that her government was "studying situations under which peaceful nuclear explosions carried out underground can be of economic benefit to India without causing environmental hazards". Earlier in 1970, India had rejected a US demarche against conducting any nuclear explosion.

1. For a detailed discussion, see Chapter 1 in Brahma Chellaney, Nuclear Proliferation: The US-Indian Conflict (New Delhi: Orient Longman, 1993).

By conducting the 1974 test, Indira Gandhi gave India a tangible nuclear option. The country broke no legal commitment and had the sovereign right to continue the testing programme. As National Security Adviser Henry Kissinger told U.S. Congress after the Pokharan test, "We objected strongly, but since there was no violation of U.S. agreements involved, we has no specific leverage on which to bring our objections to bear". The test shook the NPT to its very foundation. Had India continued to test, the NPT regime probably would have disintegrated or been seriously damaged. Instead, the U.S.-led regime emerged stronger and with fangs because India, to the great surprise of the rest of the world and its own public, did not go beyond that one single test. It will remain a riddle of history why Mrs. Gandhi did not carry out another test.

One key constraint on India going overtly nuclear was its lack of missile capability. Mrs. Gandhi sought to remedy this by formally instituting a programme in 1983 to develop ballistic missiles. An Indian nuclear-deterrent force has to be centered on missiles since bomber-aircraft cannot reach the heartland of China. India's nuclear-weapons option really opened up in an operational sense only after the Agni intermediate- range missile was flight-tested in February 1994, completing its triumphant three-test development phase.

The first Agni test in 1989 was carried out despite, in the words of Prime Minister Rajiv Gandhi, " ambassadors of certain foreign powers" threatening punitive sanctions. "I told them clearly that India would carry out the launch and we would not change our decision under pressure", the son of Mrs. Gandhi and grandson of Nehru said, Agni-type missiles make strategic sense only if they carry a nuclear weapon. While India had demonstrated its delivery capability, it had not demonstrated its ability to build a nuclear warhead for the Agni, That demanded nuclear testing.

India's turning point came when an openly pro-nuclear government took office in March 1998. The new coalition elected to power pledged, in the words of Vajpayee, to "exercise all options, including the nuclear option". The Vajpayee Government was determined not to miss India's closing opportunity to test. After legitimizing their nuclear monopoly through the NPT's permanent extension, the five traditional powers (the United

States in particular) had begun targeting India through the test-ban treaty and the proposed fissile cut-off treaty.

The point to be noted is that it took India almost a quarter century to shed its nuclear restraint and go overtly nuclear. No nation has exercised such restraint on an indigenously developed country. More importantly, India is the only democracy which debated the issue for years of whether to go nuclear. The other nuclear-armed democracies (the United States, Britain, France and Israel) went down the nuclear path quietly without any real public debate. Even in more recent times, these democracies have tested, produced and deployed new nuclear weapons without hardly any national debate prior to the decisions being made.

INDIA AND ARMS CONTROL

In announcing its first round of three simultaneous nuclear tests on May 11, 1998, New Delhi set forth nuclear India's arms-control policy. The statement, dealing with India's stance on nuclear disarmament, export controls, CTBT and the proposed Fissile Material Cut-Off Treaty, said:

"It is necessary to highlight today that India was in the vanguard of nations which ushered in the Partial Test Ban Treaty in 1963 due to environmental concerns. Indian representatives have worked in various international forums, including the Conference on Disarmament, for universal, non-discriminatory and verifiable arrangements for the elimination of weapons of mass destruction. The Government would like to reiterate its support to efforts to realise the goal of a truly comprehensive international arrangement which would prohibit underground nuclear testing of all weapons as well as related experiments described as 'subcritical' of 'hydronuclear'.

"India would be prepared to consider being an adherent to some of the undertakings in the Comprehensive Test Ban Treaty. But this cannot obviously be done in a vacuum. It would necessarily be an evolutionary process from concept to commitment and would depend on a number of reciprocal activities.

"We would like to reaffirm categorically that we will continue to exercise the most stringent control on the export of sensitive technologies, equipment and commodities—especially those related to weapons of mass destruction. Out track record has been impeccable in this regard. Therefore, we expect recognition of our responsible policy by the international community.

"India remains committed to a speedy process of nuclear disarmament leading to total and global elimination of nuclear weapons. Our adherence to the Chemical Weapons Convention and the Biological Weapons Convention is evidence of our commitment to any global disarmament regime which is non-discriminatory and verifiable. We shall also be happy to participate in the negotiations for the conclusion of a fissile material cut-off treaty in the Geneva-based Conference on Disarmament".[2]

The statement signified a major shift in the Indian position on the much-despised CTBT as well as the MFCT. India has abruptly changed its position on these two arms-control measures at the centre of U.S. non-proliferation diplomacy.

India's nuclear-weapons tests have been widely seen as a major blow to the non-proliferation regime. But those defiant tests could lead to its acquiscence in the very regime it has traditionally opposed tooth-and-nail. New Delhi now supports FMCT negotiations at the Conference on Disarmament (CD) and is ready to consider signing the CTBT if its concerns are met. Moreover, it is not insisting on being formally recognised as a nuclear-weapons state, although it has declared itself to be one. This avoids a potential clash with the 1970 Nuclear Non-proliferation Treaty (NPT) that defines the nuclear club as consisting of states that exploded a nuclear device before 1967.

It is remarkable that in a very short time India—the principal opponent of the U.S.-led non-proliferation regime—has jettisoned its unyielding opposition to the FMCT and CTBT. In mid-1996, India single-handedly blocked the approval of the CTBT at Geneva, forcing the treaty's proponents to take it to the United Nations General Assembly for endorsement though a back-door route. India vowed never to sign the treaty, which awakened the country to the technical imperatives of its long-held nuclear-weapons option.

India also opposed the FMCT, leading the group of non-aligned nations to successfully block the start of negotiations because of the refusal of the great powers to discuss nuclear

2. Government of India, Press Statement on the May 11, 1998, nuclear tests (New Delhi: Press Information Bureau, May 11, 1998), pp. 2-3.

disarmament. But after its tests, India, without consulting its non-aligned partners, announced it would participate in the FMCT negotiations, forcing Pakistan and Israel to fall in line and opening the way for negotiations to start. India was in the group of 28 non-aligned member-states at the CD that had linked the start of FMCT talks to the "immediate and concurrent commencement of negotiations and early conclusion of", among other things, a binding non-use convention on a treaty to eliminate all nuclear weapons. India reckons that by the time and FMCT is negotiated and takes effect, it would have produced enough fissile material for a minimum deterrent against its two closely aligned nuclear neighbours, China and Pakistan.

If India and Pakistan accept the CTBT and FMCT, the United States and other traditional nuclear powers could contend that far from being undermined, the non-proliferation regime has actually been strengthened. Current diplomacy efforts are thus focused on winning Indian and Pakistan support. India's new willingness to discuss the CTBT and FMCT, however, does not mean a breakthrough is near.

The need to bring India into at least the outer ring of non-proliferation regime has armed New Delhi with important leverage. India knows that unless it co-operates, non-proliferation cannot credibly remain a global norm. Without wrestling important concessions, no government in New Delhi will be able to sell the CTBT or FMCT to Parliament. Before it agrees to extent support to the non-proliferation regime, India can be expected to drive a hard bargain.

It will seek to use its leverage to beat back the rising tide of technology sanctions it has faced since conducting its first nuclear test in 1974. The technology controls were an important consideration in its decision to end almost a quarter century of nuclear restraint and resume testing. After the 1998 tests, it was slapped with limited economic sanctions, some of which have since been suspended by the United States.

New Delhi will also insist on the same privileges enjoyed by the traditional nuclear powers. These powers, for example, have defined for themselves what weapons-related activities they can engage in under (or despite) the CTBT. These include high-tech underground tests at subcritical level and other activities at nuclear test sites, simulated nuclear explosions in laboratories, and the sharing of some data and technologies among themselves. India has reserved the right to begin subcritical tests of its own. New Delhi also intends to push ahead with the necessary steps, including missile tests, to build a credible but petite nuclear deterrent against China, still modernizing its nuclear and missile forces.

It is inconceivable, however, that any Indian Government, whatever its political complexion, would unconditionally sign the CTBT. The mounting international pressure does raise some troubling questions for India. Before India's tests under its so-called Operation Shakti (Strength), the great powers had given up hope of persuading India to sign and ratify a treaty which it had rejected with a decisive "not now, not ever" pledge. These powers were looking at legal ways that would bring the CTBT—an important pillar of their NPT regime—into international effect without India. Then the Indian tests occurred, sounding what seemed to be the death knell of the CTBT. Yet, within months, these powers not only resuscitated the treaty but became confident of bringing India to heel.

What happened? Principally, India failed to resist the temptation to use the much-depised CTBT for tactical gains in negotiations. As part of negotiating tactics, the country's two most highly respected scientists were expressly ordered to publicly speak out for the treaty to help convince the Americans, before the Washington round of negotiations in September 1998, that India could live up to its side of the bargain. That a bad precedent was being set of using scientists for narrow, short-term political goals was overlooked. Some media analysts, ever so eager to take their cue from officials, were also directed to spread the word that India was "inching towards" signing the CTBT. Such signalling resulted in the Indian Government finding itself prematurely pushed towards a path for which it had not built the necessary support within the ruling coalition or the opposition.

Prime Minister Vajpayee's September 25 UN statement that India was prepared to bring its negotiations with key interlocutors "to a successful conclusion so that the entry into force of the CTBT is not delayed beyond September 1999", was thus immediately read as an Indian commitment to sign. But the speech did not contain a promise to sign, but rather a promise not to block the treaty from taking effect. The "not to block" promise was itself pivoted on two conditions set out explicitly in the speech: "Successful conclusion" of negotiations, and signature and ratification "without conditions" by all the other countries on the 44-nation list in Annex 2 of CTBT's Article XIV.[3]

Despite India being one of the 44 countries listed in Annex 2, the CTBT can take effect without India. The 1990 conventional Armed Forces in Europe (CFE) Treaty provides an example of how a pact can be brought into force provisionally without the specified set of ratifications coming in. In fact, Article 79 of the Geneva Convention on international law specifically permits that. A protocol signed by ratifying states can further ensure that the CTBT stays in force permanently.

After Mr. Vajpayee made his UN speech, Indian's negotiations with the United States got badly stuck. Also, there has been no progress in securing the ratification of key states like China, the United States and North Korea. Nor is it certain that the US Senate, even if it were to ratify the treaty, would resist the appeal to impose conditions of the kind it did while ratifying the Chemical Weapons Convention. The Prime Minister has for months been saying that India's bipartisan consensus against the CTBT cannot change unless the United States is willing to make important concessions. Only "matching steps by our interlocutors", in his words, "will help the process of shaping a new consensus in favour of accepting the CTBT". But instead of working to accommodate India's core concerns, Washington has engaged in a series of provocations. New Delhi has during the course of negotiations with the United States been subjected to international dignity, with the Americans blackballing scores of leading Indian private and state-run enterprises and many of their affiliates, restoring Pakistan's but not India's access to long-term development loans from the World Bank, vowing not to

3. In his UN speech, the Prime Minister said: "Mr. President, India having harmonised its national imperatives and security obligations and desirous of continuing to cooperate with the international community is now engaged in discussion with key interlocutors on a range of issues, including the CTBT. We are prepared to bring these discussions to a successful conclusion so that the entry into force of the CTBT is not delayed beyond September 1999. We expect that other countries, as indicated in Article XIV of the CTBT, will adhere to this Treaty without conditions". "Prime Minister's Address to the UN General Assembly", September 25, 1998, Official Transcript (New Delhi: Press Information Bureau), p. 4.

recognise India's nuclear-weapons status "even by implication" and publishing a long list of humiliating demands.[4]

The Indian prime minister has no option but to act by political consensus on the CTBT, which is not so much about testing as about verification, including technical espionage by national intelligence assets. If India accedes, it will come under the rigours of the treaty's highly discriminatory verification regime. The 'Comprehensive Test Ban Treaty', by Washington's own admission, is not a comprehensive ban on testing. With the United States set to conduct yet another underground nuclear test at professedly subcritical level, it is not even a 'test ban' since testing by technologically more advanced methods is taking place at the very site of earlier tests. It can, however, become a 'treaty', for whatever its worth, without India being dragged into it.

If most other states want the CTBT to enter into force (EIF), India could keep quiet and not campaign against it. This is what India did in 1995, although under US pressure, when the NPT was permanently extended. Mr Vajpayee's not-to-block promise prompted the Vienna-based CTBT Organisation to state at its October 1998 briefing that the CTBT's "provisional EIF" can be worked out without India. A special conference of ratifying states, as provided by Article XIV, is likely to be held in New York or Vienna in September or October 1999. This conference, if it wants, could give immediate legal effect to the CTBT, although some key states (including China) are unlikely to favour the treaty's implementation without India being on board.

On FMCT, India is likely to adopt a cautious negotiating strategy. It is not in India's interest to see an early FMCT as it will adversely impact on its development of a nuclear deterrent against China. Also, India is aware that it will be affected the most by the FMCT. India is the sole non-NPT nuclear state with a varied and growing fissile material production complex.

The only states capable of and legally not encumbered from producing fissile material for weapons use are the NPT nuclear

---

4. See, for example, William J. Clinton, Text of a Letter from the President of the Speaker of the House of Representatives and the President of the Senate, November 12, 1998 (Washington, DC: The White House); and Strobe Talbott, Deputy Secretary of State, "U.S. Diplomacy in South Asia: A Progress Report", Speech at the Brookings Institution, November 12, 1998 (Washington, DC: Department of State).

powers and the three non-NPT nuclear states. Britain, France, Russia and the United States, so awash with plutonium and highly enriched uranium (HEU) that these materials are coming out of their ears, have been forced by their circumstances to declare unilateral moratoria on production. China claims unofficially that it, too, has increased production, although there is little evidence of that. The five can turn their averred cut-off into a legal undertaking by signing an interim fissban pact among themselves. Predictably, none is interested. With all the five still qualitatively improving their nuclear arsenals, each is eager to preserve its future options and sees merit in a fissban treaty only if it can bring the non-NPT nuclear states under broad international safeguards.

Israel, the only non-NPT state without a nuclear foe in its regional backyard, has already achieved its goal of deploying a nuclear force of a desired size and has no need to sire more fissile material. The fact that its sole fissile-material source, the ageing Dimona reactor, needs to be shut down gives it more room than India to support an FMCT. Pakistan can count on China for continued clandestine transfers, even of fissile material despite an FMCT.[5] Pakistan's new Khushab plutonium-production reactor has been designed, built and outfitted by China to checkmate India through its client-state. A compact, missile deliverable nuclear warhead cannot be fabricated with HEU, which Pakistan produces at Kahuta. Khushab seeks to plug the missing link in Pakistan's deterrent plans by arming its Chinese-fathered Ghauri intermediate-range ballistic missile (IRBM) with a warhead made from plutonium, the material of choice for modern weapons.[6]

5. Tim Weiner, "U.S. and Chinese Aid Was Essential as Pakistan Built Bomb", International Herald Tribune, 2 June, 1998; Leslie H. Gelb, "Pakistan Link Perils U.S.-China Nuclear Pact", New York Times, 22 June, 1984; Nuclear Control Institute, Information Pertaining to Possible Chinese Test of a Pakistani Nuclear Device, Backgrounder (Washington, DC: Nuclear Control Institute, September, 1985); Jack Anderson and Dale Van Atta, "Nuclear Exports to China?", Washington Post, 3 November, 1985; and Leonard S. Spector with Jacqueline R. Smith, Nuclear Ambitions (Boulder, Colo.: Westview Press, 1990).
6. The Ghauri bears a striking resemblance to China's DF-21. Beijing is finding ingenious ways to circumvent the latest in its long series of broken non-proliferation promises. For the Ghauri, China has been directly providing finance, soft technology and engineering but routing the hardware through North Korea. Joseph S. Bermudez Jr., "DPRK-Pakistan Ghauri Missile cooperation", Federation of American Scientists (FAS), WWW.fas.org.

China values Pakistan as a military counterweight to India to such an extent that, FMCT or no FMCT, it could give Islamabad any critically needed fissile material to complete its FMCT thus will not lessen proliferation threats to Indian security. Rather, an early FMCT would force India's deterrent posture to be vasectomised under the treaty's sterilisation provisions even before New "Delhi has procreated enough fissile material to secure itself against its two neighbouring nuclear adversaries. Thus, India is unlikely to support an FMCT coming into effect without its bare deterrent needs having been met.

NUCLEAR PLANNING

Months after its tests, India established a National Security Council (NSC) to deal with its complex security challenges in a structured, long-term manner. India's emergence as a nuclear-weapons state has deepened its need for an effective policy-making institution such as the NSC. National security demands integrated planning and co-ordinated use of India's political, military, diplomatic and scientific resources.

An NSC thus is an essential institution for the assertive promotion of national interests. India is the only major democracy to lack institutional mechanisms to chart its long-term strategy and blend its various policies into a harmonious national vision. Even some of India's small neighbours in Southeast Asia have set up NSC-type institutions. The growing technology sanctions against India had only increased the need for an NSC.

The newly established NSC will play an important role in nuclear planning, force posture and doctrine. It sill, however, be imprudent to see the NSC as a panacea for the ills that afflict India's policy-making system. The NSC cannot overcome the limitations of the country's policy-making progress. Those limitations are rooted in India's colonial-style system of governance. They can be overcome only through radical changes in the way the Indian administration functions and acts. The setting up of the NSC constitutes only a modest effort to deal with some of the challenges India policy-making faces. Similarly, the NSC cannot by itself promote inter-ministerial and inter-departmental co-ordination. It can provide a broad and comprehensive approach to national security, but the lack of institutionalised inter-ministerial co-ordination is a problem beyond its possible reach.

Unless the NSC continually evolves a long-term national security strategy, India cannot pursue farsighted programmes and projects. Even on defence policy, new thinking is required as the ongoing global revolution in military affairs (RMA), spawned by the Information Age, is changing the very nature of war. India has to give up its outdated threat-scenario approach to defence (which puts it in a 'reactive' mode) and shift to capacity-building approach linked to clearly-defined vital interests. Capacities take a long time to be built but threats can emerge suddenly and unexpectedly. Capacity-building, tied to the evolving geostrategic landscape, can effectively daunt the rise of major threats. Lamentably, most of the key Indian ministries lack long-term planning, while the defence policy planning process is warped. The Defence Ministry makes the five-year plan, but the 15-year perspective plan and the annual are drawn by the military.

So, while the NSC cannot eliminate all the functional inefficiencies that burden Indian policy-making, it can help manage national security in a more prudent, cost-effective and result-oriented manner. It can help establish institutionalised methods so that policy-making is driven less by personalities and more by long-term integrated planning. The NSC should help blend various policies so that there is overall harmony and commonality of objectives. And in a country where governments come and go in rapid succession, the NSC should be able to ensure policy continuity.

Successive Indian governments in the 1990s have recognised that the country, with its limited resources, needs an NSC that can maximise efficiency an minimise wastage through systematic, long-term strategic planning. But in seeking to create an NSC as an essential pillar of policy-making, they ran into averred constitutional difficulties as well as the bureaucracy's reluctance to surrender some of its prerogatives. Faced with the same bureaucratic hurdles, the Vajpayee Government simply united existing arrangements and dressed them up in the form of an NSC. The formation of the NSC nonetheless is a welcome development, and improvements in its structure are likely to be part of an ongoing process at better managing security challenges.

The heart of the NSC will be its multi-disciplinary secretariat, comprising the best in-house analysts drawn from various services and departments. The secretariat, which will deal with both military as well as non-military dimensions of national security, will create ideas and proposals for approval by the NSC's Strategic Planning Group, headed by the Cabinet Secretary. The NSC's third wing, the high-sounding National Security Advisory Board comprising outside analysts, will play a role only on the margins of policy making. The existing NSC structure shows the cult Indian career bureaucrats are going to continue to wield in national security planning, including nuclear planning. Scientists will also play a crucial role in nuclear force planning and doctrine.

The Indian Military, in contrast, may remain marginalised in policy-making. India is the only major democracy that has shut out its military from defence policy-making. Uniformed officers are excluded form the Indian defence ministry, which is run by non-specialists from an elite civil service whose members move every few years to a new job, stretching from urban development and social welfare to public health and defence. As a consequence, India has never clearly defined its vital interests.

A tight civilian control over the armed forces was intended to preclude any possibility of a military coup. Over the decades, however, the country's mammoth bureaucracy, whose red tape is legendary, has pushed the military further away from the policy-making process. The backslide in the military's role began with the elimination of the Defence Committee of the Cabinet, set up at independence in 1947, and its successor, the Emergency Committee of the Cabinet, which came into being after India's humiliating 1962 defeat at the hands of the invading Chinese army. Those committees, headed by the prime minister, included the chiefs of India's three military services.

In the new NSC, the chiefs of the three military services find a place in the Strategic Policy Group, packed with senior bureaucrats ranging from the Home, Foreign, Defence, Finance Secretaries to the Reserve Bank Governor. The service chiefs have been complaining that "civilian control" over the military has come to mean "civil-services control", with the control exercised not by elected political functionaries accountable to Parliament but by bureaucrats answerable only to other bureaucrats.

Many months after conducting multiple nuclear tests, India has still to bring the military into the nuclear picture. This despite the fact it has declared itself a nuclear-weapon state. The paradox of a country proclaiming it has a nuclear deterrent without the necessary military underpinnings has created an inherently dangerous situation in which a potential adversary could be tempted to try and call India's bluff. It also highlights the marginalisation of one of the world's largest militaries.

Sooner than later, however, India will have to bring its military into nuclear planning. Without the military's involvement, it will not be possible for India to devise and to put into operation a nuclear deterrent, which would involve targeting and deployment practices.

NUCLEAR DOCTRINE

India's nuclear doctrine will determine how it copes with its fundamental external-security challenges. Given that its scientists have worked on nuclear-weapons designs for at least a quarter century and that its 1998 tests only unveiled capabilities it already had, India should have followed Israel and developed a nuclear doctrine long ago. That it did not do so is hardly a surprise. India has no strategic doctrine of long-term national security planning, lacks institutional mechanisms to develop a strategic vision or to mould its various policies into a coherent whole, and has yet to enunciate well-defined vital interests. Against this background, it would have been too much to expect India to already have a nuclear doctrine.

Rather than doctrine shaping capabilities in the classical sense, India's capabilities will dictate its nuclear doctrine. Like much else in India, developing this doctrine will be a slow, laborious process. After all, it took India 34 years after acquiring a plutonium-production capability to make up its mind on a nuclear military posture! Many months after declaring itself a nuclear-weapons state, India has yet to take the preliminary steps to integrate nuclear weapons into its defence structure and make its 'deterrent' operational in a doctrinal or military sense.

The failure to involve the military in nuclear planning has resulted in India still being vague about its nuclear doctrine. The only elements of the doctrine made public are that India will practise, in French-style terminology, "credible minimum deterrence", and not be the first to use nuclear weapons. While those objects are commendable, they only seek to make a virtue out of necessity: India does not have the plutonium or financial resources to exercise more than the barest of minimum deterrence, and is far from having the capacity to carry out a disabling first strike against an opponent. India's minimum deterrence, is likely to look in the initial years as no more than counter-city deterrence. While India is going to have a diversified nuclear dyad made up of ballistic missiles and bomber-aircraft, it is still distant from an invulnerable second-strike capability with submarine-launched missiles.

Similarly, the only aspect of India's command-and-control system discernible is that it will be firmly controlled by civilians, with the Prime Minister as the ultimate decision-maker at the head of a yet-to-be-established Strategic Nuclear Command. An effective command-and-control system, of course, can only emerge over a period of time. After all, it took the traditional nuclear powers many years (in the case of the United States more than 15 years) to develop a command-and-control system that provided a degree of self-assurance. Still, shouldn't India by now have initiated the process of designing a command, control, communications, computer and intelligence (C41) system? Such a system will come neither cheap nor fast.

While it underpins deterrence by meeting the technical-operational requirements, India has to define the central purpose of its nuclear harmony. Is it to deter nuclear threats and blackmail, or aggression in general, including conventional aggression of the kind India has suffered in the past? If it is the latter, French nuclear structure and doctrine based on 'proportional deterrence' will be more appropriate for India. It should not be simplistically assumed that nuclear weapons are only for deterring a nuclear threat, as if it is okay for others to employ conventional force against vital Indian interests. Given the decades-old technology sanctions it has suffered and the way its foreign policy has been seriously constricted, India has had to pay an extraordinarily heavy political price for its nuclear weapons. It should seek to derive the full security value from them, the way the wealthy nuclear powers do.

Indian security planners will also have to decide whether the country should have 'tactical' (theatre or battlefield) nuclear arms, or only 'strategic' (long-range) weapons. According to Pakistan, its nuclear tests involved tactical weapons. China also has such weapons in its nuclear arsenal, including tactical nuclear warheads on short-range missiles and nuclear-tipped artillery shells and mortars. Can India do without tactical nuclear arms by ignoring the fact that its neighbours have them? A pointer to the direction in which India is headed has been given by the 'father' of India's nuclear weapons, Dr. R. Chidambaram who has publicly stated that a tactical weapon is one of three "robust warhead designs" already developed. With capabilities to inspire doctrine, scientists are going to considerably influence nuclear thinking and decision-making.

Indian doctrine has to credibly take on the joint nuclear challenge posed by the countrys two linked foes, China and Pakistan. These two countries have a long history of covert nuclear and missile collaboration and anti-India collusion. Minimum deterrence in such a situation intrinsically demands a very high level of confidence in the reliability and survivability of the arsenal. Such confidence can come about only through a vigorous doctrine and rugged capabilities.

Indian doctrine also cannot ignore the hazards posed by a troubled Pakistan, which is beginning to look like a dangerously failing state. If central political authority starts to slip away, there will be a risk that renegade military elements or Taliban-type Islamic militias may try to gain control of nuclear weapons. The Indian doctrine will have to recognise that strategic weapons can have no role when renegade or terrorist elements threaten to use tactical arms.

The building of a long-term Indian nuclear doctrine demands broad political consensus not only on the value of nuclear deterrence as the country's life-insurance policy, but on the modalities of deterrence. There has been hardly any discussion on the latter. Consensus-building has been made difficult by India's raucous and splintered politics in which every issue assumes partisan colours.

India's emergence as a declared nuclear-weapons state has only brought its challenges to build a dependable nuclear deterrent into sharper focus. The national debate in India cannot shy away from the though nuts-and-bolts issues. India's intellectual Jezebels have conveniently dismissed these issues as products of Western theological thought. They had in the same manner rejected India's testing imperatives until the tests occurred, when they instantly found justification for the detonations! Thinking rationally cannot be made synonymous with thinking Western. The core tasks of nuclear deterrence are not race-specific but common to all.

For years India agonised over whether to build and test nuclear weapons. Having built and tested such weapons, it may now agonise over their deployment and the doctrine. That is why those pressuring India never call it quits but change their goal as soon as it gets overtaken by events. Earlier the U.S. goal was to prevent India from building and testing nuclear weapons; now the goal is to dissuade India from deploying nuclear warheads or mating them with missiles. India has to stop agonising or getting distracted by diplomacy and take on its challenges.

FORCE POSTURE

In the popular relation over their country's emergence as a declared nuclear-weapons state, it has been forgotten by most Indians that the tests were only the first (although most important) step to counter rising regional and extra-regional threats to Indian security from weapons of mass destruction. A lot more has to be done before India can justifiably contend that it can terminate any nuclear threat—a claim prematurely made right after the two rounds of tests in 1998.

India faces three major challenges in achieving its deterrence goals. First, the country's small reserves of weapons-grade plutonium and limited financial resources dictate that it can only have puny nuclear forces. The predetermined diminutive size of the Indian nuclear arsenal imposes high reliability and security standards on New Delhi. Without the luxury of allowing numbers to compensate for quality, India will have to rigorously meet the technical imperative of its deterrent. Unlike Pakistan and Israel which can draw on China and the United States respectively for continued nuclear-warhead design assistance, India is the only non-NPT nuclear state without an external benefactor.

Two, India has to build a credible minimal deterrent in a global environment openly hostile to further overt weaponisation, particularly by a new nuclear-weapons state. Each new action, such as an Agni 2 test or the commencement of serial production of warheads, will draw adverse international reaction. It is thus vital that the follow-up steps to erect an appropriate deterrent are taken quietly and without the official braggadocio so characteristic of every Indian advance, however, small. Classified projects in other democracies are keep under tight wraps, but in India inside information has been leaked out on the nuclear propulsion project and various missiles. Transparency is important in terms of strategic posture, not strategic projects. The United States, for example, released little information on its 1998 missile tests, testing and deployment of the new earth-penetrating B61-11 nuclear warhead, and top-classified Advanced Vortex Satellite that blew up atop a failed Titan-4A rocket.

Three, India is still some distance from acquiring the technical capacity to terminate a Chinese nuclear threat and needs to plug this vulnerability on a war footing. A deterrent against Pakistan was never the central mission of Indian nuclear strategy. There is not a single political or military reason for India to build a nuclear deterrent against Pakistan. Nuclearisation of the subcontinent only blunts India's conventional-military advantage over Pakistan, allowing the latter to employ nuclear forces as an equaliser. An Indian deterrent makes strategic sense only in context of China.

India can live with nuclear disparity with China, in the same way Beijing has done with the U.S. But India cannot live with a nuclear capability that does not provide basic cover against China. Because of a very late start, India will lag far behind China in the foreseeable future. Even if Beijing were to dramatically halt its ongoing nuclear expansion and India were to build nuclear weapons at the maximum rate possible with its present plutonium-production capacity, New Delhi would still not be able to match China's nuclear might even in a quarter century from now. The nuclear test-ban treaty and the planned FMCT will ensure that.

However, since deterrence does not demand qualitative or quantitative parity, India can still effectively deter China, which has up to 500 nuclear weapons, with less than 100 warheads—the likely maximum size of an Indian armoury if an FMCT is negotiated and brought into effect in the next five years. To ensure this, India needs to do three things. One, focus on building a hardly nuclear force, mobile and widely dispersed, that can survive an enemy first strike and be able to retaliate even if principal Indian command-and-control centres get wiped out. Two, ensure Indian capabilities are not subject to retarding arms-control fetters before they fully control to enhance, not undercut, their security.

Three, India should strive to prevent the gap with China from continuing to grow. China, with a nuclear arsenal now already larger than that of France and Britain, is engaged in the largest expansion of nuclear and missile capabilities by any state. One way for India to build international pressure on Beijing would be to tie its support to any arms-control venture to visible signs that China has stopped or slowed down its weapons modernisation. By directly raising China's strategic costs, India should also dissuade Beijing from continuing its nuclear and missile aid to Pakistan. Those costs could be increased through strategic assistance to countries that share Indian concerns over China, such as Taiwan, Vietnam and the Philippines.

The central challenge for India is that, unlike deterrence relationships elsewhere in the world, it has to adequately deter two hand-in-glove nuclear adversaries, China and Pakistan. This is no mean task, considering the fact that India has to reckon with the possibility that one of them could act as a proxy of the other and that the two would closely co-operate in any situation where one seeks to take on India.

India's nuclear self-adulation thus is misplaced. Five tests do not translate into an instant deterrent. If past mistakes are to be avoided, India should not allow its rhetoric to outstrip its capacity or its claims to overtake the reality. India has to put its head down and work with determination and humility on its tasks. While Indian scientists need to be commended for demonstrating in one go a whole range of nuclear-weapons capabilities that no other nation has similarly done, India has still important challenges to meet. Time is not on India's side. Projects that have not delivered results, such as the Advanced Technology Vehicle (ATV), need to be revamped under new leadership.

The basic reality is that India, with its modest capabilities and resources, can only pursue the modest path of building a petite, affordable but survivable nuclear force, with a doctrine that eschews both a war-fighting approach and the course of recessed or non-deployed deterrence advocated by the United States and its friends. Since independence, India has been a soft state, Nuclear weapons, however, demand hard decisions. Deterrence requires not only the technical capacity but an unmistakable national will to severely punish a potential aggressor. By keeping the military shut out from the nuclear picture months after proclaiming a "deterrent" and by engaging in a dialogue with the United States on naively optimistic premises, India has not shed its soft-state image. Rather, it has presented itself as willing to enter into several acts of compliance after one act of defiance.

One glaring mismatch India has to tackle is between its economic capacity and security needs. Without economic power, India can have no security, even with nuclear weapons. The way sanctions-hit Pakistan sunk economically after conducting its tests should serve as a lesson to India to accord the highest priority to putting its economic house in order. India should also be cognizant of the danger that nuclear weapons in the hands of a soft state can invite greater insecurity.

Address

BRAHMA CHELLANEY
Cenre for Policy Research
New Delhi 110021
India